Physics of Atoms and Molecules
An Introduction to the Structure of Matter

U. FANO AND L. FANO

THE UNIVERSITY OF CHICAGO PRESS
CHICAGO AND LONDON

THE UNIVERSITY OF CHICAGO PRESS, CHICAGO 60637
The University of Chicago Press, Ltd., London
 Published 1972
Printed in the United States of America
International Standard Book Number: 0-226-23782-6
Library of Congress Catalog Card Number: 76-184808

Contents

Preface

The study of the structure of matter tries to account for the properties of materials in terms of their atomic constitution. This book deals with isolated atoms and molecules to provide basic information for explaining and predicting properties of bulk matter. It is focused on the problems of the size and stability of atoms, their ability to aggregate, and their electric, magnetic, and spectroscopic properties. The treatment of these problems is necessarily couched in the language of quantum mechanics. The development of this language and of the underlying concepts absorbs much space and attention; yet quantum mechanics is not the subject of this book.

In our study, as in many others, one may distinguish three steps:

1. Describing instructive phenomena and drawing inferences from them. Historically, adequate appreciation of the quantum properties of atomic systems required insight of the highest order and followed long and tortuous research. With hindsight one can now identify a set of phenomena that lead more directly to the basic concepts of quantum physics.

2. Formulating basic concepts in physicomathematical language suitable for a precise description of the initial set of phenomena and for the later prediction of additional phenomena.

3. Actually applying the resulting theory and extending it to an increasing range of systems.

In this book only as much theory is developed as is required to treat the atomic and molecular properties which seem most important. We have attempted to emphasize key experiments and to describe a large variety of phenomena. Conversely, we have de-emphasized formalism, regarding it for this course as a tool rather than an aim. We seek to separate two levels in the teaching of quantum mechanics, as one usually does for classical mechanics. That is, we have approached quantum mechanics much as classical mechanics is approached in lower-division courses, for example in the *Berkeley Physics Course*, volumes 1 and 3. The formal treatment of quantum mechanics is left for later courses analogous in intent to the courses of analytical mechanics.

Much effort and space is nevertheless taken in this book by theory. This is due first to our endeavor to relate each step of abstraction explicitly to an experimental situation, instead of compressing it into mathematical language. Second, we have avoided provisional concepts (such as electron orbits, matter waves, or the direction of angular momenta) which do not actually hold in atomic physics. Instead, we have tried to abstract from the experiments concepts and formalisms that will remain valid upon deeper analysis in later courses. This endeavor places a heavy burden on student and teacher. Yet it seemed essential to minimize the danger of unwarranted implications, because misconceptions implanted in the student at the time of his first exposure to quantum physics often prove persistent and insidious.

We previously followed the same guidelines in another book (*Basic Physics of*

Atoms and Molecules, Wiley, 1959) which was aimed primarily at nonphysicists and had the character of an essay. What we are presenting now is a textbook developed for upper-division physics students in the first half of the Structure of Matter sequence at the University of Chicago.

The earlier book approached quantum mechanics through the concept of eigenstate. This approach, followed in several modern introductions to quantum physics, is strengthened in the present version by utilizing more fully the simplicity of two-level systems. All along, we stress the consistency of macroscopic and atomic physics rather than their contrasting aspects. To this end attention is drawn repeatedly to the continuous variations of nonstationary states of atoms. We have outlined the elements of the quantum treatment of radiation. Conversely, relativity effects are barely mentioned, because they are of minor importance for the structure of ordinary matter.

This book is designed for students with a solid knowledge of lower-division physics, but it does not require any previous acquaintance with quantum or "modern" physics. As a background reference we have chosen the *Berkeley Physics Course*, volumes 1, 2, 3, and 5, because of their modern approach, wide coverage, and elegance of treatment. The reader is expected to be familiar with the mathematics usually included in lower-division physics curricula. However, we utilize Fourier analysis and normal mode analysis heavily and from a physical point of view; therefore this material has been summarized in a series of appendixes.

A course covering the material in this book might extend over a year, though nearly all the material has actually been crammed into 15 weeks. In selecting the properties of atoms and molecules to be discussed, we had in mind that many students may take just one course in atomic and molecular physics. The later chapters of the book may also serve as an introduction to specialized treatises. At Chicago this material is followed by solid state and nuclear physics courses and by graduate school quantum mechanics. It might also lead into plasma physics, physical chemistry, or quantum chemistry.

We gratefully acknowledge the opportunity afforded us by the University of Chicago and by several other schools to develop and try out our new course material, as well as the advice and help provided by numerous colleagues. Mrs. Natalie W. Holzworth has contributed greatly by critically reviewing the whole manuscript and checking all the problems. Mrs. Helen Barileau has done an outstanding job of intelligent and careful typing for many successive revisions of the manuscript.

Repeated footnote references to basic textbooks are abbreviated. A selected bibliography of reference works, divided by type, is given at the end of the book.

Part I

BASIC EVIDENCE ON THE MACROSCOPIC AND QUANTUM PROPERTIES OF ATOMS

1. initial evidence

1.1 MACROSCOPIC ANALYSIS OF THE CONSTITUENTS OF MATTER

That matter is an aggregate of very small atoms was surmised by ancient philosophers. This idea remained a speculation, like most ideas about the physical world, until the advent of experimental science, but it acquired a solid foundation very rapidly once the quantitative study of chemical transformations began in the late 1700s. The main implications of this study were already organized in the minds of leaders like Dalton, Gay-Lussac, and Avogadro as early as 1810. Let us summarize the steps of this familiar development.

First, chemical substances are identified whose quantitative properties have fixed values under any given set of external conditions. Most substances can be broken down into other substances, but this process of chemical analysis reaches an ultimate limit. Substances that cannot be analyzed further are called chemical elements.

Analysis shows that each compound substance contains two or more elements in *constant proportions* of mass. Different substances exist which contain the same elements in various but still constant mass proportions bearing simple ratios to one another (law of *multiple proportions*). It follows that a unit of each compound substance, a molecule, consists of a fixed number of units of its several constituent elements, that is, of a fixed number of atoms of each element. A change in the number or species of atoms in a molecule yields a molecule of a different substance and constitutes a chemical transformation. Complementary evidence was obtained through studies of the volumes of gases taking part in chemical transformations (law of Gay-Lussac).

Experimentation on the weight of input and output substances for a large number of chemical transformations led to systematic determination of the *relative* weights of all known atoms and molecules. The actual weight of each atom or molecule could not be measured and remained unknown for a long time. The relative weight of each atom or molecule was expressed in an arbitrary scale for which the unknown weight of hydrogen was initially taken as unity. The number assigned to each atom in this scale was called its *atomic weight*.

The actual number of atoms or molecules of each species taking part in a chemical transformation cannot be counted and would anyhow be unpracticably large. The relative numbers can, however, be stated by expressing the weighed amount of each participating substance in terms of a standard quantity which contains a fixed number of atoms or molecules. This standard, called a *mole*, weighs a number of grams equal to the atomic or molecular weight of interest. The number N_A of atoms (or molecules) in a mole is called *Avogadro's number* after the man who focused attention on the problem or also *Loschmidt's number* after the man who in the 1850s first determined the value of the number by a method outlined in the next section.[1]

The volume of a given mass of liquid or solid material depends but little on external variables, such as pressure or temperature. Nor does this volume change by a large factor in chemical transformations that do not vaporize the material. It follows that each molecule, indeed each atom, occupies a moderately well-defined volume and withstands compression. (Gases differ from liquids or solids, of course, because their molecules separate from one another under the influence of thermal agitation; the average spacing of their molecules then responds readily to changes of pressure or temperature.)

Physico-chemical evidence of various kinds also shows that atoms or small molecules are neither flat nor threadlike but have comparable thickness in different directions. This evidence on the gross morphological properties of atoms should be kept in mind and combined with the more detailed evidence that will emerge in the course of our study.

Atoms had been conceived originally as units of matter that could not be further subdivided. Evidence of their having an electric structure was provided by a sequence of quantitative studies of the transport of electricity by matter, beginning in the mid-19th century.

1.1.1. *The Atomic Unit of Electricity.* Electricity flows through aqueous solutions, together with some of the dissolved substances, in the phenomenon of electrolysis. Chemical substances (e.g. hydrogen and oxygen or metals) separate out of the solution at the surface of the electrodes. The mass of each substance separated during a certain interval of time can be determined and compared with the quantity of electricity transported during the same time.

Faraday's experiments showed that the quantity q of electric charge transported is proportional to the mass m, and therefore to the number of moles, of each substance separated. He measured the ratio q/m and found it to take only a few alternative values represented by the formula

$$\frac{q}{m} = \frac{n F}{M}, \tag{1.1}$$

where n is a small integer, M the molecular weight of the substance and

$$F = 9.649 \times 10^4 \text{ coulombs/mole} \tag{1.2}$$

is called the *Faraday constant.* This constant, expressed in terms of current flowing through a cell, amounts to approximately 27 ampere-hours per mole. The quantity of charge F carried by one mole is actually so large that one could not conceive of collecting it in any laboratory volume. Consider, for

1. Actually, in common practice, "Loschmidt's number" is defined as the number of molecules per cm^3 at normal pressure and temperature (1 atmosphere and 0° C).

purpose of orientation, that if a mole of gas at normal pressure were given this charge the potential of the gas would be of the order of 10^{17} volts with respect to the ground.

Faraday's result (1.1) indicates that each atom or molecule can carry only a few alternative quantities of charge, ne, where the basic quantity

$$e = F/N_A \tag{1.3}$$

may be called an "atom of electricity" or, more commonly, an *elementary charge*. An atom or molecule carrying a nonzero charge is called an *ion*. If the elementary charge e is taken to be positive, the small integer n of equation (1.1) may have either sign and indicates the electrochemical valence of the species of ion.

Note that the experiments on electrolysis determine the ratio q/m of the total charge to mass transported by macroscopic amounts of matter and hence do not bear directly on the charge-to-mass ratio of individual ions. Experiments on the transport of electric charge through gases prove far more instructive in this respect.

1.1.2. *Transport of Charge through Gases.* Gases have very low electrical conductivity under ordinary conditions. However, a luminous discharge takes place with high flow of electricity when a sufficiently high potential is applied to electrodes separated by a rarefied gas. Electricity thus appears to be carried by gas molecules turned into ions.

The low conductivity of ordinary gases indicates that a minimal fraction of molecules carries any charge. Application of a strong potential difference imparts a sharp acceleration to these few existing ions. Lowering of the gas pressure permits each ion to build up a high velocity before colliding with another molecule. A violent collision is likely to cause additional ionization, that is, separation of charges. (The study of ionization by collision is in itself an important part of atomic physics.) An intense flow of electricity may thus result through the multiplicative process of ion production.

The carriers of electricity in a gas discharge may be studied by confining the discharge and by leaking some of the carriers through collimating pinholes in the electrodes. The gas pressure may be reduced further in the space behind the pinholes as shown schematically in Figure 1.1. A luminous beam is then observed to emerge from the negative electrode. A glowing spot appears where this beam strikes the container's wall if the wall has been coated with a suitable material. Symmetrically, a glowing spot may also appear on the side of the positive electrode, even though no luminous beam is observed here.

The properties of whatever leaks through the pinholes are studied by applying electric and magnetic fields in the regions behind the pinholes. In particular, shifts of the glowing spots confirm that carriers of positive charge emerge from the negative electrode and carriers of negative charge from the positive one (see Figure 1.2). The observation of a luminous beam only behind the negative electrode reveals a difference between the carriers of

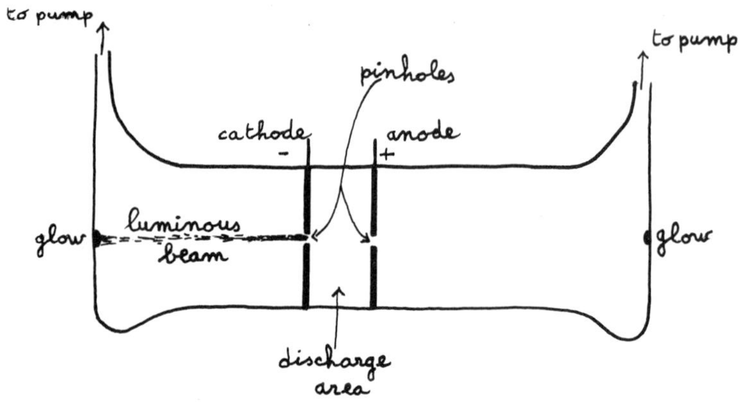

Fig. 1. 1. Sketch of apparatus for canal and cathode rays.

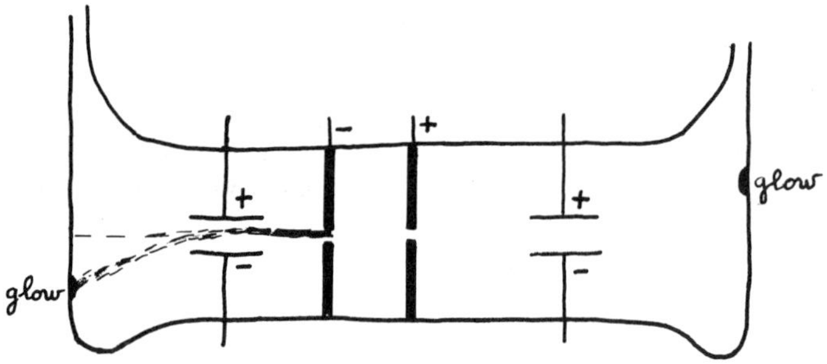

Fig. 1. 2. Deflection of canal and cathode rays.

positive and negative electricity. We will deal first with the positive carriers, which constitute the *canal rays*.

Deflection of canal rays by electric or magnetic fields is opposed by the inertia of their constituent particles. The deflection is inversely proportional to the mass and, of course, directly proportional to the charge of the particles. Therefore, measurements of deflection due to given electric and magnetic fields serve to determine the charge-to-mass ratio, or *specific charge,* of the particles. For particles of known velocity, measurement of the deflection in a single field, electric or magnetic, would suffice to determine the specific charge. Concurrent measurements of deflections in electric and magnetic fields enabled J. J. Thomson to determine the specific charge of particles, irrespective of their velocity.[2]

2. A description of typical experiments on charge-to-mass ratio is given in C. P. Harnwell and J. J. Livingood, *Experimental Atomic Physics* (New York: McGraw-Hill, 1933), chap. 4.

Since each particle in a beam responds to electromagnetic fields according to its own specific charge, the deflection method analyzes the beam into components homogeneous with respect to specific charge. Therefore, this method provides an important and sharp analytic tool. (Recall that experiments with electrolysis provide only an average specific charge for all transported particles.) This method is called *mass spectroscopy* because the charge of each particle is often regarded as known so that a measurement of specific charge amounts to a measurement of mass.[3]

The main early result of mass spectroscopy was the discovery that most elements are inhomogeneous with regard to mass, that is, they consist of different *isotopes,* indistinguishable by ordinary chemical analysis. For instance, canal rays emerging from a discharge in neon are resolved into three components with atomic weights 20.0, 21.0, and 22.0, whereas chemically pure neon appears to have atomic weight 20.2.

Canal rays emerging from discharge in a monatomic gas or vapor are also generally inhomogeneous with respect to charge. Their ions may carry different multiples of the elementary charge e, that is, charges $ne = nF/N_A$. The distribution of values of the charge multiplicity n depends on the potential difference applied to create the discharge, higher voltage favoring higher n values. In modern experiments the charge multiplicity can be increased by accelerating the ions to very high energy and then passing them through a thin metal foil. The highest possible value of n for each element coincides with its *atomic number* Z, that is, with the number of the element in the periodic system.

Passing now to the negative-charge carriers leaking from the positive electrode, one finds that they are generally a complex mixture which depends sensitively upon voltage and pressure. Beams of singly charged negative ions can be extracted under suitable circumstances. Much more important is the discovery at low pressure, of the order of 10^{-5} atmospheres, of negative-charge carriers whose specific charge exceeds that of any ion by orders of magnitude. The study of these carriers is performed most conveniently with the experimental arrangement described below.

1.1.3. *Cathode Rays.* Transport of electricity occurs between two electrodes at different potential even when the gas pressure in the intervening space is so low that the number of gas molecules is insufficient to account for the transport. The transport, which constitutes an electric current,is enhanced when the negative electrode (*cathode*) is heated or has a pointed shape as in the diagram in Figure 1.3. In this arrangement a glowing spot appears at point A behind the anode, but no similar spot appears behind the cathode. It follows that the charge carriers orginate from the cathode, are negatively charged and hence are accelerated by the anode's attraction. The flowing negative

---

3. Mass spectroscopy is very accurate; measurements can be made to better that one part in 10^7.

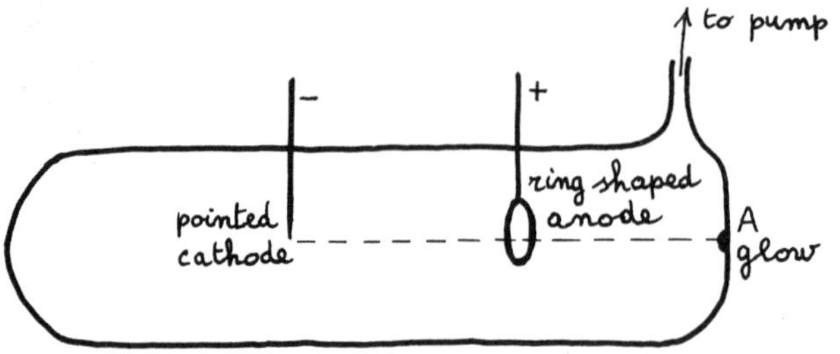

Fig. 1. 3. Sketch of apparatus for cathode rays.

carriers are called *cathode rays*. No analogous transport of charge is observed if the potential is reversed.

Analysis of cathode rays by mass spectroscopy was carried out initially by J. J. Thomson. He found cathode rays to be altogether homogeneous and to have always the same specific charge, no matter what materials were used in the experiment. Thomsons's value for the specific charge was 1.7×10^8 coulomb/gram. He called the carriers *electrons*. A recent value of the specific charge of electrons is

$$q/m = 1.7588 \times 10^8 \text{ coulombs/gram.} \qquad (1.4)$$

The simplest interpretation of these results, confirmed by all subsequent evidence, is that each electron has an elementary unit of negative charge $e = F/N_A$. The large value of q/m implies that the electron's mass m_e is orders of magnitude smaller than the mass of any ion. Quantitatively, one obtains the "atomic weight" of electrons by dividing Faraday's constant in coulombs per mole by the specific charge of electrons

$$\frac{F}{q/m} = \frac{9.649 \times 10^4 \text{ coulombs/mole}}{1.7588 \times 10^8 \text{ coulombs/gram}} = \frac{1}{1823} \frac{\text{gram}}{\text{mole}} \qquad (1.5)$$

1.1.4. *Summary of Evidence on the Constituents of Atoms*. Whereas in electrolysis atoms or molecules carry charges of either sign, in a gas discharge the negative ions tend to be replaced by free electrons. The low mass of the electrons and the unspecificity of their origin suggest that they are constituents of all species of atoms. The characteristic part of an atom of each element appears then to carry most of its mass and a positive charge; this charge is ordinarily neutralized by a complement of negatively charged electrons. Loss of one or more electrons changes a neutral atom into a positive ion; capture of an electron changes it into a negative ion. The positive charge of the characteristic part of an atom, expressed in elementary units e, equals the atomic number Z, since this number coincides with the highest multiplicity of ionization observed for that atom. Consequently, Z electrons constitute a normal complement for a neutral atom.

Picturing positive ions as incomplete atoms and negative ions as overloaded atoms accounts for the common occurrence of positive ions and the scarcity of negative ions among the violent collisions of a gas discharge. Both kinds of ions exist commonly in solutions, where violent collisions do not occur.

1.2 EVALUATION OF ATOMIC SIZES AND ENERGIES

The atomic interpretation of chemical transformations and the discovery of the electric properties of atoms are largely independent of any knowledge of the absolute magnitude of atoms, that is, of the number of atoms in a mole. Indeed, Avogadro's number was not evaluated for half a century after its conceptual introduction.

Actually, Avogadro's number is so fundamental for such a variety of properties of matter that its value can be derived by a large number of different methods. We shall outline here one method based on the properties of matter in bulk, one based on the direct measurement of the elementary charge e, and one based on a theoretical interpretation of chemical energies. Of these, the measurement of the elementary charge provides a quantitative determination of N_A, whereas the other methods provide only order-of-magnitude estimates.

1.2.1 *Loschmidt's Method.* The first determination of N_A rests on the molecular theory of the transport properties of gases, and followed rapidly Maxwell's development of the kinetic theory of gases.[4]

The transport properties of gases, such as self-diffusion, viscosity, and thermal conductivity, depend on the rate of intermixing of molecules over macroscopic distances. This rate is in turn proportional to the mean free path l of the molecules and to their root mean square velocity c. Specifically, the self-diffusion coefficient of a gas is given by

$$D = \tfrac{1}{3}\, l c. \tag{1.6}$$

This coefficient can be measured, for example, by injecting a small amount of radioactive or otherwise labeled material at one point in a gas volume and observing how rapidly it spreads throughout the volume. However, it is easier to observe the transfer of momentum or of kinetic energy (heat) from one small gas volume to another one. Indeed the viscosity η of a gas, which represents the trend to equalize the momenta of adjacent gas layers, is the product of D and the gas density ρ:

$$\eta = \rho D. \tag{1.7}$$

---

4. *Berkeley Physics Course* (New York: McGraw-Hill, 1965-71), vol. 5, chap. 8; (hereafter quoted as *BPC*).

The thermal conductivity k, which represents the trend to equalize the kinetic energy of molecules, is the product of the viscosity and of the specific heat C_V:

$$k = C_V \eta = C_V \rho D. \tag{1.8}$$

All the quantities in equations (1.6)–(1.8), except l, can be measured by experiments on matter in bulk. (The molecular velocity c is independent of molecular size and simply related to the sound velocity in the gas.) Therefore, measurements of transport properties yield a value of l. This value in turn determines N_A through the following considerations.

If a gas were condensed so as to pack the molecules close to each other, the mean free path l would be approximately equal to the diameter of a molecule. If we call v the volume of a mole of condensed gas, then l is somewhat shorter than $(v/N_A)^{1/3}$; for purposes of orientation we take $l \sim \frac{1}{3}(v/N_A)^{1/3}$. As the gas expands and a mole occupies a volume V, l extends in proportion to the expansion ratio V/v and we have

$$l \sim \frac{1}{3}\left(\frac{v}{N_A}\right)^{1/3} \frac{V}{v}. \tag{1.9}$$

Since V and v can be measured, a determination of l yields a value of N_A. An order-of-magnitude estimate of the value of N_A is obtained by entering in this formula the following values: $l \sim 10^{-5}$ cm, $V/v \sim 10^3$, $V \sim 2 \times 10^4$ cm^3. The result is 7×10^{23}.

1.2.2 *Millikan's Oil-Drop Experiment.* An absolute measurement of the elementary charge e provides directly a value of Avogadro's number N_A since 1 mole of elementary charges equals 9.649×10^4 coulombs ($F = N_A e = 9.649 \times 10^4$ coulombs/mole). The measurement of e is made possible by the enormous concentration of charge on electrons and ions, indicated by the large value of F. When a macroscopic body is charged, for example with negative electricity, it acquires a number of extra electrons which is but an infinitesimal fraction of the number of its atoms. That is, a very small percent increase in the number of electrons yields a macroscopically measurable effect. The addition or loss of a single electron may then affect the electric force acting on a particle large enough to be observed visually.

Millikan observed visually the effect of a single electron and thereby demonstrated the discrete nature of electricity and at the same time measured the elementary charge e. His method is described in detail in general textbooks[5] and briefly outlined here.

---

5. See, for instance, Physical Science Study Committee, *Physics* (Boston: D. C. Heath, 1960). The full theory of the measurement may be found in Harnwell and Livingood, *Experimental Atomic Physics*.

A spray of oil droplets is introduced between the plates of a condenser (Figure 1.4). The motion of a single droplet may be observed through a microscope. When the condenser is grounded and disturbances are avoided, a falling droplet quickly attains a uniform velocity which is limited by the viscosity of air and depends on the weight of the droplet. Measurement of this velocity serves to determine the weight of the droplet, by utilizing data on air viscosity.

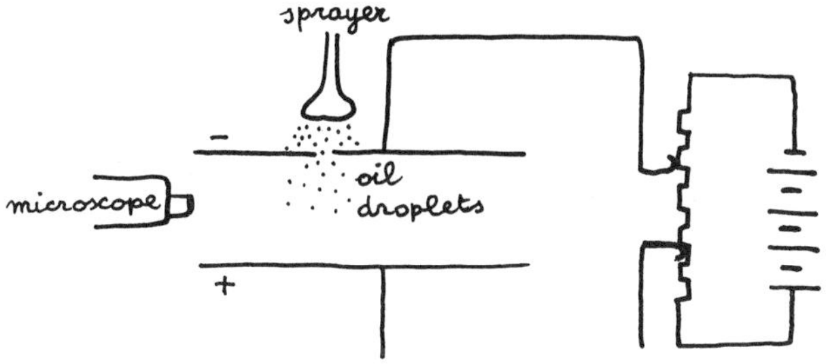

Fig. 1.4. Sketch of Millikan's apparatus for measuring the elementary charge.

Most droplets acquire an electric charge. Therefore, application of a suitable voltage between the condenser plates can apply to the droplet a force which balances its weight and keeps the droplet at rest. Knowledge of the field strength between the condenser plates and of the droplet's weight then yields the charge of the droplet.[6] Millikan saw that the charge on a droplet varies from time to time but varies always by the same amount or by a small multiple of that amount.

The elementary charge variation thus measured by Millikan, corrected by use of a recent value of air viscosity, is

$$e = 4.803 \times 10^{-10} \text{ esu} = 1.602 \times 10^{-19} \text{ coulombs.} \tag{1.10}$$

The resulting value of Avogadro's number is

$$N_A = \frac{F}{e} = \frac{9.649 \times 10^{4} \text{ coulombs/mole}}{1.602 \times 10^{-19} \text{ coulombs/electron}} = 6.022 \times 10^{23} \text{ electrons/mole.} \tag{1.11}$$

1.2.3. *Electrical Interpretation of Chemical Energies.* Start from the premise that atoms contain elementary charges of opposite sign. One may then surmise that the energy of bonds holding atoms together—chemical bonds—is

---

6. The droplet acts somewhat as the needle of an extremely sensitive electrometer. The sensitivity derives from the small capacity of the droplet, of the order of 10^{-17} farads.

of the order of magnitude of the energy between elementary charges at a distance of the order of atomic diameters. This energy can be expressed in terms of known macroscopic quantities and of Avogadro's number N_A. One may then ask, What is the value of N_A that makes the electrostatic and bond energies comparable to one another?

The electrostatic energy of two elementary charges at distance d is e^2/d in cgs units, with e expressed in esu.[7] In terms of macroscopic quantities e is given by

$$e = \frac{F}{N_A} = \frac{3 \times 10^9 \times 9.6 \times 10^4}{N_A} \approx \frac{3 \times 10^{14}}{N_A} \text{ esu.} \tag{1.12}$$

The distance d will be taken as $(v/N_A)^{1/3}$ as in equation (1.9), where v is the volume of one mole of condensed material. The electrostatic energy E per mole of charged pairs is then

$$E = \frac{N_A(3 \times 10^{14}/N_A)^2}{(v/N_A)^{1/3}} = 9 \times 10^{28} N_A{}^{-2/3}\, v^{-1/3} \text{ergs/mole.} \tag{1.13}$$

For comparison, we consider the bond energy per mole of nickel atoms in their solid state, namely,

$$E = 10^2 \text{ kcal/mole} = 4 \times 10^{12} \text{ ergs/mole.} \tag{1.14}$$

The volume of one mole of nickel is 6.6 cm^3. Equating the energies (1.13) and (1.14) gives

$$N_A = \left(\frac{9 \times 10^{28}}{4 \times 10^{12}}\right)^{3/2} v^{-1/2} = 10^{24} \text{ atoms/mole.} \tag{1.15}$$

1.2.4. *Avogadro's Number and the Size of Atoms.* A good value of Avogadro's number, obtained by simultaneous fitting of numerous experiments, is

$$N_A = 6.022 \times 10^{23} \text{ mole}^{-1}. \tag{1.16}$$

This value is adjusted to the 1961 definition of a mole, i.e., the number of atoms of the carbon-12 isotope whose total mass equals 12 grams.

Since the order of magnitude of N_A is 10^{24} and the volume of one mole of condensed atoms is of the order of 1 cm^3, the volume of one atom is of the order 10^{-24} cm^3 and its linear dimensions are of the order of 10^{-8} cm. For this reason atomic dimensions are conveniently expressed in angstroms (Å), that is, in units of 10^{-8} cm. The diameters of all atoms vary within the range 1 to 4 Å. In the example of nickel mentioned above, the volume per atom is about 10 $Å^3$.

---

7. One coulomb equals 2.998×10^9 esu.

From the value (1.16) of Avogadro's number, one can obtain the mass of any atom and the mass of an electron which is

$$m_e = 9.110 \times 10^{-28} \text{ grams.} \tag{1.17}$$

Having evaluated the order of magnitude of atomic diameters, it is of interest to evaluate also the electric potential and energies of elementary charges at atomic distances. The electrostatic potential V at a point 2 Å from a positive elementary charge is 7.20 volts.[8] Therefore, the energy necessary to rip off an electron from this point to infinity is 1.60×10^{-19} coulombs $\times$ 7.20 volts = 11.5×10^{-19} joules = 11.5×10^{-12} ergs.

To avoid dealing with the small numbers which represent atomic energies in standard units, one usually takes as a unit the potential energy of an electron at a point where the electric potential is 1 volt. This unit is called an *electron volt* and is indicated by eV.

$$\begin{aligned} 1 \text{ eV} &= 1.602 \times 10^{-19} \text{ coulombs} \times 1 \text{ volt} \\ &= 1.602 \times 10^{-19} \text{ joules} \\ &= 1.602 \times 10^{-12} \text{ ergs.} \end{aligned} \tag{1.18}$$

The energies involved in individual atomic processes and those involved in the chemical transformations of matter in bulk are related by the equation 1 eV/molecule = 23.06 kcal/mole.

1.3 OBSERVATION OF ELEMENTARY ATOMIC PROCESSES

The study of atomic and subatomic systems has been greatly advanced by the observation of isolated collisions between particles. The tools and methodology for these observations originate from the work done in the early 1900s by Rutherford and his collaborators on the radiations from radioactive substances.[9] This section reviews the principles underlying both the tools and the methods, partly as an introduction to Rutherford's main discovery and partly because of their importance throughout atomic physics.

---

8. The volt is the unit of electric potential in the mkqs system; the unit of potential in the cgs system equals 299.8 volts. The calculation of atomic potentials is simplest in the cgs system, where $V = e/r$; with $e = 4.80 \times 10^{-10}$ esu and $r = 2 \times 10^{-8}$ cm, V equals 2.40×10^{-2} cgs units, that is, 7.20 volts. The joule is the unit of energy in the mkqs system and equals 1 coulomb $\times$ 1 volt; the cgs unit of energy is the erg, equal to 10^{-7} joules; the thermal unit of energy is the calorie, equal to 4.184 joules.

9. The classic report on this work is the book by E. Rutherford, J. Chadwick, and C. Ellis, *Radiations from Radioactive Substances* (New York: Macmillan Co., 1930).

Rutherford's work utilized particularly radiations consisting of charged particles with kinetic energy of the order of millions of eV. These are the α-rays consisting of He^{++} ions and the β-rays consisting of electrons. Their high kinetic energy combined with their high specific charge enables single α- and β-particles to produce macroscopically detectable effects when passing through suitable materials. Reliance on radioactive substances for the production of high-energy particles ceased in the 1930s with the advent of accelerating machines. The technical capabilities for particle detection have also expanded greatly so that now even low-energy single particles can be detected.

1.3.1. *Particle Counting and Track Visualization.* Fast charged particles are detected through the effects of their energy dissipation in suitable materials. The most important of these effects is ionization, that is, the ejection of an electron from an atom or molecule of the material. A single ionization can be amplified by the presence of a strong electric field which accelerates the ejected electron. This mechanism can start a branching chain of successive ionizations and thus cause a spark or even a lasting discharge. Another useful effect is the efficient conversion of the kinetic energy of a particle into visible light by a luminescent material. This is what happens on the screens of oscilloscopes and television tubes, and it is the cause of the glowing spot in cathode- and canal-ray tubes. Rutherford observed and counted visually, though with difficulty, the scintillations produced by individual particles. Nowadays even faint scintillations are detected and amplified by photomultiplier tubes. Various less immediate effects of the passage of fast particles, presumably initiated by ionization, have also proven very useful. Among these are the activation of photographic emulsion grains, the formation of droplets in the supersaturated gases of cloud chambers, and the formation of bubbles in the superheated liquids of bubble chambers.

Ionization and scintillation serve in the first place to trigger counters which register the number of particles passing through well-defined portions of materials. Besides counting, various devices can also measure the energy dissipated by a particle along its track.

Counting and energy measurements are complemented by an increasing variety of "track visualization" devices. Instead of merely registering the passage of a particle or its dissipation of energy, these devices (cloud chambers, bubble chambers, spark chambers, etc.) display the tracks of all particles that have traversed a given volume within a given time. The resulting pictures show all major events occurring along the tracks, such as deflections or formations of secondary particles (Figure 1.5).

1.3.2. *Elementary Collision Processes.* Many properties of particle tracks result from a multitude of interactions between the particles and the material they traverse. Such are, for instance, the total amount of ionization produced by an α-particle in a gas, the total length of a particle track, which depends on the rate of energy dissipation, and the progressive curvature of a track caused by successive collisions. On the other hand, there are singular events occurring infrequently along tracks, such as sharp deflections and forks. These singular events appear to originate from collisions with single atoms

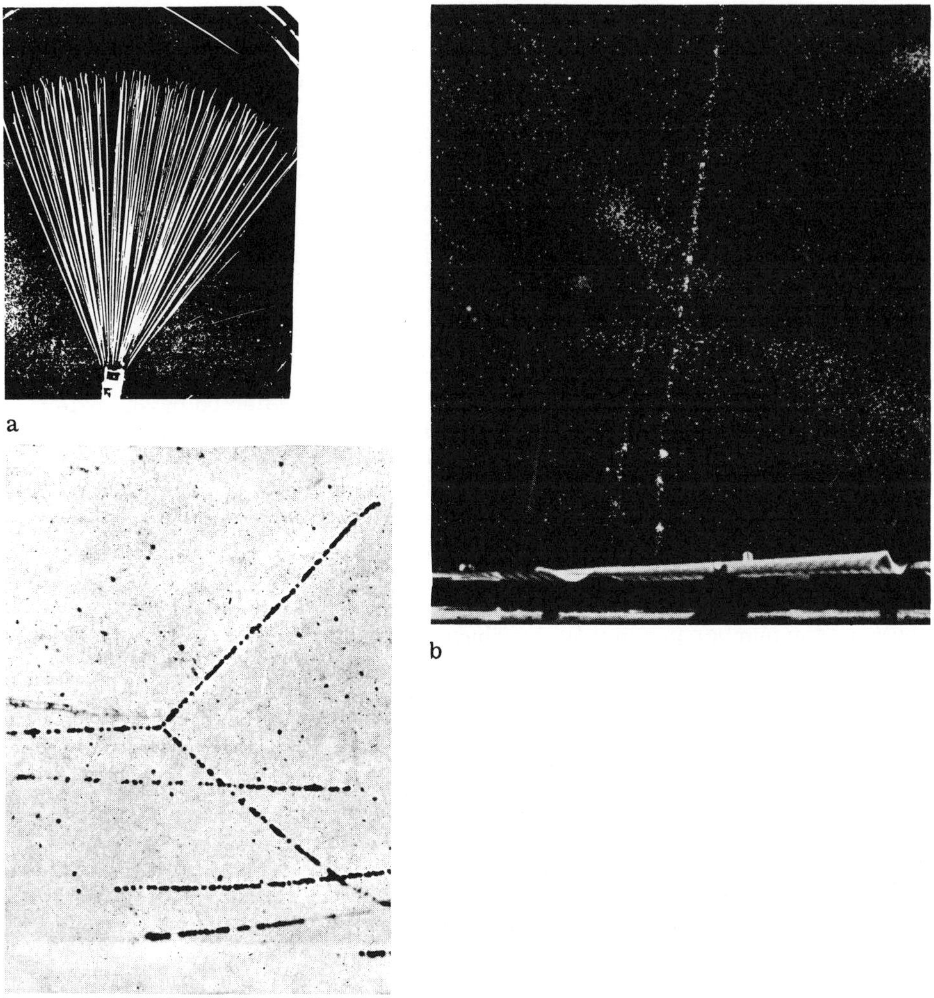

Fig. 1. 5. (a) Cloud chamber picture of α rays in fluorine showing a sharp de-deflection and recoil of a fluorine atom. Courtesy of J. K. Bøggild, from *Atlas of Typical Expansion Chamber Photographs* (London: Pergamon Press, 1954).

(b) Cloud chamber picture of β-ray tracks resolved into separate droplets. Courtesy of E. V. Hayward, University of California.

(c) Tracks of fast hydrogen ions in emulsion. One ion collides with a hydrogen atom in the emulsion and knocks it off, sharing its energy with it in nearly equal parts. Courtesy of C. F. Powell, from Powell and Occhialini, *Nuclear Physics in Photographs* (Oxford: Clarendon Press, 1947).

or molecules of the material. They are called *elementary processes* and provide direct evidence on atomic interactions. Multiple processes are also instructive, of course, but their study is more indirect and laborious.

The identification of elementary processes is sometimes trivial but a precise test is required for quantitative analysis. For example, Figures 1.5a and 1.5c show tracks which are obviously forked, but if one arm of the fork were very short it would be hard to decide whether there is a real forking or just an irregularity in the main track. Quantitative criteria for the identification of elementary processes require a sharp classification of events along the tracks. For example, it might be appropriate to classify as a sharp deflection one in which the track bends by 10° or more within 0.1 mm.

Actually, large-scale studies of elementary processes are preferentially carried out by means of counters, as illustrated schematically in Figure 1.6. Here, the counter scores only particles that have been deflected by an angle θ. The location of the deflection event and the value of θ are bracketed in this case by a system of collimating diaphragms in the apparatus. Each event registered by the counter is classified in accordance with the position of the counter, diaphragms, and foil.

The quantitative study of the events thus classified proceeds then by *counting* the number of events in each class. Specifically, one may count the number of events of a given class per incident particle, or the number per unit track length, or the number per unit time. These numbers are commonly referred to as the *frequencies* of relevant events. (Here the word "frequency" is used loosely in a statistical sense and is unrelated, of course, to the frequency of a periodic phenomenon.)

An important procedural step consists of testing whether the events of a particular class actually originate from collision of an incident particle with a single atom (or groups of atoms), that is, whether they are elementary processes or whether instead they originate from the cumulative interactions with two or more unrelated atoms. Consider that the criteria for defining the class of an event imply that the event has occurred within a specified volume of space. Events involving interaction with a single atom (or group of atoms) must occur with a frequency proportional to the number of atoms in that volume. Events involving two unrelated atoms have a frequency proportional to the number of pairs of atoms in the volume, and thus to the square of the number of atoms. Events involving n unrelated atoms depend on the nth power of the number of atoms. The test which identifies events originating from elementary processes is, therefore, whether their frequency is proportional to the number of atoms in the relevant volume. Hence the frequency of elementary processes must be proportional to the density of material traversed. For events observed on track pictures, the frequency must also be proportional to the total length of track examined; for events within a well-defined layer of material, for instance a metal foil, the frequency must be proportional to the thickness of the layer.

1.3.3. *Cross-sections.* The frequency of elementary processes in a given length of track is proportional to this length and to the density of the material, quantities which are incidental to the elementary process itself. Therefore, the significant quantity to be derived from the observation of elementary processes is their frequency per unit track length and per unit density of material. This frequency is then independent of density and track length.

Consider, for example, Rutherford's experiment for the study of large deflection of α-particles traversing a metal foil. A schematic diagram of the apparatus is shown in Figure 1.6. Tests with foils of different thicknesses show that large deflections (of ~5° or more) originate from elementary collision processes, provided the foil is sufficiently thin, of the order of 1 micron.

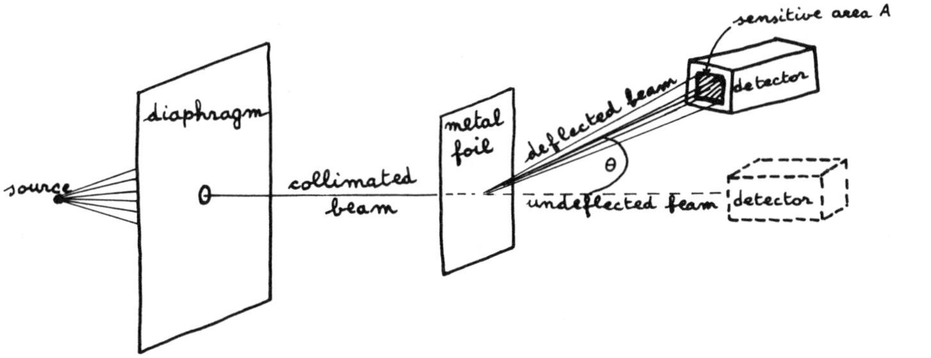

Fig. 1.6. Sketch of apparatus for observation of particles which have undergone large-angle deflection. After insertion of the metal foil the detector has been rotated by an angle θ in the horizontal plane.

As a preliminary, one determines the number N of α-particles per unit time in the incident beam by removing the foil and placing the detector in line with the undeflected beam. One then inserts the foil, places the detector at the desired angle θ, and counts the number n of α-particles received by the detector per unit time. The frequency of elementary processes thus scored is n/N and is proportional to the thickness t of the foil and to its density ρ expressed as number of atoms per unit volume. The ratio

$$\sigma = \frac{n/N}{t\rho} \tag{1.19}$$

represents the frequency of deflections per incident α-particle.

The significance of the frequency σ may be visualized by the following geometrical model. Suppose that each atom of the foil carries a target whose area is equal to σ and is much smaller than the atom itself. The aggregate area of all targets would constitute a fraction $\sigma t\rho$ of the area of the foil. If the foil is so thin that no target is screened by another one, this fraction is much smaller than unity. In this case the expected number of hits on target by a beam of N particles equals $N\sigma t\rho$, that is, the number n of counts in the detector. For this reason one talks of the frequency of collision processes *as though* it were the frequency of hits on targets of area σ. This figure of speech has proved convenient, and the ratio σ as defined by (1.19) is called the *effective target area* or, more commonly, the *effective cross-section* of an atom for the occurrence of the elementary process of interest.

The cross-section σ still depends on incidental details of the experiment, namely the sensitive area A of the detector and its distance r from the point of collision. In fact, σ must be proportional to the solid angle $\Omega = A/r^2$, if this angle is sufficiently small. The limiting ratio

$$\frac{d\sigma}{d\Omega} = \lim_{\Omega=0} \frac{\sigma}{\Omega} = \lim_{\Omega=0} \frac{n}{Nt\rho} \frac{r^2}{A} \tag{1.20}$$

is the parameter of greatest intrinsic significance to be derived from experimental data. This ratio represents the frequency—per unit track of the incident particle and per unit density of the material (one atom per unit volume)—of elementary processes from which the particle emerges traveling within a unit solid angle about a given direction. The frequency $d\sigma/d\Omega$ is called the *effective cross-section per unit solid angle* or also, briefly, the *differential cross-section* of the collision process under consideration. It depends, of course, on such variables as the angle between the directions of observation and of incidence, or the energy of the incident particles, but it is independent of N, t, ρ, A, and r, because any change of these variables is compensated by a change of the observed frequency n.

An alternative type of experiment, different from Rutherford's prototype, scores residual undeflected particles in a beam rather than those that have experienced specified deflections. This alternative experiment is performed as shown schematically in Figure 1.7. The diaphragm close to the detector serves to stop particles that have experienced small deflections.

Increasing the foil thickness increases the number of particle deflections and reduces the number of counts scored by the detector. More specifically, addition to the foil of successive layers of *equal thickness* reduces the counts by *equal fractions*, because the frequency of deflection within a layer is indepen-

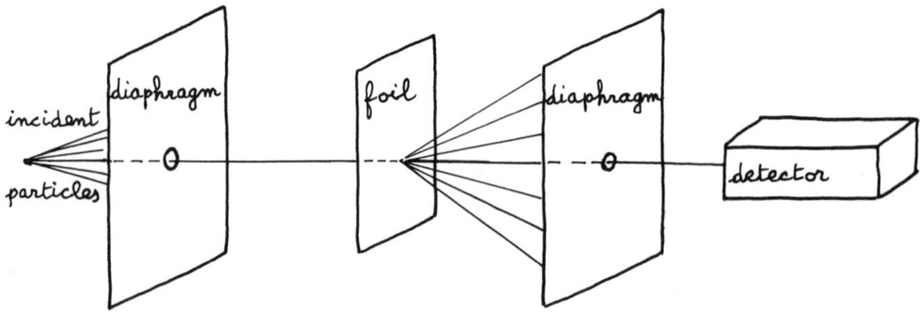

Fig. 1.7. Diagram of attenuation experiment.

dent of the existence of previous layers. It follows that the number N of undeflected particles in the beam decreases in each layer of infinitesimal thickness dt by an amount dN which is proportional to dt, to the density of atoms in the foil, and to N itself. The number of particles in the beam is now a function N(t) of the thickness of foil traversed, N(0) being the number of incident particles indicated previously by N. According to our reasoning the function N(t) obeys the differential equation

$$dN = -\sigma_T N\rho dt, \tag{1.21}$$

where σ_T is a proportionality coefficient discussed below. Integration of equation (1.21) yields the very important exponential *law of attenuation* of undeflected beams,

$$N(t) = N(0) \exp(-\sigma_T \rho t). \tag{1.22}$$

Experimental check of this law provides a sensitive test of whether the observable deflections actually constitute elementary processes.

When the density ρ is expressed as number of atoms per unit volume of the foil, the coefficient σ_T has the dimension of an area. It is then called the *total effective cross-section* for deflection of a particle by an atom or, more appropriately, for removal of a particle from the beam owing to collision with a single atom. When deflection is the only possible outcome of collision with an atom, σ_T must equal the integral of the deflection cross-section $d\sigma/d\Omega$ over all directions toward which deflection occurs. Measurements of σ_T are very accurate when the exponential attenuation can be measured until very few particles remain—for instance, until $N(t)/N(0) \lesssim 10^{-3}$. Note that this kind of experiment is not practicable with α-particles scattered by a foil because of the excessive number of small deflections which cannot be screened off the detector.

1.4. RUTHERFORD'S ATOM: A BODY OF ELECTRONS WITH A POINTLIKE NUCLEUS

Rutherford and his co-workers observed that when a beam of α-particles traverses a thin metal foil, a small fraction of the particles is sprayed, that is, *scattered* in all directions, even backward. Quantitative experimentation with this scattering process led Rutherford to a previously unsuspected conception of atomic structure. We review here the qualitative implications of this famous work by considering simultaneously the evidence provided by the observation of α-particle tracks in cloud chambers. (The metal foil in Rutherford's experiments usually consisted of a heavy element, such as gold, while cloud chambers are usually filled with gases of light elements.)

A large majority of α-particles travel in approximately straight lines; in a gas at nearly atmospheric pressure the track of a particle is several centimeters long and nearly straight until the very last few millimeters. The fraction of α-particles scattered with a deflection of the order of 25° depends on the material traversed and ranges from about 1 in 10^3 for light elements to 1 in 10 for heavy elements; deflections of the order of 90° occur about 100 times less frequently. Since each α-particle traverses as many as 100, 000 atoms, the interaction between an α-particle and an atom appears capable of causing a deflection of 90° only in 1 out of 10^{10} to 10^8 traversals, depending on the atomic weight of the element traversed.

All tracks, upon close inspection, appear to consist of multiple ionizations accompanied by small-angle deflections which build up to ~1°. About 10^5 ionizations may be observed along a track in a gas; they presumably dissipate the α-particle energy. Energy dissipation in numerous small steps and with very small deflections is just the effect to be expected from interactions between α-particles and atomic electrons in view of the energy and momentum conservation laws.[10] One concludes that the interaction between an α-particle—a He^{++} ion—and the massive, positively charged constituents of the gas atoms must be responsible for the rare large deflections.

Rutherford inferred from these facts that the massive, positively charged portions of an atom and of an incident He^{++} ion interact very seldom because both of them are extremely small. In fact, these *nuclei* of all atoms are so small that they may be considered pointlike for most purposes of atomic physics. The main interaction causing the rare large deflections should then be the electrostatic repulsion of the nuclei.

On this basis one can readily account semiquantitatively for the very low frequency of large deflections. It suffices to consider that the electric repulsion between the α-particle and the atomic nucleus can deflect a particle track sharply only if its potential energy attains a magnitude comparable to the kinetic energy of the incident particle, that is, several MeV. To estimate the distance at which this magnitude is attained, recall that the potential energy of two elementary charges separated by an atomic diameter is of the order of 10 eV. The α-particle carries two elementary charges, and the nucleus of a scatterer with atomic number Z carries Z charges. Hence the potential energy of the α-particle and nucleus at one atomic diameter amounts to $2Z \times 10$ eV, that is, to ~100 eV for a light atom and ~1, 000 eV for a heavy atom. Since the potential energy increases inversely as the distance, approach to within $\sim 1/10^5$ or $\sim 1/10^4$ of an atomic diameter will be necessary to produce a ~90° deflection by light or heavy atoms, respectively. This estimate accounts roughly for the observations on large-angle scattering.

1.4.1. *Rutherford's Calculation and Its Test.* Rutherford carried out a detailed calculation of α-particle scattering which yields the actual value of the cross-section for scattering as a function of the deflection angle and which lends itself to very extensive and successful testing against experimental data.

---

10. In a collision between two bodies with masses M and m, with $M \gg m$, and with the light body initially at rest, conservation of energy and momentum requires that the fractional change of momentum of the heavy body $\Delta p/p$ cannot exceed the order of magnitude m/M, which is about 1/7300 for an α-particle and an electron.

He regarded both α-particles and scattering nuclei as pointlike bodies and assumed that the deflection of α-particles is due to the electrostatic repulsion by nuclei.[11]

The force between an α-particle and a nucleus influences their motion with respect to one another. Its effect is conveniently calculated in a frame of reference attached to the center of mass of the two bodies. If the atomic nucleus is much heavier than the α-particle (an atom of Au is 50 times heavier than a He^{++} ion), the center of mass coincides approximately with the position of the nucleus; one may then calculate approximately as though the nucleus were at a fixed position and only the α-particle moved, whereas in practice the nucleus recoils. Here, we shall regard the nucleus as fixed, but the exact calculation differs only by a minor adjustment.[12]

The repulsion by the nucleus forces the α-particle to follow a path of the type indicated in Figure 1.8a. The Coulomb law of force between electric charges (whether attraction or repulsion) has the same analytical form as the law of gravitational attraction. In either case of repulsion or attraction, the incident particle follows a Kepler orbit which is a hyperbola.

This orbit depends on certain constants of the α-particle and of the nucleus, namely, the magnitudes ze and Ze of their charges and the mass m of the α-particle. (The charge of the α-particle is indicated here by ze, rather than 2e, so that the calculation may apply also to collisions of other particles; also, m represents actually the reduced mass of the colliding particles.) The orbit also depends on the conditions of incidence, namely, on the initial velocity v of the particle and on the distance b shown in Figure 1.8 and called the *impact parameter*. This is the distance of the initial line of flight of the α-particle from the atomic nucleus. The smaller b is, the larger becomes the repulsion and the larger the deflection. In the event of a head-on collision ($b = 0$) the α-particle continues in a straight line until its kinetic energy is spent and the repulsion forces it back in the direction from which it came ($\theta = 180°$).

The central part of Rutherford's calculation consists of determining the relation between deflection and impact parameter. Consider, for example, the two equations that give the equality of energy and of angular momentum of the α-particle at infinite distance from the nucleus and at the minimum distance, ρ, from it:

$$E = \tfrac{1}{2}mv^2 = \frac{zZe^2}{\rho} + \tfrac{1}{2}mv_{min}^2, \qquad (1.23)$$

$$J = mvb = mv_{min}\rho. \qquad (1.24)$$

---

11. The attraction exerted by atomic electrons on an α-particle counteracts the nuclear repulsion and is called the *screening effect*. It may, however, be disregarded in a calculation of large-angle scattering.

12. The adjustment consists of replacing the mass of the α-particle by the reduced mass of particle and nucleus (*BPC*, 1: 275) and of transforming the deflection angle from the center-of-mass frame to the laboratory frame (ibid., p. 172).

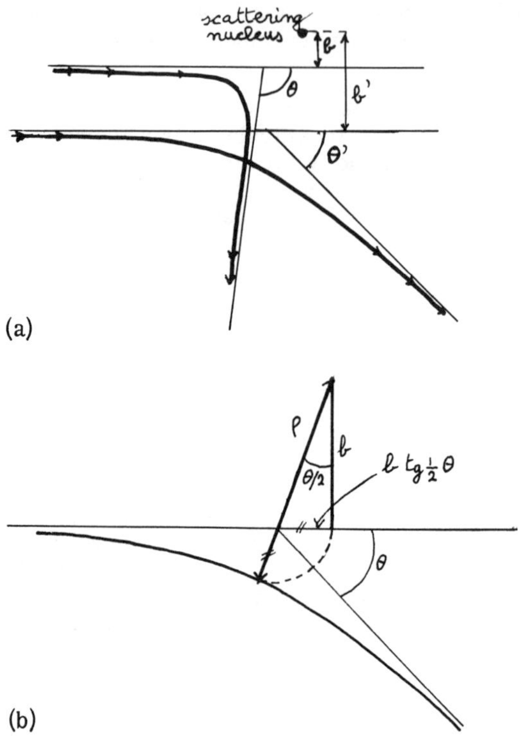

Fig. 1. 8. (a) Scattering of α-particles by a nucleus. The track with small impact parameter b is deflected by the large angle θ; the track with impact parameter b′ by θ'.
(b) Illustration of the relation between the impact parameter b and the distance of closest approach ρ.

Elimination of the velocity at ρ, v_{min}, between these two equation yields the quadratic relation between ρ and b

$$\rho^2 - 2\,\frac{zZe^2}{mv^2}\,\rho - b^2 = 0. \tag{1.25}$$

Solving for ρ gives

$$\rho = \frac{zZe^2}{mv^2} + \left[\left(\frac{zZe^2}{mv^2}\right)^2 + b^2\right]^{1/2}. \tag{1.26}$$

On the other hand, knowledge that the particle follows a hyperbolic path under the influence of Coulomb-law repulsion gives the relationship between ρ and b illustrated in Figure 1.8b:[13]

$$\rho = b \tan \tfrac{1}{2}\theta + b(1 + \tan^2 \tfrac{1}{2}\theta)^{1/2}. \tag{1.27}$$

Comparison of equations (1.26) and (1.27) establishes finally the connection between the deflection and the impact parameter

$$\tan \tfrac{1}{2}\theta = \frac{zZe^2}{mv^2 b}. \tag{1.28}$$

For α-particles of a given velocity scattered by atoms of a given kind, the deflection depends only on the impact parameter. All particles whose initial line of flight passes *within* a distance b from the nucleus are deflected by an angle larger than the value θ related to b by equation (1.28). Therefore, a circle of radius b, perpendicular to the line of flight and centered on the nucleus, constitutes an aiming target for collision processes with deflection larger than θ. Rutherford's calculation associates an *actual target* of definite size and shape to deflection by a given angle. The calculation and the underlying hypotheses are to be tested by comparing the calculated target sizes with the experimental cross-sections derived from the frequencies of large-angle scattering.

The comparison is performed conveniently for the differential cross-section given by equation (1.20) which pertains to scattering per unit solid angle with deflection θ. Consider scattering in all directions for which the deflection lies between θ and $\theta + \Delta\theta$. These directions fill a gap between two cones shown in Figure 1.9. The solid angle subtended by this gap is $\Delta\Omega = 2\pi \sin\theta\, \Delta\theta$.

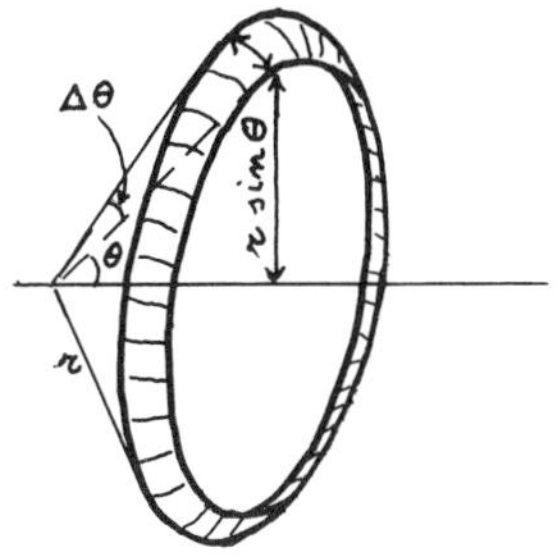

Fig. 1.9. Area $A = 2\pi r \sin\theta\, r\Delta\theta$ of a ring-shaped detector which intercepts, at a distance r from the point of scattering, all particles deflected by angles between θ and $\theta + \Delta\theta$. The detector covers a solid angle $\Omega = A/r^2 = 2\pi \sin\theta\, \Delta\theta$.

13. See, for instance, *BPC*, vol. 1, chap. 9.

Consider now the range of impact parameters of the particles that are scattered into the solid angle $\Delta\Omega$. According to equation (1.28) this range extends from

$$b = \frac{zZe^2}{mv^2} \frac{1}{\tan \frac{1}{2}\theta} \tag{1.29}$$

to $b - \Delta b$ where, by differentiation of equation (1.29),

$$\Delta b = \frac{zZe^2}{mv^2} \frac{1}{\sin^2 \frac{1}{2}\theta} \tfrac{1}{2}\Delta\theta. \tag{1.30}$$

The initial lines of flight of the particles with impact parameters in this range are aimed to fill an annular target of area $\Delta\sigma = 2\pi b\Delta b$. Therefore, the measurable differential cross-section (1.20) is calculated to be

$$\frac{d\sigma}{d\Omega} = \lim_{\Delta\Omega = 0} \frac{\Delta\sigma}{\Delta\Omega} = \lim_{\Delta\theta = 0} \frac{2\pi b\Delta b}{2\pi \sin\theta\, \Delta\theta}. \tag{1.31}$$

Upon substitution of b and Δb from equations (1.29) and (1.30) and replacement of $\sin\theta$ by $2 \sin \frac{1}{2}\theta \cos \frac{1}{2}\theta$, the theoretical cross-section (1.31) reduces to the Rutherford formula

$$\frac{d\sigma}{d\Omega} = \left(\frac{zZe^2}{2mv^2}\right)^2 \frac{1}{\sin^4 \frac{1}{2}\theta}. \tag{1.32}$$

An alternative form of this expression, namely,

$$\frac{d\sigma}{d\Omega} = \frac{(2mzZe^2)^2}{|\vec{p} - \vec{p}\,'|^4}, \tag{1.33}$$

emphasizes the dependence of the cross-section on the difference between the initial and final momenta, $\vec{p}$ and $\vec{p}\,'$, of the α-particle, that is, on the recoil momentum $|\vec{p} - \vec{p}\,'| = 2mv \sin \frac{1}{2}\theta$ received by the scatterer nucleus.

Notice in equation (1.32) that the factor $zZe^2/2mv^2$ equals one-quarter of the distance r between α-particle and nucleus at which the potential energy zZe^2/r equals the kinetic energy $\frac{1}{2}mv^2$. Notice also the very rapid decrease of $d\sigma/d\Omega$ as the deflection angle θ increases. This rapid variation permits a very critical test of the Rutherford formula.

The test was provided in 1913 with results shown in Table 1.1. Note in the table how the product $n \sin^4 \frac{1}{2}\theta$ remains nearly constant while the number of counts varies by a factor of 5,000. The slight increase in the product for small scattering angles may be due to the cumulative effect of small deflections.

In other experiments the number of observed particles scattered by small angles θ may fall below the value predicted by equation (1.32) owing to the screening effect of the attraction exerted by the atomic electrons on each

TABLE 1.1

Variation of Scattering with Angle

		Silver Foil		Gold Foil	
θ (degrees)	$(\sin^4 \frac{1}{2}\theta)^{-1}$	Particle Count, n	$n \sin^4 \frac{1}{2}\theta$	Particle Count, n	$n \sin^4 \frac{1}{2}\theta$
150	1.15	22.2	19.3	33.1	28.8
135	1.38	27.4	19.8	43.0	31.2
120	1.79	33.0	18.4	51.9	29.0
105	2.53	47.3	18.7	69.5	27.5
75	7.25	136	18.8	211	29.1
60	16.0	320	20.0	477	29.8
45	46.6	989	21.2	1435	30.8
37.5	93.7	1760	18.8	3300	35.3
30	223	5260	23.6	7800	35.0
22.5	690	20300	29.4	27300	39.6
15	3445	105400	30.6	132000	38.4

NOTE Data from Geiger and Mardsen, *Phil. Mag.* 25 (1913): 604.

α-particle. Departures are also observed for very large deflections when the impact parameter drops below 10^{-12} cm. These departures are attributed to nonelectric forces between the α-particle and the nucleus of an atom, which become appreciable for sufficiently close approach. A different type of departure from Rutherford's result, which occurs only in the scattering of α-particles by helium atoms, will be described in chapter 15.

1.4.2. *What Determines the Size, Shape, and Rigidity of Atoms?* Rutherford's discovery of the very small size of atomic nuclei set the problem of atomic structure and stability into a much more definite but altogether novel frame. Practically the whole volume of each atom, or indeed of any material, contains only electrons. Therefore, the size, stability, and impenetrability of any atom or aggregate of atoms should result from the physics of electrons subjected to electric attraction by positively charged nuclei. The main goal of our study consists then of formulating the physics of electrons so as to account for those gross properties of atoms.

The electrons that constitute the "body" of an atom are subject to their mutual repulsion and to the attraction by the nucleus, which thereby tends to contract the volume of the atom. Each kind of atom maintains, nevertheless, a rather well-defined volume, which shrinks hardly at all even under the influence of strong pressures. These facts suggest that the "body" of an atom has a tendency to expand and that its volume adjusts itself so that the tendency to expand balances the inward pull exerted by the nucleus. External pressures would then cause only small adjustments of this balance.[14]

This surmise is confirmed by data on the size of *isoelectronic sequences* of ions, that is, of sequences of ions of different elements but with equal numbers of electrons. (For example, F^-, Na^+, Mg^{++}, Al^{+++} constitute a sequence of ions with ten electrons.) Data on ion sizes are derived from the density and structure of crystals containing ions and show that the size of the "body" shrinks as the nuclear charge increases from one ion to the next in the sequence. The radii of F^-, Na^+, Mg^{++}, and Al^{+++} are, respectively, 1.36, 0.95, 0.65, and 0.50 Å. On the other hand, in the sequence of neutral atoms along the periodic system the nuclear attraction increases but the number of electrons also increases and the atomic volume shows no strong trend of uniform increase or decrease.

Notice that atoms of hydrogen, containing a single electron, have the same qualitative properties as other atoms. Hence the expansive tendency of the "body" of electrons appears not to result from interactions among different electrons but instead to be a property of even a single electron. The evidence considered thus far does not suffice to define this property more precisely or to relate it to other, more general, phenomena. This will be done progressively in the following chapters.

Following Rutherford's success in calculating the α-particle scattering by treating particles as pointlike bodies according to the ordinary methods of mechanics, it was a natural step to calculate the motion of electrons by a similar treatment. According to this point of view, an electron could move around a nucleus, under the influence of its electric attraction, on a stable circular or elliptical orbit. The centrifugal force of the motion on such an orbit would exactly balance the attraction and prevent the electron from "falling" onto the nucleus, just as it prevents planets from falling into the sun. This treatment of the mechanics of atomic electrons is called the *planetary model* of atomic structure.

The planetary model fails, however, in that it provides no reason why the radii of the orbits should be of any particular order of magnitude or why all atoms of the same element should have the same size. Moreover, the stability of orbits in the planetary model is deceptive, in that the model presupposes the absence of any substantial interaction with other physical systems. In fact, the solar system itself would collapse with a release of energy if the planets

---

14. Pressures available in the laboratory are small compared with the pull of nuclear attraction. For example, a pressure of 1,000 atmospheres/cm^2 performs work equal to 10^9 ergs per cm^3 of contraction of a material. Translated into atomic units, 10^9 ergs/cm^3 equals 6×10^{-4} eV/$Å^3$. The data in section 1.2 show that the nuclear attraction performs work at the rate of at least 10 eV/$Å^3$ of contraction of atomic volume.

were pushed in by an external pressure. Still further, the motion of an electron around a nucleus constitutes a high-frequency variable current which should act as an antenna and radiate energy away; this energy dissipation should lead by itself to rapid collapse of the electron onto the nucleus. Finally, atoms would be flat according to the planetary model, at least in the case of hydrogen, contrary to evidence.

The concepts and laws derived from macroscopic experiments thus appear inadequate to account for the properties of atoms, particularly for their stability. Bohr realized this failure very soon after Rutherford's discovery. In time, the interpretation of new experiments revealed physical relationships which are not brought out by macroscopic investigations. New concepts and laws were then formulated which account for atomic as well as macroscopic phenomena.

CHAPTER 1 PROBLEMS

1.1. Calculate the number of molecules per cm^2 if one gram of water were spread uniformly over the earth's surface.

1.2. A layer of BF_3 gas contains 0.003 grams per cm^2 of area. Calculate (a) its thickness, assuming normal pressure and temperature, (b) the number of B and F atoms per cm^2 of the layer, (c) the number of electrons per cm^2.

1.3. The isothermal compressibility coefficient of copper is

$$\kappa = -\frac{1}{V}\left(\frac{\partial V}{\partial P}\right)_T = 7.5 \times 10^{-7}\ \text{atm}^{-1}.$$

When a copper bar is compressed, its volume shrinks by a very small fraction and the work performed can be calculated, by disregarding second-order variations of V, as $W = -\int P dV \approx \kappa V \int P dP$. Calculate the work performed on each atom as the pressure over the bar increases from 1 atmosphere $= 1.01 \times 10^6$ dyne/cm^2 to 1000 atm.

1.4 A beam of particles of unit charge e, mass m, and velocity v travels along the x axis. Electric and magnetic fields of strength E and B and parallel to the z axis act over the interval $0 \leq x \leq d$, imparting to each particle small transverse momentum components δp_z and δp_y, respectively (see *BPC* vol. 1, chap. 4). (a) Calculate the magnitudes of δp_z and δp_y. (b) The particles are collected on a screen perpendicular to the x axis at $x = X >> d$. Show that particles with equal values of e/m and different v intersect the screen at points (X, Y, Z) lying on a parabola and obtain the equation of the parabola $Z = Z(Y)$, $X =$ constant.

1.5. The α-particles of a very thin source of polonium produce 1.2 scintillations per second per cm^2 of a screen at 1 m from the source in an evacuated vessel. Calculate the number of α-particles emitted per second by the source.

1.6. Calculate the current carried to the electrodes of a large ionization chamber when the source of problem 1.5 is placed in it. Consider that each α-particle dissipates an energy of 5.2×10^6 eV and that one positive and one negative ion are produced, on the average, per every 34 eV of energy dissipation.

1.7. In a hydrogen bubble chamber traversed by high-energy particles (π^- mesons) 60 events were recorded in which a track stopped abruptly and a pair of separate V-shaped tracks (due to Λ^0 and K^0 disintegrations) appeared as secondary effects. The events were observed over a track length of 21×10^3 m. Calculate the cross-section for the production of these events in the collision of the track particles with H atoms. (The hydrogen density in the chamber was 0.057 g/cm^3.)

1.8. A beam of monoenergetic electrons traverses a gas cell 5 cm long containing a gas at 27°C and 10^{-2} mm Hg pressure. (a) If 5% of the electrons fail to emerge in the initial direction, calculate the collision cross-section. (b) What fraction of electrons would fail to emerge in the initial direction if the gas density were raised by a factor of 10? (c) Keeping the gas density raised as above, calculate the length of the cell necessary to prevent 90% of the electrons from emerging in the initial direction.

1.9. A neutron beam from a nuclear reactor experiences a 2% loss of intensity when it traverses the BF_3 layer of problem 1.2 which contains 2.7×10^{19} molecules/cm^2. This loss increases to 5% when the concentration of the boron isotope of atomic weight 10 in the BF_3 is enriched from its normal proportion of 18% to 45%. Show that the neutron loss is due almost entirely to the isotope ^{10}B and calculate the cross-section of its collision with neutrons.

1.10. A beam of protons is incident on a gold foil of thickness 1 micron. (a) For a given incident proton energy, what scattering angle corresponds to the closest approach of the proton to the Au nucleus? (b) Determine the incident proton energy necessary for the proton to come within 2×10^{-12} cm of the nuclear center. (c) Using the incident energy derived in (b), calculate the cross-section for deflection by $\theta \geqslant 60°$ and the fraction of the beam deflected by $\theta \geqslant 60°$. (d) Using the incident energy derived in (b), calculate by means of the Rutherford formula the impact parameter for which the proton beam would be deflected by 1 second of arc. Why is this unreasonable?

1.11. A beam of α-particles of 5 MeV kinetic energy traverses a gold foil. One particle in 8×10^5 is scattered so as to hit a surface of 0.5 cm^2 at 10 cm from the point of traversal of the foil and in a direction at 60° from the beam axis. Calculate (a) the foil thickness t; (b) the distance ρ of closest approach to a nucleus by the particles which hit the detector's surface; (c) the change in the number of particles hitting the detector when the gold foil is replaced by a silver foil with an equal number of atoms per unit area.

SOLUTIONS TO CHAPTER 1 PROBLEMS

1.1. $N_A/(M_{H_2O}4\pi R^2) = 6.0 \times 10^{23}/[18 \times 4\pi \times (6.4 \times 10^8)^2] = 6500$ molecules/cm^2.

1.2 (a) $(0.003\ g/cm^2) \times (2.24 \times 10^4\ cm^3/mole)/M_{BF_3} = 1.0$ cm; (b) $n_B = 0.003\ N_A/M_{BF_3} = 2.7 \times 10^{19}$; $n_F = 3n_B = 8.0 \times 10^{19}$; (c) $5n_B + 9n_F = 8.5 \times 10^{20}$.

1.3 $V_{atom} = A_{Cu}/\rho_{Cu}N_A = 63.5\ g\ mole^{-1}/[(8.9\ g\ cm^{-3}) \times (6.0 \times 10^{23}\ atoms\ mole^{-1})] = 1.2 \times 10^{-23}\ cm^3/atom$; $W = \frac{1}{2}\ \kappa\ V_{atom}[P_2{}^2 - P_1{}^2] = 4.5 \times 10^{-18}$ erg/atom $= 2.8 \times 10^{-6}$ eV/atom.

1.4 (a) $\delta p_x = eEd/v$, $\delta p_y = -e(v/c)Bd/v = -eBd/c$; (b) $Y \sim X\delta p_y/mv = -XeBd/mvc$; $Z \sim X\delta p_z/mv = XeEd/mv^2$. Eliminate $v = -(eBd/mc)X/Y$; $Z = (m/e)\ (Ec^2/B^2dX)Y^2$.

1.5. $n = (1.2\ sec^{-1}\ cm^{-2}) \times 4\pi(10^2\ cm)^2 = 1.5 \times 10^5\ sec^{-1}$.

1.6. $i = ne \times 5.2 \times 10^6\ eV/(34\ eV) = (1.5 \times 10^5\ sec^{-1}) \times (1.6 \times 10^{-19}\ coulombs) \times [(5.2 \times 10^6)/34] = 3.7 \times 10^{-9}$ amperes.

1.7. Apply equation (1.19), considering that here Nt represents the observed track length. $\sigma = 60/[(2.1 \times 10^6\ cm) \times (0.057\ g\ cm^{-3}) \times (6.0 \times 10^{23}\ atoms\ g^{-1})] = 8 \times 10^{-28}\ cm^2/atom$.

1.8. Apply equation (1.22). (a) $\rho = N_A\ (2.24 \times 10^4\ cm^3\ mole^{-1})^{-1}[(10^{-2}\ mm\ Hg)/(760\ mm\ Hg)] \times [(300°K)/(273°K)] = 3.2 \times 10^{14}$ molecules/cm^3; $\sigma = -(\rho t)^{-1} \ln 0.95 \sim 0.051\ (3.2 \times 10^{14} \times 5)^{-1} = 3.2 \times 10^{-17}\ cm^2/atom$. (b) $N(t)/N(0) = \exp(-3.0 \times 10^{-17} \times 3.2 \times 10^{15} \times 5) = \exp(-0.51) = 0.61$; 40% of the electrons fail to emerge. (c) $t = -(\sigma_T\rho)^{-1} \ln(0.10) = 2.3(3.1 \times 10^{-17} \times 3.2 \times 10^{15})^{-1} = 22$ cm.

1.9. ^{10}B must be responsible because the neutron absorption is proportional to the ^{10}B concentration (2: 5 = 18: 45). ^{10}B cross-section equals fractional absorption divided by ^{10}B absolute concentration, i.e., $(0.02)/(0.18 \times 2.7 \times 10^{19}) = 4 \times 10^{-21}\ cm^2/atom$. (Note that this cross-section for a nuclear process is much larger than the nucleus itself.)

1.10. (a) Closest approach, ρ_{min}, occurs when the particle comes to rest in a head-on collision with $\theta = 180°$. When it comes to rest all of its initial kinetic energy has been transformed into potential energy. (b) $E = \frac{1}{2}mv^2 = zZe^2/\rho_{min} = 1 \times 79 \times (4.8 \times 10^{-10}\ esu)^2/2 \times 10^{-12}\ cm = 9.1 \times 10^{-6}\ ergs = 5.7$ MeV. (c) Use equation (1.28) and the preceding formulae. $\sigma(\theta \geqslant 60°) = \pi[b(\theta = 60°)]^2 = \pi[zZe^2/mv^2 \tan 30°]^2 = \pi[(\frac{1}{2}\rho_{min})/\tan 30°]^2 = 9.4 \times 10^{-24}$ cm^2/atom. From equation (1.19), $n/N = \sigma\rho_{Au}t = 9.4 \times 10^{-24}\ [6.0 \times 10^{23} \times 19.3/197] \times 10^{-4} = 5.5 \times 10^{-5}$. (d) $b(\theta = 1'') = \frac{1}{2}\rho_{min}/\tan 0.5'' = 10^{-12}/$

$(2.4 \times 10^{-6}) = 4 \times 10^{-7}$ cm. Rutherford formula is not applicable at this distance larger than the atomic radius.

1.11. (a) Combine equations (1.20) and (1.32), substituting $n/N = 1/(8 \times 10^5)$, density $\rho = (6.0 \times 10^{23} \times 19.3/197)$ atoms/cm^3, $A/r^2 = 5 \times 10^{-3}$, $z = 2$, $Z = 79$, $e = 4.8 \times 10^{-10}$ esu, $\frac{1}{2}mv^2 = 5 \times 10^6$ eV $= 8.0 \times 10^{-6}$ ergs, $\sin \frac{1}{2}\theta = \frac{1}{2}$. This gives $t = 2\mu$. (b) Combine equations (1.27) and (1.29) to find the distance of closest approach $\rho = 6.8 \times 10^{-12}$ cm. (c) According to equation (1.32) the number is reduced in the ratio $Z_{Ag}^2/Z_{Au}^2 = (47/79)^2 = 1/2.8$.

2. macroscopic interaction of electromagnetic radiation with matter

The propagation of light, X-rays, and radiofrequency radiation through matter—that is, through assemblies of atoms—provides many clues to atomic mechanics. So does the absorption and emission of these radiations. We discuss in this chapter some main facts emerging from macroscopic observations, and their interpretation. Later, we shall deal with the essential additional facts and concepts that emerge from atomistic observations.

Much of the information to be discussed here was gathered and organized about the beginning of this century within the frame of the *electron theory* of matter, due primarily to H. A. Lorentz. The theory relates to the atomic model of J. J. Thomson, according to which electrons are held elastically at equilibrium positions in a continuous distribution of positive charge. Even though this model did not prove fruitful for studies of atomic structure, most results on the macroscopic electromagnetic behavior of matter derived from it stand today unchanged. This means that the basic assumptions of the electron theory of matter are independent of the unwarranted features of the Thomson–Lorentz model.

2.1. MATTER AS AN ELECTROMAGNETIC MEDIUM[1]

Common experience shows that light and radio waves propagate through air, glass, and many other materials much in the same way as they do in empty space. These materials thus appear macroscopically as continuous and homogeneous in spite of their atomic structure and even though the electric charges within each atom must interact with electromagnetic disturbances. They appear homogeneous because the phenomena under study depend not on the reaction of individual atoms or atomic particles to disturbances but only on the collective reaction of large numbers of atomic systems. To estimate these numbers, consider that the electric and magnetic fields of visible light remain nearly uniform over distances of the order of $\lambda/2\pi \sim 10^{-5}$ cm; a volume of $(10^{-5})^3$ cm^3 contains about 10^8 atoms of a dense material and about 30,000 molecules of any gas at normal pressure.[2] In this chapter we deal with information on atoms provided by radiation processes in which matter appears as a homogeneous medium; effects of departures from homogeneity will be the subject of chapter 4.

---

1. *BPC*, vol. 3, supplementary topic 9 covers much of the subject of this section.

2. The upper atmosphere has low density and therefore does not constitute a very homogeneous medium, especially for the propagation of blue light whose wavelength is short. This lack of homogeneity causes the upper atmosphere to scatter blue light; hence, the sky looks blue.

The properties of matter that behaves as a homogeneous medium bear directly only on populations of atoms or molecules, not on the behavior of individual atomic systems. Yet they are quite relevant to individual atoms or molecules in a statistical sense, particularly so because all molecules of a pure substance are identical.[3] (By contrast the statistics of, for example, a human population bear on an idealized "average man," individuals being actually different.) Reliance on the observation of statistical properties of atoms is, in fact, a systematic feature of atomic physics rather than an incidental limitation imposed here by our focusing on macroscopic properties of matter.

The study of atomic systems through the propagation of radiation in matter is helped by several circumstances. First, the large mass difference between electrons and nuclei causes them to react preferentially to radiations in different frequency ranges. Here we shall deal primarily with visible and ultraviolet light, i.e., with radiation of frequency so high that nuclei are unable to respond appreciably to it owing to their inertia. Hence we deal with effects traceable to atomic electrons; the response of nuclei, primarily to infrared and lower-frequency radiation, will be considered in chapter 20.

Second, the electric fields within atoms, due to the charges of nuclei and electrons and evaluated in section 1.2, have a strength $\gtrsim 1$ volt/Å $= 10^8$ volt/cm which greatly exceeds the strength of radiation fields.[4] Therefore, radiation causes only small disturbances of the equilibrium of electrons within matter. These disturbances can thus be properly treated in a linear approximation.

Finally, the speed v of atomic electrons is normally much lower than the speed c of light propagation. Therefore, the magnetic effects of radiation, being proportional to v/c, remain negligible compared with the electric effects. Thus we deal, in effect, only with the linear responses of electrons to high-frequency electric fields.

The homogeneity of matter and the linearity of its response to radiation permit a treatment of this response by Fourier analysis. This means that it is sufficient to consider the response of matter to plane monochromatic waves of various frequencies. It also means that the differential equations pertaining to the propagation of radiation in matter reduce to algebraic form. Concepts and techniques of Fourier analysis, as used in this section and throughout this book, are outlined in Appendix A. The outline emphasizes the systematic exploitation of linearity and of invariances in space and time.

2.1.1. *Electric Polarization and Light Propagation.* The force exerted by a field $\vec{E}$ upon the electrons within the molecules of a material causes a relative displacement of the negative and positive charges. Thereby a unit volume of

---

3. The small mass differences among isotopes are negligible here.

4. High-intensity laser light is exceptional in this respect, but we do not consider its propagation here.

material acquires an electric dipole *polarization*[5] with moment

$$\vec{P} = N\vec{p} = \chi_e \vec{E}, \tag{2.1}$$

where N indicates the number of molecules per unit volume, $\vec{p}$ the average moment acquired by a molecule, and the proportionality coefficient χ_e is called the *electric susceptibility* of the material.

For the purpose of studying the propagation of plane monochromatic waves, it suffices to consider the polarization $\vec{P}$ produced by an electric field $\vec{E}$ which varies sinusoidally in time and can, therefore, be represented as $\vec{E} = \vec{E}_0 \times \exp(-i\omega t)$.[6] The time dependent polarization $\vec{P}$ performs then driven oscillations represented by

$$\vec{P} = \vec{P}_0 e^{-i\omega t} = N\vec{p}_0 e^{-i\omega t} = \chi_e(\omega)\vec{E}_0 e^{-i\omega t}. \tag{2.2}$$

The amplitude $\vec{P}_0$ of these oscillations is proportional to $\vec{E}_0$:

$$\vec{P}_0 = N\vec{p}_0 = \chi_e(\omega)\vec{E}_0. \tag{2.3}$$

The susceptibility function $\chi_e(\omega)$ is a characteristic property of a material, dependent on its atomic structure. It affects the propagation of radiation and can be determined experimentally by observing the propagation of monochromatic light of various frequencies. Note that it has not been necessary to consider any variation in space of $\vec{E}_0$ because the material is taken to be homogeneous. For the same reason $\chi_e(\omega)$ is independent of space variables.

As in all phenomena of driven oscillations (see p. 530-31 and *BPC*, vol. 1, pp. 211 ff.) the phase of the polarization $\vec{P}$ lags behind the phase of the driving field $\vec{E}$. The lag is represented by the fact that $\chi_e(\omega)$ is complex and is given by the phase of this complex number.

The connection between polarizability and light propagation emerges from analysis of the Maxwell equation

$$\vec{\nabla} \times \vec{B} = \frac{1}{c}\frac{\partial \vec{E}}{\partial t} + \frac{4\pi}{c}\vec{j}. \tag{2.4}$$

Here the current density $\vec{j}$ is regarded as consisting of two parts, namely, externally driven currents, called $\vec{j}_{ext}$, which constitute a source of radiation in the region of interest and the current $\vec{j}_{matter}$ induced by the action of the radiation upon the material. This induced current density is the rate of change of the polarization $\vec{P}$ in the course of time. Accordingly, for the case of mono-

---

5. See, e.g., *BPC*, vol. 2, chap. 9.

6. When a complex exponential formulation is used in this chapter, it is understood that physical significance is attributed to the real part of each time-dependent equation (see p. 529). For instance, in equation (2.2) the significant quantity is the real part of the product $\chi_e(\omega)\exp(-i\omega t)$ rather than the product of the real parts of the factors.

chromatic radiation with time variation expressed as in equation (2. 2), we have

$$\vec{j}_{\text{matter}} = \frac{\partial \vec{P}}{\partial t} = \chi_e(\omega)\frac{\partial \vec{E}}{\partial t}\,. \tag{2.5}$$

Substitution of $\vec{j} = \vec{j}_{\text{matter}} + \vec{j}_{\text{ext}}$ in equation (2. 4) yields now

$$\vec{\nabla} \times \vec{B} = \frac{1}{c}\frac{\partial \vec{E}}{\partial t} + \frac{4\pi}{c}\left[\chi_e(\omega)\frac{\partial \vec{E}}{\partial t} + \vec{j}_{\text{ext}}\right]; \tag{2.6}$$

that is,

$$\vec{\nabla} \times \vec{B} = \frac{1}{c}\,\epsilon(\omega)\frac{\partial \vec{E}}{\partial t} + \frac{4\pi}{c}\,\vec{j}_{\text{ext}}\,, \tag{2.7}$$

where

$$\epsilon(\omega) = \epsilon'(\omega) + i\epsilon''(\omega) = 1 + 4\pi\chi_e(\omega) \tag{2.8}$$

indicates the dielectric constant with real part $\epsilon'(\omega)$ and imaginary part $\epsilon''(\omega)$. Formally, the transformation of the Maxwell equation (2. 4) into equation (2. 7) has replaced any explicit consideration of the current $\partial\vec{P}/\partial t$, induced in the material and proportional to $\partial\vec{E}/\partial t$, by the introduction of the dielectric constant ϵ.

The effect of the polarization current $\partial\vec{P}/\partial t$ upon the propagation of light can be established by observing how the insertion of $\epsilon(\omega)$ into equation (2. 7) modifies the derivation of the properties of plane monochromatic waves. (This derivation is given on p. 530). Briefly, in the treatment for empty space the entire current density $\vec{j}$ is set at zero, with the understanding that the radiation originates from an infinitely distant source. Here, the analogous condition sets $\vec{j}_{\text{ext}} = 0$ in equation (2. 7). The propagation of light is described by the interplay of equation (2. 7) and of its companion equation $\vec{\nabla} \times \vec{E} = -(1/c)\partial\vec{B}/\partial t$, which is unaffected by polarization currents. In empty space, where $\epsilon = 1$, the two companion equations require the wavenumber k and the frequency ω of the plane wave,

$$\vec{E} = \vec{E}_0 e^{i(\vec{k}\cdot\vec{r} - \omega t)}, \quad \vec{B} = \vec{B}_0 e^{i(\vec{k}\cdot\vec{r} - \omega t)} \tag{2.9}$$

to obey the basic relationship $k^2 = \omega^2/c^2$. For $\epsilon \neq 1$, one finds instead

$$k^2 = \epsilon(\omega)\frac{\omega^2}{c^2} = [\epsilon'(\omega) + i\epsilon''(\omega)]\frac{\omega^2}{c^2}\,. \tag{2.10}$$

The introduction of $\epsilon \neq 1$, which changes the ratio k/ω, also modifies the relative strengths of the magnetic and electric fields, namely, the ratio

$$\frac{E_0}{B_0} = \frac{\omega}{ck} = \left(\frac{1}{\epsilon(\omega)}\right)^{1/2}. \tag{2.11}$$

This ratio of field strengths is called the *characteristic impedance* of the material.[7]

Recall that we are dealing with the propagation of monochromatic waves with real frequencies ω. Accordingly, we interpret equation (2.10) as showing that polarization currents affect the propagation of a plane monochromatic wave by changing its wavenumber k and making it complex unless $\chi_e(\omega)$ and $\epsilon(\omega)$ happen to be real. The wavenumber k is then expressed as

$$k = k' + i\kappa,\ k'^2 - \kappa^2 = \epsilon'(\omega)\frac{\omega^2}{c^2},\ 2k'\kappa = \epsilon''(\omega)\frac{\omega^2}{c^2}. \qquad (2.12)$$

The wavelength is accordingly related to the frequency[8] $\nu = \omega/2\pi$ by

$$\lambda = \frac{2\pi}{k'} = \frac{c}{\nu n} = \frac{c}{\nu \mathrm{Re}[\epsilon(\omega)]^{1/2}}, \qquad (2.13)^9$$

where $n = \mathrm{Re}[\epsilon(\omega)]^{1/2}$ represents the refractive index and equals $\epsilon'^{1/2}$ approximately, unless $\epsilon''(\omega)$ is unusually large. The imaginary part of k, κ, represents an *attenuation coefficient* because the strength of the radiation fields $\vec{E}$ and $\vec{B}$ is seen to decrease along the direction of $\vec{k}$ upon insertion of the complex $\vec{k} = (k' + i\kappa)\hat{k}$ in the plane-wave representation (2.9).

To describe the propagation of a plane wave in homogeneous matter realistically, we write the real part of the plane-wave representation (2.9) substituting in it for k its expression $k' + i\kappa$, and the value of B_0/E_0 from equation (2.11). For simplicity, we also chose the x coordinate axis parallel to the wave vector $\vec{k}$ and the y and z axes along the directions of the electric and magnetic fields, respectively. The plane monochromatic wave is then represented by

$$\mathrm{Re}(\vec{E}) = \{0,\ E_0 e^{-\kappa x}\cos[k'x - \omega t],\ 0\},$$
$$\mathrm{Re}(\vec{B}) = \{0,\ 0,\ |\epsilon(\omega)|^{1/2}\, E_0 e^{-\kappa x}\cos[k'x - \omega t + \arctan(\kappa/k')]\}. \qquad (2.14)$$

The radiation is attenuated in the direction of $\vec{k}$, as indicated by the decrease of its field strengths and of the energy flux (Poynting vector) $c\vec{E} \times \vec{B}/4\pi$.[10] This attenuation is due to the absorption of energy and is rapid when $\epsilon''(\omega)$ is large. (The energy absorption can be calculated as work performed by the

---

7. This chapter deals with nonmagnetic materials; that is, it assumes that the magnetic permeability $\mu = 1$. In general the impedance is $(\mu/\epsilon)^{1/2}$ (see *BPC*, 3: 205, 575).

8. For brevity the word "frequency" will refer indiscriminately to the quantity indicated by ω and expressed in radians/sec and to the quantity indicated by ν and expressed in cycles/sec.

9. The symbol "Re" stands, as usual, for "real part of".

10. *BPC*, 3: 360.

field $\vec{E}$ on the current density $\vec{j}_{matter}$.) In addition, equations (2.12) and (2.14) show that the attenuation coefficient κ influences the relationship between k′, ω, and ϵ', and the ratio of the strengths of $\vec{B}$ and $\vec{E}$. It also introduces a phase lag in the oscillations of $\vec{B}$ with respect to those of $\vec{E}$. These effects are often minor, insofar as $\kappa/k' \ll 1$, but the attenuation $\exp(-\kappa x)$ becomes always important for sufficiently large values of x. Of course, one deals frequently with combinations of materials and frequencies for which attenuation is minimal, as, for example, for visible light in glass. However, there are wide frequency bands in which all dense materials absorb radiation strongly.

For materials and for frequency ranges such that ϵ' is negative, the plane wave (2.14) represents a phenomenon quite different from propagation in empty space. According to equations (2.12) the attenuation κ becomes larger than k′ under these circumstances, irrespective of the value of ϵ''. There results a strong attenuation over a single wavelength, since $\kappa/k' = \kappa\lambda/2\pi > 1$, that is, inability of the radiation to penetrate the material.[11] The occurrence of this phenomenon will be discussed below.

In the absence of strong attenuation, a main feature of radiation propagation through a material is the dependence of the velocity of propagation ω/k' on the frequency of a sinusoidal wave. This dependence is called *dispersion*, because it enables an experimentalist to sort out monochromatic components with different frequencies by dispersing them into different directions as radiation passes from one material to another.[12] Apart from this effect of refraction on an interface, dispersion has a profound influence even on the rectilinear propagation of nonmonochromatic radiation in homogeneous materials. The various monochromatic components fall out of step with one another in the course of propagation. The resulting changes of the radiation are difficult to survey in general, but can be established easily for nearly monochromatic radiation, that is, for radiation whose frequency spectrum is confined to a narrow bandwidth. In this case, the superposition of plane waves with slightly different frequencies yields a *wave packet* represented by a plane wave with amplitude modulation. The plane wave has frequency ω, wavenumber k′, and attenuation parameter κ, each of which equals an average over the corresponding values for the superposed monochromatic components. However, the amplitude-modulation factor propagates with the *group velocity* $d\omega/dk'$ which differs from the phase velocity ω/k' (see pp. 536-37 and references quoted there).

2.1.2. *Connection with Single Atom Properties*. The remainder of this chapter deals with low-density monatomic gases or vapors, i.e., with the type of material whose study provides data on the atoms of each element. In this case, specifically in the low-density limit where the polarization $\vec{P}$ in equation (2.1) is proportional to the density N of atoms, the interaction between driven oscillations of different atoms is negligible and the average dipole moment $\vec{p}$ may

---

11. *BPC*, 3: 136, 574.

12. This involves the concept of refraction as described by Snell's law and related equations (*BPC*, vol. 3, chap. 7, section 11.)

be regarded as the response of a single atom to the field $\vec{E}$, at least on a statistical basis.

Insofar as an atom has a number of internal variables, its elastic distortions should be represented appropriately as superpositions of alternative *normal modes*, 1, 2, ..., r, ...[13] Each normal mode is capable of free harmonic oscillation with a characteristic (resonance) frequency ω_r. Accordingly, the driven oscillations $\vec{p}(t) = \vec{p}_0 \exp(-i\omega t)$ considered in equation (2.2) will be expressed in the form

$$\vec{p}(t) = \Sigma_r \vec{p}_r(t) = \Sigma_r \vec{p}_{0r} e^{-i\omega t}. \tag{2.15}$$

The equation for the driven oscillation of the rth normal mode is

$$\frac{d^2\vec{p}_r}{dt^2} + \Gamma_r \frac{d\vec{p}_r}{dt} + \omega_r^2 \vec{p}_r(t) = b_r \vec{E}_0 e^{-i\omega t}, \tag{2.16}$$

where the significance of the *damping parameter* Γ_r and of b_r remains to be discussed. The solution of this equation, as outlined, for example, in Appendix A, shows that each amplitude p_{0r} of equation (2.15) is proportional to the field amplitude E_0 in accordance with

$$\vec{p}_{0r} = \frac{b_r}{\omega_r^2 - \omega^2 - i\Gamma_r\omega} \vec{E}_0. \tag{2.17}$$

Combination of equation (2.17) with equations (2.15), (2.3), and (2.8) yields

$$\epsilon(\omega) = 1 + 4\pi N \Sigma_r \frac{b_r}{\omega_r^2 - \omega^2 - i\Gamma_r\omega}, \tag{2.18}$$

$$\begin{aligned} \epsilon'(\omega) &= 1 + 4\pi N \Sigma_r b_r \frac{\omega_r^2 - \omega^2}{(\omega_r^2 - \omega^2)^2 + \Gamma_r^2\omega^2}, \\ \epsilon''(\omega) &= 4\pi N \Sigma_r \frac{b_r \Gamma_r \omega}{(\omega_r^2 - \omega^2)^2 + \Gamma_r^2\omega^2} \end{aligned} \tag{2.19}$$

One can actually establish the dielectric constant representation (2.18) utilizing only general analytic properties of its frequency dependence. However, the consideration utilized here—namely, that each atom has well-defined normal modes of oscillation—implies that individual terms in the Σ_r give physically identifiable contributions to equation (2.18). We shall see in the next section that this expectation is indeed verified to a substantial extent. Experimental measurements, of the refractive index ck'/ω and of the attenuation coefficient κ as functions of ω, yield plots of $\epsilon'(\omega)$ and $\epsilon''(\omega)$ which can be fitted by the analytical expressions (2.19), thereby providing values of the atomic parameters ω_r, b_r, and Γ_r for each species of atom.

---

13. See Appendix B.

The parameter b_r in equation (2.17) is indicated as a scalar rather than a tensor— so that $\vec{p}$ is parallel to $\vec{E}$—on the assumption that atoms are isotropic, at least on the average over all their possible orientations. We have no plausible basis at this point for interpreting b_r and Γ_r significantly, but it is instructive to consider what they would represent if we adopted one or another schematic model of atomic structure. In an extreme Thomson–Lorentz model, each normal mode corresponds to the oscillation of a single electron of mass m_e which is held to its equilibrium position by an elastic force with force constant k_r and which is subject to friction with coefficient η_r. The driven oscillations of the electron's position $\vec{x}_r$ obey an equation that coincides with equation (2.16) if one sets $\vec{p}_r = -e\vec{x}_r$ and

$$\Gamma_r = \frac{\eta_r}{m_e}, \quad \omega_r^2 = \frac{k_r}{m_e}, \quad b_r = \frac{e^2}{m_e}. \tag{2.20}$$

Alternatively, one might regard atoms as containing oscillating electric circuits with induction L_r, capacity C_r, resistance R_r, and with antennas of length ℓ_r. Here again the equation for the oscillating current i_r in the circuit coincides with equation (2.16) if one sets $i_r\ell_r = d\vec{p}_r/dt$ and

$$\Gamma_r = \frac{R_r}{L_r}, \quad \omega_r^2 = \frac{1}{L_rC_r}, \quad b_r = \frac{\ell_r^2}{L_r}. \tag{2.21}$$

Irrespective of models, the coefficient Γ_r includes a contribution due to the fact that the oscillating dipole $\vec{p}_r(t)$ acts as a radio antenna and dissipates energy into the radiation field surrounding it. If the dipole oscillates with a frequency of ω radian/sec, this contribution is[14]

$$\Gamma_{r,rad} = \frac{2\omega^2}{3c^3} b_r. \tag{2.22}$$

Additional contributions to Γ_r may arise from dissipation of energy to other mechanical degrees of freedom within the material.

In the high-frequency limit the response of atoms to radiation must be limited by the inertia of electrons, irrespective of any forces that hold them within atoms.[15] That is, any material should behave in the high-ω limit as a gas containing NZ electrons per unit volume (N atoms with Z electrons each). In this case the position $\vec{x}_r$ of each electron obeys the equation

$$m_e \frac{d^2\vec{x}_r}{dt^2} = e\vec{E}_0 \exp(-i\omega t), \tag{2.23}$$

---

14. See *BPC*, 3: 377.

15. This requirement played a major role in Heisenberg's development of quantum mechanics.

from which we have

$$\vec{p}_0 = -\frac{Ze^2}{m_e\omega^2}\vec{E}_0 \tag{2.24}$$

for each atom and

$$\epsilon(\omega) = 1 - \frac{4\pi NZe^2}{m_e\omega^2}\,. \tag{2.25}$$

Note that the ratio $\vec{p}_0/\vec{E}_0$ (and therefore the susceptibility χ_e) is negative for a system of free charged particles (see *BPC*, 1: 217). On the other hand, we have from equation (2.18)

$$\epsilon(\omega) \xrightarrow[\omega \to \infty]{} 1 - 4\pi N\Sigma_r \frac{b_r}{\omega^2}\,. \tag{2.26}$$

Consistency of expressions (2.25) and (2.26) establishes the requirement

$$\Sigma_r b_r = Ze^2/m_e, \tag{2.27}$$

irrespective of any details of atomic mechanics. This circumstance suggests that we write

$$b_r = \frac{e^2}{m_e} f_r, \quad \Sigma_r f_r = Z, \tag{2.28}$$

where the dimensionless coefficients f_r—called *oscillator strengths*—replace the coefficients b_r conveniently.[16]

In the opposite limit of low frequency, equation (2.18) reduces to

$$\epsilon(\omega) \xrightarrow[\omega \to 0]{} 1 + 4\pi N\Sigma_r \frac{b_r}{\omega_r^2}\,. \tag{2.29}$$

Thus we see that the measurement of the static ("zero-frequency") dielectric constant determines the value of the static *atomic polarizability*

$$\alpha = \frac{\Sigma_r \vec{p}_{0r}}{\vec{E}_0} \xrightarrow[\omega \to 0]{} \Sigma_r \frac{b_r}{\omega_r^2} = \frac{e^2}{m_e} \Sigma_r \frac{f_r}{\omega_r^2}\,. \tag{2.29a}$$

Finally let us point out that $\epsilon(\omega)$ varies sharply whenever ω approaches one of the ω_r (see Fig. 2.1), with a behavior characteristic of all oscillations driven near resonance. The striking variations of $\epsilon'(\omega)$, with alternate sign for $\omega \lessgtr \omega_r$, constitute the optical phenomenon of *anomalous dispersion.* The nega-

---

16. The condition (2.27), especially in the form $\Sigma_r f_r = Z$, is called a *sum rule*.

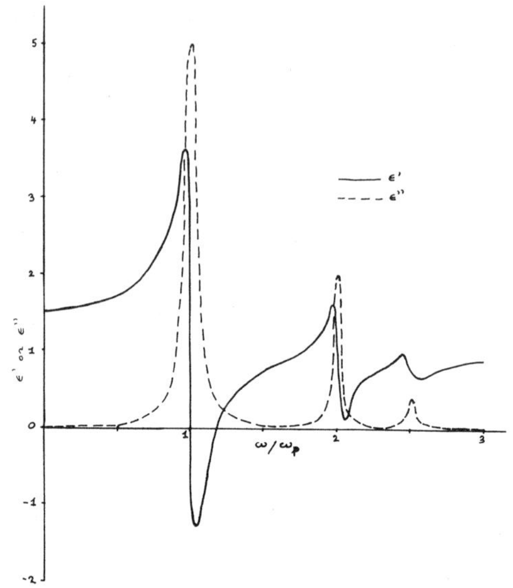

Fig. 2. 1. Plot of $\epsilon'(\omega)$ and $\epsilon''(\omega)$ for a model with three frequencies ω_r: $\omega_1 = \omega_p, \omega_2 = 2\omega_p, \omega_3 = 2.5\omega_p$, where $\omega_p = (4\pi Ne^2/m_e)^{1/2}$.

tive values of $\epsilon'(\omega)$ for a range of $\omega > \omega_r$ prevent radiation of these frequencies from entering the material, as noted above. The range of occurrence of this phenomenon is extremely narrow in gases or vapors but extends through the visible spectrum for solid metals whose lowest resonance frequency ω_1 equals zero and whose coefficient b_1 is large. Hence this phenomenon is called *metallic reflection.* On the other hand, $\epsilon''(\omega)$ is generally small owing to the smallness of the Γ_r but peaks sharply near each resonance, where $|\omega - \omega_r| \lesssim \Gamma_r$; here light absorption becomes intense, as we now proceed to describe.

2.2. ABSORPTION SPECTRA

Experimental study of the propagation of monochromatic electromagnetic radiation through a material permits one to determine the attenuation coefficient κ and the wavenumber k' of the radiation. Equation (2.12) yields then from the values of κ and k' those of $\epsilon'(\omega)$ and $\epsilon''(\omega)$ for each frequency ω utilized in the experiment. Since equation (2.19) provides an analytic representation of the functions $\epsilon'(\omega)$ and $\epsilon''(\omega)$ in terms of the sets of atomic parameters ω_r, Γ_r, and b_r, these parameters can be derived by fitting the analytic representation to numerical values of $\epsilon'(\omega)$ and $\epsilon''(\omega)$. Values of ω_r are identified quite easily when they correspond to sharp maxima of $\epsilon''(\omega)$ and $\kappa(\omega)$.

The attenuation coefficient κ is measured by passing a radiation beam through a layer of thickness t of the material of interest. According to equations (2.14) the ratio of the time averaged energy fluxes $c\vec{E} \times \vec{B}/4\pi$ in the beam after and before traversing the material is

$$\frac{|\vec{E} \times \vec{B}|_{x=t}}{|\vec{E} \times \vec{B}|_{x=0}} = e^{-2\kappa t}. \tag{2.30}$$

This attenuation law, analogous to the law (1.22) for particle beams, can be verified by carrying out measurements of energy flux for different values of t. The value of k′ in a given material is generally obtained from measurement of the refractive index $n = k'_{matter}/k'_{vacuum}$. The value of k'_{vacuum} is measured by diffraction or interferometry.

A practical method to determine κ as a function of ω consists of producing a beam of radiation whose spectral components cover a broad range of frequencies ω with nearly uniform intensity. This beam is made to traverse a layer of the material of interest after which its spectral components are separated by means of a spectroscope, so as to fan out in different directions and to be received at different positions of a detector (Fig. 2.2). The pattern of intensity received along the detector reproduces the variations of $\exp(-2\kappa t)$ and therefore of κ as a function of ω. Peaks of this function $\kappa(\omega)$ are called *absorption lines* because they appear as dark lines across the radiation pattern received by the detector in figure 2.2. Each line corresponds also to a peak of $\epsilon''(\omega)$ and therefore to one of the values ω_r in equations (2.19).

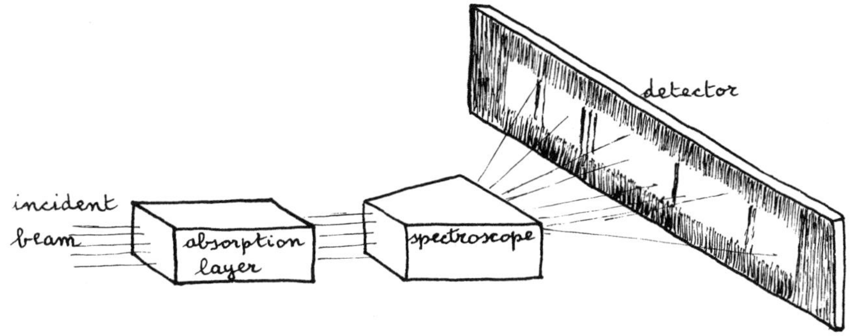

Fig. 2.2. Schematic experiment for absorption spectroscopy.

The width of each peak of $\epsilon''(\omega)$ depends, according to equations (2.19), on the parameter Γ_r which represents the friction effect for the rth mode of charge oscillation. Specifically, equations (2.19) show that the magnitude of each term in the summation is reduced to half of its maximum at $\omega = \omega_r \pm \frac{1}{2}\Gamma_r$

(Fig. 2.1).[17] Disregarding other terms of the sum, one can write

$$\epsilon''(\omega_r \pm \tfrac{1}{2}\Gamma_r) = \tfrac{1}{2}\epsilon''(\omega_r). \tag{2.31}$$

For this reason the parameter Γ_r is called the *line width at half-maximum* or, simply, the *width* of the absorption line at frequency ω_r.

The loss of radiation intensity displayed on the detector and integrated over the width of a line determines, though indirectly, the coefficient b_r for the rth mode of oscillation. The equivalent, but more meaningful, oscillator strength f_r can be determined from equation (2.28).

The collection of data on all parameters ω_r, Γ_r, and f_r of a material characterizes its *absorption spectrum*. The word "spectrum" is used widely and loosely with various meanings. For example, sometimes it refers to a set of characteristic frequencies ω_r, sometimes to a distribution of intensity as a function of frequency (such as the pattern received at the surface of a detector), and sometimes to the Fourier analysis of a mathematical function.

2.2.1. *Characteristics of Experimental Findings.* The absorption spectra of the atoms of almost all known elements have been observed extensively, though not quite sufficiently to determine all the relevant parameters over the spectral range where appreciable absorption occurs. Certain aspects of the results fit our initial expectations very well, but others could not be understood prior to the advent of atomistic rather than macroscopic observations of radiation effects.

To begin with, it is actually possible to achieve a fit of the experimental values

---

17. The behavior of ϵ near a resonance, that is, for $\omega \sim \omega_r$, is obtained by simplifying the rth term in the summations of equations (2.18) and (2.19) through the approximation $\omega_r^2 - \omega^2 = (\omega_r - \omega)(\omega_r + \omega) \sim (\omega_r - \omega)2\omega$. This simplification yields

$$\frac{b_r}{(\omega_r^2 - \omega^2) - i\Gamma_r\omega} \sim \frac{b_r}{2\omega}\,\frac{1}{\omega_r - \omega - i\tfrac{1}{2}\Gamma_r},$$

$$b_r\frac{\omega_r^2 - \omega^2}{(\omega_r^2 - \omega^2)^2 + \Gamma_r^2\omega^2} \sim \frac{b_r}{2\omega}\,\frac{\omega_r - \omega}{(\omega_r - \omega)^2 + \tfrac{1}{4}\Gamma_r^2},$$

$$\frac{b_r\Gamma_r\omega}{(\omega_r^2 - \omega^2)^2 + \Gamma_r^2\omega^2} \sim \frac{b_r}{2\omega}\,\frac{\tfrac{1}{2}\Gamma_r}{(\omega_r - \omega)^2 + \tfrac{1}{4}\Gamma_r^2}.$$

The last of these expressions represents the variation of energy absorption as a function of frequency near a resonance, the so-called *Lorentz profile* of an absorption line. This profile is a general feature of driven harmonic oscillations (*BPC*, 1: 217).

of $\epsilon'(\omega)$ and $\epsilon''(\omega)$ with the analytic expression (2.19).[18] This fit implies verification of the linear dependence of $\epsilon(\omega)$ on the number N of atoms per unit volume, assumed in expression (2.19) for monatomic low-density gases or vapors. Moreover, the relationship (2.22) between the parameters Γ_r and b_r—applicable in the absence of dissipation mechanisms other than interaction with radiation—is verified after elimination of contributions to the line widths by thermal agitation in the vapor (*Doppler broadening* and *collision broadening* of lines).

The lowest ω_r of each element is of the order of magnitude of 10^{16} radians/sec, corresponding to a vacuum wavelength of the order of 10^3 Å.[19] (Variations of the lowest ω_r from one element to another range over one factor of 10). The frequency 10^{16} radians/sec coincides with that of a harmonic oscillator having the mass of an electron and held elastically near its equilibrium position with a force constant of 6 eV/Å^2. This force is comparable to the electric forces expected to act on an electron at the periphery of an atom. Other observed values of ω_r range upward of 10^{16} radians/sec by several orders of magnitude, particularly so for heavier atoms. This finding also appears quite reasonable because electrons are presumably subject to stronger forces near the nucleus of an atom than at its periphery, particularly so in atoms with a high nuclear charge.

The observed values of the oscillator strength parameters f_r, defined by equations (2.28), are not easily interpreted. Some of them are of the order of unity, others are exceedingly small. Their total value $\Sigma_r f_r$ appears compatible with the sum rule (2.28), according to which it should equal the number of electrons; but only for a few elements are the available data adequate to test this law to within a few percent.

Features of absorption spectra that are unexpected on the basis of experience with macroscopic mechanics emerge from the large number of resonance frequencies ω_r and from the distribution of their values. Recall that a macroscopic system consisting of n bodies held by elastic forces near equilibrium positions has a *finite* number of normal modes of oscillation, of the order of n. However, the absorption spectra of atoms contain *series* of lines, such that the intervals in ω between successive lines decrease rapidly with increasing r until the lines can no longer be distinguished from one another. At still higher frequencies, absorption occurs throughout an extended continuous range

---

18. This statement must be qualified, since it holds only in the limit where the Doppler line broadening due to thermal agitation has been made negligible or corrected for.

19. Actually, some absorption of radiowaves or of microwave radiation by single atoms can often be detected. However, the corresponding frequencies ω_r correspond not to deformations of the electronic structure of an atom but to changes of its orientation with respect to the orientation of the nucleus or of an external field. This absorption is not relevant to the present analysis, but will be described in section 7.3.2.

of frequencies instead of being concentrated in lines. *In this respect* atoms behave altogether differently from a system of macroscopic particles. The origin of this behavior of atoms will begin to emerge from the study of radiation effects on single atoms. We note here only that infinite sets of oscillation frequencies are characteristic of classes of macroscopic systems which appear quite different from atoms. For example, whereas a finite chain of discrete particles has a finite number of characteristic frequencies, a finite but continuous string has an infinite set of discrete frequencies of vibration.[20] Further, a continuous spectrum of vibrational frequencies is characteristic of mechanical systems which are continuous and of infinite size.

Historically, groundwork for further study was laid by the observation of quantitative regularities in the absorption spectra and in the emission spectra to be discussed in the next section. The frequencies ω_r of numerous successive lines of a series are often represented by an empirical formula with a few constant parameters and a variable integer n which indicates the sequential number of each line in the series. Indicating for this purpose the sequence of frequencies of a series by ω_n, one observes that

$$\omega_n \approx \omega_t - \frac{\Omega}{(n-\sigma)^2}, \tag{2.32}$$

where ω_t, Ω, and σ are constants. Moreover, the constant Ω has a universal value for all regular series of all neutral atoms, namely,

$$\Omega = 2.07 \times 10^{16} \text{ radians/sec.} \tag{2.33}$$

When expressed as a wavenumber, this constant is called the *Rydberg constant*,

$$R = \Omega/2\pi c = 1.10 \times 10^5 \text{ cm}^{-1}. \tag{2.33a}$$

A still higher regularity is observed in the absorption spectrum of hydrogen atoms, whose structure is presumably simplest; this spectrum consists of a single series with

$$\omega_t = \Omega, \sigma = 0 \text{ (H atom)}. \tag{2.34}$$

Accounting for the empirical laws (2.32), (2.33), and (2.34), together with another law described in the next section, constituted one of the main challenges in the early development of atomic physics.

As noted above, absorption lines with larger and larger values of n in equation (2.32) become so close to one another that each of them is not readily distinguished from the next one. [This limitation stems mostly from incidental

---

20. *BPC*, vol. 3, chap. 2.

circumstances, because the intrinsic line width $\Gamma_{n,\,rad} = (2\omega^2 e^2/3m_e c^3)f_n$ keeps decreasing with increasing n; lines with n ~ 100 have been observed distinctly.] Nevertheless, it is now understood that each series includes in principle an infinity of lines whose frequencies ω_n converge for $n \to \infty$ to the *series limit* ω_t. A *continuous spectrum* of frequencies ω_r is said to lie beyond each limit ω_t. This limit constitutes the threshold beyond which a spectrum becomes continuous.

The absorption spectra of atoms other than H often include different series converging to the same limit ω_t. Numerous series are also observed which converge to different limits ω_t. The values of the limits ω_t range within one order of magnitude for the atoms of elements with very low Z values but extend over four orders of magnitude for high-Z elements in keeping with our earlier remarks about the range of values of ω_r.

2.3. EMISSION SPECTRA. COMBINATION PRINCIPLE

The absorption spectra discussed in the preceding section provide extensive evidence on current oscillations within atoms. Still more extensive evidence is provided by the spectral analysis of radiation emitted by atoms. Typically, one observes light emission by a hot low-density gas or vapor or by a gas or vapor through which passes an electric discharge. In a discharge, as in a hot gas, atoms are subjected to collisions by other neutral atoms, ions or electrons. The internal equilibrium of an atom is disturbed briefly by each collision. It is plausible that in atoms—as in other stable mechanical systems—any moderate, brief breach of equilibrium is followed by internal oscillations with frequencies that are characteristic of the elastic properties of the system. Since atoms contain electric charges, their mechanical oscillations constitute presumably also charge and current oscillations which are sources of radiation.

These oscillations, represented by variable dipole moments $\vec{p}(t)$, should consist of superpositions of normal modes, each of them governed by equation (2.16) with the driving force set at zero:

$$\frac{d^2\vec{p}_r}{dt^2} + \Gamma_r \frac{d\vec{p}_r}{dt} + \omega_r^2 \vec{p}_r(t) = 0 \tag{2.35}$$

Each normal mode of oscillation is characterized not only by the appropriate values of the frequency ω_r and of the coefficient Γ_r, but also by the direction of its dipole moment. This direction would be characteristic for each mode of a single anistropic atom, but becomes random for a population of randomly oriented atoms. The time dependence of $\vec{p}_r$, obtained by solving equation (2.35) by the method of page 527, is

$$\vec{p}_r(t) = \vec{p}_{r0} \exp(-\tfrac{1}{2}\Gamma_r t) \cos\left[(\omega_r^2 - \tfrac{1}{4}\Gamma_r^2)^{1/2} t + \psi\right], \tag{2.36}$$

where the peak value $\vec{p}_{r0}$ and phase constant ψ depend on the conditions under which the oscillation is started. Note that usually $\Gamma_r \ll \omega_r$, so that

$$(\omega_r^2 - \tfrac{1}{4}\Gamma_r^2)^{1/2} \sim \omega_r. \tag{2.37}$$

The light emitted by atoms is analyzed by a spectroscope that directs sinusoidal components with different ω_r to different positions of a detector. The pattern of radiation energy received by the detector reproduces the strengths and frequencies of the different components. The strengths of the different components depend on the temperature and other conditions of the gas, but the set of frequencies ω_r (the *emission spectrum*) is characteristic of the gas atoms.

The set of frequencies ω_r in the emission spectrum of the atoms of each element is similar in structure to that of its absorption spectrum, in that it consists of isolated values, which constitute line series, and of values that cover certain ranges continuously and constitute continuous spectra. However, emission spectra include a large number of series and of continuous spectra besides those found normally in absorption spectra. This observation indicates that the collisions experienced by atoms have somehow the effect of multiplying the number of their observable modes of oscillation. The origin of this multiplication is not readily apparent at this point of our study.

Some order is brought into the multiplicity of spectral frequencies by the following remark. If one lists the differences between frequencies ω_r of the emission spectrum of an element, one finds certain values recurring repeatedly. Guided by this observation, spectroscopists have succeeded in expressing the set of frequencies ω_r of each element in terms of two smaller sets of frequency values. By trial and error one constructs two sets of numbers $\tau_1, \tau_2, \ldots, \tau_\alpha, \ldots$, and $\bar{\tau}_1, \bar{\tau}_2, \ldots, \bar{\tau}_\beta, \ldots$, such that each frequency ω_r equals the difference between one element of the set τ and one element of the set $\bar{\tau}$. Accordingly, one can relabel each frequency ω_r by a pair of indices $(\alpha\beta)$ so that

$$\omega_r \equiv \omega_{\alpha\beta} = |\tau_\alpha - \bar{\tau}_\beta|. \tag{2.38}$$

The elements of the sets τ and $\bar{\tau}$ are called *spectral terms*. Clearly, the two sets τ and $\bar{\tau}$ are not unique but are defined to within a common additive constant which cancels upon taking the difference $\tau_\alpha - \bar{\tau}_\beta$. One often utilizes this indeterminacy by setting one spectral term to zero.[21]

As an illustration consider the following set of 16 frequencies (expressed in cm^{-1}) drawn from the emission spectrum of La^{++}: 10885, 12114, 15748, 16977, 28424, 30043, 31272, 31520, 37236, 40332, 65097, 68193, 79393, 79641, 80870,

21. According to usual practice a spectral term is expressed in cm^{-1}—more properly, in number of waves per cm—and is obtained from τ defined in equation (2.38) by division by $2\pi c$, where c is the light velocity in cm/sec.

82489. Two sets of four spectral terms, τ and $\overline{\tau}$, are found, namely, {13591, 82347, 110210, 124504} and {42015, 45111, 93232, 94461}, whose differences reproduce the spectral frequencies ω_r as shown in Table 2.1.

TABLE 2.1

Selected Spectral Terms and Frequencies of La^{++} Spectrum (cm^{-1})*
(Courtesy J. Sugar)

τ \ $\overline{\tau}$	42015	45111	93232	94461
13591	28424	31520	79641	80870
82347	40332	37236	10885	12114
110210	68193	65097	16977	15748
124504	82489	79393	31272	30043

* Discrepancies in the last digit between entries in this table and the given frequencies are due to round-off errors.

Naturally, if we had wanted to represent a larger set of frequencies from the lanthanum spectrum, we would have needed sets of spectral terms with larger numbers of elements. In fact, since every spectrum consists of an infinite number of frequencies, the complete sets of spectral terms would also be infinite.

In practice, one can construct very large but finite arrays similar to Table 2.1. However, not every term difference in the array corresponds to a frequency actually observed in the emission spectrum. This is not surprising because one conceives readily of oscillation modes of atomic electrons that are not readily detectable by radiation emission or absorption. For example, half of the electronic charge may move upward and half simultaneously downward so that the center of mass of all electrons remains fixed at the nucleus; the resulting dipole moment $\vec{p}_r$ remains then equal to zero.[22] (The great majority of possible oscillation frequencies actually fail to appear in the optical spectra for the very reason that their electric dipole moment vanishes.)

The idea of representing frequencies as differences of spectral terms is called the *Rydberg-Ritz combination principle.* Its empirical discovery laid a foundation for the further progress of atomic physics. Its various features will be interpreted in later chapters.

---

22. This type of oscillation is called a *quadrupole oscillation.*

All frequencies of the absorption spectrum of each element normally belong to a single column (or row) of the array of emission frequencies, the one pertaining to the lowest spectral term.[23] In the case of atomic hydrogen, all spectral terms of both sets, τ_α and $\overline{\tau}_\beta$, are expressed by the single formula

$$\tau_n = -\frac{2.07 \times 10^{16}}{n^2} \text{ radians/sec} = -\frac{1.10 \times 10^5}{n^2} \text{ cm}^{-1}, \tag{2.39}$$

where n is an integer. Actually, the combination principle originates from the much earlier (1884) remark by Balmer that the frequencies of the four prominent lines in the visible emission spectrum of atomic hydrogen could be represented by equations (2.38) and (2.39) with $n = 6, 5, 4, 3$ for τ_α, and $n = 2$ for $\overline{\tau}_\beta$. The series of hydrogen lines with $n = 2$ for $\overline{\tau}_\beta$ became known as the *Balmer series*. The analogous series of lines with $n = 1$ for $\overline{\tau}_\beta$ was discovered 20 years later by Lyman in the far-ultraviolet. The series with $n \geq 3$ for $\overline{\tau}_\beta$ lie in the infrared and were discovered still later.

2.4. INFLUENCE OF A CONSTANT MAGNETIC FIELD

A magnetic field exerts a deflecting action on any current $\vec{j}_{matter}$. Thereby it influences the emission, propagation, and absorption of radiation. The influence of a constant magnetic field $\vec{B}$ on absorption and emission spectra is called *Zeeman effect;* its study has played an important role in the development of atomic physics. The influence of $\vec{B}$ on propagation is called the *Faraday effect*. A field $\vec{B}$ also influences steady intra-atomic currents which neither emit nor absorb radiation.

We confine our attention to the action of a constant uniform field upon a population of separated, randomly oriented atoms. We also carry out the treatment to lowest order in the field strength B, disregarding quantities proportional to B^2.

A constant "external" magnetic field exerts upon any current $\vec{j}_{matter} = N d\vec{p}/dt$ the force

$$\frac{1}{c}\vec{j}_{matter} \times \vec{B} = N\frac{1}{c}\frac{d\vec{p}}{dt} \times \vec{B} \tag{2.40}$$

per unit volume of matter. This force tends to deflect the current carriers in a direction perpendicular to both $\vec{j}_{matter}$ and $\vec{B}$.

We begin by inserting a term proportional to equation (2.40) into equation (2.35) which governs the free oscillations of dipole moments $\vec{p}_r(t)$. The magnetic force is proportional to the current $d\vec{p}_r/dt$ and should contribute a rate

23. Lowest is intended here in the algebraic sense, such that, e.g., −1235 is lower than 272.

of *variation* of the current proportional to the force itself. We indicate the proportionality constant by 2γ, for later convenience, and write the modified equation (2.35) as

$$\frac{d^2\vec{p}_r}{dt^2} + \Gamma_r \frac{d\vec{p}_r}{dt} - 2\gamma \frac{d\vec{p}_r}{dt} \times \vec{B} + \omega_r^2 \vec{p}_r(t) = 0. \tag{2.41}$$

For purposes of orientation we derive the value of γ for the extreme Thomson-Lorentz model utilized to establish equation (2.20). According to this model the equation of motion of an electron at position $\vec{x}$, subject to an elastic force, to friction, and to the action of a magnetic field $\vec{B}$ is

$$m_e \frac{d^2\vec{x}}{dt^2} = -k\vec{x} - \eta \frac{d\vec{x}}{dt} - \frac{e}{c}\frac{d\vec{x}}{dt} \times \vec{B}. \tag{2.42}$$

This equation reduces to the form (2.41) upon multiplication by the electron's charge $-e$ and division by the electron's mass m_e. The product $-e\vec{x}$ represents the dipole moment $\vec{p}_r$. The equation then coincides with equation (2.41) if $\Gamma_r = \eta/m_e$ and $\omega_r^2 = k/m_e$ in accordance with equation (2.20), and if

$$\gamma = \frac{-e}{2m_e c} = -8.8 \times 10^6 \frac{\text{radians}}{\text{sec gauss}} . \tag{2.43}$$

Even for $B = 10^4$ gauss, a rather strong field, the resulting value of $\gamma B \sim 10^{11}$ radians/sec is many orders of magnitude lower than the frequencies of atomic oscillations. For this reason it is sufficient to treat γB as a small quantity.

In the absence of a magnetic field, equation (2.41) reduces to equation (2.35) and has solutions with $\vec{p}_r$, $d\vec{p}_r/dt$, and $d^2\vec{p}_r/dt^2$ all parallel to one another and of arbitrary direction. This means that there exists an arbitrary set of three mutually orthogonal normal mode oscillations with equal frequencies ω_r; as noted in section B.2, any superposition of these *degenerate* normal modes is still a normal mode with the same frequency. The equivalence of all directions no longer exists in the presence of $\vec{B}$; the complete equation (2.41) contains a term perpendicular to the fixed direction of $\vec{B}$. We expect then that a particular set of three mutually orthogonal normal modes, which are degenerate for $\vec{B} = 0$, will become a nondegenerate set for $\vec{B} \neq 0$.

One normal mode of this set clearly has $\vec{p}_r(t)$ and $d\vec{p}_r/dt$ parallel to $\vec{B}$, since in this case the term of equation (2.41) containing $\vec{B}$ vanishes. The oscillation frequency of this mode remains ω_r. If we choose coordinates with the z axis parallel to $\vec{B}$, this solution of equation (2.41) is represented by

$$\vec{p}_{r\parallel}(t) = [0,\ 0, p_0 \exp(-\tfrac{1}{2}\Gamma_r t) \cos(\omega_r t + \psi)], \tag{2.44}$$

where the subscript $\parallel$ means "parallel to $\vec{B}$,", p_0 and ψ are arbitrary, and Γ_r^2 has been disregarded in comparison with ω_r^2, in accordance with equation (2.37).

The other two normal modes will be perpendicular to $\vec{B}$, but their direction cannot be constant owing to the magnetic field. The dipole $\vec{p}_r$ keeps a constant magnitude in these normal modes but its direction rotates about $\vec{B}$ at a uniform rate; such a rotation can be represented as the superposition of two oscillations with fixed directions, respectively, along the x and y axes and 90° out of phase with respect to one another. We indicate these normal modes by $\vec{p}_{r+}(t)$ and $\vec{p}_{r-}(t)$, where the ± sign corresponds to the sense of rotation about the direction of $\vec{B}$, and we represent them by their Cartesian components as

$$\vec{p}_{r+}(t) = [p_0 \exp(-\tfrac{1}{2}\Gamma_r t) \cos(\omega_{r+}t + \psi), p_0 \exp(-\tfrac{1}{2}\Gamma_r t) \sin(\omega_{r+}t + \psi), 0], \tag{2.45}$$

$$\vec{p}_{r-}(t) = [p_0 \exp(-\tfrac{1}{2}\Gamma_r t) \cos(\omega_{r-}t + \psi), -p_0 \exp(-\tfrac{1}{2}\Gamma_r t) \sin(\omega_{r-}t + \psi), 0]. \tag{2.46}$$

These definitions of $\vec{p}_{r+}$ and $\vec{p}_{r-}$ have been so chosen that, upon substitution in (2.41), the two terms $-2\gamma(d\vec{p}_r/dt) \times \vec{B}$ and $\omega_r^2\vec{p}_r$ become antiparallel to each other for $\vec{p}_{r+}$ and parallel for $\vec{p}_{r-}$. Accordingly, the magnetic force has merely the effect of decreasing or increasing ω_r slightly. One verifies that equations (2.45) and (2.46) fulfill condition (2.41), to within terms of order Γ_r^2, $(\gamma B)^2$, or $\Gamma_r\gamma B$ which we disregard, provided

$$\omega_{r+} = \omega_r - \gamma B, \tag{2.47}$$

$$\omega_{r-} = \omega_r + \gamma B. \tag{2.48}$$

Thus we see that free oscillations $\vec{p}_r(t)$, which can occur in arbitrary directions and with a single frequency ω_r when $\vec{B} = 0$, are replaced for $\vec{B} \neq 0$ by three alternative types of normal modes with specified orientation in space and with *different* frequencies ω_r, ω_{r+}, and ω_{r-}. On the basis of the present analysis one would then expect each line of the emission spectrum of an atom to *split*, upon the onset of a magnetic field, into *three* lines separated by intervals equal to the *Larmor frequency*

$$\omega_L = \gamma B. \tag{2.49}$$

The presence of a magnetic field also has an important effect upon the intensity and polarization of radiation emitted in various directions. The currents within atoms of a hot gas which emits light are initiated by collisions in random directions. In the absence of a magnetic field these currents can oscillate with each frequency ω_r in any direction, so that light is emitted uniformly in all directions and with *natural* (i.e., random) polarization. In the presence of a magnetic field each current started by a collision resolves into three normal mode components, each having a different orientation with respect to $\vec{B}$ and a slightly different frequency, ω_r, ω_{r+}, or ω_{r-}. Light emitted by each component current, recognizable on the basis of its frequency, has a well-defined pattern of polarization and of intensity distribution in various directions, as outlined next.

Light of unshifted frequency ω_r has the familiar pattern pertaining to a

Hertzian dipole[24] parallel to $\vec{B}$, namely, intensity proportional to $\sin^2 \theta$ (where θ is the angle between the propagation direction $\vec{k}$ and $\vec{B}$) and linear polarization with the oscillating field $\vec{E}(t)$ in the plane of $\vec{k}$ and $\vec{B}$. Light of frequency ω_{r+} or ω_{r-} has the pattern pertaining to a dipole rotating in the (x, y) plane; its intensity is proportional to $1 + \cos^2 \theta$, and its polarization is elliptical; i.e., its electric field $\vec{E}(t)$ describes an elliptical pattern in the plane perpendicular to $\vec{k}$. To the frequencies ω_{r+} and ω_{r-} correspond opposite directions of rotation of $\vec{E}$ along this pattern. When $\vec{k}$ is parallel to the constant field $\vec{B}$, the ellipse rounds into a circle; and when $\vec{k}$ is perpendicular to $\vec{B}$, the ellipse flattens into a line perpendicular to both $\vec{B}$ and $\vec{k}$. Thus the constant field $\vec{B}$ effectively splits monochromatic light emitted with random direction and polarization into components with different frequencies and with different well defined characteristic patterns of intensity and polarization.

Experimentally, the modification of emission spectra by a constant magnetic field was discovered by Zeeman and was partially interpreted shortly thereafter by Lorentz essentially in the manner outlined above. Some series of lines fit the predictions of this theory exactly under typical laboratory conditions. These lines are said to display a *normal Zeeman effect*. Moreover, the experimental value of the constant γ coincides in these cases with the value (2.43) corresponding to the behavior of isolated electrons. On the other hand, numerous series of spectral lines display an *anomalous Zeeman effect*, namely, lines split into more than three components with different values of γ. This anomalous effect was explained much later and will be discussed in sections 7.3.1 and 14.3.

The influence of the magnetic field on radiation propagation and on absorption spectra can be predicted readily from a knowledge of its influence on the emission spectra. This influence is rather simple in two cases: (a) propagation along the constant field $\vec{B}$, and (b) propagation perpendicular to $\vec{B}$ with electric field $\vec{E}(t)$ parallel to $\vec{B}$. In the latter case $\vec{B}$ has no effect. For propagation parallel to $\vec{B}$, one should resolve $\vec{E}(t)$ into two components which rotate about $\vec{B}$ with positive and negative circular polarization, respectively. The propagation and absorption of these two components proceeds as described in sections 2.1 and 2.2 except that the frequencies ω_r are replaced by ω_{r+} and ω_{r-}. This shift in frequencies is observed as a Zeeman effect in the absorption spectrum. The shift of frequencies also changes $\epsilon(\omega)$ and thus changes the velocity of propagation differently for the alternative circular polarizations; i.e., it makes the material optically active (*Faraday effect*).

2.4.1. *Larmor Precession*. As noted above, the presence of a field $\vec{B}$ splits the current oscillation started within an atom by a random collision into components with different frequencies and with specified orientations. In order to keep the whole phenomenon in view (see end of section A.2), it is of interest to follow the complete process of current oscillation in the course of time. This can be done analytically by superposing the monochromatic components at

24. *BPC*, vol. 3, chap. 5, advanced topic 2, also chap. 6.

different successive times after the collision. The superposition of components with three closely spaced frequencies (ω_{r+}, ω_r, and ω_{r-}) yields, in general, an oscillation with a single average frequency ω_r modified by beats. In our particular stituation the effect of beats is unusually simple as shown by the following considerations.

The dipole rotation $\vec{p}_{r+}(t)$ represented by equation (2.45) proceeds in *positive* direction at a rate of $\omega_{r+} = \omega_r - \omega_L$ radians/sec (eqs. [2.48] and [2.49]). This rotation has an apparent rate ω_r radians/sec when observed from a frame of reference rotating in a *negative* direction at the rate ω_L radians/sec. Now, observe that the other dipole rotation $\vec{p}_{r-}(t)$, represented by equation (2.46) proceeds in a *negative* direction at a rate $\omega_{r-} = \omega_r + \omega_L$ radians/sec, which also appears to coincide with ω_r in the rotating frame. Finally, the oscillation (2.44), $\vec{p}_{r\parallel}(t)$, has the same appearance in the rotating frame as in the laboratory frame.

We conclude that *all three components* appear to proceed at the same, unshifted, rate ω_r in the rotating frame. That is, no influence of the constant field $\vec{B}$ upon the current oscillations is apparent in the rotating frame. In fact, equation (2.41) could have been reduced to its original form (2.35) with $\vec{B} = 0$, by transforming the coordinates of $\vec{p}_r(t)$ to a rotating frame.[25] Note, however, that all these statements hold only to within terms of order ω_L^2 which have also been disregarded in the solution of equation (2.41).

If, then, the effect of $\vec{B}$ appears null to an observer rotating with frequency ω_L in a negative direction, any oscillation $\vec{p}_r(t)$ under the influence of $\vec{B}$ can be described as the combination of two variations in time: (a) a free oscillation in an arbitrary direction in accordance with equation (2.35) (as though $\vec{B} = 0$) and (b) a rotation about the direction of $\vec{B}$ at the rate of $-\omega_L = -\gamma B$ radians/sec. This rotation is called the *Larmor precession.* The origin of Larmor precession can be traced, for a simple atomic model, to the electromagnetic induction which accompanies the onset of a magnetic field and drives electric charges around the field direction.[26]

The connection between Larmor precession and Zeeman splitting of spectral lines affords an example of the complementarity relation between frequency resolution and time resolution in the analysis of a phenomenon (see section A.3.1). Description of the Larmor precession implies successive observations at time intervals shorter than the precession period. On the other hand, the resolution of a spectral line into its Zeeman components implies separation of radiation into components whose frequency differences are equal to the Larmor frequency ω_L. As noted in section A.3.1, sharp time resolution is achieved by superposition of frequencies covering a range reciprocal to the

---

25. The Coriolis term (see *BPC*, 1: 85) introduced in equation (2.41) by coordinate rotation with velocity $\vec{\omega}_L = -\gamma\vec{B}$ cancels the magnetic term of equation (2.41). This simplification of equation (2.41) by coordinate transformation could have been surmised by inspection.

26. *BPC*, 2: 372.

time resolution. Therefore, high time resolution and high frequency resolution are mutually exclusive.

2.4.2. *Larmor Precession of Nonradiating Atoms; Gyroscopic Effect.* Larmor precession occurs not only for atoms that carry oscillating currents and therefore interact with radiation, but also for other atoms. We saw at the end of chapter 1 that atoms undisturbed by collisions do not emit radiation, thus showing the absence of internal variable currents. It is nevertheless conceivable and is often verified in practice (chap. 6) that undisturbed atoms carry steady electric currents subject to influence by a constant magnetic field $\vec{B}$. Such steady (time constant) currents must, of course, flow in a closed path which constitutes a current loop.

The forces exerted by a field $\vec{B}$ on a closed loop are conveniently studied by regarding the loop as an integrated whole rather than as a sum of current elements.[27] The magnetic properties of the loop are represented in condensed fashion by a magnetic *dipole moment* $\vec{\mu}$. The vector $\vec{\mu}$ is defined as having a direction perpendicular to the area A enclosed by the loop and a magnitude equal to the product of A and of the current intensity i expressed in electromagnetic units. Representing the area by a vector perpendicular to the loop, one writes

$$\vec{\mu} = i\vec{A}. \tag{2.50}$$

The forces exerted upon different portions of the loop have different directions and cancel out in the sense that they yield no net lateral pull on the loop. However, they produce a torque which strives to align the loop perpendicularly to the magnetic field with the (positive) current flowing in the positive direction about the direction of $\vec{B}$. This torque is usually expressed, analogously to the torque on a magnetic needle, as the vector product of the field $\vec{B}$ and the magnetic moment $\vec{\mu}$:

$$\vec{M} = \vec{\mu} \times \vec{B}. \tag{2.51}$$

As stated above, the torque $\vec{M}$ strives to turn $\vec{\mu}$ in the direction of $\vec{B}$, but its actual effect on $\vec{\mu}$ depends on the inertia of the loop. Since the field $\vec{B}$ has been assumed to be too weak to disturb the internal structure of an atom, the effect of $\vec{B}$ on $\vec{\mu}$ depends on the inertia of the whole atom. More specifically, the action of a torque on an isolated system such as a gas atom gives rise to a gyroscopic effect in which the inertia of the atom appears as an angular momentum. That is, the effect of the field $\vec{B}$ on the atomic current loop is similar to the effect of gravity on a spinning top and is described by a similar equation.

The weight of a spinning top applies to it a torque $\vec{M}$ which strives to drop the

---

27. *BPC*, 2: 361–79.

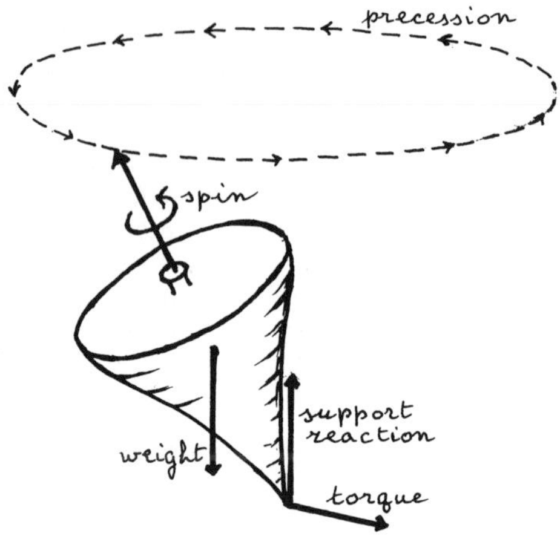

Fig. 2. 3. Precession of spinning top.

top's center of mass, turning the spin axis about the point of support[28] (Fig. 2.3). Similarly, for an atom with a circulating current, the torque $\vec{M}$ given by equation (2.51) strives to turn the atom about its center of mass. In either case the combined effect of the torque and of the inertia changes the angular momentum $\vec{J}$ which is due to the spin of the top or to the circulation of current-carriers in an atom. The equation of motion is

$$\frac{d\vec{J}}{dt} = \vec{M}. \tag{2.52}$$

Vector addition of $d\vec{J}/dt$ to the instantaneous angular momentum $\vec{J}$ causes the direction of $\vec{J}$ to precess about an axis which is vertical for the top and parallel to $\vec{B}$ for the atom. In the course of precession $\vec{J}$ maintains a constant angle with the precession axis. The angular velocity of precession is usually represented by a vector $\vec{\omega}$ parallel to the axis

$$\frac{d\vec{J}}{dt} = \vec{\omega} \times \vec{J} = \vec{M}. \tag{2.53}$$

Thus we see that atoms with a steady circulation of current precess about the direction of an applied magnetic field, in the same way as atoms with an oscillating electric moment.

The substantial identity between the precession of atoms with oscillating and with steady currents emerges by considering that a "current loop" actually means a distribution of circulating electric charges. Each element of this charge distribution has an electric dipole moment $\vec{p}$ with respect to the atomic

---

28. *BPC*, 1: 252-55.

nucleus even though the whole distribution is understood to have no net dipole moment, or, at least, no rapidly variable dipole moment. That is, the circulating current consists, by implication, of an aggregate of rotating electric dipole moments which have no net resultant but whose motion yields a nonzero resultant *magnetic moment*. Since the action of the magnetic field is known to impress a Larmor precession with velocity ω_L upon any rotating dipole, the aggregate distribution of dipoles experiences the same precession. Accordingly, we substitute for $\vec{\omega}$ in equation (2.53) its expression as a Larmor rotation velocity,

$$\vec{\omega} = \vec{\omega}_L = -\gamma \vec{B}. \tag{2.54}$$

This result establishes a most important link between the magnetic moment of a circulating current and the angular momentum of an atom. Substituting equations (2.51) and (2.54) into the precession equation (2.53) yields

$$\frac{d\vec{J}}{dt} = -\gamma \vec{B} \times \vec{J} = \vec{\mu} \times \vec{B}. \tag{2.55}$$

This equation shows that $\vec{\mu}$ is proportional to $\vec{J}$; the constant ratio of these vectors, namely,

$$\gamma = \vec{\mu}/\vec{J}, \tag{2.56}$$

is accordingly called the *gyromagnetic ratio* or, more appropriately, the *magnetomechanical ratio* of the atom. Notice that the precession depends only on this ratio but not on the separate values of $\vec{\mu}$ and $\vec{J}$. A theoretical value of γ may be obtained by starting from a specific model of the current circulation in atoms. The model of an electron moving along a circular path yields the same value $\gamma = -e/2m_e c$ as was found in equation (2.43).[29] In fact, the same dependence of γ on the charge-to-mass ratio results from any model in which charge and mass travel together.

Experimental measurements of γ thus provide evidence on the association of charge and mass circulation within atoms. Experiments on single atoms and other particles, which demonstrate Larmor precession without intervention of light emission or absorption and which measure γ values, will be described in chapter 7. In some cases the experimental values agree with the predictions of a simple model, but in other cases they do not, just as is the case for the Zeeman effect.

CHAPTER 2 PROBLEMS

2.1. Calculate and plot the real and imaginary parts of $\epsilon(\omega)$ and of the wavenumber $k = \omega[\epsilon(\omega)]^{1/2}/c$ for copper metal, using equations (2.18) and (2.12) with parameters $N = 10^{23}$ cm^{-3}, $b_0 = e^2/m_e$, $\omega_0 = 0$, $\Gamma_0 = 5 \times 10^{13}$ sec^{-1}.

---

29. *BPC*, 2: 370-71.

Consider only the $r = 0$ term of the sum, which represents the contribution to ϵ of conduction electrons.

2.2. A beam of X-rays in vacuum strikes an aluminum surface at grazing incidence, forming with the surface an angle of 10^{-3} radians ($\sim 3'$). Consider the refraction or total reflection of the beam on the basis of Snell's law and of the high-frequency limit form of $\epsilon(\omega)$ as given in equation (2.25). (a) Show that total reflection occurs for $\lambda = 0.5$ Å. (b) Will the same occur for $\lambda =$ 0.3 Å?

2.3. Consider an electron held near an equilibrium position by an elastic bond with force constant k_r such that a displacement of 0.5 Å requires an energy expenditure of 4 eV. Oscillation of the electron dissipates energy by radiation emission. The power loss equals the product of the energy of oscillation and of the damping constant $\Gamma_{r,rad}$ given by equations (2.20) and (2.22). (This result is contained in eqs. [156] and [157] on p. 377 of *BPC*, Vol. 3. It also follows from the equation of an oscillator with friction constant chosen in accordance with eq. [2.20], that is, $\eta = \Gamma_{r,rad}\, m_e$.) Calculate (a) the value of k_r expressed in terms of eV and Å and also in mks units; (b) the oscillation frequency; (c) the average radiation power loss for oscillation with 0.5 Å amplitude; (d) the energy flux at 10 m from a source consisting of 10^{16} electrons oscillating with the above amplitude in random directions.

2.4. The following absorption lines are observed in potassium vapor:

$$\nu_3 = 1.301 \times 10^4\ \text{cm}^{-1}, \qquad \nu_5 = 2.900 \times 10^4\ \text{cm}^{-1},$$
$$\nu_4 = 2.471 \times 10^4\ \text{cm}^{-1}, \qquad \nu_6 = 3.107 \times 10^4\ \text{cm}^{-1}.$$

Represent these wavenumbers by the relation $\nu_n = \nu_t - R/(n-\sigma)^2$, with $R = 1.097 \times 10^5\ \text{cm}^{-1}$, by fitting the parameters ν_t and σ. Note that the differences between any two frequencies are independent of ν_t. Determine σ graphically to fit the differences.

2.5. The spectrum of La^{++} includes, besides the lines given in Table 2.1, lines of the following frequencies: 42015, 43508, 45111, 92858, 93232, and 94461 cm^{-1}. Extend Table 2.1 so as to include these frequencies and additional spectral terms as required. Recall that not all boxes in a table need correspond to observed lines.

2.6. The main absorption spectrum of calcium vapor has a normal Zeeman effect. State how many lines will result from each line of this spectrum under the influence of a magnetic field when the field direction and the direction of observation form an angle of (a) 0°, (b) 30°.

2.7. The saturation magnetization, M_s, of a material is its maximum dipole moment per unit volume. For Fe at room temperature $M_s = 1700$ gauss. (a) Calculate the saturation mean magnetic moment of one Fe atom. (b) Assuming that this moment arises from a circular current loop 1 Å in diameter, calculate the current intensity in the loop. (c) Assuming that the current is carried by one electron moving along the loop, calculate the electron's velocity. (d) Assuming that the gyromagnetic ratio γ of the electron has the value given in equation (2.43), calculate the total angular momentum of all the electrons in 1 cm^3 of metal, in cgs units.

SOLUTIONS TO CHAPTER 2 PROBLEMS

2.1. $\epsilon(\omega) = 1 - \omega_p^2/(\omega^2 + i\Gamma_0\omega)$; $\omega_p = [4\pi Ne^2/m_e]^{1/2} = 1.8 \times 10^{16}$ radians/sec. Re $\epsilon(\omega) = 1 - \omega_p^2/(\omega^2 + \Gamma_0^2)$. Im $\epsilon(\omega) = (\Gamma_0/\omega)\omega_p^2/(\omega^2 + \Gamma_0^2)$.

$$k = \epsilon^{1/2}\omega/c = \frac{[(\omega^2 + \Gamma_0^2 - \omega_p^2)^2 + \omega_p^4\Gamma_0^2/\omega^2]^{1/4}}{(\omega^2 + \Gamma_0^2)^{1/2}} \frac{\omega}{c} (\cos \tfrac{1}{2}\varphi + i \sin \tfrac{1}{2}\varphi).$$

$$\varphi = \arctan \frac{\omega_p^2\Gamma_0/\omega}{\omega^2 + \Gamma_0^2 - \omega_p^2}$$

2.2. $\epsilon(\omega) = 1 - 4\pi NZe^2/m_e\omega^2$; $N_{Al} = 6.0 \times 10^{23} \times 2.7/27 = 6.0 \times 10^{22}$ atoms/cm^3; $Z_{Al} = 13$. $\omega = 2\pi c/\lambda$; $\epsilon = 1 - (NZe^2/\pi m_e c^2)\lambda^2$
$= 1 - (2.64 \times 10^5\ cm^{-1}\lambda)^2$. $n_{Al} = \sqrt{\epsilon} \sim 1 - \tfrac{1}{2}(2.64 \times 10^5\ cm^{-1}\lambda)^2$.
Snell's law — $\sin\theta_{vac} = n_{Al} \sin\theta_{Al}$ — implies that refraction cannot occur and therefore total reflection takes place when $n_{Al} < \sin\theta_{vac}$. Now $\theta_{vac} \sim \tfrac{1}{2}\pi - 10^{-3}$, $\sin\theta_{vac} \sim 1 - \tfrac{1}{2}10^{-6}$. Therefore, no refraction occurs when $2.64 \times 10^5\ cm^{-1}\lambda > 10^3$, i.e. $\lambda > 0.37 \times 10^{-8}$ cm $= 0.37$ Å. (a) Total reflection. (b) No.

2.3. (a) Potential energy of oscillator is $\tfrac{1}{2}k_r x^2$. Set $\tfrac{1}{2}k_r(0.5\ \text{Å})^2 = 4$ eV; then $k_r = 32\ \text{eV/Å}^2 = 32 \times 1.6 \times 10^{-19}/(10^{-10})^2 = 512$ newtons/meter.
(b) $\omega = (k_r/m_e)^{1/2} = [512\ \text{newtons/meter}/9.11 \times 10^{-31}\ \text{kg}]^{1/2} = 2.4 \times 10^{16}$ radians/sec. (c) Substitute in equation (2.22) the value of b_r from equation (2.20) and of ω_r from part (b) above to find $\Gamma_{r,rad} = 3.6 \times 10^9\ sec^{-1}$. Power loss equals $3.6 \times 10^9\ sec^{-1} \times 4$ eV $= 1.4 \times 10^{10}$ eV sec^{-1}. (Note that this model oscillator dissipates most of its energy in less than 10^{-9} sec.)
(d) Energy flux equals $(10^{16} \times 1.4 \times 10^{10}\ \text{eV sec}^{-1})/(4\pi \times 10^6\ cm^2) = 1.1 \times 10^{19}\ \text{eV cm}^{-2}\ \text{sec}^{-1} = 0.7\ \text{watts cm}^{-2}$.

2.4. $\sigma = 0.77 \pm 0.01$, $\nu_t = (3.51 \pm 0.02) \times 10^4\ cm^{-1}$.

2.5. TABLE 2.1 is extended by adding two rows which read:

$\tau(cm^{-1})$				
0	42015	45111	93232	94461
1603		43508		92858

2.6 (a) Light propagating along the magnetic field is absorbed only at two frequencies, $\omega_r \pm \omega_L$. (b) Light propagating at 30° from $\vec{B}$ is absorbed at three frequencies, ω_r, and $\omega_r \pm \omega_L$.

2.7. (a) density $\rho_{Fe} = 6.0 \times 10^{23} \times 7.8/55.9 = 8.5 \times 10^{22}$ atoms/cm^3; $\langle \mu_{Fe} \rangle = (M_S)_{Fe}/\rho_{Fe} = 2.0 \times 10^{-20}$ gauss cm^3 $= 2.0 \times 10^{-20}$ ergs/gauss. (b) $i = \langle \mu_{Fe} \rangle / (\frac{1}{4}\pi \times 10^{-16}\ \text{cm}^2) = 2.5 \times 10^{-4}$ emu $= 2.5 \times 10^{-3}$ amperes. (c) $i = ev/(\pi \times 10^{-8}\ \text{cm})$; $v = (i\pi \times 10^{-8}\ \text{cm})/e = 5.0 \times 10^8$ cm/sec. (d) $J = [(M_S)_{Fe}/\gamma] \times 1\ \text{cm}^3 = [(1700\ \text{gauss})/(-8.8 \times 10^6\ \text{sec}^{-1}\ \text{gauss}^{-1})] \times 1\ \text{cm}^3$ $= -1.93 \times 10^{-4}$ erg sec $= -1.93 \times 10^{-4}$ g cm^2 sec^{-1}.

3. energy levels of atoms and radiation

This chapter deals with a group of phenomena involving elementary processes of energy transfer between atoms, light, and electrons. The study of these energy transfers reveals a discrete structure of energy levels of both atoms and electromagnetic radiation. It also reveals a fundamental relationship between the magnitude of an elementary energy transfer and the rate of variation in time of currents, fields, or other variables.

The discrete structure of energy levels escapes detection in macroscopic experiments where one observes the aggregate amount of energy transferred in a multitude of atomic processes. The discontinuities of energy and, in general, the novel characteristics of atomic phenomena are called *quantum effects*. Quantum physics is the study of phenomena where quantum effects are apparent; macroscopic, or classical, physics is the study of phenomena where quantum effects fail to stand out.

Two basic types of experiments are considered in this chapter because their interpretation is particularly direct. They involve, respectively, the action of light and of low-energy electrons upon single atoms. The evidence concerning quantum effects provided by these experiments complements and clarifies greatly the evidence from emission and absorption spectra.

A large number of phenomena involving radiation, electrons, atoms, and their interactions fits qualitatively and semiquantitatively within the framework of concepts provided in this chapter. Phenomena of this kind are briefly described in supplementary sections; they were selected for instructive, historic, or informational value. An outline of techniques for detailed measurements of elementary processes is also given.

3.1 THE PHOTOELECTRIC EFFECT AND ITS IMPLICATIONS

The photoelectric effect is the release of electrons from matter under the influence of light or of other electromagnetic radiation. Visible light produces it at the surface of a number of substances, particularly alkaline metals. Light in the far ultraviolet produces it in most substances. Since light delivers energy to any material by which it is absorbed, it is understandable that some of this energy serves to remove electrons, as thermal energy does in a hot wire.

The photoelectric effect appears, in the first place, as a release of negative electricity. Mass spectroscopy verifies that the negative electricity consists of electrons. Procedures analogous to mass spectroscopy determine the kinetic energy of the *photoelectrons* ejected from the material.

Figure 3.1 shows a schematic photocell for the analysis of photoelectrons. Light is directed through a window onto the surface of the material under study in an evacuated vessel. Electrons, if any are released from the surface, are sucked up by a positive grid and may pass through its holes to reach a collecting electrode. A current meter measures the rate at which the photo-

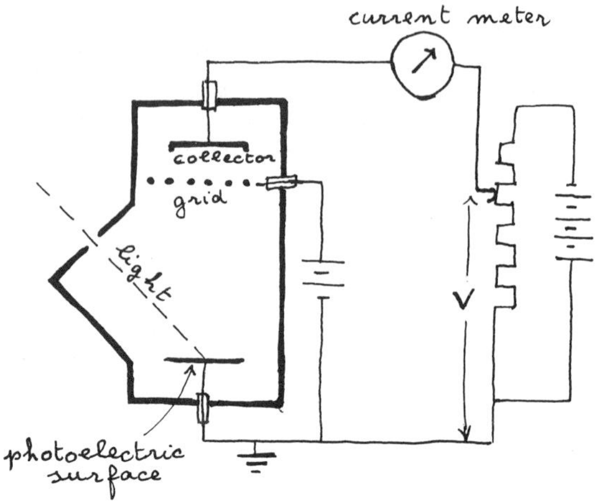

Fig. 3. 1. Diagram of cell for measurement of photoelectron energies.

electrons reach the collector. A negative potential difference of V volts is applied between the collector and the emitting surface. The collector repels the electrons and will receive only those which have left the surface with a kinetic energy of at least V electron volts, whereas lower energy electrons are turned back. Measuring the collector current i as a function of the potential V yields thereby an analysis of the energy distribution of the photoelectrons.

The experiments show that the number and the energy of the photoelectrons depend, respectively, on the intensity and on the frequency of the incident light. Other conditions being equal, the intensity of the photoelectric current is proportional to the intensity of the incident light. On the other hand, the energy distribution of the photoelectrons depends on the frequency ν (or wavelength λ) of the light.

In the first place, no photoelectric effect whatsoever is observed unless the incident light contains spectral components of frequency sufficiently far in the direction of ultraviolet. The light may be filtered preliminarily through a monochromator, which lets through radiation with frequency very close to any desired value of ν. A photoelectric current is observed only if ν exceeds a threshold value ν_0 which depends exclusively on the physico-chemical nature of the photoelectric surface. For example, the threshold frequency is 6.0×10^{14} cycles/sec for a clean sodium crystal surface, 12.1×10^{14} for nickel, and 11.7×10^{14} for gold. The threshold is wholly independent of the light intensity. Electromagnetic radiation is thus shown to have a potency which depends on its frequency. High-frequency radiation achieves effects which lower-frequency radiation cannot achieve, no matter how intense it is.[1]

1. Experiments on the chemical and biological actions of electromagnetic radiation also show very clearly that short-wave (high-frequency) radiation achieves effects that lower-frequency radiation does not achieve even upon delivery of much larger amounts of energy.

This concept is refined by the experimental analysis of photoelectron energies. One measures the photoelectric current i as a function of the counter-potential V. For incident light of any given frequency ν, above the threshold ν_0, i(V) decreases as the counter-potential V increases, and vanishes at, and above, a definite potential V_s which stops all photoelectrons (see Fig. 3. 2).

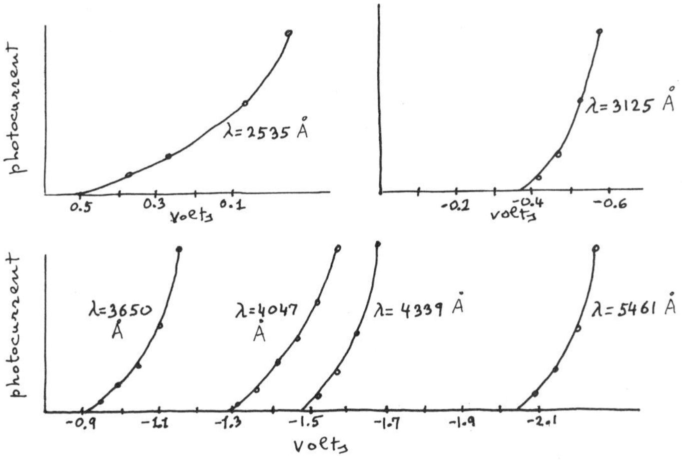

Fig. 3. 2. Photoelectric current intensity versus counter-potential for light of different wavelengths incident on sodium. Vertical scale shows relative current intensities. Horizontal scale shows counter-potential uncorrected for contact potential between sample and circuitry. Note extrapolation from experimental points to zero current. [Adapted from R. A. Millikan, *Phys. Rev.* vol. 7 (1916).]

The value of V_s may be measured for various frequencies and plotted against ν. The result, shown in Figure 3. 3, is a straight line, whose slope has the same value, 4.136×10^{-15} volt sec, [2] for all photoelectric surfaces. That is, plots corresponding to different surface materials yield parallel lines. The threshold frequency ν_0 is found by extrapolating the experimental plot to its intercept, as shown in Figure 3. 4.

For $\nu < \nu_0$ there are, of course, no photoelectrons and hence no measurable value of V_s. If the plot of Figure 3. 4 is nevertheless extrapolated to the axis $\nu = 0$, one finds a negative intercept $V_0 = 4.136 \times 10^{-15} \nu_0$. The value V_0 thus determined coincides with a characteristic of the surface material which is known from other phenomena, namely, the *extraction potential* or *work func-*

---

2. When frequencies are measured in cycles/sec, the slope, which represents the ratio potential/frequency, is expressed in volts/sec^{-1} = volt sec.

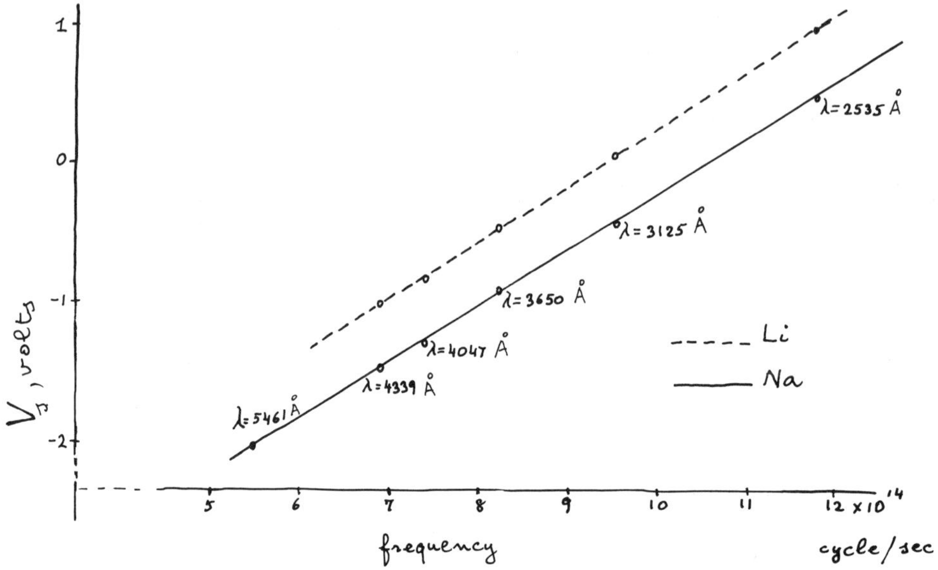

Fig. 3. 3. Plot of potential V_s which stops all photoelectrons versus light frequency. Values of V_s for Na are taken from the intercepts of figure 3. 2. [Adapted from R. A. Millikan, *Phys. Rev.* vol. 7 (1916).]

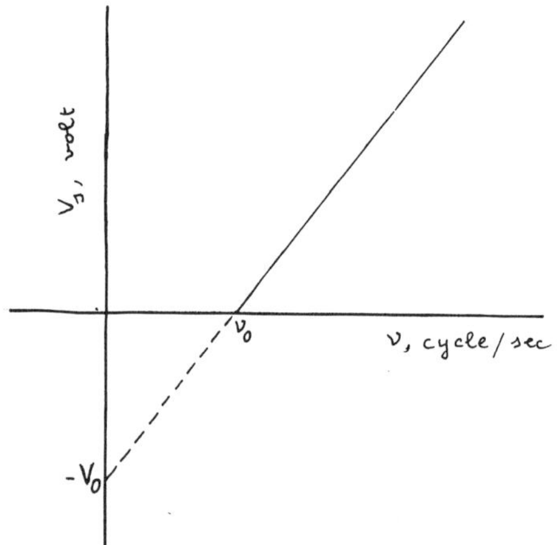

Fig. 3. 4. Plot of counter-potential V_s necessary to stop all photoelectrons emitted by a given material under the influence of light of frequency ν. For light of frequency smaller than ν_0 no photoelectrons are emitted. The slope of the plots is the same for all materials while ν_0 is a characteristic of the material and the extraction potential V_0 equals $4.136 \times 10^{-15}\ \nu_0$.

tion. This is the electric potential which holds electrons within a material and prevents them from escaping out of its surface.[3]

The energy which each photoelectron receives from the radiation consists then of two parts. One portion is spent in extracting the electron from the surface and amounts to *at least* eV_0, that is, to at least V_0 electron volts. The other portion becomes kinetic energy of the electron and amounts to *no more* than eV_S that is, than V_S electron volts, because no photoelectron surmounts a counter-potential greater than V_S. Einstein surmised in 1905 that each electron receives an amount of energy equal to $eV_0 + eV_S$; those electrons which were bound within the material by more than the minimum energy eV_0 emerge with correspondingly lower kinetic energy.

This surmise is confirmed by studying the photoelectric effect produced in monatomic gases and metal vapors rather than on solid surfaces. The threshold frequency ν_0 lies in the ultraviolet, often in the far-ultraviolet, for gases and vapors. Here, all photoelectrons ejected by monochromatic light with a given frequency have equal kinetic energy K.[4] This means that all emerging photoelectrons were bound within their atom with the same energy eV_0. Measurement of the single energy K provides, therefore, the value of eV_S directly, whereas in the case of solids V_S is obtained as the extrapolated intercept of a curve in Figure 3.2. Consequently, the interpretation of the photoelectric effect in gases is more compelling than for solids. Historically, the effect was studied primarily for metals because the experiments are easier. The difficulties of photoelectric experimentation in gases, outlined in section S3.8, have been surmounted only recently. It is now also verified that the photoelectric effect in a gas results from elementary processes, since the number of photoelectrons is proportional to the number of atoms exposed to light (see section 1.3). The number of photoelectrons produced in a gas can be counted, thus completing the quantitative study of the effect.

According to Einstein's interpretation, the total energy received by each electron is proportional to the radiation frequency, because $eV_0 + eV_S = e(V_0 + V_S)$ and $V_0 + V_S$ equals the frequency ν multiplied by 4.136×10^{-15} volt sec, as shown in Figure 3.3. This relationship is expressed by Einstein's equation

$$E = eV_0 + eV_S = h\nu, \tag{3.1}$$

---

3. The difference between the value of V_0 for different metals causes electrons to pour through the contact surface between two metals in the Volta effect. It also causes hot surfaces of different metals to release electrons at different rates in the thermionic effect.

4. To be exact, photoelectric processes in which electrons require different energies for extraction occur also in gases. Here, there is a number of different possible extraction energies $V_0, V_1, V_2, \ldots$, and only the lowest one, V_0, is relevant as long as $h\nu < eV_1$. In solids the different possible extraction potentials are not separated because of interaction among atoms.

where the proportionality factor

$$h = 4.136 \times 10^{-15} \text{ eV sec} = 6.626 \times 10^{-27} \text{ erg sec} \tag{3.2}$$

is called the *Planck constant*.

All subsequent evidence has confirmed that energy exchanges between atomic systems and electromagnetic radiation obey equation (3.1). That is, electromagnetic radiation energy is always absorbed or emitted in discrete amounts of magnitude proportional to the radiation frequency with the proportionality constant h. The elementary quantity of radiation energy transferred in an elementary process is called a *photon* or *light quantum*.

One commonly says that radiation energy is subdivided into photons. The meaning of this expression is qualified by the fact that radiation energy is observed only through its absorption and emission by matter. In fact, "radiation energy" is a name that designates energy traveling from one to another portion of matter or otherwise stored in space.

The subdivision of radiation energy into finite units escapes detection in all macroscopic experiments that fail to bring out the effects of interaction between radiation and single atomic particles (10 watts of visible light emitted by a lamp amount to about 3×10^{19} photons/sec). The discontinuous structure of radiation energy had been inferred by Planck in 1900, before Einstein's interpretation of the photoelectric effect, from an analysis of the thermal equilibrium between matter and radiation (section S3.5). His discovery showed that discontinuities akin to the atomic structure of matter are more widespread than had been expected.

The discontinuous or "atomistic" structure of radiation energy and of matter are similar insofar as photons, like atomic particles, enter elementary processes or emerge from them as single units. Other properties of matter and radiation are quite different. An atom occupies a certain portion of space and no other atom can be crowded into that space without profound disruption to both atoms; therefore, one cannot raise the density of a solid material much above its normal value without crushing its atomic structure. The intensity of electromagnetic radiation in any region of space is subject to no analogous limitation.

Another analogy between radiation and matter emerges when their resistance to compression is examined from an atomistic point of view. Consider an electromagnetic standing wave confined between opposite reflecting walls,[5] for example, microwave radiation in a cavity resonator or light between the two mirrors of a laser.[6] Pushing the walls together against the radiation pressure shortens the wavelength and increases the frequency of the standing wave. (Displacement of the walls serves, in fact, to tune the radiation frequency.) The work performed by pushing increases the radiation energy; from an

---

5 *BPC*, vol. 3, chap. 7.1.

6. B. A. Lengyel, *Introduction to Laser Physics* (New York: John Wiley, 1966).

atomistic point of view what increases is the energy of each photon rather than the number of photons. This statement can be verified by an exact calculation of the frequency shift and of the work performed. One may then say that each photon resists compression, much as each atom does.

The discontinuous structure of radiation energy permits relative, or even absolute, measurements of radiation intensity by counting. Since the photoelectric current is proportional to the intensity of incident light, a count of the number of ejected photoelectrons—which is practical at low intensity—constitutes a relative measurement of light intensity. This procedure is commonly called *photon counting*. When the cross-section for the production of photoelectrons is known, photon counting provides an absolute measurement of radiation intensity.

The relationship between energy transfer and oscillation frequency revealed by the photoelectric effect links the energy and time variation of phenomena in a previously unsuspected manner. The energy transferred in an elementary process is proportional to the oscillation frequency of the incident light and of the intraatomic currents that absorb the light. This type of relationship prevails throughout quantum physics. The delivery of a large amount of energy in a single process requires a rapidly varying action; conversely, any rapidly varying action is capable of delivering a large amount of energy in an elementary process. This relationship will be made more precise in the course of our study.

3.2 INELASTIC COLLISIONS OF ELECTRONS WITH ATOMS

An *elastic collision* between two bodies is one in which the internal energy of each body remains unchanged, whereas in an *inelastic collision* the internal energy usually increases at the expense of kinetic energy. If an electron collides with an atom at rest and the collision is elastic, the atom hardly recoils because it is much heavier than the electron and takes up a negligible fraction of the electron's kinetic energy. Therefore, if an electron emerges from a collision with a substantial energy loss, the collision must have been inelastic and the energy must have been absorbed by the atom. Observation of the energy losses experienced by electrons serves to study the energy absorption of atoms.

The first fundamental experiment was performed by Franck and Hertz in 1914. Their main result was that incident electrons experience no inelastic collisions unless their energy exceeds a threshold value E_1 characteristic for the species of atom. At somewhat higher energy many electrons appear to have lost just the energy E_1. Here we shall describe the results obtained by modern experiments.

The conceptual elements of the experimental apparatus are indicated most schematically in Figure 3.5. An outline of actual procedures is given in section S3.8. The equipment consists of three main parts:

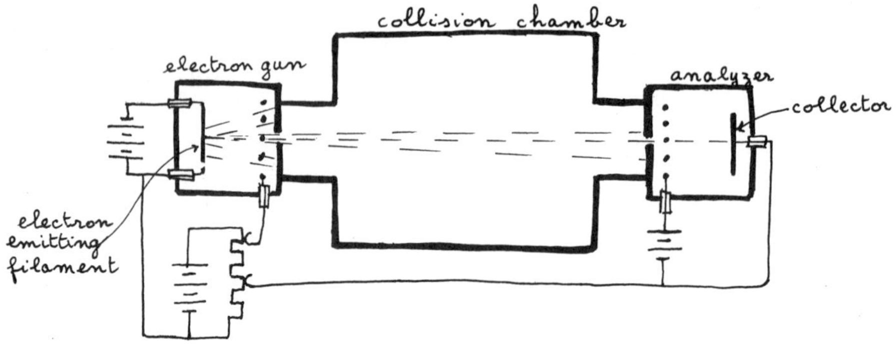

Fig. 3. 5. Diagram of apparatus for the measurement of energy losses by electrons in a gas.

(1) A device to produce a beam of electrons with uniform kinetic energy. This device, called an *electron gun,* may consist of a hot wire (filament) which releases large numbers of electrons with low velocity, of a grid or other electrode at a positive potential which attracts the electrons and lets them through, and of slits or pinholes in diaphragms which collimate the electrons. The energy of the electrons in the beam is determined by the potential of the grid. The electron gun is evacuated to avoid collisions of the electrons with gas atoms.

(2) A collision chamber containing a rarefied monatomic gas or vapor. At a pressure of 0. 01 mm Hg, electrons travel on the average about 1 cm before colliding with gas atoms; the occurrence of repeated collisions by an electron traversing the chamber may thus be minimized by keeping the pressure low.

(3) A device to analyze the kinetic energy of electrons which have traversed the chamber. This device might be of the simple type indicated in figure 3. 1 for a photocell, that is, it may consist only of a collector at a variable negative potential.

Additional devices may be attached to the collision chamber to analyze any by-products of the collisions.

3. 2. 1. *The Energy Levels of Atoms.* The experiments show, first of all, that electrons with kinetic energies up to a few electron volts experience only elastic collisions with atoms.[7] As the electrons' energy is increased, all collisions remain elastic until the energy reaches a threshold value E_1 which is characteristic of the chemical species of atom in the collision chamber, for example, 4. 9 eV for Hg, 1. 4 eV for Cs, 16. 6 eV for Ne.

7. This result holds only for collisions against isolated atoms; inelastic collisions occur at lower energies when atoms are grouped in diatomic or polyatomic molecules.

When the energy exceeds the threshold E_1, many electrons emerge from the chamber with kinetic energy reduced by an amount just equal to E_1, whereas the others lose no significant amount of energy (Fig.3. 6). This means that many electrons have experienced an inelastic collision in which a fixed amount of energy E_1 is transferred from an electron to the internal energy of an atom. The number of electrons that have lost energy is proportional to the density of atoms in the collision chamber provided this density is sufficiently low; according to the criteria of section 1. 3, the inelastic collisions are thus shown to be elementary processes.

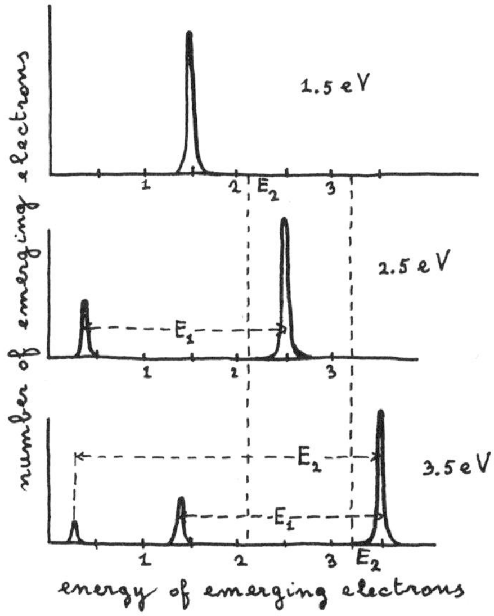

Fig. 3. 6. Diagram of energy distribution of electrons emerging after collision with Na atoms. The incident electron energies are 1. 5 eV, 2. 5 eV, and 3. 5 eV. The first two excited levels of Na lie at $E_1 = 2.1$ and $E_2 = 3.2$ eV.. For an example of actual energy distribution see Fig. 3. 12.

New types of inelastic collisions set in successively, when the energy E of the bombarding electrons is raised above successive thresholds E_2, E_3, ... Electrons emerge from the chamber with energies reduced to $E - E_2$, $E - E_3$, . . . , showing that collisions have occurred with energy transfers E_2, E_3, . . . (Fig. 3. 6).

The lack of inelastic collisions at subthreshold bombarding energies shows that the atoms are left after collisions in unchanged states of internal energy. When inelastic collisions take place, at bombarding energies above threshold, the energy lost by an electron must have been retained by an atom which is thereby left in a new state of internal energy. To each of the discrete amounts of energy that may be lost by electrons must correspond a specific state of internal energy of the atoms after collision. All atoms of the same element must accordingly possess a characteristic discrete set of states with different *levels of internal energy* in which they can exist after collisions.

These states are called *stationary states*. The name applies to all states of isolated quantum systems (atoms, electrons, radiation, etc.) which have a definite energy. This energy changes only through interaction with another system. For example, an atom which has absorbed energy in a collision remains in a stationary state only insofar as it retains that energy, that is, as long as it is effectively isolated. Any emission of light by the atom after collision involves a transfer of energy; it implies that the atom is no longer isolated, in a stationary state, but that it interacts with the radiation fields.

The existence of stationary states with well-defined, clearly separated energy levels is a typical quantum effect. By contrast, the internal energy of macroscopic bodies appears to be a continuous variable which may change ever so little.

The stationary state of lowest energy, which is the normal, stable state of a free atom, is called the *ground state*; the others are called *excited states* because their higher level of internal energy implies a more active internal motion.

The threshold energies E_1, E_2, E_3, ... observed in electron collisions experiments are also called *critical potentials*.[8] The differences between successive critical potentials decrease rapidly to a point where they can no longer be distinguished by analyzing the energies of electrons emerging from the chamber. The analysis of energy levels can, however, be continued by observing by-products of inelastic collisions.

3.2.2. *Ionization*. Positive ions appear in the collision chamber of the experiment of Figure 3.5, when the bombarding electron energy exceeds a new critical potential, somewhat higher than the potentials considered thus far. The presence of positive ions is demonstrated by inserting a negatively charged grid in a wall of the chamber and analyzing by mass spectroscopy the particles flowing through the grid. That electrons have been ejected from the atoms in the chamber is confirmed by an increase in the flow of electrons out of the chamber.

As the energy E of the bombarding electrons is increased, starting from a low value, no ions are detected until E reaches a threshold E_I, but thereafter the yield of ions rises sharply. The threshold E_I, expressed in volts of accelerating potential, is called the *ionization potential* of the element and coincides

---

8. The term "potential" refers to the accelerating potential applied to the grid of the electron gun; a critical potential should accordingly be expressed in volts, rather than in electron volts or in other energy units. In practice this distinction is seldom maintained and critical potentials are often expressed in eV.

with the extraction potential V_0 determined from the photoelectric effect in free atoms of the same element. The energy E_I is also called the binding energy of the electron which is ejected, because it represents the energy required to release the electron by breaking the bond which was holding it within the atom.

The variations of the ionization potential from one chemical element to another are indicated in Figure 3.7 and follow a periodic pattern in accordance with chemical classification. The lowest ionization potentials are observed in alkali atoms (3 to 6 volts), the highest in rare gases (up to 24 volts). On the whole, the ionization potentials decrease slightly from one row to the next of the periodic system.

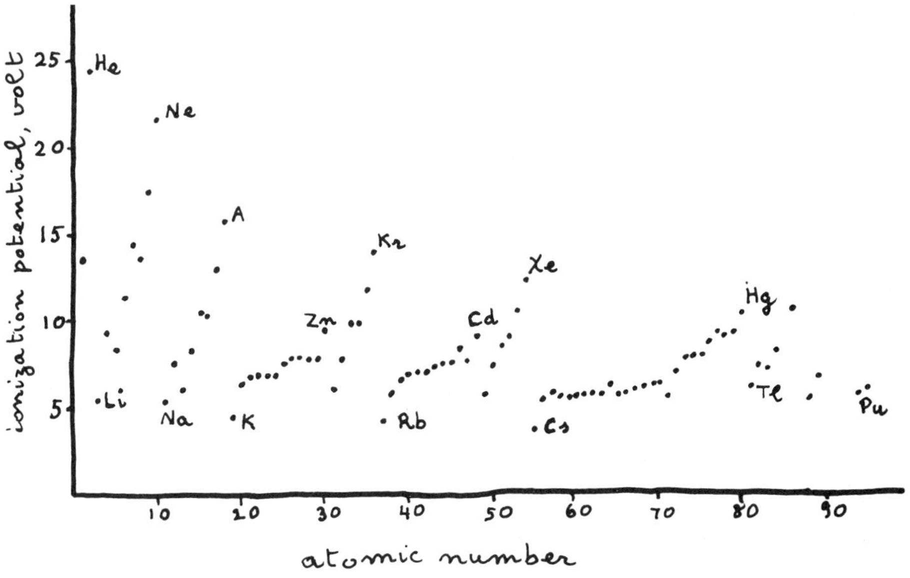

Fig. 3.7. Ionization potentials of the elements in vapor form.

When ionization takes place, electrons emerge from the collision chamber with a continuous distribution of energies. Two electrons emerge from each ionizing collision: one had arrived from the electron gun and one is released from the atom. These electrons may share in any proportion the energy $E - E_I$ available after the ionization requirement.

If the energy of the bombarding electrons is raised further above the ionization threshold E_I, the yield of ions from any element but hydrogen shows successive rises at higher critical potentials. These rises are often accompanied by the onset of light emission with new frequencies, or by the appearance or doubly or multiply charged ions.

3.2.3. *Radiation Emission.* Light usually emerges from the collision chamber of Figure 3.5 when the incident electron energy exceeds the first threshold E_1 and inelastic collisions take place. The light is monochromatic, and its intensity is proportional to the current of bombarding electrons. The outstanding experimental fact is that the light frequency is proportional to the energy E_1 in accordance with the photoelectric equation (3.1), that is, $\nu_1 = E_1/h$. For example, when sodium atoms are hit inelastically by electrons with energy in excess of $E_1 = 2.1$ eV, they emit the characteristic yellow light with wavelength 5.9×10^3 Å. The frequency of this light is $\nu = c/\lambda = 5.1 \times 10^{14}$ cycles/sec, and coincides with the ratio $E_1/h = 2.1$ eV$/4.1 \times 10^{-15}$ eV sec.

We have here a process reciprocal to the photoelectric effect: in the photoelectric effect radiation of frequency ν delivers to matter amounts of energy $E = h\nu$; here atoms which have an excitation energy E_1 emit radiation of frequency $\nu_1 = E_1/h$. It appears that in the course of light emission the electromagnetic field receives from the atoms photons of energy $h\nu_1$. The radiation emission consists of elementary processes; in each process an atom is understood to return from the excited state to its ground state and to release the excitation energy E_1 as a radiation photon.

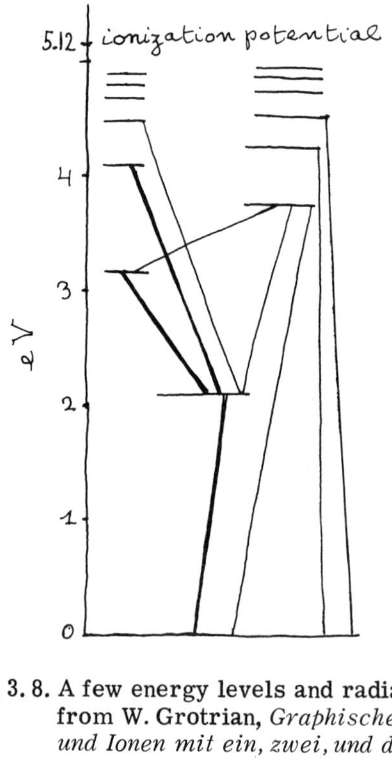

Fig. 3.8. A few energy levels and radiative transitions for sodium. Adapted from W. Grotrian, *Graphische Darstellung der Spektren von Atomen und Ionen mit ein, zwei, und drei Valenzelektronen* (Springer, 1928).

When the energy of the bombarding electrons is raised progressively above the successive thresholds E_2, E_3, . . . , one observes, in general, the emission of light with new frequencies, first, $\nu_2 = E_2/h$ or $\nu_{21} = (E_2 - E_1)/h$, then, $\nu_3 = E_3/h$, $\nu_{31} = (E_3 - E_1)/h$, etc. Thus atoms return from excited states to the ground state either in a single emission process or in successive steps through intermediate levels. In each elementary process the photon energy emitted, $h\nu$, equals the energy difference between the initial and final state of the atom (Fig. 3. 8).[9]

When a critical potential is detected through the onset of light emission of a specific frequency, it is often called a *radiation potential.* The measurement of radiation potentials permits a better analysis of atomic energy levels than the measurement of electron energies after collisions. However, the precision of this measurement is still limited by inhomogeneity of the energies of bombarding electrons for a given potential setting of the electron gun. In general, the best analysis of levels is provided by the study of light emission irrespective of the method of excitation, as discussed in the next section.

3.3 ENERGY LEVELS AND OPTICAL SPECTRA

The experiments on radiation potentials of monatomic gases establish a one-to-one quantitative correspondence between pairs of energy levels E_n $E_{n'}$ of an atom and the frequencies $\nu_{nn'}$ of emitted light. Since each frequency is itself associated with a pair of spectral terms τ_n, $\bar{\tau}_{n'}$ according to the combination principle (section 2. 3), there is actually a one-to-one correspondence between energy levels and spectral terms. Quantitatively, we have

$$\nu_{nn'} = \frac{E_n - E_{n'}}{h}; \tag{3.3}$$

that is,

$$\omega_{nn'} = \frac{E_n - E_{n'}}{\hbar}, \tag{3.3a}$$

where

$$\hbar = \frac{h}{2\pi}. \tag{3.4}$$

Comparison of equations (3. 3a) and (2. 38) establishes the relationship

$$E_n = \hbar\tau_n, \tag{3.5}$$

provided that the origins (zero points) of the scales of energies and of spectral term values are suitably chosen.

It was stressed in section 2. 3 that spectral terms are defined to within an

---

9. All emission processes consistent with this rule do not take place with comparable intensity. Many of them are hardly observable.

arbitrary additive constant which fixes the zero point of their scale of values. The same holds for energies in general since experiments always measure energy transfers or energy differences. In particular, the Franck-Hertz experiment measures *energy transfers* between electrons and atoms. The introduction of absolute spectral-term values and energy-level values constitutes a convenient bookkeeping device which requires adoption of an arbitrary zero point for their scales.

It was also pointed out in section 2.3 that there exist two separate sets of spectral terms, indicated by τ_α and $\bar{\tau}_\beta$. Observed spectral frequencies are normally represented as differences between one term of one set and one term of the other set. The correspondence (3.5) implies that energy levels are also classified into two sets. The origin of this classification will become apparent in chapters 13 and 16.[10]

The correspondence between discrete energy levels and spectral terms is matched by a correspondence between ionization potentials and threshold frequencies of continuous spectra. (Recall that these frequencies also represent photoelectric thresholds). The equation analogous to (3.5) is

$$E_I = \hbar\omega_t, \tag{3.6}$$

where ω_t is the threshold frequency introduced in section 2.2.1 and E_I is the ionization potential expressed in energy units.

Owing to the law (3.5) the names "energy level" and "spectral term" have become essentially synonymous and are generally employed interchangeably[11] being expressed either in eV or in frequency units or, more frequently, as wavenumbers (i.e., in cm^{-1}). The most extensive and accurate information on energy levels is obtained from emission or absorption spectra rather than from electron-collision experiments. Equations (3.5) and (3.6) have been verified, directly or indirectly, very widely and very accurately. Historically, the quantum interpretation of spectroscopic data resulted from Einstein's theory of the photoelectric effect and from the combination principle (section 2.3) but preceded the Franck-Hertz and related experiments which were designed to verify the existence of energy levels.

The occurrence in absorption and emission spectra of line series, whose frequencies converge to a limit on the high-frequency side, corresponds to the existence of energy levels in which a single electron is understood to be more and more loosely bound to the rest of the atom. The binding vanishes at the series limit which corresponds to an energy sufficient for the electron's escape from the atom, i.e., for ionization. The continuous absorption spectra,

---

10. The two sets are said to have *opposite parity*.

11. However, these two names have acquired again a different meaning among professional spectroscopists, in connection with the occurrence of a fine structure, i.e., of groups of levels with very small separation. In this case the word "term" refers usually to a whole group of levels or to their average value.

for frequencies extending above a threshold ω_t, correspond to absorption of radiation energy spent in the photoelectric effect, that is, in elementary processes of *photoionization*. Following this process the atom is left in an ionized state. This state may still be regarded as a stationary state of the atom even though one of its electrons is detached. The energy of this state is equal to the energy of the ground state of the atom plus the energy of the absorbed photon. The continuous emission spectra consist mostly of radiation emitted in a process inverse to photoionization, namely, in the attachment of an electron to a preexisting ion.

In the highly ionized gases called *plasmas* one also observes continuous absorption and emission spectra associated with *free-free transitions*. In these processes an unattached electron is briefly subjected to attraction by an ion, thus forming with it a temporary compound. Free-free transitions take place between different stationary states of this fleeting compound with energy transfers to or from the radiation field. They also occur in the temporary compounds formed by an electron with neutral atoms.

We have seen in section 3. 2. 3 how excitation of an atom to a high level by electron bombardment can be followed by light emission of different frequencies. For example, Figure 3. 8 indicates how excitation to the level E_3 by electron collision can be followed by radiation emission with frequencies $\omega_{30} = (E_3 - E_0)/\hbar$ or $\omega_{31} = (E_3 - E_1)/\hbar$. The same emissions also result from excitation of E_3 by absorption of light with frequency ω_{30}. Light emission with frequencies ω' which follows absorption of a frequency $\omega \geqslant \omega'$ is called *fluorescence*.

The connection between continuous absorption spectra and photoelectric emission can be illustrated by considerations of Fourier analysis. In chapter 2 we had started from the point of view that absorption of radiation of any one frequency is due to an intra-atomic dipole oscillation with that frequency. Recall now from section A. 3 that superposition of oscillations with a continuous set of frequencies can result in a nonperiodic process. The escape of an electron from an atom constitutes a nonperiodic variation of dipole moment. Description of photoemission as occurring during a limited time interval requires that the incident radiation be represented as a traveling wave packet rather than as a strictly monochromatic wave. Accordingly, neither the photon energy nor the energy of the ionized state of the atom is sharply defined and this state is no longer regarded as stationary. The same point of view regards an atom as expanding to infinite size when an electron is ejected by photoionization. In this phenomenon the infinite size of the atom is associated with its continuous spectrum of free oscillations; the association of infinite size and continuous spectrum is a general feature of mechanical systems such as a string of infinite length, as was noted in section 2. 2. 1.

3. 4. PRELIMINARY SUMMARY OF QUANTUM EFFECTS

The experiments on energy exchanges between atoms and radiation or electrons suggest certain conclusions whose general validity is demonstrated by countless additional experiments. First, an isolated atom exchanges energy with other systems only in discrete (*quantized*) amounts characteristic of each species of atom, unless the atom is ionized. An elementary process of energy exchange involves a transition of the atom from one stationary state (or energy level) to another. The radiation or the electron which exchanges energy with the atom also experiences a transition between two stationary states. In particular, monochromatic radiation receives energy from—or imparts energy to—atoms in discrete amounts (photons) whose magnitude is proportional to the radiation frequency. Here again one speaks of transitions between stationary states of the radiation, whose energy levels differ by $h\nu$.

Second, the macroscopic concept that monochromatic radiation is emitted and absorbed by electric currents oscillating with the radiation frequency is now complemented by an additional finding. This is that each monochromatic oscillation of radiation or atom corresponds to a *pair of energy levels* separated by the interval $h\nu$.

The existence of discrete energy levels and the correspondence of oscillation frequencies to pairs of levels are both foreign to macroscopic physics. Yet they are not inconsistent with it because they emerge only from experiments bearing on single atoms or particles. A theoretical formulation of these quantum phenomena and of their relation to macroscopic physics remains to be worked out in later chapters.

We anticipate here that the correspondence between oscillations and energy levels is not confined to harmonic or approximately harmonic oscillations. When any variable describing the evolution of a phenomenon in the course of time is represented by Fourier analysis, each Fourier component of frequency ω is associated with a pair of stationary states separated by energy $\hbar\omega$.

Notice how quantum effects limit the degradation of energy in a chain of elementary processes. Consider an electron which strikes an atom—for instance, in a star—and excites it with transfer of a definite amount of energy. The atom may start oscillating with a correspondingly definite frequency and emit one photon of radiation of that frequency. The radiation will eventually drive an oscillating current in another atom "tuned" to that frequency, perhaps in another star, and will be absorbed by this atom which thereby recovers the very amount of energy lost by the initial electron. Thus the energy originally transferred to a single atom is *not* degraded or dispersed in different directions, even though, macroscopically, the energy emitted by an assembly of atoms is distributed in all directions.

Compare also the macroscopic and atomistic aspects of the distribution in time of the energy emitted by a source. When oscillating currents of a single frequency are started in an assembly of atoms, progressive loss of energy by radiation emission causes the current intensity to decrease as an exponential function of time. Equation (2. 36) shows how the amplitude of dipole oscillations of frequency ω_r decreases in proportion to $\exp(-\frac{1}{2}\Gamma_r t)$; the intensity of light emission decreases in proportion to the squared amplitude, that is, to $\exp(-\Gamma_r t)$. This exponential decay is observed experimentally as a decay of luminosity of canal rays which travel away from the discharge. Various types of decay experiments have been designed for the direct measurement of Γ_r. For instance, beams of ions are accelerated and then passed through a foil from which they emerge with oscillating dipole moments. The luminosity of the beam is measured as a function of the distance from the foil. In other experiments a rarefied gas is hit by a very brief pulse of fast electrons. The induced luminosity of the gas is then measured as a function of time after the pulse. In these macroscopic experiments excitation energy appears to be dissipated gradually. On the other hand, detection by photon counting of the light emitted along a beam reveals absorption of photons of a single energy even though the counter is collimated to receive light only from a very short section of the beam. Of course, the rate of photon detection decreases exponentially as the spot observed through the collimator is moved down the beam.

The association between frequency and energy levels of radiation also has striking effects on electromagnetic standing waves confined between reflecting walls, as noted briefly in section 3.1. Narrowing of the space between the walls increases the radiation frequency and, hence, its photon energy, that is, the separation between its energy levels. This connection between confinement and energy levels is but a first example of a general phenomenon. It will be seen in chapter 11 that increased confinement of an electron or other particle increases its kinetic energy; therein lies the source of the stability of atoms and their resistance to compression.

SUPPLEMENTARY SECTIONS

S3.5. BLACKBODY RADIATION

The idea that radiation energy has discrete levels separated by intervals $h\nu$ originates historically from the analysis of the light spectrum emerging from an oven (blackbody radiation). This spectrum is continuous and does not depend on the material used in the oven walls because it results from repeated energy exchanges between the oven walls and radiation within the empty[12] space in the oven under conditions of thermal equilibrium.[13]

The very idea that empty space contains thermal energy in the form of radiation emitted by matter appeared to lead to unacceptable results in the absence of quantum concepts. The concept of thermal energy in the classical kinetic theory of gases implies that each degree of freedom of matter has an energy $\sim k_B T$ at temperature T (k_B = Boltzmann constant). If this concept is extended to radiation in space, each component of the electric and magnetic field vectors at each point of space constitutes one degree of freedom. Space, having

---

12. Air in the oven does not matter here.

13. *BPC*, 4, section 1: 35-40.

infinitely many points, would have infinitely many degrees of freedom per unit volume and, therefore, an infinite heat capacity.

Planck realized as early as 1900 that this paradox would be removed if, somehow, the exchanges of energy between radiation and the material of the oven occur in discrete amounts $h\nu$. This assumption has, of course, been confirmed by the photoelectric effect and other evidence. Here, we develop its consequences by a method of Fourier analysis.

The radiation in the oven is regarded as a superposition of monochromatic waves of all frequencies ν. The total energy of the radiation is the sum of the energies of all monochromatic components each of which counts as one degree of freedom. There are infinitely many monochromatic waves, but those of high frequency do not contribute appreciably to the heat capacity for the following reason. Adding to the radiation in the oven only one photon of frequency ν requires an amount of energy $h\nu$. When $h\nu \gg k_B T$, it is unlikely that this much energy would be concentrated on a single degree of freedom.

Quantitatively, statistical mechanics teaches that excitation of a state of energy $h\nu$ has, at temperature T, a relative probability equal to the Boltzmann factor $\exp(-h\nu/k_B T)$.[14] The relative probability of an n-photon excitation equals the Boltzmann factor $\exp(-nh\nu/k_B T)$. Accordingly, the mean energy of a monochromatic wave with frequency ν is obtained by averaging its energy levels weighted by their Boltzmann factors

$$\langle E \rangle_\nu = \frac{h\nu \exp(-h\nu/k_B T) + 2h\nu \exp(-2h\nu/k_B T) + \ldots}{1 + \exp(-h\nu/k_B T) + \exp(-2h\nu/k_B T) + \ldots} = \frac{h\nu}{\exp(h\nu/k_B T) - 1}. \qquad (3.7)$$

This equation shows that $\langle E \rangle_\nu \sim k_B T$ for $h\nu \ll k_B T$, as expected, but $\langle E \rangle_\nu \sim 0$ for $h\nu \gg k_B T$.

To obtain the total amount of radiation energy in the oven and its frequency distribution, one must combine equation (3. 7) with a knowledge of the number, $\rho(\nu)d\nu$, of different monochromatic waves in the frequency interval between ν and $\nu + d\nu$. This number can be counted if one considers a finite volume of space analogous to the finite interval of time 2T in the Fourier expansion of section A3. We take without proof that this number is[15]

$$\rho(\nu)d\nu = 8\pi\nu^2 d\nu/c^3 \qquad (3.8)$$

per unit volume. The spectral energy density of radiation at temperature T is then

---

14. See, for example, F. Reif, *Fundamentals of Statistical and Thermal Physics* (New York: McGraw-Hill, 1965), section 6. 2, or, for a brief summary, Appendix E.

15. For a derivation of this formula see, e.g., F. K. Richtmyer, E. H. Kennard, and J. N. Cooper, *Introduction to Modern Physics* (New York: McGraw-Hill, 1969), section 5. 3.

$$\langle E \rangle_\nu \rho(\nu) = \frac{h\nu}{\exp(h\nu/k_B T) - 1} \frac{8\pi\nu^2}{c^3}. \quad (3.9)$$

This is the famous Planck function. The peak of this function occurs at a characteristic value of the numerical ratio $h\nu/k_B T$. Prior to Planck's work it had been observed that the high-frequency portion of the spectrum is a function of the ratio ν/T (Wien's law), but no constant of nature was known that would establish a scale on which ν and T would be commensurable, that is, would have the same physical dimensions.

The total energy of radiation per unit volume of the oven is obtained by integrating equation (3.9) over the whole frequency spectrum. Setting $x = h\nu/k_B T$ one obtains

$$E = \int_0^\infty \langle E \rangle_\nu \rho(\nu) d\nu = \frac{8\pi (k_B T)^4}{h^3 c^3} \int_0^\infty \frac{x^3 dx}{\exp(x) - 1} = \frac{8\pi^5 (k_B T)^4}{15 h^3 c^3}. \quad (3.10)$$

This equation shows not only the proportionality of E to T^4 (Stefan-Boltzmann law) but also the value of the proportionality coefficient. The theoretical value of E obtained by Planck agreed with the result of earlier experimental studies. The derivative dE/dT represents the finite heat capacity of space per unit volume.

S3.6. BOHR'S THEORY OF THE HYDROGEN ATOM

Bohr performed in 1913 a limited but unbelievably fruitful graft of quantum concepts upon the classical mechanics of an electron moving in the Coulomb field of a nucleus. His points of departure were Rutherford's model of atomic structure and the experimental connection between energy levels and spectral frequencies (section 3.3).

Characteristically, Bohr focused at the outset on the motion of an electron which is almost detached from an atom and thus follows a very large orbit in the field of an ion—not necessarily of a bare nucleus. Classical mechanics must be applicable to this large orbit, at least as a very good approximation; its consistency with quantum theory is apparent for the following reason. Even though energy levels of the electron are discrete, their separation may be so small as to remain undetected. This separation is, nevertheless, known to exist from quantum physics; it equals $h\nu$, where ν is the frequency of the radiation emitted by the orbital motion, that is, the frequency of revolution of the electron on a stable orbit. Accordingly, the level separation $E_n - E_{n-1}$ is set equal to $h\nu_n$, where ν_n is the orbital frequency of an electron with energy E_n. The relation between ν_n and E_n is given, in essence, by Kepler's law of orbital periods.[16] The level separation is thus given as a function of E_n itself and Bohr could reconstruct the whole spectrum of high excitation levels in the following manner.

---

16. *BPC*, 1: 288-89.

A circular orbit is stable if the centrifugal force on the electron balances the electric attraction exerted on it by a nucleus with charge Ze, that is, if

$$\frac{m_e v^2}{r} = \frac{Ze^2}{r^2}, \tag{3.11}$$

where m_e and v are the mass and velocity of the electron and r is the radius of the orbit. The electron energy consists of kinetic energy and of a potential energy which represents the attraction by the nucleus and is, accordingly, negative:

$$E = \tfrac{1}{2} m_e v^2 - Ze^2/r. \tag{3.12}$$

Combination of (3.12) and (3.11) yields[17]

$$E = -\tfrac{1}{2} m_e v^2 = -\tfrac{1}{2} Ze^2/r. \tag{3.13}$$

The frequency of revolution equals the velocity divided by the length $2\pi r$ of the orbit and can be expressed in terms of E by means of equation (3.13):

$$\nu = \frac{v}{2\pi r} = \left(-\frac{2E}{m_e}\right)^{1/2} \frac{1}{2\pi} \left(-\frac{2E}{Ze^2}\right) = \frac{(-2E)^{3/2}}{2\pi m_e^{1/2} Ze^2}. \tag{3.14}$$

The level separation must then be

$$E_n - E_{n-1} = \frac{h}{2\pi m_e^{1/2} Ze^2} (-2E_n)^{3/2} = \frac{\hbar}{m_e^{1/2} Ze^2} (-2E_n)^{3/2}, \tag{3.15}$$

where, as usual, $\hbar = h/2\pi$.

For large values of n, that is, for large orbits, the difference $E_n - E_{n-1}$ may be represented approximately as a derivative, dE/dn, and thus equation (3.15) is replaced by the differential equation

$$\frac{dE_n}{dn} = \frac{\hbar}{m_e^{1/2} Ze^2} (-2E_n)^{3/2}. \tag{3.16}$$

In the straightforward integration of this equation, zero energy corresponds to an orbit of infinite radius; accordingly we take lim $E_n = 0$ for $n = \infty$. There remains an undetermined constant σ arising from the integration over n. The result of the integration is

$$E_n = -\tfrac{1}{2} \frac{m_e Z^2 e^4}{\hbar^2 (n - \sigma)^2}. \tag{3.17}$$

---

17. This equation between total, kinetic and potential energies holds for any system of particles interacting with $1/r^2$ force laws and is called the *virial theorem* (*BPC*, 1: 296-99).

Substituting in this formula the numerical values of m_e, e, and $\hbar$ yields

$$E_n = -\frac{13.6}{(n-\sigma)^2} Z^2 \text{ eV}. \tag{3.18}$$

This theoretical result is brought to coincide with the experimental values $\tau_n = E_n/\hbar$ of the spectral terms of hydrogen, given by equation (2.39), by setting the constant σ at zero and, of course, $Z = 1$. Equation (3.18) also accounts for the spectral terms of all ions with a single electron, namely He^+, Li^{++}, . . . , if Z is given the appropriate value. Finally, when σ takes appropriate nonzero values, equation (3.18) yields the sequence of highly excited levels of most atoms, as determined from the experimental absorption frequencies (2.32).

The comparison between the theoretical result (3.18) and experimental data can be pressed to high precision. However, for this purpose the constant m_e in equation (3.17) should be replaced by the reduced mass of the electron and of the nucleus (or other ion) about which the electron revolves. Accordingly, different mass values must be entered in equation (3.17) when applied to the spectra of different elements or of different isotopes of an element. These values differ by less than 1/1,000 but such differences loom large in view of the high precision of spectroscopic measurements. Agreement with experiment is outstanding. In fact, the agreement has been consistently so good that the theoretical formula is taken for granted and is combined with the experimental value of the Rydberg constant R to provide a high-accuracy input for the determination of e, m_e, h, and c. An accurate value of R, to be used in conjunction with the electron mass m_e, is

$$R = \frac{2\pi^2 m_e e^4}{h^3 c} = 109,737.31 \text{ cm}^{-1}. \tag{3.19}$$

Bohr's theory is thus remarkably successful in accounting for the highly excited levels of atoms. However, its main impact derived from heuristic extrapolation to low n values. The initial basis of the theory was replaced by the introduction of an ad hoc postulate which leads to the result (3.18) with $\sigma = 0$ and affords a broad generalization. The postulate requires the electron's momentum, integrated over the circumference of a circular orbit, to equal a multiple of the Planck constant

$$m_e v 2\pi r = nh. \tag{3.20}$$

Application of equation (3.20) dominated quantum theory until the advent of quantum mechanics. The "postulate" was then shown to follow from an approximation of the complete theory which will be encountered in chapter 12.

The application of Bohr's theory to low values of n, particularly to the ground state $n = 1$, provided not only a correct theoretical value of the ionization potential of the hydrogen atom but also a determination of its size. Equation (3.13) gives the radius of a circular orbit as a function of the electron's energy, $r = -Ze^2/2E$. Substituting in this formula the energy E_n from equation (3.17), with $n = 1$ and $\sigma = 0$, and setting $Z = 1$, one finds

$$r = \frac{\hbar^2}{m_e e^2} = a_B = 0.529\ \text{Å}. \tag{3.21}$$

This quantity, called the *Bohr radius*, serves as the basic unit for atomic dimensions.

S3.7. ATOMIC PROCESSES AT HIGHER ENERGIES

The phenomena of electron and light interaction with atoms discussed in earlier sections extend to higher electron collision energies and to higher-frequency radiation. We survey here the main processes of this class briefly; each of them forms the subject of a sizable field of study.[18]

S3.7.1. *Elastic Collisions of Electrons.* Electrons traversing atoms may experience sharp deflections without energy loss, not unlike α-particle deflections (see section 1.4). In fact, the Rutherford cross-section formula (1.32) remains approximately valid, with due allowance for the screening effect of atomic electrons.[19] Since Rutherford's cross-section is inversely proportional to the squared energy of the incident particle, one observes, for instance, that deflections of a 50-keV electron are 10,000 times more likely than those of a 5-MeV α-particle.

S3.7.2. *Inelastic Collisions of Electrons.* Collisions of electrons can transfer energy to atoms through the repulsion between the incident and the atomic electrons. For collisions with energy $\gtrsim 1$ keV, most of the inelastic processes can be classed into either of two extreme types depending on the magnitude of the incident electron's deflection (more accurately, of its momentum change).

a) *Small deflections.* This process is analogous to the collisions in a Franck-Hertz experiment, in that it leads usually either to a low excitation of the target atom or to its ionization with ejection of a low-energy electron. The incident electron thus loses a very small fraction of its energy but this amount can nevertheless be measured accurately.

---

18. The collision of atoms and ions with atoms is more complicated than electron collisions or interactions with radiation owing to the complexity of the incident particle. Basically, two colliding atoms form an unstable molecule during their interaction; therefore, the study of these collisions forms a chapter of molecular physics. The phenomenon becomes rather simple only when the incident atom is faster than all its electrons, in which case its electrons are stripped off; one deals then, in effect, with the collision of a bare nucleus, as in the example of α-particles.

19. The screening effect is evaluated in section S4.4.

These energy transfers have low probability in the sense that they seldom occur more often than once per hundred atoms traversed even for low-energy incident electrons. For incident energies above 1 keV this probability decreases approximately in inverse proportion to the energy.

b) *Large deflections.* This process is analogous to Rutherford scattering of the incident electron by an atomic electron. The atomic electron recoils sharply and thus absorbs an appreciable fraction of the incident electron's energy E. The cross-section for this process is essentially the same as Rutherford's. According to equation (1. 33) this cross-section is inversely proportional to the fourth power of the recoil momentum and hence to the square of the recoil energy Q imparted to the ejected electron and lost by the incident electron. The result of interest is the differential cross-section for collisions with energy transfers between Q and Q + dQ rather than for collisions with a given deflection. The value of this cross-section, obtained from equation (1. 33) with some adjustments, is

$$d\sigma = \frac{\pi e^4}{Q^2} \frac{dQ}{E} \tag{3. 22}$$

per electron in the target atom.

S3. 7. 3. *Bremsstrahlung (Continuous X-ray Emission).* As mentioned in section 3. 3, an electron traversing an atom forms with it, for a fleeting moment, a compound system and thus can experience a free-free transition with emission of electromagnetic radiation. This emission is predicted by the macroscopic theory of chapter 2, since the electron's motion accelerated by the atomic field constitutes a variable current. The emission becomes increasingly intense and its spectrum extends further toward high frequencies as the incident electron's energy increases. This phenomenon becomes conspicuous when the electron energy exceeds ~10 keV and the emitted spectrum extends into the X-ray range. For energies above 10-100 MeV, bremsstrahlung becomes the main mechanism of energy dissipation.

The bremsstrahlung spectrum is continuous since the variation in the time of the electron current is nonperiodic and its Fourier representation contains in principle components with all frequencies. Quantum physics sets an obvious upper limit ν_{max} to the frequency spectrum, since the photon energy $h\nu$ radiated by an electron in a free-free transition cannot exceed its initial energy eV (V = accelerating potential). Indeed, measurement of ν_{max} or, rather, of the wavelength $\lambda_{min} = c/\nu_{max}$, provides an accurate determination of the combined constant $hc/e = V\lambda_{min}$.

S3. 7. 4. *X-ray Photoelectric Absorption.* Electromagnetic radiation with frequency in the far ultraviolet and near X-ray range is absorbed very strongly by nearly all materials (so strongly, in fact, that it can be handled only in vacuum). All materials are, however, increasingly transparent to X-rays of increasing frequency. The variation of the attenuation coefficient $\kappa(\omega)$ as a function of the photon energy $\hbar\omega$ follows a rather simple pattern for $\hbar\omega \gtrsim 1$ keV whereas it is complicated at lower photon energies.[20]

The basic trend of $\kappa(\omega)$ is to decrease approximately as ω^{-3} for $\hbar\omega$ between ~1 and ~100 keV. Upon this smooth decrease are superposed sudden rises, called *absorption edges* which cause plots of $\kappa(\omega)$ to have a typical sawtooth aspect (Fig. 3.9). The set of edge frequencies forms a characteristic pattern for each absorbing element. Moreover, this pattern varies smoothly as a function of the atomic number (see Fig. 3.10). Detection and energy measurement of electrons from absorbing atoms show that the X-ray absorption edges constitute photoelectric thresholds for the ejection of strongly bound electrons.

According to this evidence, some electrons are held strongly within atoms, presumably near their nuclei where the attraction is strongest. The rapid increase of the threshold energies with increasing Z, observed in Figure 3.10, confirms that the thresholds depend on the strength of the nuclear attraction. Moreover, strong attraction is required for electrons to perform normal-mode oscillations with high frequencies ω_r and thus to absorb high-frequency X-rays.

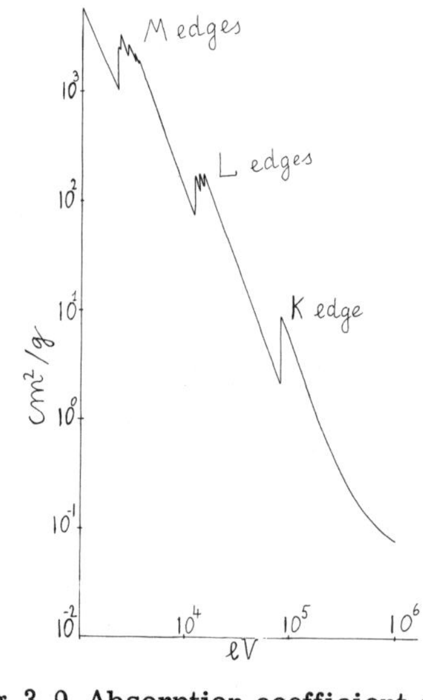

Fig. 3.9. Absorption coefficient of Au for X-rays as a function of photon energy. This coefficient is the quantity indicated by $\rho\sigma_T$ in equation (1.21); it is expressed in cm^2/gram, implying that the density ρ is given in atoms/gram. This expression is convenient because the thickness t of an absorption foil is determined by weighing and is given in grams/cm^2.

---

20. The measurement of $\kappa(\omega)$ for X-rays proceeds in principle as for light (section 2.2). Different monochromatic components of an X-ray beam are separated by crystal diffraction (see chap. 4).

Finally, the fact that the edge energies depend but little on the chemical properties of an element or on its state of chemical combination confirms that they arise from inner-atom phenomena.

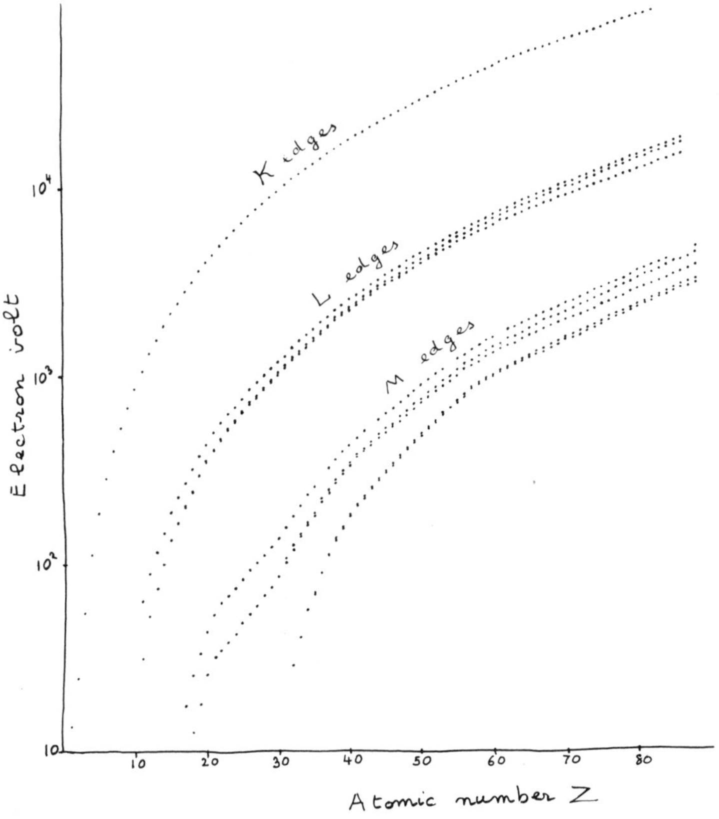

Fig. 3. 10. Energy of X-ray absorption edges. Data from J. A. Bearden and A. F. Burr, *Rev. Mod. Phys.*, **39** (1967): 78.

Figure 3.10 also shows the occurrence in each element of groups of absorption edges at widely separated frequencies, and hence the existence in an atom of electrons whose binding energies differ by almost one order of magnitude. These energy differences are attributed to differences of the electrons' distances from the nucleus. The successive groups of absorption edges, called K, L, M, . . . , are regarded as photoelectric thresholds for ejection of electrons from successive concentric shells of the atomic structure, the shells being labeled by the same letters K, L, M, ... The theoretical analysis of atomic structure in chapter 16 confirms this interpretation.

S3.7.5. *Characteristic X-ray Emission.* X-ray absorption is generally accompanied by fluorescence, that is, by emission of radiation of lower frequency (section 3.3). Since the absorption is accompanied by ejection of an electron, the fluorescence is characteristic of the resulting ion and should be accompanied by transitions of the ion from a higher to a lower level of excitation. The different levels of excitation originate from ejection of a photoelectron from different internal shells. Accordingly, the photon energies of the fluorescent X-rays should coincide with the energy differences between the absorption edges. This expectation is confirmed with great accuracy.

The spectrum of the *characteristic X-ray* emission of each element consists of a number of series with widely different photon energies. The K series consists of lines with photon energies equal to the edge differences $E_L - E_K$, $E_M - E_K$, etc.; the L series corresponds to differences $E_M - E_L$, $E_N - E_L$, etc.

The characteristic X-rays are usually produced by electron bombardment of each element rather than by prior absorption of X-rays, the main reason being the availability of intense electron beams. The various X-ray series appear in the succession M, L, K, . . . , as the incident electron energy is increased; as one would expect, they appear at critical potentials coinciding with the X-ray absorption edges.

The structure of X-ray line spectra and their dependence on the atomic number was predicted by Bohr by a daring extrapolation of his results on the planetary model described in section S3.6. He considered that any electron near the nucleus is subjected primarily to the nuclear attraction and is only secondarily disturbed by other electrons. Accordingly, he anticipated spectral emissions similar to those of an ion stripped of all but one of its electrons, with frequencies which are proportional to Z^2 according to equation (3.18), and lie in different spectral ranges as they do in the spectrum of atomic hydrogen.

S3.7.6. *Auger Effect.* X-ray emission is not the only mechanism for energy release by a highly excited ion (or neutral atom). The energy may be picked up by an electron of the same atom; this electron is thereby ejected with kinetic energy $h\nu - E_I$, where $h\nu$ is the energy of the photon that might have been emitted and E_I is the energy required to detach the electron. The earmark of this process (the Auger effect) is the ejection of electrons with this characteristic energy. The fraction of ion de-excitations that proceeds by X-ray emission rather than by ejection of Auger electrons is called *fluorescence yield* and is small for $h\nu \lesssim 10^4$ eV. For $h\nu \lesssim 100$ eV the Auger process is usually called *autoionization.*

S3.8. EXPERIMENTAL METHODS

The quantitative study of elementary processes involving electrons, radiation, and atoms requires considerable sophistication for electron energies below ~1 keV and photon energies between ~10 eV and several hundred eV. Electrons and radiation in this energy range interact with all materials strongly. Hence it is difficult to get them to collide with an adequate sample of atoms of interest without their being exposed to disturbing multiple or stray interactions.

Such contrasting requirements make it generally necessary to maintain large pressure differences between different parts of the apparatus without intervening windows. High-performance pumps are normal apparatus components.

Moreover, the field due to any accumulation of electric charges on the walls of the apparatus distorts the trajectories of low-energy electrons. Therefore, metal apparatus, bakeable in high vacuum to remove any insulating films, has come into general use.

Conceptually, the apparatus consists of the three components shown in Figure 3.5: a source, a collision chamber, and analyzers to study what emerges from collisions. Figure 3.11 illustrates a recent model of electron spectrometer. Figure 3.12 shows results obtained with this instrument. Experimental arrangements for other types of studies are illustrated in figures 3.13 and 3.14.

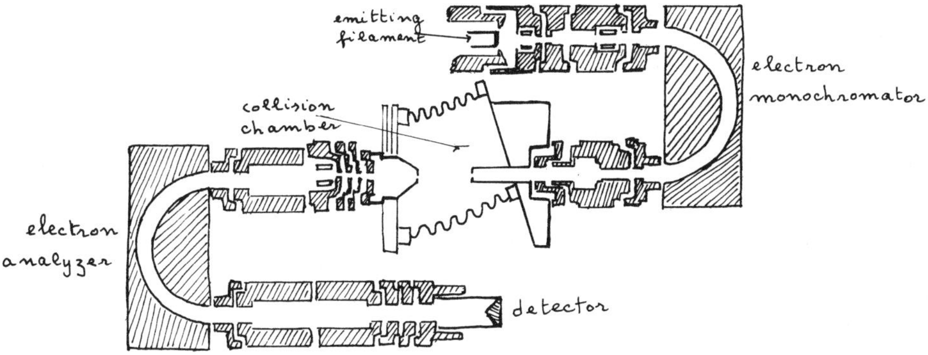

Fig. 3.11. Sketch of electron spectrometer. The cross hatched parts represent electrodes whose different potentials and shapes are designed to collimate and focus electrons of desired energies. [Adapted from C. E. Kuyatt and J. A. Simpson, *Review of Scientific Instruments* 38 (1967): 103].

S3.8.1. *Collision Chamber.* The volume within which the collisions of interest occur must be well identified by adequate collimation of the incident beam and of the entrance to analyzers. Yet, the collimators must transmit a sufficient number of electrons or sufficient radiation. The collimator edges should not deflect or diffract particles or radiation into or out of the beams.

The density of atoms in the collision volume must be known, must be sufficient to produce an adequate number of collisions, yet sufficiently small for multiple collisions to remain infrequent. Pressures of the order of 10^{-2} mm Hg or less are usually required for electron-collision studies. Since much lower pressures are required in other parts of the apparatus, the gas leaks steadily through all chamber exits. Therefore, the pressure tends to be nonuniform and difficult to determine.

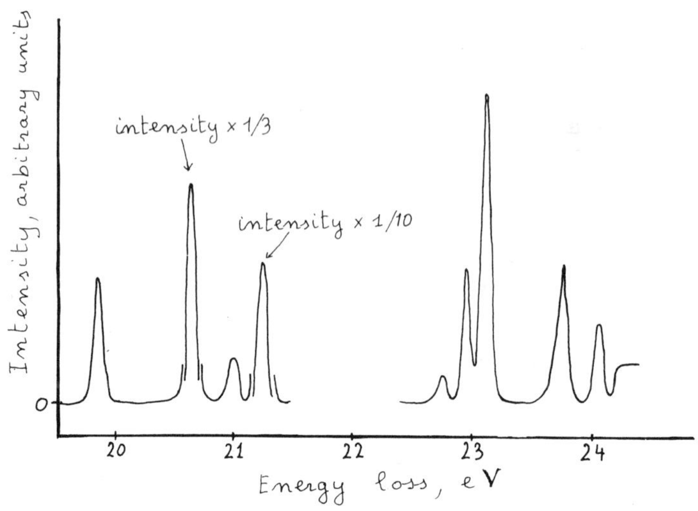

Fig. 3. 12. Intensity of electrons scattered by helium atoms with different energy losses. Each peak corresponds to a level of excitation of helium atoms. Courtesy of C. E. Kuyatt.

Easier definition of the collision volume and reduction of unwanted collisions are achieved by the cross-beam arrangement, in which the target atoms merely traverse the chamber as a gas beam, as shown in Figure 3. 14. However, the density of atoms in the beam is not easily measurable.

S3. 8. 2. *Electron Sources.* Electrons are normally obtained from a hot filament, from which they emerge with a distribution of kinetic energies ranging from zero to ~0. 3 eV. When a smaller energy spread is required, the electrons are filtered through a monochromator prior to ejection into the collision chamber. The monochromator consists of electric or magnetic lenses which focus electrons of desired energy onto an exit diaphragm. A narrow energy selection implies, of course, a lower transmission of electron current. Careful electron-optical design helps in optimizing the performance of monochromators.

S3. 8. 3. *Electron Analyzers.* The energy analysis performed by varying the potential of a collector or of a grid, shown in Figures 3. 1, 3. 5, and 3. 13, yields a current versus voltage response such as that of Figure 3. 2, which constitutes the "integral spectrum" of electron energies. This is the number of electrons with energies in excess of a given value. What is usually wanted, of course, is the "differential spectrum," that is, the number of electrons with energy in any given small interval E to E + ΔE.

The differential spectrum can be obtained by measuring the slope of the integral

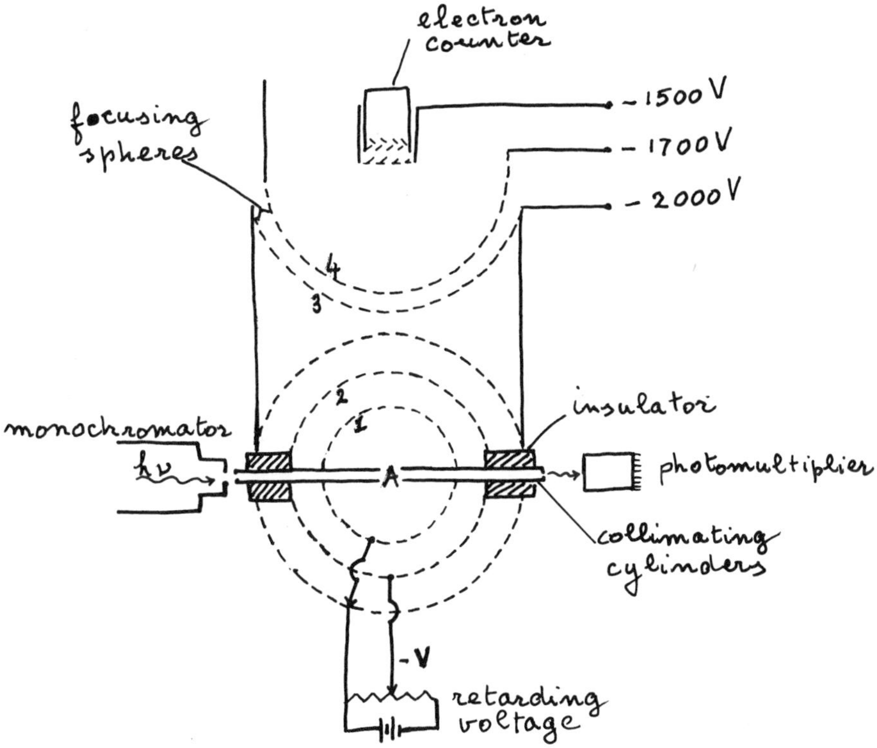

Fig. 3. 13. Diagram of apparatus for analysis of photoelectron energy. Photoelectrons released in region A can traverse the gap between the grids 1 and 2 only if they have sufficient energy. In this case they are focused and accelerated by the focusing spheres (grids 3 and 4) and accelerated further to hit the counter with ~ 2000 eV kinetic energy. [Adapted from J. R. Samson and R. B. Cairns, *Phys. Rev.* 173 (1968): 80].

spectrum curve, though this implies a loss of accuracy. Automatic differentiation can be achieved by impressing a small a.c. voltage to a control grid and measuring the Fourier component of the transmitted current with the same frequency as the a.c. voltage. A more rigorous energy selection is achieved by filtering the electrons once more through a monochromator such as those used in electron sources (see Fig. 3. 11).

Sensitivity of the electron scoring device is generally essential owing to the large losses of current through successive selections. Low-capacity current meters can detect 10^{-17} to 10^{-16} amperes, that is, a few hundred electrons per second. Electron counters measure even smaller electron flows, the limit being set by background counts due to stray particles.

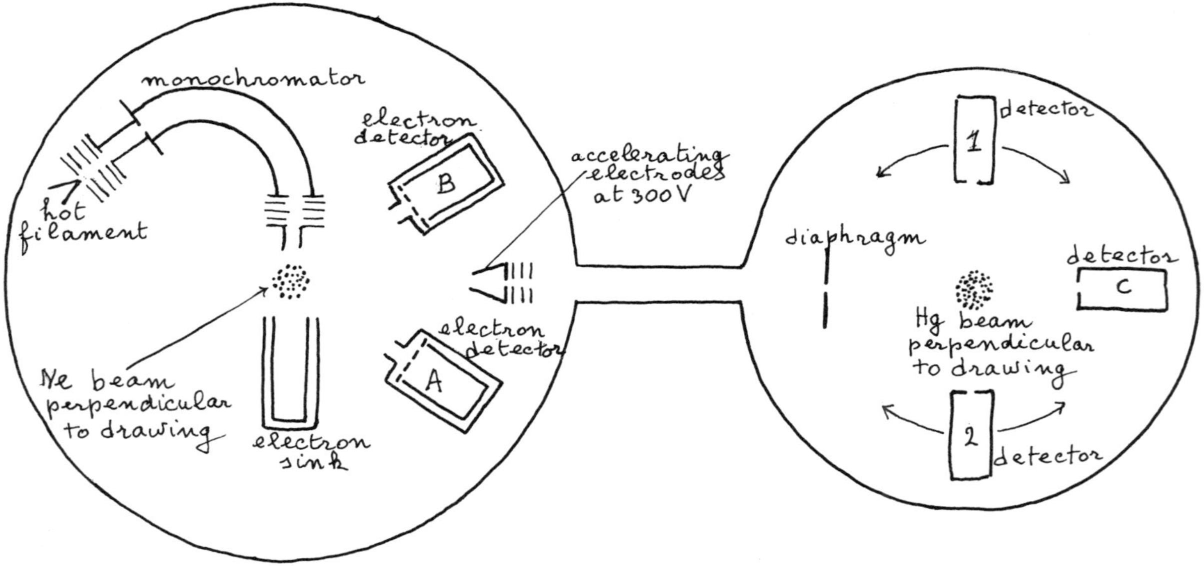

Fig. 3. 14. Diagram of apparatus for double electron scattering. Electrons from the incident beam are scattered first by atoms of a Ne beam, then they are accelerated and scattered again by atoms of a Hg beam. Any significant difference in the counts registered by detectors 1 and 2 indicates that the properties of the electrons after the first scattering in Ne depend on the direction of the incident beam. Note the use of the additional detectors A, B, and C to monitor the stability and symmetry of all beams. This diagram illustrates the sophistication of current experiments; the purpose of this particular experiment will be explained in chapter 14. [Adapted from E. Reichert and H. Deichsel, *Z. Phys.* 185 (1965): 169].

CHAPTER 3 PROBLEMS

3.1. The threshold wavelength λ_0 for the photoelectric effect in tungsten is 2700 Å. Calculate the maximum kinetic energy, K_{max}, of electrons ejected from tungsten by light of $\lambda = 1700$ Å.

3.2. Radiation with frequency 100 megahertz from a very distant antenna has an electric field that oscillates with amplitude 10^{-6} V/m. Calculate its energy flux in photons/m^2sec.

3.3. An electron with 4 eV kinetic energy collides with a Hg atom at rest and is deflected by 90°. Calculate its loss of kinetic energy, taking into account the conservation of momentum in the collision.

3.4. An electron with negligible initial energy falls through a difference of potential and produces radiation when it hits a target. Find the minimum potential difference required to produce (a) X-rays of 0.6 Å wavelength, (b) light of 600 Å, (c) microwaves of 6 cm.

3.5. Can the emission of 5890 Å light from sodium vapor result from the impact of electrons with 2.3 eV kinetic energy?

3.6. Excited Hg atoms emit light of wavelength $\lambda = 1850$ Å. In a strong magnetic field the intensity distribution of the light with undisturbed wavelength is the same as the intensity distribution of an antenna parallel to the magnetic field (as predicted by the theory of the normal Zeeman effect). Consider a gas containing 10^{-9} g of Hg in a strong magnetic field. During a discharge in the gas 10 photons/sec of 1850 Å light are detected by a photon counter placed at 10 m from the discharge in a direction at 60° from that of the magnetic field. The counter has a window area of 10^{-3} cm^2 and a detection efficiency of 10%. From these data we can obtain information on the dipole oscillations of each Hg atom, using the formula for the power dP received over a solid angle dΩ at an angle θ with respect to a dipole antenna,

$$dP = \frac{\omega_r^4}{c^3} \langle p_r^2 \rangle \sin^2\theta \frac{d\Omega}{4\pi}$$

(see *BPC*, vol. 3, p. 376, eq. [171]). In this formula $\langle p_r^2 \rangle$ indicates the average over an oscillation of the dipole moment p_r considered in chapter 2, and ω_r its frequency of oscillation. Calculate (a) the total rate R of 1850 Å photon emission by the gas; (b) the total power P of 1850 Å radiation emitted by the gas; (c) the average rate of photon emission by each Hg atom; (d) the effective amplitude p_{r0} of dipole oscillation by each atom and p_{r0}/e, the effective displacement of one electron per atom.

3.7. Atomic hydrogen is bombarded with electrons in a Franck-Hertz experiment. Emission of the red (lowest frequency) line of the Balmer series is observed, but the other lines of that series do not appear. This observation fixes an upper and a lower limit to the kinetic energy K of the bombarding electrons. Calculate these limits, assuming the atoms to be initially in their ground state and disregarding their recoil.

3.8. Mercury vapor is bombarded by a beam of monoenergetic electrons. As the electron energy is increased progressively, light emission by the vapor is observed spectroscopically. Assume that initially all Hg atoms are in their ground state. At a succession of threshold electron energies, different lines appear in the spectrum as follows:

threshold [eV]	line, λ [Å]
4.89	2537
6.70	1850
7.73	4047, 4358, 5461.

From this evidence construct as much of the Hg energy level diagram as you can, indicating the observed transitions.

3.9. A beam of Hg atoms travels down an evacuated tube at 2.5×10^7 cm/sec and is crossed by a beam of electrons of 5.3 eV kinetic energy. The Hg atoms are thus excited and emit light of 2537 Å. Three photomultipliers measure the light intensity emerging through each of three windows shown in the figure.

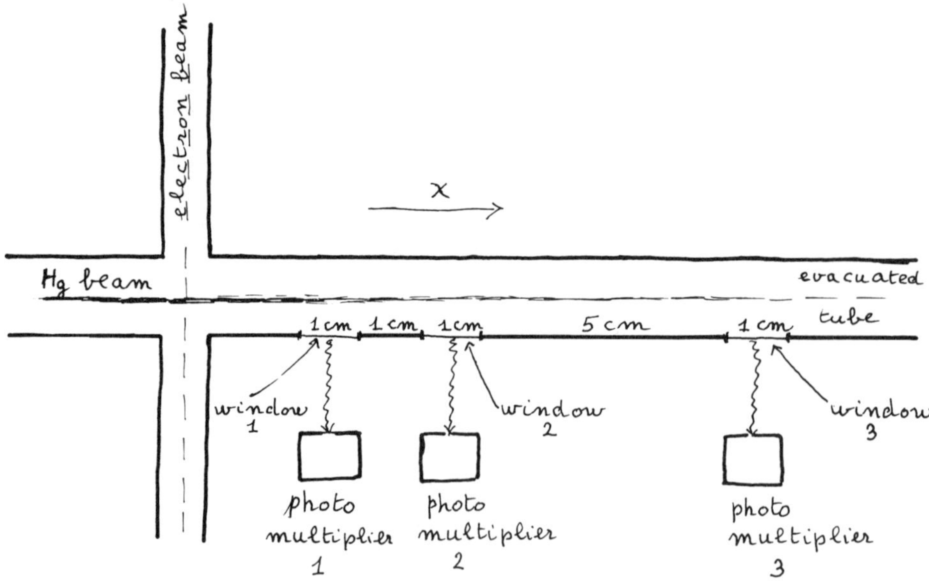

The current intensities in photomultipliers 1 and 2 are in the ratio 2:1.
(a) Calculate the ratio of current intensities in photomultipliers 1 and 3.
(b) Calculate the decay constant (or damping parameter) Γ for the light emit-

ted by the Hg beam. (c) Calculate the kinetic energy of an electron after it has excited a Hg atom.

3.10. Consider the stationary state of a hydrogen atom with n = 50, according to Bohr's model. Calculate (a) the binding energy of the electron in eV; (b) the radius of the electron orbit; (c) the frequency of revolution; (d) the wavelength of radiation emitted in transition from the n = 50 to the n = 49 stationary state.

3.11. Calculate the ratio of energy levels with the same quantum number in the spectra of the two isotopes of Li^{++} with atomic weights 6.0 and 7.0.

3.12. Electrons with kinetic energy of 200 keV traverse a cloud chamber filled with nitrogen at standard pressure and temperature. Calculate the number N_r of recoil electrons ejected from nitrogen molecules with energy in excess of 1 keV, per cm track of the high-energy electrons. (These recoil electrons appear as branches in the main track and are called δ-rays.) Disregard the energy required to ionize the nitrogen molecule.

SOLUTIONS TO CHAPTER 3 PROBLEMS

3.1. $K_{max} = h\nu - h\nu_0 = hc/\lambda - hc/\lambda_0 = 6.6 \times 10^{-27} \times 3 \times 10^{10}$
$\times(1/1.7 \times 10^{-5} - 1/2.7 \times 10^{-5}) = 4.3 \times 10^{-12}$ ergs = 2.7 eV.

3.2 For a very distant antenna, radiation is a plane wave, with $\vec{E} = \text{Re}\{E_0 \exp[i(\vec{k}\cdot\vec{r} - \omega t)]\}$ and $|\vec{E}| = |\vec{B}|$. Energy flux is the average magnitude of the Poynting vector (see p. 35, that is $c\langle|\vec{E} \times \vec{B}|\rangle/4\pi = cE_0{}^2/8\pi$. Set this equal to $nh\nu$, where n is number of photons per cm^2 per sec. Since 1 V/m = (1/3) $\times 10^{-4}$ dynes/esu, $n = [3 \times 10^{10} \times (0.33 \times 10^{-10})^2]/(8\pi \times 6.62 \times 10^{-27} \times 10^8) cm^{-2} sec^{-1} \sim 2 \times 10^{10}$ photons/m^2 sec.

3.3. The collision is elastic since the electron energy is below the threshold of excitation of Hg. Conservation of energy and momentum requires the fraction of energy lost by the electron to be $2m_e/(M_{Hg} + m_e) = 5.5 \times 10^{-6}$; the energy loss is 2.2×10^{-5} eV.

3.4. Minimum potential difference in volts equals photon energy $h\nu = hc/\lambda$, expressed in eV. (a) 2.1×10^4 V; (b) 2.1 V; (c) 2.1×10^{-5} V.

3.5. For 5890 Å, $h\nu = hc/\lambda = 2.1$ eV; therefore, the incident energy is sufficient.

3.6. (a) $R = 10 \times 10 \times [4\pi \times 10^6/10^{-3}][\langle\sin^2 \theta\rangle/\sin^2 60°] = 1.1 \times 10^{12}\ sec^{-1}$, where the first factor represents the counting rate, the second represents the inverse of counter efficiency, the third represents the solid-angle ratio, and the fourth corrects for anisotropy of emission. (b) $P = R(hc/\lambda) = 1.1 \times 10^{12} \times 6.7$ eV $sec^{-1} = 7.4 \times 10^{12}$ eV sec^{-1} = 12 ergs/sec^{-1}. (c) Number of Hg atoms $N_{Hg} = 10^{-9}$ g $N_A/M_{Hg} = 10^{-9} \times 6.0 \times 10^{23}/200 = 3 \times 10^{12}$. $R/N_{Hg} = 1.1 \times 10^{12}/3 \times 10^{12} = 0.37$ photons sec^{-1}. (d) From the formula given in the problem we have $(2/3)(\omega_r^4/c^3)\langle p_r^2\rangle = P/N_{Hg} = 4 \times 10^{-12}$ ergs sec^{-1}, $\omega_r = 2\pi c/1850$ Å $= 1.0 \times 10^{16}$ radians/sec^{-1}, $\langle p_r^2\rangle = 4 \times 10^{-12} \times (3/2) \times (3 \times 10^{10})^3/10^{64} = 1.6 \times 10^{-44}(\text{esu-cm})^2$, $p_{r0} = [2\langle p_r^2\rangle]^{1/2} = 1.8 \times 10^{-22}$ esu-cm, $p_{r0}/e = 1.8 \times 10^{-22}/4.8 \times 10^{-10} = 3.8 \times 10^{-13}$ cm. The effective amplitude of oscillation of electrons is thus very small when averaged over the whole Hg vapor.

3.7. According to the description at the end of section 2.3 the emitting atoms have been raised to their third energy level E_3 from which the red line is emitted, but they have not been raised to the fourth level. Therefore $E_3 - E_1 < K < E_4 - E_1$, that is, $12.1 < K < 12.7$ eV.

3.8. The spectral lines observed correspond to photon energies hc/λ which are, in eV, 4.89 (2537 Å), 6.70 (1850 Å), 3.06 (4047 Å), 2.84 (4358 Å), and 2.27 (5461 Å). The first two of them coincide with threshold energies and thus correspond to transitions from excited levels $E_1 = 4.89$ eV and $E_2 = 6.70$ eV to the ground state. Another energy, $2.84 = 7.73 - 4.89$ eV, corresponds to a transition from the third threshold, $E_3 = 7.73$ eV, to the level E_1. The two remaining lines must be emitted in transitions from their threshold level E_3 to lower levels $E_4 = E_3 - 3.06$ eV $= 4.67$ eV and $E_5 = E_3 - 2.27 = 5.46$ eV, which are not detected as radiation thresholds.

3.9. The light emission from the excited Hg decays exponentially in time, $I = I_0 \exp(-\Gamma t)$ and also in space, $I = I_0 \exp(-\Gamma x/v)$. The range of observation of photomultiplier 2 is shifted by 2 cm with respect to photomultiplier 1, photomultiplier 3 is shifted by 8 cm with respect to 1. Therefore, $I_1 : I_2 : I_3 = 1 : \exp(-\Gamma\, 2\text{ cm}/v) : \exp(-\Gamma\, 8\text{ cm}/v)$. (a) $I_1 : I_3 = (I_1/I_2)^4 = 16$. (b) $\Gamma = (2.5 \times 10^7 \text{ cm sec}^{-1}/2\text{ cm})\ln 2 = 8.7 \times 10^6\ \text{sec}^{-1}$. (c) $K = 5.3$ eV $- hc/2537$Å $= 0.41$ eV.

3.10. (a) Energy level $E_n = -13.6/n^2$ eV; binding energy $E_{n=\infty} - E_n = 0 - (-13.6/50^2) = 5.4 \times 10^{-3}$ eV. (b) $E_n = -e^2/2r_n$, $r_n = -e^2/2E_n = 1.3 \times 10^3$ Å $= 0.13\ \mu$. (c) Frequency of revolution $v/2\pi r = (-2E_n)^{3/2}/2\pi m_e^{1/2}e^2 = 5.3 \times 10^{10}\ \text{sec}^{-1}$. (d) $\lambda_{50\to49} = c/\nu_{50\to49} = hc/(E_{50} - E_{49}) = 0.55$ cm.

3.11. The Rydberg constants for the isotopes are in the ratio of the reduced masses $R_{Li6} : R_{Li7} = m_{Li6} : m_{Li7} = [1 + (7.0 \times 1823)^{-1}] : [1 + (6.0 \times 1823)^{-1}] \sim 1 - (42 \times 1823)^{-1} = 1 - 1.3 \times 10^{-5}$.

3.12. Integration of equation (3.22) gives $\sigma(Q > Q_{min}) = \pi e^4/Q_{min}E$. Since each N_2 molecule has 14 electrons, the electron density in the chamber is $\rho_{el} = 14 \times N_A/2.24 \times 10^4 = 3.8 \times 10^{20}\ \text{cm}^{-3}$. $N_r = \rho_{el}\sigma(Q > 1\text{ keV}) = 3.8 \times 10^{20} \times \pi \times e^4/(1\text{ keV} \times 200\text{ keV}) = 0.12\ \text{cm}^{-1}$.

4. quantum properties of momentum transfers

The amounts of energy transferred in elementary processes consist of discrete quanta; the magnitude of a quantum depends on the *variations in time* of a relevant quantity—e.g., of an intra-atomic current (chapter 3). In this chapter we shall see that momentum transfers proceed similarly by discrete quanta; the magnitude and direction of a momentum quantum depend on the *variation in space* of a relevant quantity—e.g., of the electron density in a material.

The momentum of a particle remains constant, by definition, when the potential energy is uniform along the particle's path, and it varies when the potential is nonuniform. Classical mechanics establishes relationships between the variations of the total momentum of a macroscopic body and those of its potential energy. Observations on atomic particles, suggested by analogy with radiation phenomena, establish additional relations.

Radiation propagates in a straight line through a homogeneous medium but is deflected or scattered by inhomogeneities. Changes in the direction of propagation imply changes in momentum, since a plane wave is known to carry a vector momentum of magnitude equal to its energy flux divided by the light velocity.[1]

We begin this chapter by studying how a periodic inhomogeneity of a medium influences the propagation of radiation. Thereafter section 4.3 describes experiments showing that the results are applicable to the deflection (momentum change) of particle beams as well as of radiation beams. Supplementary sections describe extensions of these results and phenomena involving the simultaneous transfer of energy and momentum.

4.1. PROPAGATION IN A SINUSOIDALLY VARIABLE MEDIUM

Crystals are materials whose internal atomic structure is highly ordered. In particular, the distribution of nuclei and electrons throughout a crystal is understood to be a periodic function of position. Accordingly, the properties of a crystal which determine the propagation of radiation, namely the refractive index and the dielectric constant ϵ, will be regarded here as functions of position which vary periodically from point to point. This assumption is introduced for brevity on plausibility grounds. Equivalent results follow from a more realistic treatment which extends our analysis of the relation between applied field and induced current (section 2.1) to the case of an inhomogeneous material.

As a preliminary to a description and discussion of experiments, this section treats the propagation of radiation in a medium whose dielectric constant

---

1. This result of Maxwell's theory is verified by the recoil of mirrors that reflect light (see *BPC*, 3: 362).

varies from point to point by a small fraction according to a simple sinusoidal law. This sinusoidal variation impresses its periodicity upon a propagating plane wave by modulating the wave's amplitude. Even though the variation of the medium is small, the resulting modulation of the propagating wave is dominant under special (resonance) conditions, as will be shown in the following.

We return to the Maxwell equation (2.7) setting $\epsilon(\omega) = 1 + \gamma(\omega) \cos \vec{q}\cdot\vec{r}$, where the coefficient γ represents the small amplitude of variation of ϵ and $\vec{q}$ indicates the direction and pitch of this variation. We also set $\vec{j}_{ext} = 0$, implying that the radiation source lies at infinity, as in section 2.1. The equation is then

$$\vec{\nabla} \times \vec{B} = \frac{1}{c}(1 + \gamma \cos \vec{q}\cdot\vec{r}) \frac{\partial \vec{E}}{\partial t}. \tag{4.1a}$$

Among the other Maxwell equations, $\vec{\nabla}\cdot\epsilon\vec{E} = 0$ becomes

$$\vec{\nabla}\cdot(1 + \gamma \cos \vec{q}\cdot\vec{r})\vec{E} = 0 \tag{4.1b}$$

while the others are independent of ϵ.

Differential equations of the type of (4.1) can be treated by Fourier analysis in spite of the variation of their coefficients. This treatment is developed in some detail in Appendix C with reference to physical phenomena involving a minimal number of variables. Here the procedure for solving the Maxwell equations is given only in broad outline.

First, the electric and magnetic fields of radiation can still be represented by plane monochromatic waves—as in empty space—though with an amplitude modulation which is a periodic function of the quantity $\vec{q}\cdot\vec{r}$. Accordingly, we represent the radiation fields by

$$\begin{aligned}\vec{E}(\vec{r}, t) &= [\Sigma_n \vec{E}_n \exp(in\, \vec{q}\cdot\vec{r})] \exp[i(\vec{k}\cdot\vec{r} - \omega t)]\\ &= \Sigma_n \vec{E}_n \exp\{i[(k + n\vec{q})\cdot\vec{r} - \omega t]\}\\ \vec{B}(\vec{r}, t) &= [\Sigma_n \vec{B}_n \exp(in\, \vec{q}\cdot\vec{r})] \exp[i(\vec{k}\cdot\vec{r} - \omega t)]\\ &= \Sigma_n \vec{B}_n \exp\{i[(\vec{k} + n\vec{q})\cdot\vec{r} - \omega t]\}.\end{aligned} \tag{4.2}$$

The next step is to reduce the differential Maxwell equations to algebraic equations in the amplitudes E_n and B_n of the Fourier expansions. The plane-wave factors of (4.2), $\exp[i(\vec{k}\cdot\vec{r} - \omega t)]$, still factor out when $\vec{E}(\vec{r}, t)$ and $\vec{B}(\vec{r}, t)$ are substituted in the Maxwell equations (4.1). However, the amplitude modulation factors require special handling. In particular, the variable coefficient $\gamma \cos \vec{q}\cdot\vec{r}$ can be combined with the amplitude modulation of the fields, to yield

$$\begin{aligned}&\gamma \cos \vec{q}\cdot\vec{r}\ \Sigma_n \vec{E}_n \exp[in(\vec{q}\cdot\vec{r})]\\ &= \tfrac{1}{2}\gamma \Sigma_n \vec{E}_n \{\exp[i(n+1)(\vec{q}\cdot\vec{r})] + \exp[i(n-1)(\vec{q}\cdot\vec{r})]\}\\ &= \tfrac{1}{2}\gamma \Sigma_n (\vec{E}_{n-1} + \vec{E}_{n+1}) \exp[in(\vec{q}\cdot\vec{r})],\end{aligned} \tag{4.3}$$

and a similar formula for $\vec{B}(\vec{r}, t)$. By utilizing these formulae, the residual space dependence of the Maxwell equations can be represented by a Fourier series, the procedure being analogous to the one used in equation (C.4) for a time-dependent problem. Since the Maxwell equations must hold at each point of space, they must be satisfied by the coefficients of each term of the series. Thus one obtains the algebraic form of the equations

$$i(\vec{k} + n\vec{q}) \times \vec{E}_n = i\,\frac{\omega}{c}\,\vec{B}_n, \tag{4.4a}$$

$$i(\vec{k} + n\vec{q}) \times \vec{B}_n = -i\,\frac{\omega}{c}\,[\vec{E}_n + \tfrac{1}{2}\gamma(\vec{E}_{n+1} + \vec{E}_{n-1})], \tag{4.4b}$$

$$i(\vec{k} + n\vec{q})\cdot[\vec{E}_n + \tfrac{1}{2}\gamma(\vec{E}_{n+1} + \vec{E}_{n-1})] = 0, \tag{4.4c}$$

$$i(\vec{k} + n\vec{q})\cdot\vec{B}_n = 0. \tag{4.4d}$$

These equations differ from equations (A.11) for empty space by the terms with the coefficient γ, which interlink the Fourier coefficient $\vec{E}_n$ with $\vec{E}_{n+1}$ and $\vec{E}_{n-1}$. Therefore, equations (4.4) constitute a system which consists of infinitely many equations. However, the system can be unraveled by successive approximations in powers of γ, as indicated in Appendix C. Note that the system (4.4) can be reduced to a system of two equations in the electric field $\vec{E}_n$, since the magnetic coefficients $\vec{B}_n$ can be obtained from (4.4a). Formally, substitution of $\vec{B}_n$ from equations (4.4a) into equation (4.4b) serves to eliminate $\vec{B}_n$, yielding the new equation

$$(\vec{k} + n\vec{q}) \times [(\vec{k} + n\vec{q}) \times \vec{E}_n] = -\frac{\omega^2}{c^2}[\vec{E}_n + \tfrac{1}{2}\gamma(\vec{E}_{n+1} + \vec{E}_{n-1})]. \tag{4.5}$$

The coefficients $\vec{E}_n$ can now be calculated by utilizing equations (4.5) and (4.4c) as a system.

To zeroth order in γ, that is, for empty space, we consider a plane monochromatic wave whose wave vector $\vec{k}$ has arbitrary direction and magnitude $k = \omega/c$, and whose amplitude $\vec{E}_0$ has arbitrary magnitude and direction perpendicular to $\vec{k}$. All $\vec{E}_n$ with $n \neq 0$ vanish in this approximation. On this basis, successive approximations determine the values of all $\vec{E}_n$.

To first order, one needs to consider only the equations in which γ is multiplied by $\vec{E}_0$, namely, those with $n = \pm 1$,

$$(\vec{k} \pm \vec{q}) \times [(\vec{k} \pm \vec{q}) \times \vec{E}_{\pm 1}] = -\frac{\omega^2}{c^2}[\vec{E}_{\pm 1} + \tfrac{1}{2}\gamma\vec{E}_0], \tag{4.6a}$$

$$(\vec{k} \pm \vec{q})\cdot[\vec{E}_{\pm 1} + \tfrac{1}{2}\gamma\vec{E}_0] = 0. \tag{4.6b}$$

These equations determine respectively the components of $\vec{E}_{\pm 1}$ transverse and parallel to $\vec{k} \pm \vec{q}$. Indicating components transverse and parallel to $\vec{k} \pm \vec{q}$ by $\perp$ and $\|$, and replacing ω/c by k, one finds

$$(E_{\pm 1})_\perp = \frac{k^2}{|\vec{k} \pm \vec{q}|^2 - k^2} \tfrac{1}{2}\gamma(\vec{E}_0)_\perp \tag{4.7a}$$

$$(\vec{E}_{\pm 1})_\parallel = -\tfrac{1}{2}\gamma(\vec{E}_0)_\parallel . \tag{4.7b}$$

Substitution of this result in equations (4.2) shows that the variation of the dielectric constant generates two secondary waves with wave vectors $\vec{k} \pm \vec{q}$ which accompany the zero-order wave $\vec{E}_0 \exp[i(\vec{k}\cdot\vec{r} - \omega t)]$. The amplitude of the secondary waves is generally so small that they remain undetected.[2]

However, the variation of ϵ has a large effect when either one of the secondary waves can propagate freely with the same frequency as the zero-order wave. This happens when one of the wave vectors $\vec{k} \pm \vec{q}$ obeys the condition

$$|\vec{k} \pm \vec{q}| \approx k \tag{4.8}$$

and thus ensures that the denominator of one of the transverse amplitudes (4.7a) approximately vanishes. Under this condition the method of successive approximations breaks down, as discussed in greater detail on p.555. Appendix C also develops a modified method of first-order approximation which yields a non divergent solution $\vec{E}_{\pm 1}$, of magnitude comparable to that of $\vec{E}_0$. We do not pursue this solution here but merely note the occurrence of an intense secondary wave for special values and directions of the incident wave vector $\vec{k}$. The superposition of the intense secondary wave with the zero-order (incident) wave constitutes the dominant effect of modulation anticipated at the beginning of this section.

Notice that equations (4.8) can be satisfied only for radiation of sufficiently large k and short wavelength, with $k \geqslant \tfrac{1}{2}q$. No secondary wave of appreciable intensity can occur unless either k or the amplitude γ of the variations of material properties are sufficiently large. This remark provides a formal justification for the assumption in chapter 2 that ϵ may be regarded as constant in the study of light propagation through a macroscopically homogeneous material. Since Fourier representation of any variation of ϵ from point to point may be expressed as a superposition of terms of the type $\gamma \cos \vec{q}\cdot\vec{r}$; and since macroscopic homogeneity implies that γ is negligible whenever q^{-1} is much larger than atoms, equation (4.8) is satisfied only by radiation with wavelengths shorter than that of visible light.

4.1.1. *Bragg Reflection.* In the experimental study of propagation of radiation in crystals, one wants to search for conditions under which an incident (zero-order) wave gives rise to an intense secondary wave. For this purpose the

---

2. Higher-order approximations show that $\vec{E}_n$ is of order γ^n, so that the number of secondary waves increases with the order of approximation. These higher-order secondary waves can become intense under a condition analogous to equation (4.8), namely, $|\vec{k} + m\vec{q}| \sim k$, where m is an integer other than ±1. However, intense secondary waves with wave vectors $\vec{k} + m\vec{q}$ arise more readily from another effect, described in section 4.2.

condition (4.8) is conveniently expressed in the form

$$2k \sin \tfrac{1}{2}\theta = q, \tag{4.9}$$

where θ is the deflection angle between $\vec{k}$ and $\vec{k} + \vec{q}$ (or $\vec{k} - \vec{q}$, as the case may be).[3] As shown in Figure 4.1, the vector $\vec{q}$ is perpendicular to the bisectrix of the angle θ and the secondary wave appears to arise by reflection from planes of constant ϵ which lie perpendicular to $\vec{q}$. The occurrence of an intense secondary beam under these conditions, first observed by W. L. and W. H. Bragg, is called *Bragg reflection*. A diagram of the experimental arrangement is shown in Figure 4.2.

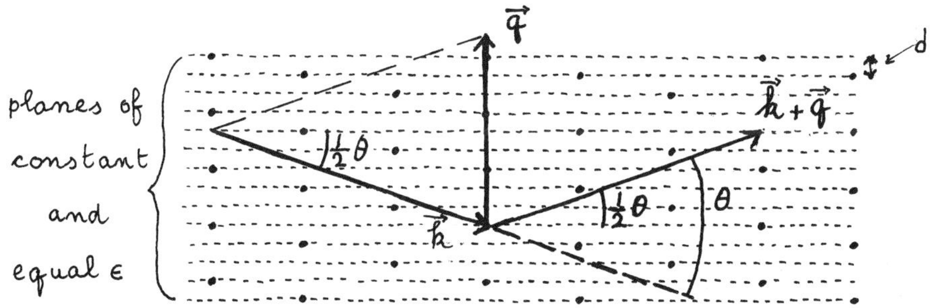

Fig. 4.1. Angle of Bragg reflection. Dots represent atoms in crystal.

The condition (4.9) for Bragg reflection is often expressed in terms of the radiation wavelength λ and the distance d between planes of constant and equal ϵ (see Fig. 4.1). The lengths λ and d are reciprocal to k and q, respectively, to within a factor 2π. Substituting $k = 2\pi/\lambda$ and $q = 2\pi/d$ in equation (4.9) gives the *Bragg law*

$$\lambda = 2d \sin \tfrac{1}{2}\theta. \tag{4.9a}$$

Bragg reflection and related diffraction and interference phenomena are usually derived by a theoretical procedure different from that described here and akin to the application of Huygens's principle in optics.[4] One considers first that various volume elements of a crystal (or other system) become the seat of oscillating currents under the action of the incident radiation with wave vector $\vec{k}$ and thereby act as secondary sources of radiation. Then one considers the superposition of the radiation components emitted by all secondary sources in any single direction with propagation vector $\vec{k}'$. The resultant of

---

3. This angle is often called 2θ, rather than θ.

4. See, for instance, C. Kittel, *Introduction to Solid State Physics* (New York: John Wiley, 1967), chap. 2; also, *BPC*, vol. 3, chap. 9.

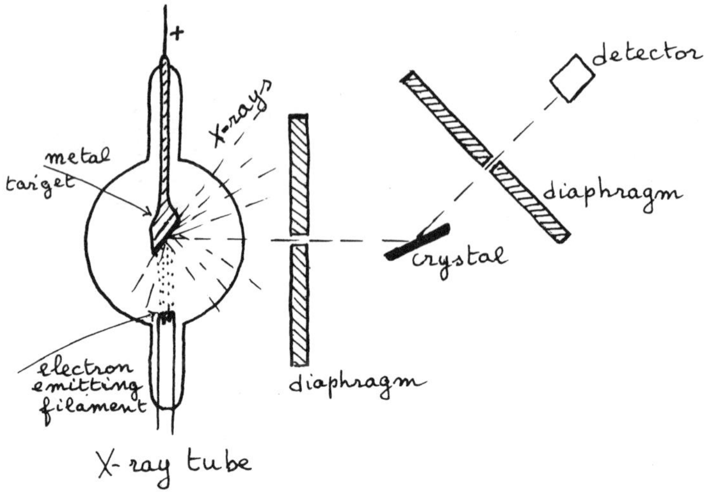

Fig. 4. 2. Diagram of apparatus for Bragg reflection.

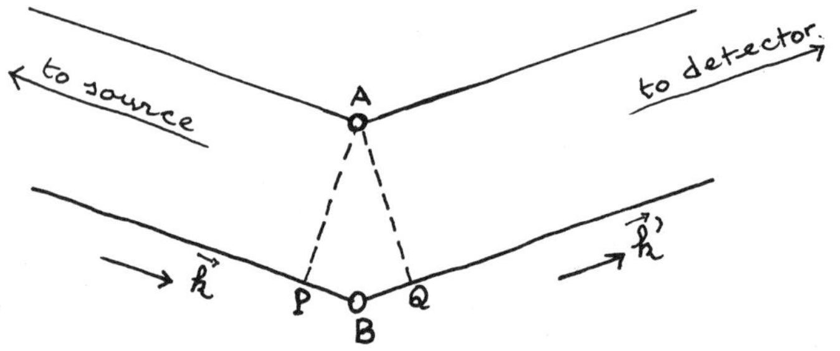

Fig. 4. 3. Diagram showing the excess path length PBQ traveled by radiation scattered by the secondary source B over the path length traveled through the secondary source A.

these components depends on their differences in path length traveled from the distant source to the respective secondary sources and on to a distant detector (see Fig. 4. 3). The distribution in space of the secondary sources causes the radiation emitted by them to interfere constructively only in a certain direction $\vec{k}'$ and destructively in all others. Equivalent results are obtained by this method and by the Fourier analysis procedure described above. In particular, the interference point of view leads directly to the Bragg law (4. 9a) whereas the Fourier analysis method of section 4. 1 leads to the vector relation (4. 8).

The method of secondary sources is familiar from ordinary optics and follows the sequence of events in space and time. On the other hand, it involves almost invariably the assumption of unrealistic features of the secondary sources; for example, each crystal cell is often regarded as a single pointlike secondary source. The method of Fourier analysis emphasizes that only the space distribution of a semimacroscopic property of the material, the dielectric constant in our problem, is relevant.

4.2. X-RAY DIFFRACTION BY CRYSTALS

Since crystals are homogeneous on a macroscopic scale, and presumably inhomogeneous but well ordered on the atomic scale, the Fourier expansion of their dielectric constant should contain terms of the type $\gamma \cos \vec{q}\cdot\vec{r}$ considered in the preceding section. The magnitude of the vectors $\vec{q}$ should be of the order of the inverse of atomic size. Accordingly, Bragg reflection should be expected when a beam of radiation of wavelength comparable to atomic size traverses a crystal. Radiation of this wavelength (X-rays) has a photon energy of the order of 10 keV and is emitted, with a convenient spectrum of monochromatic lines, by elements of medium atomic weight bombarded by 10-50 keV electrons (section S3.7).

4.2.1. *Evidence from Bragg Reflection Experiments.* Bragg reflection is observed by letting a monochromatic X-ray beam fall on a single crystal and by searching for a secondary (*diffracted*) beam. The mutual orientations of source, crystal, and detector are changed simultaneously until a diffracted beam appears. The instrumentation requires good collimation of source and detector and high mechanical precision in the control of the orientation angles.

Observation of Bragg reflection with a measured deflection angle θ determines the direction of a vector $\vec{q}$, as shown in figure 4.1. Easily observed reflection usually pertains to a $\vec{q}$ vector perpendicular to an outer face of the crystal. This circumstance facilitates the search for reflection. The measured value of θ determines the ratio $q/k = \lambda/d$ through equation (4.9) or (4.9a). If the wavenumber k of the X-rays is known, the value of the ratio q/k thus measured determines the value of q, that is, the pitch of a periodicity of atomic structure within the crystal. Conversely, if q is known from other experiments, measurement of q/k determines the value of k.

These determinations are often of high precision. An X-ray beam emerging by Bragg reflection on a known crystal is highly monochromatic with known wavenumber. Accordingly, Bragg reflection is the central element of X-ray monochromators.

The search for Bragg reflection generally shows a whole set of reflections by the same crystal with different orientations. Each reflection identifies a $\vec{q}$ vector with definite magnitude and orientation with respect to the crystal. The set of $\vec{q}$ vectors serves to determine the main structure of the crystal through the analysis in section 4.2.2. The intensities of reflected beams also provide structural information, as indicated in section 4.2.4.

4.2.2. *Lattice Structure.* Crystal structure analysis is a branch of physics covered in special treatises; here we merely point out some of its conceptual steps. We assume that a crystal is a three-dimensional stack of identical *primitive cells.*[5] Each cell has the shape of a parallelepiped whose edges are represented by *fundamental translation vectors* $\vec{a}$, $\vec{b}$, and $\vec{c}$ which do not lie in a plane and may or may not be mutually orthogonal. The set of all translation vectors $m\vec{a} + n\vec{b} + p\vec{c}$, where m, n, and p are arbitrary integers, constitutes the *lattice of* the crystal. (More properly, the lattice consists of the set of points reached from an origin by the translation vectors.)

Because of the identity of the cells, any property of the crystal that varies from point to point, specifically the dielectric constant ϵ, has the same value at identical positions within different cells. This characteristic is represented by the periodicity condition

$$\epsilon(\vec{r} + m\vec{a} + n\vec{b} + p\vec{c}) = \epsilon(\vec{r}). \tag{4.10}$$

On the other hand, $\epsilon(\vec{r})$ can also be represented by a Fourier expansion

$$\epsilon(\vec{r}) = 1 + \Sigma_j \ \gamma_j \exp(i\vec{q}_j\cdot\vec{r}), \tag{4.11}$$

a generalization of the expression $1 + \gamma \cos \vec{q}\cdot\vec{r}$ utilized in (4.1). The expansion (4.11) fulfills the periodicity condition (4.10) if and only if for each vector $\vec{q}_j$

$$\vec{q}_j\cdot\vec{a} = 2\pi h, \quad \vec{q}_j\cdot\vec{b} = 2\pi k, \quad \vec{q}_j\cdot\vec{c} = 2\pi l, \tag{4.12}$$

where h, k, and l are integers. This condition implies that each set of values $\{h, k, l\}$ identifies a vector $\vec{q}_j$, and is therefore equivalent to the label j.

The set of vectors $\vec{q}_j$ in the expansion (4.11) is, of course, the same set of vectors $\vec{q}$ obtained through the experimental observation of Bragg reflections. Verification of the existence of three vectors $\vec{a}$, $\vec{b}$, and $\vec{c}$ which satisfy the condition (4.12) for all experimental vectors $\vec{q}$ tests the cell model of a crystal; their actual determination establishes the cell size and shape.

As one verifies readily, the periodicity condition also implies that each vector $\vec{q}_j$ can be represented by the vector formula

$$\vec{q}_j = h\vec{A} + k\vec{B} + l\vec{C}, \tag{4.13}$$

where

$$\vec{A} = 2\pi \frac{\vec{b} \times \vec{c}}{\vec{a}\cdot\vec{b} \times \vec{c}}, \quad \vec{B} = 2\pi \frac{\vec{c} \times \vec{a}}{\vec{b}\cdot\vec{c} \times \vec{a}}, \quad \vec{C} = 2\pi \frac{\vec{a} \times \vec{b}}{\vec{c}\cdot\vec{a} \times \vec{b}}. \tag{4.13a}$$

The vector $\vec{A}$ is perpendicular to the plane of $\vec{b}$ and $\vec{c}$ and has magnitude reciprocal to that of $\vec{a}$; corresponding properties hold for $\vec{B}$ and $\vec{C}$. For this reason

---

5. See Kittel, *Introduction to Solid State Physics.* Chapters 1 and 2 treat crystal structure and diffraction by crystals extensively.

the vectors $\vec{A}$, $\vec{B}$, and $\vec{C}$ are called the fundamental vectors of the *reciprocal lattice* of the crystal. Just as the crystal lattice consists of all the vectors $m\vec{a} + n\vec{b} + p\vec{c}$, the reciprocal lattice consists of the set of vectors $\vec{q}_j$ defined by equation (4.13).

4.2.3. *Laue and Debye-Scherrer Methods.* For purposes of crystal analysis, experimental procedures are available to detect Bragg reflection without searching for it by varying the orientation of a crystal. The Laue procedure, utilized in the original discovery of X-ray diffraction, employs a beam of non-monochromatic X-rays incident on a crystal of fixed orientation. Bragg reflection usually takes place for several monochromatic components, irrespective of the mutual orientation of beam and crystal. A set of diffracted beams emerges then from the crystal, each of them corresponding to a vector $\vec{q}_j$ of the reciprocal lattice. This set forms a characteristic *Laue pattern* on a screen or photographic plate (see Figs. 4.4a and b). The deflection θ_j experienced by each diffracted beam identifies the direction of the corresponding vector $\vec{q}_j$ through the construction shown in Figure 4.1. Thereby examination of a Laue pattern provides direct evidence on the lattice structure.

The *powder method* utilizes instead a monochromatic beam and a crystal powder sample of the material under study. Among the randomly oriented crystals in the sample, some have an orientation appropriate to produce Bragg reflection on planes perpendicular to a vector $\vec{q}_j$ of the reciprocal lattice. The diffracted beam emerges then with a deflection $\theta_j = 2 \arcsin (q_j/2k)$. The same deflection results from diffraction by all crystals whose vectors $\vec{q}_j$ form the same angle with the direction of the incident beam. The beams reflected by these crystals travel in directions lying on a cone with its vertex at the sample, with its axis along the incident beam, and with aperture θ_j. The cone has a circular intersection with a screen or photographic plate perpendicular to the incident beam. Accordingly, the set of Bragg reflections pertaining to a set of reciprocal lattice vectors $\vec{q}_j$ of a crystalline powder produces a set of concentric circles on a screen or plate. This set of circles is called a Debye-Scherrer pattern (see Fig. 4.4c).

4.2.4. *Structure of Crystal Cells.* The experimental study of X-ray diffraction provides information not only on the arrangement of crystal cells in a periodic lattice but also on the structure of each cell. This is achieved, in essence, by reconstructing in detail the periodic variation of properties of the crystalline medium which are responsible for the diffraction.

The outline of the theory of Bragg reflection in section 4.1 suggests that the intensity of the secondary (reflected) wave is related to the magnitude of the coefficient γ. Without entering into detail of the theory and its applications, which can be very laborious, we state that intensity measurements of a large set of Bragg reflections can, in fact, determine a large number of coefficients γ_j of the expansion (4.11).

Experimental values of these coefficients find a most interesting semimacroscopic interpretation when one utilizes the expression (2.25) of the dielectric constant in the high frequency limit, $\epsilon(\omega) \to 1 - 4\pi e^2 NZ/m_e\omega^2$. (This limit is

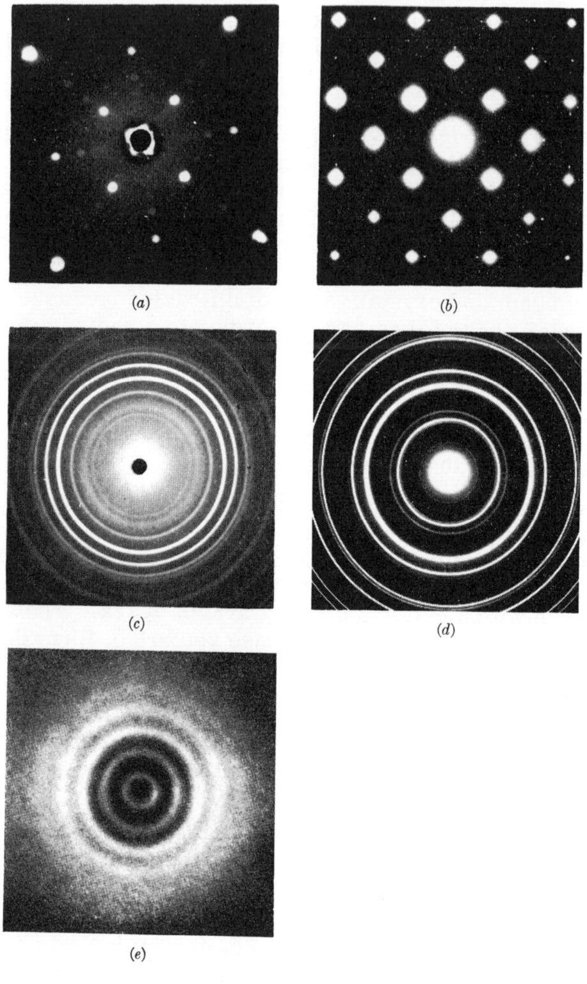

Fig. 4. 4. Typical diffraction patterns. (*a*) Laue pattern of X-ray diffraction by NaCl single crystal, (*b*) Laue pattern of electron diffraction by single gold crystal, (*c*) Debye-Scherrer pattern of X-ray diffraction by ZrO_2 crystalline powder, (*d*) Debye-Scherrer pattern of electron diffraction by a polycrystalline gold film, (*e*) Pattern of electron diffraction by As_4 vapor.

(*a*) Courtesy of G. R. Clark, University of Illinois;
(*b*) and (*d*) courtesy of J. A. Suddeth, National Bureau of Standards;
(*c*) courtesy of H. Swanson, National Bureau of Standards;
(*e*) courtesy of L. R. Maxwell, *J. Chem. Physics*.

approximated fairly well under the conditions of X-ray diffraction experiments.) The product NZ, which represents in equation (2. 25) the average electron density in a macroscopically homogeneous material, will be indicated here as a variable effective density $N\rho(\vec{r})$, where, in this case, N is the number of cells per unit volume. Introducing the Fourier expansion

$$N\rho(\vec{r}) = N\Sigma_j \rho_j \exp(i\vec{q}_j \cdot \vec{r}) \tag{4.14}$$

into the expression of ϵ (2. 25), and this formula into equation (4. 11), we have

$$\epsilon(\vec{r}) = 1 + \Sigma_j \gamma_j \exp(i\vec{q}_j \cdot \vec{r}) \rightarrow 1 - \frac{4\pi e^2 N}{m_e \omega^2} \Sigma_j \rho_j \exp(i\vec{q}_j \cdot \vec{r}); \tag{4.15}$$

that is,

$$\gamma_j \rightarrow - \frac{4\pi e^2 N}{m_e \omega^2} \rho_j. \tag{4.16}$$

Experimental determination of the coefficients γ_j provides the coefficients of the Fourier representation of the effective electron density $\rho(\vec{r})$. Since the reciprocal lattice vectors $\vec{q}_j$ are also determined experimentally, the Fourier expansion (4. 14) of $\rho(\vec{r})$ can be evaluated numerically. Thereby one constructs a map of the effective electron density $\rho(\vec{r})$. Figure 4. 5 shows such a map constructed by X-ray diffraction analysis of an organic crystal; this map is accompanied by a diagram of the chemical structure of each crystal molecule.

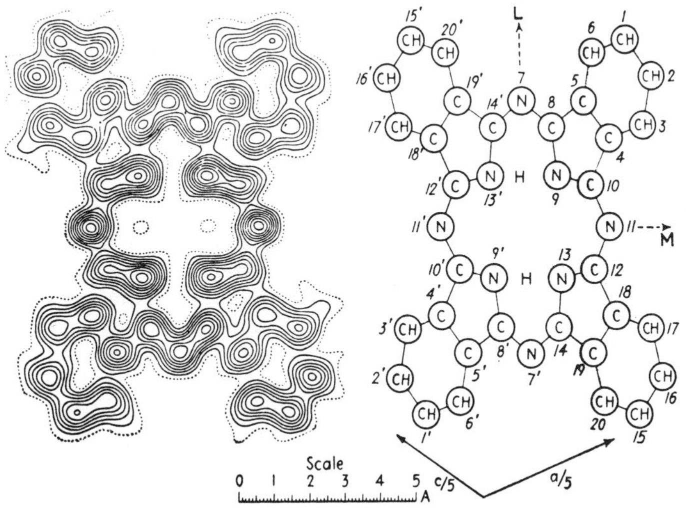

Fig. 4. 5. Electron density projection of the phthalocyanine molecule, from G. L. Clark, *Applied X-Rays* (New York: McGraw-Hill Book Co., 1940), p. 457.

The name "effective" electron density has been used advisedly to qualify the significance of $\rho(\vec{r})$. The whole development of this chapter in terms of a dielectric constant ϵ variable from point to point within single crystal cells and the further representation of ϵ in terms of $\rho(\vec{r})$ constitutes a tentative extrapolation of the semimacroscopic approach of chapter 2. There, the constant ϵ was intended to represent a property of volumes of matter containing large numbers of atoms. The successful construction of maps such as that of Figure 4.5, whose details prove consistent with evidence from chemistry and from other sources, leads to the following conclusion. The dielectric constant $\epsilon(\vec{r})$ and the electron density $\rho(\vec{r})$ can be defined beyond our original intent, on *a subatomic scale,* as properties of a material determined by X-ray diffraction experiments. This conclusion is an important point of our study of the structure of matter. The precise theoretical meaning of $\rho(\vec{r})$ remains to be developed in later chapters.

4.2.5. *Mechanical Interpretation of Bragg Reflection.* A radiation beam with propagation vector $\vec{k}$ and frequency $\omega = ck$ can deliver photons with energy $\hbar\omega$ ($\hbar = h/2\pi$) and momentum $\hbar\vec{k}$. Diffracted radiation with propagation vector $\vec{k} + \vec{q}_j$ such that $|\vec{k} + \vec{q}_j| = |\vec{k}|$ can deliver photons with the same energy $\hbar\omega$ but with momentum $\hbar(\vec{k} + \vec{q}_j)$. Accordingly, one may say that Bragg reflection changes the momentum of each photon by $\hbar\vec{q}_j$. This momentum transfer is presumably balanced by a recoil of the crystal. Observation of photons which have experienced this momentum transfer implies that the expansion (4.11) of $\epsilon(\vec{r})$ contains a Fourier component with propagation vector $\vec{q}_j$. This connection between sinusoidal *oscillations* of a variable—the parameter ϵ in our case—*from point to point* in space and *momentum transfers* was surmised at the beginning of this chapter as an analog of the connection between oscillations in time and energy transfers. The connection rests now on the results of a semimacroscopic theory of radiation diffraction combined with the knowledge of the quantum properties of radiation energy.

4.3. DIFFRACTION OF PARTICLE BEAMS BY CRYSTALS

If momentum exchanges at the atomic level depend quite generally on the occurrence of sinusoidal oscillations in space, an analogous dependence should exist for momentum changes of atomic particles as well as of radiation. This surmise has indeed been verified extensively, beginning with famous experiments on electron beams by Davisson and Germer and by G. P. Thomson. Any beam of particles with momentum $\vec{p}$ hitting a crystalline material is diffracted, in essence, like an X-ray beam with photon momentum $\vec{p} = \hbar\vec{k}$ (see Figs. 4.4d and 4.4e). That is, some of the particles are so deflected that their new momentum is

$$\vec{p} + \hbar\vec{q}, \tag{4.17}$$

where $\vec{q}$ is a characteristic vector of the crystal. In the case of particles that achieve substantial penetration into the crystal, such as electrons and neutrons, the vector $\vec{q}$ fulfills the same conditions as for X-ray diffraction, i.e., it must be one of the vectors of the reciprocal lattice represented by equation (4.13).

Experiments have also been performed with beams of atoms of thermal energy, whose momenta are equal to those of X-ray photons but which do not penetrate beyond the surface of the crystal. In this case the vector $\vec{q}$ fulfills only the condition that $\vec{q}\cdot\vec{a}$ and $\vec{q}\cdot\vec{b}$ be multiples of 2π, where $\vec{a}$ and $\vec{b}$ indicate a pair of lattice vectors parallel to the crystal surface; the component of $\vec{q}$ perpendicular to this surface is subject to no condition.

The diffraction of particle beams establishes the general law of momentum transfers at the atomic level. It also proves valuable for purposes of structural analysis of materials. Diffraction of particles often complements or replaces X-ray diffraction effectively because it reveals structural characteristics that are not easily detected by X-ray analysis. For example, the position of H atoms in a crystal coincides with no strong concentration of electrons because H nuclei have a small charge. Therefore, hydrogen atoms tend to elude X-ray analysis (recall Fig. 4.5) whereas they are easily located by neutron diffraction owing to the strong interaction between neutrons and H nuclei.

The experiments on particle diffraction and their theoretical interpretation indicated above provide, of course, only a preliminary concept of the connection between momentum transfers and oscillations of variables in space. We shall return to this connection, as well as to the connection between energy transfers and oscillations in time, in later chapters.

SUPPLEMENTARY SECTIONS

S4.4. DIFFRACTION BY ISOLATED ATOMS

The semimacroscopic theory developed in sections 4.1 and 4.2 for the diffraction of X-rays by crystals can also be applied to other materials. To this end one can formally attribute to any material a dielectric constant $\epsilon(\vec{r})$ varying from point to point. In this case, the Fourier expansion of $\epsilon(\vec{r})$, analogous to expansion (4.11), has the form of an integral over a continuous set of vectors $\vec{q}$:

$$\epsilon(\vec{r}) = 1 + \int d\vec{q}\, \gamma(\vec{q}) \exp(i\vec{q}\cdot\vec{r}). \tag{4.18}$$

This integral replaces the Σ_j over the vectors $\vec{q}_j$ of a crystal's reciprocal lattice.

The substantive point of the extension to noncrystalline materials occurs in the calculation of the Fourier coefficients $\gamma(\vec{q})$. Consider, in particular, the example of a monatomic gas with N atoms per unit volume lying at uncorrelated positions. Owing to the lack of correlation, each atom contributes independently to $\gamma(\vec{q})$. When this contribution is calculated from the electron density, as in equation (4.16), $\rho(\vec{r})$ itself must be considered separately for each atom. The coefficient $\gamma(\vec{q})$ is then represented in the form, equivalent to expression (4.16)

$$\gamma(\vec{q}) = -\frac{4\pi e^2 N}{m_e \omega^2}\,\rho(\vec{q}). \tag{4.19}$$

The Fourier coefficient $\rho(\vec{q})$ is obtained from the density function within each atom, $\rho(\vec{r})$, by a generalization of equation (A.33), namely,[6]

$$\rho(\vec{q}) = \frac{1}{(2\pi)^3}\int \exp(-i\vec{q}\cdot\vec{r})\rho(\vec{r})d\vec{r}. \tag{4.20}$$

Since the integral of equation (4.20) is to be evaluated separately for each atom, the origin of the coordinates $\vec{r}$ is conveniently taken at the nucleus. Consider further that we deal with average properties of atoms of random orientation. The density function $\rho(\vec{r})$ may then be regarded as spherically symmetric, that is, as a function only of the distance r from the nucleus. By the same token, the Fourier coefficient $\rho(\vec{q})$ also depends only on the magnitude of $\vec{q}$, and the integration over the direction of $\vec{r}$ in equation (4.20) can be carried out analytically to yield

$$\rho(q) = \frac{1}{(2\pi)^3}\int_0^\infty r^2 dr \int_{-1}^{1} d(\cos\theta) \int_0^{2\pi} d\phi\, \exp(-iqr\cos\theta)\,\rho(r)$$

$$= \frac{1}{(2\pi)^3}\int_0^\infty 4\pi r^2 dr\, \frac{\sin qr}{qr}\,\rho(r) = \frac{1}{(2\pi)^3}\,F(q). \tag{4.21}$$

In this formula we have introduced the symbol F(q) for the integral; the quantity F(q), called the *form factor* of the atom, is used more commonly than $\rho(q)$ itself. The form factor has the useful property that its value for $q = 0$ equals the integral of the electron density over the whole atom, and therefore equals the atomic number Z.

The preceding equations (4.18)-(4.21) relate the property $\epsilon(\vec{r})$—or rather the Fourier coefficients $\gamma(\vec{r})$—of a monatomic gas to the electron density $\rho(\vec{r})$ within an atom. The procedure can be extended to polyatomic molecular gases and even to amorphous condensed materials. The form factor F(q) must then be defined so as to allow for the influence of correlations between the positions of adjacent atoms. For each material F(q) constitutes an experimentally accessible parameter, because it is related to $\gamma(\vec{q})$ by a known factor. The coefficients $\gamma(\vec{q})$ themselves are determined through measurements of the intensity distribution $I(\theta)$ of X-rays scattered with various deflections θ.

---

6. The factor $(2\pi)^{-3}$ appears in equation (4.20) because we have utilized Fourier expansions in plane waves $\exp(i\vec{q}\cdot\vec{r})$ which are periodic in $\vec{q}\cdot\vec{r}$ with the period 2π, whereas the expansion in Appendix A utilizes functions $\exp(i\pi nt/T)$ which are periodic in $nt/2T$ with period 1. This difference introduces one factor $(2\pi)^{-1}$ in the Fourier coefficient formula for each variable of integration.

X-rays are scattered by noncrystalline materials with all deflections θ because the Fourier expansion (4.18) of $\epsilon(\vec{r})$ includes components with a continuous set of vectors $\vec{q}$. Vector addition of the propagation vector $\vec{k}$ of the incident X-rays and of a vector $\vec{q}$ of the material yields the propagation vector $\vec{k} + \vec{q}$ of scattered X-rays. The scattered X-rays have appreciable intensity only if the magnitude $|\vec{k} + \vec{q}|$ of their propagation vector has the same value k as that of the incident X-rays, as required by condition (4.8). This requirement leads to the Bragg relationship (4.9) between $\vec{q}$ and the deflection angle θ,

$$q = 2k \sin \tfrac{1}{2}\theta = \frac{4\pi \sin \frac{1}{2}\theta}{\lambda}. \tag{4.22}$$

Section 4.1 also shows that the amplitude of a secondary wave is proportional to the coefficient γ; hence its intensity is proportional to $|\gamma|^2$. The intensity distribution of scattered X-rays obeys the proportionality relationship

$$I(\theta) \propto |\gamma(q)|^2 \propto \left| F\left(\frac{4\pi \sin \frac{1}{2}\theta}{\gamma}\right)\right|^2. \tag{4.23}$$

Since we know that $F(0) = Z$ for neutral atoms, relative measurements of $I(\theta)$ at various values of θ yield the complete function $F(q)$, from which it is possible to determine the electron density $\rho(r)$ numerically. [A test of the accuracy of the whole theory is afforded by the prediction that F depends on the two variables θ and λ only through their combination $q = (4\pi \sin \frac{1}{2}\theta)/\lambda$; that is, one should find the same value of F for different pairs (θ, λ) which yield the same q]. Reasonably successful early experimental determinations of $\rho(r)$ by this method (Figure 4.6) have been continued only sporadically because $\rho(r)$ can now be determined theoretically with greater ease and accuracy. However, it is important for our treatment that one can map the effective electron density throughout an atom by means of experiments and of an approximate semimacroscopic interpretation.

S4.4.1. *Electron scattering.* From the experiments on electron diffraction by crystals mentioned in section 4.3, transfer of momentum $\hbar\vec{q}_j$ to an electron appears to depend on the Fourier expansion coefficient γ_j of some unspecified property which varies from point to point with the periodicity (4.10) of the crystal. It also appears that crystals exert a weak action on the electrons, as they do on the X-rays.[7] This indicates that the momentum changes occur as single elementary processes, if they occur at all. Scattering of electrons *by single atoms* (also called *electron diffraction*) should similarly depend on some coefficients analogous to the $\gamma(\vec{q})$ of equation (4.18); however, no prediction whatever can be made on these coefficients unless one knows which property of the atom is relevant to electron scattering. We anticipate here that the relevant property is the electrostatic potential energy $-eV(\vec{r})$ of an electron at each point of the atom, under the conditions of weak action which

---

7. The action on electrons is not quite as weak as that on X-rays.

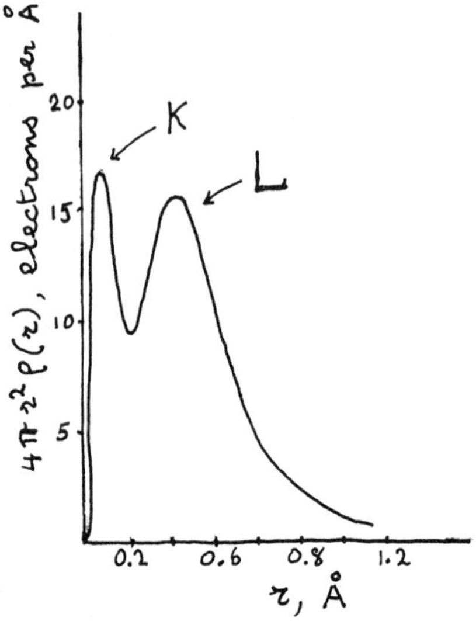

Fig. 4. 6. Radial electron distribution in neon from X-ray diffraction experiments, showing the shell structure. [Adapted from E. O. Wollan, *Rev. Mod. Phys.* 4 (1932): 205.]

obtain approximately for electrons of sufficiently high energy (e.g., $\gtrsim$ 1 keV). This assumption is plausible because the electron's momentum must remain constant in a region of space where $V(\vec{r})$ is uniform.

Under this assumption a connection between the scattering of electrons and that of X-rays is provided by the Poisson equation $\nabla^2 V = -4\pi\rho$.[8] The density of electric charge ρ in this equation includes the positive charge Ze of the nucleus as well as the negative charge of the electrons, whereas only the electron density appears in the calculation for X-rays.[9] The combined charge density is represented by

$$\rho(\vec{r}) = Ze\delta(\vec{r}) - e\rho_e(\vec{r}), \tag{4.24}$$

where $\delta(r)$ is the Dirac function defined on p. 541 and ρ_e is the same as ρ in equation (4.21). The Fourier coefficients of equation (4.24) are

$$\rho(q) = (2\pi)^{-3}[Z - F(q)]e, \tag{4.25}$$

---

8. *BPC*, 2: 61

9. The symbol ρ in this section indicates a charge density rather than a particle density.

where F(q) is the form factor which determines X-ray scattering. Utilizing the Fourier expansion formula

$$\nabla^2 V(\vec{r}) = \nabla^2 \int d\vec{q} V(\vec{q}) \exp(i\vec{q}\cdot\vec{r}) = - \int d\vec{q} q^2 V(\vec{q}) \exp(i\vec{q}\cdot\vec{r}) \tag{4.26}$$

and comparing it to the expansion of $-4\pi\rho(\vec{r})$, one sees that the Poisson equation $\nabla^2 V(\vec{r}) = 4\pi\rho$ is equivalent to the equation among Fourier coefficients

$$V(q) = \frac{Z - F(q)}{2\pi^2 q^2} e. \tag{4.27}$$

This relationship between the properties that determine the scattering of electrons and X-rays is verified experimentally.

Equation (4.27) verifies and complements remarks made in section S3.7 concerning the elastic scattering of electrons by atoms. As the intensity of scattered X-rays is proportional to the squared Fourier coefficients, $|\gamma(\vec{q})|^2$, so should the cross-section for electron scattering be proportional to $|V(\vec{q})|^2$. According to equation (4.27), this cross-section should then be inversely proportional to q^4 and hence to the fourth power of the momentum transfer $\hbar q$, in agreement with the Rutherford formula (1.33). Furthermore, the cross-section should be proportional to $|Z - F(q)|^2$; equation (1.33) shows proportionality to Z^2 but does not take into account the screening effect of atomic electrons. This effect was mentioned but not evaluated in sections 1.4 and S3.7. Thus equation (4.27) provides an evaluation of the screening effect through subtraction of the form factor F(q) from the atomic number. These predictions will be verified by quantum-mechanical calculations in section 17.4.

S4.5. SIMULTANEOUS TRANSFER OF MOMENTUM AND ENERGY. RAMAN AND COMPTON EFFECTS

The property of a material that affects the propagation of radiation, namely ϵ, may vary not only from point to point in space but also from moment to moment in time. (We mean here not the variations in time driven by the radiation itself, which may cause radiation absorption, but independent variations. For example, the electron density in a material always fluctuates owing to the thermal agitation of the atoms, an effect that was disregarded above in the first approximation.) A radiation beam traversing the material may then experience a simultaneous change of *direction and frequency*, that is, a change of the *momentum and energy* of its photons.

This effect, which appears plausible on the basis of analogies, emerges from a suitably redeveloped semimacroscopic treatment of radiation propagation. We do not carry out the redevelopment but merely indicate where changes occur in the previous equations. Let us return to the propagation equation (4.1), replacing cos $\vec{q}\cdot\vec{r}$ by cos $(\vec{q}\cdot\vec{r} - \chi t)$. The field representation (4.2) should be modified by replacing the frequency ω by $\omega + m\chi$ and the index n of the field amplitudes $\vec{E}_n$, $\vec{B}_n$ and of the sum Σ_n by a pair of indices nm. The development proceeds then as in section 4.1, the basic condition (4.8) for Bragg reflection being now replaced by

$$|k \pm q| = k \pm \frac{\chi}{c}. \tag{4.28}$$

The radiation diffracted with frequency shift χ may be said to result from beats of the incident radiation and of the oscillatory variation of ϵ.

Energy balance must be preserved in the process of diffraction with frequency shift by an accompanying transition of the scatterer atom (or other portion of matter) from a stationary state A to another stationary state B. This process may then also be described as inelastic scattering of radiation. Energy conservation requires that

$$\hbar\omega + E_A = \hbar(\omega + \chi) + E_B \tag{4.29}$$

and, hence, that

$$\hbar(\omega + \chi) - \hbar\omega = \hbar\chi = E_A - E_B. \tag{4.30}$$

When A is the ground state, E_B is larger than E_A and χ is negative.[10]

Experimentally, the simultaneous transfer of momentum and energy between radiation and matter is observed under various circumstances. Much of Bragg reflection itself is accompanied by very small frequency shifts associated with thermal vibrations of the crystal lattice. In the optical frequency range one observes the *Raman effect,* i.e., a substantial frequency shift of light scattered by molecules; the levels A and B correspond to states with different vibrations of atoms with respect to one another, but with the same electronic motion. This effect has been very important for molecular studies because it affords a convenient experimental observation by optical techniques of vibrational frequencies ω_{BA} lying in the very far infrared (see chap. 19). Another very important phenomenon involving a frequency shift is the scattering of X-rays in which the transition A→B involves ionization (the Compton effect).

S4.5.1. *Compton Effect.* This process is comparatively intense for large values of the momentum transfer $\hbar\vec{q}$. An electron receiving this large momen-

---

10. A similar energy balance holds in the fluorescence process, where an atom absorbs a photon of radiation energy $\hbar\omega$, while being excited from an initial state A to a state C, and then emits radiation while passing to a lower energy state B. However, in fluorescence the photon energy $\hbar\omega$ must equal the energy difference $E_C - E_A$ of two stationary states, but no such requirement exists in our case. Radiation scattering, with or without frequency shift, occurs for any frequency of the incident radiation and results from atomic current oscillations driven off-resonance.

tum usually recoils out of the atomic structure to which it belonged.[11] The energy transfer $-\hbar\chi = E_B - E_A$ is much larger than the original binding energy of the electron within its atom so that the electron recoil can be treated approximately as though the rest of the material were absent. This process becomes likely when the recoil momentum amounts at least to one-tenth of the product of the electron's mass and the light velocity, i.e., when $\hbar q \gtrsim m_e c/10$. Indeed, the Compton effect is the dominant process of radiation interaction with matter for photon energies above approximately 50 keV, when absorption by driven currents is negligible.[12] (See Figure 4. 7.). (The oscillator strengths f_r considered in chapter 2 are negligible for resonance frequencies with $\hbar\omega_r \gtrsim 50$ keV.) The predominance of the Compton effect extends, however, only to photon energies $\lesssim$ 10-100 MeV, after which another process, called *pair production*, becomes important.

Equation (4. 28) determines the balance of energy and of momentum in a Compton process. For a given propagation vector $\vec{k}$ of incident X-rays, it

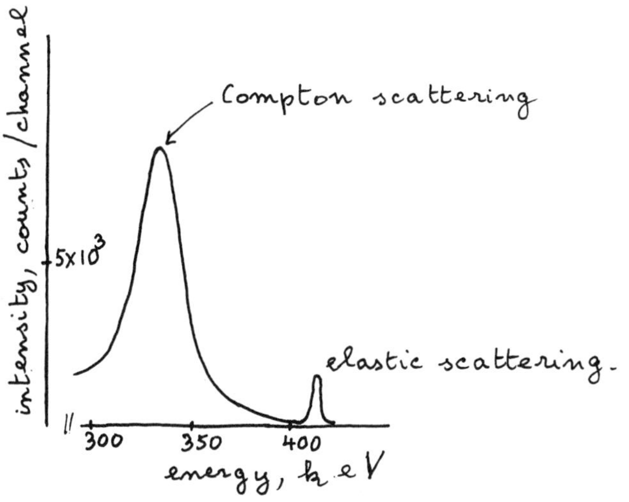

Fig. 4. 7. Spectrum of X rays scattered by lead at 45°; the energy of the incident radiation is 412 keV. The peak of the spectrum of Compton-scattered photons is not quite sharp, because the scattering electrons exchange some momentum with the lead nuclei; this exchange is disregarded in the simple theory of section 4. 5. 1. [Adapted from M. Schumaker, *Phys. Rev.* 182 (1969): 7].

---

11. The recoil electron and scattered photon were first detected simultaneously in a coincidence experiment of Bothe and Geiger. See M. Born, *Atomic Physics* 7th ed. (Darien, Conn.: Hafner, 1962), p. 90.

12. Ibid.

serves to determine the energy transfer either for a given angle θ of X-ray deflection or for a given direction of electron recoil. We derive here the most familiar formula which yields the wavelength increase of the X-rays as a function of θ. To this end we write equations (4.28) in terms of a new vector $\vec{k}'$:

$$\vec{k}' = \vec{k} + \vec{q}, \tag{4.31}$$

$$k' = k + \frac{\chi}{c}. \tag{4.32}$$

The energy $-\hbar\chi$ transferred to the electron together with the momentum $-\hbar\vec{q}$ must be represented by a relativistic formula,[13] since the recoil velocity may approach c,

$$-\hbar\chi = (m^2c^4 + \hbar^2c^2q^2)^{1/2} - m_ec^2. \tag{4.33}$$

Substitution into equation (4.32) yields

$$k' = k - (q_c{}^2 + q^2)^{1/2} + q_c, \tag{4.34}$$

where $q_c = m_ec/\hbar$. Solving equation (4.34) for q^2, one finds

$$q^2 = (k - k')(k - k' + 2q_c). \tag{4.35}$$

On the other hand, equation (4.31) serves to represent q^2 in terms of k', k, and θ by solving the vector triangle $(\vec{k}', \vec{k}, \vec{q})$:

$$q^2 = k^2 + k'^2 - 2kk' \cos\theta = (k - k')^2 + 2kk'(1 - \cos\theta). \tag{4.36}$$

Equating the right-hand sides of equations (4.35) and (4.36) yields

$$q_c(k - k') = kk'(1 - \cos\theta),$$

which, transformed to wavelength scale, turns into the Compton relation

$$\lambda' - \lambda = (h/m_ec)(1 - \cos\theta) = \lambda_c(1 - \cos\theta). \tag{4.37}$$

The quantity $\lambda_c \equiv h/m_ec = 2.42 \times 10^{-10}$ cm is known as the *Compton wavelength.*

CHAPTER 4 PROBLEMS

4.1. Deduce mathematically that the expansion (4.11) fulfills the periodicity condition (4.10) if and only if each vector $\vec{q}_j$ fulfills the condition (4.12).

---

13. *BPC*, 1: 397.

4.2. Suppose that the fundamental vectors of the reciprocal lattice of a crystal, $\vec{A}$, $\vec{B}$, and $\vec{C}$, have been determined by X-ray diffraction analysis. Show that the fundamental translation vectors of the crystal lattice are reciprocal to $\vec{A}$, $\vec{B}$, and $\vec{C}$, that is, they are given by formulae analogous to (4.13a), $\vec{a} = 2\pi(\vec{B} \times \vec{C})/(\vec{A}\cdot\vec{B} \times \vec{C})$, etc. This means that the reciprocal of the reciprocal lattice is the original lattice. Use the vector identity $(\vec{\alpha} \times \vec{\beta}) \times (\vec{\gamma} \times \vec{\delta}) = (\vec{\alpha}\cdot\vec{\beta} \times \vec{\delta})\,\vec{\gamma} - (\vec{\alpha}\cdot\vec{\beta} \times \vec{\gamma})\vec{\delta}$.

4.3. The fundamental vectors of the reciprocal lattice of crystalline NaCl are $\vec{A} = (2\pi/a_0)\,(\hat{x} + \hat{y} - \hat{z})$, $\vec{B} = (2\pi/a_0)\,(-\hat{x} + \hat{y} + \hat{z})$. $\vec{C} = (2\pi/a_0)\,(\hat{x} - \hat{y} + \hat{z})$, where $a_0 = 5.63$ Å. Calculate the fundamental translation vectors of the crystal lattice by applying the statement of problem 4.2.

4.4. A beam of X-rays with wavelength 0.4 Å passing through a crystal powder is diffracted into a cone with half-aperture $\theta = 12°$. Calculate (a) the spacing of planes of equal dielectric constant which have produced the observed diffraction; (b) the energy of electrons that would be diffracted by the same angle.

4.5. Graphite crystals consist of practically flat layers of carbon atoms in hexagonal arrangement, as shown in the figure. Two fundamental translation vectors are shown. The third fundamental vector $\vec{c}$ is perpendicular to the drawing and has length 6.70 Å. Determine the fundamental vectors of the

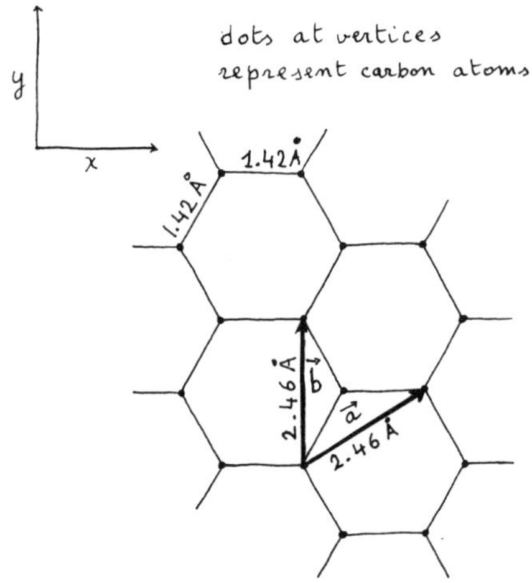

reciprocal lattice. Draw the vectors $\vec{A}$ and $\vec{B}$ in the plane parallel to the layer and the set of vectors $h\vec{A} + k\vec{B}$, with (h, k) = (1, 1), (2, 1), (−1, −1), and (1, −1).

4.6. Consider an experiment of Laue diffraction of X-rays on a graphite crystal. The X-ray beam has a continuous bremsstrahlung spectrum produced by

10-keV electrons and is directed along the positive z axis which is perpendicular to the graphite layers in the coordinate system of problem 4.5. We seek the diffracted beam which corresponds to a vector $\vec{q}_j$ identified by the indices $h = 1, k = 0$, and $l = -1$ of equation (4.13). Utilizing the solution of problem 4.5 and the Bragg law, (a) show that the given conditions identify the direction of the desired diffracted beam and the wave vector for which diffraction takes place; (b) calculate the angle θ between the diffracted beam and the z axis; (c) calculate the wavelength and photon energy; (d) determine the direction, if any, of other diffracted beams forming the same angle θ with the z axis; (e) calculate the kinetic energy of electrons and of neutrons that would be diffracted in the same direction as the X-rays.

4.7. The layers of a graphite crystal are stacked with successive layers staggered with respect to one another, while the atoms of every second layer are on top of one another (see figure). Show by a qualitative argument that the Fourier coefficients γ_j of equation (4.11) are zero for all vectors $\vec{q}_j$ that are odd multiples of the fundamental vector $\vec{C}$ of the reciprocal lattice.

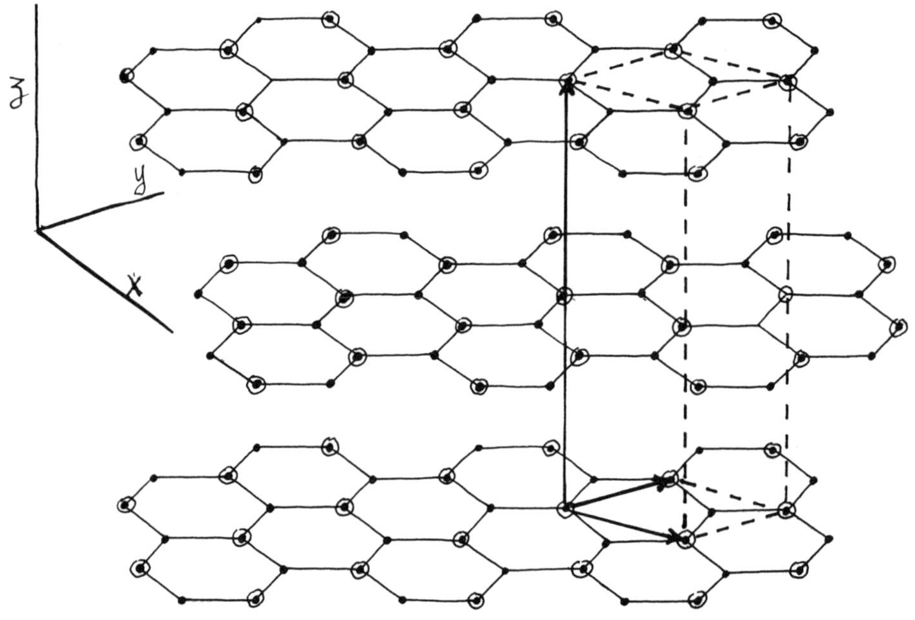

Stacking of graphite layers. Dots represent carbon atoms. Circled dots are atoms which have neighbors in the z direction in the layers immediately above and below them. Uncircled dots are atoms with no near neighbors in the z direction. On the right of the figure are shown the three fundamental translation vectors and the unit cell.

4.8. Calculate the form factor F(q) (see eq. [4.21]) for the following fictional atom. The atom contains 10 electrons distributed about its center with density

$(5/4\pi a^3)\exp(-r/a)$, where $a = 1$ Å. Plot this form factor as a function of X-ray scattering angle θ, for an X-ray wavelength $\lambda = 1.33$ Å.

4.9. Monochromatic X-rays with 125 keV photon energy are Compton-scattered by water and emerge with different energies in different directions. Determine the direction of emergence of X-rays with 100 keV photon energy.

SOLUTIONS TO CHAPTER 4 PROBLEMS

4.1. The functions $\epsilon(\vec{r})$ and $\epsilon(\vec{r} + m\vec{a} + n\vec{b} + p\vec{c})$ coincide if and only if all pairs of corresponding terms in their Fourier expansions coincide. The jth terms of the two expansions differ by a factor $\exp[i\vec{q}_j \cdot (m\vec{a} + n\vec{b} + p\vec{c})]$. This factor is unity if and only if the scalar product in the exponent is a multiple of 2π. This scalar product is a linear combination of the scalar products (4.12) with the arbitrary integer coefficients m, n, and p; it is necessarily a multiple of 2π if and only if each of the scalar products (4.12) is itself a multiple of 2π.

4.2. Substitute in the given expression of $\vec{a}$ the expressions of $\vec{A}$, $\vec{B}$, and $\vec{C}$ given by equation (4.13a) and carry out the simplifications. The result is $\vec{a}$ itself.

4.3. $\vec{a} = \frac{1}{2}a_0(\hat{x} + \hat{y})$, $\vec{b} = \frac{1}{2}a_0(\hat{y} + \hat{z})$, $\vec{c} = \frac{1}{2}a_0(\hat{z} + \hat{x})$.

4.4. (a) From equation (4.9a), $d = \lambda/(2\sin\frac{1}{2}\theta) = 0.4$ Å$/(2\sin 6°) = 1.9$ Å; (b) $E = p^2/2m_e = h^2/2m_e\lambda^2 = (6.62 \times 10^{-27})^2/[2 \times 9.1 \times 10^{-28} \times (0.4 \times 10^{-8})^2] = 1.51 \times 10^{-9}$ ergs $= 0.94$ keV.

4.5. The vector $\vec{A}$ forms an angle of 30° with $\vec{a}$ and the vector $\vec{b}$ is at 30° from $\vec{B}$; $\vec{A}$ and $\vec{B}$ have equal length, 2.95 Å^{-1}. The vector $\vec{C}$ is perpendicular to the plane of $\vec{A}$ and $\vec{B}$ and has length 0.94 Å^{-1} (see next page).

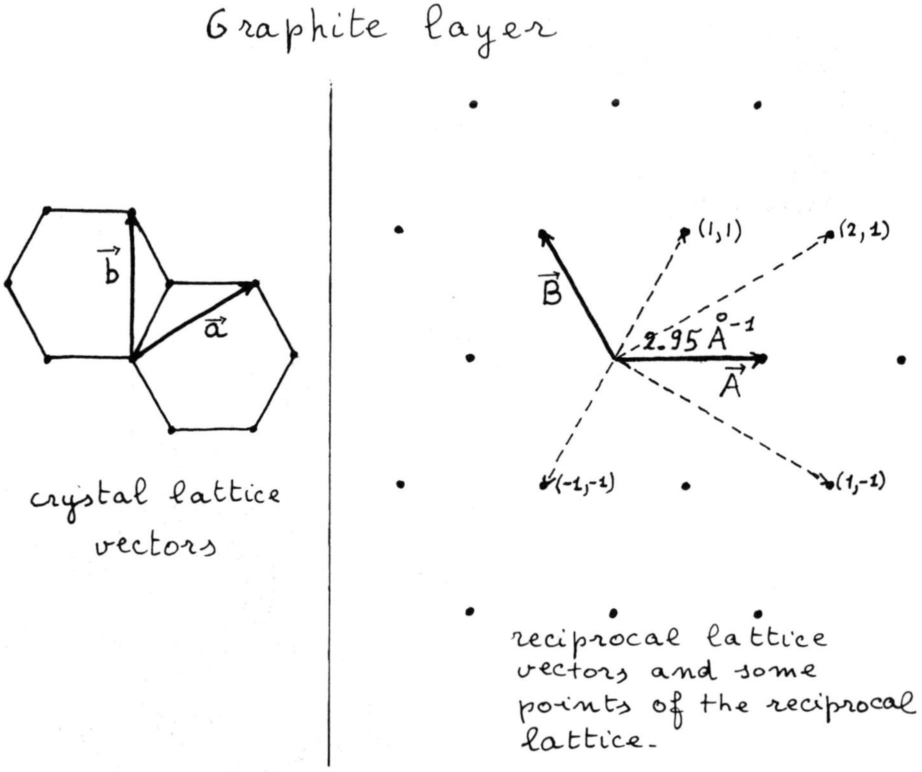
Graphite layer
b
a
crystal lattice vectors
B
2.95 Å^{-1}
A
(1,1)
(2,1)
(-1,-1)
(1,-1)
reciprocal lattice vectors and some points of the reciprocal lattice.

4.6. (a) The solution of problem 4.5 yields vectors $\vec{A}$ and $\vec{C}$ lying in the (x, z) plane as shown on page 118. The vector $\vec{q}_j = \vec{A} - \vec{C}$ has the magnitude 3.09 Å^{-1} and forms an angle of 107° 38′ with the positive z axis. The planes of Bragg reflection are perpendicular to $\vec{q}_j$ and thus form an angle of 17° 38′ with the z axis. The vector $\vec{k}$, parallel to the z axis and such that it is reflected into $\vec{k}' = \vec{k} + \vec{q}_j$, is thus uniquely determined. (b) The diffraction angle $\theta = 2 \times (17°38') = 35°16'$. (c) From the Bragg law (4.9) we have k = 5.11 Å^{-1}. The photon energy is $h\nu = \hbar ck = 10$ keV, and $\lambda = 1.2$ Å. (d) In the (x, y)-plane of the reciprocal lattice there are, besides $\vec{A}$, five additional vectors with the same magnitude as $\vec{A}$, namely, $\vec{A} + \vec{B}$, $\vec{B}$, $-\vec{A}$, $-\vec{A} - \vec{B}$, and $-\vec{B}$ (see the diagram in the solution of problem 4.5). The set of these six vectors has hexagonal symmetry. To the six vectors correspond six beams diffracted symmetrically about the z axis. (e) For equal diffraction by the same crystal, particles of mass m must have the same momentum $\hbar k$ as the X-ray photons. Their energy is $(\hbar ck)^2/2mc^2 = h\nu(h\nu/2mc^2)$, which yields 100 eV for electrons and 0.054 eV for neutrons.

4.7. For $\vec{q}_j = (2t + 1)\vec{C} = (2t + 1)C\hat{z}$, the Fourier coefficients are given by

$$\gamma_j = (2\pi)^{-3} \int d\vec{r} \exp(-i\vec{q}_j \cdot \vec{r})\epsilon(\vec{r}) = (2\pi)^{-3} \int dz \exp[-i(2t + 1)Cz][\iint dxdy\, \epsilon(\vec{r})].$$

Even though the function $\epsilon(\vec{r})$ is periodic in the z direction with the period $c_0 = 6.70$ Å of the double layer spacing, its integral over x and y is periodic with the periodicity $\frac{1}{2}c_0$ equal to a single spacing, because integration cancels the effect of layer staggering. On the other hand, $\exp[-i(2t + 1)Cz] = -\exp[-i(2t + 1)C(z + \frac{1}{2}c_0)]$ because $C\frac{1}{2}c_0 = \pi$. Hence $\gamma_j = 0$.

4.8. Integration of equation (4.21) and substitution of the given parameters yield $F = 10\,[1 + 9\pi^2 \sin^2 \frac{1}{2}\theta]^{-2}$. Note that F equals the number of electrons for $\theta = 0$, but it decreases rapidly for $\theta \gtrsim 2/3\pi = 12°$.

4.9. The Compton relation (4.37) gives the angle between incident and scattered X-rays as $\cos\theta = 1 - (\lambda' - \lambda)/\lambda_C$. The wavelength ratio λ'/λ is 5/4, being reciprocal to the ratio of photon energies. The ratio $\lambda/\lambda_C = m_e c^2/h\nu = 4.08$. Therefore, $\cos\theta = -0.02$ and $\theta = 91°$.

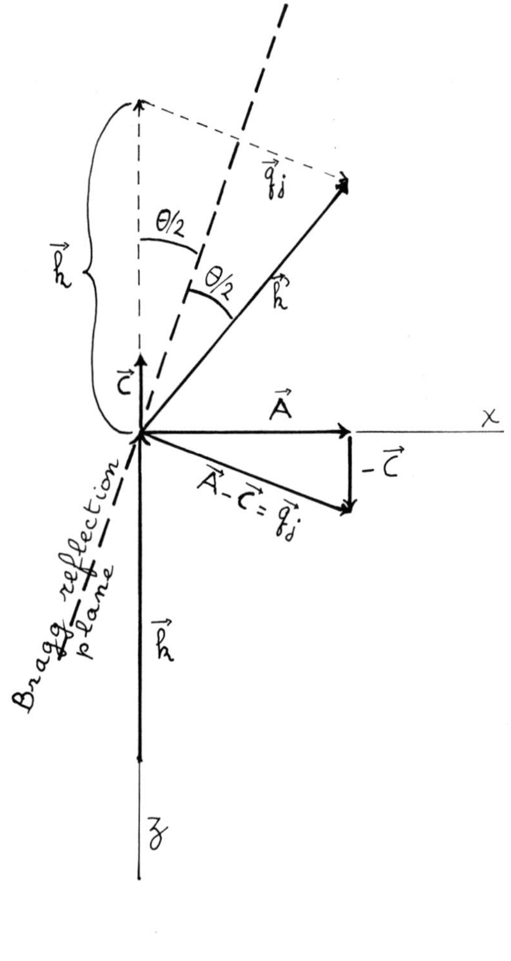

$\vec{q}_j$
θ/2
θ/2
$\vec{k}$
$\vec{k}$
$\vec{C}$
$\vec{A}$
x
$-\vec{C}$
$\vec{A}-\vec{C}=\vec{q}_j$
Bragg reflection plane
$\vec{k}$
z

5. the statistical aspects of atomic physics

This chapter begins to consider how the macroscopic and atomistic properties of radiation and matter hang together and how the macroscopic laws should be adjusted to obtain a comprehensive formulation of atomic phenomena. Note at the outset that the macroscopic description of a system need not *a priori* remain meaningful at the atomic level. For instance, macroscopic variables such as pressure and temperature are meaningless with regard to one or a few molecules.

However, concepts and laws of macroscopic electromagnetism remain relevant at such low intensities that the elementary processes of emission and absorption occur independently of one another in separate intervals of time. Specifically, experiments on polarization, interference, etc., verify the predictions of macroscopic electromagnetism even at extremely low intensity, provided the duration of an experiment permits the accumulation of a sufficient number of elementary processes, for example, on a photographic plate.[1] Accordingly, the theory of electromagnetic fields should not be abandoned at the atomic scale but should be adjusted to incorporate the evidence on the discontinuity of energy levels and of energy transfers.

On the other hand, application of concepts and laws of macroscopic mechanics to the motion of pointlike electrons fails to account for the properties of atoms. Remarkably, however, concepts pertaining to matter as a continuum, such as the density and flow of charge carriers, remain meaningful at the atomic scale. Indeed, the map in Figure 4.5 shows subatomic details of electron density. It appears then that particle mechanics should be adjusted to include evidence on the continuous aspects of electron density and flow.

On the atomic scale the predictions of electromagnetic theory hold in a statistical sense. For example, a hot gas normally emits light uniformly in all directions on a macroscopic scale. However, when this radiation is detected at low intensity by photon counters lying in various directions, first one counter and than another responds. Thus, elementary processes of light absorption are localized sharply in space and time, but their aggregate distribution is uniform in all directions.

Similarly, the continuum properties of matter that appear to remain meaningful at the atomic scale—electron density and current density—were intended to represent statistical properties of populations of atoms (section 2.1). Detection of single electrons ejected from atoms occurs, on the other hand, at localized and randomly distributed times and places.

The fact that atoms and radiation display discrete structural properties as well as continuous properties has been emphasized throughout the preceding chapters. (Recall that each atom appears to occupy a rather well-defined

---

1. An early experiment of this type was reported by G. I. Taylor, *Proc. Cambridge Phil. Soc.* 15 (1909): 114, reprinted in *Scientific Papers* (Cambridge: Cambridge University Press, 1958-71), 4:1.

volume of space.) Now we direct attention to the key point that the *continuous properties have a statistical character* whereas the *discrete properties manifest themselves with random aspects*. This fundamental phenomenon of quantum physics will be studied systematically in the following chapters, but some of its highlights are emphasized here.

Randomness means, in our context, that in different experimental runs under identical conditions elementary processes have different distributions in space and time. That is, the *detailed distribution* of individual observations *does not reproduce* from one experimental run to the next. However, the *statistical features*, such as averages, indices of fluctuation, etc., *do reproduce*.

For example, experiments of the Franck-Hertz type on inelastic collisions of electrons with atoms have been performed so as to detect individual electrons emerging from the collision chamber with se cted deflections and selected energy losses. The time intervals between successive detections and the sequence in which electrons emerge with various deflections and energy losses are quite random. The statistical features of the distribution of time intervals, deflections, and energy losses are, however, quite reproducible. In fact, reproducible values of cross-sections are measured by the procedure outlined in section 1.3.2.

Randomness of single elementary processes means that the time and place of their occurrence is unpredictable though remaining *within the constraints* of relevant statistical laws. An additional constraint derives from energy and momentum conservation in each elementary process. Numerous successful experimental tests of this randomness have been made since the early days of radioactivity and of atomic physics. Some of these tests are described in section 5.1.

The concept of different experimental runs performed under identical conditions is essential to our whole study and warrants careful statement. We refer for this purpose to the introduction of statistical concepts on pp. 50-57 of *BPC*, vol. 5. Randomness of individual results and reproducibility of statistical features hold for experiments performed on *ensembles* of "similar"—or, we would rather say, "identical"—systems. The *mean values* that will be considered throughout this book are also called *ensemble averages* and represent averages performed over ensembles of identical systems.

Historically, the randomness of elementary processes was not regarded at first as a remarkable feature of atomic and subatomic physics. For example, physicists knew that α-particles originate from the nuclei of radioactive substances but had no inkling of the circumstances leading to their emission. Hence the striking and well-documented randomness of successive emissions could be attributed to the cumulative effect of numerous unknown accidents. Quantum theory took a sharp and decisive turn at a later time, through the realization that randomness in atomic processes need not be attributed to unknown but presumably well-defined circumstances, but can be regarded as a *fundamental primitive law*. Thus—beginning with Born, Bohr, and their coworkers in 1926-27—randomness came to be regarded as a point of departure of atomic theory rather than as a phenomenon to be explained. This view proved extremely fruitful, but its novelty started widespread discussion. Section 5.2 outlines differences between the classical and quantum concepts of randomness.

Acceptance of randomness as a basic phenomenon resolved apparent inconsistencies which had stood in the way of atomic theory. For example, the fact that atomic electrons in a stationary state do not radiate, even though they are neither at rest nor in uniform motion, appeared inconsistent with electromagnetism, as long as theory was expected to assign a definite position and velocity to each electron at each instant. However, if a stationary state of the electrons is described by the probability distribution of their possible positions and velocities, the average current may vanish (no radiation emission) while the velocities nevertheless have a nonzero mean square value (positive kinetic energy).

Historically, this point of view developed from a combination of partial experimental evidence, fragmentary methods of calculation, and intuition. Its soundness was proved when consistent theoretical methods were developed from it which accounted in detail for the mass of experimental data on atoms and radiation available at the time.[2]

From this point of view, a major objective of quantum theory is to predict the probabilities of various possible outcomes of any experiment in which single events are detected. For many problems of radiation emission, propagation, and absorption, macroscopic electromagnetism provides the answer. For instance, if I indicates the macroscopic intensity of light of frequency ν at a certain position (expressed as incident energy per unit area per unit time), the probability of recording one photon with an ideal detector of unit area equals $I/h\nu$ per unit time. However, the treatment of most problems of quantum theory requires a formulation of pertinent laws and development of methods of calculation. Macroscopic laws serve as a basis for this development insofar as they remain valid as relationships among statistical averages of physical quantities.

5.1. TESTS OF RANDOMNESS OF ATOMIC EVENTS

Randomness in the time sequence of atomic events is qualitatively familiar to anyone who has listened to the irregular clicking of a Geiger counter. Quantitative tests are carried out by statistical methods, on the basis outlined in Appendix D. Some tests on the time distribution of elementary processes are described in this section; similar procedures can be applied to distributions in space, direction, etc.

---

2. Phenomena taking place within regions of space no larger that 10^{-13} cm, and involving the nature of particles, have still not been organized successfully by theory. These phenomena do not appear to have any essential influence on the subject matter of this book.

A typical laboratory experiment consists of scoring the detection of α-particles from a fixed radioactive source by a fixed detector. The scoring can be done at low intensity by recording the time of each counter response. At higher intensity the scoring may consist of registering the total number of counter responses within fixed intervals of time. Extensive experiments of this type were conducted by Rutherford and collaborators in the early studies of radioactivity.

Meyer and Gerlach carried out in 1913 an analogous experiment bearing directly on the elementary process of photoeffect. Their apparatus was similar to that used in Millikan's oil-drop experiment (Fig. 1.4). The oil drops were replaced by metal grains of low photoelectric threshold, exposed to ultraviolet light. Emission of a photoelectron by a grain causes a sudden change of its motion, so that each grain viewed by the microscope acts as a photon counter. Again, one can record the time of each photoemission event or, instead, the number of events within fixed intervals of time.

The procedure of recording the time of occurrence of individual events is slow but permits a direct test of randomness. If the average frequency of events is fixed but events happen otherwise at random, the length of the time interval between any two events should be totally uncorrelated to the lengths of preceding and successive intervals. For instance, if the average interval between counts is 10 seconds, then an unusually long interval of 30 seconds is equally likely whether the preceding interval has also been 30 seconds or only 2 seconds. For a quantitative analysis, each event can be characterized by the pair of time intervals T and T' which separate it from the previous count and from the following count. After obtaining a sufficient sample of events, one calculates the parameters $\langle T \rangle$, $\langle T' \rangle$, $\Delta T = \langle (T - \langle T \rangle)^2 \rangle^{1/2}$, $\Delta T'$ and $\Delta(T, T') = \langle (T - \langle T \rangle)(T' - \langle T' \rangle) \rangle$ defined in Appendix D. The correlation coefficient $r = \Delta(T, T')/\Delta T \Delta T'$ does not depart significantly from zero if the time intervals T and T' are random, that is, uncorrelated.[3] Figure 5.1 shows a randomness test of this type.

The experimental record of time intervals between successive counts in a long sequence also serves to test the basic assumption that events occur with equal probability between time instants t and $t + \Delta t$ for fixed Δt irrespective of t. Under this assumption, the number n(T) of intervals in which the time T has elapsed without any count decreases by a fixed fraction whenever T increases by ΔT. Therefore, n(T) decreases in accordance with the exponential attenuation law (1.22)

$$n(T) = N \exp(-T/\tau), \tag{5.1}$$

where N is the total number of intervals in a sample and τ is the mean length

3. The correlation coefficient is discussed in appendix D. Tests of significance of its departures from zero are discussed in statistics textbooks, e.g., G. H. Weinberg and J. A. Schumaker, *Statistics: An Intuitive Approach* (Belmont, Calif.: Wadsworth, 1962), p. 287.

of all intervals. Figure 5.2 shows plots of the quantity $n(T + \Delta T) - n(T)$ against T for two typical experiments.

A record of the number of counts scored in each of a large number of equal time intervals of duration T serves to test the Poisson distribution formula which is derived on the same assumption as equation (5.1) (See Appendix D).

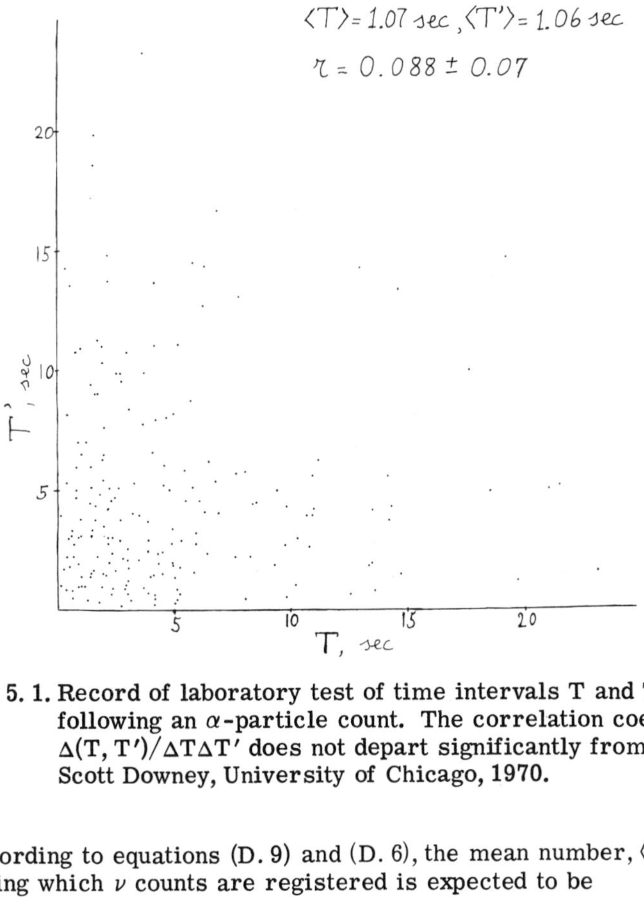

Fig. 5.1. Record of laboratory test of time intervals T and T′ preceding and following an α-particle count. The correlation coefficient $r = \Delta(T, T')/\Delta T \Delta T'$ does not depart significantly from zero. Courtesy of Scott Downey, University of Chicago, 1970.

According to equations (D.9) and (D.6), the mean number, $\langle n_\nu \rangle$ of intervals during which ν counts are registered is expected to be

$$\langle n_\nu \rangle = NP(\nu) = Ne^{-T/\tau} \frac{(T/\tau)^\nu}{\nu\,!}. \tag{5.2}$$

Here, N indicates again the number of intervals in the sample, τ is the mean time elapsed between counts, and $P(\nu)$ is the probability that exactly ν counts occur in the interval. Figure 5.3 shows the results of a very extensive test of this law.

A completely different demonstration of the equiprobability of elementary

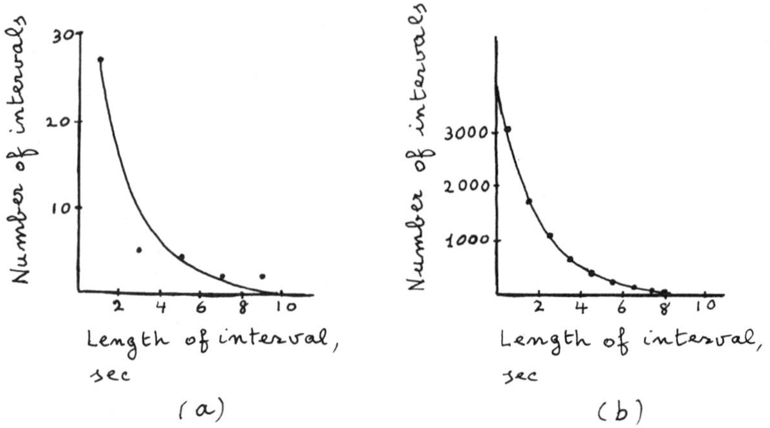

Fig. 5. 2. Distribution of time intervals between successive elementary processes. Solid curves from equation (5. 1); experimental points: (*a*) photoelectric emission [adapted from E. Meyer and W. Gerlach, *Ann. Physik*, 45 (1914): 215]; (*b*) α-particle emission [adapted from E. Rutherford, *Radioactive Substances and Their Radiations* (Cambridge University Press, 1913), p. 190].

processes during equal time intervals is afforded by the following analysis. Consider, for instance, that photoemission may occur in the Meyer-Gerlach experiment immediately after the light has been switched on. Within a finite, however short, time period ΔT there is a finite probability $\Delta T/\tau$ that an electron will be emitted from a given particle. This is true even if the period

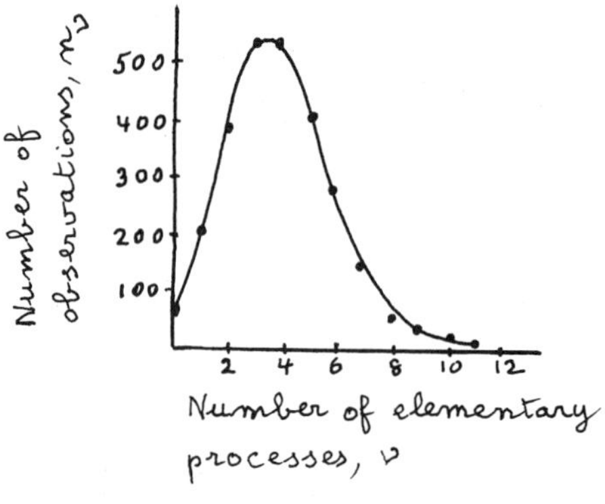

Fig. 5. 3. Distribution of numbers of elementary processes observed in equal time periods. Solid curve from equation (5. 2) with $T/\tau = 3.87$. Experimental points from E. Rutherford, *Radioactive Substances and their Radiations* (Cambridge University Press, 1913), p. 189.

ΔT is so short that the energy delivered to the particle by a weak light amounts to less than one photon $h\nu$ according to macroscopic laws. The macroscopic expression for the incident energy is $IS\Delta T$, where I is the light intensity and S the cross-sectional area of the particle. If, for example, $I = 10^{-10}$ watts/cm^2, $S = 0.1\ \mu^2 = 10^{-9}$ cm^2, and $\Delta T = 1$ sec, the energy delivered to the particle is $IS\Delta T = 10^{-19}$ joules $= 0.62$ eV. If the light wavelength is, for example, 3000 Å, corresponding to a photon energy of 4 eV, then the average number of photons available for absorption at the surface is 0.16 per second. If the metal grains produce on the average 1 photoeffect per 10 photons available, the average rate of photoemissions is 0.016 per second. Therefore, one photoelectron may have been emitted by the end of the first second after switching on, with a probability of 0.016, corresponding, on the average, to one success in 64 trails. Thus photoemission occurs with this low probability, even though the macroscopic value of the energy incident on the particle in each trial amounts to only 0.6 eV; this amount is lower than the work function and is therefore inadequate to eject an electron.

One concludes that the macroscopic expression of the light energy incident on a surface represents a statistical average. The actual arrival of energy experiences random fluctuations about the average. The metal particles in the Meyer-Gerlach experiment act as probe indicators of these fluctuations.

5.2. THE CONCEPT OF PROBABILITY IN MACROSCOPIC AND IN QUANTUM PHYSICS

Probability plays a role in various branches of physics, such as the statistical theory of gas mechanics, and in many familiar phenomena, such as dice or card games. Before discussing its role in quantum physics, let us consider how probability is introduced in the analysis of macroscopic phenomena.

The mathematical theory of probability derives logical consequences from an initial statement that a certain number of possible, mutually exclusive events are equally probable. We deal here not with the mathematical theory but with the physical considerations leading to an initial statement of equal probability.

The theory of a fair game of cards assumes that all possible initial distributions of cards among the players are equally probable. The card distribution derives from a multitude of accidental events in the course of shuffling the cards. One believes it plausible that adequate shuffling erases any correlation between the sequences of cards in a deck before and after shuffling, and consequently between the card distribution in successive games. Common experience supports this belief.

On the other hand, if one starts with a deck of cards in a given arrangement and if one controls and reproduces accurately the procedure of shuffling, one expects to obtain again and again the same card distribution among players. Accordingly, the equal probability of card distributions in a fair game rests on our intentional failure to control and reproduce the conditions affecting the operation of shuffling.

Similarly, the statistical mechanics of gases rests on the assumption that all possible positions and velocities of each gas molecule are equally probable, subject to suitable restrictions—for example, that all molecules be within a given container and that their total energy be fixed. The equiprobability is assumed to result from a multitude of collisions among molecules. Nevertheless, classical statistical mechanics also regards it as possible, in principle, to know the position and velocity of each molecule at a given time; in this event, statistical mechanics would not be applicable, just as the theory of a card game does not apply when shuffling is rigged.

In other words, the distribution of playing cards or of molecules is regarded as unpredictable and unreproducible because of failure to control the numerous variables that govern the distribution.

With regard to the time distribution of the emission of photoelectrons—for instance, in the Meyer-Gerlach experiment—these authors sought to identify the source of unreproducibility in the process of escape of electrons through the air surrounding each metal particle, but their explanation is not convincing. In other equivalent phenomena—for example, in the ejection of high-energy photoelectrons by X-rays in gases—one knows of no variables which are responsible for the random distribution of the events. The same holds for all elementary processes of atomic physics, that is for all processes involving individual particles or photons.

As no variables manifest themselves whose control would eliminate or even reduce the randomness of elementary processes, their existence need not be assumed *a priori*. In other words, the possibility that observed randomness reflects the influence of uncontrolled "hidden variables" can be disregarded when no such variable is detected. Accordingly, quantum mechanics regards the equiprobability of alternative events as a physical fact which need not be a consequence of other facts, that is, as a primitive fact.[4] More generally, it regards as a primitive fact the limited reproducibility of observations of atomic events performed under identical conditions.

Physics can make predictions only about those features of the outcome of experiments which are reproducible. In atomic experiments statistical parameters of the results, such as the average rates of occurrence of certain events or the mean square values of departures from averages, are indeed reproducible. The variables of macroscopic phenomena, such as the intensity of a photoelectric current, take reproducible values because they are statistical parameters of the same phenomena regarded from an atomistic point of view.

---

4. We take it here that the task of science consists of assembling experimental evidence and organizing it in our minds economically. Economy requires us to regard a minimum number of facts as fundamental, or primitive, and the others as their logical consequences. During any period in the history of science, the facts we regard as primitive depend on the evidence available at the time, on our analytical ability and also to some extent on preference.

CHAPTER 5 PROBLEMS

5.1 A counter registers the radiation from a radioactive source. The counts are recorded in 10-second intervals, and it is found that in 10% of the intervals no counts are registered. Utilize the Poisson distribution law, equation (5.2), to calculate (a) the mean time between successive counts; (b) the fraction of 10-second intervals in which three counts are registered.

5.2. In an experiment similar to that of problem 5.1 two counts are recorded in 10% of the 10-second intervals. Utilize the Poisson distribution law to calculate (a) the mean time between successive counts; (b) the root mean square time between successive counts; (c) the fraction of intervals in which three counts are registered. The solutions of this problem are not unique.

5.3. A radioactive source is placed halfway between two identical counters A and B so that the average counting rates of A and B are equal. Six successive counts are registered. Calculate the probabilities of the following distributions of counts: (a) the first three in one counter and the next three in the other, i.e., in either sequence AAABBB or BBBAAA; (b) alternation of successive counts in the two counters, i.e., in either sequence ABABAB or BABABA; (c) all counts in the same counter, i.e., AAAAAA or BBBBBB; (d) three counts in one counter and three in the other, in any sequence.

5.4. A counter detects radiation with a mean interval between counts $\tau = 5$ sec. The counter registers the number of counts ν in each 15-second interval. The observations extend over N such intervals. Calculate (a) the mean number of counts, $\langle \nu \rangle$, per interval; (b) the mean square $\langle \nu^2 \rangle$; (c) the mean square deviation $\Delta\nu^2$. Utilize equation (5.2); note that $\nu \langle n_\nu \rangle = (T/\tau) \langle n_{\nu-1} \rangle$ and that $\sum_{\nu=0}^{\infty} \langle n_\nu \rangle = N$.

5.5. A counter detects radiation from a distant radioactive source at the mean rate of $\mathcal{N}$ counts per second. The counter is moved 2% closer to the source; thus the new counting rate, $\mathcal{N}'$, should exceed $\mathcal{N}$ by 4%. Estimate the time t required to verify the increase, taking as a criterion of significance that $\mathcal{N}'t$ should exceed $\mathcal{N}t$ by 3 times the root mean square deviation of $\mathcal{N}t$.

5.6. In a Compton effect experiment it is desired to reduce the counting rate to approximately one per 10 sec in order to observe individual pulses on an

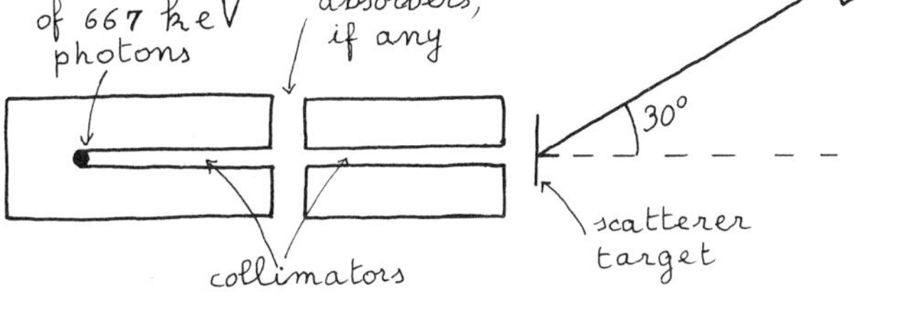

oscilloscope screen and to study counting statistics. The experimental layout is shown in the figure. Without absorbers the counter registers approximately 100 counts per second. Absorber foils are available, each of which reduces the average counting rate by 50%. (a) How many foils should one insert to reduce the average counting rate to approximately 1 per 10 sec? (b) With the number of foils chosen in (a) above, what is the probability of observing 3 counts in a 20-second interval? (c) Still with the same number of foils as in (a) and (b), what is the mean square time interval between counts? (d) In the absence of foils, what is the wavelength and photon energy of the X-rays reaching the detector?

SOLUTIONS TO CHAPTER 5 PROBLEMS

5.1. (a) Equation (5.2) for $\nu = 0$ gives $\langle n_0 \rangle = N \exp(-T/\tau)$, that is, $0.10 = \exp(-10/\tau)$ and $\tau = -10/\ln(0.10) = 4.3$ sec. (b) $\langle n_3 \rangle / N = 0.21$, from equation (5.2) with $\nu = 3$.

5.2. (a) Equation (5.2) with $\nu = 2$ yields $0.10 = \frac{1}{2}(10/\tau)^2 \exp(-10/\tau)$, which reduces to $0.2\tau \ln \tau = 0.62\,\tau - 1$. This equation can be solved graphically by plotting its two sides as functions of τ and finding the intersections of the plots. Two intersections are found at $\tau = 2$ sec and $\tau = 16$ sec. (b) From equation (5.1) the mean square interval is given by

$$\langle T^2 \rangle = \int_0^\infty T^2 \frac{1}{N}\left(-\frac{dn}{dT}\right) dt = \int_0^\infty T^2 \exp(-T/\tau)\, dT/\tau = 2\tau^2.$$

Therefore, the root mean square, $2^{1/2}\tau$, has the two alternative values 3 sec and 23 sec. (c) Equation (5.2) with $\nu = 3$ gives the two alternative values $\langle n_3 \rangle / N = 0.14$ and $\langle n_3 \rangle / N = 0.02$.

5.3. Any specified sequence of counts has the same probability $(\frac{1}{2})^6 = 1/64$ (a) 1/32, (b) 1/32, (c) 1/32, (d) 5/16 because there are 20 distinct sequences of the specified kind.

5.4. (a) $\langle \nu \rangle = \Sigma_{\nu=0}\, \nu \langle n_\nu \rangle / N = (T/\tau) \sum_{\nu=1}^{\infty} \langle n_{\nu-1} \rangle / N = T/\tau = 15/5 = 3.$

(b) $\langle \nu^2 \rangle = \sum_{\nu=0}^{\infty} \nu^2 \langle n_\nu \rangle / N = (T/\tau) \sum_{\nu=1}^{\infty} \nu \langle n_{\nu-1} \rangle / N$

$= (T/\tau) \sum_{\nu=1}^{\infty} [(\nu - 1) + 1] \langle n_{\nu-1} \rangle / N = (T/\tau)^2 + (T/\tau)$

$= \langle \nu \rangle^2 + \langle \nu \rangle = 9 + 3 = 12$

(c) $\Delta\nu^2 = \langle \nu^2 \rangle - \langle \nu \rangle^2 = \langle \nu \rangle = T/\tau = 3.$

5.5. The expected number of counts is $\langle \nu \rangle = \mathcal{N}'t = 1.04\, \mathcal{N}t$. The expected increase of counts is $\mathcal{N}'t - \mathcal{N}t = 0.04\, \mathcal{N}t$. From problem 5.4c we know that the mean square deviation $\Delta\nu^2 = \langle \nu \rangle$. Here $\langle \nu \rangle = 1.04\, \mathcal{N}t$, therefore, $\Delta\nu = 1.02 \sqrt{(\mathcal{N}t)} \sim \sqrt{(\mathcal{N}t)}$. The criterion of significance is then that $0.04\, \mathcal{N}t \geqslant 3\sqrt{(\mathcal{N}t)}$, $t \geqslant (5625/\mathcal{N})$ sec.

5.6 (a) The desired reduction in counting rate, from 100 per sec to 1 per 10 sec, amounts to a factor 1/1000. Ten foils achieve this effect very closely, since $(\frac{1}{2})^{10} = 1/1024$. (b) Assuming now $\tau = 10$ sec, and setting $T = 20$ sec and $\nu = 3$ in equation (5.2), one finds $\langle n_3 \rangle / N = 0.18$. (c) From the solution of problem 5.2b, $\langle T^2 \rangle = 2\tau^2 = 200$ sec. (d) From equation (4.37) the wavelength increase is $\lambda_c(1 - \cos\theta) = 0.0243 \times 0.134 = 3.24 \times 10^{-3}$ Å. For the source X-rays we have $\lambda = hc/(667 \text{ keV}) = 1.86 \times 10^{-2}$ Å, and for the scattered X-rays $(1.85 + 0.32) \times 10^{-2} = 2.18 \times 10^{-2}$ Å; the scattered photon energy is 570 keV.

Part II

CONCEPTS AND PHENOMENA OF QUANTUM PHYSICS

introduction

The survey of atomic properties in chapter 1 has raised a fundamental question, that is, why don't atomic electrons collapse onto the nucleus instead of occupying a fairly well-defined volume? Answering this question is the main objective of the second part of the book.

Two fundamental quantum effects have also emerged in Part I. One is the relationship between energy transfers in elementary processes and the frequencies of atomic and field oscillations. This relationship ties together previously unrelated variables. Conversely, the other effect relaxes previously rigid causal relationships. Atomic systems prepared and treated in identical ways may experience different elementary processes subject only to statistical regularities. Macroscopic laws remain valid as relations among mean values; the laws thus interpreted provide a main link between classical and quantum physics.

The first goal of Part II is to define more precisely the extent and character of the statistical aspects of atomic phenomena. It will thus become possible to develop a mathematical language which represents both rigid and statistical relationships between physical variables. To this end we will describe additional types of experiments on elementary processes and analyze them at length, step by step.

The mathematical formulation to be abstracted from these experiments not only contains the relationship between frequency and energy levels, but also expresses quite generally the time variations of atomic systems in terms of their energy levels. A precise expression of the statistical relationship between the position of an electron and its kinetic energy leads directly to an understanding of the stability of atoms and to an evaluation of their size. An analogous relationship between the electric and magnetic fields of radiation determines the photon structure of radiation energy.

6. eigenstates of atoms and radiation. incompatible quantities

This chapter describes selected experiments whose analysis leads to basic concepts of quantum theory. Experiments on radiation, familiar from macroscopic optics, will be reviewed because their analysis points the way to concepts applicable throughout quantum physics. All experiments to be considered involve the preparation and processing of radiation and particle beams through successive analyzing and measuring devices.

6.1. ENERGY AND MOMENTUM MEASUREMENTS

Given a beam of nonmonochromatic light, one extracts from it a monochromatic component of the desired frequency by filtration through a monochromator, that is, through a spectroscope provided with an exit slit at a suitable point of its screen. When absorbed, the monochromatic component delivers to matter photons of uniform energy.

Radiation filtered through a monochromator may be tested by filtration through a second monochromator identical to the first one; all of the radiation goes through the second filter if its slit is set in the same position as in the first filter, and none goes through if the setting is different. For equal settings of the slit, a photon counter scores at equal rates whether it is placed behind the first or behind the second filter; all photons received in both cases have the same energy.

This somewhat trivial retest procedure provides an operational definition of a state of radiation, the state whose analysis gives the same results as for radiation emerging from a monochromator. Radiation that meets this requirement is said to be in an *eigenstate* (that is, a characteristic, or proper, state) of photon energy. The value of the photon energy, or the corresponding value of the light frequency is called the *eigenvalue* for the eigenstate thus identified.

A nonmonochromatic beam is not in an eigenstate of photon energy; spectroscopic analysis is said to resolve the state of the beam into *component eigenstates* of photon energy. The intensities of the different components separated by the spectroscope provide a statistical analysis of photon energies in the original beam. If the light intensity on the spectroscope screen is I_1 and I_2 at positions corresponding to frequencies ν_1 and ν_2, the relative probabilities of detecting photon absorptions at these positions are $I_1/h\nu_1$ and $I_2/h\nu_2$. The theoretical counterpart of this experimental analysis consists of the Fourier analysis of the electric and magnetic fields of radiation into monochromatic components. (Recall that the sum of the intensities of all monochromatic components equals the total intensity of the radiation.)

Selection of particle beams according to kinetic energy is performed by the same conceptual procedure as for radiation, that is, by filtering the beam through a device that transmits only particles of a given energy. The retest procedure is equally relevant, and the same nomenclature (eigenstate and eigenvalue) applies. For charged particles, the main techniques of energy analysis utilize deflection in magnetic and electric fields and have been indicated in section S3.8. Beams of neutral atoms or molecules are also used in atomic physics. These beams are usually formed by leaking some gas or vapor out of an oven into a high-vacuum vessel and then selecting particles that travel in a desired direction by collimation through pinholes in diaphragms. Velocity (and hence energy) selection can then be performed for particles of thermal energy, for instance by filtering the beam through additional diaphragms in rotating wheels, as indicated in figure 6.1. The same technique is applicable to neutrons of thermal energies. Electronic techniques serve to "chop" beams of faster neutral particles, performing a function equivalent to the pair of rotating wheels.

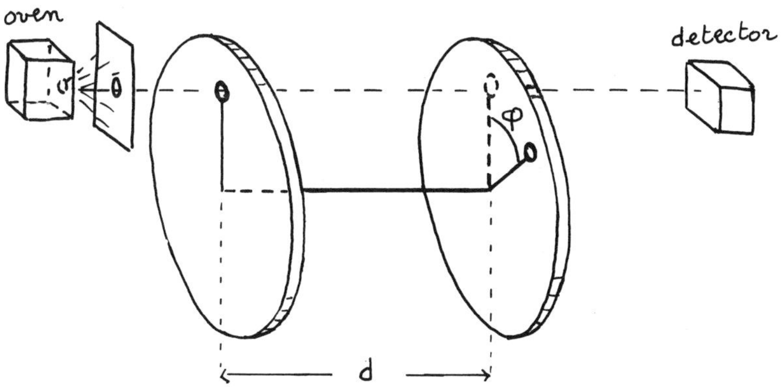

Fig. 6.1. Velocity selection of atomic beam by rotating wheels. The second wheel slit transmits only particles of velocity equal to $\omega d/\varphi$, where ω is the angular speed of the wheels and d and φ are the distance and angle shown in the figure.

The photon energy eigenstates of radiation and the kinetic energy eigenstates of particles are stationary states in the sense explained in section 3.2 for atoms with well-defined internal energies. (In fact, the names "energy eigenstate" and "stationary state" are synonymous: one of them emphasizes that the state has a well-defined energy, and the other emphasizes that the state is invariant in time.) A beam of radiation or particles in an energy eigenstate remains in that state until it interacts with a material or device with which it can exchange energy. Upon emerging from a region of interaction—for instance, the collision chamber in an experiment of the Franck-Hertz type—the beam can be energy-analyzed once more. Its state will then usually be found to contain different component eigenstates of energy; the results obtained with a counter will be analogous to those obtained by analysis of a nonmonoenergetic beam.

As implied above, preparation of a beam in an energy eigenstate often makes the beam also monodirectional. In electron optical devices, selection of particles of uniform energy is achieved by bringing them to a focus from which they diverge; use of an additional lens can make the beam parallel without disturbing its energy characteristics. Thus an eigenstate of particle kinetic energy can also be eigenstate of the vector momentum $\vec{p}$ of the particles.

For radiation, as mentioned in chapter 4, a monodirectional beam in an eigenstate of photon energy delivers photons of uniform momentum $\vec{p}$ in each elementary absorption process; the magnitude of $\vec{p}$ is equal to $h\nu/c$, and its direction is parallel to the beam.

6.1.1. *Bandwidth and Complementarity*. For radiation in an eigenstate of photon energy and momentum, the electric and magnetic fields are represented by a plane monochromatic wave. The sinusoidal dependence of the fields on space and time coordinates is understood to extend with full regularity to infinite values of the coordinates. Accordingly, a plane wave describes an idealized physical phenomenon, as emphasized in section A.2. By the same token, the very concept of an energy, or momentum, eigenstate of radiation is an idealization.

Radiation whose state is nearly an eigenstate of photon energy and momentum is represented by a superposition of plane waves of nearly equal frequencies and wave vectors. In fact, a spectroscope, or any other monochromator, never sorts out ideally monochromatic radiation. The design performance of the instrument is limited by the dimensions of slits, gratings, etc., which set a lower limit to the range (bandwidth) of frequencies and wavenumbers of the transmitted radiation. The bandwidth in the Fourier representation of a phenomenon with limited extension in time (or space) is discussed in section A.3.

The essential point here is that frequency and wave vector (and hence photon energy and momentum) are *defined* as parameters of a phenomenon with infinite extension in time and space. Any limitation in time or space introduces a spread of the frequencies or wave vectors required to represent the phenomenon. Conversely, precise localization in time or space is represented by a Fourier superposition with infinite bandwidth.

This type of relation between a variable x and the parameters of the Fourier expansion of f(x) is fundamental and well known in the theory of Fourier analysis. The experimental discovery of the proportionality of frequency and photon energy forces the same relationship upon the pairs of variables time and photon energy, and space coordinates and photon momentum. Since definition of one variable in a pair implies a spread of values in the other one, Bohr has called such pairs of variables complementary and has emphasized *complementarity* as a key phenomenon of quantum physics.

That energy and momentum eigenstates of particles are also idealizations is suggested by the equivalent behavior of particle and radiation beams in diffraction experiments (see section 4.3). It will be seen in chapter 11 that complementarity applies to particles as well as to radiation, but we defer this subject.

We also defer any further sharpening and generalization of the concept of complementarity for energy and for momentum in favor of studying simpler systems. Energy and momentum have infinite and continuous sets of eigenvalues to which correspond infinite sets of eigenstates; furthermore, the analysis of energy and momentum measurements involves incidental instrumental uncertainties besides those due to complementarity. We concentrate here on polarization variables which have a small finite set of discrete eigenstates. This study will show complementarity to be a particular case of a more general phenomenon of incompatibility of variables.

6.2. ANALYSIS OF LIGHT POLARIZATION[1]

According to macroscopic electromagnetism, the electric and magnetic fields of radiation traveling in empty space or in an isotropic material are perpendicular to one another and to the direction of propagation. These limitations leave the direction of the electric field undetermined within the plane perpendicular to the direction of propagation. The field direction need not remain constant at any one point in the course of time. Radiation whose electric field maintains a constant direction in the course of time is said to be polarized[2] linearly in that direction.

Light detectors, such as photographic plates, photocells, or photon counters, usually give a response independent of polarization. However, the polarization of light can be analyzed, like its frequency. A polarizer filter consists of a device that separates light with different polarizations, followed by a diaphragm placed so as to let through only light in a state of specified polarization. A polarizer filter can be utilized much as a monochromator. Separation of different polarizations is achieved by means of materials whose properties are anisotropic, that is, not uniform in all directions.

We consider here in particular crystals of Iceland spar, whose electrons yield to the pull of an electric field by a displacement skew to the pull unless the field is either perpendicular or parallel to an axis of symmetry of the crystal (optical axis). (In the language of section 2.1, the induced current $\vec{j}_{matter}$ is not parallel to the field $\vec{E}$.) Figure 6.2a shows a spar crystal, its optical axis, and a principal section, that is, a plane through the optical axis and perpendicular to a crystal face. A light beam travels undisturbed only if it enters the crystal in a direction perpendicular to a crystal face and lying on a principal section, and if it has polarization perpendicular to the principal section (Fig. 6.2b). If the direction of polarization lies on the principal section, the light experiences an anomalous refraction; it follows the path marked in Figure 6.2c and emerges in a direction parallel to the direction of incidence but shifted sideways. If the direction of the electric field is oblique to the principal section, the incident light beam splits into two components which follow respectively

---

1. General background on polarization is provided by *BPC*, vol. 3, chap. 8.

2. The word "polarization" applies generally to any physical action that singles out a particular direction of space and is accordingly used in various unrelated contexts.

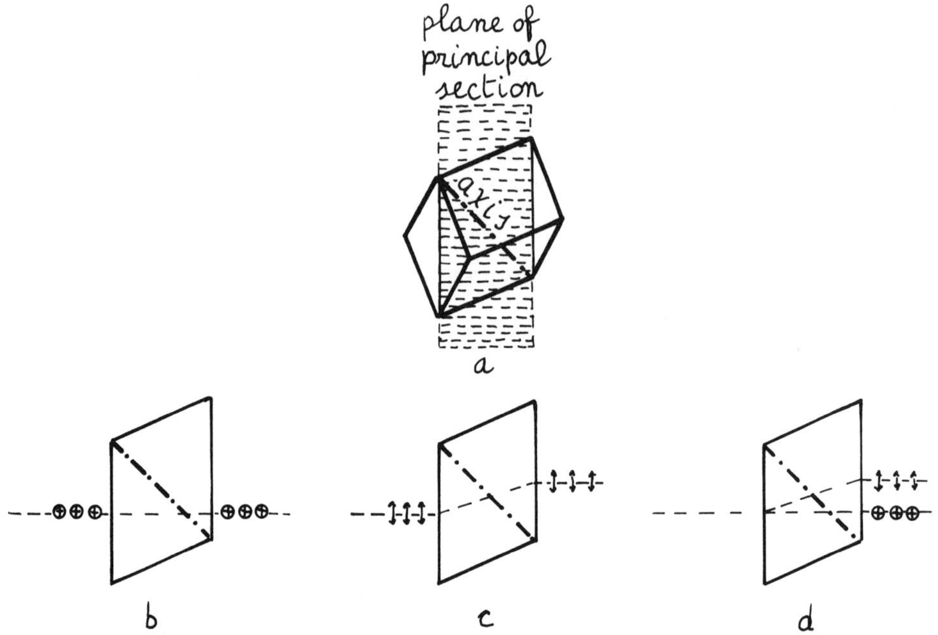

Fig. 6. 2. (*a*) Iceland spar crystal with optical axis and principal section in the plane of the paper. In b, c, and d the crystal is represented by its principal section.
(*b*) Light polarized with the electric field perpendicular to the principal section travels undisturbed—ordinary ray.
(*c*) Light with the electric field in the principal section is shifted—extraordinary ray.
(*d*) Light in any other state of polarization is split into an ordinary and an extraordinary ray.

the two paths of Figure 6. 2d. The component which follows the straight path is called the *ordinary ray;* the other one, the *extraordinary ray.*

When a diaphragm is placed behind a spar crystal so as to let through either the ordinary ray only or the extraordinary ray only, the crystal acts as a polarizer filter. The transmitted light is linearly polarized in a direction perpendicular or parallel to the principal section, independently of whether the incident light was linearly polarized or in what direction.

The transmitted light may be tested with a second filter identical with the first one and equally oriented. All of the light goes through the second filter if its diaphragm is set in the same position as in the first filter; none of the light goes through if the settings are opposite. For equal settings of the filter a photon counter scores equally whether placed behind the first or behind the second filter.

On the basis of these experiments, a state of linear polarization can be identified operationally without explicit reference to the direction of the electric field. The identifying observation is that all of the light intensity passes *as a single beam* through a spar crystal of definite orientation about the beam direction. The light is then said to be in an eigenstate of linear polarization characterized by the orientation of the crystal. For any given orientation there are two eigenstates *orthogonal* to one another, transmitted as an ordinary ray or as an extraordinary ray. A state of linear polarization is completely identified by specifying the orientation of the spar crystal and whether the light is transmitted as the ordinary or the extraordinary ray.

Consider now what happens when a light beam in a definite state of linear polarization traverses a second spar crystal which serves as analyzer. For instance, the beam may have been prepared by filtration as an ordinary ray through a polarizer, and may traverse an analyzer of orientation different from that of the polarizer. The beam is then split by the analyzer into an ordinary and an extraordinary ray; its state was thus *not* an eigenstate of the analyzer's orientation. The fractional intensities of the two rays depend upon the angle α between the principal section planes of the polarizer and analyzer crystals. According to both experimental results and electromagnetic theory, the two fractions are $\cos^2 \alpha$ and $\sin^2 \alpha$ for the ordinary and extraordinary ray, respectively. If the ordinary (or extraordinary) ray proceeds through a diaphragm, it displays all the characteristics of an eigenstate of the analyzer orientation. Its previous polarization, impressed by the initial polarizer, has been wiped out by transmission through the analyzer filter. In particular, the ordinary ray of the analyzer is sorted out once more into fractions $\cos^2 \alpha$ and $\sin^2 \alpha$ by a further analyzer parallel to the initial polarizer.

The processes of analysis of a light beam through spar crystals of different orientations[3] are said to be *incompatible*, because the polarization of the ordinary ray of one crystal depends on the orientation of that crystal and bears no trace of previous transmission through a crystal of different orientation. More formally and more generally, two processes of analysis are said to be incompatible when they select different sets of eigenstates.

6.2.1. *Survey of Polarization States.* Radiation whose electric field vector $\vec{E}$ at any one point changes direction periodically is said to be elliptically or circularly polarized, depending on whether the tip of the vector describes an ellipse or a circle. This radiation may be represented mathematically by superposition of two monochromatic linearly polarized components with their electric fields $\vec{E}_1$ and $\vec{E}_2$ orthogonal to one another and 90° out of phase. The fields $\vec{E}_1$ and $\vec{E}_2$ are directed along the axes of the ellipse, and their magnitudes are equal if the polarization is circular. From this point of view, linear polarization is a special case of elliptical polarization in which one of the vectors $\vec{E}_1$ or $\vec{E}_2$ vanishes.

3. Spar crystals with principal sections orthogonal to one another perform equivalent analyses, since each of them transmits as an ordinary ray the extraordinary ray of the other one.

Operationally, elliptical polarization can be analyzed by passing the light through a *quarter wavelength* ($\lambda/4$) plate and a spar crystal. A $\lambda/4$ plate is a slice of crystal with two characteristic orthogonal directions. When these two directions are set parallel to the component fields $\vec{E}_1$ and $\vec{E}_2$, the phase of one of the two linearly polarized components of the radiation is retarded by 90°. The two components emerge in phase and yield a resultant linear polarization. This linear polarization is then identified by adjusting the orientation of the spar crystal until only one ray emerges. The essential point is that each state of polarized light[4] is identified completely by the *two* angles of orientation of the $\lambda/4$ plate and of the spar crystal about the beam direction, and by indicating whether the radiation emerges as an ordinary or an extraordinary ray. Thus the crystal settings alone identify a pair of eigenstates of elliptical polarization (just as the setting of a single spar crystal identifies a pair of eigenstates of linear polarization). The two eigenstates are said to be orthogonal in the same sense as the ordinary and extraordinary rays of a spar crystal; for instance, right-circular and left-circular polarization are orthogonal to one another.

6.2.2. *Atomistic Aspects.* The preceding discussion has been couched in the language of macroscopic phenomena, but slanted for application at the atomic level. Low-intensity experiments may be conducted by placing two photon counters so that one receives the ordinary ray and the other the extraordinary ray of the spar crystal. For arbitrary orientation of the analyzer crystal (or crystals), now one counter scores, now the other, in random sequence though with average frequencies in a fixed ratio. When only one of the counters scores, the incident radiation is in an eigenstate characterized by the crystal orientation. Light polarization can also be detected through the effects of its absorption by atoms. For example, photoelectrons ejected by linearly polarized light are distributed anisotropically about the incident beam.

6.3. MAGNETIC ANALYSIS OF ATOMS

This section deals with the identification of atomic states which differ in the distribution of internal currents and are analogous in many respects to states of light polarization. In the case of atoms the task must be carried out on the basis of typical quantum phenomena and therefore, unlike the study of light polarization, without the benefit of a preexisting macroscopic theory.

Light polarization is studied by passing a beam of light through an anisotropic material; atomic current distribution is studied by passing a beam of atoms through a region of space made anisotropic by a magnetic field $\vec{B}$. As mentioned in section 2.4.2, atoms may be the seat of steady nonradiating internal currents whose magnetic properties are represented by a magnetic moment $\vec{\mu}$. A magnetic field $\vec{B}$ exerts on an atom a torque $\vec{\mu} \times \vec{B}$, and the atom in the

---

4. Partially polarized and unpolarized light will be mentioned briefly in section 6.4.

field acquires a potential energy $E_{magn} = -\vec{\mu}\cdot\vec{B}$. We choose the z axis in the direction of $\vec{B}$, whereby this energy is represented by

$$E_{magn} = -\mu_z B. \tag{6.1}$$

Stern and Gerlach developed in 1921 a famous experiment to determine the value of μ_z by measuring the deflection of atoms which traverse a magnetic field. To obtain a deflection, the field strength B is made nonuniform in a direction perpendicular to the line of flight of the atoms; the atoms then experience a deflecting force

$$\vec{F} = -\vec{\nabla} E_{magn} = \mu_z \vec{\nabla} B. \tag{6.2}$$

Call m the mass of the atom, v its velocity through the magnet, and l the length of travel through the magnetic field. The acceleration due to the force (6. 2) builds up a transverse velocity whose value at the exit of the magnet is $v_t = \mu_z |\vec{\nabla} B| l/mv$. Assuming that the deflection angle θ is small, its value is then

$$\theta = \frac{v_t}{v} = \frac{\mu_z |\vec{\nabla} B| l}{mv^2}. \tag{6.3}$$

Measurement of θ determines μ_z when all other quantities are known.

An incident beam of atoms[5] is prepared as indicated in section 6. 1, and the velocity of the atoms may be made uniform by the method shown in Figure 6. 1. The magnet has pole faces shaped approximately as shown in Figure 6. 3 to produce large values of $|\vec{\nabla} B|$ near the central wedge. The arrival of atoms at different positions of a screen was detected in early experiments through the accumulation of material deposited. More sensitive detection methods are restricted by the fact that the incoming atoms carry no charge and little kinetic energy. However, atoms of some elements, particularly alkalis, are easily detected since they readily yield an electron to a hot tungsten wire. The wire is displaced to scan the point of arrival of atoms with various deflections, as if on a screen.

6. 3. 1. *Basic Results*. The following fundamental result is observed: A collimated monoenergetic beam of atoms which passes through a Stern-Gerlach magnet either remains undeflected or, depending on the nature of the atoms, is split into a small number of components that experience different deflections. For example, a beam of He atoms remains undeflected, whereas a beam of H or Ag atoms splits into two components, a beam of N atoms into four, a suitably prepared beam of O atoms into five components.[6] A beam of neutrons is

---

5. These atoms may be assumed to be in their ground state of internal energy, unless otherwise stated.

6. These statements must be qualified because the pertinent properties of the internal structure of the incident atoms depend to some extent on the strength of the magnetic field to which they are subjected (see section 14. 3. 1).

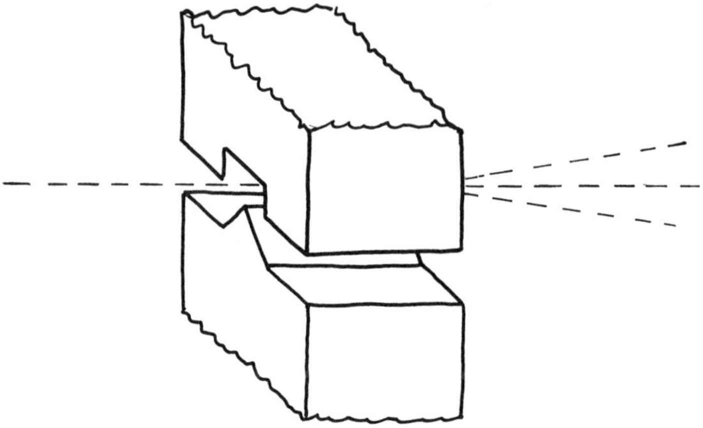

Fig. 6. 3. Splitting of a particle beam after traversing a Stern-Gerlach magnet.

split into two components whose deflection is much smaller than in the other examples indicated here.

The component beams hit the screen at *equally spaced* intervals symmetrically distributed with respect to a point of no deflection,[7] that is, to the point of arrival in the absence of a magnetic field. The symmetry implies that the middle component beam remains undeflected if there is an odd number of components. If the number of components is even, there is no "middle" component, and the point of no deflection lies midway between the two components nearest to it.

Finally, the *intensities* of all component beams are *equal* provided the incident beam has not been subjected to previous magnetic actions.

The fact that all atoms in any one component beam undergo equal deflections shows that their state is characterized by a specific value of μ_z and, therefore, by a specific value of the magnetic energy (6. 1) in a field of given strength. The different values of μ_z for the different beam components correspond to discrete energy levels $-\mu_z B$ of the atoms in a magnetic field. In other words, atoms which are all presumably in their ground state of internal energy in the absence of a magnetic field are separated by the field into groups with somewhat different energy levels. The differences between measured values of $-\mu_z B$, converted to a frequency scale, are of the order of magnitude of the Zeeman splitting of spectral lines (section 2. 4) in a field of the same strength.

7. This result holds insofar as the deflections are so small that each atom is subjected to the same net force throughout the region between the pole faces, that is, insofar as the gradient of the magnetic field is uniform throughout the region traversed by the atoms.

The connection between the Stern-Gerlach results and the Zeeman effect will be discussed in section 7.3.

The results of the Stern-Gerlach experiment indicate that all atoms in a component beam are in an eigenstate of μ_z and therefore of E_{magn}. Confirmation is provided by a retest procedure analogous to those of sections 6.1 and 6.2. To this end, a diaphragm may be inserted at the exit of a Stern-Gerlach magnet so as to let through only one among the component beams. This component beam is allowed to travel on for a distance and is then analyzed with a second Stern-Gerlach magnet equal to the first one and equally oriented (Fig. 6.4). The beam experiences *no further splitting* in the second magnet and is deflected to the same extent as in the first magnet.

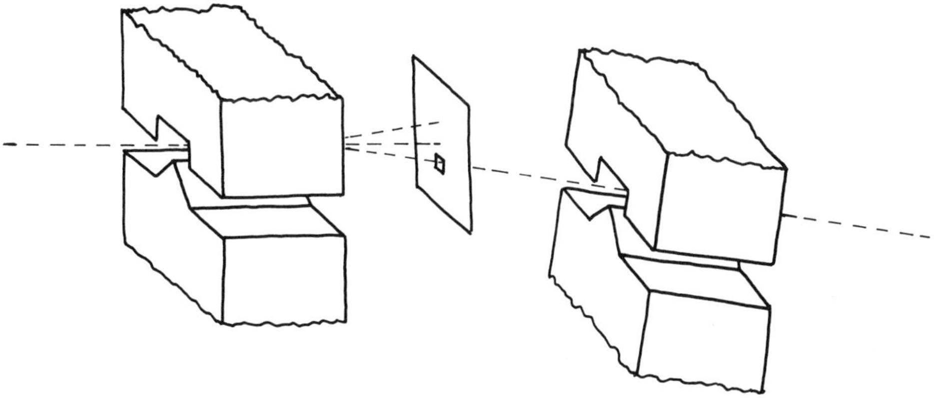

Fig. 6.4. Retest of beam with a second Stern-Gerlach magnet. A component beam is not resplit by a second magnet parallel to the first.

A Stern-Gerlach experiment on atoms of a given element thus determines a set of eigenvalues of μ_z, one for each eigenstate, that is, one for each beam component. Each eigenvalue is commonly represented in terms of three parameters which together indicate its position in the set, the number of eigenvalues in the set, and the difference between two successive eigenvalues. The choice of parameters originates from historical reasons and from the relationship with angular momentum eigenvalues described in section 7.2. One of the parameters is the value μ of the largest eigenvalue of μ_z; μ is called the *magnetic moment* of the atom.[8] The difference between successive eigenvalues of the set is indicated by μ/j. This implies that the number of eigenvalues of μ_z

---

8. The name "magnetic moment" was applied to the largest eigenvalue in the belief that it represents the magnitude of the vector magnetic moment; the smaller eigenvalues were believed to represent projections of this vector from various directions oblique to the z axis.

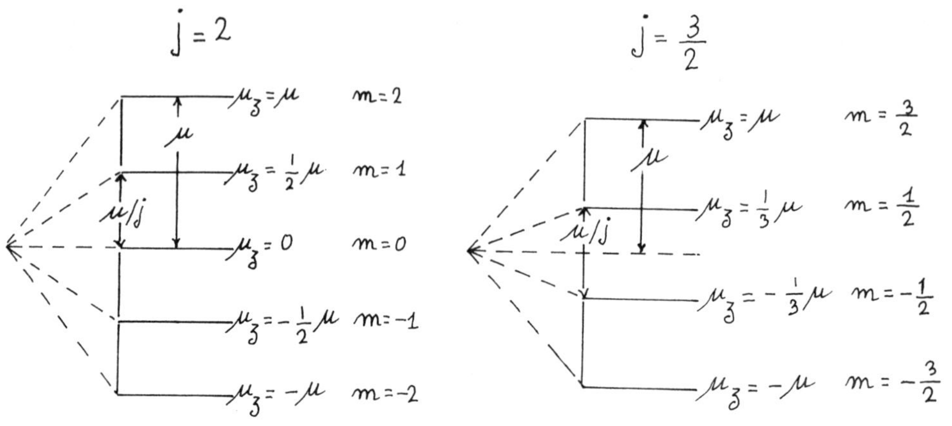

Fig. 6. 5. Diagram of eigenvalues of μ_z as determined by the separation of beam components in a Stern-Gerlach experiment.

is equal to 2j + 1 (see Fig. 6. 5). Finally, the position of one particular eigenvalue in the set is indicated by a number m which varies in steps of one from −j to j. Thus the whole set of eigenvalues is represented by

$$\mu_z = \mu \frac{m}{j}, \tag{6.4}$$

for j ≠ 0. Whereas μ has the physical dimensions of a magnetic moment (measured in ergs/gauss or in eV/gauss), m and j are pure numbers; they are called *quantum numbers* like all indices which identify discontinuous features of quantum phenomena. The number m is called the *magnetic quantum number* and j the *angular momentum quantum number* for reasons that will become apparent later. Note that m and j are integers when the number of beam components is odd, and half-integers when it is even.

6.3.2. *Repeated Analysis with Magnets of Different Orientation.* The study of states of polarized light centers on experiments with a polarizer and an analyzer of different orientations. By analogy we consider here the following schematic problem: One component beam has been filtered out of an initial atomic beam by means of a magnetic deflection followed by a selecting slit (as in Fig. 6. 4). This component beam now enters "suddenly"[9] into another

---

9. The state of an atom changes progressively if the atom moves slowly on a path along which the magnetic field changes direction. This change is effectively "slow" or "fast" depending on the promptness of response of the intra-atomic currents, that is, depending on the frequency of Larmor precession (section 7. 1).

A clear-cut study of the effect of two successive different magnetic fields upon

deflecting magnet whose field direction forms an angle θ with the direction of the field in the first magnet (Fig. 6. 6). We call the first magnet a polarizer and the second an analyzer.

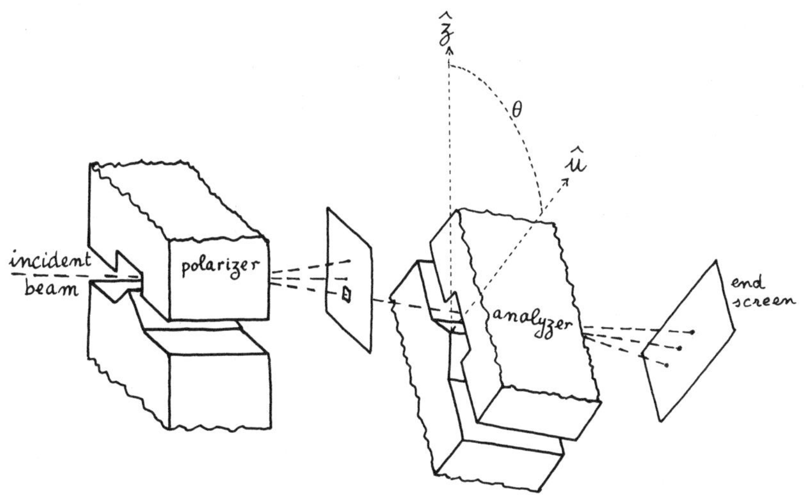

Fig. 6. 6. Retest of a beam with a second Stern-Gerlach magnet. A component beam is resplit by a second magnet oriented differently from the first one.

The analyzer magnet splits the polarized beam into a new set of components. These component beams reach the end screen at a series of points aligned in a direction parallel to the magnetic field of the analyzer. The analyzer resolves as many component beams as the polarizer, and the deflections are equal in magnitude if the two instruments are identical except for orientation. However, the *intensities* of the component beams resolved by the analyzer are in general *unequal*, and depend upon the angle θ between the directions $\hat{z}$ and $\hat{u}$ of the polarizer and analyzer magnets.

Qualitatively this result is quite analogous to the corresponding result for light polarization. A beam of atoms in an eigenstate of μ_z can be resolved into component beams of atoms in different eigenstates of μ_u, where μ_u is the

---

atoms requires that the transition from the first to the second field be sufficiently sudden. Experimental arrangements to achieve this goal were developed by Stern and Phipps and especially by Frisch and Segrè. Without going into detail we assume that one can achieve in practice an ideally sudden transition from field-free space into the space between the poles of a magnet, or from one magnet into another one with different orientation.

component of the magnetic moment along the analyzer's axis $\hat{u}$. A component beam separated by the analyzer is split once more if retested with a further magnet of orientation different from the analyzer—for instance, parallel to the polarizer. Magnets with different orientations perform *incompatible types of analysis*.

The quantitative dependence of the component intensities upon the relative orientations of a polarizer and an analyzer is different for light and for atoms. In the analysis of linearly polarized light with an Iceland spar crystal, the relative intensities of the ordinary and the extraordinary beam are $\cos^2 \alpha$ and $\sin^2 \alpha$ and are accounted for by macroscopic theory. For atomic beams there is a different intensity law depending on whether the beam is split into two, three, four, etc., components.

Each of the relative intensities of component beams represents the probability that an atom, having emerged from the polarizer magnet in the component beam identified by a quantum number m, will emerge from the analyzer magnet in the component beam identified by m'. This probability may be indicated as $P^{(j)}{}_{m'm}$. It is a function of the angle θ between the polarizer and analyzer axes and it depends on m and m' and on the number of beam components, that is, on the quantum number j of equation (6.4) which equals the largest value of m (or of m'). For purposes of illustration, Table 6.1 gives analytical expressions that represent the probabilities $P^{(j)}{}_{m'm}(\theta)$ for cases of two-, three-, and four-component beams ($j = \frac{1}{2}$, $j = 1$, and $j = \frac{3}{2}$). The entries for $j = \frac{1}{2}$ are derived theoretically below; the others are discussed in section 10.3. In general, $P^{(j)}{}_{m'm}$ is a polynomial function of $\cos \theta$ of degree 2j. The table shows equal probability for an atom in an eigenstate with $m = a$ to emerge in an eigenstate with $m' = b$, and for an atom with $m = b$ to emerge with $m' = a$ (*reciprocity law*). The table also shows complete symmetry between deflections in opposite directions, characterized by values of m and m' equal but with opposite sign.

6.3.3. *Survey of Magnetic Energy Eigenstates*. The Stern-Gerlach experiments identify finite sets of eigenstates and eigenvalues of the magnetic energy $E_{magn} = -\vec{\mu}\cdot\vec{B}$ for atoms in a field $\vec{B}$. We introduce now a symbol that serves to identify quantum-mechanical states. It consists of a vertical bar and an angular bracket, $|\ldots\rangle$, between which one writes sufficient data to identify the state. Specifically, we indicate by $|\hat{B}\ m\rangle$ an eigenstate of magnetic energy with quantum number m in a field of direction $\hat{B}$. Additional notations might be inserted to indicate, for instance, the quantum number j, the species of atom under consideration, or its level of nonmagnetic energy. This symbol is called a "Dirac *ket*" for historical reasons to be indicated later on.

Eigenstates with the same $\hat{B}$ and different quantum numbers m and m' are said to be *orthogonal* in the same sense as, for instance, states of right-circular and left-circular polarization are orthogonal. The operational definition of orthogonality of quantum-mechanical states, a very important concept, is illustrated by the diagram of Figure 6.4. Atoms emerging from a polarizer magnet in a beam component with quantum number m have zero probability of emerging from the parallel retest magnet in a component beam with $m' \neq m$.

TABLE 6.1

Probability Formulae for Two-, Three-, and Four-Component Beam Splitting

$P^{(1/2)}_{m'm}(\theta)$

m' \ m	$\frac{1}{2}$	$-\frac{1}{2}$
$\frac{1}{2}$	$\frac{1}{2}(1+\cos\theta)=\cos^2\frac{1}{2}\theta$	$\frac{1}{2}(1-\cos\theta)=\sin^2\frac{1}{2}\theta$
$-\frac{1}{2}$	$\frac{1}{2}(1-\cos\theta)=\sin^2\frac{1}{2}\theta$	$\frac{1}{2}(1+\cos\theta)=\cos^2\frac{1}{2}\theta$

$P^{(1)}_{m'm}(\theta)$

m' \ m	1	0	−1
1	$\frac{1}{4}(1+\cos\theta)^2$ $=\cos^4\frac{1}{2}\theta$	$\frac{1}{2}(1-\cos^2\theta)$ $=2\sin^2\frac{1}{2}\theta\cos^2\frac{1}{2}\theta$	$\frac{1}{4}(1-\cos\theta)^2$ $=\sin^4\frac{1}{2}\theta$
0	$\frac{1}{2}(1-\cos^2\theta)$ $=2\sin^2\frac{1}{2}\theta\cos^2\frac{1}{2}\theta$	$\cos^2\theta$ $=(\cos^2\frac{1}{2}\theta-\sin^2\frac{1}{2}\theta)^2$	$\frac{1}{2}(1-\cos^2\theta)$ $=2\sin^2\frac{1}{2}\theta\cos^2\frac{1}{2}\theta$
−1	$\frac{1}{4}(1-\cos\theta)^2$ $=\sin^4\frac{1}{2}\theta$	$\frac{1}{2}(1-\cos^2\theta)$ $=2\sin^2\frac{1}{2}\theta\cos^2\frac{1}{2}\theta$	$\frac{1}{4}(1+\cos\theta)^2$ $=\cos^4\frac{1}{2}\theta$

$P^{(3/2)}_{m'm}(\theta)$

m' \ m	$\frac{3}{2}$	$\frac{1}{2}$	$-\frac{1}{2}$	$-\frac{3}{2}$
$\frac{3}{2}$	$\frac{1}{8}(1+\cos\theta)^3$ $=\cos^6\frac{1}{2}\theta$	$\frac{3}{8}(1+\cos\theta)(1-\cos^2\theta)$ $=3\cos^4\frac{1}{2}\theta\sin^2\frac{1}{2}\theta$	$\frac{3}{8}(1-\cos^2\theta)(1-\cos\theta)$ $=3\cos^2\frac{1}{2}\theta\sin^4\frac{1}{2}\theta$	$\frac{1}{8}(1-\cos\theta)^3$ $=\sin^6\frac{1}{2}\theta$
$\frac{1}{2}$	$\frac{3}{8}(1+\cos\theta)(1-\cos^2\theta)$ $=3\cos^4\frac{1}{2}\theta\sin^2\frac{1}{2}\theta$	$\frac{1}{8}(1+\cos\theta)(1-3\cos\theta)^2$ $=\cos^2\frac{1}{2}\theta(1-3\sin^2\frac{1}{2}\theta)^2$	$\frac{1}{8}(1-\cos\theta)(1+3\cos\theta)^2$ $=\sin^2\frac{1}{2}\theta(3\cos^2\frac{1}{2}\theta-1)^2$	$\frac{3}{8}(1-\cos^2\theta)(1-\cos\theta)$ $=3\cos^2\frac{1}{2}\theta\sin^4\frac{1}{2}\theta$
$-\frac{1}{2}$	$\frac{3}{8}(1-\cos^2\theta)(1-\cos\theta)$ $=3\cos^2\frac{1}{2}\theta\sin^4\frac{1}{2}\theta$	$\frac{1}{8}(1-\cos\theta)(1+3\cos\theta)^2$ $=\sin^2\frac{1}{2}\theta(3\cos^2\frac{1}{2}\theta-1)^2$	$\frac{1}{8}(1+\cos\theta)(1-3\cos\theta)^2$ $=\cos^2\frac{1}{2}\theta(1-3\sin^2\frac{1}{2}\theta)^2$	$\frac{3}{8}(1+\cos\theta)(1-\cos^2\theta)$ $=3\cos^4\frac{1}{2}\theta\sin^2\frac{1}{2}\theta$
$-\frac{3}{2}$	$\frac{1}{8}(1-\cos\theta)^3$ $=\sin^6\frac{1}{2}\theta$	$\frac{3}{8}(1-\cos^2\theta)(1-\cos\theta)$ $=3\cos^2\frac{1}{2}\theta\sin^4\frac{1}{2}\theta$	$\frac{3}{8}(1+\cos\theta)(1-\cos^2\theta)$ $=3\cos^4\frac{1}{2}\theta\sin^2\frac{1}{2}\theta$	$\frac{1}{8}(1+\cos\theta)^3$ $=\cos^6\frac{1}{2}\theta$

As stated above, magnets with different orientations $\hat{z}$ and $\hat{u}$ perform incompatible selections of states. Accordingly, the alternative magnetic energies of atoms in fields of different orientation are also said to be *incompatible quantities*. Circular and linear polarization are incompatible in the same sense, but are not often regarded as "quantities" or as "physical variables" because they have no measurable magnitude.

A set of mutually orthogonal states $|\hat{B}\ m\rangle$, with $2j + 1$ alternative values m, exists for each direction of space $\hat{B}$. Stern-Gerlach magnets are normally designed to select atoms in these states only if the atoms travel in a direction perpendicular to $\hat{B}$. However, different instruments could serve to select states without this restriction. When the field strength B vanishes, the magnetic energies $-\mu_z B$ of all states $|\hat{B}\ m\rangle$ converge to the same value, namely zero; the states are then said to be *degenerate*.

The application of a magnetic field sorts out and identifies degenerate eigenstates of internal energy, that is, states which have the same internal energy in the absence of the field. For atomic states with $j = 0$, there is no splitting of the beam in a Stern-Gerlach magnet, the magnetic energy is zero, and there is, in fact, no degeneracy to be removed. For $j = \frac{1}{2}$, application of a nonzero magnetic field in various directions $\hat{B}$ of space identifies the whole variety of states $|\hat{B}\ \pm\frac{1}{2}\rangle$ which are degenerate for zero field. For $j > \frac{1}{2}$ it is not sufficient to vary the direction of a magnetic field in order to identify each possible state as a stationary state in that field; suitable combinations of nonuniform electric and magnetic fields would be necessary for this purpose.

6.3.4. *Statistical Analysis*. Striking quantum properties of atomic systems emerge from a discussion of Stern-Gerlach experiments. These properties have a statistical character as expected from the remarks in chapter 5. We start here from two basic concepts: the incompatibility of experimental analysis by magnets of different orientation and the symmetry of space about the direction of a magnetic field; we shall also utilize macroscopic vector properties of the magnetic moment which remain valid on a statistical basis.

Macroscopically, the magnetic moment $\vec{\mu}$ of a system has a well-defined magnitude and direction. These are determined, for example, from simultaneously measured values of three Cartesian components μ_x, μ_y, and μ_z. For an atom, these three components are incompatible. Thus, for example, atoms selected by a Stern-Gerlach magnet in the z direction have a definite value of μ_z, but further analysis of these atoms by a magnet in the x (or y) direction yields nonunique values of μ_x and μ_y. In this sense, the direction and magnitude of an atom's magnetic moment are undefined. Well-defined quantities, however, are the *mean value* of the vector $\vec{\mu}$ and its *squared magnitude*, $|\vec{\mu}|^2$.

Statistically, one can define the mean magnetic moment vector $\langle\vec{\mu}\rangle$ of atoms emerging from a given source. This is the vector whose components along three Cartesian axes equal the mean values of μ_x, μ_y, and μ_z determined by three Stern-Gerlach experiments performed separately on beams emerging from identical sources.[10] For atoms in an eigenstate $|\hat{z}\ m\rangle$, the mean value of the component μ_z equals the eigenvalue itself, that is, $\mu m/j$. On the other hand, the mean values of the components μ_x and μ_y vanish. This is because eigenvalues of μ_x (or μ_y) of equal magnitude and opposite sign, $\mu m'/j$ and $-\mu m'j$, have equal probabilities for atoms in the state $|\hat{z}\ m\rangle$, since the axes +x and −x (or +y and −y) are symmetrical with respect to $\hat{z}$ (Fig. 6.7). Therefore, we write

$$\langle\vec{\mu}\rangle_{\hat{z}m} = \left(0, 0, \mu\frac{m}{j}\right) \equiv \mu\frac{m}{j}\hat{z}. \tag{6.5}$$

Consider next the mean square values of the three components μ_x, μ_y, and μ_z,

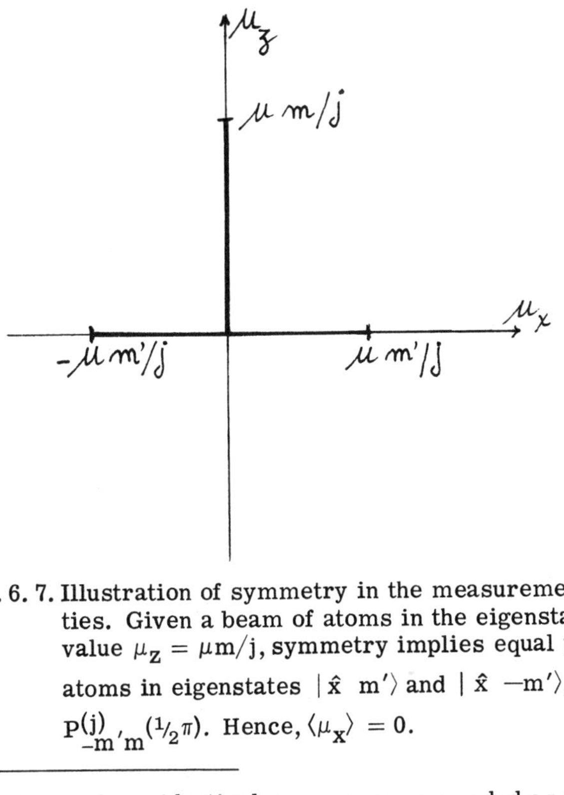

Fig. 6.7. Illustration of symmetry in the measurement of incompatible quantities. Given a beam of atoms in the eigenstate $|\hat{z}\ m\rangle$ with the eigenvalue $\mu_z = \mu m/j$, symmetry implies equal probability of detecting atoms in eigenstates $|\hat{x}\ m'\rangle$ and $|\hat{x}\ -m'\rangle$, that is $P^{(j)}_{m'm}(\frac{1}{2}\pi) = P^{(j)}_{-m'm}(\frac{1}{2}\pi)$. Hence, $\langle\mu_x\rangle = 0$.

10. Atoms from identical sources are regarded as having identical probability of passing through identical analyzers that select a single component. This is an important example of the statistical point of view of quantum mechanics discussed in section 5.2. In quantum mechanics, separate experiments on atoms from identical sources replace the simultaneous measurements of different variables which are deemed possible in macroscopic physics.

or rather, their sum which is the mean squared moment,

$$\langle |\vec{\mu}|^2 \rangle = \langle \mu_x^2 \rangle + \langle \mu_y^2 \rangle + \langle \mu_z^2 \rangle. \tag{6.6}$$

Obtaining this quantity is straightforward for atoms or other particles with $j = \frac{1}{2}$, because in this case each component of $\vec{\mu}$ has only the two eigenvalues $+\mu$ and $-\mu$, and its square has the single eigenvalue μ^2. Therefore, the expression (6.6) takes a value independent of the state of the atom, namely,

$$\mu^2 + \mu^2 + \mu^2 = 3\mu^2. \tag{6.7}$$

Hence, in this case each state $|\hat{B}\ m\rangle$ is an eigenstate of $|\vec{\mu}|^2$ with the eigenvalue $3\mu^2$. This was to be expected because $|\vec{\mu}|^2$ is a scalar and, therefore, independent of the orientation in space of any magnet that may have selected the atoms. Note particularly that for the state $|\hat{z}\ \frac{1}{2}\rangle$ the components μ_x and μ_y have mean value zero but contribute to $\langle|\vec{\mu}|^2\rangle$ amounts $\langle\mu_x^2\rangle$ and $\langle\mu_y^2\rangle$ equal to μ^2. This is why the value of $|\vec{\mu}|^2$ exceeds the squared value, μ^2, of the largest component of $\vec{\mu}$.

Irrespective of the value of j, the fact that $|\vec{\mu}|^2$ is a scalar ensures that its mean value is independent of the orientation in space of the polarizer through which an atomic state has been selected. In fact, the mean value of $|\vec{\mu}|^2$ is altogether independent of any method of selection that may separate one state out of a set of degenerate states. We accept here this result even though its generality does not follow from the evidence we have presented. In other words, all atomic states of a degenerate set are normally eigenstates of $|\vec{\mu}|^2$ with the same eigenvalue.[11] This eigenvalue is usually obtained as a function of j by a procedure beyond the scope of this book. For the particular cases of $j = 1$ and $j = 3/2$, the eigenvalues may be calculated by evaluating equation (6.6) for a particular state $|\hat{z}\ m\rangle$, using the probability Table 6.1. One has, by definition,

$$\langle\mu_z^2\rangle_{\hat{z}m} = \left(\mu\frac{m}{j}\right)^2,$$

$$\langle\mu_x^2\rangle_{\hat{z}m} = \langle\mu_y^2\rangle_{\hat{z}m} = \Sigma_{m'}\left(\frac{\mu m'}{j}\right)^2 P^{(j)}_{m'm}\left(\frac{1}{2}\pi\right). \tag{6.8}$$

By use of the expression for $P^{(j)}$ from Table 6.1, the $\Sigma_{m'}$ yields $\frac{1}{2}(\mu/j)^2 \times [j(j+1) - m^2]$, and, hence, the complete result

$$\langle|\vec{\mu}|^2\rangle_{\hat{z}m} = \mu^2\frac{1}{j^2}\,j(j+1). \tag{6.9}$$

The right-hand side of this equation has been obtained as the mean value of $|\vec{\mu}|^2$ for the particular state $|\hat{z}\ m\rangle$; the fact that it is independent of both $\hat{z}$ and m confirms that equation (6.9) represents the eigenvalue common to all degenerate states. In equation (6.9) as well as in equation (6.7), even though the

---

11. An exceptional case will be considered in chap. 13.

mean values of μ_x and μ_y vanish, their mean square values contribute to $\langle |\vec{\mu}|^2 \rangle$ equal amounts which, added to $\langle \mu_z{}^2 \rangle$, make the result independent of m.

For the simple case $j = \frac{1}{2}$, it is possible to calculate the probabilities $P^{(1/2)}_{m'm}$ in Table 6.1 with the theoretical means developed in this section. We require only that the mean magnetic moment (6.5) have the same vector properties as the corresponding macroscopic vector $\vec{\mu}$. Specifically, we require that, given a beam of atoms in the state $|\hat{z}\ m\rangle$, its analysis by a Stern-Gerlach magnet with field direction $\hat{u}$ at an angle θ with $\hat{z}$ yield a mean moment component

$$\langle \mu_u \rangle_{\hat{z}m} = \langle \vec{\mu} \rangle_{\hat{z}m} \cdot \hat{u} = \frac{\mu m}{j}\ \hat{z} \cdot \hat{u} = \frac{\mu m}{j} \cos\theta. \tag{6.10}$$

Consider now the alternative expression of $\langle \mu_u \rangle_{\hat{z}m}$ in terms of the probabilities $P^{(j)}_{m'm}(\theta)$, which we want to calculate, namely

$$\langle \mu_u \rangle_{\hat{z}m} = \Sigma_{m'} \frac{\mu m'}{j} P^{(j)}_{m'm}(\theta). \tag{6.11}$$

For $j = \frac{1}{2}$ and $m = \frac{1}{2}$, m' equals $\pm\frac{1}{2}$, that is, $\pm j$; therefore, equation (6.11) reduces to

$$\langle \mu_u \rangle_{\hat{z}\frac{1}{2}} = \mu \left[P^{(1/2)}_{\frac{1}{2}\frac{1}{2}}(\theta) - P^{(1/2)}_{-\frac{1}{2}\frac{1}{2}}(\theta) \right], \tag{6.12}$$

while equation (6.10) reduces to $\mu \cos\theta$. Comparison of these two formulae yields

$$P^{(1/2)}_{\frac{1}{2}\frac{1}{2}}(\theta) - P^{(1/2)}_{-\frac{1}{2}\frac{1}{2}}(\theta) = \cos\theta. \tag{6.13}$$

Since, on the other hand, the total probability

$$P^{(1/2)}_{\frac{1}{2}\frac{1}{2}}(\theta) + P^{(1/2)}_{-\frac{1}{2}\frac{1}{2}}(\theta) = 1, \tag{6.14}$$

solution of the pair of equations (6.13) and (6.14) yields

$$P^{(1/2)}_{\pm\frac{1}{2}\frac{1}{2}} = \tfrac{1}{2}(1 \pm \cos\theta), \tag{6.15}$$

in agreement with the entries in Table 6.1.

Extension of this calculation to higher values of j involves considerable complications. The theoretical calculation of all functions $P^{(j)}_{m'm}$ will be discussed briefly in section 10.3.

6.4. USE AND PROPERTIES OF EIGENSTATES

The concepts and language drawn from the study of particular experiments in this chapter serve throughout quantum physics, regardless of whether the test and retest procedures in their operational definitions are applicable in practice. To each eigenvalue of a variable of an atom, for instance, its energy in a magnetic field parallel to $\hat{z}$, one associates the eigenstate[12] for which that value obtains. The incompatibility of variables (for instance, of the energy in fields with different directions) is then expressed by the existence of different sets of eigenstates for incompatible variables. The probability P_{ba} that atoms having a value a of a variable A will be found to have a value b of the variable B is expressed by saying that a fraction P_{ba} of the eigenstate $|A\ \ a\rangle$ consists of the eigenstate $|B\ \ b\rangle$. Physical problems are thus expressed in terms of relationships between alternative sets of eigenstates; such relationships will be defined quantitatively and evaluated in later chapters as we have just done for a simple problem.

Problems are reduced to this general form even when they appear different at first. For example, X-ray diffraction by atoms has been related in section S4.4 to the average density distribution of electrons throughout atoms. We deal, of course, with undisturbed atoms in their stationary state of lowest energy E_0, that is, in their ground state. The average density of electrons at any one point of coordinates (x, y, z) with respect to the nucleus, $\rho(x, y, z)$, equals the probability of finding an electron at that point by a hypothetical experiment designed for this purpose. Accordingly, one considers an electron "position eigenstate" $|x, y, z\rangle$ even in the absence of any practical device for selecting or "filtering out" electrons at that position. The density $\rho(x, y, z)$ is then calculated by resolving the energy eigenstate $|E_0\rangle$ into component eigenstates $|x, y, z\rangle$. A procedure for selecting atoms (or other systems) in a particular eigenstate of a variable by filtering them out of a beam of atoms in an initial state is described in mathematical language as the projection of the initial state onto the particular eigenstate.

The method of analysis into eigenstates applies also to atomic systems whose state varies in the course of time. The analysis of the state into eigenstates of any one variable is then meant to provide a snapshot of the state at a particular instant of time. At any different time, the characterization would differ in the values of the probabilities of finding the system in any one component eigenstate. The component eigenstates themselves would remain the same because they are singled out by application of the *same operation*, even though applied at different times. For example, the state of polarization of a light beam can be analyzed by a spar crystal after the beam has traversed different thicknesses of an optically active medium. The component eigenstates selected by the crystal are always the same, but their intensities vary. Analysis into component eigenstates is a theoretical construct which is used regardless of whether the analysis is performed experimentally. Note that experimental application of an analysis changes the conditions of the system under study.

The theoretical analysis characterizes the state of a system in terms of the

---

12. Various eigenstates often correspond to a single value of the variable, as, for instance, degenerate states may correspond to a single energy level of an atom in the absence of a magnetic field.

components that would be detected if an experimental analysis were applied at a given time.

An electron-atom collision process is described in eigenstate language as follows. One considers first an eigenstate of kinetic energy of an incident electron and the stationary ground state of a target atom. One then considers the combined system electron + atom whose initial state is simply a combination of the states of its initially independent parts. The electric forces between the electron and the constituents of the atom transform this initial state into a final state. (The crucial part of the calculation is to work out this transformation.) The final state is no longer an eigenstate of the separate energies and momenta of the electron and atom. Resolving the final state into component eigenstates of the energy and momentum of the escaping electron determines the probability of detecting it with a specified analyzer.

Once we consider eigenstates of position of a system as we consider eigenstates of any other variable, the complementarity of position and momentum pointed out in section 6.1 appears as an example of the incompatibility of physical variables. The representation of incompatibility as a relation between sets of eigenstates is essentially the same for pairs of analogous variables, such as the components μ_x and μ_z of a magnetic moment, as for entirely different variables, such as position and momentum. A more precise definition of complementarity will be given in section 11.4.

The probabilities of alternative events for an atomic system in an arbitrary state must add up to unity, the alternative events being, of course, mutually exclusive. In order that the probabilities be represented by resolving the state into a set of component eigenstates, the set must be *orthogonal* and *complete*. In an orthogonal set any two states must be orthogonal, that is mutually exclusive; this means, as noted before, that if a system is found in one of the states, the probability of finding it in the other one is zero. Mathematically, states are orthogonal when their projections onto one another is zero. A set is complete when there exists no further state orthogonal to all states of the set. The physical analysis of a state into a complete set of orthogonal states corresponds to well-known mathematical methods of analysis. For instance, the analysis of a light beam into component eigenstates of photon energy corresponds to Fourier analysis of the electromagnetic field into monochromatic components. (The orthogonality and completeness of Fourier components are discussed in section A.3.)

Returning now to experiments with Stern-Gerlach magnets, a reciprocity property was noticed in Table 6.1, namely, the probabilities are equal for an atom in a state $|\hat{z}\ m\rangle$ to be found in a state $|\hat{u}\ m'\rangle$ as for an atom in $|\hat{u}\ m'\rangle$ to be found in $|\hat{z}\ m\rangle$. In this case the reciprocity property has an *a priori* plausibility because the reciprocal experiments are performed by exchanging the role of the polarizer and the analyzer, which are identical except for orientation. However, the reciprocity property of probabilities is quite general; it holds not only when the probability links eigenstates of variables of the same kind, but also for variables of entirely different kinds. For instance, the probability of finding an electron of an atom at one point P if the atom is known to be in its ground state equals the probability of finding the atom in its ground state if the electron is assumed to be at P.

The states of atoms and radiation considered in this chapter are meant to belong to a set of eigenstates. That is, there exists, at least in principle, an analyzer which may be set to transmit fully the system in a given state; or alternatively, it may be set to reject that system with certainty. However, this requirement is not met, for example, by states of atoms emerging from an oven, or of light with natural—or even partial—polarization. The states of these systems which do not fit in any complete orthogonal set belong to a more general class which is not considered in this book or in most introductory courses of quantum mechanics.

CHAPTER 6 PROBLEMS

6.1. A light beam traverses in sequence three spar crystals A, B, and C, each of them fitted with a screen that stops the extraordinary ray. A and C are parallel to each other, B is rotated by 60° with respect to them. A photon counter behind C monitors the transmitted light. By what factor will the counting rate increase upon removal of B? (Regard each crystal as perfectly transparent to the ordinary beam.)

6.2. A beam of atoms emerging from an oven is split into three components by a Stern-Gerlach magnet. Show that the component of the mean magnetic moment in the direction of the magnetic field, $\langle \mu_z \rangle$, is zero for the incident beam while the mean square component $\langle \mu_z{}^2 \rangle$ is $\frac{2}{3}\,\mu^2$.

6.3. A beam of atoms with $j = \frac{1}{2}$ is passed initially through a Stern-Gerlach polarizer P whose field is parallel to the z axis and which transmits only the beam component with $m = \frac{1}{2}$. The emerging beam is filtered again by a second Stern-Gerlach magnet B whose field forms an angle of 60° with the z axis and which transmits only the $m = \frac{1}{2}$ component. The beam emerging from B is finally filtered by a third magnet A with orientation and setting identical to that of P. What fraction of the beam emerging from P will finally emerge from A?

6.4. A beam of nitrogen atoms is split by a Stern-Gerlach magnet into four components of which only the one with $m = \frac{1}{2}$ is allowed to proceed. The field in the magnet makes angles θ of 90°, 120°, and 30° with the x, y, and z coordinate axes. The magnetic moment μ of the atoms is 2.8×10^{-20} ergs/gauss. For the emerging beam with $m = \frac{1}{2}$ calculate (a) the mean values of μ_x, μ_y, and μ_z; (b) the square of the mean moment $|\langle \vec{\mu} \rangle|^2$; (c) the mean square deviations of μ_x, μ_y, and μ_z; (d) the mean square moment $\langle |\vec{\mu}|^2 \rangle$ (use results of question 6.4c); (e) the mean magnetic energy of the atoms when introduced in a uniform magnetic field of 1000 gauss directed along the z axis; (f) the energy levels of the atoms in the uniform field.

6.5. Consider two incompatible variables A and B of a system and indicate their sets of eigenstates by $|A\ n\rangle$ and $|B\ m\rangle$, where n and m are quantum numbers that take as many values as there are eigenstates in a complete set. Call P_{mn} the probability that a system in the state $|A\ n\rangle$ will be found in the state $|B\ m\rangle$ when transmitted through an analyzer that selects eigenstates of B. The set of probabilities P_{mn} can be organized into a square array, that is, into a matrix whose rows are labeled by m and columns are labeled by n. (Note that the first index m of the matrix P_{mn} indicates the eigenstate selected by the analyzer.) (a) What is the sum of the elements of each column of the matrix? (b) What is the relationship of the matrix P_{mn} to the matrix P_{nm}? (The matrix P_{nm} consists of the array of probabilities that a system in the state $|B\ m\rangle$ will be found in the state $|A\ n\rangle$.) (c) What is the sum of the elements of each row of P_{mn}? Verify that the matrices in Table 6.1 comply with your answers.

SOLUTIONS TO CHAPTER 6 PROBLEMS

6.1. The fraction of the output of A transmitted by B is $\cos^2 60° = 1/4$. The same fraction of the output of B is transmitted by C; in the absence of B, C transmits the output of A fully. Therefore, removal of B increases the counting rate by 16.

6.2. Since the component beams have equal intensities and form a symmetric pattern with respect to the point of no deflection, the value of $\langle\mu_z\rangle$ determined by the mean deflection is zero. The eigenvalues of μ_z^2 equal μ^2, 0, and μ^2 for the three beam components of equal intensity. The probability of each atom to land in any one component is 1/3, hence $\langle\mu_z^2\rangle = 2/3\ \mu^2$.

6.3. The probability that an atom arriving from P will emerge from B is $P^{(1/2)}_{1/2,1/2}(60°) = 1/2(1 + 1/2) = 3/4$, according to Table 1, or to equation (6.15). The same probability applies to an atom arriving from B to emerge from A. The combined probability of passage through B and A is $(3/4)^2$.

6.4. The desired mean values would be determined experimentally by passing the beam emerging from the polarizer magnet through a second magnet parallel to one of the coordinate axes. We can predict the result by using the probability Table 6.1 for $j = 3/2$ (four beam components). Specifically, take $P^{(3/2)}_{m'1/2}(\theta)$ for the given values of θ. The eigenvalues of μ_x, μ_y, and μ_z are represented by $2/3\ \mu m'$, according to equation (6.4) with $j = 3/2$. (a) $\langle\mu_x\rangle = 2/3\ \mu\Sigma_{m'}m'P^{(3/2)}_{m'1/2}(90°) = 0$; $\langle\mu_y\rangle = -1/6\ \mu$; $\langle\mu_z\rangle = (1/6\sqrt{3})\mu$. (Note that these mean values coincide with $1/3\mu\cos\theta$ in accordance with the vector property of $\langle\vec{\mu}\rangle$ represented by eq. [6.10].) (b) $|\langle\vec{\mu}\rangle|^2 = \langle\mu_x\rangle^2 + \langle\mu_y\rangle^2 + \langle\mu_z\rangle^2 = (0 + 1/36 + 3/36)\mu^2 = \mu^2/9$. (This is the square of the eigenvalue $\mu/3$ for $m = 1/2$.) (c) $\Delta\mu_x^2 = \langle\mu_x^2\rangle - \langle\mu_x\rangle^2 = (4/9)\mu^2\Sigma_{m'}m'^2\,P^{(3/2)}_{m'1/2}(90°) - 0 = 7/9\mu^2$; $\Delta\mu_y^2 = 4/9\mu^2\Sigma_{m'}m'^2\,P^{(3/2)}_{m'1/2}(120°) - \mu^2/36 = 21/36\mu^2$; $\Delta\mu_z^2 = 7/36\mu^2$. (d) $\langle|\vec{\mu}|^2\rangle = \langle\mu_x^2\rangle + \langle\mu_y^2\rangle + \langle\mu_z^2\rangle = \mu^2[7/9 + 11/18 + 5/18] = 5/3\mu^2$. (This result verifies eq. [6.9] for $j = 3/2$.) (e) $\langle E_{magn} \rangle = -\langle \vec{\mu} \cdot \vec{B} \rangle = -\langle \mu_z \rangle B =$ $-\frac{1}{6}\sqrt{3} \times 2.8 \times 10^{-20}$ ergs/gauss $\times 10^3$ gauss $= -8.0 \times 10^{-18}$ ergs $=$ -5.0×10^{-6} eV. (f) The eigenstate with $\mu_z = \frac{2}{3}\mu m'$ has the energy level $E_{magn} = -\mu_z B = -\frac{2}{3} \times 2.8 \times 10^{-20}\, m' \times 10^3$ ergs. Therefore, $E_{magn} =$ $\mp 2.8 \times 10^{-17}$ ergs for $m' = \pm 3/2$ and $E_{magn} = \mp 0.9 \times 10^{-17}$ ergs for $m' =$ $\pm 1/2$. These levels are evenly spaced.

6.5. (a) $\Sigma_m P_{mn} = 1$ since it represents the total probability of finding the system in any eigenstate of the complete set $|B\ m\rangle$. (b) The matrix P_{nm} is the transpose of P_{mn} owing to the reciprocity property of quantum-mechanical probabilities. (c) $\Sigma_n P_{mn} = \Sigma_n P_{nm} = 1$.

7. Larmor precession and angular momentum

Experiments on the response of atoms to magnetic fields provide key items of information on atomic mechanics, in addition to the identification of eigenstates of magnetic energy described in section 6.3. The main experiments described in this chapter demonstrate the Larmor precession of atoms and other particles in a magnetic field. In this phenomenon, anticipated on a macroscropic basis in section 2.4, the mean magnetic moment of atoms changes direction periodically. The description of these changes will serve as a prototype for the description of nonstationary states of atomic systems. The study of nonstationary states deals directly with the evolution of atomic phenomena in the course of time, whereas the study of stationary states emphasizes properties that remain constant.

Since Larmor precession involves a dynamic balance between the magnetic force upon circulating currents and the inertia represented by the angular momentum of the charge carriers, its experimental study also determines angular momentum eigenvalues, as will be seen in section 7.2. The simple mathematical form of the eigenvalues thus determined suggests that they could have been obtained by a more direct method; however, such a direct derivation would involve a systematic study of symmetry under space rotations and is not attempted here.

The existence of magnetic energy levels influences the atomic spectra, owing to the connection between energy levels and spectral lines (section 3.3). Section 7.3 develops the connection between magnetic levels and the Zeeman effect, and describes other spectroscopic phenomena related to Larmor precession.

7.1. EXPERIMENTS ON LARMOR PRECESSION

Consider a beam of atoms, or other particles, with velocity v which traverse in succession three magnetic field devices. The first device is a Stern-Gerlach magnet with field orientation $\hat{P}$, which prefilters (polarizes) the beam by selecting particles in the eigenstate $|\hat{P}\ m\rangle$. The second device is an ordinary magnet which produces a uniform field $\vec{B}$ over a path length l; the particles remain in this field for a time interval l/v which can be varied by changing either l or v. The third and last device is a Stern-Gerlach analyzer magnet, with field orientation $\hat{A}$. This analyzer separates component beams in the various eigenstates $|\hat{A}\ m'\rangle$ and measures their respective intensities $I_{m'}$. These intensities depend on the time interval l/v and on the field strength B. The mutual orientation of $\hat{P}$, $\hat{B}$, and $\hat{A}$ can be controlled, as well as the channel selection of the polarizer P, that is, the value of m (see Fig. 7.1). The quantum number j and the magnetic moment μ are fixed characteristics of the particles.

The experiment reduces to that shown in Figure 6.6 if the uniform field strength B is set at zero. In this case the intensity $I_{m'}$ is proportional to the probability $P^{(j)}_{m'm}(\theta)$, where θ is the angle between $\hat{P}$ and $\hat{A}$. For a nonzero value of B, a first result is that the relative intensities $I_{m'}$ are the same as

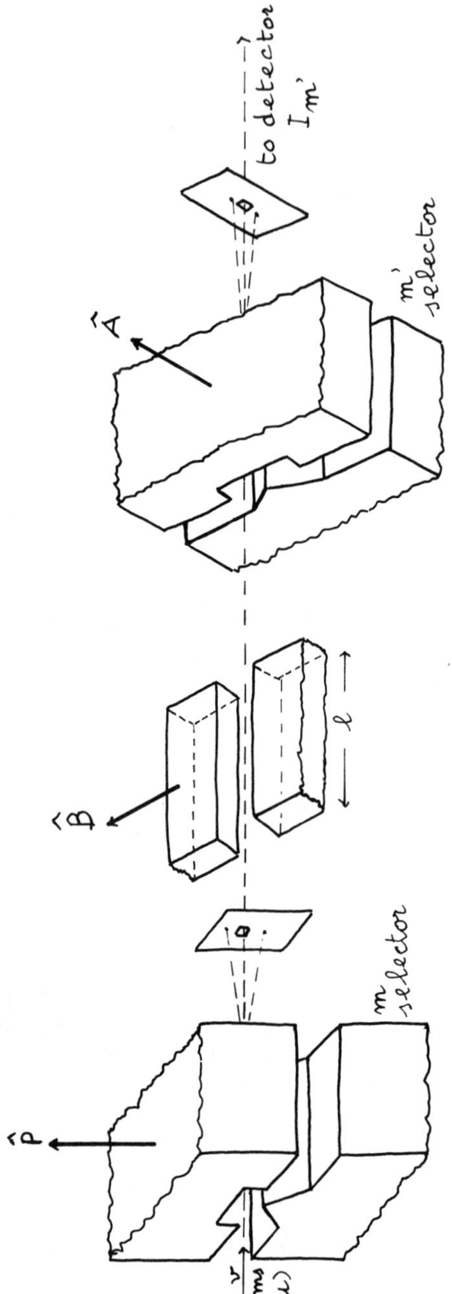

Fig. 7. 1. Atomic beam analysis by two Stern-Gerlach magnets of different orientations, P̂ and Â, and one ordinary magnet of orientation B̂.

though B were still zero but the orientation $\hat{P}$ of the polarizer had been changed. In other words, the intensities $I_{m'}$ can be represented as proportional to probabilities $P^{(j)}_{m'm}(\theta')$, where θ' has taken a value different from θ. This result indicates that the state $|\hat{P}\ \ m\rangle$ of the particles emerging from the polarizer has been changed by the action of the uniform field $\hat{B}$ into a state which we call $|\hat{Q}\ \ m\rangle$; here $\hat{Q}$ indicates a direction of space that forms an angle θ' with $\hat{A}$.

Experiments with a single analyzer orientation $\hat{A}$ do not suffice to determine the direction of $\hat{Q}$ uniquely, since a whole cone of directions forms with $\hat{A}$ the same angle θ'. However, the direction of $\hat{Q}$ can be identified by experimentation with the analyzer in two or three different orientations.

The direction $\hat{Q}$ thus identified depends on the field strength B and on the time l/v spent by the particles in the uniform field. A main result of the experiments is that $\hat{Q}$ forms a constant angle with the direction $\hat{B}$ of the uniform field irrespective of the values of B and l/v. This angle is the same as the angle between $\hat{P}$ and $\hat{B}$ since $\hat{Q}$ reduces to $\hat{P}$ in the limit $B = 0$ or $l = 0$. In other words the direction $\hat{Q}$ describes a cone of axis $\hat{B}$ as the parameters B, l, and v are varied. A further important result is that $\hat{Q}$ returns to coincide with $\hat{P}$ whenever the combination of parameters $(\mu B/jh)(l/v)$ equals a positive or negative integer. A positive integer indicates the number of turns performed by $\hat{Q}$ in a clockwise direction; a negative integer, that in a counter-clockwise direction. More specifically, $\hat{Q}$ precesses uniformly about the direction $\hat{B}$ as the time l/v spent in the magnet increases, for constant field strength B.

The analysis performed by the magnet A bears directly on the state $|\hat{Q}\ \ m\rangle$ of the particles as they emerge from the uniform field $\vec{B}$. However, we infer that the state of the particles has changed with continuity while particles were subjected to the field $\vec{B}$. Accordingly, we introduce the symbol $|\hat{Q}(t)\ \ m\rangle$ to represent a nonstationary state of particles while they are in the field $\vec{B}$. The function $\hat{Q}(t)$ is studied by determining $\hat{Q}$ experimentally for different values of the exit time $t = l/v$.

All these results can be summarized by formulating an equation of motion for the vector $\hat{Q}(t)$. Since $\hat{Q}$ precesses about $\hat{B}$ at a uniform rate in a clockwise (that is, negative) direction for positive values of $\mu B/jh$ and has period one in the variable $(\mu B/jh)t$, it obeys the equation

$$\frac{d\hat{Q}}{dt} = -\frac{\mu}{j\hbar}\,\vec{B} \times \hat{Q}. \tag{7.1}$$

Recall now from equation (6.5) that the mean magnetic moment of particles in the state $|\hat{z}\ \ m\rangle$ is parallel to the axis z. Accordingly, the mean magnetic moment $\langle\vec{\mu}\rangle$[1] of the particles in the nonstationary state $|\hat{Q}(t)\ \ m\rangle$ is parallel to

---

1. Recall that $\langle\vec{\mu}\rangle$ refers to the average of measurements on many atoms, not to a time average. Experiments performed with the analyzer magnet in various orientations effectively determine $\langle\vec{\mu}\rangle$ for any given value of the time l/v spent in the magnetic field $\vec{B}$.

$\hat{Q}(t)$. Equation (7.1) is thus equivalent to

$$\frac{d\langle\vec{\mu}\rangle}{dt} = -\frac{\mu}{j\hbar}\vec{B} \times \langle\vec{\mu}\rangle, \tag{7.1a}$$

which represents the precession of the mean magnetic moment. The phenomenon described here constitutes an example of the Larmor precession introduced in section 2.4.2 on a macroscopic basis. According to equation (7.1) the Larmor precession has the angular velocity

$$\omega_L = \frac{\mu B}{j\hbar} \text{ radians/sec.} \tag{7.1b}$$

The significance of this result with regard to the angular momentum of atomic particles and to the gyromagnetic ratio of their internal currents will be discussed in section 7.2.

Notice that equation (7.1b) represents the Larmor frequency ω_L as the ratio of $\mu B/j$ to the Planck constant $\hbar$. Now, $\mu B/j$ is just the separation of successive magnetic energy levels of the particles in the field $\vec{B}$. We have thus a new example of the relationship between frequency and energy difference introduced in chapter 3. In the Larmor precession the frequency ω_L pertains to the variation of a nonstationary state in the field $\vec{B}$; the energy separation $\mu B/j$ pertains to pairs of stationary states in the same field.

7.1.1. *Observation of Larmor Precession in Muon Decay.* The experiment described above with three magnets of orientations $\hat{P}$, $\hat{B}$, and $\hat{A}$ is not performed easily or accurately for the reason indicated in the note on p. 144. However, a variety of experiments provide direct and striking demonstrations of Larmor precession of atoms and other particles. Among these are observations on the decay of positive muons, unstable particles about 200 times heavier than an electron. These particles emerge as secondary products of high-energy collisions in a state with nonzero mean magnetic moment $\langle\vec{\mu}\rangle$. In the absence of an external field the direction of this moment is opposite to the direction of motion, i.e., it points toward the particles' source. This property of the initial state removes the need for a polarizer magnet $\hat{P}$. Muons can be stopped in a block of matter. Here they decay, that is, they experience a radioactive transformation with a mean delay $\tau = 2.20 \times 10^{-6}$ sec; the decay yields a high-energy positron (positive electron) which easily escapes from the stopping material and can be detected by a counter. The number of positrons detected is known to depend on the angle θ shown in Figure 7.2 and therefore on the angle between the counter's direction and the direction of $\langle\vec{\mu}\rangle$ for the decaying muon. Accordingly, positron counting in various directions provides information on the orientation of $\langle\vec{\mu}\rangle$, equivalent to that which could be provided by an analyzer magnet $\hat{A}$. Specifically, the rate of positron counting is represented by

$$N(t) \propto (1 + a\cos\theta)\exp(-t/\tau), \tag{7.2}$$

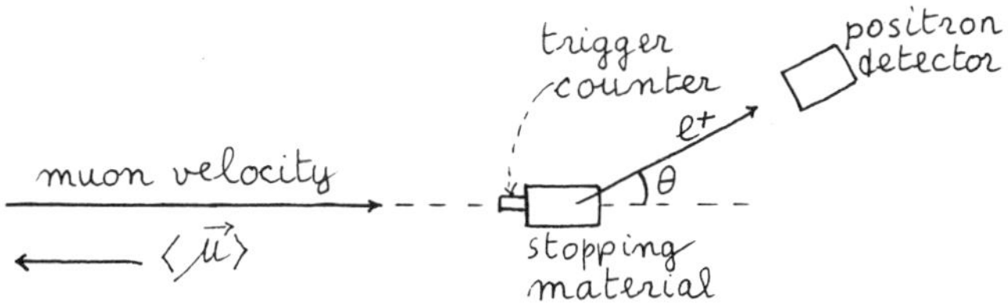

Fig. 7. 2. Diagram of arrangement for observing decay of positive muons.

where a is a numerical coefficient irrelevant to us and t indicates the time interval between the arrival of a muon at the stopping material and the the positron detection. (The muon arrival is marked by the discharge of a trigger counter from which the time delay t is measured.)

To induce Larmor precession of $\vec{\mu}$ between the arrival of a muon and its decay, one applies to the stopping material a magnetic field $\vec{B}$ perpendicular to the plane of the figure. The time interval t between the arrival and the decay of the muon plays the same role as the time l/v spent by atoms in the center magnet of Figure 7. 1. The rate of counting (7. 2) becomes

$$N(t, B) \propto [1 + a \cos(\theta + \gamma Bt)] \exp(-t/\tau), \tag{7.3}$$

where the proportionality coefficient γ represents the muon's gyromagnetic ratio. That is, the positrons decay preferentially in a direction which precesses about $\hat{B}$ at a rate proportional to the field strength.[2]

7. 1. 2. *Nonstationary States and Quantum Jumps.* We have seen how the nonstationary state of particles in Larmor precession can be described in terms of the orientation of a vector $\hat{Q}$ which varies in the course of time. In general, however, no such direct characterization of nonstationary states is readily available. As noted in section 6. 4, quantum-mechanical states are ordinarily identified indirectly through their analysis into component eigenstates of some variable. In the case of Larmor precession the complete operational identification of the state $|\hat{Q}(t) \; m\rangle$ requires analysis into two sets of eigenstates corresponding to different orientations of the analyzer $\hat{A}$. The mathematical identification of a state in terms of a single set of component eigenstates will be developed in following chapters.

Here we point out how the incompleteness of the description of $|\hat{Q}(t) \; m\rangle$ by a single analyzer $\hat{A}$ is reflected in the language commonly used for describing changes of quantum states. Consider, for example, the case of an analyzer $\hat{A}$ parallel to the polarizer $\hat{P}$ of Figure 7. 1. At $t = 0$ (or in the absence of the field $\hat{B}$), analysis by $\hat{A}$ yields a single component beam, with $m' = m$. At later

---

2. For details of this experiment see E. Segrè, *Nuclei and Particles* (New York: Benjamin, 1964), pp. 606, 607.

times, the same analysis yields components of nonzero intensity with $m' \neq m$. On this basis, the action of the field $\vec{B}$ is said to have induced *transitions*, or *quantum jumps*, between eigenstates of the analyzer A, that is, between discrete energy levels of the particles in a magnetic field parallel to $\hat{A}$. This type of statement seems to imply a discontinuity in the evolution of the state $|\hat{Q}(t)\ m\rangle$. In fact, the discontinuity is a characteristic of the process of analysis into a set of discrete eigenstates, and thus belongs to the analyzer $\hat{A}$ itself. The intensities $I_{m'}$ of the component beams separated by the analyzer vary continuously as functions of time, and thus serve appropriately to characterize the continuous evolution of the state $|\hat{Q}(t)\ m\rangle$.

7. 2. ANGULAR MOMENTUM

The macroscopic treatment of the Larmor precession in section 2. 4. 2 has introduced the macroscopic proportionality relation $\vec{\mu} = \gamma\vec{J}$ between the magnetic moment and the angular momentum of the atom (see eq. [2. 56]). This relation must remain valid in a quantum treatment provided the macroscopic vectors $\vec{\mu}$ and $\vec{J}$ are interpreted as mean values. Consequently, the evidence on the magnetic moment from Stern-Gerlach experiments can be reinterpreted as evidence on the angular momentum of atoms. This transfer of information is quantitative if the value of the gyromagnetic ratio is known.

This value is actually obtained from the experimental results (7. 1a) on Larmor precession. The equation represents the mean magnetic moment $\langle\vec{\mu}\rangle$ as precessing about $\hat{B}$ with angular velocity $\omega_L = \mu B/j\hbar$ radians/sec. Since the mean magnetic moment $\langle\vec{\mu}\rangle$ equals $\gamma\langle\vec{J}\rangle$ according to the macroscopic treatment, its velocity of precession coincides with that of $\langle\vec{J}\rangle$. On the other hand, we know that $\langle\vec{J}\rangle$ precesses at γB radians/sec, according to equation (2. 54); therefore, we have

$$\gamma = \frac{\omega_L}{B} = \frac{\mu}{j\hbar}. \tag{7. 4}$$

From the macroscopic proportionality of $\vec{\mu}$ and $\vec{J}$ we take it that in atomic problems the mean values of these vectors are proportional, with the same ratio (7. 4), for all states of an atom that are degenerate at $B = 0$. It follows that the eigenstates of any component of $\vec{\mu}$, for instance the z component are also eigenstates of the corresponding component of $\vec{J}$ and have eigenvalues in the ratio γ. Hence, the angular momentum component J_z has the eigenvalues

$$J_z = \frac{\mu_z}{\gamma} = \mu\frac{m}{j}\frac{j\hbar}{\mu} = m\hbar. \tag{7. 5}$$

Here we have found the main significance of the quantum number m, namely, that for the eigenstate $|\hat{z}\ m\rangle$ it equals the eigenvalue of J_z expressed in units of $\hbar$. The quantum number j represents the largest eigenvalue of J_z. By the same procedure one obtains from equations (6.5) and (6.9) the corresponding formulae

$$\langle\vec{J}\rangle_{\hat{z}m} = m\hbar\hat{z} \tag{7.6}$$

$$|\vec{J}|^2 = j(j+1)\hbar^2. \tag{7.7}$$

Note that while $\langle\vec{J}\rangle$ and $|\vec{J}|^2$ have well-defined values, $\vec{J}$ itself, like $\vec{\mu}$, has no well-defined direction in quantum physics.

The observation that the eigenvalues of J_z differ by multiples of $\hbar$ parallels the key observations on discrete exchanges of energy and momentum described in chapters 3 and 4. Any phenomenon that varies in time with period $2\pi/\omega$ is associated with discrete energy exchanges $\hbar\omega$ or multiples thereof. Any variation of properties of a material from point to point with period $2\pi/q$ produces momentum transfers by multiples of $\hbar q$. Notice now that any variation of physical conditions which is a function of the azimuth angle about an axis has necessarily a period 2π; accordingly, such variations transfer angular momentum in discrete multiples of $\hbar$.

7.3. SPECTRA AND MAGNETIC MOMENTS

7.3.1. *Magnetic Energy Levels and the Zeeman Effect.* As outlined in section 6.3.1, many energy eigenstates of atoms are degenerate in the absence of a magnetic field but acquire different energies upon application of a field. This increase in the number of energy levels leads to an increase in the number of spectral lines observed in a magnetic field; that is, it leads to the splitting of spectral lines which constitutes the Zeeman effect described macroscopically in section 2.4.

We propose here to discuss the Zeeman effect from the point of view of transitions between energy levels. To this end we assume that application of a magnetic field splits all energy levels of atoms—except those with quantum number $j = 0$—even though direct evidence from Stern-Gerlach experiments bears mostly on atoms in their ground state. Accordingly, the eigenstate symbol $|\hat{B}\ m\rangle$ will be generalized, when appropriate, to $|n\ B\ m\rangle$, where the index n identifies the energy eigenvalue E_n at field strength $B = 0$. For $B \neq 0$ the energy of this state is

$$E_n - (\mu_n m/j_n)B = E_n - m\hbar\gamma_n B, \tag{7.8}$$

where the subscript n attached to μ, j, and γ reminds us that these parameters may have different values for different states of excitation.

From equation (7.8) we now try to predict the pattern of spectral line splitting in the Zeeman effect. For $B = 0$ a spectral line of frequency $\omega_{nn'}$ is associated with a pair of energy levels $(E_n, E_{n'})$ by equation (3.3a), $\omega_{nn'} = (E_n - E_{n'})/\hbar$. For $B \neq 0$ we shall have

$$\omega_{nm,n'm'} = (E_n - E_{n'})/\hbar - (m\gamma_n - m'\gamma_{n'})B = \omega_{nn'} - (m\gamma_n - m'\gamma_{n'})B. \tag{7.9}$$

This equation should be compared with the results of the Zeeman effect theory in section 2.4, which predicts the occurrence of three lines where a single one occurs at $B = 0$, namely, one with undisplaced frequency ω_r and two with $\omega_{r\pm} = \omega_r \mp \gamma B$. The two results coincide only if one restricts equation (7.9) by the conditions

$$\gamma_n = \gamma_{n'} = \gamma, \tag{7.10}$$

$$m - m' = 0 \text{ or } \pm 1. \tag{7.11}$$

Condition (7.11) is, in fact, verified quite generally by spectroscopic experiments, in that pairs of energy levels associated with observed spectral lines have quantum numbers that satisfy equation (7.11).

This condition is called a *selection rule.* As noted in section 2.3, atomic currents interact strongly with the electromagnetic field only if the center of mass of the electrons departs from the nucleus. It will be shown in chapter 13 for the simple case of the H atom that the center of mass of the electron departs from the nucleus in a transition between two levels only if condition (7.11) is fulfilled. Therefore, violation of condition (7.11) suppresses the interaction between atomic currents and radiation. Transitions between two levels which satisfy (7.11) are said to be *allowed*; transitions which violate it are called *forbidden.* It will also be shown in chapter 13 that if condition (7.11) is fulfilled, the three alternative values of $m' - m$ in equation (7.11) correspond to shifts of the electron center of mass represented by the three vectors $\vec{p}_r$, $\vec{p}_{r+}$, and $\vec{p}_{r-}$ of section 2.4.

Condition (7.10) is met for a certain class of levels of even-Z atoms. In this case $\gamma = \mu/j\hbar$ has the value $-e/2m_ec$ predicted by macroscopic theory in section 2.4. Violation of condition (7.10) in other cases leads to the occurrence of the anomalous Zeeman effect. Equation (7.9) suggests that experimental data on the anomalous Zeeman effect be reduced to the form of values of the gyromagnetic ratios $\gamma_n, \gamma_{n'}, \ldots$ for the various atomic levels. Most values of these ratios have in fact been obtained from spectroscopic evidence. The origin of the departures of γ values from $-e/2m_ec$ will be discussed in chapter 14, but it may be noted here that the values of γ obtained by Stern-Gerlach measurements of the ground states of many odd-Z atoms (for instance, H, Na, Ag, N) are very approximately equal to $-e/m_ec$ rather than to $-e/2m_ec$.

7.3.2. *Radiofrequency Spectra.* The separation of magnetic energy levels in the presence of an external field can be studied by spectroscopic techniques. One observes transitions between levels which would be degenerate in the absence of a magnetic field and whose separation is extremely small under laboratory conditions; for example, for a field of 1000 gauss it may be of the order of 10^{-5} eV. Photon energies of this order of magnitude belong to electromagnetic fields in the range of radiofrequencies or microwaves. Quantitatively, the frequencies of transitions between otherwise degenerate magnetic levels are represented by equation (7.9) with $n = n'$ and $\omega_{nn'} = 0$.

The transitions occur when atoms are subjected simultaneously to a uniform magnetic field and to an oscillating electromagnetic field. In this case an atom interacts with the oscillating field through oscillations of its mean magnetic moment rather than of its electric dipole moment. This magnetic interaction is much weaker than electric interactions, but at microwave or lower frequencies the applied oscillating magnetic fields can attain strengths that are orders of magnitude higher than those they attain in optical beams. Radio and microwave frequencies can also be controlled to extremely high accuracy. Therefore, spectroscopy in this range has proven a high-precision and broadly useful tool for the study of properties of matter. The main approach for this study by magnetic resonance techniques will be outlined in section S18.5. Here we mention an earlier and very fruitful method for the measurement of magnetic moments which utilizes atomic beams and which was developed by Rabi and his collaborators in the 1930's.

In this method, beam components with different quantum numbers m and m′ are not actually separated out. Thereby the available beam intensity is increased and velocity selection of atoms is no longer required, even for precision work. Instead, an atomic beam is passed through a polarizer Stern-Gerlach magnet designed to separate and then refocus all beam components in a region of space where the magnetic field is uniform and parallel to the field in the polarizer. Here a radiofrequency electromagnetic field is applied. The beam passes then through a second (analyzer) Stern-Gerlach magnet, identical to the first one but with reverse orientation, which brings again all beam components to a second focus on a detector. The refocusing works, however, only for atoms that traverse both magnets in the same beam component, that is, in the same eigenstate m. Any transition with a change of m which takes place between polarizer and analyzer causes a loss of beam intensity at the second focus. Such transitions occur with appreciable intensity only when the photon energy of the radiofrequency field matches the separation of magnetic energy levels. Very extensive and accurate measurements of magnetic moments have been performed by this method.

Notice that spontaneous emission of radiofrequency radiation by atoms while they traverse the apparatus is altogether unlikely, particularly because the rate of spontaneous emission decreases as the cube of the frequency of oscillation. Transitions of atoms between different levels of magnetic energy occur under the influence of the applied oscillating field, not only with absorption but also with emission of energy by the atoms. The applied field, of low frequency and high strength, acts as a macroscopic drive and may be discussed usefully from such a point of view.

7.3.3. *Representation of Energy Transfers as Effects of Larmor Precession.* Consider a beam of atoms subjected to a constant and uniform magnetic field $\vec{B}_0$ and initially in the energy eigenstate $|\hat{B}_0\ m\rangle$. At the time $t = 0$ these atoms enter a region where they are also subjected to a much weaker oscillating field $\vec{B}_1$, which is not parallel to $\vec{B}_0$. The state $|\hat{B}_0\ m\rangle$ is now nonstationary and may be indicated by $|\hat{Q}(t)\ m\rangle$, where $\hat{Q}(0)$ coincides with $\hat{B}_0$ and the variations of $\hat{Q}(t)$ in the course of time obey the Larmor equation (7.1) with $\vec{B} = \vec{B}_0 + \vec{B}_1(t)$. Note that equation (7.1) holds regardless of whether $\vec{B}$ is constant and can be solved without difficulty as a first-order linear differential equation, if necessary by numerical procedure.[3]

The precession does not generally drive $\hat{Q}(t)$ appreciably away from its initial direction $\hat{B}_0$ because the direction of the field $\vec{B}_0 + \vec{B}_1(t)$ is nearly parallel to $\hat{B}_0$ and also oscillates rapidly. The important exception occurs at resonance when the frequency ω of oscillation of $\vec{B}_1$ coincides with the Larmor precession frequency of the atoms, namely, with $\gamma|\vec{B}_0 + \vec{B}_1| \sim \gamma B_0$. In this case the actions of $\vec{B}_1$ in successive half-cycles of its oscillation add up to drive $\hat{Q}(t)$ further and further away from $\hat{B}_0$.[4] When the atoms leave the oscillating field region at the time t, the orientation vector $\hat{Q}(t)$ will normally be, and remain, far from $\hat{B}_0$ under the resonance condition $\omega \sim \gamma B_0$. If the state $|\hat{Q}(t)\ m\rangle$ is then resolved by an analyzer magnet into component energy eigenstates $|\hat{B}_0\ m'\rangle$, different components will generally be found, with $m' \neq m$. We say, then, that transitions from $|\hat{B}_0\ m\rangle$ to $|\hat{B}_0\ m'\rangle$ have occurred with absorption or emission of energy depending upon the sign of $m' - m$ and of γ. The statement that interaction with radiation causes transitions between stationary states must be interpreted in the context of section 7.1.2.

The description of transitions between atomic levels with energy transfer to or from a radiofrequency electromagnetic field provides a physico-mathematical model for the description of all atomic transitions with absorption or emission of radiation. An initial stationary state of an atom is no longer stationary when its interaction with another system is taken into account. The state thus departs from its initial characteristics in the course of time. When the nonstationary state is reanalyzed at a later time into component stationary states of the isolated atom, components different from the initial state are generally found and one says that the atom has performed a transition with emission or absorption of radiation energy. This sketchy description of the process is for orientation purposes only. Detailed implementation of a calculation involves numerous details, many of which lie beyond the scope of this book.

7.3.4. *Zeeman Effect and Larmor Precession.* Equation (7.9) represents the Zeeman frequencies of light emitted by an atom in a magnetic field as resulting from the zero-field frequency $\omega_{nn'}$ by addition or subtraction of multiples of the Larmor precession frequencies $\gamma_n B$ and $\gamma_{n'} B$. (This combination of frequencies is of the type discussed in section 4.5 and Appendix C.) Conversely, the Larmor precession was itself introduced initially in section 2.4.1 as a "beat" due to the superposition of the three different normal-mode frequencies (ω_r, $\omega_r + \omega_L$, $\omega_r - \omega_L$) which appear in a normal Zeeman effect. Thus one may either regard the individual Zeeman frequencies as basic, or emphasize the amplitude modulation of a single frequency by the effect of

---

3. The solution is facilitated by recalling from section 2.4.1 that transformation to coordinates rotating about $\hat{B}_0$ with velocity $-\gamma B_0$ cancels the effect of the stronger field $\vec{B}_0$.

4. This phenomenon will be treated analytically in section S18.5.

Larmor precession. Which point of view is appropriate depends upon experimental circumstances.

The frequencies of the several lines observed in a Zeeman effect are resolved and measured by spectroscopes of high resolving power. The necessary resolving power is attained only if the observation lasts under steady conditions for a sufficiently long time, owing to bandwidth considerations. On the other hand, if one observes the emitted intensity at successive closely spaced instants, and with correspondingly lower frequency resolution, he can observe directly the Larmor precession of the radiating atom. To perform this kind of observation, atoms are excited at a well-defined time. This may be achieved, for example, by excitation through a very brief burst ($\leqslant 10^{-7}$ sec) of light. If the ground state of atoms has quantum number $j = 0$ and the exciting light is linearly polarized, the atoms would normally become the seat of oscillating currents parallel to the direction of the light polarization. These currents in turn emit radiation with the pattern of direction and polarization characteristic of a linear antenna. Specifically, a radiation detector placed at an angle θ from the antenna direction receives an intensity proportional to $\sin^2 \theta$. In the presence of an external uniform magnetic field $\vec{B}$ the atomic oscillating current experiences a Larmor precession about the direction $\hat{B}$ with frequency γB. The precession begins at the time of excitation $t = 0$. If the external field $\vec{B}$ is perpendicular to the plane of the angle θ, the intensity received by the detector is no longer simply proportional to $\sin^2 \theta$ but oscillates with the Larmor frequency and is proportional to $\sin^2 (\theta + \gamma Bt)$.

CHAPTER 7 PROBLEMS

7.1. When the experiment illustrated in Figure 7.1 is performed on nitrogen atoms the analyzer A resolves the beam into four components ($j = 3/2$). The intensity I of each component depends on the time $t = l/v$ spent by the atoms in the field B, and can be represented by the Fourier formula $I = \Sigma_{n=0}^{3} I_n \cos (n\omega_L t + \varphi_n)$. Explain (a) why the frequencies $n\omega_L$ are multiples of a single frequency; (b) why the highest frequency has $n = 3$.

7.2. Write equation (7.1) in the form $d\hat{Q}/dt = -\omega_L \hat{B} \times \hat{Q}$, and set $\hat{Q}(0) = \hat{P}$.
(a) Verify that $\hat{Q}(t) = (\hat{P}\cdot\hat{B})\hat{B} + [\hat{P} - (\hat{P}\cdot\hat{B})\hat{B}] \cos \omega_L t - \hat{B} \times \hat{P} \sin \omega_L t$.
(b) Refer to Figure 7.1 and assume $j = \frac{1}{2}$ and $m = m' = \frac{1}{2}$. Show that the intensity $I_{1/2}$ received by the detector is proportional to $\frac{1}{2}\{1 + (\hat{P}\cdot\hat{B})(\hat{B}\cdot\hat{A}) + [\hat{P}\cdot\hat{A} - (\hat{P}\cdot\hat{B})(\hat{B}\cdot\hat{A})] \cos \omega_L t - (\hat{B} \times \hat{P}\cdot\hat{A}) \sin \omega_L t\}$.

7.3. With reference to the experimental arrangement of Figure 7.1 show that the uniform field $\vec{B}$ has no influence on the transmission through the analyzer if (a) $\hat{B}$ is parallel to $\hat{P}$, (b) $\hat{B}$ is parallel to $\hat{A}$.

7.4. Consider the Larmor precession of Ag atoms ($j = \frac{1}{2}$, $\mu = 0.9 \times 10^{-20}$ ergs/gauss) in an experimental arrangement of the type of Figure 7.1 with the uniform field strength $B = 100$ gauss. Assume $\hat{P}$ and $\hat{A}$ parallel and $\hat{B}$ at 45° from them. The polarizer P transmits only atoms with $m = \frac{1}{2}$. Calculate the probabilities of transmission through the two channels of $\hat{A}$ as functions of the time t spent in the field $\hat{B}$.

7.5. Consider the Larmor precession of nitrogen atoms ($j = 3/2, \mu = 2.8 \times 10^{-20}$ ergs/gauss) in an experimental arrangement of the type of Figure 7.1 for a field strength B = 100 gauss. Assume that $\hat{B}$ forms an angle of 60° with $\hat{P}$ and $\hat{A}$ forms an angle of 30° with $\hat{P}$, in the plane of $\hat{P}$ and $\hat{B}$, and is orthogonal to $\hat{B}$. Assume that the component $m = \frac{1}{2}$ is transmitted by P. Plot the fraction of this component which is transmitted in the channel $m' = -\frac{1}{2}$ of A, as a function of the time t spent in the uniform field $\vec{B}$.

7.6. The green line of the Hg spectrum ($\lambda = 5461$ Å) is emitted in a transition between levels n and n′ with quantum numbers $j_n = 1$ and $j_{n'} = 2$ and with gyromagnetic ratios $\gamma_n \sim 2\gamma$ and $\gamma_{n'} \sim \frac{3}{2}\gamma$. Calculate, in terms of the normal Zeeman ratio γ, (a) the number and separation of magnetic energy levels that result from the splitting of E_n and $E_{n'}$ by a field $\vec{B}$; (b) the number and frequency shifts of the anomalous Zeeman components of the green line which are consistent with the selection rule, equation (7.11).

SOLUTIONS TO CHAPTER 7 PROBLEMS

7.1. (a) The properties of a nonstationary state oscillate with frequencies proportional to the energy differences of the stationary states. Since the energy levels of the atoms in the field B are evenly spaced (see section 6.3), their differences are multiples of a single difference. (b) Since atoms with $j = \frac{3}{2}$ have four energy levels in a magnetic field, the largest difference between two levels is 3 times the smallest. (For an alternative approach see end of solution 7.2)

7.2. (a) Carry out the operations on $\hat{Q}(t)$ indicated on both sides of the differential equation. Simplify by means of the vector identities $\hat{B} \times \hat{B} = 0$ and $\hat{B} \times (\hat{B} \times \hat{P}) = (\hat{B}\cdot\hat{P})\hat{B} - \hat{P}$. Note that $\hat{Q}(t)$ reduces to $\hat{P}$ at $t = 0$. (b) $I_{1/2}$ is proportional to the probability given by equation (6.15) $P^{(1/2)}_{1/2\,1/2}(\theta) = \frac{1}{2}[1 + \hat{Q}(t)\cdot\hat{A}]$. Substitute the expression of $\hat{Q}(t)$ to obtain the given solution. With reference to problem 7.1, note that $P^{(3/2)}_{m'm}(\theta)$ (given in Table 6.1) is a cubic function of $\cos\theta = \hat{Q}(t)\cdot\hat{A}$, where $\hat{Q}(t)$ is a linear function of $\cos\omega_L t$ and $\sin\omega_L t$. The Fourier expansion of a cubic function of these variables has terms $\cos n\omega_L t$ and $\sin n\omega_L t$ up to $n = 3$.

7.3. (a) Since $\hat{Q}(0) = \hat{P}$, it would coincide with the precession axis $\hat{B} \equiv \hat{P}$ and would thus remain fixed. (b) $\hat{Q}(t)$ precesses about $\hat{B} \equiv \hat{A}$ and thus maintains a constant angle with $\hat{A}$. Since the probability of transmission through any channel of A depends only on the angle between $\hat{Q}$ and $\hat{A}$, it remains unaffected by the precession.

7.4. If we call $|\hat{Q}(t)\ \frac{1}{2}\rangle$ the state of the atoms as they precess in the magnet B, the desired probabilities are $P^{(1/2)}_{\mp 1/2\,1/2} = \frac{1}{2}(1 \mp \hat{Q}(t)\cdot\hat{A})$. The dot product of two unit vectors may be calculated by the formula of spherical trigonometry $\hat{u}_1 \cdot \hat{u}_2 = \cos\theta_1 \cos\theta_2 + \sin\theta_1 \sin\theta_2 \cos\varphi$, where θ_1 and θ_2 are the angles of $\hat{u}_1$ and $\hat{u}_2$ with an axis of polar coordinates z and φ is the azimuth angle between the planes $\hat{u}_1\hat{z}$ and $\hat{u}_2\hat{z}$. In our problem we take $\hat{z}$ parallel to the direction of the uniform field $\hat{B}$, so that $\theta_1 = \theta_2 = 45°$ and φ is the angle of Larmor precession $\omega_L t = (\mu B/j\hbar)t = (0.9 \times 10^{-20} \times 100)\,(\frac{1}{2} \times 1.05 \times 10^{-27})^{-1} t = 1.7 \times 10^9 t$, with t given in seconds. Therefore, $\hat{Q}(t)\cdot\hat{A} = \frac{1}{2} + \frac{1}{2}\cos(1.7 \times 10^9 t)$. Thus $P^{(1/2)}_{1/2\;1/2} = \frac{3}{4} + \frac{1}{4}\cos(1.7 \times 10^9 t)$, $P^{(1/2)}_{-1/2\;1/2} = \frac{1}{4} - \frac{1}{4}\cos(1.7 \times 10^9 t)$.

The value of $\hat{Q}(t)\cdot\hat{A}$ can also be obtained by using the expression of $\hat{Q}(t)$ given in problem 7.2

7.5. If we call $|\hat{Q}(t)\ \frac{1}{2}\rangle$ the state of the atoms as they precess in the magnet B, the desired fraction is given according to Table 6.1 by $P^{(3/2)}_{-1/2\;1/2} = \frac{1}{8} \times (1 - \hat{Q}(t)\cdot\hat{A})(1 + 3\hat{Q}(t)\cdot\hat{A})^2$. From problem 7.2a, we have $\hat{Q}(t)\cdot\hat{A} = (\hat{P}\cdot\hat{B})(\hat{B}\cdot\hat{A}) + [\hat{P}\cdot\hat{A} - (\hat{P}\cdot\hat{B})(\hat{B}\cdot\hat{A})]\cos\omega_L t - \hat{B}\times\hat{P}\cdot\hat{A}\sin\omega_L t$, which reduces to $\hat{Q}(t)\cdot\hat{A} = \frac{1}{2}\,3^{1/2}\cos\omega_L t$ with $\omega_L = 1.8 \times 10^9\ \text{sec}^{-1}$.

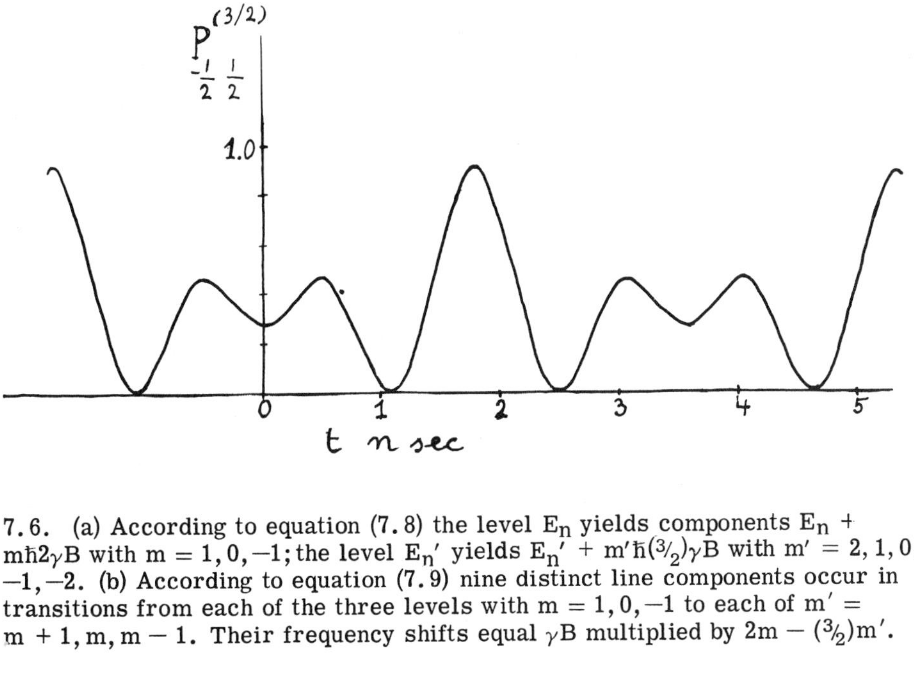

7.6. (a) According to equation (7.8) the level E_n yields components $E_n + m\hbar 2\gamma B$ with m = 1, 0, −1; the level E_n' yields $E_n' + m'\hbar(\frac{3}{2})\gamma B$ with m′ = 2, 1, 0, −1, −2. (b) According to equation (7.9) nine distinct line components occur in transitions from each of the three levels with m = 1, 0, −1 to each of m′ = m + 1, m, m − 1. Their frequency shifts equal γB multiplied by $2m - (\frac{3}{2})m'$.

8. quantum mechanics of atomic states with $j = \frac{1}{2}$

8.1 INTRODUCTION

In quantum physics a functional relationship between two quantities involves not only their magnitudes but also their eigenstates because the quantities may be incompatible. Consider a pair of incompatible variables A and A′ of an atomic system and their respective sets of eigenstates, $|A\ n\rangle$ and $|A'\ n'\rangle$. The relationship between A and A′ is characterized by the set of probabilities $P(A\ n, A'\ n')$ that a system preselected in the state $|A'\ n'\rangle$ will be found in the state $|A\ n\rangle$ by a second selection process. For example, the sets of eigenstates of magnetic moment components $\vec{\mu}\cdot\hat{u}$ and $\vec{\mu}\cdot\hat{z}$ have been indicated by $|\hat{u}\ m'\rangle$ and $|\hat{z}\ m\rangle$, and the corresponding probabilities by $P_{mm'}(\theta)$ with $\cos\theta = \hat{z}\cdot\hat{u}$.

There is one example of quantum systems about which we already possess full knowledge, namely, an atom with quantum number $j = \frac{1}{2}$ if we consider only variations of its magnetic energy and disregard any other form of excitation. Full knowledge of this system was readily achieved because the state of the system depends only on orientation in space and thus any physical variable of the system is proportional to its angular momentum. For this system, a single pair of orthogonal states—for example, $|\hat{z}\ \frac{1}{2}\rangle$ and $|\hat{z}\ -\frac{1}{2}\rangle$—constitutes a complete set. The totality of states of the system can be represented by $|\hat{u}\ \frac{1}{2}\rangle$ for all orientations of the vector $\hat{u}$ since the state $|\hat{u}\ -\frac{1}{2}\rangle$ coincides with $|-\hat{u}\ \frac{1}{2}\rangle$. The whole set of probabilities $P(A\ n, A'\ n')$ is represented for this system by

$$P(\hat{u}\ \tfrac{1}{2}, \hat{u}'\ \tfrac{1}{2}) = P^{(1/2)}_{1/2\,1/2}(\theta) = \tfrac{1}{2}(1 + \hat{u}\cdot\hat{u}'), \tag{8.1}$$

with $\cos\theta = \hat{u}\cdot\hat{u}'$, and is thus fully known.

Even the kinematics and dynamics of this system are fully known, since an initially given state $|\hat{u}\ \frac{1}{2}\rangle$ can only vary in the course of time by a change in direction of the vector $\hat{u}$. Any progressive change of direction is described by the differential equation

$$\frac{d\hat{u}}{dt} = \vec{\omega}(t) \times \hat{u}. \tag{8.2}$$

This equation coincides with the Larmor precession equation (7.1) if one sets $\vec{\omega} = -\gamma\vec{B}$. Any change of state of our system in the course of time may then be attributed to the action of a variable magnetic field $-\vec{\omega}(t)/\gamma$.

Thus, for an atom with $j = \frac{1}{2}$ we can identify all states, the probability relations among them, and the form of the equation of motion. This knowledge actually applies to all *two-level systems,* for which two orthogonal states form a complete set. These systems are all equivalent in that they have the same mathematical representation, as will be shown in chapter 9.

If one progresses to consider systems with more than two, say with N, orthogonal states in a complete set, the basic equations (8.1) and (8.2) can be generalized. In fact, they might keep essentially the same form with a suitable redefinition of the vectors $\hat{u}$ and $\vec{\omega}(t)$ in a model space. Complications emerge, however, from the fact that the dimensionality of the model space increases quadratically with N while the components of the vectors are not independent. Hence we shall not follow this approach to the generalization of the quantum laws (8.1) and (8.2).

A clue to a more practical extension of these laws is provided by the remark that the probabilities $P(A\ n, A'\ n')$ are nonnegative quantities, like the light intensity. In the macroscopic theory of radiation the nonnegative character of the intensity emerges from its representation as a quadratic function of the electric and magnetic fields. These fields obey linear equations, and their calculation is simpler than that of intensities. Similarly, we shall see that quantum mechanics is simplified formally by representing probabilities as quadratic expressions in terms of suitably defined amplitudes. The dimensionality of such amplitudes increases only linearly with the number of orthogonal states in a set; the amplitudes themselves obey linear equations.

As the number N of states in a complete set increases, the manifold of complete orthogonal sets of a system, and hence the manifold of its different physical variables, expands quadratically with N. To master the relationships among all these variables one may express the relation between any two states of the system in terms of the relation of each state to all the eigenstates of a single base set. Such a base set serves much as a base set of coordinate axes serves in analytical geometry. In fact, each state of an atom or other physical system is often identified through its relation to the states of the base set, rather than through directly observable physical characteristics. This approach will be utilized here.

Notice at the outset that, given a base set of states $|A\ n\rangle$ and given the probabilities $P(A\ n, A'\ n')$ and $P(A\ n, A''\ n'')$ which relate two sets $|A'\ n'\rangle$ and $|A''\ n''\rangle$ to it, the relationships between the variables A' and A'' and between their eigenstates are not yet identified. In the example of angular momentum components $\vec{J}\cdot\hat{u}$, the probabilities given in Table 6.1 depend only on the angle θ between $\hat{u}$ and $\hat{z}$. Thereby, they do not identify the direction of $\hat{u}$ with respect to all coordinate axes, nor does one know the angle between two directions $\hat{u}$ and $\hat{u}'$ from data on $\hat{z}\cdot\hat{u}$ and $\hat{z}\cdot\hat{u}'$. Accordingly, full identification of a state $|A'\ n'\rangle$ in terms of a complete base set $|A\ n\rangle$ requires more information than is contained in the set of probabilities $P(A\ n, A'\ n')$.

Proceeding from this background, the objective of this chapter is to recast our knowledge of two-level systems in a form suitable for extension to other systems, even though the new formulae will appear far more complicated than (8.1) and (8.2). This will be done initially by expressing a probability formula for the Larmor precession of an atom with $j = 1/2$ in terms of a base set of energy eigenstates (section 8.2). The new expression of the probability will be found to have a quadratic structure (section 8.3). This remark will lead by successive steps to represent the relationships between quantum-mechanical states by probability amplitudes (section 8.4). These amplitudes are handled conveniently in the language of matrix algebra (section 8.5).

8.2. LARMOR PRECESSION FORMULA FOR $j = \frac{1}{2}$

Consider again the precession phenomenon studied with the setup of Figure 7.1. The nonstationary state of particles precessing in the uniform field $\vec{B}$ was indicated by $|\hat{Q}(t)\ m\rangle$ in section 7.1. Here we can set $m = \frac{1}{2}$ without loss of generality since $|\hat{u}\ -\frac{1}{2}\rangle = |-\hat{u}\ \frac{1}{2}\rangle$. The orientation $\hat{Q}$ coincides with that of the polarizer $\hat{P}$ at $t = 0$. After $\hat{Q}$ has precessed about $\hat{B}$ for a given time interval t, the particles pass through an analyzer magnet of orientation $\hat{A}$ which lets through only the component with $m' = \frac{1}{2}$. (Here again the selection of $m' = -\frac{1}{2}$ would correspond to $m' = \frac{1}{2}$ and reversal of $\hat{A}$.) The fraction of the particles found in this component eigenstate is then, according to equation (8.1),

$$P(\hat{A}\ \tfrac{1}{2}, \hat{Q}(t)\ \tfrac{1}{2}) = P^{(1/2)}_{1/2\,1/2}(\theta_{AQ}) = \tfrac{1}{2}[1 + \hat{A}\cdot\hat{Q}(t)]. \tag{8.3}$$

Since $\hat{Q}$ precesses about the field $\vec{B}$, the variation of the product $\hat{A}\cdot\hat{Q}$ in the course of time is described most conveniently utilizing $\hat{B}$ itself as an axis of polar coordinates. Both $\hat{A}$ and $\hat{Q}$ form fixed angles, θ_A and θ_Q, with this axis. The azimuth of $\hat{A}$, measured from the half-plane $\hat{B}\hat{P}$, will be indicated by ϕ_A. The variable azimuth of $\hat{Q}$, measured from the same half-plane, is indicated by $-\gamma Bt = -\omega_L t$ (see sections 7.1 and 2.4). The probability (8.3) depends on time through the angle ϕ between the projections of $\hat{A}$ and of $\hat{Q}$ on a plane perpendicular to $\hat{B}$, that is, through the difference between the azimuths of $\hat{A}$ and $\hat{Q}$. Accordingly, we rewrite equation (8.3) in the form

$$P(\hat{A}\ \tfrac{1}{2}, \hat{Q}(t)\ \tfrac{1}{2}) = \tfrac{1}{2}[1 + \cos\theta_A \cos\theta_Q + \sin\theta_A \sin\theta_Q \cos\phi], \tag{8.4}$$

with $$\phi = \phi_A + \omega_L t. \tag{8.5}$$

Notice that since $\hat{Q}(0) = \hat{P}$, the constant angle θ_Q may be called θ_P; ϕ_A usually equals 0 or π in a realistic application and equals π in Figure 8.1. Equations (8.4) and (8.5) show explicitly how the probability P varies sinusoidally in the course of time.

At this point we turn to the idea outlined in section 8.1, of relating eigenstates of different variables to those of a base set, instead of relating the eigenstates of components of $\vec{J}$ in the directions $\hat{A}$, $\hat{Q}$ (or $\hat{P}$) directly to one another. The set of eigenstates of the component $\vec{J}\cdot\hat{B} = J_z$ affords a natural choice for a base set because the probability (8.4) is already expressed in terms of the angles θ_A and $\theta_Q = \theta_P$, and because the eigenstates of $\vec{J}\cdot\hat{B}$ are stationary in the center magnet. Accordingly, we express in equation (8.4) the functions of θ_A and of $\theta_Q = \theta_P$ in terms of probabilities that relate eigenstates of $\vec{J}\cdot\hat{A}$ and of $\vec{J}\cdot\hat{P}$ to the base set of eigenstates of $\vec{J}\cdot\hat{B}$. To this end, since the probabilities have the form $\frac{1}{2}(1 \pm \cos\theta)$, one may rewrite equation (8.4) as an explicit

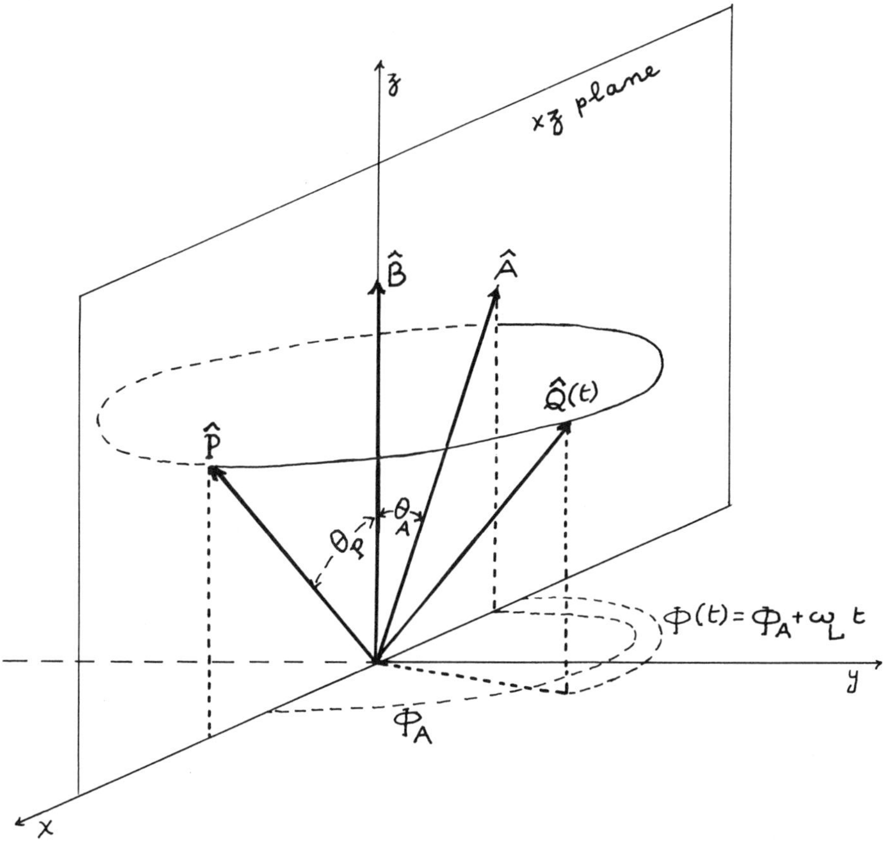

Fig. 8. 1. Relative orientations of $\hat{P}$, $\hat{Q}$, and $\hat{A}$ and their projections onto the (x, y)-plane. $\hat{B}$ coincides with the z axis; $\hat{A}$ is coplanar with $\hat{z}$ and $\hat{P}$.

function of $\frac{1}{2}(1 \pm \cos\theta_A)$ and $\frac{1}{2}(1 \pm \cos\theta_P)$. This is done through the identities

$$\tfrac{1}{2}(1 + \cos\theta_A\cos\theta_P) = \tfrac{1}{2}(1 + \cos\theta_A)\tfrac{1}{2}(1 + \cos\theta_P) + \tfrac{1}{2}(1 - \cos\theta_A)\tfrac{1}{2}(1 - \cos\theta_P), \tag{8.6}$$

$$\sin\theta_A = [1 - \cos^2\theta_A]^{1/2} = 2[\tfrac{1}{2}(1 + \cos\theta_A)\tfrac{1}{2}(1 - \cos\theta_A)]^{1/2}, \tag{8.7}$$

with the analogous formula for $\sin\theta_P$. Substituting θ_Q by θ_P and [1]

---

1. For simplicity in this chapter we omit the superscript index of $P^{(1/2)}$ since it has always the same value.

$$\tfrac{1}{2}(1 \pm \cos\theta_A) = P_{1/2\ \pm 1/2}(\theta_A),\ \tfrac{1}{2}(1 \pm \cos\theta_P) = P_{\pm 1/2\ 1/2}(\theta_P), \tag{8.8}$$

reduces equation (8.4) to the form

$$P(\hat{A}\ \tfrac{1}{2}, \hat{Q}(t)\ \tfrac{1}{2}) = P_{1/2\ 1/2}(\theta_A)\, P_{1/2\ 1/2}(\theta_P) + P_{1/2\ -1/2}(\theta_A)\, P_{-1/2\ 1/2}(\theta_P) \tag{8.9}$$

$$+\ 2[P_{1/2\ 1/2}(\theta_A) P_{1/2\ 1/2}(\theta_P) P_{1/2\ -1/2}(\theta_A) P_{-1/2\ 1/2}(\theta_P)]^{1/2} \cos(\phi_A + \omega_L t).$$

The first two terms on the right of equation (8.9) represent the result that would be obtained by a variant of the Larmor precession experiment. In this variant the energy eigenstates of particles in the uniform field B would be actually separated by inserting ahead of $\vec{B}$ an additional Stern-Gerlach magnet with field parallel to $\vec{B}$. Particles could then traverse the sequence of devices through two alternative sequences of states, namely,

$$\begin{array}{l} \xrightarrow{\text{polarizer } \hat{P}} |\hat{P}\ \tfrac{1}{2}\rangle \\ \xrightarrow{\text{selector } \hat{B}} \begin{cases} |\hat{B}\ \tfrac{1}{2}\rangle \xrightarrow[\text{(stationary)}]{\text{constant } \vec{B}} |\hat{B}\ \tfrac{1}{2}\rangle \xrightarrow{\text{analyzer } \hat{A}} |\hat{A}\ \tfrac{1}{2}\rangle \\ |\hat{B}\ -\tfrac{1}{2}\rangle \xrightarrow[\text{(stationary)}]{\text{constant } \vec{B}} |\hat{B}\ -\tfrac{1}{2}\rangle \xrightarrow{\text{analyzer } \hat{A}} |\hat{A}\ \tfrac{1}{2}\rangle. \end{cases} \end{array} \tag{8.10}$$

The first term on the right of equation (8.9) represents the "compound" probability of following the upper path in expression (8.10), after transmission through $\hat{P}$. This is the product of the probability of clearing selector $\hat{B}$ in the $m = \tfrac{1}{2}$ component and the probability of clearing $\hat{A}$ in the $m' = \tfrac{1}{2}$ component. The second term of equation (8.9) represents the corresponding probability for the lower path in expression (8.10). The sum of these probabilities represents the "total" probability of reaching the final state $|\hat{A}\ \tfrac{1}{2}\rangle$ through *either path* calculated on the assumption that the two paths afford distinct *mutually exclusive* alternatives. The occurrence of the last term of equation (8.9) implies that the two paths in expression (8.10) do *not* constitute mutually exclusive alternatives in the actual Larmor precession experiment, where the selector $\hat{B}$ is absent. The fact that the complete probability $P(\hat{A}\ \tfrac{1}{2}, \hat{Q}(t)\ \tfrac{1}{2})$ given by equation (8.9) depends on t *only through this last term* indicates that time variations occur through the *unresolved coexistence* of alternative stationary states. We shall return to this fundamental point later on.

The last term of equation (8.9) consists of twice the geometric mean of the first two terms times the cosine of the angle $\phi(t)$. The occurrence of this term confirms the expectation indicated in section 8.1 that the network of connections between all states $|A'\ n'\rangle$ of a system could not be represented completely by the set of probabilities $P(A\ n, A'\ n')$ relating each state to the various states $|A\ n\rangle$ of a single base set; the angle $\phi(t)$ in equation (8.9) constitutes additional information required to connect $|\hat{A}\ \tfrac{1}{2}\rangle$ to $|\hat{Q}(t)\ \tfrac{1}{2}\rangle$.

8.3. VECTOR DIAGRAMS AND COMPLEX NUMBERS.

Equation (8.9) has the structure of the equation that relates the sides of a triangle with one angle ϕ. A diagram of this triangle is shown in Figure 8.2.

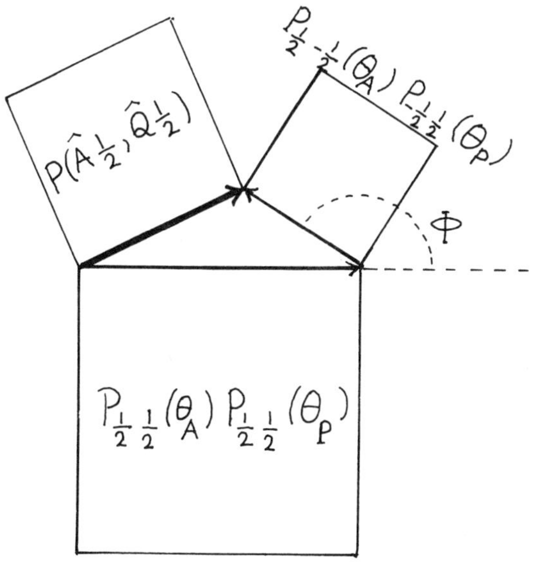

Fig. 8.2. Combination of probabilities in the precession experiment.

An equivalent analytical representation gives $P(\hat{A}\ \tfrac{1}{2},\ \hat{Q}(t)\ \tfrac{1}{2})$ as the square of the sum of two vectors whose directions form an angle ϕ. That is, we rewrite (8.9) in the form

$$P(\hat{A}\ \tfrac{1}{2},\ \hat{Q}(t)\ \tfrac{1}{2}) = |\vec{V}_{1/2} + \vec{V}_{-1/2}|^2, \tag{8.11}$$

where

$$|\vec{V}_{1/2}|^2 = P_{1/2\ 1/2}(\theta_A)P_{1/2\ 1/2}(\theta_P),$$

$$|\vec{V}_{-1/2}|^2 = P_{1/2\ -1/2}(\theta_A)P_{-1/2\ 1/2}(\theta_P), \tag{8.11a}$$

$$\hat{V}_{1/2} \circ \hat{V}_{-1/2} = \cos\phi(t) = \cos(\phi_A + \omega_L t). \tag{8.11b}$$

Thereby the time dependence of $P(\hat{A}\ \tfrac{1}{2},\ \hat{Q}(t)\ \tfrac{1}{2})$ due to the Larmor precession is described in terms of the rotation of two vectors with respect to one another, with uniform angular velocity ω_L.

Equation (8.11) represents the probability $P(\hat{A}\ \frac{1}{2}, \hat{Q}\ \frac{1}{2})$ through the vector sum of two terms, which correspond respectively to the two eigenstates of the base set consisting of $|\hat{B}\ \frac{1}{2}\rangle$ and $|\hat{B}\ -\frac{1}{2}\rangle$. (This formulation provides a clue for the extension to systems with larger sets of orthogonal eigenstates.) However, the squared magnitude of each term of equation (8.11) is defined by (8.11a) as a compound probability linking three states, namely: $|\hat{Q}\ \frac{1}{2}\rangle$, one of the states $|\hat{B}\ \pm\frac{1}{2}\rangle$ and $|\hat{A}\ \frac{1}{2}\rangle$. Our intention was to relate separately each of the states $|\hat{Q}\ \frac{1}{2}\rangle$ and $|\hat{A}\ \frac{1}{2}\rangle$ to the base set of states $|\hat{B}\ \frac{1}{2}\rangle$ and $|\hat{B}\ -\frac{1}{2}\rangle$. The vector symbols in equation (8.11) do not lend themselves to this kind of factorization; however, they can be replaced by complex numbers whose algebra includes operations of multiplication and factorization besides addition by vector rules.

Let us then replace the pair of vectors $\vec{V}_{1/2}$ and $\vec{V}_{-1/2}$ by a pair of complex numbers $V_{1/2}$ and $V_{-1/2}$ whose squared magnitudes are given by equation (8.11a) and whose imaginary phases *differ* by ϕ. *No physical basis exists* for assigning a specified phase to *each* of these complex numbers (or to the direction of each vector in eq. [8.11] or in Fig. 8.2). A specific assignment must be made in practice solely *on the basis of analytical* ("bookkeeping") *convenience*. We set

$$P(\hat{A}\ \tfrac{1}{2}, \hat{Q}(t)\ \tfrac{1}{2}) = |V_{1/2} + V_{-1/2}|^2 = (V_{1/2} + V_{-1/2})(V_{1/2}{}^* + V_{-1/2}{}^*), \tag{8.12}$$

$$V_{1/2} = [P_{1/2\ 1/2}(\theta_A)P_{1/2\ 1/2}(\theta_P)]^{1/2} \exp[i\tfrac{1}{2}\phi(t)],$$

$$V_{-1/2} = [P_{1/2\ -1/2}(\theta_A)\ P_{-1/2\ 1/2}(\theta_P)]^{1/2} \exp[-i\tfrac{1}{2}\phi(t)], \tag{8.12a}$$

where the asterisk denotes complex conjugate and the vertical bars indicate absolute magnitude of a complex number rather than of a vector.

It is important to remember that an arbitrary convention has been introduced in our formulation through the specific allotment of the phase difference ϕ in equation (8.12a). Additional conventions will be made. Conventions involving the introduction of unphysical quantities are a standard feature of quantum theory. They are the price to be paid for analytical simplifications, in our case for the simplification of equation (8.9) to its form (8.12). Notice that the squared magnitude in equation (8.12) is itself independent of such conventions though $V_{1/2}$ and $V_{-1/2}$ are not.

Equation (8.12a) involves the square roots of probabilities $P_{m'm}(\theta)$ besides the newly introduced complex phase factors. These square roots are easily obtained because the expression (8.1) or $P_{m'm}(\theta)$ can also be represented as the square of a trigonometric function of the angle θ between $\hat{z}$ and $\hat{u}$,

$$P_{mm}(\theta) = \cos^2 \tfrac{1}{2}\theta,\ P_{-mm}(\theta) = \sin^2 \tfrac{1}{2}\theta \tag{8.13}$$

for $m = \pm\frac{1}{2}$.

8.4. PROBABILITY AMPLITUDES

We are now in a position to complete our task of representing the probability $P(\hat{A}\ \tfrac{1}{2}, \hat{Q}(t)\ \tfrac{1}{2})$ in a form that displays the separate relationships between the state $|\hat{A}\ \tfrac{1}{2}\rangle$ and the two states $|\hat{B}\ \pm\tfrac{1}{2}\rangle$ of the base set and between $|\hat{Q}(t)\ \tfrac{1}{2}\rangle$ and the same base set. Thereby we shall also formulate each of these separate relationships completely and in a form suitable for further applications. As pointed out before, the probability $P(\hat{A}\ \tfrac{1}{2}, \hat{B}\ \tfrac{1}{2})$—a real number—does not specify the relationship of $|\hat{A}\ \tfrac{1}{2}\rangle$ and $|\hat{B}\ \tfrac{1}{2}\rangle$ completely; a complex number, which summarizes two pieces of information, will prove adequate.

Let us examine, then, equation (8.12a) from the point of view of sorting out factors that depend either on the relationship between $|\hat{A}\ \tfrac{1}{2}\rangle$ and $|\hat{B}\ \pm\tfrac{1}{2}\rangle$ or on the relationship between $|\hat{Q}\ \tfrac{1}{2}\rangle$ and $|\hat{B}\ \pm\tfrac{1}{2}\rangle$. The magnitude of $V_{1/2}$ is already given as the product of two factors, $[P_{1/2\ 1/2}(\theta_A)]^{1/2}$ and $[P_{1/2\ 1/2}(\theta_P)]^{1/2}$, which depend respectively on $\hat{A}\cdot\hat{B}$ and on $\hat{B}\cdot\hat{P} = \hat{B}\cdot\hat{Q}$; the same holds for the magnitude of $V_{-1/2}$. Equation (8.12a) also contains the phase factors $\exp[\pm i\tfrac{1}{2}\phi(t)]$, which provide the additional information required to determine completely the mutual orientation of $\hat{A}$ and of $\hat{Q}(t)$. Each of these factors is also readily split into one factor that depends on $\hat{A}$ and one that depends on $\hat{Q}(t)$; this is done by setting $\phi(t) = \phi_A - \phi_Q = \phi_A + \omega_L t$, as in equation (8.5). (Note that this treatment of phases again involves an arbitrary choice, namely the earlier choice of a zero-azimuth for the polar coordinates attached to the base set of states $|\hat{B}\ \pm\tfrac{1}{2}\rangle$.) We are thus led to rewrite equation (8.12a) in the form

$$V_{1/2} = \{[P_{1/2\ 1/2}(\theta_A)]^{1/2} \exp(i\tfrac{1}{2}\phi_A)\}\{[P_{1/2\ 1/2}(\theta_P)]^{1/2} \exp(i\tfrac{1}{2}\omega_L t)\}, \tag{8.14a}$$

$$V_{-1/2} = \{[P_{1/2\ -1/2}(\theta_A)]^{1/2} \exp(-i\tfrac{1}{2}\phi_A)\}\{[P_{-1/2\ 1/2}(\theta_P)]^{1/2} \exp(-i\tfrac{1}{2}\omega_L t)\}. \tag{8.14b}$$

By including the time-dependent exponential in the second factors of (8.14) we represent the rotation of $\hat{Q}(t)$ with respect to the zero-azimuth plane of polar coordinates attached to $\hat{B}$.[2]

The quantity in each of the braces in equation (8.14) is called a *probability amplitude* because its squared magnitude represents a probability, much as the squared amplitude of a wave phenomenon represents the intensity of the

---

2. An alternative convention is sometimes used in which the variation in time is attributed to the energy eigenstate. In our example this convention implies that the plane of zero-azimuth attached to $\hat{B}$ rotates about $\hat{B}$ so as to follow the motion of $\hat{Q}$. In this case the time-dependent exponential is placed in the first braces of equation (8.14).

phenomenon. Probability amplitudes are the basic quantities that serve to relate any two states in quantum mechanics. Each probability amplitude is indicated by a two-part symbol into which are written letters or numbers that identify the two states. For example, the quantity in the first braces of equation (8.14a), indicated by

$$\langle \hat{A}\ \tfrac{1}{2} | \hat{B}\ \tfrac{1}{2} \rangle = [P_{1/2\ 1/2}(\theta_A)]^{1/2} \exp(i\tfrac{1}{2}\phi_A), \tag{8.15}$$

is the probability amplitude that connects the state $|\hat{B}\ \tfrac{1}{2}\rangle$ of the base set to the analyzer state $|\hat{A}\ \tfrac{1}{2}\rangle$. The ordering of the two states in the symbol and of successive symbols in more extensive formulae is essential for an orderly and consistent performance of calculations. It will be seen later that interchange of the two states in the symbol amounts to changing its value into its complex conjugate. Specifically, the analyzer state in which the particles are observed is entered on the *left* of the state $|\hat{B}\ \tfrac{1}{2}\rangle$ pertaining to an earlier stage of the experiment.[3]

Similarly, the quantity in the second braces of equation (8.14a) is the probability amplitude connecting the nonstationary state $|\hat{Q}(t)\ \tfrac{1}{2}\rangle$ to the stationary state $|\hat{B}\ \tfrac{1}{2}\rangle$ of the base set. Its expression is brought to a more general and standard form by recalling that $\theta_P = \theta_Q$ and especially that

$$\tfrac{1}{2}\omega_L = \tfrac{1}{2}\gamma B = \mu B/\hbar \equiv -E_{1/2}/\hbar, \tag{8.16}$$

since $\gamma = \mu/\tfrac{1}{2}\hbar$ and $E_{1/2} = -\mu B$. Thus we write

$$\langle \hat{B}\ \tfrac{1}{2} | \hat{Q}(t)\ \tfrac{1}{2} \rangle = [P_{1/2\ 1/2}(\theta_Q)]^{1/2} \exp(-iE_{1/2}t/\hbar) \tag{8.17a}$$

$$= \exp(-iE_{1/2}t/\hbar)\langle \hat{B}\ \tfrac{1}{2} | \hat{Q}(0)\ \tfrac{1}{2} \rangle.$$

The exponential factor has been brought to the left of the symbol $\langle \hat{B}\ \tfrac{1}{2} | \hat{Q}(0)\ \tfrac{1}{2} \rangle$ so as to be next to the state $|\hat{B}\ \tfrac{1}{2}\rangle$ to which it pertains. The probability amplitude connecting $|\hat{Q}(t)\ \tfrac{1}{2}\rangle$ with $|\hat{B}\ -\tfrac{1}{2}\rangle$ is given by

$$\langle \hat{B}\ -\tfrac{1}{2} | \hat{Q}(t)\ \tfrac{1}{2} \rangle = \exp(-iE_{-1/2}t/\hbar)\langle \hat{B}\ -\tfrac{1}{2} | \hat{Q}(0)\ \tfrac{1}{2} \rangle, \tag{8.17b}$$

where $E_{-1/2} = \mu B = \tfrac{1}{2}\hbar\omega_L$.

Equations (8.12) and (8.12a) can now be reformulated once again in terms of probability amplitudes by writing

$$P(\hat{A}\ \tfrac{1}{2}, \hat{Q}(t)\ \tfrac{1}{2}) = |\langle \hat{A}\ \tfrac{1}{2} | \hat{Q}(t)\ \tfrac{1}{2} \rangle|^2, \tag{8.18}$$

---

3. Notice that the symbols relating to the later steps of analysis are placed on the left of those relating to earlier ones. Mathematical symbols indicating successive steps of a process are usually ordered from the right to the left. Mathematics, like the English language, places the indications of operations and modifications before the indication of the object; one says "two times a" and one writes "2a" rather than "a2."

$$
\begin{aligned}
&\langle \hat{A}\ \tfrac{1}{2} | \hat{Q}(t)\ \tfrac{1}{2} \rangle \\
&= \langle \hat{A}\ \tfrac{1}{2} | \hat{B}\ \tfrac{1}{2} \rangle \langle \hat{B}\ \tfrac{1}{2} | \hat{Q}(t)\ \tfrac{1}{2} \rangle + \langle \hat{A}\ \tfrac{1}{2} | \hat{B}\ -\tfrac{1}{2} \rangle \langle \hat{B}\ -\tfrac{1}{2} | \hat{Q}(t)\ \tfrac{1}{2} \rangle \\
&= \langle \hat{A}\ \tfrac{1}{2} | \hat{B}\ \tfrac{1}{2} \rangle \exp(-iE_{1/2}t/\hbar) \langle \hat{B}\ \tfrac{1}{2} | \hat{Q}(0)\ \tfrac{1}{2} \rangle \qquad (8.18a) \\
&+ \langle \hat{A}\ \tfrac{1}{2} | \hat{B}\ -\tfrac{1}{2} \rangle \exp(-iE_{-1/2}t/\hbar) \langle \hat{B}\ -\tfrac{1}{2} | \hat{Q}(0)\ \tfrac{1}{2} \rangle .
\end{aligned}
$$

This pair of equations, with its generalization outlined in chapter 10, is a prototype for the analytical description and calculation of quantum phenomena. It solves the problem outlined in section 8.1, since it represents a probability pertaining to an arbitrary pair of states $P(A''\ n'', A'\ n')$ in terms of probability amplitudes $\langle A''\ n'' | A\ n \rangle$ and $\langle A\ n | A'\ n' \rangle$ which relate each of these states to the eigenstates of a single base set $|A\ n\rangle$. The choice of energy eigenstates for the base set is often convenient because, as we shall see, it permits factorization of the time variation of probability amplitudes as it does in equation (8.17).

Equations (8.18) and (8.18a) also solve the problem of relating the oscillation frequency ω_L of the probability $P(\hat{A}\ \tfrac{1}{2}, \hat{Q}(t)\ \tfrac{1}{2})$ to the difference between the energy eigenvalues of the two stationary states $|\hat{B}\ \tfrac{1}{2}\rangle$ and $|\hat{B}\ -\tfrac{1}{2}\rangle$. The association of the nonstationary states $|\hat{Q}(t)\ \tfrac{1}{2}\rangle$ with the energy eigenvalues is described by the pair of probability amplitudes (8.17a) and (8.17b). The nonzero magnitude of these probability amplitudes represents the fact that analysis of the nonstationary state $|\hat{Q}(t)\ \tfrac{1}{2}\rangle$ into energy eigenstates yields two components whose eigenvalues $E_{\pm 1/2}$ differ by $\hbar\omega_L$. The phase of each of the two probability amplitudes (8.17) varies linearly with time at a rate proportional to the energy eigenvalue $E_{\pm 1/2}$; the difference between the phases varies as $\omega_L t$.

8.5. MATRICES

The treatment of physical problems by means of probability amplitudes is made more powerful by expressing the properties of entire sets of probability amplitudes in the language of matrix algebra. This language helps to clarify physical relationships among different variables even though actual calculations be means of matrix operations are cumbersome. For this reason we introduce here the essentials of matrix concepts and language, while, on the other hand, we will use matrix techniques sparingly in the following chapters.

8.5.1. *Transformation Matrices.* We begin by considering together all the probability amplitudes which connect each state of a complete set $\{|\hat{Q}(t)\ \tfrac{1}{2}\rangle, |\hat{Q}(t)\ -\tfrac{1}{2}\rangle\}$ to each state of another set, $\{|\hat{B}\ \tfrac{1}{2}\rangle, |\hat{B}\ -\tfrac{1}{2}\rangle\}$. These four probability amplitudes are conveniently arranged in a square array, that is, in a matrix

$$
\begin{vmatrix}
\langle \hat{B}\ \tfrac{1}{2} | \hat{Q}(t)\ \tfrac{1}{2} \rangle & \langle \hat{B}\ \tfrac{1}{2} | \hat{Q}(t)\ -\tfrac{1}{2} \rangle \\
\langle \hat{B}\ -\tfrac{1}{2} | \hat{Q}(t)\ \tfrac{1}{2} \rangle & \langle \hat{B}\ -\tfrac{1}{2} | \hat{Q}(t)\ -\tfrac{1}{2} \rangle
\end{vmatrix} .
$$

Each probability amplitude is called an *element* of the matrix. All elements in one column pertain to the same state of the set $\{|\hat{Q}(t) \ \pm\frac{1}{2}\rangle\}$, that is, have the same value of the quantum number m. All elements in one row pertain to the same state of the set $\{|\hat{B} \ \pm\frac{1}{2}\rangle\}$, that is, they have the same value of the quantum number which we call $\overline{m}$. (We must use different symbols for quantum numbers of states of different sets.) The matrix of all probability amplitudes connecting states $|\hat{Q}(t) \ m\rangle$ with $|\hat{B} \ \overline{m}\rangle$ is indicated by $\|\langle\hat{B} \ \overline{m}|\hat{Q}(t) \ m\rangle\|$, that is, by the symbol of a generic probability amplitude placed between heavy vertical bars.[4] The probability amplitude (8.17a) is the element of this matrix with $\overline{m} = m = \frac{1}{2}$.

To calculate the complete matrix $\|\langle\hat{B} \ \overline{m}|\hat{Q}(t) \ m\rangle\|$ one must extend the considerations of the previous sections to the state $|\hat{Q}(t) \ -\frac{1}{2}\rangle$. One must also utilize the expression (8.13) of $P_{m'm}$ for $m' \neq m$.[5] Thus one finds

$$\|\langle\hat{B} \ \overline{m}|\hat{Q}(t) \ m\rangle\| = \begin{array}{c|cc} \overline{m} \backslash m = & \frac{1}{2} & -\frac{1}{2} \\ \hline \frac{1}{2} & \exp(-iE_{1/2}t/\hbar)\cos\frac{1}{2}\theta_Q & -\exp(-iE_{1/2}t/\hbar)\sin\frac{1}{2}\theta_Q \\ -\frac{1}{2} & \exp(-iE_{-1/2}t/\hbar)\sin\frac{1}{2}\theta_Q & \exp(-iE_{-1/2}t/\hbar)\cos\frac{1}{2}\theta_Q \end{array} . \tag{8.19}$$

Similarly, one obtains the matrix connecting the set of states $|\hat{B} \ \overline{m}\rangle$ with the set $|\hat{A} \ m'\rangle$,

$$\|\langle\hat{A} \ m'|\hat{B} \ \overline{m}\rangle\| = \begin{array}{c|cc} m' \backslash \overline{m} = & \frac{1}{2} & -\frac{1}{2} \\ \hline \frac{1}{2} & \cos\frac{1}{2}\theta_A \exp(i\frac{1}{2}\phi_A) & \sin\frac{1}{2}\theta_A \exp(-i\frac{1}{2}\phi_A) \\ -\frac{1}{2} & -\sin\frac{1}{2}\theta_A \exp(i\frac{1}{2}\phi_A) & \cos\frac{1}{2}\theta_A \exp(-i\frac{1}{2}\phi_A) \end{array} . \tag{8.20}$$

Use of matrix symbols is particularly appropriate because the structure of equation (8.18a) coincides with that of the general formula which defines a matrix product. In fact, one may rewrite the expression (8.18a) for a single probability amplitude in the more general form

$$\langle\hat{A} \ m'|\hat{Q}(t) \ m\rangle = \Sigma_{\overline{m}}\langle\hat{A} \ m'|\hat{B} \ \overline{m}\rangle\langle\hat{B} \ \overline{m}|\hat{Q}(t) \ m\rangle . \tag{8.21}$$

The same relation is represented for complete matrices by the equation

---

4. Light bars mean "absolute magnitude of," as usual.

5. The determination of the sign of each matrix element rests on an analysis of the relations between different base sets, which is rather complicated and is given partly in section 8.5.2 and partly in section S10.3.

$$|\langle\hat{A}\ \ m'|\hat{Q}(t)\ \ m\rangle| = |\langle\hat{A}\ \ m'|\hat{B}\ \ \overline{m}\rangle| \times |\langle\hat{B}\ \ \overline{m}|\hat{Q}(t)\ \ m\rangle|. \tag{8.21a}$$

The matrices (8.19) and (8.20) are called transformation matrices because they serve much like the coordinate transformation matrices in analytical geometry. An analogy between a set of eigenstates and a set of coordinate axes in geometry was pointed out when the idea of utilizing a base set of reference eigenstates emerged in section 8.1. This analogy extends to the idea of treating all sets of eigenstates as alternative sets of coordinate axes in geometry.[6] The matrix (8.19) may then be regarded as representing a "coordinate" transformation from the set of eigenstates $\{|\hat{Q}(t)\ \ \pm\frac{1}{2}\rangle\}$ to the set $\{|\hat{B}\ \ \pm\frac{1}{2}\rangle\}$, while (8.20) connects sets $\{|\hat{B}\ \ \pm\frac{1}{2}\rangle\}$ and $\{|\hat{A}\ \ \pm\frac{1}{2}\rangle\}$. The matrix (8.21a) which connects the sets $\{|\hat{Q}(t)\ \ \pm\frac{1}{2}\rangle\}$ and $\{|\hat{A}\ \ \pm\frac{1}{2}\rangle\}$ directly is then obtained, as in geometry, by multiplication of the successive transformation matrices (8.19) and (8.20).

8.5.2. *Reciprocity and Unitarity.* It was pointed out in chapter 6 with reference to Table 6.1, and also in section 6.4, that the probabilities $P_{m'm}$ obey a reciprocity law, that is, they do not change if one interchanges the input and output channels in a procedure of eigenstate analysis. This symmetry property is included in a more comprehensive property of probability amplitudes.

Consider the special case where the two transformation matrices (8.19) and (8.20) pertain to inverse transformations. Equation (8.19) pertains to transformation from the eigenstates of $\hat{Q}(t)$ to those of $\hat{B}$, equation (8.20) to transformation from $\hat{B}$ to $\hat{A}$. The two transformations are inverse when $\hat{Q}$ and $\hat{A}$ coincide, that is, when $\theta_Q = \theta_A$ and $E_{1/2}t/\hbar = \frac{1}{2}\phi_A$. In this case the probability $P(\hat{A}\ \ m', \ \hat{Q}\ \ m) = |\langle\hat{A}\ \ m'|\hat{Q}\ \ m\rangle|^2$ must, of course, equal unity for $m' = m$ and must vanish for $m' \neq m$. In addition, phase conventions are usually established so that the probability amplitude itself, $\langle\hat{A}\ \ m\ |\hat{Q}\ \ m\rangle$—not merely its magnitude—equals unity when $\hat{A}$ and $\hat{Q}$ coincide. These two requirements are summarized by stating that $\langle\hat{A}\ \ m'|\hat{Q}\ \ m\rangle$ equals $\delta_{mm'}$ for $\hat{A} = \hat{Q}$, that is, that the matrices (8.19) and (8.20) are *inverse* of each other when $\hat{A}$ and $\hat{Q}$ coincide. That the matrices are inverse is verified by setting $\theta_A = \theta_Q$ and $\frac{1}{2}\phi_A = E_{1/2}t/\hbar$ in the matrix (8.20) and working out the matrix multiplication indicated in (8.21a); one obtains the unit matrix.

Indeed, the very correspondence between sets of eigenstates and sets of coordinate axes implies the reciprocity of probabilities, $P(A\ \ n,\ A'\ \ n') = P(A'\ \ n', A\ \ n)$, since in geometry the projection of an axis vector $\hat{x}$ on a vector $\hat{x}'$ of another coordinate system coincides with the projection of $\hat{x}'$ on $\hat{x}$. Another important property of the matrices (8.19) and (8.20) emerges by setting $\theta_A = \theta_Q$ and $\frac{1}{2}\phi_A = E_{1/2}t/\hbar$ in the matrix (8.20). The two matrices are then seen to be Hermitian conjugate of each other.[7] A matrix whose inverse is its

---

6. The concepts of orthogonality and projections of eigenstates introduced in section 6.4 fit into this geometrical picture. The model space in which a set of eigenstates constitutes the coordinate axes is an example of *Hilbert space*.

7. Hermitian conjugate means transposed and complex conjugate.

Hermitian conjugate is unitary. All probability amplitude matrices are unitary. Unitarity of these matrices is represented by the formula

$$\langle\hat{Q}\ \ m|\hat{B}\ \ \bar{m}\rangle = \langle\hat{B}\ \ \bar{m}|\hat{Q}\ \ m\rangle^*, \tag{8.22}$$

which relates corresponding elements of a matrix and of its inverse. This formula implies the reciprocity property of probabilities, $P(\hat{B}\ \ \bar{m},\ \hat{Q}\ \ m) = P(\hat{Q}\ \ m,\ \hat{B}\ \ \bar{m})$. In addition, it requires that the probability amplitudes $\langle\hat{Q}\ \ m|\hat{B}\ \ \bar{m}\rangle$ and $\langle\hat{B}\ \ \bar{m}|\hat{Q}\ \ m\rangle$ have phases of equal magnitude and opposite signs.

The unitarity of transformation matrices makes it possible to represent a probability as a product of a probability amplitude and of the amplitude of the inverse transformation:

$$P_{\bar{m}m}(\theta_Q) = \langle\hat{Q}\ \ m|\hat{B}\ \ \bar{m}\rangle\langle\hat{B}\ \ \bar{m}|\hat{Q}\ \ m\rangle. \tag{8.23}$$

The sum of these probabilities gives then directly the expected identity

$$\textstyle\sum_{\bar{m}} P_{\bar{m}m}(\theta_Q) = \sum_{\bar{m}}\langle\hat{Q}\ \ m|\hat{B}\ \ \bar{m}\rangle\langle\hat{B}\ \ \bar{m}|\hat{Q}\ \ m\rangle = \langle\hat{Q}\ \ m|\hat{Q}\ \ m\rangle = 1. \tag{8.24}$$

Equation (8.23) proves convenient for representing mean values.

8.5.3. *Mean Value Calculations and Matrix Representation of Physical Quantities.* We have seen how transformation matrices represent the relationships between complete sets of eigenstates of different variables. As a next step, one can combine information on both eigenvalues and eigenstates of physical quantities in a unified matrix representation. This representation serves directly to formulate mean values, as we shall show for a particular example; its power will become apparent in later chapters.

As an example of matrix formulation we shall develop the mathematical construction of the mean values of the angular momentum components introduced in section 7.2. Specifically, we will calculate the Cartesian components of the time-dependent mean angular momentum $\langle\vec{J}\rangle_{\hat{Q}(t)1/2}$ of a particle with $j = \frac{1}{2}$ which performs Larmor precession as described in section 8.2. The definition (7.6) of $\langle\vec{J}\rangle$ and the description of the precession of $\hat{Q}$ about the z axis, that is, about $\hat{B}$, identify the vector and its Cartesian components as

$$\langle\vec{J}\rangle_{\hat{Q}(t)1/2} = \tfrac{1}{2}\hbar\,\hat{Q}(t) \equiv \tfrac{1}{2}\hbar\,(\sin\theta_Q \sin\omega_L t,\ -\sin\theta_Q\sin\omega_L t,\ \cos\theta_Q). \tag{8.25}$$

We shall now rederive the mean value of each component of $\langle\vec{J}\rangle$ by the alternative method of multiplying each eigenvalue by its probability and summing over these products. This procedure is lengthy, but it introduces concepts and methods which are generally applicable when no analog of the direct formulation (8.25) is available.

When each probability is expressed by the product of two matrix elements, in the form (8.23), the eigenvalue may be written conveniently between the two probability amplitudes, that is, next to the indications of its corresponding eigenstate. Thus we write the definitions of our mean values in the form

$$\langle J_x\rangle_{\hat{Q}(t)1/2} = \sum_{m_x} \langle\hat{Q}(t)\ \tfrac{1}{2}|\hat{x}\ m_x\rangle\ m_x\hbar\ \langle\hat{x}\ m_x|\hat{Q}(t)\ \tfrac{1}{2}\rangle, \tag{8.26a}$$

$$\langle J_y\rangle_{\hat{Q}(t)1/2} = \sum_{m_y} \langle\hat{Q}(t)\ \tfrac{1}{2}|\hat{y}\ m_y\rangle\ m_y\hbar\ \langle\hat{y}\ m_y|\hat{Q}(t)\ \tfrac{1}{2}\rangle, \tag{8.26b}$$

$$\langle J_z\rangle_{\hat{Q}(t)1/2} = \sum_{m_z} \langle\hat{Q}(t)\ \tfrac{1}{2}|\hat{z}\ m_z\rangle\ m_z\hbar\ \langle\hat{z}\ m_z|\hat{Q}(t)\ \tfrac{1}{2}\rangle, \tag{8.26c}$$

Our aim in the following is to show that each of these expressions may be regarded as an element of a matrix built as a product of matrices.

The equations (8. 26a) and (8. 26b) utilize probability amplitudes that relate eigenstates of $\hat{x}$ and $\hat{Q}$ (or of $\hat{y}$ and $\hat{Q}$) whereas we have been striving to represent all characteristics of the state $|\hat{Q}(t)\ \tfrac{1}{2}\rangle$ in terms of its relationship to a single base set $|\hat{B}\ \overline{m}\rangle$, that is, $|\hat{z}\ m_z\rangle$. However, the probability amplitudes $\langle\hat{z}\ m_z|\hat{Q}(t)\ \tfrac{1}{2}\rangle$ can be introduced in (8. 26a) (and [8. 26b]) by utilizing the basic formula (8. 21), where we now identify $\hat{A}$ with $\hat{x}$ (or with $\hat{y}$). That is, we set $\theta_A = \tfrac{1}{2}\pi$ and $\phi_A = 0$ for $\hat{A} \equiv \hat{x}$ (and $\theta_A = \tfrac{1}{2}\pi$, $\phi_A = \tfrac{1}{2}\pi$ for $\hat{A} \equiv \hat{y}$). Thus we obtain

$$\langle J_x\rangle_{\hat{Q}(t)1/2} = \sum_{m_x} [\sum_{m_z} \langle\hat{Q}(t)\ \tfrac{1}{2}|\hat{z}\ m_z\rangle\langle\hat{z}\ m_z|\hat{x}\ m_x\rangle]\ m_x\hbar \times$$
$$[\sum_{m_z'} \langle\hat{x}\ m_x|\hat{z}\ m_z'\rangle\langle\hat{z}\ m_z'|\hat{Q}(t)\ \tfrac{1}{2}\rangle], \tag{8.27a}$$

and an analogous formula for $\langle J_y\rangle$.[8]

To single out the probability amplitudes $\langle\hat{z}\ m_z|\hat{Q}(t)\ \tfrac{1}{2}\rangle$ in (8. 27a) we group together the factors that are independent of $\hat{Q}$. The grouping is done by carrying out the summation over m_x first and by representing the result of this summation by a new suitable matrix symbol. We write then

$$\langle J_x\rangle_{\hat{Q}(t)1/2} = \sum_{m_z}\sum_{m_z'} \langle\hat{Q}(t)\ \tfrac{1}{2}|\hat{z}\ m_z\rangle\langle\hat{z}\ m_z|J_x|\hat{z}\ m_z'\rangle\langle\hat{z}\ m_z'|\hat{Q}(t)\ \tfrac{1}{2}\rangle, \tag{8.28a}$$

where $\langle\hat{z}\ m_z|J_x|\hat{z}\ m'_z\rangle$ indicates an element of the matrix

$$|\langle\hat{z}\ m_z|J_x|\hat{z}\ m'_z\rangle| = |\sum_{m_x} \langle\hat{z}\ m_z|\hat{x}\ m_x\rangle\ m_x\hbar\ \langle\hat{x}\ m_x|\hat{z}\ m_z'\rangle|$$

$$\equiv \begin{array}{c|cc|} m_z \backslash m_z' & \tfrac{1}{2} & -\tfrac{1}{2} \\ \tfrac{1}{2} & 0 & \tfrac{1}{2}\hbar \\ -\tfrac{1}{2} & \tfrac{1}{2}\hbar & 0 \end{array}\ . \tag{8.29a}$$

Similarly, we write

---

8. Here the "prime" on the index m_z' indicates that the last summation in (8. 27a) is carried out independently of the preceding one.

$$\langle J_y\rangle_{\hat{Q}(t)1/2} = \sum_{m_z}\sum_{m'_z}\langle \hat{Q}(t)\ \tfrac{1}{2}|\hat{z}\ m_z\rangle\langle \hat{z}\ m_z|J_y|\hat{z}\ m'_z\rangle\langle \hat{z}\ m'_z|\hat{Q}(t)\ \tfrac{1}{2}\rangle, \tag{8.28b}$$

with the definition

$$|\langle \hat{z}\ m_z|J_y|\hat{z}\ m'_z\rangle| = |\sum_{m_y}\langle \hat{z}\ m_z|\hat{y}\ m_y\rangle\, m_y\hbar\, \langle \hat{y}\ m_y|\hat{z}\ m'_z\rangle| \equiv$$

$$\begin{array}{c|cc|} m_z \backslash m'_z & \tfrac{1}{2} & -\tfrac{1}{2} \\ \tfrac{1}{2} & 0 & -i\tfrac{1}{2}\hbar \\ -\tfrac{1}{2} & i\tfrac{1}{2}\hbar & 0 \end{array}. \tag{8.29b}$$

The matrices (8. 29a) and (8. 29b) are constructed by combining the eigenvalues of J_x (or J_y) with the probability amplitudes that relate the eigenstates of J_x (or J_y) to those of J_z. They are called the *matrix of the variable J_x* (or J_y) in the representation of eigenstates of J_z. The explicit forms of the matrices (8. 29) are obtained by utilizing the matrices of probability amplitudes (8. 20) with $\hat{x}$ (or $\hat{y}$) in place of $\hat{A}$, with $|\hat{z}\ m_z\rangle$ in place of $|\hat{B}\ \overline{m}\rangle$, and with the values $\theta_A = \tfrac{1}{2}\pi$ and $\phi_A = 0$ (or $\tfrac{1}{2}\pi$) noted above.[9]

Notice how (8. 28a) and (8. 28b) yield mean values, as desired, in terms of probability amplitudes $\langle \hat{z}\ m_z|\hat{Q}\ \tfrac{1}{2}\rangle$ but involve a double summation over quantum numbers m_z and m'_z whereas (8. 26c) involves a single sum. The double sum occurs because the states $|\hat{z}\ m_z\rangle$ of the base set are not eigenstates of the variable J_x or J_y. A double sum occurs normally in quantum-mechanical calculations involving quantities incompatible with the base set of states. Actually also (8. 26c) itself can be cast in the same form as (8. 28a) or (8. 28b), namely,

$$\langle J_z\rangle_{\hat{Q}(t)1/2} = \sum_{m_z}\sum_{m'_z}\langle \hat{Q}(t)\ \tfrac{1}{2}|\hat{z}\ m_z\rangle\langle \hat{z}\ m_z|J_z|\hat{z}\ m'_z\rangle\langle \hat{z}\ m'_z|\hat{Q}(t)\ \tfrac{1}{2}\rangle, \tag{8.28c}$$

---

9. The matrix (8. 20) with these substitutions is $|\langle \hat{x}\ m'_x|\hat{z}\ m'_z\rangle| = \begin{vmatrix} \sqrt{\tfrac{1}{2}} & \sqrt{\tfrac{1}{2}} \\ -\sqrt{\tfrac{1}{2}} & \sqrt{\tfrac{1}{2}} \end{vmatrix}$.

The matrix for the inverse transformation is its Hermitian conjugate, that is,

$|\langle \hat{z}\ m_z|\hat{x}\ m_x\rangle| = \begin{vmatrix} \sqrt{\tfrac{1}{2}} & -\sqrt{\tfrac{1}{2}} \\ \sqrt{\tfrac{1}{2}} & \sqrt{\tfrac{1}{2}} \end{vmatrix}$. The set of the two eigenvalues of J_x, $\tfrac{1}{2}\hbar$ and

$-\tfrac{1}{2}\hbar$, may be written as a diagonal matrix $|\langle \hat{x}\ m_x|J_x|\hat{x}\ m'_x\rangle| = \begin{vmatrix} \tfrac{1}{2}\hbar & 0 \\ 0 & -\tfrac{1}{2}\hbar \end{vmatrix}$.

The matrix (8. 29a) is obtained as the product of these three matrices, that is,

$$|\langle \hat{z}\ m_z|\hat{x}\ m_x\rangle| \times |\langle \hat{x}\ m_x|J_x|\hat{x}\ m'_x\rangle| \times |\langle \hat{x}\ m'_x|\hat{z}\ m'_z\rangle|$$

where

$$|\langle \hat{z}\ m_z | J_z | \hat{z}\ m_z' \rangle| = |m_z \hbar\ \delta_{m_z m_z'}| = \begin{array}{c|cc|} & m_z' \ 1/2 & -1/2 \\ m_z & & \\ 1/2 & 1/2\hbar & 0 \\ -1/2 & 0 & -1/2\hbar \end{array} \tag{8.29c}$$

is a *diagonal matrix*.

The time dependence of $\langle J_x \rangle_{\hat{Q}(t)1/2}$ (or $\langle J_y \rangle_{\hat{Q}(t)1/2}$) can be factored out explicitly from the time-dependent probability amplitudes by writing them in the form (8.17). Thus we rewrite (8.28a) in the form

$$\langle J_x \rangle_{\hat{Q}(t)1/2} = \sum_{m_z} \sum_{m_z'} \langle \hat{Q}(0)\ 1/2 | \hat{z}\ m_z \rangle \exp(-i m_z \omega_L t) \langle \hat{z}\ m_z | J_x | \hat{z}\ m_z' \rangle$$
$$\times \exp(i m_z' \omega_L t) \langle \hat{z}\ m_z' | \hat{Q}(0)\ 1/2 \rangle. \tag{8.30}$$

Substituting in this equation the probability amplitudes from equations (8.17a) and (8.17b) and the matrix of J_x from (8.29a) and carrying out the sum, one reproduces the value of $\langle J_x \rangle$ given in equation (8.25). As anticipated, this method of calculation is more complicated than the direct calculation of equation (8.25) but it is generally used for other systems; we will show an example in section 10.2 for which no more direct calculation method is available.

In view of the definition (8.29a) of the matrix of J_x in the representation of eigenstates of J_z, equation (8.26) which defines the mean value of J_x in the state $|\hat{Q}(t)\ 1/2\rangle$ may also be regarded as defining the element $m = m' = 1/2$ of the matrix $|\langle \hat{Q}(t)\ m | J_x | \hat{Q}(t)\ m' \rangle|$. Indeed, the mean value of any physical quantity in the state $|\hat{Q}(t)\ 1/2\rangle$ may be regarded as a diagonal element of the matrix of that quantity in the representation of eigenstates $|\hat{Q}(t)\ m\rangle$. This point of view is also apparent from the structure of the lengthy expression of $\langle J_x \rangle$ in equation (8.27a). This expression consists of the eigenvalues of J_x, which constitute the diagonal matrix

$$|\langle m_x | J_x | m_x' \rangle| = |m_x \hbar\ \delta_{m_x m_x'}| = \begin{vmatrix} 1/2\hbar & 0 \\ 0 & -1/2\hbar \end{vmatrix} \tag{8.31}$$

in the representation of the eigenstates of J_x, multiplied on either side by elements of successive transformation matrices.[10]

---

10. In fact, each of the various expressions of $\langle J_x \rangle_{\hat{Q}(t)1/2}$ in this section may be regarded as the element $m = m' = 1/2$ of the matrix $|\langle \hat{Q}(t)\ m | j_x | \hat{Q}(t)\ m' \rangle|$, this matrix being factored in various forms. The factoring for the different expressions is

(8.26a) : $|\langle \hat{Q}\ m | \hat{x}\ m_x \rangle| \times |m_x \hbar \delta_{m_x m_x'}| \times |\langle \hat{x}\ m_x' | \hat{Q}\ m' \rangle|$;

CHAPTER 8 PROBLEMS

8.1 Construct the complete matrix $|\langle \hat{A}\ m' | \hat{Q}(t)\ m\rangle|$ by working out explicitly the matrix product in equation (8.21a), for the case where $\hat{P}$ and $\hat{A}$ are orthogonal to $\hat{B}$, i.e, for $\theta_Q = \theta_A = \frac{1}{2}\pi$ and for $\phi_A = \pi$ as in Figure 8.1. Set in matrix (8.19) $E_{1/2}/\hbar = -\frac{1}{2}\omega_L$ and $E_{-1/2}/\hbar = \frac{1}{2}\omega_L$. Verify that the variation in time of the resulting probability $P_{1/2\,1/2} = |\langle \hat{A}\ \frac{1}{2} | \hat{Q}(t)\ \frac{1}{2}\rangle|^2$ reproduces the Larmor precession.

8.2. For particles with $j = \frac{1}{2}$ the squared magnetic moment component μ_x has the single eigenvalue μ^2 (see p. 150); similarly, $J_x{}^2$ has the single eigenvalue $(\frac{1}{2}\hbar)^2$. Verify this fact by constructing the matrix of $J_x{}^2$ as the square of the matrix of J_x given by equation (8.29a).

SOLUTIONS TO CHAPTER 8 PROBLEMS

8.1. For the given conditions we have

$$|\langle \hat{A}\ m' | \hat{B}\ \bar{m}\rangle| \times |\langle \hat{B}\ \bar{m} | \hat{Q}(t)\ m\rangle|$$

$$= \begin{vmatrix} \sqrt{\tfrac{1}{2}}\,i & -\sqrt{\tfrac{1}{2}}\,i \\ -\sqrt{\tfrac{1}{2}}\,i & -\sqrt{\tfrac{1}{2}}\,i \end{vmatrix} \times \begin{vmatrix} \exp(i\tfrac{1}{2}\omega_L t)\sqrt{\tfrac{1}{2}} & -\exp(i\tfrac{1}{2}\omega_L t)\sqrt{\tfrac{1}{2}} \\ \exp(-i\tfrac{1}{2}\omega_L t)\sqrt{\tfrac{1}{2}} & \exp(-i\tfrac{1}{2}\omega_L t)\sqrt{\tfrac{1}{2}} \end{vmatrix}$$

$$= \tfrac{1}{2}i \begin{vmatrix} 1 & -1 \\ -1 & -1 \end{vmatrix} \times \begin{vmatrix} \exp(i\tfrac{1}{2}\omega_L t) & -\exp(i\tfrac{1}{2}\omega_L t) \\ \exp(-i\tfrac{1}{2}\omega_L t) & \exp(-i\tfrac{1}{2}\omega_L t) \end{vmatrix}$$

$$= \begin{vmatrix} -\sin \tfrac{1}{2}\omega_L t & -i\cos \tfrac{1}{2}\omega_L t \\ -i\cos \tfrac{1}{2}\omega_L t & -\sin \tfrac{1}{2}\omega_L t \end{vmatrix}.$$

$P_{1/2\,1/2} = \sin^2 \frac{1}{2}\omega_L t$ equals zero at $t = 0$ when $\hat{Q} \equiv \hat{P}$ and $\hat{A}$ are parallel but in opposite directions; it oscillates for $t > 0$; and it equals 1 for $\omega_L t = \pi$ after one-half precession cycle, when $\hat{Q}$ coincides with $\hat{A}$.

(8.27a) : $|\langle \hat{Q}(t)\ m | \hat{z}\ m_z\rangle| \times |\langle \hat{z}\ m_z | \hat{x}\ m_x\rangle| \times |m_x \delta_{m_x m_x'}| \times |\langle \hat{x}\ m_x' | \hat{z}\ m_z'\rangle|$
$\times |\langle \hat{z}\ m_z' | \hat{Q}\ m'\rangle|$;

(8.28a) : $|\langle \hat{Q}(t)\ m | \hat{z}\ m_z\rangle| \times |\langle \hat{z}\ m_z | J_x | \hat{z}\ m_z'\rangle| \times |\langle \hat{z}\ m_z' | \hat{Q}(t)\ m'\rangle|$;

(8.30) : $|\langle \hat{Q}(0)\ m | \hat{z}\ m_z\rangle| \times |\exp(-im_z\omega_L t)\,\delta_{m_z m_z'}| \times |\langle \hat{z}\ m_z' | J_x | \hat{z}\ m_z''\rangle|$

$\times |\exp(im_z''\omega_L t)\,\delta_{m_z'' m_z'''}| \times |\langle \hat{z}\ m_z''' | \hat{Q}(0)\ m'\rangle|$

8.2. $|\langle \hat{z}\ m_z | J_x^2 | \hat{z}\ m_z' \rangle| = |\langle \hat{z}\ m_z | J_x | \hat{z}\ m_z'' \rangle| \times |\langle \hat{z}\ m_z'' | J_x | \hat{z}\ m_z' \rangle|$

$$= \begin{vmatrix} 0 & \frac{1}{2}\hbar \\ \frac{1}{2}\hbar & 0 \end{vmatrix} \times \begin{vmatrix} 0 & \frac{1}{2}\hbar \\ \frac{1}{2}\hbar & 0 \end{vmatrix} = \begin{vmatrix} \frac{1}{4}\hbar^2 & 0 \\ 0 & \frac{1}{4}\hbar^2 \end{vmatrix} = \frac{1}{4}\hbar^2 \begin{vmatrix} 1 & 0 \\ 0 & 1 \end{vmatrix} .$$

9. two-level systems. interference and superposition

It often happens that a pair of orthogonal states of a physical system may be regarded as a complete set, because experimental circumstances make all other states irrelevant in practice. Such circumstances have been assumed in the last chapter for states of an atom with $j = \frac{1}{2}$, in that the experiments dealt only with the magnetic energy of atoms, disregarding other types of excitation. Again, when dealing with light polarization in chapter 6, we considered only pairs of orthogonal states of polarization and disregarded the frequency of the light, which need not be relevant. The name "two-level system" has come into use particularly in connection with the study of atoms which perform rapid transitions between two levels of internal energy under the influence of an intense beam of light at resonance. Other energy levels need not be considered because the light frequency fails to meet the requirements for their excitation.

The behavior of all two-level systems is described by the very same mathematical formulation developed for atoms with $j = \frac{1}{2}$. While for these atoms $\hat{u}$ and $\vec{\omega}(t)$ in equations (8.1) and (8.2) and the vectors $\hat{Q}$, $\hat{A}$, and $\hat{B}$ represent vectors of physical space, the corresponding quantities in other problems represent vectors in a model mathematical space. The modeling is described in section 9.1 with particular reference to light polarization. The following sections develop qualitative aspects of the quantum mechanics of two-level systems which deserve particular attention.

9.1. THE MODEL SPACE

The model space is constructed by establishing a one-to-one correspondence between states of the system under consideration and states of an atom with $j = \frac{1}{2}$. For example, in the case of light polarization one may take as a base set two orthogonal states of linear polarization with electric field directions $\hat{E}_1$ and $\hat{E}_2$, as in section 6.2. These two states then correspond to the pair of atomic states $|\vec{z}\ \pm\frac{1}{2}\rangle$. To represent the pair of polarization eigenstates of an arbitrary analyzer $\hat{A}$, one may apply to the pair $(\hat{E}_1, \hat{E}_2)$ the transformation matrix (8.20). This yields the pair of electric field vectors

$$\hat{A}_1 = \cos \tfrac{1}{2}\theta_A \exp(i\tfrac{1}{2}\phi_A)\hat{E}_1 + \sin \tfrac{1}{2}\theta_A \exp(-i\tfrac{1}{2}\phi_A)\hat{E}_2, \qquad (9.1)$$

$$\hat{A}_2 = -\sin \tfrac{1}{2}\theta_A \exp(i\tfrac{1}{2}\phi_A)\hat{E}_1 + \cos \tfrac{1}{2}\theta_A \exp(-i\tfrac{1}{2}\phi_A)\hat{E}_2.$$

For $\phi_A = 0$ this formula represents linear polarizations rotated by an angle $\frac{1}{2}\theta_A$ from the direction of $\hat{E}_1$ toward $\hat{E}_2$. For $\phi_A \neq 0$ the polarization is, in general, elliptical; it is right-circular for $\frac{1}{2}\theta_A = 45°$ and $\phi_A = 90°$ and left-circular for $\frac{1}{2}\theta_A = 45°$ and $\phi_A = 270°$.[1] Equation (9.1) thus establishes a

1. Note that in equation (9.1) the phase difference of the coefficients of $\hat{E}_2$ and $\hat{E}_1$ is ϕ_A, *not* $\frac{1}{2}\phi_A$

mapping of all light polarizations on the pair of angles (θ_A, ϕ_A) which identify a direction of physical space.

An analogous mapping can be developed for each two-level system. In the example of the excited and ground states of an atom, these energy eigenstates are usually made to correspond, again, to the states $|\hat{z} \ \pm\frac{1}{2}\rangle$. Each state of the system corresponds to a particular "model direction" (θ_A, ϕ_A). The physical characteristics of each state can be derived from those of the two base states, for instance, of their electron density distribution. Examples of electron density distributions in states of a two-level system will be given in section 11.3.

The evolution of a state of any two-level system is also represented, as in chapter 8, by a rotation of the vector $\hat{u}$ defined by the angles (θ_A, ϕ_A). For example, the polarization of light propagating through a $\lambda/4$ plate with axes parallel to $\hat{E}_1$ and $\hat{E}_2$ is represented by a precession of $\hat{u}$ about $\hat{z}$, that is, by a variation of ϕ_A at constant θ_A; ϕ_A is increased by 90° at emergence from the $\lambda/4$ plate. In the example of the two-level atom which does not interact with radiation, a nonstationary state can be resolved into two component energy eigenstates $|E \ 0\rangle$ and $|E \ 1\rangle$ with probabilities $\cos^2 \frac{1}{2}\theta_A$ and $\sin^2 \frac{1}{2}\theta_A$. For this state the vector $\hat{u}$ precesses, again, about $\hat{z}$ with frequency $(E_1 - E_0)/h$. In this type of problem the geometrical interpretation affords a convenient mathematical model of a phenomenon, but it has no physical meaning.

Note that for an atom with $j = \frac{1}{2}$ all degenerate states are equivalent because each state is identified by one direction of physical space — $|\hat{z} \ -\frac{1}{2}\rangle$ coincides with $|-\hat{z} \ \frac{1}{2}\rangle$ — and the physical space is isotropic in the absence of external fields. On the other hand, in a model space different directions may represent states with different characteristics. In particular, circular and linear polarization are qualitatively different. There are but two states of circular polarization, each of them with axial symmetry about the direction of light propagation; the two are distinguished by opposite directions of rotation of the electric field vectors and they are represented in the model space by two vectors (θ_A, ϕ_A) opposite to each other, that is, $\theta_A = 90°$, and $\phi_A = 90°$ or $270°$. Conversely, for linear polarization there are infinitely many equivalent pairs of orthogonal states; each state is identified by a direction of physical space perpendicular to the direction of light propagation and in the model space by a direction (θ_A, ϕ_A) perpendicular to those that identify the circular polarization, that is, by any direction on a meridian circle with $\phi_A = 0°$ or $180°$.

9.2. INTERFERENCE AND PHASE DIFFERENCES

We return now to the discussion of the Larmor precession formula (8.9) and particularly to the comparison of that formula with the results of the "variant" two-path experiment described by equation (8.10). As emphasized there, the cross-term in equation (8.9) exists because the two paths of atoms from polarizer to analyzer "through" the two stationary states in $\vec{B}$ do not constitute mutually exclusive alternatives. Owing to these terms, the observed probability $P(\hat{A} \ \frac{1}{2}, \hat{Q} \ \frac{1}{2})$ departs from the expectation based on the unjustified assumption of mutually exclusive channels. The departure is called an *interference effect* in analogy to macroscopic wave theories. Similar interference effects occur not only for all two-level systems, but throughout quantum physics. They are easily studied in optics where their interpretation by macroscopic theory is familiar and where their probability aspects emerge when the intensities are measured by photon counting.

The experiments are particularly striking when two paths of transmission are separated in space. This separation, shown schematically by equation (8.10) for particles, may be actually performed for light beams—for example, by means of a Jamin interferometer (Fig. 9.1). This instrument is designed to separate in space an ordinary and an extraordinary ray and then to recombine them without disturbing their phase relation. At the same time the instrument permits the insertion of a screen (or of a phase shifter) on the path of either ray so as to alter its contribution to the recombined emerging beam.

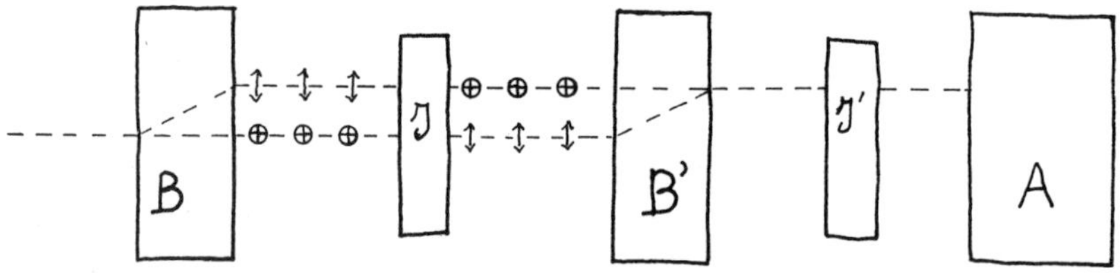

Fig. 9. 1. Light beam separation and recombination by a Jamin interferometer. The iceland spar crystal B splits incident light into ordinary and extraordinary rays which travel separately and are recombined by a second crystal B′. The inverter crystal 𝒥 makes the two path lengths equal by rotating each polarization by 90°, so that each beam component traverses one spar as ordinary ray and the other one as extraordinary ray. The action of the inverter 𝒥 is compensated by the second inverter 𝒥′ between B′ and the final spar analyzer A.

Suppose that a light beam prefiltered by an unknown polarizer P traverses the interferometer whose final analyzer A is turned at an angle α with respect to B and B' and lets through only the ordinary ray. The transmitted intensity is measured by a macroscopic device or by a photon counter. Consider the following set of three measurements:

1) With the instrument as shown in Figure 9.1, the intensity I is the same as if B and B' were removed.

2) When the upper beam in Figure 9.1 is intercepted between B and B', the measured intensity is $I_o \cos^2 \alpha$, where I_o is the intensity of the ordinary ray transmitted by crystal B.

3) When the lower component is intercepted, the intensity is $I_e \sin^2 \alpha$, where I_e is the extraordinary ray of B.

Experiments and macroscopic theory show that

$$I = I_o \cos^2 \alpha + I_e \sin^2 \alpha + 2(I_o \cos^2 \alpha \, I_e \sin^2 \alpha)^{1/2} \cos \phi, \qquad (9.2)$$

where ϕ is the phase difference between the incident beam components which are parallel and perpendicular to the principal section of the crystal B.

Notice the correspondence between equations (9.2) and (8.9). The intensities I_o and I_e correspond to the probabilities $P_{\pm 1/2\ 1/2}(\theta_P)$; $\cos^2 \alpha$ and $\sin^2 \alpha$ correspond to the probabilities $P_{1/2 \pm 1/2}(\theta_A)$; and ϕ to $(\phi_A + \omega_L t)$.

The important point here is: On the one hand $\phi_A + \omega_L t$ is known from the analysis of the precession phenomenon to be an angle in physical space, and ϕ in the interferometer experiment is known from macroscopic theory to be a phase difference between two electric field components. On the other hand, $\cos \phi$ can be determined by the three intensity measurements I, $I_o \cos^2 \alpha$, and $I_e \sin^2 \alpha$ through equation (9.2) without any reference to preexisting theory. For any two-level system the phase difference can be defined and measured by three intensity measurements of the type indicated above. This holds in particular for the precession experiment where the phase difference $\phi_A + \omega_L t$ could be determined by measuring the probability (8.9) as well as the separate probabilities indicated in (8.10).

Interference effects can be observed for all quantum systems, whether two-level or multilevel. They are detected as differences between the observed probability of an event and the sum of probabilities attributable to separate contributing processes. For example, the occurrence of diffraction of particles by crystals hinges on the fact that the different atoms of the crystal contribute together to the phenomenon rather than singly.

9.2.1. *Incoherence.* It is well known in optics that interference effects do not normally occur with beam components originating from different sources. For instance, if the spar crystal B in Figure 9.1 were removed and the upper and lower beams were piped in with the polarization shown in the figure, but from different sources, the intensity measurement behind A would yield

$$I = I_o \cos^2 \alpha + I_e \sin^2 \alpha. \qquad (9.3)$$

Analogous results are obtained with other systems. When the combination of two beams yields no interference effect, that is, no departure from equation (9.3), the component beams are said to be *incoherent*.

Incoherence occurs when contributing processes are *mutually exclusive*. Whenever any feature of the experimental conditions makes it possible—even only in principle—to associate the final detection of a particle or photon with its passage through one specified channel rather than another, the cross-term in equations (8.9) and (9.2) disappears. This association can be established, for example, by inserting counters along alternative paths and recording passage through one counter in coincidence with the final detector. More generally, any condition which permits a particle or a photon to leave any trace whatsoever of its passage through a channel destroys interference.

The distinction between coherent and incoherent combinations is of great conceptual importance. Its correct application to the analysis of any specific phenomenon requires some care, because relevant circumstances often fail to be perceived or assessed accurately.

9.3 SUPERPOSITION

The concept of resolving the state of a system into component eigenstates of a given variable has been derived in chapter 6 from experimental procedures. We can now represent mathematically a state as a combination of the eigenstates of a base set. We take as a model the representation of the electric field vector of polarized light in terms of the fields of two component beams with orthogonal polarizations. We will utilize the language developed for atoms with $j = \frac{1}{2}$.

According to equation (9.1), the electric field vector $\hat{A}_1$ is a linear combination of the unit vectors of a base set $\hat{E}_1, \hat{E}_2$. The coefficients of the combination are the probability amplitudes which relate the state with polarization $\hat{A}_1$ to the two base states with polarization $\hat{E}_1$ and $\hat{E}_2$. We return now to atoms with $j = \frac{1}{2}$, whose states have been indicated by ket symbols. A state $|\hat{u}\ \frac{1}{2}\rangle$ of these atoms will now be represented as a combination of eigenstates $|\hat{z}\ \pm\frac{1}{2}\rangle$ by

$$|\hat{u}\ \tfrac{1}{2}\rangle = |\hat{z}\ \tfrac{1}{2}\rangle\langle\hat{z}\ \tfrac{1}{2}|\hat{u}\ \tfrac{1}{2}\rangle + |\hat{z}\ -\tfrac{1}{2}\rangle\langle\hat{z}\ -\tfrac{1}{2}|\hat{u}\ \tfrac{1}{2}\rangle$$
$$= \sum_{m=-1/2}^{1/2} |\hat{z}\ m\rangle\langle\hat{z}\ m|\hat{u}\ \tfrac{1}{2}\rangle. \qquad (9.4)$$

This formula should read: "The eigenstate $|\hat{u}\ \frac{1}{2}\rangle$ is a superposition of the two eigenstates $|\hat{z}\ \pm\frac{1}{2}\rangle$ of a set incompatible with the set $|\hat{u}\ \pm\frac{1}{2}\rangle$. The coefficients of the superposition are the probability amplitudes $\langle\hat{z}\ m|\hat{u}\ \frac{1}{2}\rangle$."

An explicit expression of equation (9.4) is obtained by utilizing the probability amplitudes (8.19) for a two-level system. The amplitudes (8.19) are functions of the polar coordinate angles θ_Q and $-2E_{1/2}t/\hbar$ of the vector $\hat{Q}(t)$ which identifies the state $|\hat{Q}(t)\ \frac{1}{2}\rangle$; to represent $|\hat{u}\ \frac{1}{2}\rangle$, we replace these angles by the angles θ and ϕ which identify $\hat{u}$ in polar coordinates with axis $\hat{z}$. Thus we write

$$|\hat{u}\ \tfrac{1}{2}\rangle = |\hat{z}\ \tfrac{1}{2}\rangle \cos\tfrac{1}{2}\theta\, \exp(i\tfrac{1}{2}\phi) + |\hat{z}\ -\tfrac{1}{2}\rangle \sin\tfrac{1}{2}\theta\, \exp(-i\tfrac{1}{2}\phi). \qquad (9.5)$$

As in the vector formula (9.1), the sum of the squared magnitudes of the coefficients $|\hat{z}\ \frac{1}{2}\rangle$ and $|\hat{z}\ -\frac{1}{2}\rangle$ equals unity. The squared magnitude of each coefficient represents the fraction of $|\hat{u}\ \frac{1}{2}\rangle$ which consists of $|\hat{z}\ \frac{1}{2}\rangle$ or of $|\hat{z}\ -\frac{1}{2}\rangle$, respectively.

The correspondence between sets of eigenstates and sets of coordinate axes in geometry (see end of section 8.5.1) serves also to illustrate the structure of equations (9.4) and (9.5). These equations correspond to the representation of a space vector as a superposition of unit vectors directed along a set of coordinate axes. For this reason any ket symbol representing a quantum-mechanical state is often called a *state vector*.

Equation (9.4) has a counterpart in a basic relationship among probability amplitudes. This relationship is obtained by replacing each ket symbol of a state in equation (9.4) by the probability amplitude which connects that state, for instance, to the state $|\hat{A}\ \tfrac{1}{2}\rangle$. The replacement yields

$$\langle \hat{A}\ \tfrac{1}{2}|\hat{u}\ \tfrac{1}{2}\rangle = \sum_{m=-1/2}^{1/2} \langle \hat{A}\ \tfrac{1}{2}|\hat{z}\ m\rangle\langle \hat{z}\ m|\hat{u}\ \tfrac{1}{2}\rangle, \tag{9.6}$$

that is, the relationship (8.21) among probability amplitudes. Conversely, one may regard the representation (9.4) of the state $|\hat{u}\ \tfrac{1}{2}\rangle$ as constructed, in essence, by working backward from equation (9.6).

In other words, the representation of the state $|\hat{u}\ \tfrac{1}{2}\rangle$ as a superposition of states, $|\hat{z}\ \tfrac{1}{2}\rangle$ and $|\hat{z}\ -\tfrac{1}{2}\rangle$, is an alternative formulation of the concept of representing the probability connection between two states, such as $|\hat{u}\ \tfrac{1}{2}\rangle$ and $|\hat{A}\ \tfrac{1}{2}\rangle$, by a sum of two terms. One term represents the probability connection between $|\hat{u}\ \tfrac{1}{2}\rangle$ and $|\hat{A}\ \tfrac{1}{2}\rangle$ via $|\hat{z}\ \tfrac{1}{2}\rangle$; the other one, the connection via $|\hat{z}\ -\tfrac{1}{2}\rangle$. The essential step lies in the linearization process. The linearization occurs when one replaces the actual probability formula of the type (8.9), namely,

$$\begin{aligned} P(\hat{A}\ \tfrac{1}{2}, \hat{u}\ \tfrac{1}{2}) &= P(\hat{A}\ \tfrac{1}{2}, \hat{z}\ \tfrac{1}{2})P(\hat{z}\ \tfrac{1}{2}, \hat{u}\ \tfrac{1}{2}) \\ &+ P(\hat{A}\ \tfrac{1}{2}, \hat{z}\ -\tfrac{1}{2})P(\hat{z}\ -\tfrac{1}{2}, \hat{u}\ \tfrac{1}{2}) + 2[P(\hat{A}\ \tfrac{1}{2}, \hat{z}\ \tfrac{1}{2})P(\hat{z}\ \tfrac{1}{2}, \hat{u}\ \tfrac{1}{2}) \\ &\times P(\hat{A}\ \tfrac{1}{2}, \hat{z}\ -\tfrac{1}{2})P(\hat{z}\ -\tfrac{1}{2}, \hat{u}\ \tfrac{1}{2})]^{1/2} \cos\phi, \end{aligned} \tag{9.7}$$

by the equality of the squared magnitudes of the two sides of equation (9.6). This replacement expresses probabilities, nonnegative quantities which are not represented by sums of contributions of separate channels, as the squares of quantities represented by just such sums. Quantum physics can thus benefit from the simplifications which are inherent in linear representations.

Quantum-mechanical states can also be represented by symbols with the parentheses on their left side. An equation analogous to (9.4) and similarly related to (9.6) is

$$\langle \hat{A}\ \tfrac{1}{2}| = \sum_m \langle \hat{A}\ \tfrac{1}{2}|\hat{z}\ m\rangle\langle \hat{z}\ m|. \tag{9.8}$$

The symbol used in (9.6), $|\hat{u}\ \tfrac{1}{2}\rangle$, is employed conventionally to indicate states on the basis of their initial preparation by a "polarizer" instrument; the second type, $\langle \hat{A}\ \tfrac{1}{2}|$, to identify states on the basis of their future selection by an "analyzer" instrument. Hence the symbols are chosen to complement each other. The probability amplitude symbol $\langle \hat{A}\ \tfrac{1}{2}|\hat{u}\ \tfrac{1}{2}\rangle$ accordingly pertains to the process of analyzer selection of $\langle \hat{A}\ \tfrac{1}{2}|$ from a state prepared as indicated by $|\hat{u}\ \tfrac{1}{2}\rangle$. Because the symbol $\langle \hat{A}\ \tfrac{1}{2}|\hat{u}\ \tfrac{1}{2}\rangle$ may be called a *bracket*, the author of this system of notation (Dirac) has called $\langle \hat{A}\ \tfrac{1}{2}|$ a *bra* and $|\hat{u}\ \tfrac{1}{2}\rangle$ a *ket*. Let us repeat that (9.8) and (9.4) are symbolic equations, whereas (9.6) is a quantitative relation between (complex) numbers. Note, specifically, that the unitarity relation between numbers, (8.22), does *not* imply that the symbol

$\langle \hat{z}\ \tfrac{1}{2}|$ should be regarded as the complex conjugate of $|\hat{z}\ \tfrac{1}{2}\rangle$; indeed, a bra or ket symbol, not being a number, does not possess a complex conjugate.

The representation of a state as a superposition of eigenstates of a base set does more than formalize the concept of analysis into components. Experimental analysis into eigenstates of the base set relates only to the probabilities, that is, to the magnitudes of the probability amplitudes. The phases of the probability amplitudes provide information necessary to predict the results of experimental analysis into different sets of eigenstates. As noted in section 6.4, actual experimental analysis into component eigenstates changes the physical conditions of the system under study and thus changes its state; it normally wipes out the phase relations of the probability amplitudes of component states.

CHAPTER 9 PROBLEMS

9.1. The analysis of light into polarization eigenstates can be represented in terms of probability amplitudes $\langle \hat{B}\ \overline{m}|\hat{P}\ m\rangle$, where, for linear polarization, the vector $\hat{B}$ is perpendicular to the principal section of a spar analyzer and $\hat{P}$ is perpendicular to that of another analyzer. The index m (or $\overline{m}$) may be taken as 1 or 2 for the ordinary or extraordinary ray of each analyzer. With suitable conventions we may take

$$\|\langle \hat{B}\ \overline{m}|\hat{P}\ m\rangle\| = \begin{vmatrix} \cos\beta & \sin\beta \\ -\sin\beta & \cos\beta \end{vmatrix}$$

where β is the angle from $\hat{P}$ to $\hat{B}$. Verify that equation (8.21a) holds for the product of this matrix and of an analogous matrix $\|\langle \hat{A}\ m'|\hat{B}\ \overline{m}\rangle\|$.

9.2. Consider light of 6000 Å wavelength prepared as ordinary ray of a polarizer P and incident on the Jamin interferometer shown in Figure 9.1. On the path of the extraordinary ray after the crystal B the refractive index of air is increased by 2 parts in 100,000 over a distance of 0.5 cm, with a resultant phase retardation of $2 \times 10^{-5} \times 0.5\ \text{cm}/6 \times 10^{-5}\ \text{cm} = 1/6$ of a wavelength $= \tfrac{1}{3}\pi$ radians. Calculate the fraction f of the incident intensity which is transmitted as an ordinary ray of the analyzer A, as a function of the angles between $\hat{P}$, $\hat{B}$, and $\hat{A}$. (Apply the method of problem 9.1.)

SOLUTIONS TO CHAPTER 9 PROBLEMS

9.1. Calling α the angle from $\hat{B}$ to $\hat{A}$, we have

$$\|(\hat{A}\ m'|\hat{P}\ m)\| = \begin{vmatrix} \cos\alpha & \sin\alpha \\ -\sin\alpha & \cos\alpha \end{vmatrix} \times \begin{vmatrix} \cos\beta & \sin\beta \\ -\sin\beta & \cos\beta \end{vmatrix}$$

$$= \begin{vmatrix} \cos(\alpha+\beta) & \sin(\alpha+\beta) \\ -\sin(\alpha+\beta) & \cos(\alpha+\beta) \end{vmatrix}.$$

9.2. The phase shift introduced in the interferometer may be represented by the matrix $\begin{vmatrix} 1 & 0 \\ 0 & e^{i\pi/3} \end{vmatrix}$, to be inserted between the matrices of problem 9.1. Therefore $f = |\langle \hat{A}\ 1|\hat{P}\ 1\rangle|^2 = |\cos\alpha\cos\beta - \sin\alpha\, e^{i\pi/3}\sin\beta|^2$
$= \cos^2(\alpha + \beta) + \frac{1}{4}\sin 2\alpha \sin 2\beta$.

10. multilevel systems

Ideas and formulae for the treatment of atomic problems have been presented in chapters 8 and 9 with reference to two-level systems. This chapter indicates how they are generalized to systems with an arbitrary number of orthogonal states. The extension will not be carried out in a deductive way because experiments to verify each step of the generalization have not, in fact, been performed. Historically, quantum mechanics developed through a number of enlightened guesses and inductive leaps. That its formalism is appropriate for the description of atomic phenomena is verified by a large number of successful and consistent applications.

10.1. PROBABILITY AMPLITUDES AND PROBABILITIES

Given a complete base set of N eigenstates $|A \ \ n\rangle$ of a variable A, an eigenstate $|A' \ \ n'\rangle$ of a variable A′ incompatible with A is identified by the set of probability amplitudes $\langle A \ \ n | A' \ \ n'\rangle$ with $n = 1, 2, 3, \ldots, N$, and n′ constant. The eigenstate indicated by $|A' \ \ n'\rangle$ may also be represented by the superposition formula analogous to equation (9.4):

$$|A' \ \ n'\rangle = \sum_{n=1}^{N} |A \ \ n\rangle\langle A \ \ n | A' \ \ n'\rangle. \tag{10.1}$$

Very often it is neither practical nor actually desirable to identify a state as an eigenstate of a specific physical quantity; in other words, one does not identify an experimental device which would select atoms in that state. In such cases a state, which may be indicated by the ket symbol $|a\rangle$, is identified operationally through its analysis into component eigenstates of two or more incompatible variables A and A′. That is, the state is identified by the set of probabilities $P(A \ \ n, a)$ and $P(A' \ \ n', a)$ or by equivalent measurable quantities such as mean values.

Analytically, the state $|a\rangle$ is identified through the formula (10.1), that is, by giving the set of complex numbers

$$a_n = \langle A \ \ n | a\rangle, \tag{10.2}$$

which constitute the coefficients of the superposition[1]

$$|a\rangle = \sum_{n=1}^{N} |A \ \ n\rangle \, a_n. \tag{10.2a}$$

---

1. As an example of states represented by (10.2a) one may mention those that result from the superposition of eigenstates of a Stern-Gerlach magnet operating on particles with $j > 1/2$, that is, states $|a\rangle = \sum_m |\hat{B} \ \ m\rangle a_m$. In general, these states are not eigenstates of any angular momentum component.

The set of numbers a_n is identified in turn by its relations to measurable quantities such as

$$P(A\ \ n,\ a) = |a_n|^2, \tag{10.3a}$$

$$P(A'\ \ n',\ a) = |a_{n'}|^2 = |\textstyle\sum_n \langle A'\ n'|A\ \ n\rangle\, a_n|^2. \tag{10.3b}$$

For a system with a known complete set of eigenstates $|A\ \ n\rangle$, the whole gamut of possible states $|a\rangle$ is represented by letting the set of values of the coefficients a_n range arbitrarily subject only to the *normalization condition*

$$\textstyle\sum_n |a_n|^2 = 1. \tag{10.4}$$

This condition states that, upon experimental analysis of $|a\rangle$ into component eigenstates $|A\ \ n\rangle$, the total probability of all eigenvalues of A equals unity.

Returning now to eigenstates $|A'\ \ n'\rangle$ of a specified physical quantity A′, the complete set of probability amplitudes $\langle A\ \ n|A'\ \ n'\rangle$, with n and n′ varying from 1 to N, constitutes a transformation matrix, as indicated in section 8.5 for the case of N = 2. Usual bookkeeping requirements on the phases of probability amplitudes make the matrix unitary, for the reasons explained in section 8.5.2. This property is indicated by

$$\langle A'\ \ n'|A\ \ n\rangle = \langle A\ \ n|A'\ \ n'\rangle^*, \tag{10.5}$$

which is a generalization of equation (8.22). Here, the matrix of probability amplitudes $\|\langle A'\ \ n'|A\ \ n\rangle\|$ is the inverse of the transformation matrix $\|\langle A\ \ n|A'\ \ n'\rangle\|$ and thus has the property

$$\textstyle\sum_n \langle A'\ \ \alpha|A\ \ n\rangle\langle A\ \ n|A'\ \ \beta\rangle = \delta_{\alpha\beta} \tag{10.6}$$

where $\delta_{\alpha\beta}$ is the Kronecker delta and α and β are two values of n′.

Transformation matrices yield the solution of the problem outlined in section 8.1, that is, of representing the probability relation $P(A''\ \ n'',\ A'\ \ n')$ between any two eigenstates $|A'\ \ n'\rangle$ and $|A''\ \ n''\rangle$ of incompatible variables. This probability was to be expressed in terms of the connection between each of these states and the eigenstates $|A\ \ n\rangle$ of a base set. The problem is solved, in terms of probability amplitudes, by the matrix product equation

$$\|\langle A''\ \ n''|A'\ \ n'\rangle\| = \|\langle A''\ \ n''|A\ \ n\rangle\| \times \|\langle A\ \ n|A'\ \ n'\rangle\|, \tag{10.7}$$

where the elements of the product matrix are given by

$$\langle A''\ \ n''|A'\ \ n'\rangle = \sum_{n=1}^{N} \langle A''\ \ n''|A\ \ n\rangle\langle A\ \ n|A'\ \ n'\rangle. \tag{10.7a}$$

To pass from the probability amplitude (10.7a) to an expression of the probability itself, we write each probability amplitude in terms of its magnitude and phase, namely, for instance,

$$\langle A\ n | A'\ n' \rangle = |\langle A\ n | A'\ n' \rangle| \exp(i\phi_{nn'}), \tag{10.8}$$

$$P(A\ n,\ A'\ n') = |\langle A\ n | A'\ n' \rangle|^2. \tag{10.9}$$

The squared magnitude of (10.7a) is then the generalization of equation (8.9):

$$P(A''\ n'',\ A'\ n') = \sum_{n=1}^{N} P(A''\ n'',\ A\ n) P(A\ n,\ A'\ n')$$
$$+ 2 \sum_{n=1}^{N} \sum_{m<n} [P(A''\ n'',\ A\ m) P(A\ m,\ A'\ n') P(A''\ n'',\ A\ n)$$
$$\times P(A\ n,\ A'\ n')]^{1/2} \cos[(\phi_{n''m} + \phi_{mn'}) - (\phi_{n''n} + \phi_{nn'})]. \tag{10.10}$$

The two indices m and n, corresponding to different values of the same parameter n, serve to designate the two-dimensional array of cross-terms.

As in the case of a two-level system, discussed after equation (8.9) and again in section 9.2, the cross-terms represent interference effects. We have here one interference term for each pair of base-set states with different n, that is, for each pair $[|A\ m\rangle, |A\ n\rangle]$. Equation (10.10) may also be expressed as a squared vector sum analogous to that of (8.11), namely,

$$P(A''\ n'',\ A'\ n') = |\sum_{n=1}^{N} \vec{V}_n|^2, \tag{10.11}$$

where

$$|\vec{V}_n|^2 = P(A''\ n'', A\ n) P(A\ n, A'\ n'). \tag{10.11a}$$

An important feature of equation (10.10) is that the phase differences in the two-dimensional array of cross-terms are not independent of one another but are actually differences between pairs of phases $(\phi_{n''m} + \phi_{mn'})$ and $(\phi_{n''n} + \phi_{nn'})$; these pairs form a one-dimensional set. This fact causes the vectors $\vec{V}_n$ to be coplanar, so that the probability $P(A''\ n'',\ A'\ n')$ and the probability products (10.11a) are related to one another as the squares of the sides of the polygon in figure 10.1.

10.1.1. *Quantities with Continuous Sets of Eigenvalues.* Many quantities, such as position coordinates, momentum components, and energies, may have a continuous range of eigenvalues. If the variable A whose eigenstates $|A\ n\rangle$ are chosen as a base set belongs to this class, the discrete set of probability amplitudes $\langle A\ n | A'\ n' \rangle$ for a single value of n′ turns into a function of the continuous variable n.

If both variables A and A′ have continuous ranges of eigenvalues, each probability amplitude becomes a function of the two continuous variables n and n'. The complete function of these two variables for all values of n and n' is, nevertheless, still called a transformation matrix. Quantum mechanics extends to continuous functions the language used for discrete sets of eigenstates and dis-

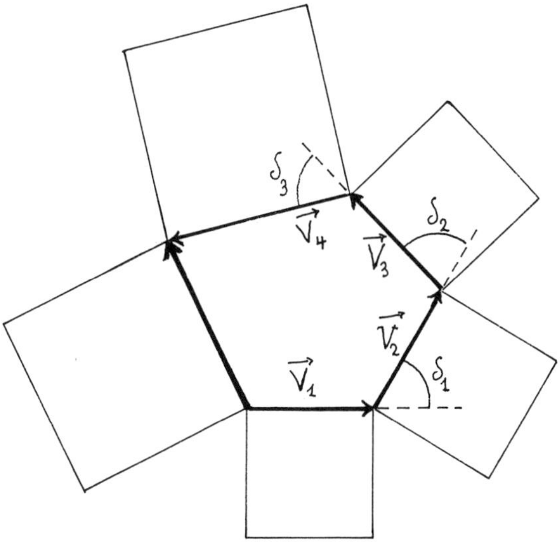

Fig. 10. 1. Diagram illustrating the determination of probability according to equations (10. 10) and (10. 11), for N = 4. The angles between vectors are $\delta_n = (\phi_{n'' \, n+1} + \phi_{n+1 \, n'}) - (\phi_{n''n} + \phi_{nn'})$.

crete arrays of matrix elements, thus departing from the usual practice of matrix algebra.[2]

All summations over an index n are then replaced by integrals. In this book we shall occasionally use the summation symbol with the understanding that it means integration whenever appropriate. In the formulation (10. 6) of the unitarity property of discrete transformation matrices, the Kronecker symbol $\delta_{\alpha\beta}$ represents the unit matrix. If the variable A′ has a continuous range of eigenvalues, the unit matrix is represented by the Dirac delta function introduced in section A. 3. 5. The condition (10. 6) then becomes

$$\sum_n \langle A' \ \alpha | A \ n \rangle \langle A \ n | A' \ \beta \rangle = \delta(\alpha - \beta). \tag{10.12}$$

This equation is equivalent to equation (A. 46a).

The position coordinates (x, y, z) of a particle are an important example of variables with continuous eigenvalues. Probability amplitudes connecting any other variable A to position coordinates, $\langle x, y, z | A \ n \rangle$, are functions of (x, y, z). These functions have often oscillatory behavior, as illustrated in the following

---

2. This generalization of the language of vector and matrix algebra to functions of continuous variables was introduced by Hilbert long before the development of quantum mechanics.

chapters. The probability amplitudes of the type $\langle x, y, z | A \quad n\rangle$ are accordingly called *wave functions* and are often indicated by $\psi_n(x, y, z)$.

10.2. TIME VARIATIONS IN ATOMIC SYSTEMS

The variations of nonstationary states of an atomic system in the course of time are described by a generalization of the formulae which describe the Larmor precession of a two-level system in chapter 8. The probability amplitudes that connect a nonstationary state $|\hat{Q}(t) \quad \tfrac{1}{2}\rangle$ to each of two stationary states are shown in equation (8.17a) to depend on time through an exponential phase factor. For a multilevel system the time variations of a nonstationary state $|a(t)\rangle$ are described by representing this state as a *superposition of stationary states*, that is, of energy eigenstates,

$$|a(t)\rangle = \sum_n |E \quad n\rangle\langle E \quad n|a(t)\rangle. \qquad (10.13)$$

The time dependence of the probability amplitudes themselves is expressed by the law, of which equations (8.17) are special cases,

$$\langle E \quad n|a(t)\rangle = \exp(-iE_n t/\hbar)\langle E \quad n|a(0)\rangle. \qquad (10.14)$$

For the calculation of mean values, following the pattern of section 8.5, one also represents the state by bra symbols,

$$\langle a(t)| = \sum_n \langle a(t)|A \quad n\rangle\langle A \quad n|. \qquad (10.13a)$$

In this case the probability amplitude $\langle a(t)|A \quad n\rangle$ is Hermitian conjugate to (10.14) and thus varies according to the complex conjugate law

$$\langle a(t)|A \quad n\rangle = \langle a(0)|A \quad n\rangle \exp(iE_n t/\hbar). \qquad (10.14a)$$

Sets of energy eigenstates serve conveniently as base sets throughout atomic physics just because the probability amplitudes which identify other states have then the simple time dependence (10.14). The probabilities $|\langle E \quad n|a(t)\rangle|^2$ must be constant in time to ensure that the mean value of the energy and of all its powers are constant for any state of an isolated system. Therefore, only the phase of the probability amplitude can vary in the course of time. That this variation is linear, with coefficient proportional to E_n, is a fundamental *experimental* law of atomic physics. Recall that in the case of atoms with $j = \tfrac{1}{2}$, the probability amplitudes (8.17) refer to the variations of a state characterized by a vector $\hat{Q}(t)$ which precesses in physical space about a fixed axis with a uniform rate. Equation (10.14) refers to the variations of a state $|a(t)\rangle$ which one might characterize by a vector precessing in a multidimensional model space.

The two equations (10.13) and (10.14) serve to express the law of time variation of any probability of interest and of the mean values of any physical quantity. Suppose, for example, that one wants to represent the time variations of the probability of finding a particle at a point (x, y, z) when the particle is in a state $|a(t)\rangle$ identified by (10.13). One begins by writing the probability amplitude

$$\langle x, y, z|a(t)\rangle = \sum_n \langle x, y, z|E\ n\rangle \exp(-iE_n t/\hbar)\langle E\ n|a(0)\rangle. \qquad (10.15)$$

Here, $\langle x, y, z|E\ n\rangle$ is the probability amplitude (wave function) which connects the *n*th stationary state of a particle to its position; the wave function is time-independent because it pertains to a stationary state. Explicit calculation of wave functions for stationary states of atomic electrons is a main task of atomic physics.

Note that the dependence on time of $|a(t)\rangle$ is represented entirely by the phases of the probability amplitudes $\langle E\ n|a(t)\rangle$.[3] In particular, the probability amplitudes $\langle x, y, z|E\ n\rangle$ are taken to be time-independent even though a state initially identified as $|x, y, z\rangle$ would be nonstationary. The point here is that the representation of a state as a superposition of component eigenstates is meant to give a "snapshot" picture of the state at a particular instant of time. As mentioned in section 6.4, the eigenstates are characteristic of a specific analyzer and are independent of the time at which the analysis is performed.

The square magnitude of (10.15) gives the probability expression, of the type of (10.10),

$$\begin{aligned} P(x, y, z;\ a(t)) &= |\langle x, y, z|a(t)\rangle|^2 = \sum_n |\langle x, y, z|E\ n\rangle|^2 |\langle E\ n|a(0)\rangle|^2 \\ &+ 2\sum_n \sum_{m<n} |\langle x, y, z|E\ n\rangle|\, |\langle E\ n|a(0)\rangle|\, |\langle x, y, z|E\ m\rangle|\, |\langle E\ m|a(0)\rangle| \\ &\times \cos\phi_{mn}(t), \end{aligned} \qquad (10.16)$$

where

$$\phi_{mn}(t) = \frac{E_m - E_n}{\hbar}\, t + \phi_{mn}(0). \qquad (10.17)$$

Since the probability $P(x, y, z;\ a(t))$ depends upon time only through the factors $\cos\phi_{mn}(t)$, and since each phase difference is a linear function of time, equation (10.16) constitutes a Fourier expansion of the probability. The frequencies $(E_m - E_n)/\hbar$ in this expansion reproduce the experimental frequencies of spectroscopy given by equation (3.3a).

The probability expression in (10.16) is a special case of the expression (10.10) and can accordingly be illustrated by a polygonal diagram of the type of figure 10.1. In expression (10.16) the phases are explicit functions of time. The variations in time of the probability result then from the variation of each angle δ_n in the diagram at the constant rate $(E_{n+1} - E_n)/\hbar$.

---

3. An alternative convention is indicated in the footnote on p. 177.

10.2.1. *Mean Values and their Time Dependence.* It was shown in section 8.5 how the mean value of a physical variable can be expressed through the matrix of the variable in the representation of a base set. If the base set consists of stationary states, the time dependence of the mean value is obtained in the form of a Fourier expansion analogous to (10.16).

We consider here the mean value of the x coordinate of an atomic electron. This value is proportional to the x component of the mean electric dipole moment, and its variation in time causes emission or absorption of radiation. To obtain this mean value we start from an expression of the probability of electron position equivalent to (10.16) but of a form analogous to (8.23), namely,

$$P(x,\ a(t)) = \iint dydz\langle a(t)\,|\,x, y, z\rangle\langle x, y, z\,|\,a(t)\rangle = \iint dydz \sum_n \langle a(0)\,|\,E\ \ n\rangle$$

$$\times \exp(iE_n t/\hbar)\langle E\ \ n|\,x, y, z\rangle \sum_{n'} \langle x, y, z\,|\,E\ \ n'\rangle \exp(-iE_{n'}t/\hbar)\langle E\ \ n'\,|\,a(0)\rangle. \quad (10.18)$$

The integration over the position coordinates y and z is required because equation (10.18) represents the probability of the value x of one coordinate irrespective of the values of the other coordinates.

The mean value of the position coordinate x can now be expressed in a form analogous to the mean value $\langle J_x \rangle$ of an angular momentum component in equation (8.30), namely,

$$\langle x\rangle_{a(t)} = \sum_n \sum_{n'} \langle a(0)\,|\,E\ \ n\rangle \exp(iE_n t/\hbar)\langle E\ \ n\,|\,x\,|\,E\ \ n'\rangle \exp(-iE_{n'}t/\hbar)$$

$$\times \langle E\ \ n'\,|\,a(0)\rangle, \quad (10.19)$$

where

$$\langle E\ \ n\,|\,x\,|\,E\ \ n'\rangle = \iiint \langle E\ \ n\,|\,x, y, z\rangle x \langle x, y, z\,|\,E\ \ n'\rangle dxdydz. \quad (10.20)$$

The matrix elements of (10.20) along the diagonal, that is those with $n = n'$, generally vanish for symmetry reasons, much as the diagonal elements of J_x vanish in (8.29a). Note the structure of equation (10.19), which represents the Fourier expansion of $\langle x\rangle_{a(t)}$ into complex exponentials with the frequencies $(E_n - E_{n'})/\hbar$. Each coefficient of this expansion contains one factor, namely the matrix element (10.20), which depends on the coordinate x and on the pair of energy eigenstates (n, n') with the particular energy difference $E_n - E_{n'}$, but does *not* depend on the specific state of interest, $|a(t)\rangle$. The other two factors of the coefficient, namely, $\langle a(0)\,|\,E\ \ n\rangle$ and $\langle E\ \ n'\,|\,a(0)\rangle$, depend instead on the identification of the state of interest at the instant $t = 0$ in terms of $\langle E\ \ n|$ and $|E\ \ n'\rangle$ but do *not* depend on the variable x under consideration.

In the example of the two-level system considered in section 9.1, all probability amplitudes $\langle E\ \ n|\,a(0)\rangle$ vanish except those with $n = 0$ or $n = 1$. The double summation in expression (10.19) reduces then to four terms of which the two time-independent terms with $n = n'$ usually vanish, as noted above. The remaining two cross-terms are complex conjugate and yield a mean value which oscillates sinusoidally in the course of time with frequency $(E_1 - E_0)/\hbar$.

This oscillation proceeds at the same rate as the precession of the vector $\hat{u}$ that represents the state in the model space, much as $\langle J_x \rangle$ oscillates in the course of the Larmor precession studied in chapter 8.

SUPPLEMENTARY SECTION

S10.3. TRANSFORMATION MATRICES FOR STERN-GERLACH EIGENSTATES WITH $j > \frac{1}{2}$.

A task of atomic theory is to calculate the transformation matrices which relate eigenstates of different analyzers and to determine the corresponding eigenvalues. For the sets of eigenstates selected by Stern-Gerlach analyzers in different directions, the different analyzers are all identical except for orientation. Therefore, the transformation matrices in this case describe how the representation of eigenstates is affected by rotations of coordinate axes attached to analyzers. The determination of these matrices requires no further information on the physics of the system and is thus reduced to a well-defined mathematical problem. The solution of this problem for $j = \frac{1}{2}$ was obtained by physical analysis of Larmor precession in chapter 8. However, for any value of j the problem can be regarded as purely mathematical.

The basic input in the mathematical problem is the law (10.7) for multiplication of the two transformation matrices which correspond to two successive rotations of the analyzer system. One matrix may represent the rotation from the orientation of a polarizer $\hat{P}$ to that of a base field $\hat{B}$, and the second matrix the rotation from $\hat{B}$ to an analyzer $\hat{A}$. The product represents the direct rotation from $\hat{P}$ to $\hat{A}$. (A first example of this kind of matrix product was given in eq. [8.21a].) The transformation matrices must fulfill the unitarity requirements (10.5) and (10.6). These conditions set the transformation matrices in one-to-one correspondence with the operations of the group of coordinate rotations; the determination of the matrices is, therefore, a problem in group theory. The solution of this problem provides at the same time the theoretical determination of the eigenvalues of components of the angular momentum $\vec{J}$ and of its squared magnitude; these eigenvalues were derived in section 7.2 from experimental information. The complete solution of the mathematical problem is not worked out in this book, but some results are given below.

We indicate here the initial procedural step which splits the matrix for any change of coordinate orientation into three separate factors. In order to identify a system of polar coordinates attached to each Stern-Gerlach analyzer one must specify not only the direction $\hat{B}$ of its magnetic field but also a reference half-plane with zero-degree azimuth from which rotations about $\hat{B}$ are measured. In the example of section 8.5 the zero-azimuth half-plane attached to the base-set axis $\hat{B}$ was taken to contain the axis of the polarizer $\hat{P}$; the zero-azimuth half-plane of the coordinate systems attached to $\hat{P}$, $\hat{Q}$, and $\hat{A}$ was implied to contain the direction of $-\hat{B}$, so that $\hat{B}$ itself has azimuth π.

The operation which brings a coordinate system attached to $\hat{B}$ to coincide with a system attached to $\hat{A}$ can be described by a sequence of three steps:

1) a rotation by ϕ_A radians about $\hat{B}$ which brings the zero-azimuth or $\hat{B}$ to coincide with the plane $\hat{B}\hat{A}$;

2) a rotation by θ_A radians which brings $\hat{B}$ itself to coincide with $\hat{A}$ (this rotation leaves the zero-azimuth half-plane with an orientation that contains $-\hat{B}$);

3) A rotation by ψ_A radians about $\hat{A}$ which brings the zero-azimuth half-plane to coincide with the preassigned azimuth of $\hat{A}$.[4]

The rotations in steps 1 and 3 of the coordinate system about the axis of the analyzer do not change either the analyzer itself or its eigenstates. Therefore, they must have no influence on probabilities; yet they influence the phases of the probability amplitudes. The matrices representing these rotations are therefore diagonal because they yield no intermixing of the analyzer's eigenstates. Since the total transformation matrix consists of a nondiagonal matrix representing step 2 multiplied by two diagonal matrices representing steps 1 and 3, each element of the complete transformation matrix splits into three factors depending on ϕ, θ, and ψ, respectively. Since, for each matrix element, the factors depending on ϕ and ψ are complex phase factors whereas the factor depending on θ can be real, we can write each transformation matrix element in the form

$$\langle \hat{A}' \ m' | \hat{B} \ m \rangle = \exp[i f_{m'}(\psi_A)] |\langle \hat{A} \ m' | \hat{B} \ m \rangle| \exp[i f_m(\phi_A)]. \qquad (10.21)$$

The next task consists of determining the form of the function $f_m(\phi)$. Since successive rotations by angles ϕ_1 and ϕ_2 must yield the same transformation matrix as a single rotation by $(\phi_1 + \phi_2)$, the function $f_m(\phi)$ must be linear. Consider now that probabilities of the type of (8.9) or (10.10) constructed with the matrix elements (10.21) include interference factors $\cos[f_m(\phi) - f_{\bar{m}}(\phi)]$, where $\bar{m}$ indicates a different value of the index m. These probabilities must, of course, be periodic functions of ϕ with period 2π. Therefore, the linear functions $f_m(\phi)$ and $f_{\bar{m}}(\phi)$ must have coefficients differing by an integer. These coefficients turn out to be the indices m and $\bar{m}$ themselves. In fact, a theoretical analysis of this type establishes the possible values of m without any resort to experiments. The same chain of arguments applies to the function $f_{m'}(\psi)$.[5]

The determination of the remaining factor of (10.21) constitutes a more complicated task which we omit. This factor is a standard function of mathematical

---

4. The breaking up of coordinate rotations into these three steps is a standard procedure of analytical geometry; the angles ϕ, θ, and ψ are called Euler angles.

5. The procedure for the determination of the phase factors in equation (10.21) is logically straightforward. It provides a consistent code for the determination of the phase conventions discussed in sections 8.3 and 8.4. However, its practical application, taking correct account of all relevant details, often proves confusing and easily open to error.

TABLE 10.1

Probability Amplitudes for Two-, Three-, and Four-Component Beam Splitting
($\psi = \phi = 0$)

$d^{(1/2)}_{m'm}(\theta)$

m' \ m	½	−½
½	$\cos \frac{1}{2}\theta$	$\sin \frac{1}{2}\theta$
−½	$-\sin \frac{1}{2}\theta$	$\cos \frac{1}{2}\theta$

$d^{(1)}_{m'm}(\theta)$

m' \ m	1	0	−1
1	$\cos^2 \frac{1}{2}\theta$	$2^{1/2} \sin \frac{1}{2}\theta \cos \frac{1}{2}\theta$	$\sin^2 \frac{1}{2}\theta$
0	$-2^{1/2} \sin \frac{1}{2}\theta \cos \frac{1}{2}\theta$	$\cos^2 \frac{1}{2}\theta - \sin^2 \frac{1}{2}\theta$	$2^{1/2} \sin \frac{1}{2}\theta \cos \frac{1}{2}\theta$
−1	$\sin^2 \frac{1}{2}\theta$	$-2^{1/2} \sin \frac{1}{2}\theta \cos \frac{1}{2}\theta$	$\cos^2 \frac{1}{2}\theta$

$d^{(3/2)}_{m'm}(\theta)$

m' \ m	3/2	½	−½	−3/2
3/2	$\cos^3 \frac{1}{2}\theta$	$3^{1/2} \cos^2 \frac{1}{2}\theta \sin \frac{1}{2}\theta$	$3^{1/2} \cos \frac{1}{2}\theta \sin^2 \frac{1}{2}\theta$	$\sin^3 \frac{1}{2}\theta$
½	$-3^{1/2} \cos^2 \frac{1}{2}\theta \sin \frac{1}{2}\theta$	$\cos \frac{1}{2}\theta\,(1-3 \sin^2 \frac{1}{2}\theta)$	$\sin \frac{1}{2}\theta\,(3 \cos^2 \frac{1}{2}\theta-1)$	$3^{1/2} \cos \frac{1}{2}\theta \sin^2 \frac{1}{2}\theta$
−½	$3^{1/2} \cos \frac{1}{2}\theta \sin^2 \frac{1}{2}\theta$	$\sin \frac{1}{2}\theta\,(1-3 \cos^2 \frac{1}{2}\theta)$	$\cos \frac{1}{2}\theta\,(1-3 \sin^2 \frac{1}{2}\theta)$	$3^{1/2} \cos^2 \frac{1}{2}\theta \sin \frac{1}{2}\theta$
−3/2	$-\sin^3 \frac{1}{2}\theta$	$3^{1/2} \cos \frac{1}{2}\theta \sin^2 \frac{1}{2}\theta$	$-3^{1/2} \cos^2 \frac{1}{2}\theta \sin \frac{1}{2}\theta$	$\cos^3 \frac{1}{2}\theta$

physics, often indicated by $d^{(j)}_{m'm}(\theta)$. The complete matrix element (10.21) has then the form

$$\langle \hat{A}\ m' | \hat{B}\ m\rangle = \exp(im'\psi_A)\, d^{(j)}_{m'm}(\theta_A)\, \exp(im\phi_A). \tag{10.22}$$

The set of functions $d(\theta)$ is given in Table 10.1 for $j \leq 3/2$. The squares of these functions are, of course, the entries in the probability Table 6.1.

CHAPTER 10 PROBLEMS

10.1. Consider a particle oscillating along the z axis whose position eigenstates are identified in dimensionless units by a parameter ξ ranging from $-\infty$ to $+\infty$. The energy eigenvalues are labeled by an integer index $n = 0, 1, 2, \ldots$ and are equal to $n\epsilon$. The wave functions of the three stationary states of lowest energy are $\langle \xi | E\ 0\rangle = \pi^{-1/4}\exp(-\tfrac{1}{2}\xi^2)$, $\langle \xi | E\ 1\rangle = 2^{1/2}\pi^{-1/4}\xi \times \exp(-\tfrac{1}{2}\xi^2)$, and $\langle \xi | E\ 2\rangle = 2^{1/2}\pi^{-1/4}(\xi^2 - \tfrac{1}{2})\exp(-\tfrac{1}{2}\xi^2)$. (a) Write the wave function of a nonstationary state $|a(t)\rangle$ identified at $t = 0$ by the probability amplitudes $\langle E\ n | a(0)\rangle = \sqrt{(1/3)}$ for $n \leq 2$ and $\langle E\ n | a(0)\rangle = 0$ for $n > 2$. (b) Calculate the probability $P(\xi, E\ n)$ of finding the particle at the position ξ for each of the three stationary states with $n = 0, 1,$ and 2. (c) Calculate the probability $P(\xi, a(t))$. (d) Calculate the mean position $\langle \xi\rangle_{a(t)}$ for the nonstationary state. This calculation involves the integral $I_p = \int_{-\infty}^{+\infty} d\xi\, \xi^p \exp(-\xi^2)$ whose values are

$$\begin{array}{llllllll} p & = 0 & 1, & 2, & 3, & 4, & 5, & 6, \ldots \\ I_p & = \pi^{1/2}, & 0, & \tfrac{1}{2}\pi^{1/2}, & 0, & \tfrac{1}{2}\tfrac{3}{2}\pi^{1/2}, & 0, & \tfrac{1}{2}\tfrac{3}{2}\tfrac{5}{2}\pi^{1/2}, \ldots . \end{array}$$

10.2. For a particle with $j = 1$, calculate the matrix of the angular momentum component J_x, $\|\langle \hat{z}\ m_z | J_x | \hat{z}\ m'_z\rangle\|$ analogous to that given in equation (8.29a) for $j = \tfrac{1}{2}$. Use the transformation matrix $\|\langle \hat{x}\ m_x | \hat{z}\ m'_z\rangle\| = \|d^{(1)}_{m_x m'_z}(\tfrac{1}{2}\pi)\|$ given in Table 10.1 and its inverse.

10.3. For particles with $j = 1$ the matrices of the angular momentum components, analogous to the matrices (8.29) for $j = \tfrac{1}{2}$, are

$$\|\langle \hat{z}\ m_z | J_x | \hat{z}\ m'_z\rangle\| = 2^{-1/2}\hbar \begin{vmatrix} 0 & 1 & 0 \\ 1 & 0 & 1 \\ 0 & 1 & 0 \end{vmatrix},\ \|\langle \hat{z}\ m_z | J_y | \hat{z}\ m'_z\rangle\|$$

$$= 2^{-1/2}\hbar \begin{vmatrix} 0 & -i & 0 \\ i & 0 & -i \\ 0 & i & 0 \end{vmatrix},\ \|\langle \hat{z}\ m_z | J_z | \hat{z}\ m'_z\rangle\| = \hbar \begin{vmatrix} 1 & 0 & 0 \\ 0 & 0 & 0 \\ 0 & 0 & -1 \end{vmatrix}.$$

Calculate the Cartesian components and the magnitude of the mean angular momentum $\langle \vec{J}\rangle_a$ for a state $|a\rangle = \Sigma_{m=-1}^{1} |\hat{z}\ m\rangle\langle \hat{z}\ m | a\rangle$, with $\langle \hat{z}\ 1 | a\rangle = \sqrt{\tfrac{1}{2}}$ and $\langle \hat{z}\ 0 | a\rangle = \langle \hat{z}\ -1 | a\rangle = \tfrac{1}{2}$.

10.4. The particles of problem 10.3 are subjected to a magnetic and an electric field parallel to the z axis. The eigenstates of J_z, $|\hat{z}\ \ m\rangle$, remain stationary because the fields apply no torque about the z axis and thus leave $\langle J_z \rangle$ constant. The energy levels E_m are not evenly spaced in the presence of an electric field. Assume the levels $E_1 = -E_{-1} = \hbar\omega_1$ and $E_0 = \hbar\omega_0$ and consider a nonstationary state $|a(t)\rangle$ which coincides with the state $|a\rangle$ of problem 10.3 at $t = 0$. Calculate the cartesian components of $\langle \vec{J} \rangle_{a(t)}$. Will this vector precess about $\hat{z}$ with constant magnitude? with constant speed?

SOLUTIONS TO CHAPTER 10 PROBLEMS

10.1. (a) Apply equation (10.15): $\langle \xi | a(t) \rangle = \Sigma_{n=0}^{2} (\xi | E\ \ n) \exp(-i\, n\epsilon t/\hbar) 3^{-1/2}$ $= 3^{-1/2}\pi^{-1/4}\exp(-\tfrac{1}{2}\xi^2)[1 + 2^{1/2}\xi\exp(-i\epsilon t/\hbar) + 2^{1/2}(\xi^2 - \tfrac{1}{2})\exp(-i2\epsilon t/\hbar)]$. (b) $P(\xi,\ E\ \ 0) = |\langle \xi | E\ \ 0\rangle|^2 = \pi^{-1/2}\exp(-\xi^2)$, $P(\xi,\ E\ \ 1) = 2\pi^{-1/2}\xi^2\exp(-\xi^2)$, $P(\xi,\ E\ \ 2) = 2\pi^{-1/2}(\xi^2 - \tfrac{1}{2})^2\exp(-\xi^2)$ (see Fig. a). (c) $P(\xi,\ a(t)) = |\langle \xi | a(t)\rangle|^2 = \tfrac{1}{3}\pi^{-1/2}\exp(-\xi^2)[1 + 2\xi^2 + 2(\xi^2 - \tfrac{1}{2})^2 + 8^{1/2}\xi\cos(2\epsilon t/\hbar) + 8^{1/2}(\xi^2 - \tfrac{1}{2}) \times \cos(2\epsilon t/\hbar) + 4\xi(\xi^2 - \tfrac{1}{2})\cos(\epsilon t/\hbar)]$ (see Fig. b). (d) $\langle \xi \rangle_{a(t)} = \int_{-\infty}^{\infty} \xi P(\xi,\ a(t))\, d\xi$.

Substitute P from (c) and carry out integrals. Note that the first, second, third, and fifth terms in brackets of P do not contribute because they give integrands with odd power ξ^p. Result is $\langle \xi \rangle_{a(t)} = \tfrac{1}{3}(\sqrt{2} + 2)\cos(\epsilon t/\hbar)$. The above solution

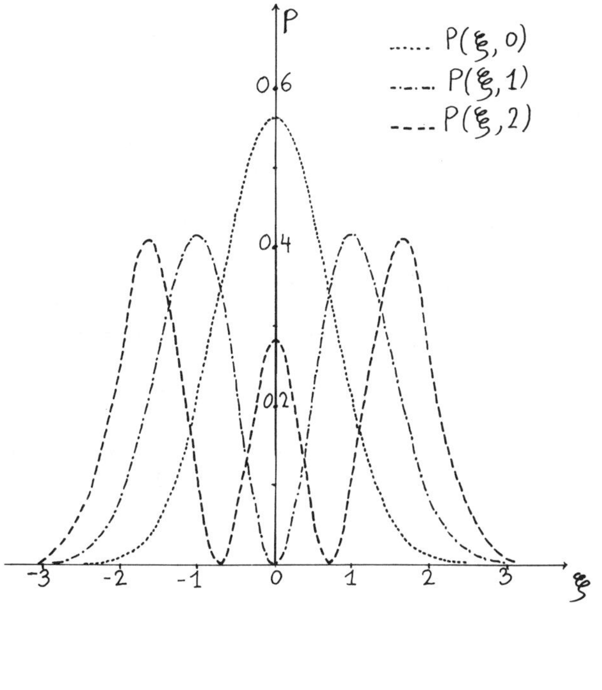

(a)

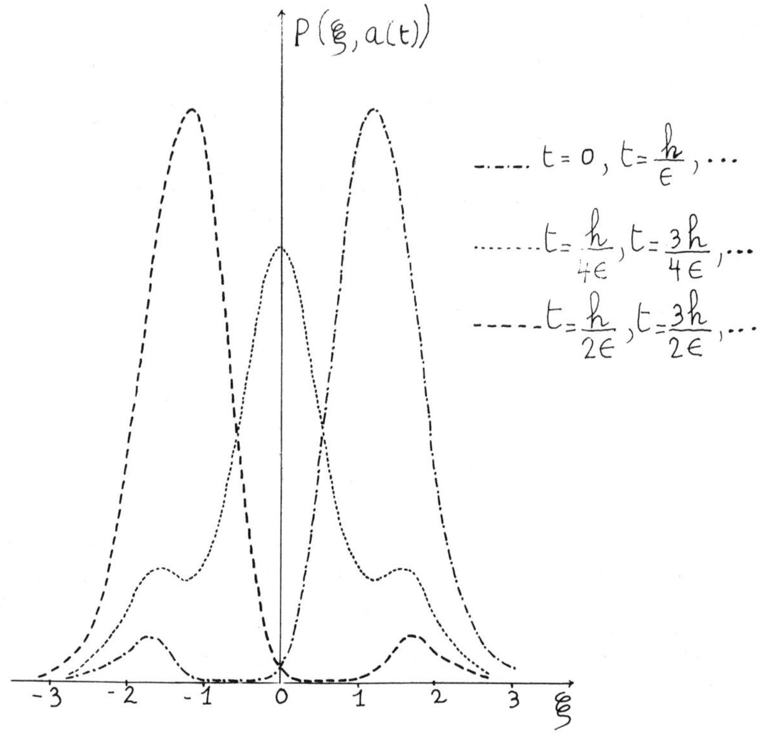

(b)

of (d) uses the previous steps in which $\langle E\ \ n|a(t)\rangle$ was first combined with $\langle \xi|E\ \ n\rangle$, then the wave function was squared to construct $P(\xi,\ a)$; integration over ξ follows for the solution of (d). To solve (d) by direct application of equation (10.19), calculate first the elements of the matrix (10.20) with n and $n' \leq 2$, i.e., $|\langle E\ \ n|\xi|E\ \ n'\rangle| = \begin{vmatrix} 0 & \sqrt{1/2} & 0 \\ \sqrt{1/2} & 0 & 1 \\ 0 & 1 & 0 \end{vmatrix}$. Then represent the expression on the right of equation (10.19) as a product of matrices. For this purpose the set of probability amplitudes $\langle E\ \ n|a(0)\rangle = 3^{-1/2}$ (for $n \leq 2$) given in part (a) may be written as a one-column matrix, and the set $\langle a(0)|n\rangle$ as a one-row matrix. Thus (10.19) is written and calculated as

$$\langle\xi\rangle_{a(t)} = \begin{vmatrix} \sqrt{1/3} & \sqrt{1/3} & \sqrt{1/3} \end{vmatrix} \times \begin{vmatrix} 1 & 0 & 0 \\ 0 & e^{i\epsilon t/\hbar} & 0 \\ 0 & 0 & e^{i2\epsilon t/\hbar} \end{vmatrix} \times \begin{vmatrix} 0 & \sqrt{1/2} & 0 \\ \sqrt{1/2} & 0 & 1 \\ 0 & 1 & 0 \end{vmatrix}$$

$$\times \begin{vmatrix} 1 & 0 & 0 \\ 0 & e^{-i\epsilon t/\hbar} & 0 \\ 0 & 0 & e^{-i2\epsilon t/\hbar} \end{vmatrix} \times \begin{vmatrix} \sqrt{1/3} \\ \sqrt{1/3} \\ \sqrt{1/3} \end{vmatrix} = 1/3\ (\sqrt{2} + 2)\cos(\epsilon t/\hbar).$$

10.2. $|\langle \hat{z}\ m_z | J_x | \hat{z}\ m'_z \rangle| = |\sum_{m_x} \langle \hat{z}\ m_z | \hat{x}\ m_x \rangle \hbar m_x \langle \hat{x}\ m_x | \hat{z}\ m'_x \rangle|$

$$= \begin{vmatrix} \frac{1}{2} & -\sqrt{\frac{1}{2}} & \frac{1}{2} \\ \sqrt{\frac{1}{2}} & 0 & -\sqrt{\frac{1}{2}} \\ \frac{1}{2} & \sqrt{\frac{1}{2}} & \frac{1}{2} \end{vmatrix} \times \begin{vmatrix} \hbar & 0 & 0 \\ 0 & 0 & 0 \\ 0 & 0 & -\hbar \end{vmatrix} \times \begin{vmatrix} \frac{1}{2} & \sqrt{\frac{1}{2}} & \frac{1}{2} \\ -\sqrt{\frac{1}{2}} & 0 & \sqrt{\frac{1}{2}} \\ \frac{1}{2} & -\sqrt{\frac{1}{2}} & \frac{1}{2} \end{vmatrix} = \begin{vmatrix} 0 & \sqrt{\frac{1}{2}}\,\hbar & 0 \\ \sqrt{\frac{1}{2}}\,\hbar & 0 & \sqrt{\frac{1}{2}}\,\hbar \\ 0 & \sqrt{\frac{1}{2}}\,\hbar & 0 \end{vmatrix}.$$

10.3. Using the same one-row and one-column matrix notation as in the solution of problem 10.1, the analog of equation (8.28a) is

$$\langle J_x \rangle_a = \left| \sqrt{\tfrac{1}{2}}\ \ \tfrac{1}{2}\ \ \tfrac{1}{2} \right| \times \sqrt{\tfrac{1}{2}}\,\hbar \begin{vmatrix} 0 & 1 & 0 \\ 1 & 0 & 1 \\ 0 & 1 & 0 \end{vmatrix} \times \begin{vmatrix} \sqrt{\frac{1}{2}} \\ \frac{1}{2} \\ \frac{1}{2} \end{vmatrix} = \tfrac{1}{2}\hbar(1 + 2^{-1/2}).$$

Similarly, one finds $\langle J_y \rangle_a = 0$ and $\langle J_z \rangle_a = \frac{1}{4}\hbar$. The vector $\langle \vec{J} \rangle_a$ lies in the (x, z) plane, and its magnitude is 0.89 $\hbar$. If the state $|a\rangle$ were the eigenstate $|\hat{u}\ m\rangle$ of the component of $\vec{J}$ along an axis $\hat{u}$, we would have $\langle \vec{J} \rangle_a \equiv \langle \vec{J} \rangle_{\hat{u}m} = m\hbar\hat{u}$; we find instead $|\langle \vec{J} \rangle_a| \neq m\hbar$.

10.4. $\langle J_x \rangle$, $\langle J_y \rangle$, and $\langle J_z \rangle$ are given by formulae having the structure of equation (10.19) with the matrix of x replaced by the matrices of J_x, J_y, and J_z given in problem 10.3. These mean values may be calculated as in the solution of problem 10.3, replacing the column matrix of probability amplitudes $\langle \hat{z}\ m | a \rangle$ by $\langle \hat{z}\ m | a(t) \rangle = \exp(-i\omega_m t) \langle \hat{z}\ m | a(0) \rangle$ and the row matrix of $\langle a | \hat{z}\ m \rangle$ by $\langle a(0) | \hat{z}\ m \rangle \exp(i\omega_m t)$. One then finds

$$\langle J_x \rangle_{a(t)} = \left| 2^{-1/2}\exp(i\omega_1 t)\ \ \tfrac{1}{2}\exp(i\omega_0 t)\ \ \tfrac{1}{2}\exp(-i\omega_1 t) \right| \times 2^{-1/2}\hbar \begin{vmatrix} 0 & 1 & 0 \\ 1 & 0 & 1 \\ 0 & 1 & 0 \end{vmatrix} \times$$

$$\begin{vmatrix} 2^{-1/2}\exp(-i\omega_1 t) \\ \frac{1}{2}\exp(-i\omega_0 t) \\ \frac{1}{2}\exp(i\omega_1 t) \end{vmatrix} = 2^{-1/2}\hbar[2^{-1/2} \cos(\omega_1 - \omega_0)t + \tfrac{1}{2} \cos(\omega_1 + \omega_0)t],$$

$$\langle J_y \rangle_{a(t)} = 2^{-1/2}\hbar[2^{-1/2} \sin(\omega_1 - \omega_0)t + \tfrac{1}{2} \sin(\omega_1 + \omega_0)t], \ \langle J_z \rangle_{a(t)} = \tfrac{1}{4}\hbar.$$

The z component of $\langle \vec{J} \rangle_{a(t)}$ remains constant as expected, but the x and y components consist of two terms which precess about $\hat{z}$ at different speeds $\omega_1 - \omega_0$ and $\omega_1 + \omega_0$. The squared magnitude $|\langle \vec{J} \rangle_{a(t)}|^2$ has a constant term and one proportional to $\cos(2\omega_0 t)$.

11. incompatibility and the motion of particles

Having formulated procedures for the theoretical description of typical quantum phenomena, we begin now to apply them to particle motion. Our main goal is still to account for the size and stability of atoms. We shall find, at the end of this chapter, that these properties stem from the incompatibility of the position coordinates with the energy of atomic electrons, to which we now direct our attention.

The motion of a particle is described in quantum mechanics through variations in the course of time of the probability of observing the particle at different positions of space. These variations are represented by interference terms in the expansion (10. 16) of the probability. The coefficient of each interference term contains the factors $|\langle x, y, z | E\ n\rangle|$ and $|\langle x, y, z | E\ m\rangle|$ which connect the position of a particle to different stationary states. Motion of a particle, that is, variation of the probability $P(x, y, z; a(t))$, requires that at least two of these probability amplitudes, corresponding to eigenvalues E_n and $E_m \neq E_n$, be nonzero. Hence, a position eigenstate does not coincide with an energy eigenstate. In other words, position and energy are incompatible. Indeed, the generality of equation (10. 16) implies that any and all quantities that may vary in time are incompatible with energy.

This chapter deals primarily with free particles, whose potential energy is zero. That these particles may move shows that it is the *kinetic* energy of each particle, and hence its velocity, which is incompatible with the particle's position. (The potential energy of a particle subject to forces is compatible with its position, since it is a function of the position.)

As a corollary, we note that the kinetic and the potential terms of the energy of a particle subject to forces are incompatible with one another. The effects of this incompatibility will be the main subject of chapter 12. Here we deal with the incompatibility of position with velocity and kinetic energy; that is, we deal mainly with the states of motion of free particles.

Our study can be narrowed further at the outset as the motion of a free particle along one Cartesian axis is independent of its motion in perpendicular directions. More precisely, the kinetic energy K of a particle of mass m may be expressed as the sum of three terms

$$K = \tfrac{1}{2}mv_x^2 + \tfrac{1}{2}mv_y^2 + \tfrac{1}{2}mv_z^2. \tag{11. 1}$$

Variations in time of the mean value of one position coordinate, for instance, of $\langle x\rangle$ depend on the value of $\tfrac{1}{2}mv_x^2$ and do not depend on the other two terms of equation (11. 1). We can, for example, consider $\langle x\rangle$ for particle states which are eigenstates of v_y and v_z with eigenvalues $v_y = v_z = 0$. This example shows that the position coordinate x is actually incompatible with the term $\tfrac{1}{2}mv_x^2$ of (11. 1) but not with the other terms. That is, we can restrict our study to a one-dimensional motion.

It is important to point out that the word "motion" is used with different meanings in quantum physics. The literal meaning implies that the position of a body, or, at least, its probability distribution, varies in the course of time. This is the common meaning in macroscopic physics, and it is also the meaning intended for the word in the preceding paragraphs. However, quantum physics considers stationary states in which no physical quantity and no probability varies in the course of time. Yet, stationary states of particles have normally nonzero kinetic energy and therefore should be fairly regarded as states of motion. That a particle may have nonzero kinetic energy while all variables are constant in time is a typical quantum effect which has no parallel in classical physics. In the following we will speak of "motion" of a particle whenever its kinetic energy is nonzero, whether its state is nonstationary or stationary. Characteristics of different types of motion will be illustrated in section 11.3.

The main step in developing the quantum mechanics of particle motion is to determine the probability amplitudes that connect eigenstates of position and eigenstates of velocity. This is done in section 11.1. The rest of the chapter develops consequences of the relationship between position and velocity, including the complementarity of particle position and momentum.

11.1. WAVE FUNCTIONS AND VELOCITY EIGENSTATES.

A nonstationary state of a particle moving freely along the x axis is identified by its wave function $\psi_a(x, t) \equiv \langle x | a(t) \rangle$. The mean velocity of a free particle is constant, which means that its velocity eigenstates are stationary. Therefore, the expansion (10.15) of the wave function takes the form

$$\psi_a(x, t) \equiv \langle x | a(t) \rangle = \int_{-\infty}^{\infty} dv_x \langle x | v_x \rangle \exp\left(-i \frac{\frac{1}{2} m v_x^2}{\hbar} t\right) \langle v_x | a(0) \rangle. \tag{11.2}$$

The macroscopic statement that a particle travels with uniform velocity v_x is translated in quantum mechanics by saying that the plot of the probability distribution $|\langle x | a(t) \rangle|^2$ travels along x with the average velocity $\langle v_x \rangle$, under nearly macroscopic conditions. These conditions require that the velocity of the particle be rather well defined, that is, that the probability $|\langle v_x | a(0) \rangle|^2$ vanish for values of v_x much different from $\langle v_x \rangle$. The question is now: What must be the mathematical form of the probability amplitudes $\langle x | v_x \rangle$ in order that the wave function $\langle x | a(t) \rangle$ travel with the expected velocity $\langle v_x \rangle$? The answer to this question is suggested and verified here; a mathematical derivation is given in supplementary section 11.6.

The suggestion is that the probability amplitudes $\langle x | v_x \rangle$ have the form of plane progressive waves exp(ikx), where the wavenumber k is a suitable function of v_x. Under this assumption the integral in equation (11.2) has the mathematical form of a Fourier integral of the type called a wave packet. If the effective range of integration over v_x is sufficiently narrow, so that both the "frequency" $\omega = \frac{1}{2} m v_x^2/\hbar$ and k depart but little from their average values, $\langle \omega \rangle$ and $\langle k \rangle$, the wave packet is known to travel with the group velocity $d\langle \omega \rangle / d\langle k \rangle$.[1] This velocity

1. See section A3.1 and especially *BPC*, vol. 3, chap. 6.

coincides with the mean velocity $\langle v_x \rangle$, as required, if we set

$$k = mv_x/\hbar \equiv p_x/\hbar. \tag{11.3}$$

The deviations of v_x from its mean value $\langle v_x \rangle$ are essential for the existence and propagation of the wave packet with the group velocity $\langle v_x \rangle = d\langle \omega \rangle / d\langle k \rangle$. However, their square magnitude has vanishing influence on the propagation in the macroscopic limit, as will be illustrated in sections S11.6 and S11.9.

To determine the values of $\langle x | v_x \rangle$ completely, a *normalization factor* multiplying exp(ikx) must be chosen such that the probability amplitudes satisfy the unitary property (10.12). The normalization factor is worked out by substituting exp(ikx) in the probability amplitudes of equation (10.12) and comparing the resulting equation with the definition (A.46) of the Dirac delta. The most common expression of the resulting probability amplitude is given in terms of the eigenstates of the momentum component p_x rather than of those of v_x, because v_x appears in equation (11.3) in the combination mv_x. The basic result is then[2]

$$\langle x | p_x \rangle = h^{-1/2} \exp(ip_x x/\hbar), \tag{11.4a}$$

with the companion complex conjugate formula

$$\langle p_x | x \rangle = h^{-1/2} \exp(-ip_x x/\hbar). \tag{11.4b}$$

This result has been suggested here by the requirement that the wave function (11.2) travel with the group velocity $\langle v_x \rangle$ under approximately macroscopic conditions. However, it has general validity, as shown in section S11.6. Using the expression (11.4a) of $\langle x | p_x \rangle$, a wave function $\psi_a(x)$ is represented as a superposition of momentum eigenfunctions by

$$\psi_a(x) \equiv \langle x | a \rangle = \int_{-\infty}^{\infty} dp_x \langle x | p_x \rangle \langle p_x | a \rangle = h^{-1/2} \int_{-\infty}^{\infty} dp_x \exp(ip_x x/\hbar) \langle p_x | a \rangle. \tag{11.5}$$

This formula, which is an application of the quantum mechanical matrix relation (10.7a), has the form of a Fourier expansion of the function $\psi_a(x)$. Because the probability amplitudes (11.4) have the form of plane waves and because of their importance in describing the motion of atomic particles, the study of this motion is called *wave mechanics*.[3]

---

2. The unitary property (10.12) requires that $\int_{-\infty}^{\infty} dx \langle p_x | x \rangle \langle x | p_x' \rangle = \delta(p_x - p_x')$. To verify that equations (11.4a) and (11.4b) meet this requirement, substitute their right-hand sides in the integral and change the variable of integration from x to $\xi = x/h$. The integral becomes $\int_{-\infty}^{\infty} d\xi \exp[i2\pi\xi(p_x - p_x')]$, which has the same form as the left-hand side of (A.46) and hence represents $\delta(p_x - p_x')$.

3. That the motion of particles exhibits wave properties was first suggested by de Broglie to provide a basis for Bohr's postulate of atomic mechanics represented by equation (3.20). Wave functions are often called "de Broglie waves" or "matter waves."

Consider now a particle moving in the three dimensions of space. Some of its states are eigenstates of the vector momentum $|\vec{p}\rangle \equiv |p_x, p_y, p_z\rangle$. The probability amplitudes connecting these states to eigenstates of the position vector $|\vec{r}\rangle \equiv |x, y, z\rangle$ are products of the probability amplitudes for independent separate motions along the three coordinate axes

$$\langle\vec{r}\,|\vec{p}\rangle = \langle x\,|p_x\rangle\,\langle y\,|p_y\rangle\,\langle z\,|p_z\rangle = h^{-3/2}\exp\,(i\vec{p}\cdot\vec{r}/\hbar). \tag{11.6}$$

11.2. PARTICLE MOTION AND WAVE FUNCTION

The state of a particle may be identified by its wave function $\langle x, y, z\,|a\rangle \equiv \psi_a(\vec{r})$. Hence, an explicit expression of the wave function, analytical or numerical, must contain all the information required to describe the particle's motion. According to general procedure one should obtain statistical predictions about the particle's momentum (or velocity), kinetic energy, etc., from the set of probabilities $|\langle\vec{p}\,|a\rangle|^2$. These are obtained, in turn, from the wave function by calculating the three-dimensional Fourier integral of a given wave function $\psi_a(\vec{r})$,

$$\langle\vec{p}\,|a\rangle = \int d\vec{r}\,\langle\vec{p}\,|\,\vec{r}\rangle\,\langle\vec{r}\,|\,a\rangle = h^{-3/2}\int d\vec{r}\,\exp\,(-i\vec{p}\cdot\vec{r}/\hbar)\psi_a(\vec{r}). \tag{11.7}$$

However, characteristics of the particle motion can also be extracted from the position derivatives of the wave function, by-passing the calculation of the integral (11.7). This simplification is made possible by the mathematical form (11.6) of the probability amplitudes $\langle\vec{p}\,|\vec{r}\rangle$. The method applies whether the particle is free or subject to forces and whether its state is stationary or not. We begin by calculating the mean momentum $\langle\vec{p}\rangle_a$.

For this purpose the main property is that $\langle x\,|p_x\rangle$ obeys the differential equation

$$\frac{\hbar}{i}\frac{d}{dx}\langle x\,|p_x\rangle = \frac{\hbar}{i}\frac{d}{dx}\,h^{-1/2}\exp\,i\,\frac{p_x x}{\hbar} = \langle x\,|p_x\rangle p_x. \tag{11.8}$$

In the language used in section A.2, $\langle x\,|p_x\rangle$ is called an eigenfunction of this equation with the eigenvalue p_x. In three dimensions equation (11.8) takes the vector form

$$\frac{\hbar}{i}\vec{\nabla}\langle\vec{r}\,|\,\vec{p}\rangle = \langle\vec{r}\,|\vec{p}\rangle\vec{p}. \tag{11.9}$$

We start from the definition of mean value,

$$\langle\vec{p}\rangle_a = \int d\vec{p}\,\langle a\,|\,\vec{p}\rangle\,\vec{p}\,\langle\vec{p}\,|\,a\rangle. \tag{11.10}$$

We then substitute for $\langle\bar{a}\,|\vec{p}\rangle$ the complex conjugate of (11.7), without intending to calculate the integral, and we write

$$\langle\vec{p}\rangle_a = \int d\vec{p}\int d\vec{r}\,\langle a\,|\,\vec{r}\rangle\,\langle\vec{r}\,|\,\vec{p}\rangle\,\vec{p}\,\langle\vec{p}\,|\,a\rangle. \tag{11.11}$$

As a second step we replace the product $\langle\vec{r}\,|\,\vec{p}\rangle\vec{p}$ by the left hand side of (11.9)

$$\langle\vec{p}\rangle_a = \int d\vec{p}\int d\vec{r}\;\langle a\,|\,\vec{r}\rangle\;(\hbar/i)\vec{\nabla}\langle\vec{r}\,|\,\vec{p}\rangle\;\langle\vec{p}\,|\,a\rangle. \tag{11.12}$$

At this point, since the eigenvalue $\vec{p}$ no longer appears explicitly in the integrand, the integral over $\vec{p}$ reduces to the inverse of the transformation formula (11.7), namely, to $\int d\vec{p}\langle\vec{r}\,|\vec{p}\rangle\langle\vec{p}\,|a\rangle = \langle\vec{r}\,|a\rangle$. Thus the mean value of the momentum (11.12) becomes

$$\langle\vec{p}\rangle_a = \int d\vec{r}\;\langle a\,|\,\vec{r}\rangle\;(\hbar/i)\vec{\nabla}\langle\vec{r}\,|\,\vec{a}\rangle. \tag{11.13}$$

This expression of the mean value of $\vec{p}$ utilizes the base set of probability amplitudes $\langle\vec{r}\,|\,a\rangle$—that is, the wave function—and its derivative $\vec{\nabla}\langle\vec{r}\,|\,a\rangle$. Comparing the right hand side of (11.10) and (11.13), we see that they have the same structure. The wave functions $\langle a\,|\vec{r}\rangle$ and $\langle\vec{r}\,|a\rangle$ appear in place of $\langle a\,|\vec{p}\rangle$ and $\langle\vec{p}\,|a\rangle$; the expression $(\hbar/i)\vec{\nabla}$ appears in place of $\vec{p}$. One says that the gradient operator $(\hbar/i)\vec{\nabla}$ "represents $\vec{p}$" in the "position representation."[4]

The dependence of $\langle\vec{p}\rangle_a$ on the gradient of the wave function, that is, on the gradient of $\langle\vec{r}\,|a\rangle = |\langle\vec{r}\,|a\rangle|\,\exp[i\phi(\vec{r})]$, can be analyzed into contributions from the gradients of the magnitude $|\langle\vec{r}\,|a\rangle|$ and of the phase $\phi(\vec{r})$. To this end, one makes the integrand in (11.13) symmetrical with respect to $\vec{\nabla}\langle\vec{r}\,|\,a\rangle$ and $\vec{\nabla}\langle a\,|\,\vec{r}\rangle$. This is done by integrating by parts one half of the integrand and considering that, for a realistic state of the particle, $\langle\vec{r}\,|a\rangle$ vanishes for $r \to \infty$.[5] One finds

$$\langle\vec{p}\rangle_a = (\hbar/2i)\int d\vec{r}[\langle a\,|\vec{r}\rangle\vec{\nabla}\langle\vec{r}\,|a\rangle - \langle\vec{r}\,|a\rangle\vec{\nabla}\langle a\,|\vec{r}\rangle]. \tag{11.14}$$

---

4. This operator method of calculating mean values and the method which utilizes matrices are related in the following way. Application of the definition (8.28) or (10.20) to the matrix which represents the momentum component p_x with reference to the base set of position coordinates yields

$$\langle x\,|p_x\,|x'\rangle = \int\langle x\,|p_x\rangle p_x\langle p_x\,|x'\rangle dp_x = \int\frac{\hbar}{i}\frac{d}{dx}\langle x\,|p_x\rangle\,\langle p_x\,|x'\rangle dp_x = \frac{\hbar}{i}\frac{d}{dx}\,\delta(x-x').$$

Substitution of this equation in the general formula $\langle p_x\rangle_a = \int\int\langle a\,|\,x\rangle\langle x\,|p_x|\,x'\rangle\langle x'\,|\,a\rangle dx dx'$ reduces it to the x component of (11.13).

5. The derivation of equation (11.14) involves a three-dimensional integral over an infinite range and need not be followed in detail here. Its principle is illustrated by the one-dimensional integral formula involving two functions u(x) and v(x) and their derivatives,

$$\int_a^b u(x)(dv/dx)dx = u(b)v(b) - u(a)v(a) - \int_a^b (du/dx)v(x)dx.$$

The integrals on the right- and left-hand sides of this equation are equal if u(x) and v(x) vanish at both limits $x = a$ and $x = b$ (at infinity in our three-dimensional problem). The integral on the left is then also equal to half the sum of itself and of the integral on the right,

$$\int_a^b u(x)(dv/dx)dx = \frac{1}{2}\int_a^b dx[u(x)(dv/dx) - (du/dx)v(x)].$$

Upon substitution of $\langle a|\vec{r}\rangle = |\langle a|\vec{r}\rangle| \exp[-i\phi(r)]$ and $\langle\vec{r}|a\rangle = |\langle\vec{r}|a\rangle| \exp[i\phi(\vec{r})]$ in the integrand of (11.14), the gradient of $|\langle\vec{r}|a\rangle|$ cancels out. Equation (11.14) reduces then to

$$\langle\vec{p}\rangle_a = \hbar \int d\vec{r}\, |\langle\vec{r}|a\rangle|^2\, \vec{\nabla}\phi(\vec{r}). \tag{11.15}$$

Thus each volume element $d\vec{r}$ contributes to the mean momentum in proportion to the probability $|\langle\vec{r}|a\rangle|^2$ of finding the particle within $d\vec{r}$ and to the gradient of the phase of the wave function.

The significance of the integrand in equation (11.15) may be brought out by regarding the probability $|\langle\vec{r}|a\rangle|^2 = |\psi_a|^2$ of finding a particle at point $\vec{r}$ as a "particle density." The product $|\psi_a|^2\, \vec{\nabla}\phi(\vec{r})$ represents then, apart from a constant factor, the "current density" or "flux" of the particle at the point $\vec{r}$. The flux at each point contributes to the mean momentum in proportion to the mass m of the particle and to the volume element $d\vec{r}$. Specifically, one defines a flux vector by

$$\vec{\Phi}(\vec{r}, a) = \frac{\hbar}{m} |\psi_a|^2\, \vec{\nabla}\phi(\vec{r}) \tag{11.16}$$

or by the equivalent formula

$$\vec{\Phi}(\vec{r}, a) = \frac{\hbar}{2im} [\langle a|\vec{r}\rangle\vec{\nabla}\langle\vec{r}|a\rangle - \langle\vec{r}|a\rangle\vec{\nabla}\langle a|\vec{r}\rangle]. \tag{11.16a}$$

The appropriateness of this definition of flux as a property of the state $|a\rangle$ at each point of space is verified by showing that the particle density and flux fulfill a continuity equation stating that there exists a conservation of probability over the whole space. Mathematically, the particle flux exiting from a unit volume, $\mathrm{div}\vec{\Phi}(\vec{r}, a)$, equals the rate of decrease of the particle density.[6] For a stationary state $|\psi_a|^2$ is constant in time and the flux must be solenoidal.

Operationally, the particle flux is defined with reference to the response of an ideal particle counter placed at $\vec{r}$, which scores a positive count when its sensitive surface is traversed in one direction and a negative count when traversed in the opposite direction. The direction of $\vec{\Phi}_a$ is perpendicular to that orientation of the counter surface which yields the highest net score; the magnitude of the flux is the average counter score per unit of sensitive surface area and per unit time.

The mean kinetic energy of a particle, $\langle K\rangle$, can also be expressed by using the derivatives of the wave function $\langle\vec{r}|\vec{p}\rangle$ instead of a Fourier integral. The prop-

---

6. To work out the conservation law, enter the right-hand side of equation (11.16a) in the expression of $\mathrm{div}\,\vec{\Phi} = \vec{\nabla}\cdot\vec{\Phi}$. After cancellation of terms $\pm[\vec{\nabla}\langle a|\vec{r}\rangle]\cdot[\vec{\nabla}\langle\vec{r}|a\rangle]$, one finds $\vec{\nabla}\cdot\vec{\Phi} = (\hbar/2im)\{\langle a|\vec{r}\rangle[\nabla^2\langle\vec{r}|a\rangle] - [\nabla^2\langle a|\vec{r}\rangle]\langle\vec{r}|a\rangle\}$. The expression $\nabla^2\langle\vec{r}|a\rangle$ is proportional to $\partial\psi_a/\partial t$ according to the wave equation (11.22). Using this equation and its complex conjugate for $\nabla^2\langle a|\vec{r}\rangle$, $\vec{\nabla}\cdot\vec{\Phi}$ is reduced to $-\partial|\psi_a|^2/\partial t$.

erty of $\langle \vec{r} | \vec{p} \rangle$ used for this purpose is that it is an eigenfunction of the second-order equation

$$-\hbar^2 \nabla^2 \langle \vec{r} | \vec{p} \rangle = \langle \vec{r} | \vec{p} \rangle \langle p_x^2 + p_y^2 + p_z^2 \rangle = \langle \vec{r} | \vec{p} \rangle p^2. \tag{11.17}$$

The same manipulations which change the expression (11.10) of $\langle \vec{p} \rangle_a$ into (11.13) yield, for the kinetic energy,

$$\langle K \rangle_a = \left\langle \frac{p^2}{2m} \right\rangle_a = \int \langle a | \vec{r} \rangle \frac{-\hbar^2}{2m} \nabla^2 \langle \vec{r} | a \rangle d\vec{r}. \tag{11.18}$$

Here again we can analyze the dependence of $\langle K \rangle_a$ on the second derivative of $\langle \vec{r} | a \rangle$ into contributions from the derivatives of $|\langle \vec{r} | a \rangle|$ and of $\phi(\vec{r})$. With some analytical work whose details do not matter here, one finds[7]

$$\langle K \rangle_a = \frac{\hbar^2}{2m} \int d\vec{r} [\, |\psi_a|^2 |\vec{\nabla}\phi|^2 - |\psi_a| \nabla^2 |\psi_a| \,]. \tag{11.19}$$

Thus the *mean kinetic energy consists of two terms:* The first one depends on $\vec{\nabla}\phi$ and hence on the particle flux; this term vanishes when the wave function is real and the flux is zero. The second term depends on $\nabla^2 |\psi_a|$ and hence on variations of the particle density from point to point; this term vanishes when the density is uniform, that is, when the state $|a\rangle$ is a momentum eigenstate. The first term is the average of $|\hbar \vec{\nabla}\phi|^2/2m$ weighted by the particle density $|\psi_a|^2$; it may be compared with $\langle \vec{p}_a \rangle$ which, according to equation (11.15), is the average of $\hbar \vec{\nabla}\phi$ with the same weight. The second term is related to the excess of $\langle |\vec{p}|^2 \rangle_a$ over $|\langle \vec{p} \rangle_a|^2$ since it gives a nonzero contribution to the kinetic energy even when $\langle \vec{p} \rangle_a$ vanishes.

We conclude that a particle may have different kinds of motion for a given value of its mean kinetic energy. The particle flux $\vec{\Phi}(\vec{r}, a)$ may contribute all, some, or none of the kinetic energy. On the other hand, as anticipated in chapter 5, an electron may have nonzero kinetic energy even when the mean current vanishes. An illustration of these different types of motion is given in the next section.

We have seen how physical characteristics of the motion of a particle—particle density and particle flux—are expressed in terms of a wave function. Conversely, a wave function could be determined in principle from data on particle density and flux. This inverse problem is seldom encountered in its general form. Most cases of free particle motion are represented either by a plane wave or by a simple superposition of plane waves.

The method of studying the motion of a particle using derivatives of its wave function serves also to describe the variations of the state of a free particle in

---

7. The $\nabla^2 \langle \vec{r} | a \rangle$ contains imaginary terms whose net contribution to the integral vanishes.

the course of time. In general, this variation is represented by means of a superposition of stationary states, that is, by the formula

$$\psi_a(\vec{r}, t) \equiv \langle \vec{r} \,|\, a(t) \rangle = \int d\vec{p}\, \langle \vec{r} \,|\, \vec{p} \rangle \exp\left(-i\, \frac{p^2/2m}{\hbar}\, t\right) \langle \vec{p} \,|\, a(0) \rangle, \tag{11.20}$$

which is analogous to (11. 2). The time derivative of this wave function is

$$\frac{\partial}{\partial t}\psi_a(\vec{r}, t) = \left(-\frac{i}{\hbar}\right) \int d\vec{p}\, \langle \vec{r} \,|\, \vec{p} \rangle\, \frac{p^2}{2m} \exp\left(-i\, \frac{p^2/2m}{\hbar}\, t\right) \langle \vec{p} \,|\, a(0) \rangle. \tag{11.21}$$

Here, as in the calculation of $\langle K \rangle_a$, the product $\langle \vec{r} \,|\, \vec{p} \rangle p^2$ may be replaced by the left-hand side of (11. 17), and derivatives ∇^2 of this expression can be taken outside the integral. After ∇^2 is taken out, the integral itself reduces to the same form as in (11. 20). Thus equation (11. 21) becomes

$$-\frac{\hbar}{i}\frac{\partial}{\partial t}\psi_a(\vec{r}, t) = -\frac{\hbar^2}{2m}\nabla^2 \psi_a(\vec{r}, t). \tag{11.22}$$

This is a wave equation which determines the variation of the wave function of a free particle in the course of time from a knowledge of its derivatives in space. Equation (11. 22) is a special form of the Schrödinger equation which will be derived in the next chapter.

11. 3. STATIONARY STATES OF ONE-DIMENSIONAL MOTION[8]

This section describes the possible stationary states of a free particle whose motion is parallel to the x axis. Some of these states have kinetic energy due entirely to particle flux, and some due entirely to variations of particle density; all others fall in between. An example of nonstationary state will be described in supplementary section 11. 9.

Momentum eigenstates whose eigenvalues are equal but of different sign

$$p_x = \pm |p_x|, \tag{11.23}$$

have equal eigenvalues of the kinetic energy

$$K = \tfrac{1}{2} m v_x^2 = p_x^2/2m. \tag{11.24}$$

These two eigenstates of p_x are orthogonal and they are degenerate since they have the same energy. If we restrict attention to degenerate states only, disregarding states with different energy, we have here another example of a two-level system. The variety of states of a two-level system has been described

---

8. Recall the remarks which qualify the meaning of the word "motion" for stationary states, in the introduction to this chapter.

in section 9.1. In the example of degenerate states with equal energy $p_x^2/2m$, the various states can be identified as superpositions of $|p_x\rangle$ and $|-p_x\rangle$.

The state $|p_x\rangle$ itself has the wave function (11.4), $\langle x|p_x\rangle = h^{-1/2} \exp(ip_x x/\hbar)$. In this state the particle density, that is, the probability $|\langle x|p_x\rangle|^2$ of finding the particle at x, has the constant value[9] h^{-1} over the whole space. The particle flux, $\Phi(x, p_x) = h^{-1} p_x/m = h^{-1} v_x$, is also uniform. The kinetic energy is due entirely to particle flux. The state $|-p_x\rangle$ has the same characteristics as $|p_x\rangle$ except for opposite flux direction.

A pair of orthogonal states with quite different properties is represented by[10]

$$\begin{aligned} |p_x^2\ +\rangle &= |p_x\rangle\sqrt{1/2} + |-p_x\rangle\sqrt{1/2}, \\ |p_x^2\ -\rangle &= |p_x\rangle i\sqrt{1/2} - |-p_x\rangle i\sqrt{1/2}. \end{aligned} \tag{11.25}$$

These two states are connected to the position eigenstates by the probability amplitudes

$$\begin{aligned} \langle x|p_x^2\ +\rangle &= \left(\frac{2}{h}\right)^{1/2} \cos\frac{p_x x}{\hbar}, \\ \langle x|p_x^2\ -\rangle &= -\left(\frac{2}{h}\right)^{1/2} \sin\frac{p_x x}{\hbar}, \end{aligned} \tag{11.26}$$

which have the analytical form of standing waves. The symbol + (or −) which distinguishes these two states may be called a *parity quantum number* because it characterizes the symmetry of the wave functions (11.26) under reflection of the x coordinate at the origin $x = 0$. In fact, the two states may be called *eigenstates of parity*. Notice that this mode of expression treats a symmetry as a physical quantity like the momentum. This way of treating symmetry is characteristic of quantum physics.

---

9. The physical dimensions of h^{-1} differ from those of a particle density which should normally be a reciprocal length in a one-dimensional problem. This departure originates from the expression (10.12) of the unitarity for probability amplitudes with continuous ranges of eigenvalues. The delta function in equation (10.12), $\delta(\alpha - \beta)$, becomes a pure number only after integration over one of the parameters α and β, which represent momentum values in our case. Actually, problems involving continuous eigenvalues should be formulated more carefully than has been done here, to avoid inconsistency. For example, one could make the momentum eigenvalues discrete by initially confining the particle to a limited volume of space, following the pattern of section A3. The limit for infinite volume is taken only at the end of a complete calculation.

10. The imaginary unit is inserted in $|p_x^2\ -\rangle$ for the purpose of making the wave function $\langle x|p_x^2\ -\rangle$ real. The two states $|p_x^2\ \pm\rangle$ are identified in the model space of section 9.1 by the angles $\theta = \pi/2$ and $\phi = 0$ or π, respectively, as detailed below. The symbol p_x^2 in the kets on the left of (11.25) has the exponent 2 simply to emphasize that negative values of p_x are now excluded; it could be replaced by $|p_x|$. The normalization remains the same as for $|p_x\rangle$.

The particle flux vanishes for the states (11. 25) since their wave functions are real. Hence the kinetic energy stems here entirely from variations of the particle density. These variations follow a typical pattern throughout space, represented by

$$P(x, p_x^2\ +) = \frac{2}{h}\cos^2\frac{p_x x}{\hbar} = h^{-1}\left(1 + \cos\frac{2p_x x}{\hbar}\right),$$
$$P(x, p_x^2\ -) = \frac{2}{h}\sin^2\frac{p_x x}{\hbar} = h^{-1}\left(1 - \cos\frac{2p_x x}{\hbar}\right). \quad (11.\ 27)$$

The two states have sinusoidal distribution of particle density. The positions of density peaks and troughs are staggered for the two states; $P(x, p_x^2\ +)$ has a peak at $x = 0$, whereas $P(x, p_x^2\ -)$ has a trough there. The particle density has thus a waving pattern with period $h/2p_x$, that is, inversely proportional to the square root of the kinetic energy. This pattern is a typical feature of stationary states which are not momentum eigenstates. This is true even when the particle is subject to forces, as will be seen in chapter 12.

Thus we see that a particle moving along the x axis with a given kinetic energy has *qualitatively different* degenerate states of motion. One characteristic type consists of momentum eigenstates with progressive wave functions $\langle x | p_x \rangle$, the other has the standing-wave functions $\langle x | p_x^2\ \pm \rangle$. The existence of qualitatively different states of a two-level system has been stressed in section 9. 1 for the example of light polarization. Light polarization may be circular or linear, particle motion may have flux or standing-wave character. A one-to-one correspondence exists between states of the two-level system of light polarization and of free particle one-dimensional motion. There is just one pair of states of circular polarization and one pair of momentum eigenstates. Circular polarization has full axial symmetry, momentum eigenstates have full symmetry under translation along x. Right and left circular polarizations are distinguished by the direction of rotation of the field, the two momentum eigenstates by the direction of particle flux. On the other hand, each state of linear polarization singles out a constant direction of space, each standing-wave state singles out a constant set of values of x at which the particle density is maximum. Pairs of orthogonal linear polarizations correspond to pairs of standing-wave states, $|p_x^2\ +\rangle$ and $|p_x^2\ -\rangle$, such that the particle density peaks of one state coincide with the density troughs of the other one. As we shall see, there are infinitely many pairs of standing-wave states, just as there are infinitely many pairs of linear polarizations.

The expression (11. 26) of $\langle x | p_x^2\ +\rangle$ may be regarded as the superposition of two terms, $\langle x | p_x \rangle$ and $\langle x | -p_x \rangle$, whose amplitudes are equal and uniform in space and whose phases are different and vary from point to point. Similarly, the electric field of linear polarization may be represented as the superposition of right and left circularly polarized fields whose strengths are equal and constant in time but whose directions are different and vary in the course of time. Accordingly, the variations in space of particle density and the variations in time of squared field strength of linearly polarized light may be regarded as effects of interference. In particular, one may interpret the first term, h^{-1}, on the right-hand side of (11. 27) as $\frac{1}{2}h^{-1} + \frac{1}{2}h^{-1}$, the analog of the first two terms on the right hand side of (8. 9); the second term, $h^{-1} \cos (2p_x x/\hbar)$, represents the interference term. In fact, one often describes the waving pattern of particle density in equations (11. 27)—and in numerous analogous situations—as the "effect of interference" between two component states with opposite particle flux. According to this optical analogy, the peaks and troughs of the density distribution (11. 27) are often called bright and dark interference fringes. The distance $h/2p_x$ between the centers of two successive bright fringes is commonly called a half-wavelength.[11] The particle density (11. 27) vanishes at the center of each dark fringe where the wave function reverses its sign.

Any superposition of the standing-wave states (11. 25) with real coefficients yields another standing-wave state with zero flux and with shifted interference fringes. On the other hand, states with nonzero flux are obtained from the superposition of $|p_x^2\ +\rangle$ and $|p_x^2\ -\rangle$ with complex coefficients or from the superposition of $|p_x\rangle$ and $|-p_x\rangle$ with coefficients of different magnitude.

All degenerate states of motion along the x axis can be mapped on the directions (θ, ϕ) of the model space for a two-level system. If we assign the direction $\theta = 0$ to the momentum eigenstate $|p_x\rangle$ and $\theta = \pi$ to $|-p_x\rangle$, an arbitrary state $|a\rangle$ may be represented by

$$|a\rangle = |p_x\rangle \cos \tfrac{1}{2}\theta \exp(i\tfrac{1}{2}\phi) + |-p_x\rangle \sin \tfrac{1}{2}\theta \exp(-i\tfrac{1}{2}\phi). \qquad (11.28)$$

For this state the particle flux is represented by

$$\Phi(x, a) = h^{-1} \frac{p_x}{m} \cos \theta \qquad (11.28a)$$

and the particle density by

$$P(x, a) = h^{-1}\left[1 + \sin \theta \cos\left(\frac{2p_x x}{\hbar} + \phi\right)\right]. \qquad (11.28b)$$

For this state the particle flux is a fraction $\cos \theta$ of the flux for the momentum eigenstate $|p_x\rangle$. The peaks and troughs of the waving particle density have a height and depth equal, respectively, to $1 \pm \sin \theta$ of the density for the momentum eigenstate.

11.4. COMPLEMENTARITY AND UNCERTAINTY RELATIONS

The sets of probability amplitudes $\langle x|a\rangle$ and $\langle p_x|a\rangle$, which connect a state $|a\rangle$ of a particle to sets of eigenstates of its x coordinate and of its momentum component p_x, are related by the Fourier expansion formula (11. 5). This formula

11. The distance h/p_x is called the de Broglie wavelength of the particle.

coincides—to within inessential details—with the relation between the electric field of radiation at a point of coordinate x, $\vec{E}(x)$, and the coefficient $\vec{E}_{k_x}$ of its Fourier expansion into plane waves. The squared magnitudes $|\langle x|a\rangle|^2$ and $|\langle p_x|a\rangle|^2$ give the probability distributions of the coordinate and of the momentum component of a particle, while $|\vec{E}(x)|^2$ and $|\vec{E}_{k_x}|^2$ provide probability distributions of observing a photon at x and of receiving from it a momentum $\hbar k_x$. This correspondence confirms the surmise of section 6.1 that the position and momentum of a particle are complementary, in the same sense as they are for photon absorption.

Complementarity of two variables is a particular case of incompatibility. It can now be identified by a specific property: The probabilities of eigenvalues of two complementary variables are given by the squared magnitudes of two functions related to one another by a Fourier transformation. In the limiting, usually unrealistic, case where one of the variables has a well-defined value, the complementary variable becomes indeterminate in the sense that all its eigenvalues have equal probability. In the realistic case where eigenvalues of one variable have significantly nonzero probability over a range, say, Δx, eigenvalues of the complementary variable have significant probability over a range Δp_x. This range Δp_x is inversely related to Δx by the bandwidth considerations illustrated in section A.3.

The example of section A.3 concerns a function of a variable t and a complementary oscillation frequency ν whose effective range of variation, $\Delta\nu$, is indicated by equation (A.26) and is approximately reciprocal to the effective range of variation of t. The variable complementary to a coordinate x in the same sense is a wave vector component k_x, to which corresponds in equation (11.4a) the ratio $p_x/\hbar$, rather than p_x itself. Therefore, the inverse relation between Δx and Δp_x is not a simple reciprocity, which would not be expected on dimensional grounds, but a reciprocity between Δx and $\Delta p_x/\hbar$.

In this book we regard the complementarity of the pairs of variables frequency-time, and wave vector-position coordinates, as a phenomenon inherent in the concepts of frequency and of wave vector and grounded in the Fourier analysis of macroscopic variables. This phenomenon carries over to energy and momentum through the quantum effects which connect energies with frequencies and momenta with wave vectors. Historically, the complementarity relations involving energy and momentum first came to Heisenberg's and Bohr's attention through considerations that were not yet clearly related to a wave-mechanical point of view. The formulation of these relations as the Heisenberg uncertainty principle combined with the appearance of statistics in a fundamental role (see chap. 5) to cause surprise and controversy.

With hindsight, Heisenberg's formulae appear as a quantitative refinement of the rough estimates of bandwidth limitations indicated in section A.3. The first step of refinement replaces for a pair of complementary variables their ill-defined range of values with nonzero probability by a precisely defined parameter. The mean square deviation of each variable from its mean value, defined in Appendix D, serves for this purpose. Thus one considers, for example, the quantities Δx^2 and Δp_x^2—that is, the averages of $(x - \langle x\rangle)^2$ and

$(p_x - \langle p_x \rangle)^2$ over the probability distributions $|\langle x|a\rangle|^2$ and $|\langle p_x|a\rangle|^2$—and the corresponding root mean square deviations Δx and Δp_x. Together with Δp_x one may consider the corresponding parameter for the wave-vector component, namely, $\Delta k_x = \Delta p_x/\hbar$. Note that these mean values can also be defined as averages over the energy density or over the intensity distribution of a macroscopic field in space, or over the corresponding variation of Fourier coefficients.

The basic uncertainty relation is derived in supplementary section 11.8. The derivation pertains, in essence, to pairs of variables directly reciprocal to each other, such as x and k_x, but is adapted to momentum components by multiplying k_x and its operator representation, $-i\nabla_x$, by $\hbar$. The Heisenberg uncertainty relations are

$$\Delta x \Delta k_x \geqslant 1/2, \tag{11.29}$$

$$\Delta x \Delta p_x \geqslant 1/2\hbar \tag{11.29a}$$

and analogous ones for the pairs Δy, Δk_y, etc. The analogous relations for time, frequency, and energy variables are

$$\Delta t \Delta \omega \geqslant 1/2, \tag{11.30}$$

$$\Delta t \Delta E \geqslant 1/2\hbar. \tag{11.30a}$$

11.5. PHYSICAL BASIS FOR THE STABILITY AND SIZE OF ATOMS

The preliminary discussion of atomic structure and properties in section 1.4 emphasized that atomic electrons occupy a fairly well-defined volume which does not readily shrink under the attraction of the electrons by the nucleus or under external pressure. It was suggested tentatively that this stability might derive from some sort of expansive tendency of electrons which counterbalances their attraction by the nucleus. The complementarity of position and momentum, as expressed by the uncertainty relations, provides not only a definite basis for that suggestion but also an evaluation of the tendency to expand and of the size of atoms.

Recall, from the beginning of this chapter, that the kinetic and potential energies of a particle are incompatible with one another and with the total energy of the particle. Therefore, an atomic electron in a stationary state does not have a well-defined kinetic energy, K; the mean value $\langle K\rangle = \langle p^2\rangle/2m_e$ is, however, well defined. The essential point to be developed here is that $\langle K\rangle$ has a minimum value, which is determined by the uncertainty relations and is inversely related to the volume within which the electron is confined by the nuclear attraction. Any decrease of this volume requires expenditure of energy to increase $\langle K\rangle$; in this sense one may well say that the electron resists compression, which amounts to the same as its having a tendency to expand.

Even without explicit reference to the uncertainty relations, a minimum value of $\langle K\rangle$ can be estimated from the second term of its expression (11.19). If a particle is confined within a limited volume, the magnitude of its wave function is sharply peaked within that volume, and, therefore, has a large negative second derivative. In general, the mean kinetic energy can be expressed in terms of the square of the mean momentum and of the mean square deviations of the momentum components p_x, p_y, and p_z,

$$\begin{aligned}\langle K\rangle &= \frac{1}{2m_e}\langle |\vec{p}|^2\rangle = \frac{1}{2m_e}\langle\, |\langle\vec{p}\rangle + (\vec{p} - \langle\vec{p}\rangle)\,|^2\rangle \\ &= \frac{1}{2m_e}[\,|\langle\vec{p}\rangle|^2 + \langle|\vec{p} - \langle\vec{p}\rangle|^2\rangle] \\ &= \frac{1}{2m_e}|\langle\vec{p}\rangle|^2 + \frac{1}{2m_e}[\Delta p_x^2 + \Delta p_y^2 + \Delta p_z^2].\end{aligned} \tag{11.31}$$

Using now the uncertainty relation (11.29a) and its analogs for Δp_y and Δp_z, we may set a lower limit to $\langle K\rangle$ even when $|\langle\vec{p}\rangle|^2$ is zero:

$$\langle K\rangle \geqslant {}^1\!/_4 \frac{\hbar^2}{2m_e}\left[\frac{1}{\Delta x^2} + \frac{1}{\Delta y^2} + \frac{1}{\Delta z^2}\right]. \tag{11.32}$$

This relationship means that the mean kinetic energy of an electron cannot vanish whenever the electron is so confined that the mean square deviation of even one of its position coordinates remains finite rather than infinitely large. Assuming a roughly spherical shape for the atom, we may consider a mean square distance, a^2, of the electron from the nucleus such that $1/a^2 = (1/\Delta x^2 + 1/\Delta y^2 + 1/\Delta z^2)/4$, from which follows

$$\langle K\rangle \gtrsim \hbar^2/2m_e a^2. \tag{11.33}$$

The physical basis for this formula is, of course, that if the probability distribution of electron positions is nonzero only within a region of dimensions $\sim a$, values $\sim\hbar/a$ of each momentum component, p_x, p_y, or p_z, have significant probability.

As a measure of the expansive force of the electron we may take the rate of decrease of $\langle K\rangle$ with increasing a, namely,

$$-\frac{d\langle K\rangle}{da} \sim \frac{\hbar^2}{m_e a^3}. \tag{11.34}$$

This force should balance the attraction of the electrons by the nucleus for a value of a of the actual atomic size, according to the suggestion of section 1.4. Note that the nuclear attraction is inversely proportional to the squared distance of the electron from the nucleus and, hence, to $1/a^2$. This consideration shows that the expansive force, being proportional to $1/a^3$, would certainly prevail for sufficiently small a, while the attraction would prevail for large a. An intermediate value of a should then exist for which the two forces balance.

If the electric attraction for the electron of a hydrogen atom is taken to be e^2/a^2, it balances the expansive force (11.34) for

$$a \sim \frac{\hbar^2}{m_e e^2} = 0.53\ \text{Å}. \tag{11.35}$$

This value of a coincides with the Bohr radius a_B which was defined by equation (3.21); it represents the size of the hydrogen atom and happens to represent it exactly, as will be seen in chapter 13.

The discussion of the balance between the expansive force of the electron and its attraction by the nucleus may be formulated more precisely if one assumes a specific form of the electron wave function; this will be done in the following.

11.5.1. *Illustration of Atomic Stability.* The minimum estimate of the mean kinetic energy of an atomic electron, $\langle K \rangle$, provided by the uncertainty relations may be refined by actually calculating $\langle K \rangle$ for an electron state characterized by a tentative wave function of simple analytic form. We choose for this purpose a wave function whose value depends only on the distance r of the electron from the nucleus namely,

$$\langle \vec{r} | a \rangle = \psi_a(\vec{r}) = (\pi a^3)^{-1/2}\, e^{-r/a}. \tag{11.36}$$

Here, the parameter a indicates approximately the mean distance of the electron from the nucleus and the factor in front of the exponential is chosen so that the integral over the probability equals unity. Equation (11.18) may now be used to evaluate $\langle K \rangle$. The calculation is simplified by the fact that $\langle \vec{r} | a \rangle$ does not depend on the direction of $\vec{r}$. In this case ∇^2 takes the form $r^{-2}(d/dr)r^2(d/dr)$, and one finds[12]

$$\langle K \rangle = \frac{\hbar^2}{2m_e a^2}. \tag{11.37}$$

Utilizing the same wave function (11.36), we can also calculate the mean potential energy of the electron due to the nuclear attraction. For the electron of a hydrogen atom, the potential energy is $V(r) = -e^2/r$ and its mean value is

$$\langle V \rangle = \int_0^\infty 4\pi r^2 dr \left(-\frac{e^2}{r}\right) \frac{1}{\pi a^3}\, e^{-2r/a} = -\frac{e^2}{a}. \tag{11.38}$$

Figure 11.1 shows plots of $\langle K \rangle$ and $\langle V \rangle$ as well as of their sum which equals the mean total energy, as function of a. Note how $\langle K \rangle$ predominates over $\langle V \rangle$

---

12. This result appears to coincide formally with the lower bound of $\langle K \rangle$ given by (11.33), but the parameter a does not have quite the same meaning in the two formulae.

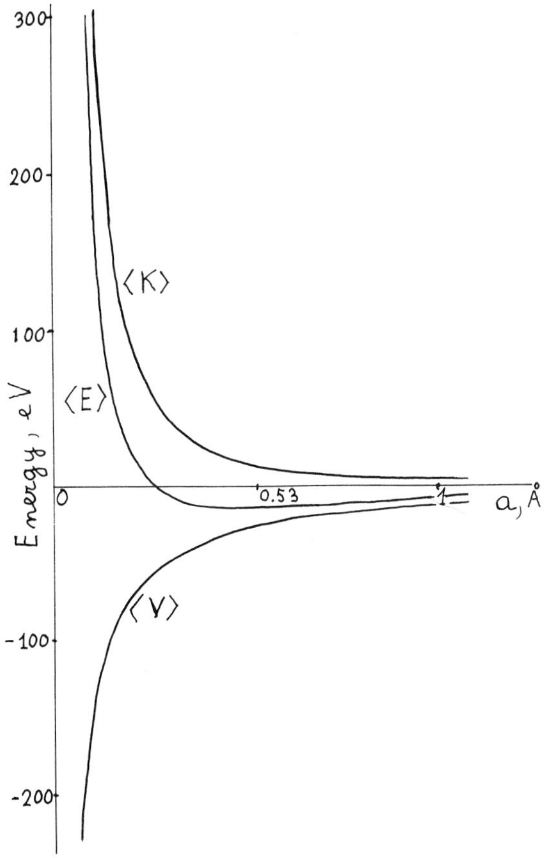

Fig. 11. 1. Dependence of the kinetic, potential, and total energy upon the atomic "radius" a.

for small a and how the situation is reversed for large a. The mean energy $\langle E \rangle$ has a minimum for the Bohr radius value $a = a_B$ given by (11.35).[13]

It will be shown in the next chapter that one may legitimately determine the

---

13. For the value $a = a_B$, the mean kinetic and potential energies also satisfy the *virial theorem*

$$\langle K \rangle = -\tfrac{1}{2} \langle V \rangle$$

which holds in classical mechanics for the kinetic and potential energy of any system of particles interacting with $1/r^2$ forces (see *BPC*, 1: 296-299). The application of this theorem to the hydrogen atom in Bohr's semi-classical calculation was mentioned in the footnote on p. 78.

approximate wave function of a stationary state by assuming its analytical form, estimating $\langle K\rangle$, $\langle V\rangle$, and $\langle E\rangle$ as we have done, and choosing the value of any undetermined parameters for which $\langle E\rangle$ is a minimum. In our example the wave function (11.36) happens to have the correct form of the wave function for the ground state of the hydrogen atom, and the minimum value of $\langle E\rangle$ in Figure 11.1, namely, -13.6 eV, yields the correct eigenvalue.

SUPPLEMENTARY SECTIONS

S11.6. CALCULATION OF PROBABILITY AMPLITUDES FOR ONE-DIMENSIONAL MOTION

The sets of eigenstates of the position and of the velocity of a particle moving on the x axis are connected by a transformation matrix $|\langle v_x|x\rangle|$. To calculate this matrix we utilize the general method of translating a macroscopic law into quantum-mechanical language. In this case the relationship $dx/dt = v_x$ is translated into the equation between mean values

$$\frac{d\langle x\rangle}{dt} = \langle v_x\rangle. \tag{11.39}$$

We apply this law initially to the mean values of x and v_x for an arbitrary non-stationary state $|a(t)\rangle$. However, our aim is really to derive from equation (11.39) an equation which holds independently of this arbitrary state and indeed is expressed without reference to it.

Recall, to this end, that mean values have been expressed in section 8.5 and again in chapter 10 by means of two separate types of quantities, namely, (a) probability amplitudes which identify the state under consideration as a superposition of states of a base set (these amplitudes are independent of the variable whose mean value is being calculated) and (b) matrices which represent the variable of interest in the base set representation and are altogether independent of the state $|a(t)\rangle$. Our first objective is then to extract from equation (11.39) an equation which relates the matrices of x and v_x in the same base-set representation. From this equation we will obtain the transformation matrix $|\langle v_x|x\rangle|$. Since the time dependence of mean values appears explicitly in equation (11.39), it is convenient to use a base set of stationary states; the eigenstates of v_x are stationary in a free-particle problem.

The mean value $\langle x\rangle$ in (11.39) is represented, using the base set of eigenstates of v_x, by an adaptation of (10.19) and (10.20),

$$\langle x\rangle = \int_{-\infty}^{\infty}\int_{-\infty}^{\infty} dv_x\, dv_x'\, \langle a(t)|v_x\rangle\langle v_x|x|v_x'\rangle\langle v_x'|a(t)\rangle, \tag{11.40}$$

where the matrix elements of x are defined by

$$\langle v_x|x|v_x'\rangle = \int_{-\infty}^{\infty} dx \langle v_x|x\rangle\, x\,\langle x|v_x'\rangle. \tag{11.40a}$$

The fact that the definition (11.40a) depends on the unknown amplitudes $\langle v_x | x \rangle$ will serve to obtain these amplitudes after $\langle v_x | x | v'_x \rangle$ itself has been calculated. From the expression (11.40) of $\langle x \rangle$ we must now calculate $d\langle x \rangle/dt$, which is to be entered in equation (11.39).

In the expression of $\langle x \rangle$ only the probability amplitudes depend on time. Their variation, prescribed by equations (10.14) and (10.14a), is

$$\begin{aligned} \langle a(t) | v_x \rangle &= \langle a(0) | v_x \rangle \exp (i \tfrac{1}{2} m v_x^2 \, t/\hbar), \\ \langle v'_x | a(t) \rangle &= \exp(-i \tfrac{1}{2} m v'^2_x \, t/\hbar) \langle v'_x | a(0) \rangle . \end{aligned} \tag{11.41}$$

These functions obey the differential equations

$$\begin{aligned} \frac{d}{dt} \langle a(t) | v_x \rangle &= \langle a(t) | v_x \rangle \, i \, \frac{\tfrac{1}{2} m v_x^2}{\hbar}, \\ \frac{d}{dt} \langle v'_x | a(t) \rangle &= -i \, \frac{\tfrac{1}{2} m v'^2_x}{\hbar} \langle v'_x | a(t) \rangle . \end{aligned} \tag{11.42}$$

Accordingly, the derivative of equation (11.40) is

$$\begin{aligned} \frac{d}{dt} \langle x \rangle = \int_{-\infty}^{\infty} \int_{-\infty}^{\infty} dv_x \, dv'_x \, \langle a(t) | v_x \rangle \Bigg[i \, \frac{\tfrac{1}{2} m v_x^2}{\hbar} \langle v_x | x | v'_x \rangle \\ + \langle v_x | x | v'_x \rangle \left(-i \, \frac{\tfrac{1}{2} m v'^2_x}{\hbar} \right) \Bigg] \langle v'_x | a(t) \rangle . \end{aligned} \tag{11.43}$$

The mean value of v_x,

$$\langle v_x \rangle = \int_{-\infty}^{\infty} dv_x \langle a(t) | v_x \rangle \; v_x \; \langle v_x | a(t) \rangle , \tag{11.44}$$

is expressed by means of the same probability amplitudes which appear in equation (11.43) but has the form of a simple integral whereas equation (11.43) is a double integral. An analogous difference was found in section 8.5 between the expression (8.26c) of $\langle J_z \rangle$ as a single sum and the expression (8.28a) of $\langle J_x \rangle$ as a double sum. To eliminate this difference we introduced there the alternative form (8.28c) of J_z in which the eigenvalues $m_z \hbar$ of J_z were cast in the form (8.29c), that is, as the diagonal matrix $m_z \hbar \delta_{m_z m'_z}$. The same method will be applied here to v_x. However, since the eigenvalues of v_x form a continuous set, the Kronecker symbol is now replaced by the Dirac delta function. Thus we cast the eigenvalues of v_x in the form of a diagonal matrix $v_x \delta(v_x - v'_x)$, so that (11.44) takes the equivalent form

$$\langle v_x \rangle = \int_{-\infty}^{\infty} \int_{-\infty}^{\infty} dv_x dv'_x \langle a(t) | v_x \rangle \; v_x \; \delta(v_x - v'_x) \langle v'_x | a(t) \rangle . \tag{11.45}$$

Our initial equation (11.39) requires equations (11.43) and (11.45) to be equal. This equality holds for all states $|a(t)\rangle$, that is, irrespective of the values of the probability amplitudes $\langle v'_x | a(t) \rangle$, if and only if the factors multiplying the probability amplitudes in the integrands of both formulae are equal. Thus we obtain the desired equation between matrix elements

$$\frac{i}{\hbar}\,\tfrac{1}{2}m\,(v_x^2 - v_x'^2)\langle v_x|x|v_x'\rangle = v_x\delta(v_x - v_x'). \tag{11.46}$$

A simplification of this formula is obtained by noticing that the factor v_x on the right-hand side can be replaced by $\tfrac{1}{2}(v_x + v_x')$ since the other factor, $\delta(v_x - v_x')$, vanishes for $v_x' \neq v_x$. This factor, $\tfrac{1}{2}(v_x + v_x')$, then cancels on the two sides of equation (11.46), whereby

$$\frac{i}{\hbar}\,m(v_x - v_x')\langle v_x|x|v_x'\rangle = \delta(v_x - v_x'). \tag{11.47}$$

Equation (11.47) determines the matrix element $\langle v_x|x|v_x'\rangle$ as the ratio of Dirac's "generalized function" $\delta(v_x - v_x')$ and of $(i/\hbar)m(v_x - v_x')$. This ratio is another generalized function of $v_x - v_x'$ indicated by

$$\langle v_x|x|v_x'\rangle = -\frac{\hbar}{mi}\,\delta'(v_x - v_x'), \tag{11.48}$$

where δ' is the derivative of the Dirac delta.[14]

From this initial result we want to determine the probability amplitudes. An equation for this purpose is obtained by multiplying both sides of equation (11.48) by $\langle v_x'|x\rangle$ and integrating over v_x':

$$\int\langle v_x|x|v_x'\rangle\langle v_x'|x\rangle dv_x' = -\frac{\hbar}{mi}\int\delta'(v_x - v_x')\,\langle v_x'|x\rangle dv_x'. \tag{11.49}$$

The integral on the left-hand side of equation (11.49) is worked out by utilizing the definition (11.40a) of $\langle v_x|x|v_x'\rangle$ and the unitary property (10.12) of the probability amplitudes,

$$\int_{-\infty}^{\infty}\langle v_x|x|v_x'\rangle\langle v_x'|x\rangle dv_x' = \int_{-\infty}^{\infty}\left[\int_{-\infty}^{\infty}\langle v_x|x'\rangle\,x'\langle x'|v_x'\rangle dx'\right]\langle v_x'|x\rangle dv_x'$$

$$= \int_{-\infty}^{\infty}\langle v_x|x'\rangle\,x'\;\delta(x' - x)dx' = \langle v_x|x\rangle x. \tag{11.50}$$

The right-hand side of equation (11.49) is given by the basic property of

---

14. The Dirac delta is identified by the property (A.47), $\int_{-\infty}^{\infty} f(t')\delta(t - t')dt' = f(t)$, its derivative by $\int_{-\infty}^{\infty} f(t')\delta'(t - t')dt' = df(t)/dt$. The two functions δ' and δ are also related by the property, utilized in obtaining equation (11.48) from (11.47),

$$\delta(t - t') = -(t - t')\delta'(t - t').$$

This property is verified by multiplying both sides by $f(t')$ and integrating over t'. The left hand side yields $f(t)$. The right hand side yields $-\{d[f(t')(t - t')]/dt'\}_{t'=t}$ which coincides with $f(t)$.

$\delta'(v_x - v_x')$, whereby (11.50) reduces to

$$\langle v_x | x \rangle\, x = -\frac{\hbar}{mi}\frac{d}{dv_x}\langle v_x | x \rangle. \tag{11.51}$$

The solution of this equation, with an arbitrary factor chosen to make the matrix unitary, is

$$\langle v_x | x \rangle = \left(\frac{m}{h}\right)^{1/2} \exp(-imv_x x/\hbar). \tag{11.52}$$

and is equivalent to expression (11.4b).

S11.7. THE LANGUAGE OF OPERATORS

In the preceding section we have disentangled laboriously the equation (11.46) between matrix elements from the initial compact equation between mean values $d\langle x \rangle/dt = \langle v_x \rangle$. Section 11.2 has developed alternative expressions of mean values by transformations of differential and integral calculus. There exists, however, a formalism that provides unified synthetic statements of the relations between different physical variables, that is, between their eigenstates and eigenvalues.

This formalism may be derived from the representation of states of a system as vectors of a model space (see section 8.5.1). A main feature of vector notation is that it represents relations among vectors without specifying any set of coordinate axes. In our development thus far we have used specific "representations," that is, coordinate axes, to express mean values or relations between matrices. Quantum mechanics uses an *operator* notation which avoids reference to any specific representation.

Consider, for example, the expression on the right of equation (11.44) for the mean value $\langle v_x \rangle$. It consists of an integral over products of three factors: (a) the set of probability amplitudes $\langle v_x | a(t) \rangle$ which identify the state $|a(t)\rangle$ in relation to the eigenstates $|v_x\rangle$; (b) the set of eigenvalues v_x; (c) the set of complex-conjugate probability amplitudes $\langle a(t) | v_x \rangle$. In operator notation this equation is written as

$$\langle v_x \rangle = \psi_a{}^\dagger\, \nu_x\, \psi_a, \tag{11.53}$$

where ν_x indicates the velocity operator which "multiplies" the *state vector* ψ_a on its right or the *Hermitian conjugate* $\psi_a{}^\dagger$ on its left. Equation (11.44) is regarded as a particular form of (11.53); (11.45) as another form. Still another form would be given in "position representation" by $\langle v_x \rangle = \int dx \int dx' \langle a(t) | x \rangle \langle x | v_x | x' \rangle \langle x' | a(t) \rangle$; here the operator ν_x is represented by the matrix $\langle x | v_x | x' \rangle$. "Application" of an operator changes a state much as multiplication by a tensor changes a vector. The symbols $\psi_a{}^\dagger$ and ψ_a are equivalent to a bra, $\langle a |$, and ket, $|a\rangle$, respectively; in fact, (11.53) could be expressed as $\langle v_x \rangle_a = \langle a | v_x | a \rangle$.

Equation (11.43) involves on its right-hand side two physical quantities, namely, the kinetic energy and the position; they can be represented by the two operators $\mathcal{K}$ and x. The equation takes then the form

$$\frac{d\langle x\rangle}{dt} = \psi_a{}^\dagger \frac{i}{\hbar} (\mathcal{K}x - x\mathcal{K})\, \psi_a. \tag{11.54}$$

Note that an operator product does not have the commutative property, in general, just as a product of tensors does not. The requirement that the mean values (11.53) and (11.54) be equal for all states ψ_a leads to the synthetic operator equation

$$\frac{i}{\hbar} (\mathcal{K}x - x\mathcal{K}) = \nu_x, \tag{11.55}$$

of which (11.46) is the matrix form in the representation of eigenstates of v_x.

To each physical variable corresponds a quantum mechanical operator. For instance, the operator p_x corresponds to the momentum component p_x; equation (11.13) shows that $(\hbar/i)\nabla_x$ is the differential form of p_x in the position representation. Equations (11.10) and (11.13) are alternative forms of $\psi_a{}^\dagger \vec{p}\, \psi_a$. Equation (11.47) represents the key property of p_x which identifies the connection between its eigenstates and those of x, namely

$$\frac{i}{\hbar} (p_x x - x p_x) = 1 \tag{11.56}$$

Here the identity operator 1 corresponds to the delta-function which is understood to represent a unit matrix; the operator p_x is written on the right of x when it stands for the eigenvalue mv_x' which corresponds in (11.47) to the eigenstate on the right of the matrix $\langle v_x|x|v_x'\rangle$.

The operator language provides a mathematical formulation of the concept of incompatibility of physical variables. Consider the *commutator* $\mathcal{AB} - \mathcal{BA}$ constructed with the operators of two variables A and B. If these variables are compatible, they have a common set of eigenstates $|\alpha)$. In the representation of $|\alpha\rangle$, both $\mathcal{A}$ and $\mathcal{B}$ reduce to diagonal matrices $A_\alpha\delta_{\alpha\alpha'}$ and $B_\alpha\delta_{\alpha\alpha'}$, whose commutator vanishes. The commutator of $\mathcal{A}$ and $\mathcal{B}$ vanishes then in any representation. The statement that A and B are compatible is equivalent to the operator equation $\mathcal{AB} - \mathcal{BA} = 0$. From this point of view equation (11.54) shows that the incompatibility of position and kinetic energy is inherent in nonzero velocity.

S11.8. DERIVATION OF UNCERTAINTY RELATIONS

The uncertainty relations are analogous to a standard formula of statistics. The product of the two mean square derivations of two statistically distributed variables m and n, $\Delta m^2\ \Delta n^2$, cannot be smaller than the mean square deviation of their product, $(\Delta mn)^2$; otherwise the correlation coefficient of m and n, defined in Appendix D, would exceed unity. Thus one establishes the standard inequality $\Delta m^2 \Delta n^2 \geqslant (\Delta mn)^2$. However, the mean square deviation of the product of two macroscopic variables, $(\Delta mn)^2$, may vanish and does so when the two variables are uncorrelated, whereas the analogous deviation for two complementary quantum-mechanical variables has a minimum nonzero value.

The uncertainty relation (11.29) will be established here in three steps. First, Δk_x^2 will be expressed, like Δx^2, as an integral over $\vec{r}$. Then we shall obtain the analog of the statistical formula $\Delta m^2 \Delta n^2 \geqslant (\Delta mn)^2$. Finally, the analog of $(\Delta mn)^2$ will be evaluated.

The mean square deviation Δx^2 for a particle in a state $\langle \vec{r} | a \rangle$ is given by

$$\begin{aligned} \Delta x^2 &= \int d\vec{r} (x - \langle x \rangle)^2 |\langle \vec{r} | a \rangle|^2 \\ &= \int d\vec{r} |f(\vec{r})|^2 \equiv \langle |f|^2 \rangle \end{aligned} \tag{11.57}$$

with

$$f(\vec{r}) = (x - \langle x \rangle) \langle \vec{r} | a \rangle. \tag{11.57a}$$

We seek now an analogous expression for Δk_x^2.

Section 11.2 shows how to calculate the mean value and mean square deviation of a momentum component p_x utilizing the representation $(\hbar/i)\nabla_x$ of p_x (see eq. [11.13]). Since the wave-vector component k_x is one and the same thing as the ratio $p_x/\hbar$, k_x is represented by $-i\nabla_x$ for the purpose of mean value calculations. We write then

$$\begin{aligned} \Delta k_x^2 &= \langle (k_x - \langle k_x \rangle)^2 \rangle = \langle (k_x - \langle k_x \rangle)(k_x - \langle k_x \rangle) \rangle \\ &= \int d\vec{r} \langle a | \vec{r} \rangle (-i\nabla_x - \langle k_x \rangle)(-i\nabla_x - \langle k_x \rangle) \langle \vec{r} | a \rangle. \end{aligned} \tag{11.58}$$

To achieve analogy with equation (11.52), the integral in equation (11.58) is now transformed so as to be obviously nonnegative. This result is achieved by integration by parts, with respect to the first factor $-i\nabla_x$, which yields

$$\begin{aligned} \Delta k_x^2 &= \int d\vec{r} \, [(i\nabla_x - \langle k_x \rangle) \langle a | \vec{r} \rangle] \, [(-i\nabla_x - \langle k_x \rangle) \langle \vec{r} | a \rangle] \\ &= \int d\vec{r} |g(\vec{r})|^2 \equiv \langle |g|^2 \rangle, \end{aligned} \tag{11.59}$$

where

$$g(\vec{r}) = (-i\nabla_x - \langle k_x \rangle) \langle \vec{r} | a \rangle. \tag{11.59a}$$

As a next step we state and prove the inequality

$$\Delta x^2 \Delta k_x^2 = \langle |f|^2 \rangle \langle |g|^2 \rangle \geqslant |\langle f^* g \rangle|^2, \tag{11.60}$$

which is analogous to $\Delta m^2 \Delta n^2 \geqslant (\Delta mn)^2$ and holds for any pair of well-behaved functions $f(\vec{r})$, $g(\vec{r})$. Consider the expression $|\langle |g|^2 \rangle f - \langle fg^* \rangle g|^2$, which is the square of a magnitude and, therefore, certainly nonnegative. The integral of this expression over all points of space is also certainly nonnegative; therefore, we write

$$\begin{aligned} 0 &\leq \int |\langle |g|^2\rangle f - \langle fg^*\rangle g|^2 \, d\vec{r} \\ &= \langle |g|^2\rangle^2 \langle |f|^2\rangle - \langle |g|^2\rangle\langle fg^*\rangle\langle f^*g\rangle - \langle |g|^2\rangle\langle f^*g\rangle\langle fg^*\rangle + |\langle fg^*\rangle|^2\langle |g|^2\rangle \\ &= \langle |g|^2\rangle\{\langle |f|^2\rangle\langle |g|^2\rangle - |\langle f^*g\rangle|^2\}. \end{aligned} \quad (11.61)$$

The factor $\langle |g|^2\rangle$ is certainly nonnegative, and therefore the same must hold for the factor in braces, which proves the inequality (11.60).

Equation (11.60) would be equivalent to the relation $\Delta m^2 \Delta n^2 \geq (\Delta mn)^2$ of macroscopic statistics if the functions f and g were real. Actually, $\langle f^*g\rangle$ is complex, but the squared magnitude $|\langle f^*g\rangle|^2$ may be expressed as the sum of two squared real quantities, each of which may represent the mean value of a physical variable,

$$\begin{aligned} |\langle f^*g\rangle|^2 &= |\langle \tfrac{1}{2}(f^*g + fg^*)\rangle + i\langle \tfrac{1}{2}(f^*g - fg^*)/i\rangle|^2 \\ &= \langle \tfrac{1}{2}(f^*g + fg^*)\rangle^2 + \langle \tfrac{1}{2}(f^*g - fg^*)/i\rangle^2. \end{aligned} \quad (11.62)$$

The essential point of our problem lies then in the identification and evaluation of the two separate quantities whose mean values equal, respectively, the real and imaginary part of $\langle f^*g\rangle$.

Utilizing the definitions (11.57a) and (11.59a) of f and g, we have

$$\langle f^*g\rangle = \int d\vec{r}\langle a|\vec{r}\rangle (x - \langle x\rangle) (-i\nabla_x - \langle k_x\rangle)\langle \vec{r}\,|a\rangle. \quad (11.63)$$

In this formula the constant $\langle k_x\rangle$ is multiplied by $\langle x - \langle x\rangle\rangle = 0$ and hence can be deleted; thus equation (11.63) reduces to

$$\langle f^*g\rangle = \int d\vec{r}\, \langle a|\vec{r}\rangle\, (x - \langle x\rangle)[-i\nabla_x\langle \vec{r}\,|a\rangle]. \quad (11.64)$$

Similarly we have

$$\langle fg^*\rangle = \int d\vec{r}\,[i\nabla_x\langle a|\vec{r}\rangle](x - \langle x\rangle)\langle \vec{r}\,|a\rangle. \quad (11.64a)$$

Of the two mean values whose squares appear in (11. 62), one is

$$\begin{aligned} \tfrac{1}{2}[\langle f^*g\rangle + \langle fg^*\rangle] &= \int d\vec{r}\,(x - \langle x\rangle)\tfrac{1}{2}\{\langle a|\vec{r}\rangle[-i\nabla_x\langle \vec{r}\,|a\rangle] + [i\nabla_x\langle a|\vec{r}\rangle]\langle \vec{r}\,|a\rangle\} \\ &= \frac{m}{\hbar}\int d\vec{r}\,(x - \langle x\rangle)\Phi_x(\vec{r}, a), \end{aligned} \quad (11.65)$$

where $\Phi_x(\vec{r}, a)$ is the x component of the vector $\vec{\Phi}(\vec{r}, a)$ defined by equation (11.16a). Equation (11.65) represents to within a factor $m/\hbar$ the mean deviation of x weighted by the particle flux in the x direction. It constitutes the quantum-mechanical analog of the root mean square deviation of the product of x and k_x. The other mean value of interest is

$$\frac{1}{2}\frac{1}{i}[\langle f^*g\rangle - \langle fg^*\rangle] = -\frac{1}{2}\int d\vec{r}\,(x - \langle x\rangle)\,\{\langle a|\vec{r}\rangle[\nabla_x\langle\vec{r}|a\rangle] + [\nabla_x\langle a|\vec{r}\rangle]\langle\vec{r}|a\rangle\}$$
$$= -\frac{1}{2}\int d\vec{r}\,(x - \langle x\rangle)\nabla_x[\langle a|\vec{r}\rangle\langle\vec{r}|a\rangle]. \tag{11.66}$$

Integration by parts reduces the last integral to the integral of $\langle a|\vec{r}\rangle\langle\vec{r}|a\rangle$, which equals unity regardless of the characteristics of the state $|a\rangle$, so that

$$\frac{1}{2}\frac{1}{i}[\langle f^*g\rangle - \langle fg^*\rangle] = \frac{1}{2}. \tag{11.67}$$

This is the key result of the calculation, namely, that the last term of (11.62) equals ¼ irrespective of the state of the particle. Consequently, the left-hand side of (11.60) is certainly not smaller than ¼. Hence follow the uncertainty relations (11.29).

S11.9. AN EXAMPLE: THE GAUSSIAN WAVE PACKET

The motion of a free particle can be described in great detail and compared with the macroscopic limit when the initial wave function has a suitably simple analytical form. A well-known example is afforded when the wave function at $t = 0$ is represented by a plane wave with a Gaussian amplitude modulation

$$\psi_a(\vec{r}, 0) \equiv \langle\vec{r}|a(0)\rangle = (2\pi b^2)^{-3/4}\exp\left(-\frac{r^2}{4b^2} + i\,\frac{m\vec{v}_0\cdot\vec{r}}{\hbar}\right). \tag{11.68}$$

For this state the particle density $|\psi_a|^2$ and the particle flux $\vec{\Phi}(\vec{r}, a)$, defined by (11.16) are

$$|\psi_a(\vec{r}, 0)|^2 = (2\pi b^2)^{-3/2}\exp\left(-\frac{r^2}{2b^2}\right), \tag{11.68a}$$

and

$$\vec{\Phi}(\vec{r}, a) = (2\pi b^2)^{-3/2}\exp\left(-\frac{r^2}{2b^2}\right)\vec{v}_0. \tag{11.68b}$$

The analysis of this state and of its variations in the course of time are facilitated by the following mathematical properties:

a) The wave function (11.68) is the product of three functions $f_x(x)f_y(y)f_z(z)$ with the same analytical form

$$f_x(x) = (2\pi b^2)^{-1/4}\exp\left(-\frac{x^2}{4b^2} + \frac{imv_{0x}x}{\hbar}\right). \tag{11.69}$$

b) The Gaussian function has the integral

$$\int_{-\infty}^{\infty}\exp(-\alpha x^2)\,dx = (\pi/\alpha)^{1/2}; \tag{11.70}$$

this property serves to verify that the integral of the density (11.68a) over the whole space equals unity.

c) The $\int_{-\infty}^{\infty} x \exp(-\alpha x^2)\, dx$ vanishes for reasons of symmetry.

d) The Gaussian has also the integral

$$\int_{-\infty}^{\infty} x^2 \exp(-\alpha x^2)\, dx = \frac{(\pi/\alpha)^{1/2}}{2\alpha}. \tag{11.71}$$

Therefore, the root mean square deviations of the particle density (11.68a) from its center, at $\langle \vec{r} \rangle = 0$, are $\Delta x = \Delta y = \Delta z = b$.

e) The Fourier integral of the Gaussian function is another Gaussian function. This property is found by noticing that

$$\exp(-\alpha x^2 - ikx) = \exp[-\alpha (x + ik/2\alpha)^2] \exp(-k^2/4\alpha); \tag{11.72}$$

the $\int_{-\infty}^{\infty} \exp[-\alpha (x + ik/2\alpha)^2] dx$ can be evaluated as in item (a) above with a change of integration variable.

Property (b) underlies the value of the normalization coefficient $(2/\pi b^2)^{-3/4}$ in equation (11.68), and property (d) indicates the physical significance of the parameter b. Property (e) permits the explicit calculation of the probability amplitudes $\langle \vec{p}\, | a(0) \rangle$ by application of the Fourier transformation formula (11.7) and shows that $\langle \vec{p}\, | a(0) \rangle$ is represented by the Gaussian function

$$\langle \vec{p}\, | a(0) \rangle = (2\pi m^2 v_1^2)^{-3/4} \exp\left(- \frac{|\vec{p} - m\vec{v}_0|^2}{4m^2 v_1^2} \right). \tag{11.73}$$

where

$$v_1 = \tfrac{1}{2} \frac{\hbar}{mb}. \tag{11.73a}$$

It follows from this formula and from properties (c) and (d) that the momentum has the mean value $\langle \vec{p} \rangle = m\vec{v}_0$, which coincides with the most probable value, and that the possible values of $\vec{p}$ are distributed about the mean with root mean square deviations $\Delta p_x = \Delta p_y = \Delta p_z = mv_1$. The value of v_1 shows that these deviations attain the minimum consistent with the uncertainty principle (11.29a). (In other words, the contribution of eq. [11.65] to eq. [11.62] vanishes for a state with the wave function [11. 68].) The mean kinetic energy can be obtained from equation (11.73) — or also from the wave function (11.68) by using equation (11.19) — and is

$$\langle K \rangle = \frac{\langle |\vec{p}|^2 \rangle}{2m} = \tfrac{1}{2} m v_0^2 + 3 \times \tfrac{1}{2} m v_1^2 = \tfrac{1}{2} m v_0^2 + 3 \frac{\hbar^2}{8mb^2}. \tag{11.74}$$

As expected from section 11.2, $\langle K \rangle$ consists of two terms. The first represents the kinetic energy corresponding to the particle flux (11.68b); the second has the same form as (11.37) and corresponds to the expansive force of the particle. Note that the second term is inversely related to the mass of the particle and to the packet "size" b.

The variation of the state $|a(t)\rangle$ in the course of time is represented in the standard manner by multiplying equation (11.73) by $\exp(-ip^2t/2m\hbar)$, since momentum eigenstates are stationary for a free particle. The time-dependent wave function may then be obtained by inverse Fourier transformation utilizing once more the property (e). One finds

$$\psi_a(\vec{r}, t) = \int d\vec{p} \langle \vec{r} | \vec{p} \rangle \exp(-ip^2t/2m\hbar) \langle \vec{p} | a(0) \rangle$$

$$= [2\pi(b + iv_1t)^2]^{-3/4} \exp\left[-\frac{|\vec{r} - \vec{v}_0t|^2}{4b(b + iv_1t)} + i\frac{m\vec{v}_0}{\hbar} \cdot \vec{r} - i\frac{\frac{1}{2}mv_0^2}{\hbar}t\right]. \qquad (11.75)$$

(This result could have been obtained by direct integration of eq. [11.22] with the initial condition that ψ_a coincide with eq. [11.68] at $t = 0$.) The wave function (11.75) yields the particle density

$$|\psi_a(\vec{r}, t)|^2 = [2\pi(b^2 + v_1^2t^2)]^{-3/2} \exp\left[-\frac{|\vec{r} - \vec{v}_0t|^2}{2(b^2 + v_1^2t^2)}\right]. \qquad (11.75a)$$

Thus the wave packet travels with the group velocity $\vec{v}_0$; it spreads out progressively with square deviations

$$\Delta x^2 = \Delta y^2 = \Delta z^2 = b^2 + v_1^2t^2. \qquad (11.76)$$

The simultaneous displacement and spreading out of the packet are described respectively by two terms of the particle flux

$$\vec{\Phi}(\vec{r}, a(t)) = |\psi_a(\vec{r}, t)|^2 \left(\vec{v}_0 + \frac{v_1^2t}{b^2 + v_1^2t^2}(\vec{r} - \vec{v}_0t)\right). \qquad (11.75b)$$

In the macroscopic limit—that is, for a particle with large mass—the velocity v_1 is negligible, the wave packet travels without appreciable spreading, and the flux remains proportional to $\vec{v}_0$.

CHAPTER 11 PROBLEMS

11.1. A particle moves along the x axis in a nonstationary state. At the time $t = 0$ its wave function is the superposition of two wave functions $\langle x | p_x^2 + \rangle$ with different kinetic energies and is given by $\psi_a(x, 0) = h^{-1/2}(\cos kx + \cos 2kx)$. Calculate the particle density and particle flux as functions of time and identify the limits between which these quantities oscillate. (Eq. [11.16a] is convenient here for the flux calculation.) Verify the continuity equation $\partial\Phi_x/\partial x + \partial|\psi_a(x, t)|^2/\partial t = 0$.

11.2. Consider an electron whose state is represented by the wave function $(\pi a^3)^{-1/2}\exp(-r/a)$ as in section 11.5.1. Calculate the probability P(R) that the electron lies within a sphere of radius R about the origin of coordinates. Find the value $R_{1/2}$ for which $P(R_{1/2}) = \frac{1}{2}$.

11.3. Calculate the value of $\Delta x \Delta p_x$ for the particle state with the wave function (11.68), as a function of time. (a) Verify that $\Delta x \Delta p_x \geqslant \frac{1}{2}\hbar$ at all times. (b) Show that any excess of $\Delta x \Delta p_x$ over $\frac{1}{2}\hbar$ is represented by the first terms on the right side of equation (11.62).

11.4. The nonstationary state of a free particle is represented at $t = 0$ by a standing wave with Gaussian amplitude modulation, $\psi(\vec{r}, 0) = (2\pi b^2)^{-3/4} \exp(-r^2/4b^2)\, 2^{1/2} \cos(m\vec{v}_0 \cdot \vec{r}/\hbar)$. Calculate how this state evolves in the course of time, studying in particular whether and under what conditions the standing-wave pattern of particle density remains in evidence. (*Hint*: the state considered in this problem may be regarded as the superposition of states studied in section S11.9.)

SOLUTIONS TO CHAPTER 11 PROBLEMS

11.1. Application of equation (10.16) gives $|\psi_a(x,t)|^2 = h^{-1}[\cos^2 kx + \cos^2 2kx + 2\cos kx \cos 2kx \cos \omega t]$, where $\omega = (E_2 - E_1)/\hbar = [(2k\hbar)^2/2m - (k\hbar)^2/2m]/\hbar = 3\hbar k^2/2m$. The particle density has extreme limits when $\cos \omega t = 1$ or -1. For $t = 0$, $\cos \omega t = 1$, we have $|\psi_a(x)|^2 = h^{-1}[\cos kx + \cos 2kx]^2 = h^{-1}\, 4\cos^2 \frac{1}{2}kx \cos^2 \frac{3}{2}kx = h^{-1}\, 2\cos^2 \frac{1}{2}kx\,(1 + \cos 3kx)$ which represents a modulated waving pattern. For $t = 2\pi m/3hk^2$, $\cos \omega t = -1$, we have $|\psi_a(x)|^2 = h^{-1}\, 2\sin^2 \frac{1}{2}kx\,(1 - \cos 3kx)$. The pattern of particle density changes so that at time $t = 2\pi m/3hk^3$ a trough is at the position of each peak at $t = 0$. Application of equation (11.16a) gives $\Phi_x(x, a) = (\hbar/mh)[\cos 2kx\,(d \cos kx/dx) - \cos kx(d \cos 2kx/dx)]\sin \omega t = (k/4\pi m)(3\sin kx + \sin 3kx)\sin \omega t$. This flux distribution has an oscillatory pattern along the x axis and reverses its sign at each point in the course of time. To verify the continuity equation, work out the flux divergence $\partial\Phi_x/\partial x = (3k^2/2\pi m)2\cos kx \cos 2kx \sin \omega t$ and see that it equals the time derivative of the oscillating term of $|\psi_a(x,t)|^2$.

11.2. Integration of the particle density from 0 to R yields $P(R) = \int_0^R \exp(-2r/a)4\pi r^2 dr/\pi a^3 = 1 - \exp(-2R/a)\,[1 + 2R/a + 2R^2/a^2]$. Numerical evaluation of this function yields $R_{1/2} = 1.34\, a$.

11.3. (a) Equation (11.76) gives $\Delta x = (b^2 + v_1^2 t^2)^{1/2}$. Equation (11.73) and property (d) give $\Delta p_x = mv_1$ at $t = 0$; this value remains constant because the probability function $|\langle \vec{p} | a(t)\rangle|^2$ is stationary. It follows that $\Delta x \Delta p_x = \frac{1}{2}\hbar \times (1 + v_1^2 t^2/b^2)^{1/2} \geqslant \frac{1}{2}\hbar$. (b) According to section 11.8 an excess of $\Delta x \Delta p_x$ over $\frac{1}{2}\hbar$ may be contributed by the mean deviation of x weighted by the flux distribution, equation (11.65). Substitute in this equation the expression (11.75b) of the particle flux and observe that the first term in the braces of (11.75b) yields no contribution; the second term, due to the spreading out of the packet, yields

$(m/\hbar)\,[v_1^2 t/(b^2 + v_1^2 t^2)]\;\Delta x^2 = (m/\hbar) v_1^2 t$. This contribution amounts to $\frac{1}{2} v_1 t/b$ according to equation (11.73a). When its square is entered in equation (11.62) and added to the value $(\frac{1}{2})^2$ of the last term, it increases the total by a factor $(1 + v_1^2 t^2/b^2)^{1/2}$, thus accounting for the answer to question 11.3a.

11.4. Regard the initial wave function $\psi(\vec{r}, 0)$ as the superposition of the wave function ψ_a of equation (11.68) and of an analogous wave function with the velocity $\vec{v}_0$ replaced by $-\vec{v}_0$, with equal coefficients $\sqrt{\frac{1}{2}}$. At the time t, $\psi(\vec{r}, t)$ will be obtained by superposing, with the same coefficients, the wave function $\psi_a(r, t)$ given by (11.75) and a wave function obtained from it by the replacement $\vec{v}_0 \to -\vec{v}_0$. Thus $\psi(\vec{r}, t)$ consists of two wave packets traveling away from each other. The resulting particle density is

$$|\psi(\vec{r}, t)|^2 = [2\pi(b^2 + v_1{}^2 t^2)]^{-3/2}\, \tfrac{1}{2} \left\{ \exp\left[-\tfrac{1}{2} \frac{|\vec{r} - \vec{v}_0 t|^2}{b^2 + v_1^2 t^2}\right] + \exp\left[-\tfrac{1}{2} \frac{|\vec{r} + \vec{v}_0 t|^2}{b^2 + v_1^2 t^2}\right] + 2\exp\left[-\tfrac{1}{2} \frac{r^2 + v_0^2 t^2}{b^2 + v_1^2 t^2}\right] \cos\left[2 \frac{m\vec{v}_0 \cdot \vec{r}}{\hbar} \frac{b^2 + 2v_1^2 t^2}{b^2 + v_1^2 t^2}\right] \right\}.$$

The first two terms in the braces represent two Gaussian density distributions traveling away from $\vec{r} = 0$ in opposite directions. The third term in the braces represents a waving pattern with Gaussian modulation that remains near $\vec{r} = 0$. Note that this waving pattern spreads out too and that its integrated intensity decreases with increasing t in proportion to $\exp[-\frac{1}{2} v_0^2 t^2/(b^2 + v_1^2 t^2)]$; for large t this factor approaches the limit $\exp(-\frac{1}{2} v_0^2/v_1^2)$ which is vanishingly small if $v_0 \gg v_1$.

12. quantum effects of forces

This chapter concludes the second part of the book with a study of phenomena displayed by particles subject to forces. Quantum physics seldom considers forces directly, as they appear, e.g., in the equations of motion of macroscopic physics. It starts from an expression of the energy of the system under consideration. This expression represents forces indirectly, through the dependence of potential energy terms on the position of particles. We deal here only with conservative forces, which can be represented by the gradient of a potential energy function.[1]

The motion of particles is studied in atomic physics mostly by identifying and describing stationary states. Accordingly, we start by formulating an equation, the *Schrödinger equation,* whose solution yields the wave functions and the energy eigenvalues for the stationary states of a particle subject to forces. Since the Schrödinger equation has the form of a wave equation, the motion of atomic particles shows a number of features common to all wave phenomena. The principal effects described in this chapter are: the partial reflection of a particle, the formation of interference fringes when a particle is turned back by a force and the occurrence of a discrete set of stationary states when a particle is confined within a limited region of space.

The Schrödinger equation serves to determine the stationary states not only of a particle but also of other systems, such as the electromagnetic field of radiation. Thus it will be shown that monochromatic radiation has discrete energy levels separated by an interval $\hbar\omega$, in accordance with the experimental results on the photoelectric effect.

12.1. THE SCHRÖDINGER EQUATION

Consider a particle whose total energy consists of two parts, namely, of the kinetic and potential energies. The classical relation $E = K + V$ must be translated into an equation among mean values for an arbitrary state $|a\rangle$ of the particle

$$\langle E\rangle_a = \langle K\rangle_a + \langle V\rangle_a. \tag{12.1}$$

---

1. Friction and other dissipative forces, which are nonconservative and do not seem to result from variations of potential energy, do not appear at the outset in atomic problems. They result indirectly from conservative forces acting between a small system of interest and a much larger system. Quantum physics treats such interactions from a consistently atomistic point of view, requiring in principle a fully detailed, particle by particle, consideration of the large system too. In practice, effects of friction appear at later stages of the treatment when the effects of interaction between the large and the small system are represented by a few effective parameters.

In order to determine the stationary states $|E\ n\rangle$ we begin by showing that $\langle E\rangle_a$ depends on $|a\rangle$ in a characteristic way when $|a\rangle$ approximately coincides with a stationary state.

The mean total energy of a particle in a state $|a\rangle$ can also be represented by

$$\langle E\rangle_a = \sum_n E_n |\langle E\ n|a\rangle|^2, \tag{12.2}$$

and thus varies, generally, in proportion to small variations of the probability amplitudes $\langle E\ n|a\rangle$. However, an exception occurs when $|a\rangle$ nearly coincides with one of the energy eigenstates, which we call $|E\ \bar{n}\rangle$. In this case all probability amplitudes $\langle E\ n|a\rangle$ with $n \neq \bar{n}$ are small, and we indicate them by

$$\langle E\ n|a\rangle = g_n\epsilon \qquad \text{for } n \neq \bar{n}, \tag{12.3a}$$

where ϵ is a small number and g_n is an arbitrary coefficient. The condition $\Sigma_n |\langle E\ n|a\rangle|^2 = 1$ requires that

$$|\langle E\ \bar{n}|a\rangle|^2 = 1 - \Big(\sum_{n\neq\bar{n}} |g_n|^2\Big)\epsilon^2. \tag{12.3b}$$

Therefore, the mean energy,

$$\begin{aligned}\langle E\rangle_a &= E_{\bar{n}}\Big[1 - \Big(\sum_{n\neq\bar{n}} |g_n|^2\Big)\epsilon^2\Big] + \sum_{n\neq\bar{n}} E_n |g_n|^2\epsilon^2 \\ &= E_{\bar{n}} + \sum_{n\neq\bar{n}} (E_n - E_{\bar{n}})|g_n|^2\epsilon^2\end{aligned} \tag{12.4}$$

departs from $E_{\bar{n}}$ only by a second-order term proportional to ϵ^2. Equation (12.4) shows that $\langle E\rangle_a$ is insensitive to the value of ϵ when ϵ is sufficiently small. This behavior of $\langle E\rangle_a$ is stated precisely by

$$\lim_{\epsilon\to 0} \frac{d\langle E\rangle_a}{d\epsilon} = 0. \tag{12.5}$$

This is an important *variational property* of the mean energy. It serves to identify a state $|a\rangle$ as an eigenstate of the energy by the property that its mean energy is insensitive to small variations of the state.[2]

---

2. This mathematical characterization of an eigenstate has familiar analogs throughout physics. For example, a marble is in equilibrium at the bottom of a bowl where the derivative of its potential energy with respect to virtual displacements vanishes, that is, where its energy is insensitive to small displacements. One can conceive hypothetical experiments to identify eigenstates by a variational method. Suppose that a beam of particles is received by an analyzer which shows the mean particle energy $\langle E\rangle$ on a dial and suppose that the beam has been prepared by a polarizer with orientation and field settings controlled by knobs. Preparation of the beam particles in an energy eigenstate of the analyzer is achieved by adjusting the polarizer until the reading of $\langle E\rangle$ becomes insensitive to further knob adjustments.

The variational property of eigenstates permits the determination of their wave functions through the following sequence of considerations. For an arbitrary state $|a\rangle$, the dependence of $\langle E\rangle_a$ on the wave function is expressed through equation (12.1) and through the position representation of $\langle K\rangle_a$ and $\langle V\rangle_a$:

$$\langle E\rangle_a = \langle K\rangle_a + \langle V\rangle_a = \int d\vec{r}\langle a|\vec{r}\rangle[-\frac{\hbar^2}{2m}\nabla^2\langle\vec{r}|a\rangle + V(\vec{r})\langle\vec{r}|a\rangle]. \quad (12.6)$$

Here m indicates the mass of the particle, the mean value of the kinetic energy is taken from (11.18), and the potential energy function $V(\vec{r})$ is assumed to be known. The integral in equation (12.6) is a linear function of the *two* sets of values of $\langle\vec{r}|a\rangle$ and $\langle a|\vec{r}\rangle$ at the various points $\vec{r}$. The state $|a\rangle$ is an energy eigenstate if the variations of the mean energy vanish to first order for small variations of either set of probability amplitudes $\langle a|\vec{r}\rangle$ or $\langle\vec{r}|a\rangle$. That is, the derivative of the integral[3] in equation (12.6) with respect to each of these probability amplitudes must vanish when the state $|a\rangle$ coincides with an energy eigenstate.[4] Since $\langle a|\vec{r}\rangle$ and $\langle\vec{r}|a\rangle$ are complex conjugates, it is enough that the derivative of $\langle E\rangle_a$ with respect to $\langle a|\vec{r}\rangle$ vanish. Vanishing of this derivative for each point of space $\vec{r}$ yields an equation that determines $\langle\vec{r}|a\rangle$. Once $\langle\vec{r}|a\rangle$ is obtained, the value of the integral in equation (12.6) equals an energy eigenvalue.

The derivative of the integral in equation (12.6) with respect to the probability amplitude $\langle a|\vec{r}\rangle$ at any given point $\vec{r}$ is not a simple partial derivative because the probability amplitudes at different points $\vec{r}$ are not quite independent. The whole set of $\langle a|\vec{r}\rangle$ must satisfy the total probability condition

$$\mathcal{P} = \int\langle a|\vec{r}\rangle\langle\vec{r}|a\rangle d\vec{r} = 1. \quad (12.7)$$

---

3. An integral over the product of two or more functions a(r) and b(r), may be regarded as the limit of the sum, I, of values of the integrand at a set of discrete points r_n, $I = \Sigma_n a(r_n)b(r_n)$. The derivative of the integral with respect to a value of a(r) is then the limit of $\partial I/\partial a(r_n) = b(r_n)$.

4. Instead of taking the derivatives of (12.6) with respect to each probability amplitude $\langle a|\vec{r}\rangle$, one may assume an analytical form of the entire wave function $\langle a|\vec{r}\rangle$ which depends on one or more parameters $\alpha_1, \alpha_2, \ldots, \alpha_i, \ldots$ The derivatives are then taken with respect to each of these parameters and set equal to zero:

$$\frac{\partial\langle E\rangle_a}{\partial\alpha_i} = 0 \text{ for each i.}$$

Solution of this system of equations yields the values of the parameters α_i and a "variational estimate" of the energy eigenvalue and eigenfunction. We have in fact applied this method in section 11.5.1 when considering a tentative wave function (11.36) for the hydrogen atom with a parameter a; the value of a was selected which minimizes the total mean energy, that is, the sum of equations (11.37) and (11.38).

The partial derivative symbol $\partial\langle E\rangle_a/\partial\langle a|\vec{r}\rangle$ does not apply in this case because it implies that $\langle a|\vec{r}'\rangle$ is kept constant at all points $\vec{r}' \neq \vec{r}$, a condition inconsistent with condition (12.7) when $\langle a|\vec{r}\rangle$ varies. We shall then use the symbol $\delta\langle E\rangle_a/\delta\langle a|\vec{r}\rangle$ to indicate the derivative taken under the condition $\mathcal{P} = 1$. The variational condition which identifies the energy eigenstates takes therefore the form

$$\lim_{|a\rangle\to|E\ n\rangle} \frac{\delta\langle E\rangle_a}{\delta\langle a|\vec{r}\rangle} = 0. \tag{12.8}$$

Explicit forms of equation (12.8) are obtained starting from the expression of the derivative at constant $\mathcal{P}$ in terms of partial derivatives

$$\frac{\delta\langle E\rangle_a}{\delta\langle a|\vec{r}\rangle} = \frac{\partial\langle E\rangle_a}{\partial\langle a|\vec{r}\rangle} - \frac{\partial\langle E\rangle_a}{\partial\mathcal{P}}\,\frac{\partial\mathcal{P}}{\partial\langle a|\vec{r}\rangle}. \tag{12.8a}$$

Substitution of this expression changes equation (12.8) into

$$\lim_{|a\rangle\to|E\ n\rangle} \frac{\partial\langle E\rangle_a}{\partial\langle a|\vec{r}\rangle} = \lim_{|a\rangle\to|E\ n\rangle} \frac{\partial\mathcal{P}}{\partial\langle a|\vec{r}\rangle}\,\frac{\partial\langle E\rangle_a}{\partial\mathcal{P}}. \tag{12.8b}$$

The partial derivatives $\partial\langle E\rangle_a/\partial\langle a|\vec{r}\rangle$ and $\partial\mathcal{P}/\partial\langle a|\vec{r}\rangle$ are obtained respectively from equations (12.6) and (12.7) before going to the limit,

$$\frac{\partial\langle E\rangle_a}{\partial\langle a|\vec{r}\rangle} = -\frac{\hbar^2}{2m}\nabla^2\langle\vec{r}|a\rangle + V(\vec{r})\langle\vec{r}|a\rangle, \tag{12.8c}$$

$$\frac{\partial\mathcal{P}}{\partial\langle a|\vec{r}\rangle} = \langle\vec{r}|a\rangle. \tag{12.8d}$$

The derivative $\partial\langle E\rangle_a/\partial\mathcal{P}$ at $\mathcal{P} = 1$ is easily evaluated in the limit since in this case equation (12.2) reduces to $\langle E\rangle_a = E_n\mathcal{P}$ and therefore

$$\lim_{|a\rangle\to|E\ n\rangle} \frac{\partial\langle E\rangle_a}{\partial\mathcal{P}} = E_n. \tag{12.8e}$$

Substitution of the equations (12.8c-e) finally reduces (12.8b) to[5]

$$-\frac{\hbar^2}{2m}\nabla^2\langle\vec{r}|E\ n\rangle + V(\vec{r})\langle\vec{r}|E\ n\rangle = \langle\vec{r}|E\ n\rangle E_n. \tag{12.9}$$

This is the Schrödinger equation for the stationary states of a particle.

---

5. This procedure for deriving a variational equation under the "subsidiary condition" (12.7) is but an application of the general method of Lagrange multipliers used, for instance, in the mechanics of a body attached to a string.

The Schrödinger equation (12.9) and the mean energy equation $\langle E \rangle_a = \langle K \rangle_a + \langle V \rangle_a$ are derived from the classical expression of the energy in terms of particle momentum, $E = p^2/2m + V(\vec{r})$. This expression is called a *Hamiltonian* in classical analytic mechanics. By similarity, any expression of the energy that underlies a Schrödinger equation or a quantum-mechanical calculation involving energy is usually called a Hamiltonian. This name is used particularly in connection with the operator notation sketched in section S11.7. With this notation, equation (12.6) takes the form $\langle E \rangle_a = \psi_a{}^\dagger \mathcal{H}\, \psi_a$; here $\mathcal{H}$ indicates the Hamiltonian operator, which takes the form $\mathcal{H} = -(\hbar^2/2m)\nabla^2 + V(\vec{r})$ in the position representation. In the same operator notation, the Schrödinger equation (12.9) takes the compact form $\mathcal{H}\psi_n = \psi_n E_n$.

12.1.1. *Preliminary Discussion.* The Schrödinger equation belongs to the class which describes the variations in space of a wave phenomenon and has the general form

$$\nabla^2 u(\vec{r}) + k^2 u(\vec{r}) = 0. \qquad (12.10)$$

The wavenumber k is constant for a wave phenomenon in a uniform medium but is itself a function of $\vec{r}$ when the properties of the medium vary from point to point. (The ratio of the local value of k to its vacuum value, or to some other standard value, is generally called the refractive index.) An alternative form of the Schrödinger equation (12.9) is obtained as the particular case of (12.10) in which

$$k^2 = \frac{2m}{\hbar^2}\,[E_n - V(\vec{r})] \qquad (12.11)$$

and $u(\vec{r}) = \langle \vec{r}\,|E\;\;n\rangle$, namely,

$$\nabla^2 \langle \vec{r}\,|E\;\;n\rangle + \frac{2m}{\hbar^2}\,[E_n - V(\vec{r})]\langle \vec{r}\,|E\;\;n\rangle = 0. \qquad (12.12)$$

When $V(\vec{r})$ vanishes, equation (12.12) is equivalent to the free-particle wave equation (11.17).

The calculation of the wave function for a stationary state of a particle is thus mathematically equivalent to the calculation of a monochromatic wave in a medium with nonuniform refractive index. The forces acting on a particle are represented mathematically by the gradient of the refractive index. This remark accounts for the correspondence between the diffraction of radiation and of particle beams described in chapter 4. Particles and radiation incident with equal momentum are diffracted at equal angles if the potential energy $V(\vec{r})$ of the particles and the dielectric constant $\epsilon(\vec{r})$ have Fourier expansions with equal wave vectors $\vec{q}$.

From the Schrödinger equation (12.9), which determines the wave functions of stationary states, one obtains readily an equation obeyed by the wave functions $\psi_a(\vec{r}, t) \equiv \langle \vec{r}\,|a(t)\rangle$ of nonstationary states. The manipulation used in section

11.2 to establish the free-particle wave equation (11.22) now yields

$$-\frac{\hbar^2}{2m}\nabla^2\psi_a(\vec{r}, t) + V(\vec{r})\psi_a(\vec{r}, t) = -\frac{\hbar}{i}\frac{\partial}{\partial t}\psi_a(\vec{r}, t). \qquad (12.13)$$

Given the values of the wave function ψ_a at an initial time $t = 0$, this equation determines its values at later times.

In particular, equation (12.13) provides a link between wave mechanics and the classical mechanics of a macroscopic particle. For a sufficiently large value of the particle mass, equation (12.13) is obeyed by wave packets which remain rather well concentrated in space and whose centers $\langle\vec{r}\rangle$ move very approximately according to $md^2\langle\vec{r}\rangle/dt^2 = -\vec{\nabla}V(\langle\vec{r}\rangle)$. The motion of a particle represented by such a wave packet is analogous to the propagation of light which follows the laws of geometrical optics when the refractive index varies but little over a wavelength. In fact, the time-dependent Schrödinger equation (12.13) was first established as an analog of the wave equations of physical optics, by requiring that its solutions would propagate in accordance with ordinary mechanics in the limit of a large particle mass.

12.2. PROPAGATION AND REFLECTION OF PARTICLE WAVE FUNCTIONS.

The influence of forces on the motion of particles[6] is highlighted by working out a number of prototype problems. We limit this study to stationary states, that is, to the variation of the wave function of a particle from point to point. This approach makes the task simpler by eliminating the time variable, but it obscures the cause-effect relationships, as discussed at the end of section A.2. Stationary states are, of course, idealizations; but they are worth studying just as monochromatic waves are. More realistic descriptions of particle motion are obtained by constructing superpositions of stationary state wave functions.

This section considers two schematic types of potential whose effects on the motion of particles are clear-cut and easy to calculate. For both types the field of force is everywhere parallel to one coordinate axis. Thus we assume that the potential depends only on the x coordinate, $V(\vec{r}) = V(x)$.

In this case the Schrödinger equation is *separable*, that is, it resolves into three separate equations. The wave function $\langle\vec{r}\,|E\ \ n\rangle$ factors in the form, analogous to equation (11.6),

$$\langle\vec{r}\,|E\ \ n\rangle = \langle x\,|E_x\ \ n_x\rangle\langle y\,|E_y\ \ n_y\rangle\langle z\,|E_z\ \ n_z\rangle. \qquad (12.14)$$

(In the following the indices n_x, etc. are dropped for brevity.) The probability

6. Recall the remarks on the meaning of the word "motion" in the introduction to chap. 11.

amplitude $\langle x | E_x \rangle$ obeys the single-variable equation

$$-\frac{\hbar^2}{2m}\frac{d^2}{dx^2}\langle x | E_x \rangle + V(x)\langle x | E_x \rangle = \langle x | E_x \rangle E_x, \tag{12.15}$$

while the factors $\langle y | E_y \rangle$ and $\langle z | E_z \rangle$ obey free particle equations. The total energy eigenvalues are

$$E_n = E_x + E_y + E_z. \tag{12.16}$$

[The separation of variables holds under the more general condition $V(\vec{r}) = V_1(x) + V_2(y) + V_3(z)$.]

We will deal in this chapter with the motion of a particle parallel to the x axis and indicate its energy by E rather than E_x, thus assuming $E_y = E_z = 0$. Equation (12.15) is then written conveniently for our purposes in the form, corresponding to equation (12.10),

$$\frac{d^2\psi}{dx^2} + k^2(x)\psi(x) = 0, \tag{12.17}$$

where

$$k^2(x) = \frac{2m}{\hbar^2}[E - V(x)]. \tag{12.17a}$$

12.2.1. *Slowly Varying Potential: WKB Semiclassical Approximation.* When the potential energy of a particle varies but little from point to point, one conveniently represents the wave function by an adjustment of a free-particle wave function. For a free particle, the momentum eigenstates have the wave function $\exp(ikx)$ with $k = p_x/\hbar$. (For simplicity, we omit in this section the normalization factor.) A slowly varying potential causes p_x and k to vary slowly as functions of x. Similarly, light in a medium with slowly varying refractive index propagates undisturbed but for a slow change of wavelength. A slow change of refractive index permits use of geometrical optics; similarly, a slowly varying potential permits a semi classical approximation to wave mechanics.

We allow for the slow variation of k by representing the wave function in the form

$$\psi(x) \equiv \langle x | E \rangle = \exp\left[i\int^x \bar{k}(x')dx'\right], \tag{12.18}$$

which reduces to exp (ikx) if $\bar{k}$ is constant and equal to k. Equation (12.18) has the derivatives

$$\frac{d\psi}{dx} = i\bar{k}(x)\psi(x), \quad \frac{d^2\psi}{dx^2} = \left[i\frac{d\bar{k}}{dx} - \bar{k}^2(x)\right]\psi(x), \tag{12.19}$$

so that $\psi(x)$ factors out of the wave equation (12.17). Thereby this equation becomes

$$i\frac{d\bar{k}}{dx} - \bar{k}^2(x) + k^2(x) = 0, \qquad (12.20)$$

that is, a Riccati type of equation. If the potential were constant, equation (12.20) would be solved by $\bar{k} = k$, as expected. For a variable potential $V(x)$, the equation can be solved by successive approximations provided the change of k within a distance 1/k amounts to a small fraction of k itself, that is, provided

$$\left|\left(\frac{dk}{dx}\frac{1}{k}\right)\frac{1}{k}\right| = \left|\frac{1}{k^2(x)}\frac{dk}{dx}\right| << 1. \qquad (12.21)$$

This condition underlies the Wentzel-Kramers-Brillouin (WKB) method of approximation.

The solution of equation (12.20) is obtained by a method which permits successive approximations. One starts by writing $\bar{k} = k(x) + \alpha(x)$, where α represents the small departure of $\bar{k}$ from k. Substitution of this expression of $\bar{k}$ changes (12.20) into an equation for α, namely,

$$i\frac{dk}{dx} + i\frac{d\alpha}{dx} + 2k(x)\alpha(x) + \alpha^2(x) = 0. \qquad (12.20a)$$

At this point one introduces the approximation of disregarding both α^2 and $d\alpha/dx$, since $\alpha(x)$ itself is expected to be small and to be nearly constant. Thereby equation (12.20a) is solved approximately by $\alpha = \frac{1}{2}i\, d(\ln k)/dx$ and we obtain the first-order approximation to the solution of (12.20)

$$\bar{k} = k(x) + \tfrac{1}{2}i\frac{d(\ln k)}{dx}. \qquad (12.22)$$

Substitution of this expression of $\bar{k}$ into (12.18) yields

$$\psi(x) = [k(x)]^{-1/2} \exp\left[i\int^x k(x')dx'\right]. \qquad (12.23)$$

According to this equation, the particle density $|\psi|^2$ is inversely proportional to $k(x)$ and hence to the velocity of the particle at the point x. This result is obvious from a simple-minded point of view, since the probability of finding the particle at a point is the smaller the faster the particle passes through it; however, this simple result holds only in the first-order approximation. It also follows from equation (12.23) that the particle flux, defined by equation (11.16), is independent of x. This last result holds rigorously because the flux is divergenceless for any stationary state and must then be uniform if the motion is one-dimensional.

These results verify the initial surmise that the wave function of a particle propagates under the condition (12. 21) as though the particle were free, with the adjustments of amplitude and phase indicated by (12. 22). Under these conditions there exists the same variety of degenerate stationary states as was described in section 11. 3 for a free particle.

However, the simple dependence of the particle density $|\psi|^2$ on the velocity breaks down in higher approximation and with it the simple propagation akin to that of a free particle. Moreover, the approximation breaks down necessarily if the particle is slowed down by the force until it is nearly at rest and k nearly vanishes in the denominator of equation (12. 21). To study the effects disregarded by the WKB first-order approximation we consider now a type of potential with opposite characteristics.

12. 2. 2. *Potential Step.* Consider a schematic arrangement in which the electric force acting on a charged particle is concentrated in the gap between two grids at different potentials and both grids are brought infinitely close to the plane $x = 0$. The electric field becomes infinitely strong in the vanishing gap, and the potential energy of the particle is represented by

$$V(x) = 0 \text{ for } x < 0, V(x) = V' \text{ for } x > 0, \tag{12. 24}$$

where V' is the product of the particle's charge and of the potential difference of the grids. To these values of V corresponds, according to equation (12. 17a),

$$\begin{aligned} k^2(x) &= k_0^2 = \frac{2m}{\hbar^2} E & \text{for } x < 0, \\ k^2(x) &= k'^2 = \frac{2m}{\hbar^2} (E - V') & \text{for } x > 0. \end{aligned} \tag{12. 25}$$

This schematization is analogous to that usually practiced in optics at the interface between two media with different refractive index.

On either side of the potential step, the potential is constant and therefore the wave function of the particle coincides with that of a free-particle stationary state. The state is thus an eigenstate of kinetic energy on either side of the step. However, the kinetic energy eigenvalues are different since the potential energy is different and the total energy is fixed. Thus the effect of a localized force is to produce a discontinuity in the kinetic energy.

This discontinuity causes the type of motion of the particle to be different on the two sides of the potential step. Eigenstates of momentum have different fluxes on the two sides of the step, since the kinetic energy and the valocity v_x are different. Yet, the particle flux must be equal on the two sides of the step, because the flux is solenoidal, as noted in the preceding section. Therefore, a stationary state can not be a momentum eigenstate on both sides of the step. The step causes a discontinuity of the parameter θ which was introduced in equation (11. 28) to characterize the waving pattern of particle density and the flux of free-particle stationary states. This discontinuous change of type of motion takes the aspect of a partial reflection when the wave function is represented as a superposition of momentum eigenstates. This representation is the one actually used to determine the wave function.

Solutions of the wave equation (12.17), with the wavenumbers given by (12.25) for the two sides of the potential step, are represented as superpositions of momentum eigenfunctions $\langle x|p_x\rangle$ and $\langle x|-p_x\rangle$,

$$\psi(x) = A \exp(ik_0 x) + B \exp(-ik_0 x) \qquad \text{for } x \leqslant 0, \tag{12.26a}$$

$$\psi'(x) = C \exp(ik'x) + D \exp(-ik'x) \qquad \text{for } x \geqslant 0. \tag{12.26b}$$

The coefficients A, B, C, and D, replace here for brevity the probability amplitudes $\langle \pm p_0 | E\rangle$ and $\langle \pm p' | E\rangle$. They must satisfy the condition that ψ and ψ', as well as their first derivatives, form together a function continuous at $x = 0$. (The second derivative of ψ is discontinuous at the potential step; this discontinuity is taken into account by the Schrödinger equation which fixes the values [12.25] of k_0 and k'.) The continuity conditions yield

$$\psi(0) = \psi'(0) \qquad \rightarrow \ A + B = C + D, \tag{12.27a}$$

$$\left(\frac{d\psi}{dx}\right)_0 = \left(\frac{d\psi'}{dx}\right)_0 \rightarrow \ k_0(A - B) = k'(C - D). \tag{12.27b}$$

We have here two equations among four parameters. The resulting indeterminacy of the parameters implies a degeneracy of the energy eigenstates in our problem, analogous to the degeneracy described for free particles in section 11.3.

Among these degenerate states we consider one that corresponds most nearly to a realistic state. This is the state in which the particle propagates from infinity toward the step on one side only, specifically from $x = -\infty$ Thus we set $D = 0$ and, for simplicity, $A = 1$. Solution of the system (12.27) yields then $B = (k_0 - k')/(k_0 + k')$. According to equation (12.26) the value of $|B|^2$ represents a reflection coefficient because the wave with amplitude B travels back from the step. This value,

$$|B|^2 = \frac{(k_0{}^2 - k'^2)^2}{(k_0 + k')^4} = \frac{V'^2}{[E^{1/2} + (E - V')^{1/2}]^4} \tag{12.28}$$

approaches zero for $E/V' \rightarrow \infty$, showing that the potential step is ineffective in this limit. Conversely, $|B|^2$ approaches unity for $E/V' \rightarrow 1$. Since no implication has been made that V' is positive, the result (12.28) holds irrespective of the sign of V'; thus partial reflection occurs whether the potential step slows down the particle or, instead, accelerates it. The value of $(k'/k_0)|C|^2 = 1 - |B|^2$ represents a transmission coefficient.

The type of motion of the particle is also changed by a force represented by a sufficiently steep, though not steplike, potential. As in the optical analog, partial reflection and related effects occur whenever the wavenumber k(x) varies appreciably over a distance 1/k, that is, when the first WKB approximation is not applicable. The partial reflection of a particle by a field of force can thus be regarded as a general phenomenon. This phenomenon influences, for example, the ejection of electrons from atoms. Even though an electron may have acquired an energy E larger than the potential V′ at infinite distance, partial reflection by the field surrounding the residual ion influences the electron's eventual escape.

12.2.3. *Total Reflection by a Potential Step.* Returning now to the schematic example of a potential step, total reflection is expected when the particle's energy E is lower than the step's height V′. Indeed, the reflection coefficient (12.28) reaches unity at $E = V'$. However, the coefficient C of the wave function (12.26) for $x > 0$ equals 2 rather than zero when $E = V'$. The wave function thus appears to extend beyond the potential step. This situation bears investigating, the more so since electrons are held within atoms just by virtue of having insufficient energy to overcome the potential of nuclear attraction.

When $E < V'$, the wavenumber k′ defined by equation (12.25) is imaginary. We can set $k' = i\kappa$, whereby the wave function $\psi'(x)$ defined by equation (12.26b) becomes

$$\psi'(x) = Ce^{-\kappa x} + De^{\kappa x} \qquad \text{for } x \geqslant 0. \tag{12.29}$$

This generalization makes the description of the stationary state of a free particle applicable to a particle with "negative kinetic energy," $E - V' < 0$. Note that the mean kinetic energy as represented by equation (11.19) can take negative values provided $\nabla_x^2 |\psi_a(x)|$ is positive. A second derivative with the same sign as the function is characteristic of real exponentials such as those in equation (12.29) and of hyperbolic functions; the opposite sign occurs for complex exponentials and for sinusoidal functions. The particle density

$$|\psi'(x)|^2 = |C|^2 e^{-2\kappa x} + |D|^2 e^{2\kappa x} + (C^*D + D^*C) \tag{12.30}$$

accordingly follows a hyperbolic curve rather than a waving pattern.

We are now faced with the problem that the particle density (12.30) diverges as $x \to \infty$. To remove this divergence, consider initially the case where the potential does not remain equal to V′ up to $x = \infty$, but drops back to zero for $x > a > 0$. This model represents schematically the situation of an electron held within an atom by a potential V′ when a second atom lies at a distance a. The particle density ratio $|\psi(a)|^2/|\psi(0)|^2$ represents essentially the relative probability that the electron be found in the regions $x > a$ or $x < 0$ and depends on the ratio of coefficients $|D|^2/|C|^2$. If one wants to deal with the case in which the particle is definitely on the side of the barrier where $x \lesssim 0$, he selects among the degenerate states the one with $D = 0$. This selection implies that the electron is known to belong to the first atom rather than to the second one. Setting $D = 0$, with the resulting removal of degeneracy, is necessary in the limit of $a \to \infty$, which corresponds to our original schematization.

After this selection the wave function is obtained by solving the system (12.27) with $D = 0$ and $k' = i\kappa$. One finds

$$\frac{B}{A} = \frac{k_0 - k'}{k_0 + k'} = \frac{k_0 - i\kappa}{k_0 + i\kappa} = \exp\left(-2i \arctan \frac{\kappa}{k_0}\right) = \exp\left[-2i \arccos \left(\frac{E}{V'}\right)^{1/2}\right]. \tag{12.31}$$

Given this ratio, we may set the magnitude of A and B by requiring that the wave function (12.26a) be normalized like the standing-wave function (11.26). Thus we find $A = (2h)^{-1/2} \exp[i \arccos(E/V')^{1/2}]$ and $B = (2h)^{-1/2} \times \exp[-i \arccos(E/V')^{1/2}]$. With these values of A and B, and with $D = 0$, equation (12.27a) yields $C = A + B = (2E/hV')^{1/2}$. Substitution into (12.26) gives the wave functions

$$\psi(x) = \left(\frac{2}{h}\right)^{1/2} \cos\left[k_0 x + \arccos\left(\frac{E}{V'}\right)^{1/2}\right] \quad \text{for } x \leqslant 0, \tag{12.32a}$$

and

$$\psi'(x) = \left(\frac{2E}{hV'}\right)^{1/2} e^{-\kappa x} \quad \text{for } x \geqslant 0. \tag{12.32b}$$

For $x < 0$ this wave function has the characteristics of the standing waves $\langle x | p_x^2 \ \pm\rangle$ described in section 11.3. It represents a state of motion with zero particle flux, with a particle density which drops to zero at the trough centers and whose peaks are separated by a half-wavelength π/k_0.

The fact that the wave function does not vanish altogether for $x > 0$ constitutes a partial penetration of the particle into the region of high potential. This phenomenon is not observed under macroscopic conditions for which the exponent κx is very large. It is analogous to the partial penetration of light under conditions of total reflection. Note that this penetration disappears altogether for $E/V' \to 0$ in which case $\psi(x)$ has a node at $x = 0$. Conversely, for $E/V' \to 1$ the penetration stretches indefinitely and $\psi(x)$ has an antinode at $x = 0$. Figure 12.1 illustrates the behavior of the wave function and particle density.

12.2.4. *Total Reflection by a Slowly Varying Potential.* Total reflection occurs whenever the potential $V(x)$ equals the energy E of the particle at a point $x = x_0$ and rises above E throughout the range $x > x_0$, even though $V(x)$ varies slowly and the force opposing the particle's motion is weak. As we know, the WKB approximation fails at $x \sim x_0$ where $k^2 = (2m/\hbar^2)[E - V(x)] \sim 0$. As shown in more advanced textbooks, the WKB method can be used nevertheless, if $V(x)$ is approximately linear through the neighborhood of x_0 where the WKB condition (12.21) fails.[7] One obtains a complete wave function by combining an analytical solution in the linear range with a WKB treatment outside

7. See, for instance, E. Merzbacher, *Quantum Mechanics* (New York: Wiley, 1970), chap. 7.

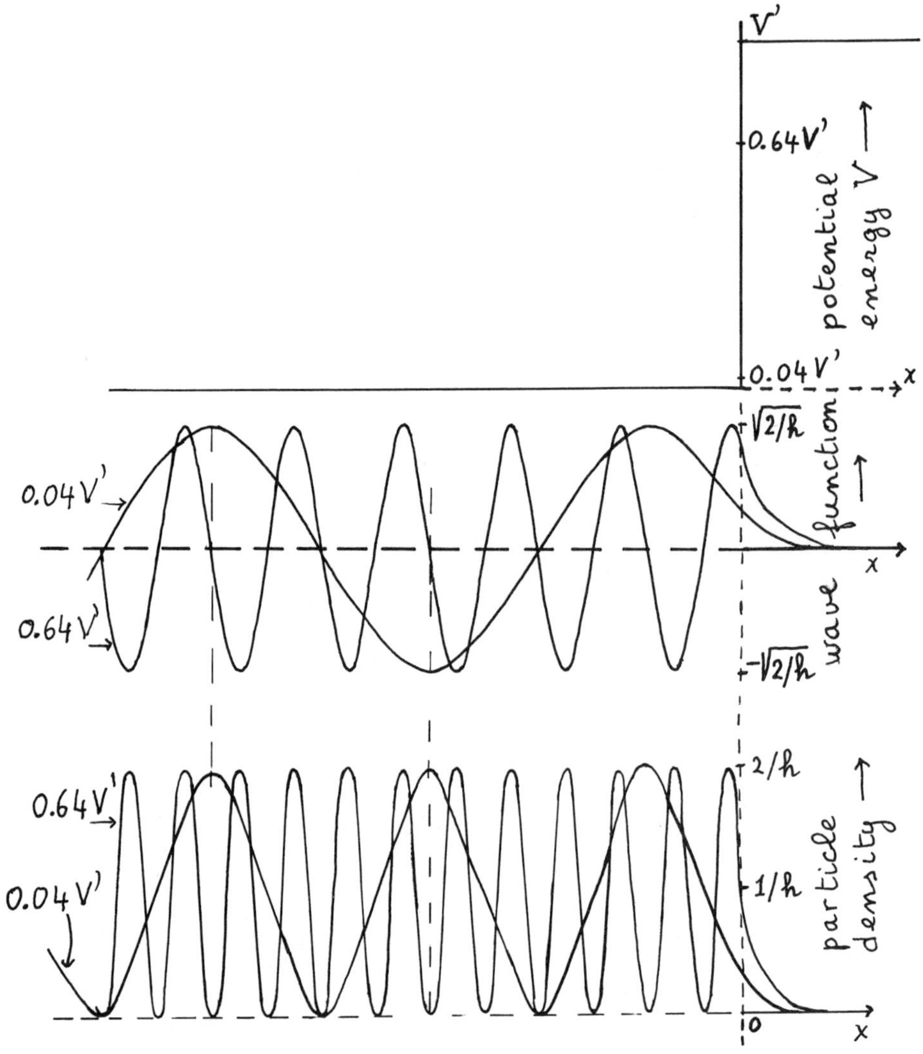

Fig. 12.1. Potential energy step, wave functions, and particle densities for energy eigenstates with $E = 0.04V'$ and $E = 0.64V'$.

of it. As in all one-dimensional cases of total reflection, the wave function is a standing wave, namely,

$$\psi(x) \sim [k(x)]^{-1/2} \cos \left[\int_{x}^{x_0} k(x')dx' + \frac{1}{4}\pi\right], \qquad \text{for } x < x_0, \tag{12.33a}$$

$$\psi(x) \sim [\kappa(x)]^{-1/2} \exp \left[-\int_{x_0}^{x} \kappa(x')dx'\right], \qquad \text{for } x > x_0, \tag{12.33b}$$

where $\kappa^2(x) = (2m/\hbar^2)[V(x) - E]$. Equation (12.33a) permits a rapid estimation of the position of the interference fringes, that is, of the peaks and troughs of the particle density, even when the WKB approximation is not very accurate. Similarly, equation (12.33b) serves to estimate the particle penetration beyond $x = x_0$.

Experimentally, the waving pattern of particle density near a reflecting barrier is not observed directly because the spacing of fringes is usually of the order of 1 Å. However, this pattern may be detected indirectly through its influence on another phenomenon. For example, the energy levels of an atom vary, sometimes rapidly, when the atom approaches another atom during a collision. A spectral line emitted by an excited atom during a collision is then broadened into a continuous band; each frequency of the band corresponds to emission by the excited atom at a particular distance from the other atom. Under these circumstances the intensity distribution across the spectral emission band shows bright and dark fringes corresponding to the alternation of particle density at various interatomic distances.[8]

12.3. TUNNELING

When a particle is reflected by a potential barrier, its wave function extends with nonzero values into the region with $x > x_0$ where the potential energy exceeds the total energy E of the particle, as illustrated by equations (12.32b) and (12.33b). Therefore, if the potential barrier is not infinitely thick but has a profile of the type shown in Figure 12.2, the wave function will still have a nonzero value at points with $x \geqslant x_1$ where E exceeds V(x) again. Under this condition a particle incident on the barrier from the direction of negative x has a finite probability of traveling toward $x = \infty$ instead of being turned back at or near $x = x_0$. This partial transmission is called *tunneling* through the barrier.

An example of tunneling calculation is sketched in supplementary section S12.7. For purposes of estimation one may take the probability of transmission to be the square of the exponential factor in equation (12.33b) with the

---

8. The fringe position is shifted when an isotope of the emitting atom is replaced by another one, in accordance with the dependence of the wavenumber on the atomic mass.

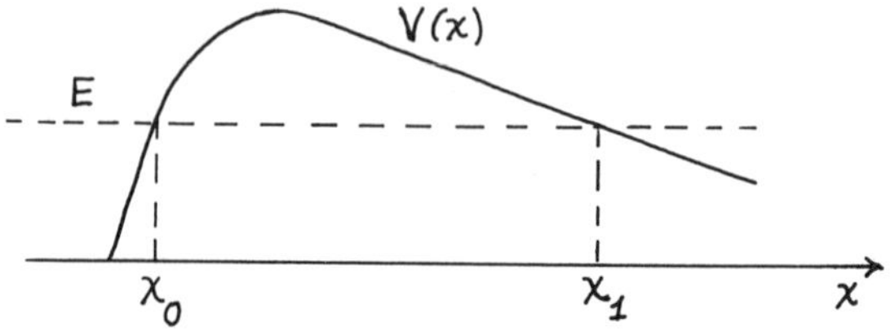

Fig. 12.2. Profile of potential barrier that permits particle tunneling.

integral extended to $x = x_1$. Note that, if the barrier does not extend to $x = \infty$, the coefficient D no longer need vanish in equation (12.29) or in the analogous formula of a WKB calculation. In the absence of this limitation, there again exists a pair of degenerate energy eigenstates for each value E of the energy. The stationary state of interest with a given energy is identified by the circumstances of the problem under consideration, usually by the direction of incidence of the particle.

Tunneling is detected experimentally in various important phenomena. Among these is the field emission of electrons from metals. The escape of electrons from a metal is prevented by the attraction of the nuclei. This attractive force is represented schematically by the rising slope of the potential near $x = x_0$ in Figure 12.2. Application of a strong electric field, for instance, of the order of 10^6 eV/cm, which strives to rip electrons off a metal surface causes the potential to slope off for $x \gtrsim x_0$, as shown in Figure 12.2. The distance $x_1 - x_0$ may amount to $\gtrsim 10^{-6}$ cm, so that the tunneling probability is quite low. However, very appreciable field emission currents result because the number of metal electrons near the surface is extremely large. Historically the first phenomenon to be interpreted as particle tunneling was the escape of α-particles from radioactive nuclei.

Phenomena analogous to particle tunneling are well known for electromagnetic radiation and for other wave propagation processes. For example, a microwave signal of frequency too high to "pass" through a narrow, infinitely long waveguide is nevertheless partially transmitted by a finite section of that waveguide. In the total reflection of light at a glass-air interface, some light leaks through an air gap if another glass surface is brought within a few wavelengths of the totally reflecting surface.

12.4. PARTICLE BOUND IN A POTENTIAL WELL

One of the striking phenomena of quantum physics is the occurrence of discrete energy levels for a particle which is confined by forces within a limited region of space. We consider here three examples of bound one-dimensional motion, in which the confining force is, respectively, very strong, very weak, and elastic.

12.4.1. *Square Potential Well.* **Consider first the motion of a particle confined between two potential steps. The potential energy function for this particle is shown in the upper part of Figure 12.3. Since each potential step is identical to that shown in Figure 12.1, we can utilize here the solution of the reflection problem. In particular, the wave function (12.32a) yields a sequence of peaks of particle density (bright fringes) at points $x < 0$ where $k_0x + \arccos (E/V')^{1/2}$**

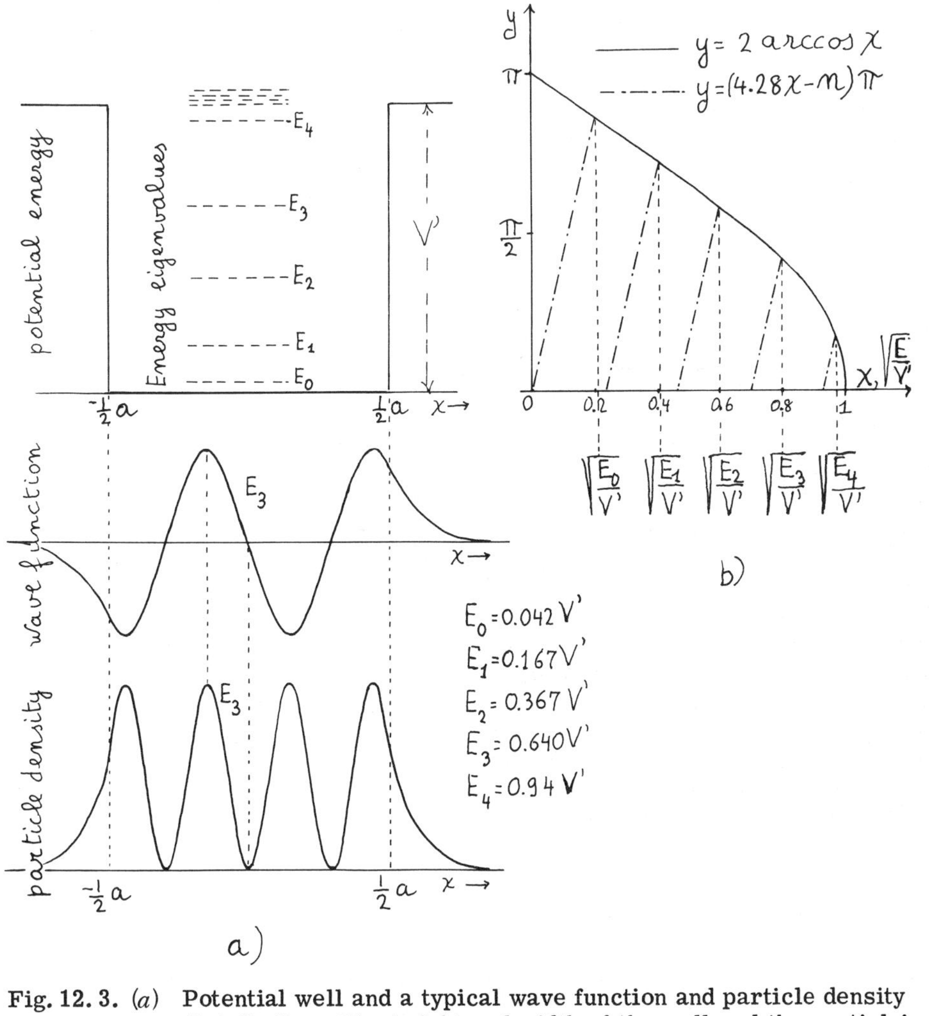

Fig. 12.3. (*a*) **Potential well and a typical wave function and particle density distribution. The height and width of the well and the particle's mass are related by $(2mV')^{1/2}a/\hbar = 13.4$.**

(*b*) **Graphical solution of equation (12.35a).**

is zero or a negative multiple of π, indicated by $-n_1\pi$. These points lie at a distance from the potential step equal to[9].

$$\frac{\hbar}{(2mE)^{1/2}}\left[n_1\pi + \arccos\left(\frac{E}{V'}\right)^{1/2}\right], \quad n_1 = 0, 1, 2, \ldots. \tag{12.34}$$

In our problem the wave function in the region $\pm\frac{1}{2}a$ is also a standing wave similar to equation (12.32a), but the position of each bright fringe must now fulfill *two conditions*. Its distance from *each* of the two potential steps, at $x = \pm\frac{1}{2}a$, must be given by expression (12.34). These two requirements are generally inconsistent. They become consistent only if the energy E of the particle is such that the sum of two terms (12.34) with integers n_1 and n_2 equals the width a of the well. Therefore, the energy must be a root E_n of the equation

$$\frac{\hbar}{(2mE_n)^{1/2}}\left[n\pi + 2\arccos\left(\frac{E_n}{V'}\right)^{1/2}\right] = a, \quad n = n_1 + n_2 = 0, 1, 2, \ldots, \tag{12.35}$$

that is,

$$E_n = \frac{\hbar^2}{2ma^2}\left[n\pi + 2\arccos\left(\frac{E_n}{V'}\right)^{1/2}\right]^2. \tag{12.35a}$$

Thus the eigenvalues E_n smaller than V' form a discrete sequence in correspondence to the successive values of n. The number of peaks of the particle density is n + 1, the number of troughs is just n. In other words, the existence of an energy eigenstate of a bound particle is conditioned by the requirement that an integral nonzero number of peaks of its particle density fit within the potential well. The peak-to-peak distance is inversely related to the particle's kinetic energy in the well, E; more accurately, it is proportional to $E^{-1/2}$. Therefore, as E increases, successive levels are found at which successive numbers of fringes fit in the well.

The well diagram in Figure 12.3 and equation (12.35) show that a particle is bound in a square well only in a finite number of discrete energy levels. This number equals the largest integer n for which the trigonometric equation (12.35) has a root $E_n < V'$. The trigonometric equation (12.35) is solved graphically in Figure 12.3b. For $E \geqslant V'$ the particle is no longer bound and its energy eigenvalues range continuously up to infinity; its eigenfunctions are obtained from a calculation described in supplementary section S12.7.

For a particle bound in a potential well the energy eigenfunctions ψ_n are constructed by adapting the wave functions (12.32) for a particle reflected by a single potential step at $x = 0$. In our problem one of the potential steps lies at $x = \frac{1}{2}a$ (see Fig. 12.3). Accordingly, the wave function for x between $-\frac{1}{2}a$ and $\frac{1}{2}a$ is obtained from equation (12.32a) by replacing x with $x - \frac{1}{2}a$. We must

---

9. In this section we replace k_0 by its definition (12.25).

also replace the wavenumber k_0 by $k_{0n} = (2mE_n/\hbar^2)^{1/2}$ and the normalization coefficient $(2/h)^{1/2}$ with a coefficient N_n which remains to be determined. Thus we write

$$\psi_n(x) = N_n \cos\left[k_{0n}(x - \tfrac{1}{2}a) + \arccos\left(\frac{E_n}{V'}\right)^{1/2}\right] \quad \text{for } -\tfrac{1}{2}a \leqslant x \leqslant \tfrac{1}{2}a \tag{12.36a}$$

and

$$\psi_n(x) = N_n\left(\frac{E_n}{V'}\right)^{1/2} \exp\left[-\kappa_n(x - \tfrac{1}{2}a)\right], \quad \text{for } x \geqslant \tfrac{1}{2}a. \tag{12.36b}$$

Equation (12.36a) is simplified by eliminating $\arccos (E_n/V')^{1/2} - \tfrac{1}{2}k_{0n}a$ through the condition (12.35), which gives

$$\psi_n(x) = N_n \cos(k_{0n}x - \tfrac{1}{2}n\pi) = \begin{cases} (-1)^{n/2} N_n \cos k_{0n}x & \text{for } n \text{ even} \\ (-1)^{(n-1)/2} N_n \sin k_{0n}x & \text{for } n \text{ odd.} \end{cases} \tag{12.37}$$

Each energy eigenstate is thus seen to be an eigenstate of parity under reflection at $x = 0$, in keeping with the symmetry of the potential well. The parity property permits us to complete the description of ψ_n for the remaining range of x

$$\psi_n = (-1)^n N_n \left(\frac{E_n}{V'}\right)^{1/2} \exp\left[\kappa_n(x + \tfrac{1}{2}a)\right] \quad \text{for } x < -\tfrac{1}{2}a. \tag{12.36c}$$

The normalization coefficient N_n is determined, for a particle in a discrete energy level, by applying the condition (10.6), whereas (10.12) applied to free particles. That is, the integral of the particle density $|\psi_n(x)|^2$ over the whole range of x must equal unity. Thus we set

$$1 = \int_{-\infty}^{\infty} \psi_n^2\, dx = N_n^2 \left\{ \int_{-\infty}^{-a/2} \frac{E_n}{V'} \exp\left[2\kappa_n(x + \tfrac{1}{2}a)\right] dx + \int_{-a/2}^{a/2} \cos^2(k_{0n}x - \tfrac{1}{2}n\pi)\, dx + \int_{a/2}^{\infty} \frac{E_n}{V'} \exp\left[-2\kappa_n(x - \tfrac{1}{2}a)\right] dx \right\}. \tag{12.38}$$

The three integrals in equation (12.38) are standard, the first and the last one are equal. The value of the middle integral is simplified by using equation (12.35) once again. Thus one finds

$$N_n = \left\{ \tfrac{1}{2}a + \frac{\hbar}{[2m(V' - E_n)]^{1/2}} \right\}^{-1/2} \tag{12.39}$$

12.4.2. *Shallow Potential Well.* The forces acting on a particle confine it within a potential well when V(x) exceeds the total energy E of the particle at all points outside a finite range $x_1 \leq x \leq x_2$. As in the case of the square well, the particle has a sequence of discrete energy eigenvalues, corresponding to successive integral numbers of peaks of the particle density. The determination of the energy eigenvalues consists, in essence, of a study of the fitting of fringes within the well, and thus involves a calculation of the fringe width as a function of energy.

This calculation can be carried out by the WKB approximation method provided the potential energy varies sufficiently slowly. For a free particle the full width of a bright fringe (trough-to-trough distance) equals one-half the wave length, that is, it equals π/k. If the potential varies slowly, one may still take the fringe width as $\pi/k(x)$, where $k(x)$ is the wavenumber defined by (12.17a) and evaluated at some point x within the fringe. This approach of approximating the fringe width by introducing a continuously varying wavenumber implies that the wave number varies slowly across a fringe and thus it is the very core of the WKB approximation.

Utilizing the WKB approximation to write a wave function in the form (12.33a) between x_1 and x_2, one finds that n + 1 bright fringes fit in the potential well if[10]

$$\int_{x_1}^{x_2} k(x)dx = n\pi + \tfrac{1}{2}\pi. \qquad (12.40)$$

This equation determines the energy eigenvalues through the dependence of k, x_1, and x_2 on E.

Equation (12.40) is often cast in the form

$$2\int_{x_1}^{x_2} \{2m[E - V(x)]\}^{1/2} dx = 2\int_{x_1}^{x_2} |p|\, dx = (n + \tfrac{1}{2})h. \qquad (12.40a)$$

Here $|p|$ coincides with $\hbar k(x)$ and represents the magnitude of the particle's momentum at the point x, evaluated in accordance to classical mechanics. Use of this classical concept is equivalent to the WKB approximation. The left-hand side of equation (12.40a) represents the integral of $|p|$ over a whole cycle of macroscopic motion of the particle from x_1 to x_2 and back to x_1. In this sense it is analogous to the expression $2\pi mvr$ for the motion of an electron on a circular orbit in equation (3.20). The requirement that this type of integral equal a multiple of h was postulated before the development of quantum mechanics and was called the *Sommerfeld quantum condition*. The value $\frac{1}{2}$ was added to the integer n later as a result of the WKB approximation to wave mechanics.

---

10. The term $\frac{1}{2}\pi$ on the right hand side of equation (12.40) consists of two equal contributions arising from accurate integration in the neighborhood of the limits x_1 and x_2; see the discussion leading to equation (12.33). The term $\frac{1}{2}\pi$ implies that extension of the wave function outside the potential well contributes altogether one-half of one bright fringe.

12.4.3. *Harmonic Oscillator*. The Schrödinger equation (12.15) for the one-dimensional motion of a particle can be solved analytically when the particle is held near an equilibrium position, $x = 0$, by an elastic force $F = -kx$. To this force corresponds the parabolic potential well

$$V(x) = \tfrac{1}{2}kx^2. \tag{12.41}$$

The Schrödinger equation with this potential is solved by a mathematical procedure which can also be applied for other potentials and is described in supplementary section S12.6. Here we give only the analytical representation of the eigenvalues and eigenfunctions and describe some of their properties.

The solutions of the problem are expressed conveniently with reference to the frequency of macroscopic oscillation of a body in the parabolic well

$$\omega = \left(\frac{k}{m}\right)^{1/2}, \tag{12.42}$$

and to a characteristic amplitude of oscillation[11]

$$a = \left(\frac{\hbar}{m\omega}\right)^{1/2} = \left(\frac{\hbar^2}{km}\right)^{1/4}. \tag{12.43}$$

The energy eigenvalues are

$$E_n = (n + \tfrac{1}{2})\hbar\omega, \quad \text{where } n = 0, 1, \ldots, \tag{12.44}$$

and the corresponding eigenfunctions are

$$\psi_n(x) = \langle x \mid E\ \ n\rangle = \frac{\exp(-x^2/2a^2)}{(2^n n!a)^{1/2}\pi^{1/4}} H_n\left(\frac{x}{a}\right); \tag{12.45}$$

here H_n indicates the Hermite polynomial of nth degree,

$$H_n\left(\frac{x}{a}\right) = \sum_{q \leq n/2} \frac{n!(-1)^q}{(n-2q)!q!} \left(2\frac{x}{a}\right)^{n-2q}. \tag{12.45a}$$

As in the preceding problems, n indicates here the number of dark fringes, that is, the number of nodes of the wave function. In the present problem, this number is determined by the degree of the Hermite polynomial. Each Hermite polynomial contains only even or odd powers of x, depending on whether n is even or odd; accordingly, each stationary state of the particle is an eigenstate of parity, like the states $|p_x{}^2\ \pm\rangle$ of a free particle represented by equations (11.26). For this reason the matrix elements of x, namely, the quantities $\langle n | x | n'\rangle$ defined by equation (10.20), vanish unless $n - n'$ is an odd number.

---

11. The potential energy for $x = a$, $\tfrac{1}{2}ka^2$, equals the lowest energy eigenvalue $E_0 = \tfrac{1}{2}\hbar\omega$.

Actually, these matrix elements vanish unless $n - n' = \pm 1$, and are given by

$$\langle n|x|n'\rangle = \begin{cases} (\tfrac{1}{2}n')^{1/2}\, a & \text{for } n' = n+1 \\ (\tfrac{1}{2}n)^{1/2}\, a & \text{for } n' = n-1 \\ 0 & \text{for } n' \neq n \pm 1. \end{cases} \tag{12.46}$$

Owing to this formula and to equation (12.40), the general expression (10.19) of the mean value of x in a state $|b\rangle$ reduces to

$$\langle x\rangle_b = a \sum_n (\tfrac{1}{2}n)^{1/2}[\langle b(0)\,|\,n\rangle e^{i\omega t}\langle n-1\,|\,b(0)\rangle + \langle b(0)\,|\,n-1\rangle e^{-i\omega t}\langle n\,|\,b(0)\rangle]. \tag{12.47}$$

Thus the mean position of the particle performs sinusoidal oscillations with the single frequency ω, for an arbitrary stationary state, in accordance with classical mechanics. These phenomena are illustrated by the solution of problem 10.1.

The oscillator problem has been formulated here as it occurs in macroscopic contexts where a particle is held near a fixed point by gravity or by a spring. In atomic physics no problem is formulated exactly in this manner, but many systems present the very same physico-mathematical features as the harmonic oscillator and have energy eigenvalues and eigenstates represented by equations (12.44) and (12.45). The essential common feature of these systems is that their energy depends on a variable x according to

$$E = \tfrac{1}{2}m\left(\frac{dx}{dt}\right)^2 + \tfrac{1}{2}kx^2. \tag{12.48}$$

The physical nature of the variable x and of the coefficients m and k may be quite different. Nevertheless, the developments of sections S11.6 and 12.1, which yield the connection between the eigenstates of $v_x = dx/dt$ and of x and the Schrödinger equation for the wave functions $\langle x\,|\,E\ \ n\rangle$, remain fully applicable, as do the calculations of E_n, of $\langle x\,|\,E\ \ n\rangle$, and of $\langle x\rangle$.

12.5. ENERGY LEVELS AND OSCILLATIONS OF THE ELECTROMAGNETIC FIELD

Radiation is a typical system with properties mathematically equivalent to those of a harmonic oscillator. The macroscopic fields of monochromatic radiation with frequency ω oscillate sinusoidally in the course of time as the mean value of an oscillator coordinate x does according to equation (12.47). The energy levels of this radiation are separated by intervals $\hbar\omega$, in accordance with equation (12.44). To establish the correspondence completely we must show how the radiation energy can actually be expressed in the form (12.48).

The energy of the electromagnetic field is represented by the integral of the energy density $[\,|\vec{E}|^2 + |\vec{B}|^2]/8\pi$ over the whole space. To sort out the energy of monochromatic components, the fields $\vec{E}$ and $\vec{B}$ can be expanded into monochromatic waves; the energy is then proportional to the sum of the squared Fourier coefficients of the expansions of $\vec{E}$ and $\vec{B}$. (This representation of the radiation energy was considered in the blackbody discussion in section S3.5.) The Fourier coefficients of one of the fields, $\vec{E}$ or $\vec{B}$, are proportional to the time derivative of the Fourier coefficients of the other field. It is this property, implied by the Maxwell equations, which causes the fields to oscillate as though they were responding to an elastic force and possessed an effective inertia. The same property causes the combined energy of the fields to take the form (12.48).

In order to avoid the complications of the Fourier expansion of unspecified radiation fields in free space, we consider here only the energy of a particular type of electromagnetic field, namely, the field in one mode of oscillation of a cavity resonator with perfectly reflecting walls. Radiation in free space can be treated as a limiting case of radiation in a very large cavity.

Call X, Y, and Z the dimensions of the cavity, and take the origin of coordinate axes at one vertex and the axes along three edges. The multiple reflection on the cavity walls restricts the possible distributions of the electric and magnetic fields to certain modes with specified orientations of the fields $\vec{E}$ and $\vec{B}$ and specified numbers of interference fringes.[12] One possible distribution, with the electric field parallel to the z edge of the cavity and with j fringes in the x direction and l fringes in the y direction is

$$E_x = 0,\ E_y = 0,\ E_z = -\frac{1}{c}\frac{dQ}{dt}\sin\left(j\pi\frac{x}{X}\right)\sin\left(l\pi\frac{y}{Y}\right), \tag{12.49}$$

$$B_x = \frac{l\pi}{Y}\,Q(t)\sin\left(j\pi\frac{x}{X}\right)\cos\left(l\pi\frac{y}{Y}\right),$$

$$B_y = -\frac{j\pi}{X}\,Q(t)\cos\left(j\pi\frac{x}{X}\right)\sin\left(l\pi\frac{y}{Y}\right),$$

$$B_z = 0.$$

In this formula, c is the light velocity and Q is a variable that indicates the amplitude and phase of the radiation. The strength of the magnetic field at any point is proportional to Q, the strength of the electric field to the time derivative dQ/dt. The macroscopic Maxwell equations require that Q(t) oscillate sinusoidally with the frequency

$$\omega = 2\pi\nu = \pi c(j^2/X^2 + l^2/Y^2)^{1/2}. \tag{12.50}$$

According to quantum mechanics, the Maxwell equations are fulfilled by the mean values $\langle\vec{E}\rangle$ and $\langle\vec{B}\rangle$. The mean values $\langle Q\rangle$ and $\langle dQ/dt\rangle$ must fulfill corresponding equations, and in particular $\langle dQ/dt\rangle$ must equal $d\langle Q\rangle/dt$. Because of

12. See *BPC*, 3 : 337 ff.

this requirement and because the radiation energy depends on both the magnetic and the electric fields, which are proportional to Q and dQ/dt, respectively, the field strengths B and E are incompatible variables (see the beginning of chapter 11).

Progressing to greater detail, we now express the energy of the radiation in the resonator as a function of Q and dQ/dt. The energy stored per unit volume of space in the form of electric (or magnetic) field equals the squared electric (or magnetic) field strength divided by 8π. We take the squared strength at each point of the cavity for each of the fields given by equations (12.49) and integrate over the volume of the cavity. We find for the radiation energy the expression

$$E = \tfrac{1}{2}\frac{XYZ}{16\pi c^2}\left(\frac{dQ}{dt}\right)^2 + \tfrac{1}{2}\frac{XYZ}{16\pi c^2}\,\omega^2 Q^2. \tag{12.51}$$

This expression is formally equivalent to equation (12.48) with the correspondence

$$Q \longleftrightarrow x, \qquad \frac{dQ}{dt} \longleftrightarrow \frac{dx}{dt}, \tag{12.52}$$

$$\frac{XYZ}{16\pi c^2} \longleftrightarrow m, \qquad \frac{XYZ}{16mc^2}\,\omega^2 \longleftrightarrow k.$$

Each of the procedural steps which have led to establish the Schrödinger equation for the wave function $\langle x | E\ \ n\rangle$ can be repeated to establish the same equation for probability amplitudes of radiation field variables $\langle Q | E\ \ n\rangle$. Thus quantum theory shows that monochromatic radiation has energy levels separated by $\hbar\omega$ and that its mean fields $\langle \vec{E}\rangle$ and $\langle \vec{B}\rangle$ oscillate sinusoidally in nonstationary states.

SUPPLEMENTARY SECTIONS

S12.6. ANALYTICAL SOLUTION OF THE OSCILLATOR EQUATION

The Schrödinger equation for the wave function $\psi(x) = \langle x | E\rangle$ of a particle with the potential energy (12.41) is

$$-\frac{\hbar^2}{2m}\frac{d^2\psi}{dx^2} + \tfrac{1}{2}kx^2\psi(x) = E\psi(x). \tag{12.53}$$

We separate the process of its solution into several steps.

Step I. Choose appropriate units for E and x which make the equation dimentionless. To this end one divides the equation by the energy unit $\hbar\omega = \hbar(k/m)^{1/2}$, which yields

$$-\tfrac{1}{2}\hbar(mk)^{-1/2}\frac{d^2\psi}{dx^2} + \tfrac{1}{2}\hbar^{-1}(mk)^{1/2}\,x^2\,\psi = \frac{E}{\hbar\omega}\,\psi. \tag{12.54}$$

Notice that $\hbar\omega$ is the natural unit for the energy; similarly, $a = \hbar^{-1/2}(mk)^{1/4}$ serves as a unit for the particle displacement x to make the equation dimensionless. Setting $\epsilon = E/\hbar\omega$ and $\xi = x/a$ reduces equation (12.54) to

$$-\tfrac{1}{2}\frac{d^2\psi}{d\xi^2} + \tfrac{1}{2}\xi^2\psi = \epsilon\psi. \tag{12.55}$$

Step II. Determine an approximate solution valid for large values of ξ, where $\tfrac{1}{2}\xi^2 >> \epsilon$, and factor this solution out of ψ. The asymptotic solution is, in this problem, $\psi \sim \exp(-\tfrac{1}{2}\xi^2)$. [The expression $\exp(\tfrac{1}{2}\xi^2)$ also obeys eq. (12.55) for large $|\xi|$ but diverges as $|\xi| \to \infty$.] Accordingly, we set

$$\psi(\xi) = \exp(-\tfrac{1}{2}\xi^2)u(\xi). \tag{12.56}$$

The $\exp(-\tfrac{1}{2}\xi^2)$ factors out of equation (12.55), which thus becomes

$$\frac{d^2u}{d\xi^2} - 2\xi\frac{du}{d\xi} + (2\epsilon - 1)u = 0. \tag{12.57}$$

Step III. Attempt a solution by power series, utilizing equation (12.57) to establish a recursion relation among its coefficients and requiring the series to remain finite by cutting off the recursion toward both low- and high-power terms. (These restrictions are necessary because $|\xi|$ ranges from 0 to ∞.) Accordingly we set

$$u(\xi) = \xi^\nu \sum_{r=0}^{s} c_r\xi^r \tag{12.58}$$

where ν indicates the lowest and $\nu + s$ the highest power of ξ, with ν, s, and c_r yet to be determined. Substitution of equation (12.58) into (12.57) yields

$$\begin{aligned}\xi^\nu &\sum_{r=0}^{s} c_r[(\nu + r)(\nu + r - 1)\xi^{r-2} - 2(\nu + r)\xi^r + (2\epsilon - 1)\xi^r]\\ &= \xi^{\nu-2}\nu(\nu - 1)c_0 + \xi^\nu \sum_{r=0}^{s-2} \xi^r[(\nu + r + 2)(\nu + r + 1)c_{r+2}\\ &+ (2\epsilon - 1 - 2\nu - 2r)c_r] + \xi^{\nu+s}(2\epsilon - 1 - 2\nu - 2s)c_s = 0. \end{aligned}\tag{12.59}$$

This polynomial equation holds for all ξ if the coefficient of each power of ξ vanishes, that is, if

$$\nu(\nu - 1)c_0 = 0; \tag{12.60a}$$

$$(\nu + r + 2)(\nu + r + 1)c_{r+2} + (2\epsilon - 1 - 2\nu - 2r)c_r = 0, \qquad 0 \leq r \leq s - 2; \tag{12.60b}$$

$$(2\epsilon - 1 - 2\nu - 2s)c_s = 0. \tag{12.60c}$$

Equation (12.60a) requires

$$\nu(\nu - 1) = 0, \tag{12.61}$$

since the expression (12.58) of $u(\xi)$ implies that $c_0 \neq 0$. This indicial equation is solved by either

$$\nu = 0 \quad \text{or} \quad \nu = 1. \tag{12.62}$$

Equation (12.60b) constitutes a recursion relation which yields c_{r+2} as a function of c_r. Since the recursion starts with $r = 0$ and proceeds in steps of 2, all coefficients c_r with odd r vanish. Therefore, $u(\xi)$ contains only even powers of ξ when $\nu = 0$ and only odd powers when $\nu = 1$. This result actually follows from the invariance of (12.53) under the *reflection symmetry operation* $x \to -x$; owing to this invariance the parity of a state (see section 11.3) is compatible with its energy. Equation (12.60c), which arises from the truncation of the series in ξ at $r = s$, relates the values of ϵ and s and thus fixes the energy eigenvalues, by setting

$$\epsilon = \nu + s + \tfrac{1}{2}, \quad \text{i.e.,} \quad E = (\nu + s + \tfrac{1}{2})\hbar\omega = (n + \tfrac{1}{2})\hbar\omega. \tag{12.63}$$

Whereas s is necessarily even, $n = \nu + s$ can be any integer; in fact, it is even or odd, depending on whether $\nu = 0$ or 1.

Step IV. Normalization of the eigenfunction $\psi_n(x)$, that is, determination of the coefficients c_0. This may be done laboriously by calculating the $\int_{-\infty}^{\infty} dx\,|\psi_n(x)|^2$ utilizing equations (12.56), (12.58), and (12.60) and thus reducing the integral to a sum of standard integrals of the type

$$\int_{-\infty}^{\infty} \exp(-\xi^2)\xi^{2m}\, d\xi = \begin{cases} \pi^{1/2} & \text{for } m = 0 \\ \pi^{1/2}\,\frac{1}{2}\,\frac{3}{2}\cdots(m - \tfrac{1}{2}) & \text{for } m > 0. \end{cases} \tag{12.64}$$

The result is proportional to $|c_0|^2$ and will equal unity provided the magnitude of c_0 is given an appropriate value. The phase of c_0 is arbitrary, and it is usually taken to be $\tfrac{1}{2}s\pi$, which makes c_0 real and c_s positive.

The wave function $\psi_n(x)$ given by equation (12.45) coincides with the $\psi_n(x)$ obtained here; specifically, the coefficients of the Hermite polynomials (12.45a) fulfill the recursion formula (12.60b).

S12.7. TRANSMISSION AND REFLECTION BY A SQUARE POTENTIAL BARRIER

Numerous problems of particle motion can be schematized and solved by representing a one-dimensional potential by a sequence of steps. The solution procedure is analogous to the one used for a single potential step in section 12.2.2. This section outlines the calculation and results for the stationary states of a particle under the influence of a square potential barrier represented by

$$V(x) = 0 \qquad \text{for } x < 0 \qquad \text{and } x > a,$$
$$V(x) = V' \qquad \text{for } 0 < x < a. \tag{12.65}$$

The particle's wave function is represented by the formula analogous to (12. 26):

$$\psi(x) = Ae^{ik_0x} + Be^{-ik_0x}, \qquad \text{for } x \leq 0; \tag{12.66a}$$

$$\psi'(x) = Ce^{ik'x} + De^{-ik'x}, \qquad \text{for } 0 \leq x \leq a; \tag{12.66b}$$

$$\psi''(x) = Ee^{ik_0(x-a)} + Fe^{-ik_0(x-a)}, \quad \text{for } a \leq x; \tag{12.66c}$$

where k_0 and k' are defined by (12. 25) and k' may be replaced by $i\kappa$, as in equation (12. 29), when $E < V'$ and $k'^2 < 0$. The replacement of k' by $i\kappa$ is deferred until we discuss the results.

The requirement of continuity of the wave function and of its first derivative at the two potential steps, that is, at $x = 0$ and $x = a$, yields four equations analogous to (12. 27),

$$A + B = C + D, \tag{12.67a}$$

$$k_0(A - B) = k'(C - D), \tag{12.67b}$$

$$Ce^{ik'a} + De^{-ik'a} = E + F, \tag{12.67c}$$

$$k'(Ce^{ik'a} - De^{-ik'a}) = k_0(E - F). \tag{12.67d}$$

Here again, we have fewer equations than parameters. The resulting indeterminacy of the parameters implies a degeneracy of the stationary states.

Among the degenerate states we consider, as in section 11. 2. 2, the one in which the particle propagates from infinity toward the barrier only from the side of $x = -\infty$. Thus we set $F = 0$ and, for simplicity, $A = 1$. The squared magnitude of E represents the transmission coefficient, that is, for $E < V'$ the probability of tunneling through the barrier, since k has the same value on both sides of the barrier. The squared magnitude of B represents the reflection coefficient, as in section 12. 2. 2. We shall calculate only B, since the uniformity of flux requires that $|E|^2 = 1 - |B|^2$.

The complete solution of the system (12. 67) is laborious, but B alone can be obtained rapidly by elimination of variables. The two equations (12. 67a, b) can be solved for C and for D as functions of B, while equations (12. 67c, d) serve to eliminate E and to provide a homogeneous relation between C and D. This relation becomes a single equation in B upon substitution of the expressions of C and D as functions of B. The result is

$$B = \frac{k_0^2 - k'^2}{k_0^2 + k'^2 + 2ik_0k'\cot(k'a)}. \tag{12.68}$$

The reflection coefficient can be cast in alternative forms

$$|B|^2 = \frac{(k_0^2 - k'^2)^2}{(k_0^2 + k'^2)^2 + 4k_0^2k'^2\cot^2(k'a)}$$
$$= \frac{(k_0^2 - k'^2)^2}{(k_0^2 - k'^2)^2 + 4k_0^2k'^2[1 + \cot^2(k'a)]} \tag{12.69}$$
$$= \left[1+4\frac{k_0^2}{(k_0^2 - k'^2)^2}\frac{k'^2}{\sin^2(k'a)}\right]^{-1} = \left[1 + 4\frac{\hbar^2}{2ma^2}\frac{E}{V'^2}\frac{(k'a)^2}{\sin^2(k'a)}\right]^{-1}.$$

The last of these forms shows that $|B|^2$ is reduced below unity by the product of two numerical factors in the denominator. One of these, $4(\hbar^2/2ma^2)E/V'^2$, increases linearly with E and thus tends to reduce $|B|^2$ to zero in the high-energy limit. The second factor, $(k'a)^2/\sin^2(k'a)$, equals 1 when $k' = 0$, that is, when $E = V'$. For $E < V'$, the wavenumber k' becomes imaginary and is replaced by $i\kappa$, as noted above; the factor $(k'a)^2/\sin^2(k'a)$ then extrapolates smoothly to $(\kappa a)^2/\sinh^2(\kappa a)$. As E drops well below V', $\sinh^2(\kappa a)$ diverges and thus causes $|B|^2$ to approach one and the tunneling probability $1 - |B|^2$ to approach zero exponentially. On the other hand, as E rises above V', $\sin^2(k'a)$ vanishes periodically whenever $k'a = n\pi$; $|B|^2$ vanishes for each of these values, indicating zero reflection and full transmission. The condition for this event, namely, that the barrier width equals a whole number of half-wavelengths π/k', indicates that full transmission results from the cumulative effect of multiple reflection at the two potential steps.

When V' is negative, that is, when the potential barrier is replaced by a well, the formulae of this section remain valid, with $k'^2 = (2m/\hbar^2)(E + |V'|)$. This result provides the solution of the problem of section 12.4.1 for values of E above the edge of the well.

S12.8. PERIODIC POTENTIAL

The one-dimensional motion of a particle subject to a periodic steplike potential serves as a prototype for particle motions in a crystal. Consider a potential consisting of the array of barriers shown in figure 12.4 and represented by

$$\begin{aligned} V &= 0 \quad \text{for} \quad m(a + b) - b \leqslant x \leqslant m(a + b). \\ V &= V' \quad \text{for} \quad m(a + b) \leqslant x \leqslant m(a + b) + a, \end{aligned} \tag{12.70}$$

where m is an arbitrary integer.

This potential consists of a sequence of barriers identical to that treated in section S12.7. Accordingly, we know the analytical form of the wave function of a stationary state for each section of the potential. For example, this wave

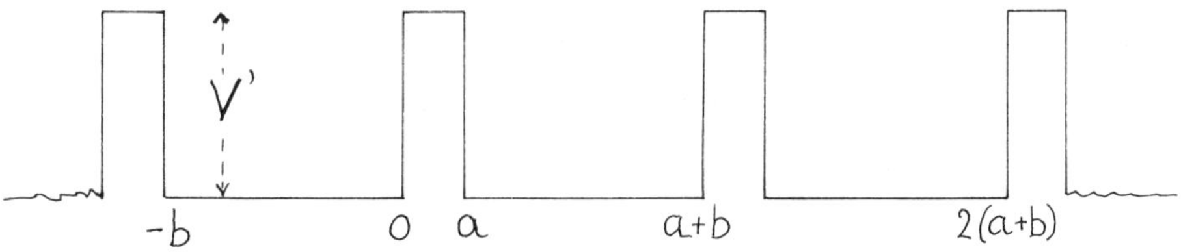

Fig. 12. 4. Periodic array of barriers.

function may be represented by the expression (12. 66) in the three intervals $-b \leqslant x \leqslant 0$, $0 \leqslant x \leqslant a$, $a \leqslant x \leqslant a + b$. The continuation of this wave function through the rest of the array is secured by applying a general result of Appendix C, namely, that the solution of a wave equation with periodic coefficients is the product of a plane wave and a periodic function. Since we have the analytical form (12. 66) for the general solution, we do not need in this problem to assume that V′ is very small, as was effectively assumed in Appendix C.

Following Appendix C, we then require the wave function to have the form $\exp(ivx)f(x)$, where v is an arbitrary wavenumber and $f(x + a + b) = f(x)$. The values of such a wave function at two points x and $x + a + b$ are in the ratio

$$\frac{e^{iv(x+a+b)}\, f(x + a + b)}{e^{ivx}\, f(x)} = e^{iv(a+b)} \tag{12.71}$$

This condition establishes a relation between the coefficient E (or F) of ψ'' in equation (12. 66c) and the coefficient A (or B) in equation (12. 66a), namely

$$Ee^{ik_0 b} = Ae^{iv(a+b)}, \quad Fe^{-ik_0 b} = Be^{iv(a+b)}. \tag{12.72}$$

Substitution of E and F from this formula reduces the equations (12. 67) to a system of four linear homogeneous equations in the coefficients A, B, C, and D. The system has nonzero solutions if and only if its determinant vanishes. This condition establishes a relation between the wavenumber v and the wavenumber k_0, that is, between v and the energy eigenvalue E. (The other parameters, a, b, and k′, are not free, k′ being determined by E and V′.) The determinant form of this relation is

$$\begin{vmatrix} 1 & 1 & -1 & -1 \\ k_0 & -k_0 & -k' & k' \\ \exp[iv(a+b) - ik_0 b] & \exp[iv(a+b) + ik_0 b] & -\exp(ik'a) & -\exp(-ik'a) \\ k_0 \exp[iv(a+b) - ik_0 b] & -k_0 \exp[iv(a+b) + ik_0 b] & -k' \exp(ik'a) & k' \exp(-ik'a) \end{vmatrix} = 0. \tag{12.73}$$

The determinant is simplified by replacing the first two columns (and the last two) by their half-sum and half-difference. Its expansion reduces equation (12.73) to

$$\cos[v(a + b)] = \cos(k_0 b)\cos(k'a) - \frac{k_0^2 + k'^2}{2k_0 k'}\sin(k_0 b)\sin(k'a). \qquad (12.74)$$

The connection between v and k_0 established by this equation is illustrated by combining $\cos(k_0 b)$ and $\sin(k_0 b)$ into a single sine function,

$$\cos[v(a + b)] = f(E)\sin[k_0 b + \delta(E)], \qquad (12.75)$$

where

$$f(E) = -\left[1 + \frac{V'^2}{4E(E - V')}\sin^2(k'a)\right]^{1/2}, \qquad (12.75a)$$

$$\tan\delta(E) = -\frac{2[E(E - V')]^{1/2}}{2E - V'}\cot(k'a). \qquad (12.75b)$$

The essential feature of equation (12.75) is that the absolute value of the coefficient f(E) is no smaller than one and drops to unity only for a few values of E. [The ratio $\sin^2(k'a)/(E - V')$ is nonnegative even when $E < V'$, since k′ is then imaginary.] Now, since $k_0 b + \delta(E)$ increases smoothly with increasing energy E and since the sine oscillates between −1 and +1, there are ranges of E for which the right-hand side of (12.75) exceeds unity in absolute magnitude, as shown in Figure 12.5. For values of E in these ranges, no value of v exists which satisfies equation (12.75), and therefore the Schrödinger equation has no eigenfunction. In the energy ranges for which the right side of equation (12.75) is smaller than one in magnitude, every value of E is a degenerate eigenvalue, because it corresponds to a pair of equal and opposite values of v which yield the same value of $\cos[v(a + b)]$.

Thus the energy eigenvalues of a particle in a periodic potential range over *continuous bands* separated by *gaps*. The gaps exist for energies E above the barrier potential V′, as well as for $E < V'$. The existence of gaps in the frequency spectrum of waves propagating in a periodic medium is discussed in section C.3. It is also pointed out there that waves with frequencies in the gaps can have a limited propagation when the periodic medium is not infinite or perfectly regular. Waves with imaginary values of $v = iw$, and values of $\cos[v(a + b)] = \cosh[w(a + b)] > 1$, then represent realistic physical phenomena. Suppose, for example, that the array of potential barriers (12.70) is semi-infinite, such that $V = 0$ throughout for $x < 0$. A wave function with $v = iw$ then represents the state of a particle which is reflected by the barriers but penetrates partially through them, being attenuated in the amplitude ratio $\exp[-w(a + b)]$ per barrier traversed. Bragg reflection is the analog of this phenomenon in a three-dimensional lattice.

The spectral width of the gaps depends on the transmission of particles by each barrier and decreases, generally, with increasing energy. Indeed, the

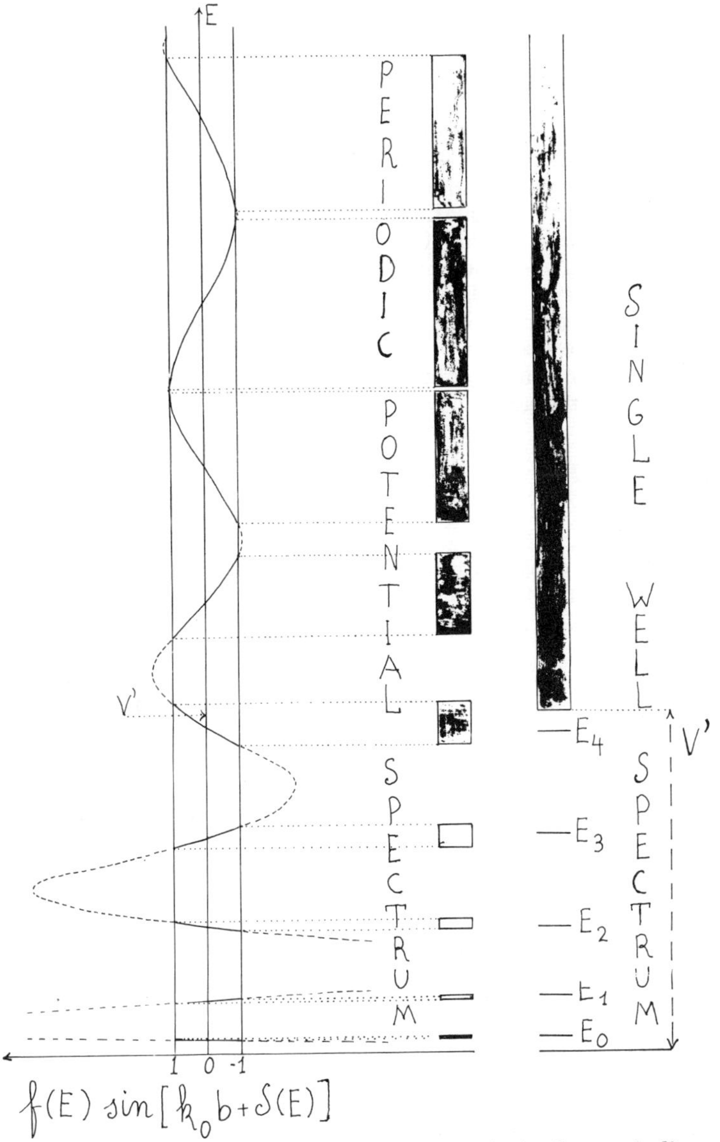

Fig. 12. 5. Spectrum of energy levels of particle in the periodic potential shown in Figure 12. 4, determined by graphical solution of $|f(E) \sin [k_0b - \delta(E)]| \leq 1$. Parameters $(2mV')^{1/2}b/\hbar = 13.4$, $b/a = 4$. Spectrum at right from single-well problem illustrated in Figure 12. 3 for the same value of well parameter.

Note gaps in spectrum, whose width depends on barrier transmission function.

square of the coefficient f(E) given by equation (12. 75a) equals the reciprocal of the barrier transmission $1 - |B|^2$ obtained in section 12. 7 and given by equation (12. 69). The phase constant $\delta(E)$ given by equation (12. 75b) can also be interpreted in the limiting cases of very low and very high barrier transmission. For very low barrier transmission, $k'b$ is imaginary and large and $\cot(k'a) \sim i$; the value of δ then reduces to the constant $-2 \arccos(E/V')^{1/2}$ which appears in the eigenvalue formula (12. 35) for a square well. In this case $|\sin^2(k'a)|$ is very large in equation (12. 75a) and so is $|f(E)|$. Therefore, the gaps are very large, and equation (12. 75) has solutions v only when $\sin[k_0 b + \delta(E)] \sim 0$, that is, when the energy nearly coincides with an eigenvalue of a single well between two barriers. For energies far above the barrier, when the transmission approaches unity, the coefficient of $\cot(k'a)$ is ~ 1 in (12. 75b); in this case we have $\delta \sim k'a - \frac{1}{2}\pi$, and the gaps occur when $k_0 b + k'a$ is a multiple of π, that is, when a whole number of half-wavelengths fits in the period a + b of the array.

CHAPTER 12 PROBLEMS

12. 1. Consider a particle confined in a potential well as in section 12. 4. Take the limit $V' \to \infty$, which is often assumed to establish simple highly schematized conditions. Give formulae for the energy eigenvalues and for the corresponding wave functions.

12. 2. Calculate the mean potential energy, $\langle \frac{1}{2}kx^2 \rangle_n$, of the oscillator particle considered in section 12. 4. 3, for the three lowest stationary states with n = 0, 1, and 2. (The wave functions of these three states are given in problem 10. 1, where ξ stands for x/a.) Take the ratio of $\langle \frac{1}{2}kx^2 \rangle_n$ to the energy eigenvalue E_n and verify that this ratio equals the ratio of potential to total energy for a macroscopic oscillator when the potential energy is averaged over a whole cycle.

12. 3. Using the WKB approximation method of section 12. 4. 2, calculate the energy eigenvalues of a particle with the oscillator potential $\frac{1}{2}kx^2$. Compare your result with the exact eigenvalues given by equation (12. 44).

12. 4. Represent the potential barrier of Figure 12. 2 schematically by V = 0 for x < 0, V = V′ − Fx for x > 0. (V′ may represent the work function for extraction of a metal electron, F the strength of an applied field.) Estimate the barrier transmission by evaluating the squared amplitude $\exp\left[-2\int_0^{x_1}\kappa(x)dx\right]$ (see eq. [12. 33b]), where x_1 is defined by $V(x_1) = E$. Assume the particle to be an electron of mass m_e, take $F = 10^6$ volts/cm and take two values of V′ − E, namely, 0. 5 and 1. 0 eV.

12. 5. A particle of mass m is confined in three dimensions by an elastic force represented by the potential energy $V(\vec{r}) = \frac{1}{2}k(x^2 + y^2 + z^2)$. Calculate its four lowest energy levels and determine the degeneracy of each of them by

finding a set of orthogonal eigenfunctions. (The four levels together have 20 orthogonal eigenfunctions.)

12.6. An electron of mass $m_e = 9.0 \times 10^{-28}$ g moves within a conduit of rectangular cross-section and infinite length represented by the potential energy function $V(r) = V_1(x) + V_2(y) + V_3(z)$, where

$$V_1(x) = 0, \quad \text{for } -\tfrac{1}{2}a < x < \tfrac{1}{2}a; \qquad V_1(x) = \infty, \quad \text{for } |x| > \tfrac{1}{2}a;$$
$$V_2(y) = 0, \quad \text{for } -\tfrac{1}{2}b < y < \tfrac{1}{2}b; \qquad V_2(y) = \infty, \quad \text{for } |y| > \tfrac{1}{2}b;$$
$$V_3(z) = 0.$$

Utilizing the results of problem 12.1, give formulae for the energy eigenvalues and eigenfunctions identified by a particle density with n_x dark fringes perpendicular to the x axis, n_y dark fringes perpendicular to y, and by the eigenvalue p_z of the momentum along the z axis. Sketch a map of the particle density $|\langle x|n_x\rangle \langle y|n_y\rangle|^2$ with $n_x = 2$ and $n_y = 1$ over a cross-section of the conduit, assuming $a = 6$ Å and $b = 3$ Å.

SOLUTIONS TO CHAPTER 12 PROBLEMS

12.1. The limit of $V' \to \infty$ implies that $(E_n/V') \to 0$ for all n. Hence we have $2\arccos(E_n/V')^{1/2} \to \pi$ in equation (12.35a) and $E_n \to \hbar^2(n + 1)^2\pi^2/2ma^2$, that is, a quadratic dependence of E_n on n. The wave function outside the potential well vanishes in the limit $V' \to \infty$, as seen from equations (12.36b) and (12.36c). Using equation (12.39) and the value of E_n, the wave function (12.37) within the well reduces to the simple form

$$\psi_n = \begin{cases} (-1)^{n/2}\left(\dfrac{2}{a}\right)^{1/2} \cos\,(n + 1)\pi\dfrac{x}{a}, & \text{for n even} \\[2ex] (-1)^{(n-1)/2}\left(\dfrac{2}{a}\right)^{1/2} \sin(n + 1)\pi\dfrac{x}{a}, & \text{for n odd.} \end{cases}$$

The ψ_n has a node at each infinitely high potential step as it does for a single potential step (see end of section 12.2.2.).

12.2. The mean values $\langle \xi^2 \rangle$, averaged over the distributions $P(\xi, E\ \ n)$ of problem 10.1b, are equal to $\frac{1}{2}, \frac{3}{2}, \frac{5}{2}$, for $n = 0, 1, 2$. Therefore, we have $\langle \tfrac{1}{2}kx^2 \rangle_n = \tfrac{1}{2}ka^2\langle \xi^2 \rangle_n = \tfrac{1}{2}ka^2(n + \tfrac{1}{2})$. Equations (12.43) and (12.42) give $\tfrac{1}{2}ka^2 = \tfrac{1}{2}k\hbar/(km)^{1/2} = \tfrac{1}{2}\hbar\omega$ and hence $\langle \tfrac{1}{2}kx^2 \rangle_n = \tfrac{1}{2}(n + \tfrac{1}{2})\hbar\omega = \tfrac{1}{2}E_n$, that is one-half of the energy eigenvalue. A macroscopic oscillator has a displacement $x = x_0 \cos \omega t$ and a potential energy $\tfrac{1}{2}kx_0^2 \cos^2 \omega t$; on the average over an oscillation period $\langle \cos^2 \omega t \rangle = 1/2$ and hence $\langle 1/2 kx^2 \rangle = 1/2 \times 1/2 kx_0^2$, that is, one-half of the total energy.

12.3. Substitution of $V = 1/2 kx^2$ in equation (12.40a) gives the condition $2\int_{x_1}^{x_2}\left[2m(E_n - 1/2 kx^2)\right]^{1/2} dx = (n + 1/2)h$. The integration limits x_1 and x_2 are defined by the condition $E_n - 1/2 kx_1^2 = E_n - 1/2 kx_2^2 = 0$. Setting $x = x_2 \xi$ reduces our condition to $2(2mE_n)^{1/2} x_2 \int_{-1}^{1}(1 - \xi^2)^{1/2} d\xi = 2(2mE_n)^{1/2} x_2 1/2 \pi = (n + 1/2)h$. Since $x_2 = (2E_n/k)^{1/2}$ and $(m/k)^{1/2} = 1/\omega$, the condition yields $E_n = (n + 1/2)\hbar\omega$, which coincides with the exact equation (12.44).

12.4. The definitions $\kappa^2 = 2m_e(V - E)/\hbar^2 = 2m_e(V' - Fx - E)/\hbar^2 = 2m_eF(x_1 - x)/\hbar^2$ give: $\exp\left[-2\int_0^{x_1}\kappa(x)dx\right] = \exp\left[-2(2m_eF)^{1/2}\int_0^{x_1}(x_1 - x)^{1/2}dx/\hbar\right]$ $= \exp\left[-4(2m_eFx_1^3)^{1/2}/3\hbar\right] = \exp\left[-4(2m_e)^{1/2}(V' - E)^{3/2}/3F\hbar\right]$

$$= \begin{cases} 2 \times 10^{-30} \text{ for } V' - E = 1 \text{ eV} \\ 3 \times 10^{-11} \text{ for } V' - E = 0.5 \text{ eV.} \end{cases}$$

12.5. The Schrödinger equation for this problem is separable, and can be treated as in equations (12.14) ff., because the potential $V(\vec{r})$ is the sum of separate terms $1/2 kx^2 + 1/2 ky^2 + 1/2 kz^2$. Separate energy levels and eigenfunctions for the motion along x, y, and z are provided in section 12.4.3. Call n_x, n_y, and n_z the quantum numbers for the separate motions. The energy of the combined motion depends only on the number $N = n_x + n_y + n_z$, according to equations (12.16) and (12.44), its eigenvalues being given by $E_N = (n_x + 1/2 + n_y + 1/2 + n_z + 1/2)\hbar\omega = (N + 3/2)\hbar\omega$. The four lowest levels are $3/2 \hbar\omega$, $5/2 \hbar\omega$, $7/2 \hbar\omega$, and $9/2 \hbar\omega$. Energy eigenfunctions with different sets of quantum numbers $\{n_x, n_y, n_z\}$ are orthogonal. For $N = 0$ there is a single (nondegenerate) eigenfunction with $n_x = 0, n_y = 0, n_z = 0$. For $N = 1$ there are three eigenfunctions identified by the sets $\{1, 0, 0\}$, $\{0, 1, 0\}$, and $\{0, 0, 1\}$. For $N = 2$ there are six eigenfunctions $[\{1, 1, 0\}, \{1, 0, 1\}, \{0, 1, 1\}, \{2, 0, 0\}, \{0, 2, 0\}, \{0, 0, 2\}]$; and for $N = 3$ there are ten $[\{1, 1, 1\}, \{2, 1, 0\}, \{2, 0, 1\}, \{1, 2, 0\}, \{0, 2, 1\}, \{1, 0, 2\}, \{0, 1, 2\}, \{3, 0, 0\}, \{0, 3, 0\}, \{0, 0, 3\}]$.

12.6. Equations (12.16) and (12.14) and problem 12.1 give $E_{n_x n_y p_z} = (\hbar^2/2m_e)[(n_x + 1)^2\pi^2/a^2 + (n_y + 1)^2\pi^2/b^2] + p_z^2/2m_e$ and $\langle x, y, z \mid n_x, n_y, p_z\rangle = \{(2/a)^{1/2} \cos[(n_x + 1)x/a - 1/2 n_x]\pi\} \times \{(2/b)^{1/2} \cos[(n_y + 1)y/b - 1/2 n_y]\pi\} \times \{(1/h)^{1/2} \exp(ip_z z/\hbar)\}$. For $n_x = 2$ and

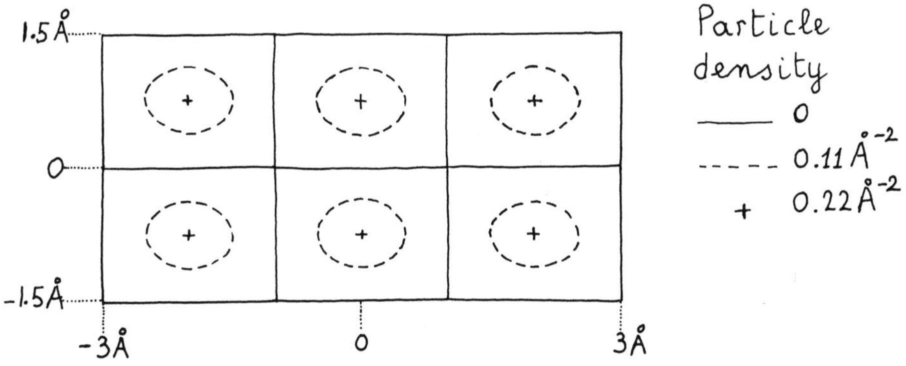

$n_y = 1$ the particle density across the conduit is $P(xy, n_x n_y) = (2/9)\cos^2(3\pi x/a)\sin^2(2\pi y/b)$ Å^{-2}.

Part III

MECHANICS OF SINGLE ATOMS

introduction

Part II has developed the main physical concepts and mathematical tools for the treatment of atomic phenomena. It has thus been possible to interpret some outstanding observations such as the stability of atoms and the occurrence of photons. Part III begins a theoretical treatment of the mechanical properties of single atoms.

The motion of atomic electrons is described through the evolution in time of the state of the electrons. This evolution is represented, in turn, by a superposition of stationary state wave functions. Therefore, the theory consists primarily of the study of stationary states of atoms, of their energy levels, and of their eigenfunctions. Most of the work goes into constructing and solving the Schrödinger equation appropriate to each problem.

The Schrödinger equation obtained in chapter 12 pertains to the motion of a single pointlike particle. Its direct application to the motion of a single electron in the hydrogen atom represents accurately many experimental results. There are, however, outstanding properties which are not explained on the basis of the motion of a single pointlike particle and which have not yet been stressed in this book. In the first place, each electron has a magnetic moment and an angular momentum which influence the magnetic properties and energy levels of atoms. A second and most important point is that the ability of matter to withstand pressure is not fully explained by the stability of each electron under nuclear attraction. The main additional mechanism underlying the incompressibility of matter will be identified as a property of systems of two or more electrons called the *Pauli exclusion principle*.

These additional phenomena will be incorporated in the theory by expanding and adapting the treatment developed in Part II. The essential tools remain the probability amplitudes which connect eigenstates of incompatible variables and the derivation of the Schrödinger equation for each system from an expression of its mean energy. It is thus possible to write, for each atomic system, a Schrödinger equation which includes all variables relevant to our study.

An analytical solution of this equation can be obtained, however, only in a few simple cases. Therefore, one proceeds in most cases by construction of a sequence of expressions for the mean energy of each system which depend on increasing numbers of variables and on increasing correlations among them. The Schrödinger equation corresponding to each of these expressions can then be solved, at least approximately, by utilizing the solutions of the previous simpler equations. Thus one accounts for an increasing fraction of the observable properties of each system.

The development of atomic theory along these lines can be condensed to some extent by following a "postulational approach." In this approach one initially draws some general ideas from experiments of the types described in Parts I and II. Such ideas suggest the postulate that each physical variable is represented by a Hermitian operator.[1] Relationships between incompatible variables are represented by the commutators of their operators. In particular, the commutator of position and momentum is postulated to be proportional to the identity operator. Again, the commutator of angular momentum components $\mathcal{J}_x$ and $\mathcal{J}_y$ is proportional to the component $\mathcal{J}_z$. Instead of obtaining the Schrödinger equation from an expression of the mean value of the energy, the postulational method leads directly to the construction of the energy operator and to the determination of its eigenvectors.

Despite the economy and elegance of the postulational approach, we shall continue to use the language and procedures of Part II. Although the actual calculations are cast in mathematical language, we wish to keep in sight as much as possible the connection between each mathematical step and its physical substrate. Maximizing the contact between mathematical symbols and the observations they represent seems worth an additional effort in an introductory study of atomic physics.

1. A brief introduction to operator language is given in supplementary section S11. 7.

13. the hydrogen atom

Having determined in Part II general laws for the motion of a pointlike particle, we can now apply them to a single atomic electron subject to attraction by a nucleus. In particular, we shall determine the energy levels of a hydrogen atom and the corresponding wave functions which represent the motion of its single electron. This will be done by solving the Schrödinger equation (12. 9) with the potential energy function $V(\vec{r}) = - e^2/r$. The eigenvalues of this equation are correctly represented by Bohr's formula derived in section S3. 6 and thus agree with spectroscopic evidence to one part in 10^5.

Results of this study of the hydrogen atom are applicable to various atomic problems. In the first place, the whole treatment is applicable, as in section S3. 6, to hydrogen-like ions consisting of a nucleus of charge Ze and of a single electron. Each formula obtained for the hydrogen atom is extended by replacing the product of charges e^2 by Ze^2 and by adjusting the reduced mass of the electron and nucleus. Second, since the potential energy $V(\vec{r})$ depends only on the distance r of the electron from the nucleus, many results obtained for the hydrogen atom apply to the motion of a particle in any potential field with spherical symmetry. In particular, the classification of the excited stationary states of hydrogen is approximately applicable to atoms with many electrons because the potential energy of each atomic electron depends primarily on its distance from the nucleus.

Nonstationary states of the hydrogen atom are constructed by superposition of stationary states. Nonstationary states are discussed in section 13. 5, where we identify the types of nonstationary states in which the mean position of the electron oscillates about the nucleus. Since these are the states that emit or absorb radiation strongly, the rules that identify them take the name of *selection rules* of spectroscopy.

13. 1. SEPARATION OF ROTATIONAL AND RADIAL MOTION

The potential energy $-e^2/r$ of the electron in a hydrogen atom depends only on its distance from the nucleus, that is, it depends on the single coordinate r. We have already treated in section 12. 2 the motion of a particle whose potential energy depends on one Cartesian coordinate only. The Schrödinger equation was then separated into three equations pertaining to motion along the axes x, y, and z. These three motions are independent of one another. The Schrödinger equation for the electron of the hydrogen atom also separates into equations pertaining to motion along the polar coordinates r, θ, ϕ. However, in this case the three types of motion are not quite independent of one another.

The relation among these three motions is brought out by partitioning the kinetic energy. The electron velocity $\vec{v}$ may be split first into two components, namely, $\vec{v}_{rad}$, parallel to the radius from the nucleus to the electron, and $\vec{v}_{rot}$, a rotation around the nucleus which is perpendicular to $\vec{v}_{rad}$. The electron energy is accordingly expressed as[1]

$$E = K + V = \tfrac{1}{2}\bar{m}v^2 - \frac{e^2}{r} = (\tfrac{1}{2}\bar{m}v_{rad}^2 - \frac{e^2}{r}) + \tfrac{1}{2}\bar{m}v_{rot}^2. \qquad (13.1)$$

The kinetic energy of rotation is not constant even though no force or torque acts in the direction of $\vec{v}_{rot}$, because the splitting of $\vec{v}$ into $\vec{v}_{rad}$ and $\vec{v}_{rot}$

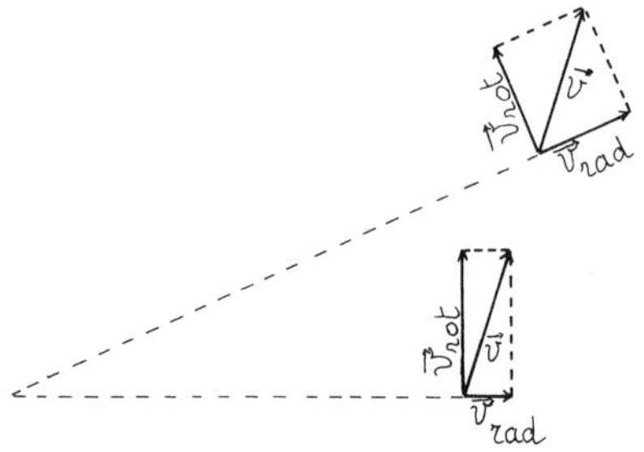

Fig. 13.1. The components $\vec{v}_{rot}$ and $\vec{v}_{rad}$ of the same velocity $\vec{v}$ depend on the position of the electron in the atom.

depends on the position of the electron, as shown in Figure 13.1. What remains constant is the *orbital angular momentum*

$$\vec{l} = \vec{r} \times \vec{p} = \bar{m}\vec{r} \times \vec{v} \qquad (13.2)$$

whose magnitude is $|\vec{l}| = \bar{m}rv_{rot}$.[2] The total energy E is then written as

---

1. In this and in the following equations in this chapter, $\bar{m}$ represents the reduced mass of the electron and of the nucleus. Note that the total kinetic energy of two particles of masses m_1 and m_2 and velocities $\vec{v}_1$ and $\vec{v}_2$ can be expressed as

$$\tfrac{1}{2}(m_1 + m_2)\left|\frac{m_1\vec{v}_1 + m_2\vec{v}_2}{m_1 + m_2}\right|^2 + \tfrac{1}{2}\frac{m_1 m_2}{m_1 + m_2}|\vec{v}_1 - \vec{v}_2|^2.$$

The first term represents the energy associated with the motion of the center of mass; the second, the energy of relative motion. Only the second term is included in equation (13.1) since the motion of the H atom as a whole is not relevant to its levels of internal energy.

2. The word "orbital" is a historic remnant of the planetary model of atomic structure. In general one still speaks, somewhat loosely, of "orbital motion" with reference to the motion of a pointlike particle in a field of force.

$$E = (\tfrac{1}{2}\bar{m}v_{rad}^2 - \frac{e^2}{r}) + \tfrac{1}{2}\bar{m}\frac{(rv_{rot})^2}{r^2} = (\frac{p_{rad}^2}{2\bar{m}} - \frac{e^2}{r}) + \frac{|\vec{l}\,|^2}{2\bar{m}r^2}. \tag{13.3}$$

Thus, for motion with constant $|\vec{l}\,|^2$, the kinetic energy of rotation is still a function of the radius r. One may then study the radial motion by regarding the rotational energy $|\vec{l}\,|^2/2\bar{m}r^2$ as an additional potential energy. Incidentally, note that the gradient of this potential energy is the centrifugal force whose strength is given by[3]

$$-\frac{d}{dr}\frac{|\vec{l}\,|^2}{2\bar{m}r^2} = \frac{|\vec{l}\,|^2}{\bar{m}r^3} = \frac{(\bar{m}rv_{rot})^2}{\bar{m}r^3} = \frac{\bar{m}v_{rot}^2}{r}. \tag{13.4}$$

The rotational motion itself is studied by partitioning in turn $\vec{v}_{rot}$ and $|\vec{l}\,|^2$ into components pertaining to motion along meridian and parallel lines. We set $\vec{v}_{rot} = \vec{v}_\theta + \vec{v}_\phi$, $v_{rot}^2 = v_\theta^2 + v_\phi^2$ and $|\vec{l}\,|^2 = (\bar{m}rv_\theta)^2 + (\bar{m}rv_\phi)^2$. Here again, v_θ and v_ϕ are not constant, nor are $(\bar{m}rv_\theta)^2$ and $(\bar{m}rv_\phi)^2$. Constant, however, is any component of the angular momentum; in particular, the component along the axis of polar coordinates, the Cartesian z axis, is constant with the value $l_z = \bar{m}(r \sin\theta)v_\phi$. The square angular momentum $|\vec{l}\,|^2$ and the kinetic energy of rotation are expressed in terms of l_z^2 by

$$|\vec{l}\,|^2 = (\bar{m}rv_\theta)^2 + \frac{(\bar{m}r\sin\theta\, v_\phi)^2}{\sin^2\theta} = (\bar{m}rv_\theta)^2 + \frac{l_z^2}{\sin^2\theta}, \tag{13.5}$$

$$\frac{|\vec{l}\,|^2}{2\bar{m}r^2} = \frac{(\bar{m}rv_\theta)^2}{2\bar{m}r^2} + \frac{l_z^2}{2\bar{m}r^2\sin^2\theta}. \tag{13.5a}$$

For a given value of l_z^2 one may then study the motion along meridians by regarding the term $l_z^2/(2\bar{m}r^2\sin^2\theta)$ as a potential energy which depends on θ. For $l_z^2 \neq 0$ this potential energy is lowest at the equator, where $\sin\theta$ is maximum, and is infinite at the poles; it corresponds to a centrifugal force which draws the electron away from the poles toward the equatorial plane.

The quantum-mechanical treatment of the electron's motion in the hydrogen atom could be carried out by rederiving the Schrödinger equation from the energy expression (13.3), with $|\vec{l}\,|^2$ in the form (13.5). However, the same result is obtained by entering in the general Schrödinger equation (12.9) the expression of ∇^2 in polar coordinates[4] and $V(\vec{r}) = -e^2/r$. The result is

---

3. See, e.g., *BPC*, 1: 76, 285.

4. The procedure for expressing differential operators in non-Cartesian coordinates is described in calculus textbooks. See, e.g., W. Kaplan, *Advanced Calculus* (Reading, Mass.: Addison Wesley, 1952), pp. 154 ff.; M. L. Boas, *Mathematical Methods in the Physical Sciences* (New York: Wiley, 1966), pp. 44 ff.; also E. Merzbacker. *Quantum Mechanics* (New York: John Wiley and Sons, 1970), pp. 178-79.

$$-\frac{\hbar^2}{2\overline{m}}\left[\left(\frac{\partial^2}{\partial r^2}+\frac{2}{r}\frac{\partial}{\partial r}\right)+\frac{1}{r^2}\left(\frac{\partial^2}{\partial\theta^2}+\cot\theta\frac{\partial}{\partial\theta}\right)+\frac{1}{r^2\sin^2\theta}\frac{\partial^2}{\partial\phi^2}\right]\langle r\ \theta\ \phi\,|\,E\ n\rangle$$

$$-\frac{e^2}{r}\langle r\ \theta\ \phi\ |\ E\ n\rangle=\langle r\ \theta\ \phi\,|\,E\ n\rangle E_n. \qquad (13.6)$$

Note the coefficient $1/(r^2\sin^2\theta)$ in the last term in the square brackets, which corresponds to the kinetic energy term $l_z^2/(2\overline{m}r^2\sin^2\theta)$ in equation (13.5a); similarly, all terms corresponding to $|\vec{l}|^2/2\overline{m}r^2$ have coefficient $1/r^2$.

The Schrödinger equation (13.6) is separated by expressing the probability amplitude $\langle r\ \theta\ \phi\,|\,E\ n\rangle$ as a product of three factors depending respectively on r, θ, and ϕ, much as was done in equation (12.14). Since states of motion along ϕ and θ are not eigenstates of the energy, we leave for the time being their identifying labels blank,

$$\langle r\ \theta\ \phi\,|\,E\ n\rangle=\langle r\,|\,E\ n\rangle\langle\theta\,|\,\rangle\langle\phi\,|\,\rangle. \qquad (13.7)$$

For each factor on the right-hand side of equation (13.7) one constructs an eigenvalue equation by sorting out of equation (12.6) the terms which contain derivatives with respect to the coordinates of that factor. The three equations are

$$-\frac{\hbar^2}{2\overline{m}r^2\sin^2\theta}\frac{d^2}{d\phi^2}\langle\phi\,|\,\rangle=\frac{l_z^2}{2\overline{m}r^2\sin^2\theta}\langle\phi\,|\,\rangle, \qquad (13.8a)$$

$$-\frac{\hbar^2}{2\overline{m}r^2}\left(\frac{d^2}{d\theta^2}+\cot\theta\frac{d}{d\theta}\right)\langle\theta\,|\,\rangle+\frac{l_z^2}{2\overline{m}r^2\sin^2\theta}\langle\theta\,|\,\rangle=\frac{|\vec{l}|^2}{2\overline{m}r^2}\langle\theta\,|\,\rangle, \qquad (13.8b)$$

$$-\frac{\hbar^2}{2\overline{m}}\left(\frac{d^2}{dr^2}+\frac{2}{r}\frac{d}{dr}\right)\langle r\,|\,E\ n\rangle+\left(\frac{|\vec{l}|^2}{2\overline{m}r^2}-\frac{e^2}{r}\right)\langle r\,|\,E\ n\rangle=\langle r\,|\,E\ n\rangle E_n. \qquad (13.8c)$$

Note that equation (13.8a) actually involves only the ϕ coordinate because its dependence on r and on θ can be factored out. Similarly, equation (13.8b) is actually independent of r, but it does involve as a parameter the eigenvalue l_z^2 of the motion along ϕ. Finally, equation (13.8c) depends only on r but involves as a parameter the eigenvalue $|\vec{l}|^2$.

The wave function (13.7) satisfies the complete Schrödinger equation (13.6) provided its separate factors satisfy the separate equations (13.8a-c), as shown by the following considerations.

Equation (13.8a) ensures that the probability amplitude $\langle\phi\,|\,\rangle$ factors out when the wave function (13.7) is entered into the complete equation (13.6). Following this first factorization, equation (13.8b) ensures that also $\langle\theta\,|\,\rangle$ factors out of (13.6). This double factorization reduces (13.6) to (13.8c).

Each of the equations (13.8) is an eigenvalue equation. The eigenvalues of (13.8a) and (13.8b) represent respectively the kinetic energy of rotation along parallels and the total kinetic energy of rotation. The two equations (13.8a) and (13.8b) determine the eigenvalues and the eigenfunctions of the squared angular momentum component l_z^2 and of the squared momentum $|\vec{l}|^2$. The solutions of these equations will be discussed in section 13. 2, and those of the radial equation (13. 8c) in section 13. 3.

13. 2. ROTATIONAL MOTION AND ORBITAL MOMENTUM

The equations (13. 8a) and (13. 8b) are simplified by dropping factors that appear on both sides. Thus they reduce to

$$-\hbar^2 \frac{d^2}{d\phi^2}\langle\phi|\rangle \equiv l_z^2\langle\phi|\rangle, \tag{13.9}$$

$$-\hbar^2\left(\frac{d^2}{d\theta^2} + \cot\theta\,\frac{d}{d\theta}\right)\langle\theta|\rangle + \frac{l_z^2}{\sin^2\theta}\langle\theta|\rangle = |\vec{l}|^2\langle\theta|\rangle. \tag{13.10}$$

These equations have been established on the grounds that the electron's potential energy depends only on its distance from the nucleus without any reference to the analytic form of the potential function V(r). Accordingly, they apply to the orbital motion of any particle subject to a central force.[5]

Equations (13. 9) and (13. 10) could actually have been derived by seeking directly the eigenvalues and eigenfunctions of l_z^2 and $|\vec{l}|^2$. The wave-mechanical procedure of section 11. 2 associates to any function of the particle momentum $\vec{p}$ an operator obtained by replacing $\vec{p}$ by $(\hbar/i)\vec{\nabla}$. To the component of the orbital angular momentum

$$l_z = (\vec{r}\times\vec{p})_z = xp_y - yp_x \tag{13.11}$$

corresponds the operator[6]

$$l_z = \frac{\hbar}{i}(x\nabla_y - y\nabla_x) = \frac{\hbar}{i}\frac{\partial}{\partial\phi}. \tag{13.12}$$

Following the procedure developed in section 12. 1, this equation could be used to express the mean value $\langle l_z\rangle_a$ for an arbitrary state $|a\rangle$ and eventually to construct a Schrödinger equation. Thus one would obtain the equation

$$\frac{\hbar}{i}\frac{d}{d\phi}\langle\phi|\rangle = \langle\phi|\rangle l_z. \tag{13.13}$$

---

5. See n. 2, p. 278

6. Equation (13. 12) utilizes the formulae

$$\nabla_x = \frac{x}{r}\frac{\partial}{\partial r} + \frac{x}{z(x^2+y^2)^{1/2}}\frac{\partial}{\partial\theta} - \frac{y}{x^2+y^2}\frac{\partial}{\partial\phi},$$

$$\nabla_y = \frac{y}{r}\frac{\partial}{\partial r} + \frac{y}{z(x^2+y^2)^{1/2}}\frac{\partial}{\partial\theta} + \frac{x}{x^2+y^2}\frac{\partial}{\partial\phi}.$$

This equation is analogous to equation (11.8) which is satisfied by the momentum eigenfunction. We regard equation (13.13) as a complement to the main equation (13.9) for the motion along ϕ. The equation (13.9) for l_z^2 could also be obtained directly, and so could the equation (13.10) for $|\vec{l}|^2$; the latter is analogous to (11.17).

13.2.1. *Motion along Parallel Circles.* States of motion of a particle along a parallel may be described as superpositions of eigenstates of the orbital angular momentum component l_z, and hence by means of joint eigenfunctions of equations (13.9) and (13.13). These eigenfunctions are complex exponentials of the form $\exp(il_z\phi/\hbar)$, analogous to the plane waves $\exp(ip_x x/\hbar)$. However, whereas all different values of the coordinate x represent different points of physical space, values of the longitude coordinate ϕ differing by any multiple of 2π radians represent the same point. Accordingly, the probability amplitude $\langle\phi|a\rangle$ for any state $|a\rangle$ must be a periodic function of ϕ with period 2π. This condition requires that in the eigenfunctions $\exp(il_z\phi/\hbar)$ the ratio $l_z/\hbar$ be a whole number m. Thus, the eigenfunctions and eigenvalues of equation (13.13) are usually represented by

$$\langle\phi|m\rangle = (2\pi)^{-1/2}\, e^{im\phi}, \tag{13.14}$$

$$l_z = m\hbar. \tag{13.15}$$

This eigenvalue formula for the orbital motion of a particle coincides with the general formula (7.5) for the eigenvalues of angular momentum components; m is accordingly called the *magnetic quantum number*. However, m is necessarily an integer for the motion of a particle, whereas half-integer values are also observed in the analysis of atomic beams (see chaps. 6 and 7).

The eigenfunctions (13.14) represent a *discrete* set of eigenstates of orbital motion even though this motion is not confined by forces. The point is that the very nature of motion along a *closed loop* provides an effective confinement. For motion along a parallel circle with given radius r sin θ, the eigenstates of l_z represented by (13.14) are also eigenstates of the kinetic energy with the discrete eigenvalues

$$K_{\phi m} = \frac{l_z^2}{2\bar{m}r^2\sin^2\theta} = \frac{m^2\hbar^2}{2\bar{m}r^2\sin^2\theta}. \tag{13.16}$$

Eigenstates represented by (13.14) with values of the quantum number m of equal magnitude and opposite sign are degenerate since they have equal eigenvalues of K_ϕ and of l_z^2. The degenerate states with these eigenvalues constitute a two-level system quite analogous to the system of eigenstates of p_x^2 described in section 11.3. The eigenfunctions (13.14) represent states with uniform particle density and nonzero particle flux. Eigenstates of l_z^2 with zero flux and nonuniform particle density are represented by wave functions analogous to (11.26):

$$\langle\phi|m^2\,+\rangle = \pi^{-1/2}\cos m\phi, \tag{13.17a}$$

$$\langle\phi|m^2\,-\rangle = \pi^{-1/2}\sin m\phi. \tag{13.17b}$$

These states have a waving pattern of particle density with dark fringes centered on m meridian planes. On each of these planes the wave function $\langle\phi|m^2 \pm\rangle$ vanishes and so does, therefore, the wave function (13.7) of the three-dimensional motion; accordingly, the meridian planes at the center of dark fringes are nodal surfaces of (13.7). The wave function (13.17a) has a bright fringe centered on the meridian plane defined by $\phi = 0$ and $\phi = \pi$; this is, instead, a nodal plane for the wave function (13.17b), as shown in Figure 13.2 for the example of $m = 2$. For $m = 0$ there is a single nondegenerate state with constant probability amplitude $\langle\phi|0\rangle = (2\pi)^{-1/2}$ and without any nodal plane.

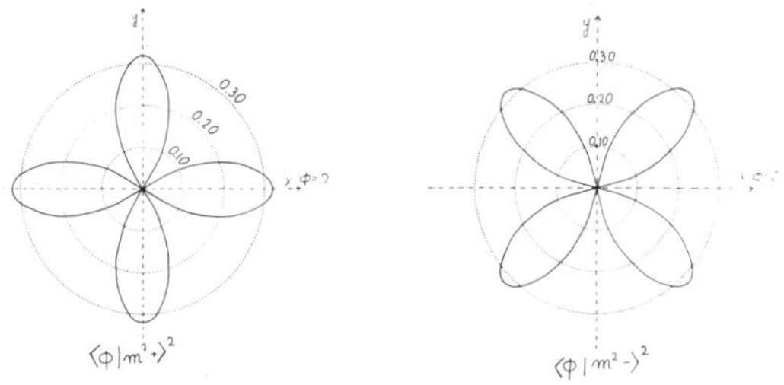

Fig. 13.2. Electron density distributions about the z axis in states $|m^2 \pm\rangle$ with $m = 2$.

13.2.2. *Motion along Meridians*. This motion is effectively confined because the θ coordinate runs only from 0 to π, that is, from one to the other pole of the coordinates. The limitation of θ to a finite range forces the particle flux to be zero in the motion along meridians. (A nonzero flux could not have zero divergence particularly at the poles, whereas the flux must be solenoidal for any stationary state.) Accordingly, the eigenfunctions of equation (13.10) are real standing waves with waving patterns of particle density. Each node of a wave function $\langle\theta|\;\rangle$, occurring at a certain value of θ, identifies a parallel circle $\theta =$ constant as a nodal line of the wave function for motion on a spherical surface of radius r. When the radial motion is taken into account, this nodal line generates a nodal cone of aperture θ (see Fig. 13.3).[7] The nodal cones serve, together with nodal meridian planes, to characterize eigenstates of rotational motion.

The motion along meridians depends on the motion along parallels, since equation (13.10) includes a term proportional to l_z^2. (Recall from the discussion of eq. [13.5a] that this term represents a centrifugal potential.) This term vanishes for the eigenstates of no motion along ϕ, when $m = 0$ and $l_z^2 = 0$. In this case the motion along meridians is subject to no force and equation (13.10) has an eigenfunction $\langle\theta|\;\rangle =$ constant, with eigenvalue $|\vec{l}\,|^2 = 0$; this eigenfunction represents a state with no rotational motion whatever. In the same case

7. A nodal cone of aperture $\frac{1}{2}\pi$ coincides with the equatorial plane and occurs, for reasons of symmetry, whenever the number of nodal cones is odd.

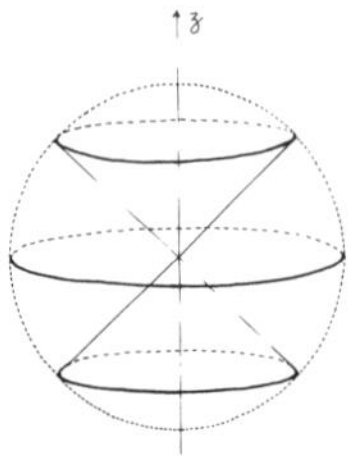

Fig. 13. 3. Nodal lines of particle density for three values of θ.

of $m = 0$, there are excited stationary states of motion along meridians. The study of these states is conducted by solving equation (13. 10) by the method of section S12. 6. The analytical development is somewhat laborious,[8] and we confine ourselves to an outline of its results.

Each stationary state is characterized by a number l of nodal cones. The wave function of this state is indicated by $\langle\theta|l\ m\rangle$, with $m = 0$, and is represented by a polynomial of degree l in the variable $\cos\theta$. For this reason, $\cos\theta$ serves conveniently as the independent variable and equation (13. 10) takes the form

$$-\left[(1-\cos^2\theta)\frac{d^2}{d(\cos\theta)^2} - 2\cos\theta\frac{d}{d\cos\theta}\right]\langle\theta|l\ m\rangle + \frac{m^2}{1-\cos^2\theta}\langle\theta|l\ m\rangle$$

$$= \frac{|\vec{l}|^2}{\hbar^2}\langle\theta|l\ m\rangle, \qquad (13.18)$$

where we have replaced l_z^2 by $m^2\hbar^2$. One must also require the solution of (13. 18) for $m = 0$ to have antinodes at $\theta = 0$ and $\theta = \pi$, to ensure smoothness at the poles. The eigenfunctions $\langle\theta|l\ 0\rangle$ are called Legendre polynomials and are given by[9]

$$\langle\theta|0\ 0\rangle = (1/2)^{1/2},$$

$$\langle\theta|1\ 0\rangle = (3/2)^{1/2}\cos\theta,$$

$$\langle\theta|2\ 0\rangle = (5/8)^{1/2}(3\cos^2\theta - 1), \qquad (13.19)$$

$$\cdots$$

$$\langle\theta|l\ 0\rangle = (l + 1/2)^{1/2}\frac{1}{2^l l!}\left[\frac{d}{d(\cos\theta)}\right]^l(\cos^2\theta - 1)^l.$$

---

8. See, for instance, Merzbacher, *Quantum Mechanics*, pp. 181 ff.

9. The eigenfunctions (13. 9) contain a normalization factor which ensures that $\int_0^\pi |\langle\theta|l\ 0\rangle|^2 \sin\theta\, d\theta = 1$. The name "Legendre polynomial" often applies to functions indicated by $P_l(\cos\theta) = (l + 1/2)^{1/2}\langle\theta|l\ 0\rangle$, for which $P_l(1) = 1$.

The corresponding eigenvalues of (13.18) are

$$\frac{|\vec{l}|^2}{\hbar^2} = l(l+1), \tag{13.20}$$

and thus agree with the more general equation (7.7). Since l identifies the eigenvalue of the squared orbital angular momentum it is called the *orbital quantum number*. The eigenvalue of $|\vec{l}|^2$ and the corresponding value of the rotational kinetic energy

$$K_{\text{rot},l} = \frac{|\vec{l}|^2}{2\bar{m}r^2} = \frac{l(l+1)\hbar^2}{2\bar{m}r^2}, \tag{13.21}$$

increase quadratically with the number l of nodal cones (dark fringes), as in the prototype equation (12.35a) for a particle confined within a square well.

For $l_z^2 \neq 0$ the centrifugal force due to the motion along parallels confines the motion along meridians to a region about the equator. This dynamical confinement causes the particle density to be nonuniform along meridians even in the ground state where the density distribution along meridians consists of a single bright fringe. The eigenvalues of $|\vec{l}|^2$ and of $K_{\text{rot},\,l}$ have thus a nonzero minimum value, which increases with increasing l_z^2 and increasing confinement.[10] As in the case of $m = 0$, excited states of rotational motion with $m \neq 0$ are characterized by their number of nodal cones. However, this number is indicated by $l - |m|$ rather than by l, so that the orbital quantum number l *represents the total number of nodal surfaces* — cones + meridian planes—for a wave function $\langle\theta|l\,m\rangle\langle\phi|m^2\,\pm\rangle$.

As in the case of $m = 0$, we give here only the results obtained by analytical solution of the equation (13.10) or of its alternative form (13.18). The eigenfunction $\langle\theta|l\;m\rangle$ with $m \neq 0$ consists of a factor $(\sin\theta)^{|m|}$, which vanishes at the poles where the centrifugal potential is infinite, and of an associate Legendre polynomial of degree $l - |m|$ in $\cos\theta$. These eigenfunctions are[11]

$$\begin{aligned} &\langle\theta|1\;\pm1\rangle = \mp(3/4)^{1/2}\sin\theta, \\ &\langle\theta|2\;\pm1\rangle = \mp(15/4)^{1/2}\sin\theta\cos\theta, \quad \langle\theta|2\;\pm2\rangle = (15/16)^{1/2}\sin^2\theta, \\ &\cdots \\ &\langle\theta|l\;\pm|m|\rangle = (\mp1)^m\left[\frac{(l-|m|)!}{(l+|m|)!}\right]^{1/2}(\sin\theta)^{|m|}\left[\frac{d}{d(\cos\theta)}\right]^{|m|}\langle\theta|l\;0\rangle. \end{aligned} \tag{13.22}$$

---

10. The existence of this minimum energy is an effect of complementarity, which was not noticed in the early development of quantum physics and emerged as an unexpected result from the solution of the Schrödinger equation.

11. The sign factors $(\mp1)^m$ are required so that the whole set of wave functions is transformed consistently by a rotation of the coordinate axes.

The eigenvalues of (13.18) are represented by the same equation (13.20) irrespective of the value of m; however, l is necessarily $\geqslant |m|$, since $l - |m|$ indicates the nonnegative number of nodal cones. In (13.22), as well as in (13.20), the numerical coefficient of each wave function is normalized so that the integrated particle density $\int_1^1 \langle \theta | l\, m \rangle^2 \, d(\cos\theta) = 1$.

13.2.3. *Degeneracy.* We have seen that equation (13.18) for the motion along meridians has the same eigenvalue $|\vec{l}|^2/\hbar^2 = l(l+1)$ for different values of m, provided only that $|m| \leqslant l$. This means that the combined rotational motion along parallels and meridians has degenerate states with equal $|\vec{l}|^2$, equal kinetic energy $K_{rot,\, l}$, and different values of the squared momentum component l_z^2. These states differ from one another in the partition of $|\vec{l}|^2$ between l_z^2 and $l_x^2 + l_y^2$;[12] the degeneracy of the eigenstates of l_z^2 has been described in section 13.2.1. An orthogonal set of degenerate eigenstates of $|\vec{l}|^2$ is represented by the wave functions

$$\langle \theta\, \phi \,|\, l\, m \rangle = \langle \theta \,|\, l\, m \rangle \langle \phi \,|\, m \rangle, \tag{13.23}$$

where $\langle \theta | l\, m \rangle$ is given by (13.22) and $\langle \phi | m \rangle$ by (13.14). There are $2l + 1$ states of this set, with values of m ranging from $-l$ to l. No further state of orbital motion exists which is degenerate with these states and orthogonal to all of them.

This set of states is an example of the sets of magnetic energy eigenstates $|\hat{B}\, m\rangle$ described on p. 148. A variety of additional states degenerate with them can be obtained by superposition of the states (13.23). A particular example is afforded by the superposition

$$\sum_m \langle \theta\, \phi \,| l\; m \rangle \langle \hat{z}\; m \,|\, \hat{z}'\; m' \rangle, \tag{13.24}$$

whose coefficients are the probability amplitudes described in section 10.3 for the case of $j = l$. These wave functions represent eigenstates of the orbital momentum component $\vec{l} \cdot \hat{z}'$ along the axis z'; accordingly, they must coincide with the function obtained from (13.23) with the following substitutions corresponding to coordinate rotation: $\hat{z} \to \hat{z}'$, $\theta \to \theta'$, and $\phi \to \phi'$. Superpositions with coefficients different from those used in (13.24) represent degenerate states which are not related to $\langle \theta\, \phi \,|\, l\, m \rangle$ by a simple rotation of coordinates.

Eigenstates of the squared orbital angular momentum $|\vec{l}|^2$ are often called

---

12. Prior to the development of quantum mechanics it was not realized that a state need not be an eigenstate of every variable. The occurrence of different values of l_z for a given value of $|\vec{l}|^2$ was interpreted on the basis of the "vector model" as reflecting different orientations of the vector $\vec{l}$. The symbol l was adopted for the orbital quantum number in the belief that this number represented the magnitude of the angular momentum $\vec{l}$ divided by $\hbar$. Here we use the symbol $|\vec{l}|^2$ instead of l^2 to avoid confusion between the magnitude of l and the quantum number l.

by standard code letters rather than by the corresponding values of l, as follows:[13]

l value:	0	1	2	3	4	5	...
code:	s	p	d	f	g	h	....

The state $|l\,m\rangle$ with $l = 1$ and $m = 0$ is often called p_z because its particle density extends in the z direction and has the nodal plane (x, y). Similarly, the two states $|l\,m^2\,\pm\rangle$ with $l = m = 1$ are called p_x and p_y, respectively. Sample maps of electron density distribution for states with $l = 2$ are shown in Figure 13.4.

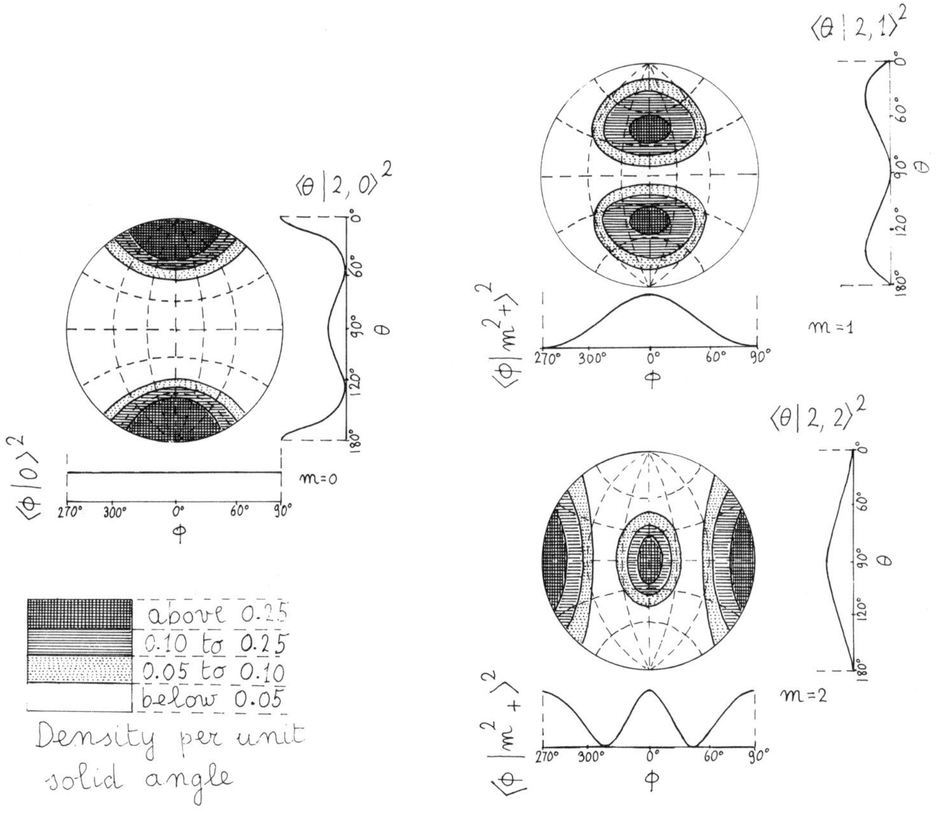

Fig. 13.4. Electron density distribution on spherical surfaces for eigenstates with quantum numbers $l = 2$, $m = 0, 1$, and 2. Profiles represent density distribution along meridians and parallels.

---

13. This code originates from early spectroscopic studies.

13.2.4. *Parity.* The maps in Figure 13.4 show that the particle density is invariant under reflection at the equatorial plane, which changes θ into $\pi - \theta$, and under reflection at the z axis, which changes ϕ into $\phi + \pi$. This invariance of the density may or may not be accompanied by a change of sign of the wave function. In fact, the wave function $\langle \theta | l\ m \rangle$ of the motion along meridians is multiplied by $(-1)^{l-m}$ when $\theta \to \pi - \theta$ while all eigenfunctions of l_z^2 are multiplied by $(-1)^m$ when $\phi \to \phi + \pi$. [The wave functions $\langle \phi | m^2 \pm \rangle$ are also multiplied by ± 1 when $\phi \to -\phi$.] Any such symmetry property of a wave function is called its parity under the specified reflection of coordinates (see section 11.3). However, the parity of the complete wave function of rotational motion $\langle \theta | l\ m \rangle \langle \phi | m \rangle$ under reflection at the origin, that is, under the combined operation $\theta \to \pi - \theta, \phi \to \phi + \pi$ is usually called "parity" without any further qualification; the multiplying factor is in this case simply $(-1)^l$.

13.3 RADIAL MOTION

The radial motion is confined whenever the electron has insufficient total energy E to escape from the nucleus, that is, when E does not exceed the potential energy $V(\infty)$ at infinite distance from the nucleus. Since we have taken $V(\infty) = 0$, the motion is confined when E is negative. Ionized states with $E > 0$ will be considered briefly in section S13.6.

Each stationary state of confined radial motion (*bound state*) is characterized by a number n_r of spherical nodal surfaces, called the *radial quantum number*. In the absence of rotational motion, that is, when $l = 0$, the radial motion of the electron is confined only by the nuclear attraction. For each stationary state with $l = 0$ the electron density is spherically symmetric with a bright fringe centered at the nucleus. The ground state of the atom has just this single bright fringe. In states with $l \neq 0$ the centrifugal force keeps the electron away from the nucleus. This force is represented in equation (13.8c) by the effective potential energy $|\vec{l}|^2/2\overline{m}r^2$.[14] The combined potential energy, centrifugal and electrostatic, has the trend shown in figure 13.5. The minimum of the total energy lies at

$$r = \frac{\hbar^2 l(l+1)}{\overline{m}e^2} = l(l+1)a_B, \qquad (13.25)$$

where

$$a_B = \frac{\hbar^2}{\overline{m}e^2} = 0.53\text{Å} \qquad (13.26)$$

is the Bohr radius.

---

14. The treatment up to this point applies to any central potential. From here on it is restricted to the potential $V(r) = -e^2/r$.

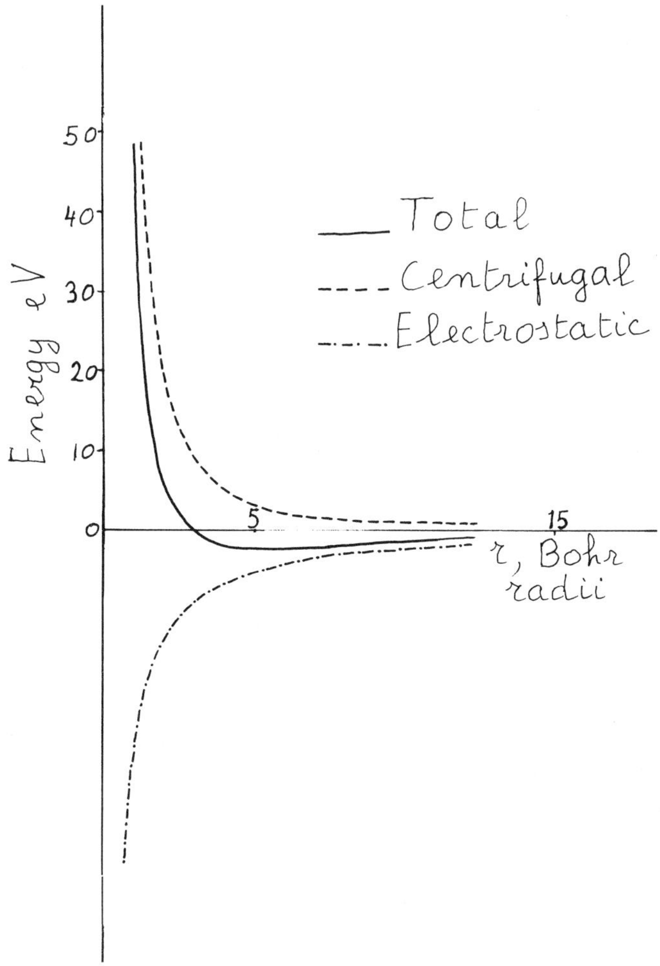

Fig. 13. 5. Dependence on the radius r of the electrostatic, centrifugal, and total potential energy for a hydrogen atom in a state with $l = 2$. Note that the centrifugal potential parallels the mean kinetic energy of Figure 11. 1.

The Schrödinger equation (13.8c) for the radial motion is solved by a procedure similar to that described in section S12.6. We sketch here a bare outline of the successive analytical steps.[15] Since the radial motion depends on the orbital quantum number l, the probability amplitude will be indicated by $\langle r | E\,n\,l \rangle$. To show the relative importance of the various terms of the equation, one may perform three changes of variables:

1) The energy eigenvalue E_n is replaced by a dimensionless parameter ν which is related to E_n through the combination of physical constants appearing in Bohr's formula

$$E_n = -\frac{\overline{m}e^4}{2\hbar^2}\frac{1}{\nu^2} = -\tfrac{1}{2}\frac{e^2}{a_B}\frac{1}{\nu^2}. \tag{13.27}$$

2) The radial coordinate r is replaced by a dimensionless parameter $\rho = r/\nu a_B$.

3) The eigenfunction of (12.8c), $\langle r | E\,n\,l \rangle$, is replaced by a new radial function $u(\rho) = r\langle r | E\,n\,l \rangle$. This new function $u(\rho)$ has the property that $u(\rho)^2 dr = |\langle r | E\,n\,l \rangle|^2 r^2\,dr$ represents the integral of the electron density over the shell between concentric spheres of radii r and r + dr. Analytically, the replacement of $\langle r | E\,n\,l \rangle$ with $u(\rho)$ has the effect of eliminating the first-derivative term in (13.8c). Thereby this equation is reduced to the standard form (12.15) for one-dimensional motion.

These three changes of variable reduce (13.8c) to the form

$$\frac{d^2u}{d\rho^2} - \left[1 - \frac{2\nu}{\rho} + \frac{l(l+1)}{\rho^2}\right] u(\rho) = 0. \tag{13.28}$$

The three terms in the brackets represent respectively the total energy of the electron (that is the energy eigenvalue which has been reduced to unity by our changes of variable), the electrostatic energy (proportional to the parameter ν), and the centrifugal energy (proportional to $l(l+1)$). For very large values of ρ the total energy term, 1, is the dominant term in the bracket. In this range the dominant factor of $u(\rho)$ is obtained by disregarding the other two terms and therefore must be $\exp(\pm\rho)$; the divergent solution $\exp(+\rho)$ must be discarded according to the discussion on p. 248. For very small values of ρ the centrifugal term is dominant. Disregarding then the terms $1 - 2\nu/\rho$, one finds that $u(\rho)$ must have a dominant factor ρ^{l+1}. The complete solution of equation (13.28) is obtained as the product of $\exp(-\rho)$, of ρ^{l+1}, and of a power series of ρ. As pointed out in section S12.6, the power series must be truncated to prevent its upsetting the trend of $u(\rho)$ for large ρ. Truncation is achieved by taking the parameter ν equal to an integer. This integer is usually expressed as

$$n = n_r + l + 1, \tag{13.29}$$

15. Details of the calculation may be found, for example, in Merzbacher, *Quantum Mechanics*, pp. 200 ff.

where n_r indicates the degree of the polynomial and is called the *radial quantum* number. The integer n itself is called the *principal quantum number*.

Stationary states with quantum numbers n and l are often indicated by a numerical index equal to n combined with one of the code letters s, p, d, ... listed at the end of section 13. 2. We return now to the original variables and give the radial wave function $\langle r | E\, n\, l\rangle$ for the first few values of n. The functions include the appropriate normalization coefficient which makes the integrated electron density equal to one,[16]

$$\langle r | E\ 1\ 0\rangle = \frac{2}{a_B{}^{3/2}} \exp(-r/a_B) \qquad \text{(1s, ground state)}$$

$$\langle r | E\ 2\ 0\rangle = \frac{2}{(2a_B)^{3/2}} \exp(-r/2a_B) \left(\frac{r}{2a_B} - 1\right) \qquad \text{(2s state),}$$

$$\langle r | E\ 2\ 1\rangle = \frac{2}{[3(2a_B)^3]^{1/2}} \exp(-r/2a_B) \frac{r}{2a_B} \qquad \text{(2p state),} \tag{13.30}$$

$$\langle r | E\ 3\ 0\rangle = \frac{2}{(3a_B)^{3/2}} \exp(-r/3a_B) \left[\tfrac{2}{3}\left(\frac{r}{3a_B}\right)^2 - 2\frac{r}{3a_B} + 1\right] \text{ (3s state),}$$

$$\langle r | E\ 3\ 1\rangle = \tfrac{4}{3}\left[\frac{2}{(3a_B)^3}\right]^{1/2} \exp(-r/3a_B) \frac{r}{3a_B} \left(\tfrac{1}{2}\frac{r}{3a_B} - 1\right) \text{ (3p state),}$$

$$\langle r | E\ 3\ 2\rangle = \frac{4}{3[10(3a_B)^3]^{1/2}} \exp(-r/3a_B) \left(\frac{r}{3a_B}\right)^2 \qquad \text{(3d state).}$$

The energy eigenvalue formula (13. 27) may now be compared with the old Bohr formula (3. 17). The requirement that ν be an integer in equation (13. 27) gives a theoretical proof of Bohr's assumption that $\sigma = 0$ in equation (3. 17). As discussed in chapter 3, the theoretical energy levels represented by these formulae agree accurately with the experimental data.

The energy eigenvalues depend on n and therefore only on the total number of nodal surfaces $n_r + l$, rather than on the separate values of n_r and l. Therefore, the combination of radial and rotational motion increases the degeneracy of each energy level with $n > 1$. The additional degeneracy will be discussed in section 13. 4. Since there are $2l + 1$ orthogonal states for a given l, and l itself ranges from zero to $n - 1$ for a given n, the total number of orthogonal degenerate states is

$$1 + 3 + 5 + \ldots + [2(n-1) + 1] = n^2. \tag{13.31}$$

---

16. The sign of the normalization coefficient is arbitrary and differs in different textbooks. Here it is chosen so as to make the wave function positive for large values of r.

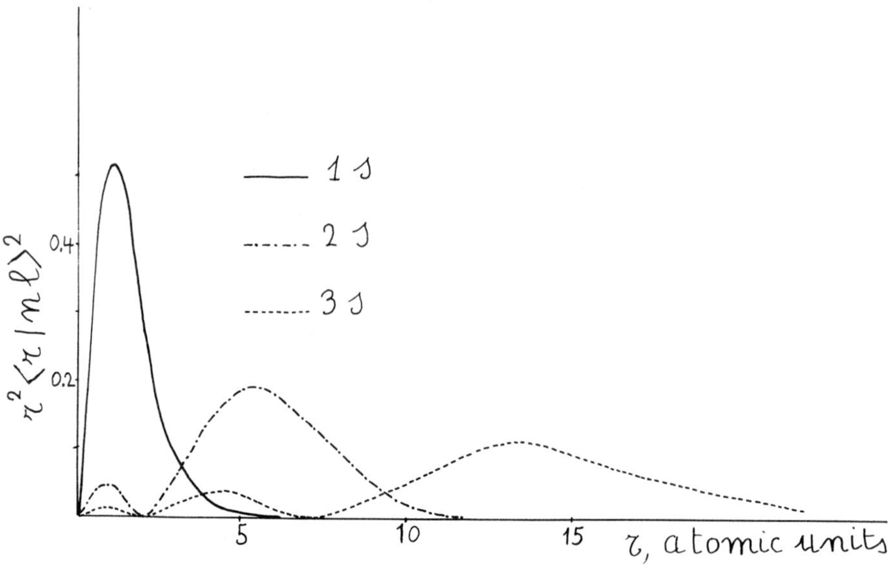

Fig. 13. 6. Electron density distribution for states with different radial excitation.

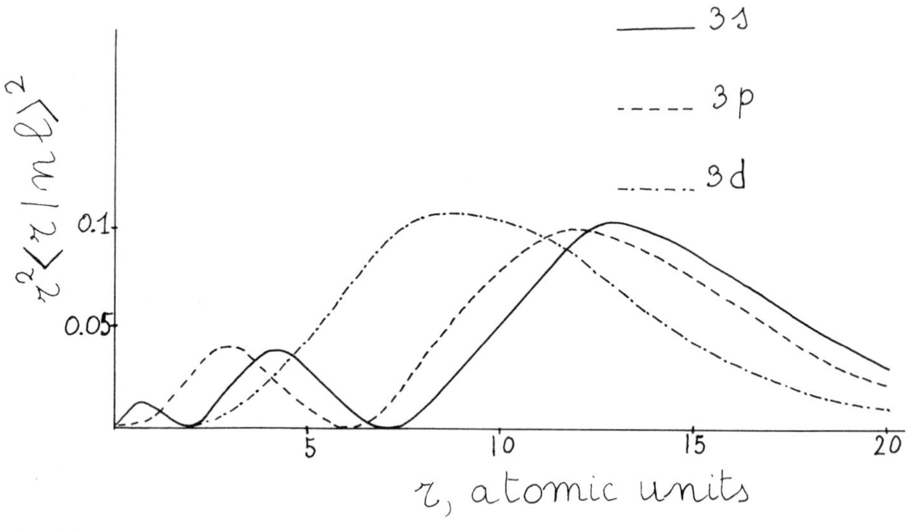

Fig. 13. 7. Electron density distribution for states with different rotational excitation.

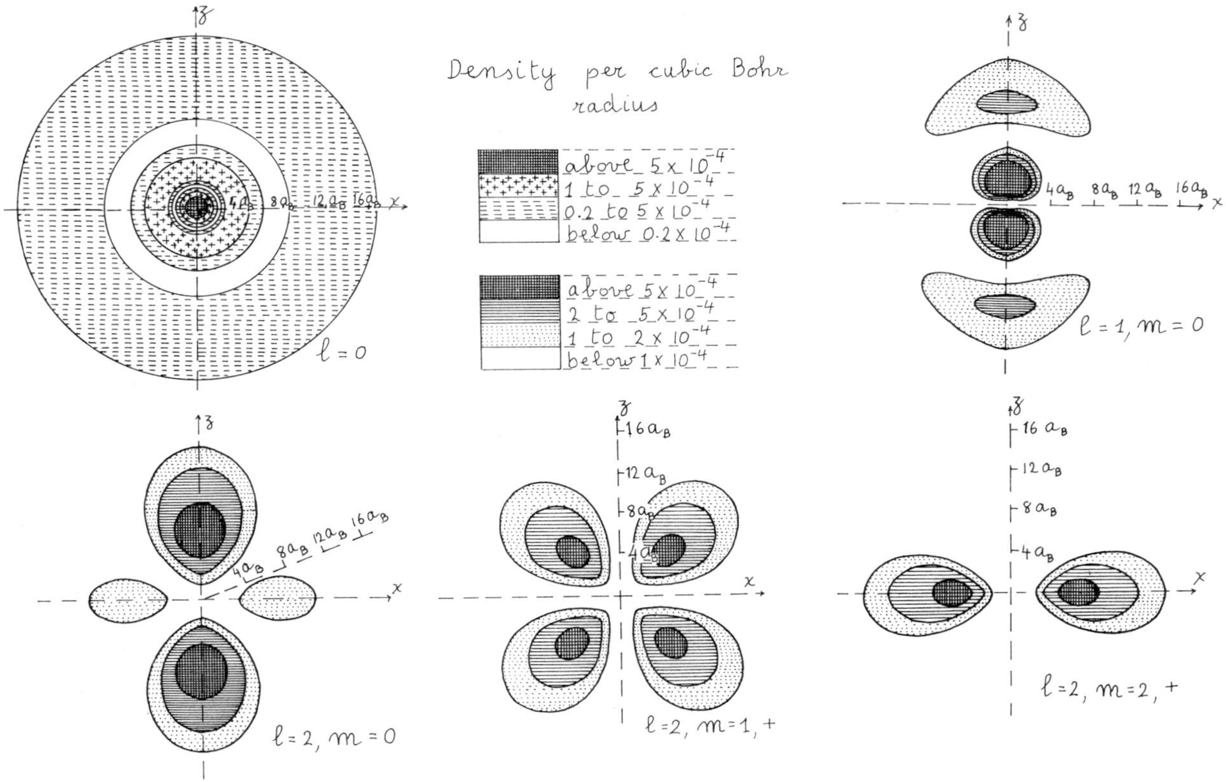

Fig. 13. 8. Electron density distribution on the (x, z) plane for eigenstates with n = 3.

This result is of fundamental importance for the structural and chemical properties of atoms, as will be seen in chapter 16.

Successive eigenstates of increasing energy differ from one another by a substantial increase of the mean distance of the electron from the nucleus. As the electron's energy increases, the electron meets a decreasing attraction by the nucleus in the farthest region it can reach. The electron density for two successive energy eigenstates differs primarily by the addition of an outer bright fringe, as shown in Figure 13.6. The radius of peak electron density in the outermost fringe is proportional to n^2.

For rotational motion with $l \neq 0$ the predominance of the centrifugal force near the nucleus has the effect of wiping out the l innermost bright fringes that would occur for $l = 0$ (see Fig. 13.7). On the other hand, for large r the nuclear attraction predominates and the radial motion is largely independent of l. For purposes of illustration, Figure 13.8 shows the electron density distribution over a plane across the atom for a few states with $n = 3$.

13.4. STATIONARY STATES WITH OFF-CENTER ELECTRON DENSITY

The stationary states of the hydrogen atom considered thus far are eigenstates of the squared angular momentum $|\vec{l}|^2$. The pattern of their electron density has the nucleus as a center of symmetry; therefore, the mean position of the electron coincides with the nucleus. There also exist stationary states of the hydrogen atom which are not eigenstates of $|\vec{l}|^2$. They can be described as superpositions of degenerate states with equal n (other than 1) and different l. The mean position of the electron differs, in general, from the position of the nucleus in any state which is not an eigenstate of the squared angular momentum. States with off-center electron density are important particularly because they favor the formation of chemical bonds (chap. 19).

The degeneracy of states with different l, which underlies the existence of off-center densities, is sometimes called "accidental" but derives in fact from properties which are characteristic of the $1/r^2$ force field (Coulomb or gravitational). In this field, a macroscopic particle may follow an elliptical orbit with a focal axis of fixed direction; any departure of the field from the $1/r^2$ law prevents the orbit from closing at every turn and causes its axis to precess about the center of attraction. Motion in a Coulomb field has then, in addition to a constant angular momentum, another constant represented by a vector $\vec{A}$. In quantum mechanics, eigenstates of any component A_z may be stationary but different components of $\vec{A}$ are incompatible with one another and with $|\vec{l}|^2$. There exists, therefore, a large variety of combinations of components of $\vec{l}$ and $\vec{A}$ whose eigenstates may be stationary, and hence a large variety of degenerate states which may be constructed and identified in various manners.

Here we confine our attention to three sets of states with $n = 2$ which are important in quantum chemistry. One of these sets consists of two states whose electron density distribution extends in opposite directions; another set has three states with densities extending toward the vertices of a triangle;

the last one has four states with densities extending toward the vertices of a tetrahedron. Prototypes of each of these three sets are constructed by superposition of states of the type 2s ($n = 2, l = 0$) and $2p_z$ ($n = 1, l = 1, m = 0$) with various coefficients.

1) Superposition of the states 2s and $2p_z$ with equal coefficients yields, according to equations (13. 14), (13. 19), and (13. 30),

$$2^{-1/2}[\langle r \ \theta \ \phi | 2s\rangle + \langle r \ \theta \ \phi | 2p_z\rangle]$$

$$= [2\pi(2a_B)^3]^{-1/2}[\exp(-r/2a_B)(\frac{r}{2a_B} - 1) + \exp(-r/2a_B)\frac{r}{2a_B}\cos\theta] \qquad (13.32)$$

$$= [2\pi(2a_B)^3]^{-1/2}\exp(-r/2a_B)(\frac{r+z}{2a_B} - 1),$$

where we have reintroduced the Cartesian coordinate $z = r\cos\theta$. This wave function has a single nodal surface, represented by the equation $r + z = 2a_B$ which is a paraboloid with the symmetry axis z.[17] Its electron density distribution is mapped in Figure 13. 9 and has mean position on the z axis at $\langle z\rangle = 3a_B$. The wave function obtained by combining 2s and $2p_z$ with coefficients of equal magnitude and opposite sign is orthogonal to the function (13. 32) and differs from it by reversal of the z axis.

2) Superposition of 2s and $2p_z$ with coefficients in the ratio $1 : \sqrt{2}$ yields the wave function

$$\sqrt{1/3}\,(r \ \theta \ \phi | 2s) + \sqrt{2/3}\,\langle r \ \theta \ \phi | 2p_z\rangle = [3\pi(2a_B)^3]^{-1/2}\exp(-r/2a_B)\left(\frac{r + \sqrt{2}z}{2a_B} - 1\right). \qquad (13.33)$$

The nodal surface of this wave function is a paraboloid somewhat wider than that of the function (13. 32), with the same symmetry about the z axis. The mean position of the electron lies at $z = 2\sqrt{2}a_B = 2.83\ a_B$. Two states orthogonal to equation (13. 22) and to one another are obtained as superpositions of 2s, $2p_z$, and $2p_x$, where $2p_x$ is the state $|n\ l\ m^2 +\rangle$ with $n = 2, l = 1, m = 1$. Their wave functions are

$$\sqrt{1/3}\,\langle r \ \theta \ \phi | 2s\rangle - \sqrt{1/6}\,\langle r \ \theta \ \phi | 2p_z\rangle \pm \sqrt{1/2}\,\langle r \ \theta \ \phi | 2p_x\rangle. \qquad (13.33a)$$

The wave functions (13. 33) and (13. 33a) differ from each other by 120° rotations of the coordinates in the zx plane.

---

17. The wave function (13. 32) belongs to a complete set of stationary states which are constructed by utilizing the parabolic coordinates $(r + z)$, $(r - z)$, and ϕ, where ϕ is the same as in polar coordinates. The Schrödinger equation with the potential $-e^2/r$ separates in terms of these coordinates as it does in terms of polar coordinates. Note that the variable r in the Coulomb potential and in the exponent of (13. 32) may be expressed as $\frac{1}{2}[(r + z) + (r - z)]$.

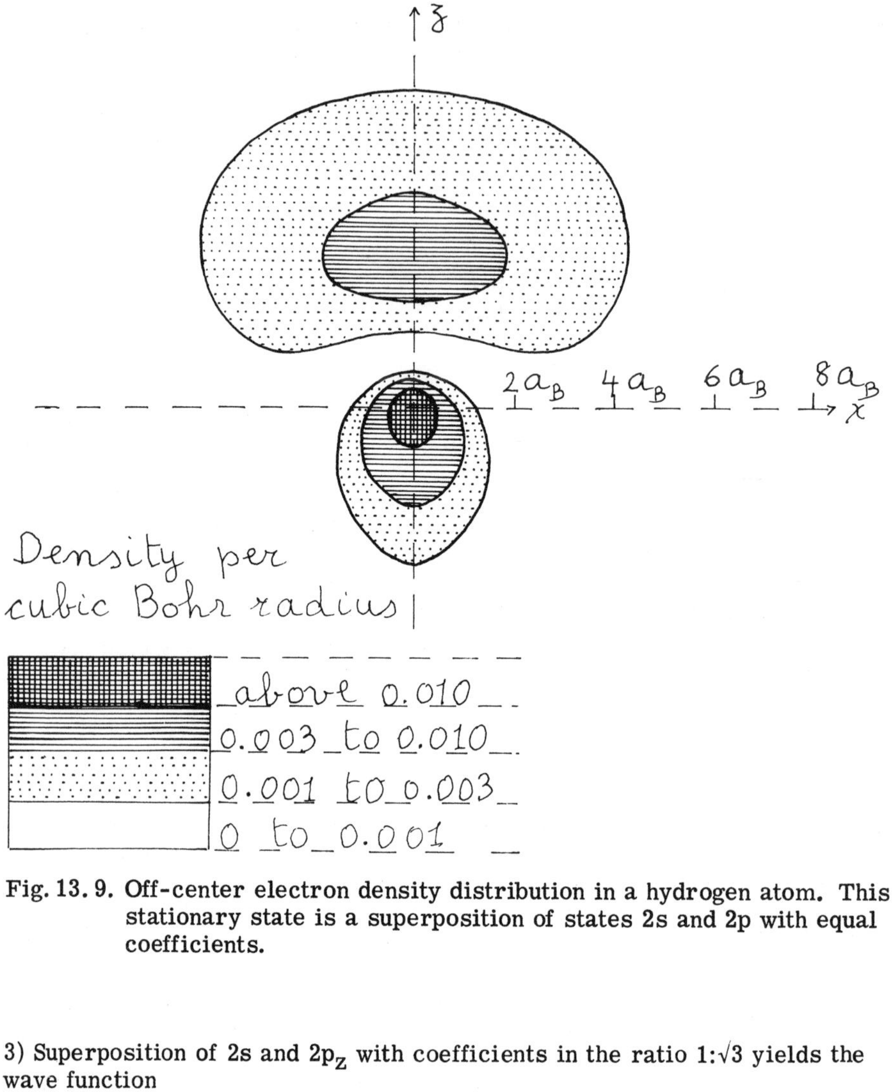

Fig. 13. 9. Off-center electron density distribution in a hydrogen atom. This stationary state is a superposition of states 2s and 2p with equal coefficients.

3) Superposition of 2s and $2p_z$ with coefficients in the ratio $1{:}\sqrt{3}$ yields the wave function

$$\tfrac{1}{2}\langle r \;\; \theta \;\; \phi\,|2s\rangle + \sqrt{3/4}\,\langle r \;\; \theta \;\; \phi\,|\,2p_z\rangle$$

$$= [4\pi(2a_B)3]^{-1/2}\exp(-r/2a_B)\left(\frac{r + \sqrt{3}z}{2a_B} - 1\right). \qquad (13.34)$$

This wave function has the same symmetry about the z axis as equations (13. 32) and (13. 33) and has a nodal surface still wider than that given by equation (13. 33). The mean position of the electron lies at $z = \tfrac{3}{2}\sqrt{3}\,a_B = 2.60\,a_B$.

Three states orthogonal to (13.34) and to one another are obtained as superpositions of 2s, $2p_z$, $2p_x$, and $2p_y$. Their wave functions are

$$\tfrac{1}{2}\langle r\ \theta\ \phi\,|\,2s\rangle - \sqrt{\tfrac{1}{12}}\,\langle r\ \theta\ \phi\,|\,2p_z\rangle + \sqrt{\tfrac{2}{3}}\,\langle r\ \theta\ \phi\,|\,2p_x\rangle, \tag{13.34a}$$

$$\tfrac{1}{2}\langle r\ \theta\ \phi\,|\,2s\rangle - \sqrt{\tfrac{1}{12}}\,\langle r\ \theta\ \phi\,|\,2p_z\rangle - \sqrt{\tfrac{1}{6}}\langle r\ \theta\ \phi\,|\,2p_x\rangle \pm \sqrt{\tfrac{1}{2}}\,\langle r\ \theta\ \phi\,|\,2p_y\rangle. \tag{13.34b}$$

The four states (13.34), (13.34a), and (13.34b) have symmetry axes pointing toward the vertices of a tetrahedron centered at the nucleus.

13.5. NONSTATIONARY STATES AND SELECTION RULES

Having studied the stationary states of the hydrogen atom, we are now in a position to discuss its nonstationary states, which occur when the atom interacts with radiation or with any other system. For these states the variations of electron density in the course of time are represented in terms of stationary-state wave functions by using the general formula (10.16). We can now substitute in this equation explicit expressions of the wave functions of the hydrogen atom and map out the resulting variations of electron density, much as we have done for the constant density distribution of stationary states.

This mapping is of particular importance because, as noted in chapter 2, an atom hardly interacts with radiation unless the mean position of the electron oscillates in the course of time. As we shall verify, the majority of the oscillating terms in equation (10.16) have such symmetry that the mean electron position remains at rest. The mean position oscillates only under conditions identified by the selection rules anticipated in section 7.3.

Consider a nonstationary state $|a(t)\rangle$ whose wave function is represented by a superposition of the stationary-state wave functions with quantum numbers n, l, m defined in equations (13.30), (13.22), and (13.14),

$$\langle r\ \theta\ \phi\,|\,a(t)\rangle = \sum_{n\,l\,m} \langle r\,|\,n\ l\rangle\langle\theta\,|\,l\ \ m\rangle\langle\phi\,|\,m\rangle \exp(-iE_n t/\hbar)\,\langle n\ \ l\ \ m\,|\,a(0)\rangle. \tag{13.35}$$

(The coefficients $\langle n\ \ l\ \ m\,|\,a(0)\rangle$ are assumed to be real for simplicity.) We represent the electron density of this state by the expansion (10.16). In this representation the quadratic terms are time independent, and therefore it is sufficient for our purpose to discuss oscillating cross-terms. The term which arises from interference between two stationary states with quantum numbers n, l, m and n', l', m' has the form

$$\langle r\,|\,n\ l\rangle\langle r\,|\,n'\ \ l'\rangle\langle\theta\,|\,l\ \ m\rangle\langle\theta\,|\,l'\ \ m'\rangle\,\frac{1}{\pi}\cos\left[\frac{E_{n'} - E_n}{\hbar}\,t - (m' - m)\phi\right]. \tag{13.36}$$

Variations in time of the electron density are represented by a sum of terms of this type with different frequencies and with coefficients $\langle a(0)\,|\,n'\ \ l'\ \ m'\rangle$ $\langle n\ \ l\ \ m\,|\,a(0)\rangle$. Note that, whereas the complete expression of the electron density must be nonnegative, the sign of individual oscillating terms like expression (13.36) generally alternates in space and in time.

Figure 13.10 shows sample maps of regions of positive and negative electron density for various terms of the type (13.36). The maps pertain to the time $t = 0$. In the course of time the electron distribution revolves about the z axis, if $m \neq m'$, according to equation (13.36). The direction of rotation is positive or negative depending on the sign of $(E_{n'} - E_n)/(m' - m)$. For $m' = m$ the electron density at each point oscillates in the course of time between positive and negative values.[18]

To determine the selection rules we must calculate the mean position of the electron with the probability distribution (13.36) and determine the conditions under which it departs from the nucleus. Calculation of the mean value of the electron's coordinate $z = r \cos\theta$ involves, in the first place, integration of expression (13.36) over ϕ. The integral vanishes unless

$$m' - m = 0, \qquad \text{for } \langle z \rangle \neq 0, \tag{13.37a}$$

as could also be seen by considering in the maps of Figure 13.10 an average of the electron density in the equatorial plane (x, y). A second condition for $\langle z \rangle \neq 0$ emerges from evaluating $\int_0^\pi \cos\theta \langle\theta | l\ m\rangle\langle\theta | l'\ m'\rangle \sin\theta\, d\theta$ with $m' = m$, using the expressions (13.22) of the wave functions. This integral is also found to vanish unless

$$l - l' = \pm 1, \qquad \text{for } \langle z \rangle \neq 0. \tag{13.37b}$$

Inspection of the maps of figure 13.10, in the meridian plane (x, z), shows that $\langle z \rangle$ vanishes for any even value of $l - l'$. (Actually, eq. [13.37b] follows from symmetry considerations that are beyond our scope.)

Consider now the mean values of the coordinates $x = r \sin\theta \cos\phi$ and $y = r \sin\theta \sin\phi$. Calculation of the mean value of x (or y) involves integration over ϕ of the product of $\cos\phi$ (or $\sin\phi$) and of expression (13.36). Both integrals vanish unless

$$m - m' = \pm 1 \qquad \text{for } \begin{cases} \langle x \rangle \neq 0 \\ \langle y \rangle \neq 0. \end{cases} \tag{13.38a}$$

This result is also apparent on inspection of the maps in Figure 13.10. A second condition emerges from evaluation of $\int_0^\pi \sin\theta \langle\theta | lm\rangle\langle\theta | l'm'\rangle \sin\theta\, d\theta$ with $m' = m \pm 1$. This integration leads to the same condition (13.37b) found in

---

18. The time variations of electron density are associated with an electron flux $\vec{\Phi}$. This flux can be calculated by equation (11.16) starting from the complete wave function (13.35). Alternatively, it may be calculated as the vector field whose divergence equals the time derivative of expression (11.36) with the opposite sign.

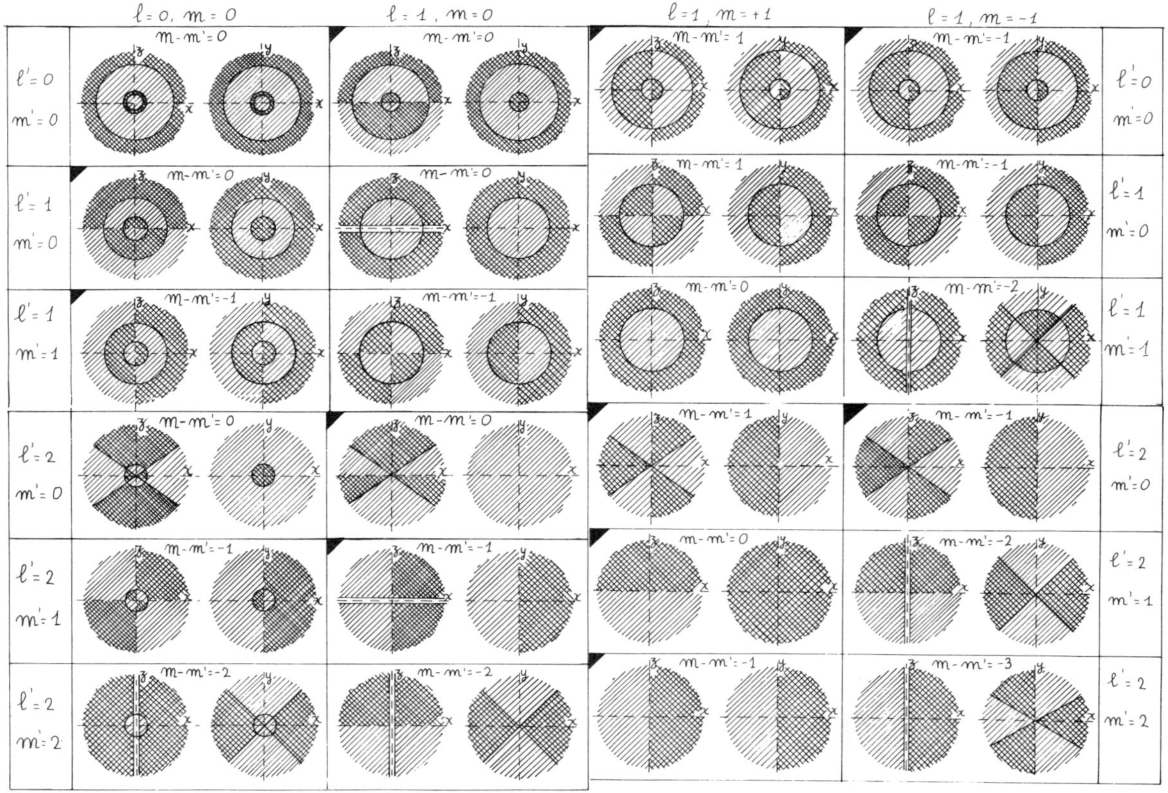

Fig. 13. 10. Effects of interference of stationary states of the hydrogen atom with $n = 2, n' = 3$. The diagrams represent cross-sectional maps of the departures of electron density distribution from its stationary components, for different pairs of interfering states $|n\ l\ m\rangle$ $|n'\ l'\ m'\rangle$. The cross-hatched areas represent positive departures in equation (13. 36) at $t = 0$; the hatched areas, negative departures. When the departures vanish in the (x, y) plane, the (x, y) map represents a cross-section just above this plane. Interference patterns with an off-center electron distribution are marked with a black corner.

the discussion of $\langle z \rangle$,

$$l - l' = \pm 1, \quad \text{for} \begin{cases} \langle x \rangle \neq 0 \\ \langle y \rangle \neq 0. \end{cases} \tag{13.38b}$$

The calculation of the mean value of each electron coordinate also involved integration over r. However, this integral does not vanish in general and therefore yields no additional selection rule. Thus the selection rules result entirely from the rotational motion of the electron; accordingly they hold for any atomic electron moving in a central field, irrespective of screening of the nuclear attraction by other electrons.

Physically, the selection rule $l - l' = \pm 1$ means that the squared orbital angular momentum of the electron changes by one unit in each allowed process of photon emission or absorption that is, in each transition for which the mean position of the electron oscillates about the nucleus. Electromagnetic theory shows that the radiation field can itself be analyzed in eigenstates of angular momentum. Without pursuing this subject, we simply note that each process of photon emission or absorption involves a transfer of angular momentum together with the transfer of energy and momentum. The selection rules derive from the conservation of angular momentum. In particular, the rules (13.37a) and (13.38a) pertaining to the magnetic quantum number derive from the conservation of the angular momentum component on the z axis.

The selection rules, together with the discussion of the oscillation and rotation of the electron density, account for the polarization characteristics of light emitted or absorbed in the Zeeman effect, which have been described in sections 2.4 and 7.3. Equation (13.36) represents the variable electron density associated with a transition between stationary states with quantum numbers (n, l, m) and (n', l', m'). A transition with $m' = m$ yields an electron oscillation parallel to the z axis and thus emits linearly polarized radiation. Transitions with $m' = m \pm 1$ yield a rotation of the electron about the z axis in a direction which depends on the sign of $m' - m$. This rotation emits circularly or elliptically polarized radiation. Notice how in this case the selection rules relate the variations of the angular momentum components l_z of the atom to the circular polarization of light.

SUPPLEMENTARY SECTIONS

S13.6 STATIONARY STATES OF THE CONTINUOUS SPECTRUM

The set of stationary states with quantum numbers (n, l, m) described in sections 13.2 and 13.3 is not complete. That is, the wave function of an electron in an arbitrary state $|a\rangle$, $\psi_a(xyz)$, cannot be represented in general as a superposition of the wave functions $\langle r | E\, n\, l \rangle \langle \theta | l\, m \rangle \langle \theta | m \rangle$. The reason is that the treatment of the Schrödinger equation for radial motion (13.8c) has been restricted in section 13.3 to negative energy eigenvalues E_n corresponding to stationary states of an electron bound within the atom. The set of wave functions is made complete by extending the treatment of the radial equation to the continuous spectrum of positive-energy eigenvalues.

A stationary state with positive energy pertains to an electron moving in the electrostatic field of a nucleus but free to move to infinity. An electron initially bound in the ground state of a hydrogen atom may be transferred to a positive-energy eigenstate by absorption of a photon with frequency in excess of the ionization threshold. Eigenfunctions with positive energy are obtained by the same procedure utilized for the bound-state eigenfunctions (13.30). Since the negative discrete eigenvalue E_n is now replaced in equation (13.27) by a positive value E, the parameter ν defined by that equation becomes imaginary. It is usually expressed as $1/ika_B$, where k represents the wavenumber of the electron at infinite distance from the nucleus. The exponential factor of the wave function $u(\rho)$ becomes thus exp (-ikr). The factor ρ^{l+1} becomes $(ikr)^{l+1}$. The series in powers of (ikr) is no longer truncated and does influence the behavior of $u(\rho)$ at large values of r. In fact, the whole wave function $u(\rho)$ becomes real; it resembles a sine function and extends to infinity. The radial eigenfunctions thus obtained are combined with eigenfunctions of the rotational motion from section 13.2.

There exist also stationary-state eigenfunctions of the continuous spectrum which are not products of functions of r, θ, and ϕ but are of the types mentioned in section 13.4. In particular, one constructs a wave function that represents the motion of a particle under the conditions of Rutherford scattering, utilizing the parabolic coordinates $r - z$ and $r + z$.[19] This wave function consists of a plane wave incident from infinity and of a spherical wave diverging from the nucleus. The intensity of the scattered spherical wave depends on the angle of scattering in full quantitative agreement with the Rutherford formula (1.27) which was obtained from classical mechanics.

S13.7. LOWERING OF ENERGY LEVELS BY RELATIVISTIC EFFECT

The energy levels calculated from the Schrödinger equation for the hydrogen atom are modified by a correction of the order of one part in 10^5 if one takes into account a relativistic effect. The correction arises from a modification of the relation between kinetic energy and momentum. The relativistic expression of the energy of a free electron of momentum p is $(m_e^2c^4 + p^2c^2)^{1/2}$. The electron's kinetic energy K is the difference between its energy at momentum p and its rest energy m_ec^2. The nonrelativistic expression of K and the relativistic correction are obtained by expansion in powers of p^2

$$K = (m_e^2c^4 + p^2c^2)^{1/2} - m_ec^2 = \frac{p^2}{2m_e} - \frac{p^4}{8m_e^3c^2} + \dots \quad (13.39)$$

19. See, e.g., Merzbacher, *Quantum Mechanics*, p. 248, also L. D. Landau and E. M. Lifschitz, *Quantum Mechanics* (London: Pergamon, 1958), pp. 516 ff.

This equation shows that the kinetic energy increases with increasing momentum less rapidly than it does in the nonrelativistic limit. The electron's reaction to confinement by the nuclear attraction is therefore slightly smaller than we have considered thus far and, correspondingly, the energy levels are somewhat depressed. The ratio of the correction to the nonrelativistic term is

$$\frac{p^4/8m_e^3c^2}{p^2/2m_e} = \frac{p^2/2m_e}{2m_ec^2}, \tag{13.40}$$

where $2m_ec^2 = 1.02$ MeV. For the hydrogen atom $p^2/2m_e$ is of the order of 10 eV and the correction is therefore of the order of one part in 10^5. For heavier atoms the effect is much larger and may exceed 10%.

The relativistic shift in energy levels can be calculated by assuming that the effect does not appreciably disturb the wave function. Accordingly, one need not solve a new Schrödinger equation, and it is sufficient to calculate the mean value of $p^4/8m_e^3c^2$ for each state under consideration. Recall that $p^4/8m_e^3c^2 \sim K^2/2m_ec^2$; in this approximation we may then take $K = E_n - V(r)$, where E_n is the energy eigenvalue derived from the Schrödinger equation. Thus one finds for the hydrogen atom the energy shift

$$\begin{aligned} \Delta E_n &= -\left\langle \frac{p^4}{8m_e^3c^2} \right\rangle_{n,l,m} = -\left\langle \frac{[E_n - V(r)]^2}{2m_ec^2} \right\rangle_{n,l,m} \\ &= E_n \left(\frac{e^2}{\hbar c}\right)^2 \left[\frac{1}{n(l + 1/2)} - \frac{3}{4n^2}\right], \end{aligned} \tag{13.41}$$

where the ratio

$$\frac{e^2}{\hbar c} = \frac{1}{137.0}. \tag{13.42}$$

The relativistic effect discussed here is often accompanied by other effects of the same order of magnitude. Therefore, equation (13.41) is incomplete and should not be used for detailed comparison with experiments.

The importance of the level shift given by equation (13.41) increases rapidly for hydrogen-like ions of increasing atomic number Z. The energy eigenvalues E_n increase in proportion to Z^2, and the level shift, which depends on E_n^2, increases in proportion to Z^4.

CHAPTER 13 PROBLEMS

13.1. The atom of "muonic hydrogen" consists of a proton and of a negative muon. The muon is similar to an electron except for its mass, which is in the ratio $m_\mu/m_e = 206.8$. Calculate the binding energy of muonic hydrogen in its ground state, in eV, and write the formula for its wave function. Take into account the proton mass.

13.2. The atom of positronium consists of a negative and a positive electron, particles of equal mass and opposite charge, held together by their Coulomb attraction. Calculate its binding energy in the ground state.

13.3. The expression $V(r) = -e^2/r$ of the potential energy for the electron in a hydrogen atom implies that the atomic nucleus (a proton) is a point charge. Assume, instead, that the proton charge $+e$ is distributed uniformly over the surface of a sphere of radius $R = 10^{-13}$ cm. (a) Calculate the "modified" potential V_m which corresponds to the nuclear charge thus distributed. (b) Assuming that the eigenfunctions of the H atom are not changed significantly by the change of nuclear charge distribution, calculate the mean energy shift $\langle \Delta V \rangle_{nlm}$ for the states $|1\ 0\ 0\rangle$, $|2\ 0\ 0\rangle$, and $|2\ 1\ 0\rangle$, to lowest order in R/a_B. (c) On the same basis, calculate $\langle \Delta V \rangle_{1,0,0}$ for the atom of muonic hydrogen considered in problem 13.1.

13.4. Calculate the mean potential energy $\langle V(r) \rangle_{nlm}$ of the electron in the stationary state of hydrogen $2p_z$ (with $n = 2, l = 1, m = 0$). Compare the result with the energy eigenvalue and thus with $\langle K \rangle_{2,1,0}$ in order to verify the virial theorem (see n.13, page 225).

13.5. The problem 12.5, concerning a particle in the three-dimensional oscillator potential $\frac{1}{2}k(x^2 + y^2 + z^2) = \frac{1}{2}kr^2$, could also be solved in polar coordinates by the methods of chapter 13. Utilizing the solutions of problem 12.5, construct the polar-coordinate eigenfunctions with the quantum numbers $n_r = 0, l = 0, m = 0$; $n_r = 0, l = 1, m = 1, 0, -1$; and $n_r = 1, l = 0, m = 0$.

13.6. Construct the hydrogen atom nonstationary wave function (13.35) with $\langle 1\ 0\ 0 \,|\, a(0)\rangle = \langle 2\ 1\ 1 \,|\, a(0)\rangle = \sqrt{\frac{1}{2}}$ and with all other $\langle n\ l\ m \,|\, a(0)\rangle = 0$. Calculate the mean position $\langle r \rangle_{a(t)}$ and the mean value $\langle z^2 \rangle$ as functions of time.

SOLUTIONS TO CHAPTER 13 PROBLEMS

13.1 The theory of the hydrogen atom applies to muonic hydrogen, with the simple replacement of the reduced mass of hydrogen, $\overline{m} = m_e M_p/(M_p + m_e)$, by $\overline{m}_\mu = m_\mu M_p/(M_p + m_\mu) = 206.8\ m_e M_p/(M_p + 206.8\ m_e) = 186\overline{m}$. According to equation (13.27) with $\nu = 1$, the binding energy increases from $\overline{m}e^4/2\hbar^2$ to $\overline{m}_\mu e^4/2\hbar^2 = 186 \times 13.6$ eV $= 2.53$ keV. The ground-state wave function is $(\pi a_{\mu B}^3)^{-1/2} \exp(-r/a_{\mu B})$ where $a_{\mu B} = a_B/186 = 2.8 \times 10^{-11}$ cm.

13.2. The theory of the hydrogen atom applies with the reduced mass for identical particles $m = \frac{1}{2}m_e$. The binding energy given by equation (13.27) with $\overline{m} = \frac{1}{2}m_e$ and $\nu = 1$ is $\frac{1}{2} \times 13.6$ eV $= 6.8$ eV.

13.3a. The potential V_m inside the sphere of radius R is uniform; outside the sphere it is the same as for a point charge. Thus we have $\Delta V = -e^2/R + e^2/r$ for $r < R$ and $\Delta V = 0$ for $r \geqslant R$.

13.3b. Since ΔV depends only on r and not on θ and ϕ, its mean value depends only on the radial distribution of electron density and we have $\langle \Delta V \rangle_{nlm} = e^2 \int_0^R (1/r - 1/R) \langle r | n\, l \rangle^2 r^2 dr$. Since $R \ll a_B$, take $\langle r | n\, l \rangle$ from equation (13.30) and expand $(r \mid n\, l)^2$ to lowest order in r/a_B, which gives $\langle r \mid 1\ 0 \rangle^2 \sim 4/a_B^3$, $\langle r \mid 2\ 0 \rangle^2 \sim 1/2a_B^3$, $\langle r \mid 2\ 1 \rangle^2 \sim r^2/24a_B^5$, and then $\langle \Delta V \rangle_{1,0,0} = 2e^2R^2/3a_B^3 = 6.4 \times 10^{-9}$ eV, $\langle \Delta V \rangle_{2,0,0} = \frac{1}{8} \langle \Delta V \rangle_{1,0,0} = 8 \times 10^{-10}$ eV, $\langle \Delta V \rangle_{2,1,0} = \frac{1}{40}(R^2/a_B^2)$, $\langle \Delta V \rangle_{2,0,0} = 7.2 \times 10^{-18}$ eV.

13.3c. Problem 1 gives $a_{\mu B} = a_B/186$. Therefore, $\langle \Delta V \rangle_{1,0,0}$ is $(186)^3 = 6.4 \times 10^6$ times larger for muonic hydrogen than for the H atom, that is, 4.1×10^{-2} eV.

13.4. Since $V(r) = -e^2/r$ depends only on r, $\langle V \rangle_{2,1,0}$ is obtained by averaging over the radial density distribution $\langle r | 2\ 1\ 0 \rangle^2 = \exp(-r/a_B) r^2/24a_B^5$. Thus we find $\langle \Delta V \rangle_{2,1,0} = e^2 \int_0^\infty (1/r) \langle r | 2\ 1\ 0 \rangle^2 r^2\, dr = -e^2 \int_0^\infty \exp(-r/a_B) r^3 dr/24a_B^5 = -e^2/4a_B$. This value is double the eigenvalue $E_2 = -e^2/8a_B$ given by equation (13.27) with $\nu = 2$. Since $E_2 = \langle K \rangle_{2,1,0} + \langle V \rangle_{2,1,0}$, we have $\langle K \rangle_{2,1,0} = -\frac{1}{2} \langle V \rangle_{2,1,0}$.

13.5. The desired eigenfunctions must be superpositions of degenerate eigenfunctions of problem 12.5. Eigenfunctions with $l = 0$, $m = 0$ depend on the r coordinate only. The ground-state wave function, called $\{0, 0, 0\}$ in problem 12.5, has the form $(\pi a^2)^{-3/4} \exp[-(x^2 + y^2 + z^2)/2a^2]$. It fits this requirement and is nodeless; therefore, it is a 1s function with $n = 1$, $l = m = 0$. The wave functions $\{1, 0, 0\}$, $\{0, 1, 0\}$, and $\{0, 0, 1\}$ have a factor x, y, or z, respectively; no linear combination of them depends on r only. However, since $z = r \cos\theta$, the function $\{0, 0, 1\}$ has the form $f(r) \langle \theta | l = 1, m = 1 \rangle$; therefore, it represents a p_z state, also its radial part is nodeless. Similarly, $\{1, 0, 0\}$ and $\{0, 1, 0\}$ depend on θ and ϕ through factors $x = r \sin\theta \cos\phi$ and $y = r \sin\theta \sin\phi$, respectively, and thus represent p_x and p_y states. The desired eigenfunctions with $l = 1$ and $m = \pm 1$ must depend on θ and ϕ through factors $\sin\theta \exp(\pm i\phi) = \sin\theta(\cos\phi \pm i \sin\phi)$ and are thus superpositions of $\{1, 0, 0\}$ and $\{0, 1, 0\}$. The remaining eigenfunctions with $n_r = 1$, $l = m = 0$, must have one radial node and therefore a polynomial form in r or r^2; r is not expressed as a rational function of x, y, z, but $r^2 = x^2 + y^2 + z^2$. Superposition of $\{2, 0, 0\}$, $\{0, 2, 0\}$, and $\{0, 0, 2\}$ with equal coefficients yields a wave function with a factor $H_2(x/a) + H_2(y/a) + H_2(z/a)$ with the required property.

13.6. Equation (13.35) with the appropriate substitution gives $\langle r\ \theta\ \phi \mid a(t)\rangle =$ $(2\pi a_B^3)^{-1/2}$ $[\exp(-r/a_B - iE_1t/\hbar) - (r/8a_B)\sin\theta\exp(-r/2a_B + i\phi - iE_2t/\hbar)]$, $|\langle r\ \theta\ \phi \mid a(t)\rangle|^2 = (2\pi a_B^3)^{-1}\{\exp(-2r/a_B) + (r/8a_B)^2\sin^2\theta\exp(-r/a_B) - (r/4a_B)\sin\theta\exp(-3r/2a_B)\cos[(E_2 - E_1)t/\hbar - \phi]\}$. The mean values, weighted with this electron density, are $\langle x\rangle = \langle r\sin\theta\cos\phi\rangle = 2^7 \times 3^{-5}\, a_B \cos[(E_2 - E_1)t/\hbar]$, $\langle y\rangle = 2^7 \times 3^{-5}\, a_B \sin[(E_2 - E_1)t/\hbar]$, $\langle z\rangle = 0, \langle z^2\rangle = (7/2)a_B^2$. Note that $\langle x\rangle$ and $\langle y\rangle$ oscillate 90° out of phase—thus representing a rotation of the vector $\langle\vec{r}\rangle$—that $\langle z\rangle$ vanishes, and that $\langle z^2\rangle$ is constant.

14. electron spin and magnetic interactions

The quantum mechanics of the electron in a hydrogen atom predicts the eigenvalues of the squared orbital angular momentum $|\vec{l}|^2$ and of its component l_z, as shown in chapter 13. It also predicts implicitly how the atom's energy levels should split in an external magnetic field $\vec{B} \equiv B\hat{z}$. According to equations (6.1) and (7.8) each energy eigenvalue E_n is modified in this field by the addition of a magnetic energy term $-\mu_z B = -\gamma l_z B$. For the electron's orbital motion the gyromagnetic ratio γ should have the value $-e/2m_e c$ predicted by equation (2.43) on the basis of macroscopic theory.[1] In particular, a beam of hydrogen atoms in their ground state, with $l = 0$, should not be split by a Stern-Gerlach magnet. The hydrogen spectrum should show a normal Zeeman effect. These predictions are contradicted by experimental results.

The experimental analysis of excited levels of the hydrogen atom is complicated by the occurrence of small effects of relativistic mechanics which are mentioned only briefly at the end of chapter 13. These effects remove the degeneracy of states with equal quantum number n and different l and are comparable in magnitude to the magnetic energy of an atom in a normal laboratory field $\vec{B}$. The same effects can, however, be disregarded for our purposes when dealing with atoms other than hydrogen whose levels with equal n and different l are well separated owing to electric forces, as discussed below.

We shall consider particularly the alkaline metals whose atoms may be regarded in good approximation as consisting of a single outer electron attached to a positive ion (see chap. 16). In this case the potential energy of the electron in the ionic field does not follow a Coulomb law but has greater magnitude near the nucleus, owing to the combined effect of nuclear charge and of screening by other electrons. Consequently, the energy levels of the outer electron are lowered to an increasing extent for decreasing values of l which permit closer approach to the nucleus. Otherwise, the orbital motion of the electron in an alkali atom follows the same laws as for hydrogen and should show the same magnetic properties. For the above reason this chapter deals with the excited states of alkalis rather than of hydrogen.

14.1. THE ELECTRON SPIN

The following qualitative results on intra-atomic currents cannot be interpreted entirely on the basis of the electron motion studied in chapter 13.

1) Beams of hydrogen atoms *in their ground state* split into two components upon analysis with a Stern-Gerlach magnet, whereas no electron flux exists in this state. That is, magnetic analysis reveals the existence of two ground states whose energies coincide in the absence of an external field. A number

---

1. The small difference between the electron mass m_e and the reduced mass $\bar{m}$ of the nucleus and electron will be disregarded in this chapter and in the following ones.

of energy levels which are *double* that expected from the study of electron motion is observed upon magnetic analysis of all systems with a single electron, notably in the spectra of hydrogen and of the alkalis.

2) Even in the absence of an external magnetic field, the spectra of alkali atoms show the existence of two closely spaced levels, rather than a single one, for any pair of quantum numbers n and $l \neq 0$. The well-known *doublet* of the sodium spectrum consisting of the two yellow lines with wavelengths of 5890 and 5896 Å is due to transitions from two such closely spaced levels to the ground state. This doublet is an example of the *fine structure* observed in most atomic spectra.

3) The observed gyromagnetic ratios depart, in general, from the value $-e/2m_ec$ expected for currents consisting of an electron flux. The gyromagnetic ratio observed by Stern-Gerlach analysis of many monovalent atoms in their ground state is approximately $-e/m_ec$, that is, double that of orbital currents.

4) The splitting of ground states into two components, and in general the splitting of any energy level in an *even* number of components, is incompatible with the assumption that the squared angular momentum $|\vec{j}|^2$ of the level arises entirely from the orbital motion of electrons. [The squared orbital angular momentum $|\vec{l}|^2$ has the eigenvalues $l(l+1)\hbar^2$, with l an integer, and it has an *odd* number, $2l+1$, of eigenstates for each eigenvalue.] In other words, the experimental occurrence of states of current circulation with *half-integer* quantum numbers m and j (see section 6.3) is not accounted for by the orbital motion of the electrons. Even numbers of components and half-integer quantum numbers are observed for all levels of atoms with an odd number of electrons and never for atoms with an even number of electrons.

These results can be sorted out empirically in terms of two types of circulating currents, one of which consists of an electron flux, as expected from the treatment of the electron motion, and another one, called the *spin current*, which requires further definition. From this standpoint one may state that:

5) Stationary states with no orbital motion—that is, with $l = 0$—and without any electron flux exhibit spin currents with gyromagnetic ratio $-e/m_ec$.

6) Stationary states with $l \neq 0$, which are expected to possess an orbital current, exhibit in general circulating currents whose gyromagnetic ratio is neither $-e/m_ec$ nor $-e/2m_ec$. This observation indicates the coexistence of orbital and spin currents.

Even though the experimental observations on circulating currents are not accounted for by the treatment of electron motion in chapter 13, that treatment proves quite successful in many respects. It yields not only the correct gross spectrum of energy levels and the size of the hydrogen atom but also many other correct results, for example, the intensity of spectral lines. Besides these successes, one must consider that currents carried by an electron flux are present, as predicted in chapter 13, even though accompanied by other currents. The treatment of chapter 13 appears thus to be incomplete rather than erroneous.

Let us recall that the objective in chapter 13 was to represent a complete set of energy eigenstates as superpositions of position eigenstates of the electron. The problem so formulated was solved completely, that is, without overlooking any combination of position eigenstates.[2] The detection by magnetic experiments of twice as many stationary states as predicted in chapter 13 means that stationary states are not described fully in terms of the electron position. The electron must then have another variable with just two eigenvalues, no more, no less; otherwise the number of stationary states would not be exactly doubled. This variable must be independent of the position coordinates and compatible with them; that is, either of its eigenvalues is consistent with any eigenvalue of the electron position. It also interacts weakly with other characteristics of the electron motion, as shown by the small separation of the energy levels in the fine structure.

The additional variable is represented by a quantum number which identifies the eigenstates of spin-current circulation about an arbitrary axis. Stationary states that exhibit only spin currents, such as the ground state of hydrogen (item 5 above), are identified by eigenvalues of the angular momentum component in the direction of the magnetic field, $j_z = m\hbar = \pm\frac{1}{2}\hbar$. One takes the magnetic quantum number $m = \pm\frac{1}{2}$ in this formula as the spin-current variable, with two eigenvalues, $\frac{1}{2}$ and $-\frac{1}{2}$. To indicate that this quantum number pertains to the spin current, a subscript s is added to the symbol m. The spin variable is then

$$m_s = \pm\tfrac{1}{2}. \tag{14.1}$$

This variable describes adequately the relevant information about the spin current, since this current is detected and defined only by its angular momentum and magnetic moment. The spin angular momentum is usually indicated by the symbol $\vec{s}$ whose component along the z axis has the eigenvalues

$$s_z = m_s\hbar = \pm\tfrac{1}{2}\hbar. \tag{14.2}$$

Since external actions influence only the orientation of the spin current but never its magnitude, the squared spin angular momentum is a constant characteristic of the electron, much like its mass. The squared spin angular momentum is indicated by $|\vec{s}|^2$ even though it has a single eigenvalue identified in accordance with equation (7.7) by a quantum number s with the single value $\frac{1}{2}$. The eigenvalue of $|\vec{s}|^2$ is therefore

$$|\vec{s}|^2 = s(s+1)\hbar^2 = \tfrac{1}{2}(\tfrac{1}{2}+1)\hbar^2 = \tfrac{3}{4}\hbar^2. \tag{14.3}$$

The spin magnetic moment has the eigenvalues[3]

---

2. A complete solution includes the eigenstates of the continuous spectrum mentioned in section S13.6.

3. A small correction amounting to one part in 1,000 is disregarded in equation (14.4) and throughout this book.

$$\mu_z = -\frac{e}{m_e c} s_z = -m_s \frac{e}{m_e c} \hbar = \pm \frac{e\hbar}{2m_e c} . \qquad (14.4)$$

The magnitude of these eigenvalues,

$$\mu_B = \frac{e\hbar}{2m_e c} = 9.27 \times 10^{-21} \text{ ergs/gauss} = 5.78 \times 10^{-9} \text{ eV/gauss}, \qquad (14.5)$$

is a convenient unit of atomic magnetic moments, called the *Bohr magneton.* The Bohr magneton serves also as a unit for the magnetic moments of orbital currents, since

$$\mu_z = -\frac{e}{2m_e c} l_z = -m \frac{e\hbar}{2m_e c} . \qquad (14.6)$$

The spin current introduced in this section and its associated angular momentum are regarded as basic properties of electrons called *electron spin.* More generally, the word "spin" refers to any angular momentum or current circulation that one fails, by inability or choice, to describe in terms of a specified structure of a particle or other system.[4]

14.1.1. *Evidence from Double Collisions.* Evidence that the spin current is an intrinsic property of electrons, rather than a property of electrons as constituents of atoms, emerges from a different line of experiments. These are experiments in which electrons are deflected twice in succession by collisions with different atoms, as illustrated in Figure 3.14. Electrons from an incident beam are deflected by 90° through collisions with neon atoms in a first scattering chamber. Then they pass into a second chamber where they can be deflected once more by collisions with mercury atoms. The second deflection is observed by the pair of counters 1 and 2 in Figure 3.14, which are placed symmetrically with respect to the path of the electrons from the first to the second chamber. Remarkably, the two counters do not score at the same rate. This observation shows that the electrons in the beam passing between the two chambers are not isotropic about the direction of motion. This means that electrons possess some structural property which becomes oriented, that is, polarized, in the course of the first collision. Additional experiments show that the property responsible for the anisotropy has the symmetry characteristics of a dipole moment. This is shown by rotating the two collision chambers with respect to one another about the axis joining their

4. Dirac has shown that the electron spin emerges from a treatment of electron quantum mechanics which is consistent from the start with the requirements of relativity theory. The treatment in this book extends to quantum physics the laws of ordinary, nonrelativistic, macroscopic physics. Dirac's theory relates the spin to a fluctuating motion of the electron's charge and mass about an average position coordinate.

centers. The difference between the counting rates of detectors 1 and 2 is found to be proportional to the cosine of the angle of rotation.

14.2 SPIN-ORBIT COUPLING

The currents associated with the orbital motion of an electron and with its spin interact with one another magnetically. Each of these currents generates a magnetic field which influences the other current. The magnetic energy of this *spin-orbit coupling* is much smaller than the differences between the energy levels of the electron motion under the influence of nuclear attraction.[5] The spin-orbit coupling has nevertheless a very substantial qualitative influence on the stationary states of atoms.

The large influence of this weak interaction is made possible by the degeneracy of levels which would exist in the absence of the interaction. First, the levels of rotational motion with orbital quantum number l are $(2l + 1)$-fold degenerate (section 13.2). This degeneracy is increased by the existence of the electron spin whose orientation does not affect the kinetic or electrostatic potential energy of an electron. Different *mutual* orientations of orbital and spin currents are no longer energetically equivalent when spin-orbit coupling is taken into account. Spin-orbit coupling thus reduces the degeneracy of states with the same quantum number l. This removal of degeneracy is the main source of the fine structure exhibited by atomic spectra in the absence of external fields.

This section deals with the problem of determining the stationary states of an atomic electron in the presence of spin-orbit coupling. These states will be represented in first and good approximation as superpositions of states that would be degenerate, with energy E_{nl}, if the spin-orbit coupling were disregarded.[6] The problem of determining the influence of a weak disturbance on a group of degenerate (or quasi-degenerate) states occurs frequently throughout quantum physics.

The first step of the procedure consists of listing the states that would be degenerate with energy E_{nl} in the absence of spin-orbit coupling. We have indicated here the energy as dependent on l, as it is for an alkali atom, for the reasons discussed in the introduction to this chapter. Stationary states with energy E_{nl} have been identified in chapter 13 by the set of quantum numbers (n, l, m) and are represented by wave function $\langle r\ \theta\ \phi | n\ l\ m\rangle$. A state of spin orientation with respect to the z axis is identified by the quantum number

---

5. The word "coupling" is used here to indicate an interaction energy. However, it is also used frequently in this book and elsewhere to indicate a type of addition of orbital and/or spin angular momenta.

6. The exact eigenstates actually include components with different energies $E_{n'l'}$; however, the probability amplitudes of these components are small in the ratio of the coupling energy to $E_{nl} - E_{n'l'}$, as will be shown in chapter 17.

$m_s = \pm\frac{1}{2}$. Therefore, a state with definite circulation of both orbital and spin currents about the z axis is represented by

$$|n\ l\ m\ m_s\rangle. \tag{14.7}$$

These quantum numbers indicate that the state is an eigenstate of the following quantities:

(a) spin angular momentum component s_z with eigenvalue $m_s\hbar$;

(b) orbital angular momentum component l_z with eigenvalue $m\hbar$;

(c) squared orbital angular momentum $|\vec{l}|^2$ with eigenvalue $l(l+1)\hbar^2$;

(d) radial motion with $n - l - 1$ dark interference fringes.

The stationary states we are trying to determine will have energy eigenvalues close to E_{nl} and will be represented by superpositions of states (14.7) with equal n, l, and different m and m_s. There are $2l + 1$ values of m for the given l and two values of m_s. Therefore we are considering a group of $2(2l + 1)$ states.

The general procedure for solving a problem of this type is to construct an appropriate form of the Schrödinger equation by the variational method of section 12.1. An example of calculation by this method is given in section S14.4. However, in our problem one may proceed more directly by a semi-qualitative analysis, sufficient to identify the eigenstates and the eigenvalues of spin-orbit coupling.

14.2.1. *Eigenstates of Total Angular Momentum.* The magnetic coupling between orbital and spin currents within an atom exerts no torque on the atom as a whole. Therefore, the coupling leaves unaffected the total angular momentum of the atom, which may be indicated as the vector sum of the orbital and spin angular momenta

$$\vec{j} = \vec{l} + \vec{s}. \tag{14.8}$$

It follows that the energy of spin-orbit coupling is compatible with $|\vec{j}|^2$ and with any one of its components, for example, with $j_z = l_z + s_z$.[7] Therefore, one can construct a set of stationary states of spin-orbit coupling each of which is an eigenstate of $|\vec{j}|^2$ and of j_z. In general, the eigenvalues of $|\vec{j}|^2$ and j_z do not identify by themselves the total energy of the state. For example, two (or more) eigenstates of $|\vec{j}|^2$ and j_z with the same pair of eigenvalues may have different radial energy; in this case superposition of the two states yields a nonstationary state. However, we are considering a system with fixed

---

7. In our problem the atom consists of a spherical core and of a single electron. The angular momentum of the atom is therefore the same as that of the electron. Accordingly, the angular momentum is indicated by the lowercase letter j as is conventional when referring to a single electron.

radial energy E_{nl}. As we shall show, this system possesses no two eigenstates of $|\vec{j}|^2$ and j_z with the same pair of eigenvalues; in this case, each eigenstate of $|\vec{j}|^2$ and j_z must be stationary.

Eigenstates of $|\vec{j}|^2$ and j_z with the same eigenvalue of $|\vec{j}|^2$ and different eigenvalues of j_z must have the same energy of spin-orbit coupling, because the component j_z of the angular momentum relates to an arbitrary coordinate axis which is irrelevant to spin-orbit coupling. Therefore, the spin-orbit energy can have only as many different eigenvalues as there are different eigenvalues of $|\vec{j}|^2$. We shall see that only two eigenvalues of $|\vec{j}|^2$ are consistent with a given value of $|\vec{l}|^2$.

Our problem is then to construct eigenstates of $|\vec{j}|^2$ and j_z by superposition of our limited set of "initial" states $|n\ l\ m\ m_s\rangle$ with the same values of n and l. (Here and in the following, "initial" means at the beginning of our treatment rather than at $t = 0$.) This problem simplifies at the start because each of these states, being an eigenstate of both l_z and s_z, is also an eigenstate of their sum j_z. The eigenvalues of j_z are expressed in terms of a *total magnetic quantum number* m_j by

$$j_z = m_j\hbar, \qquad m_j = m + m_s. \tag{14.9}$$

Each joint eigenstate of $|\vec{j}|^2$ and j_z has therefore a definite value of m_j and is a superposition of only those initial states for which $m + m_s = m_j$.

In the initial set of states $|n\ l\ m\ m_s\rangle$ the values of $m + m_s = m_j$ range from a maximum of $l + s = l + \frac{1}{2}$ to a minimum $-(l + \frac{1}{2})$. To the maximum value of m_j corresponds a *single state*, $|n\ l\ l\ \frac{1}{2}\rangle$. Since our system has no other state with the same value of m_j (and with the same n and l), this state must itself be an eigenstate of $|\vec{j}|^2$ and must be stationary in the presence of spin-orbit coupling. Physically, this means that the initial state with $m = l$ and with $m_s = \frac{1}{2}$ must remain stationary in the presence of spin-orbit coupling, because this coupling can not change the state without violating the conservation of the angular momentum component j_z. Similarly, the single state $|n\ l\ {-l}\ {-\frac{1}{2}}\rangle$, with the minimum value of m_j, must also be an eigenstate of $|\vec{j}|^2$ and must be stationary.

For each intermediate value of m_j there are two initial states with alternative spin orientation, that is, with $m_s = \pm\frac{1}{2}$ and $m = m_j \mp \frac{1}{2}$. Each joint eigenstate of j_z and $|\vec{j}|^2$ is then represented as a superposition of only two initial states with the same value of $m + m_s$. This pair of states may be treated as the base set of a two-level system; accordingly, we indicate a pair of orthogonal states of this system by[8]

$$|n\ l\ m_j{-}\tfrac{1}{2}\ \tfrac{1}{2}\rangle \cos\beta + |n\ l\ m_j{+}\tfrac{1}{2}\ {-\tfrac{1}{2}}\rangle \sin\beta, \tag{14.10a}$$

$$|n\ l\ m_j{-}\tfrac{1}{2}\ \tfrac{1}{2}\rangle (-\sin\beta) + |n\ l\ m_j{+}\tfrac{1}{2}\ {-\tfrac{1}{2}}\rangle \cos\beta. \tag{14.10b}$$

---

8. The coefficients of the superposition (14.10) are real owing to the phase conventions on the transformation matrices for eigenstates of angular momentum, which were introduced in sections 8.5 and S10.3.

The problem is to determine which values of β make these states eigenstates of $|\vec{j}|^2$ and which are the corresponding eigenvalues of $|\vec{j}|^2$.

Let us consider now the possible eigenvalues of $|\vec{j}|^2$. According to equation (7.7) these eigenvalues are represented by $j(j+1)\hbar^2$, where j is called, in this case, the *total angular momentum quantum number*. The largest value of the number j in our problem must coincide with the largest value of m_j and is thus given by

$$j = l + \tfrac{1}{2}. \tag{14.11a}$$

The state with j and m_j both equal to $l + \frac{1}{2}$ is $|n\ l\ l\ \frac{1}{2}\rangle$. For $j = l + \frac{1}{2}$ there are $2j + 1 = 2l + 2$ states which have different values of m_j and the same energy of spin-orbit coupling. (The degeneracy of these $2j + 1$ states would be removed by subjecting the atom to a weak magnetic field which sorts out different eigenstates of its magnetic moment μ_z without disturbing its spin-orbit coupling.) Each of these degenerate states with $m_j < l + \frac{1}{2}$ can be identified by different methods which are beyond the scope of this book. A simple example will be worked out in supplementary section S14.4. The general method proceeds, in essence, by studying the transformations of the wave function with $m_j = l + \frac{1}{2}$ under rotation of coordinate axes. One finds that the state with $j = l + \frac{1}{2}$ and $m_j < l + \frac{1}{2}$ has the form (14.10a), with β given by

$$\cos\beta = \left[\frac{l + m_j + \frac{1}{2}}{2l + 1}\right]^{1/2}. \tag{14.12}$$

For $m_j = -(l + \frac{1}{2})$ this formula yields $\cos\beta = 0$ and the state is represented by $|n\ l\ -l\ -\frac{1}{2}\rangle$ as expected.

There remains the set of states represented by expression (14.10b), of which there are $2l$, with m_j values ranging from $l - \frac{1}{2}$ to $-(l - \frac{1}{2})$. The largest of these values of m_j characterizes the whole set as consisting of eigenstates of $|\vec{j}|^2$ with the quantum number

$$j = l - \tfrac{1}{2}. \tag{14.11b}$$

(The value of $\cos\beta$ to be entered in expression [14.10b] is again given by eq. [14.12].)

In conclusion, the stationary states of an electron in the presence of spin-orbit coupling are eigenstates of $|\vec{j}|^2$ with the two alternative values $j = l \pm \frac{1}{2}$. These two values are loosely said to correspond to parallel and antiparallel orientations of the spin and orbital angular momenta, even though the directions of the vector angular momenta are not well defined in quantum physics. When an eigenstate of the square orbital angular momentum is labeled by one of the code letters s, p, d, . . . , etc. (see p. 287) and the state is also an eigenstate of $|\vec{j}|^2$, its j value is indicated by a subscript. For example, $p_{1/2}$ means $l = 1$ and $j = l - \frac{1}{2} = \frac{1}{2}$. The procedure followed above for combining angular momenta of orbital and spin motion is a first example of a general procedure of *addition of angular momenta*, to which we will return in section 14.2.3.

Prior to the development of quantum mechanics it was thought that each of the vectors $\vec{l}$ and $\vec{s}$ would precess about the direction of their resultant $\vec{j}$ under the influence of the magnetic fields generated by the spin and orbital currents. As usual for macroscopic models, this picture applies to the mean values of our variables. Suppose, for instance, that the orbital and spin currents were known to be at the time $t = 0$ in eigenstates of different angular momentum components $\vec{l}\cdot\hat{u}$ and $\vec{s}\cdot\hat{v}$. Under the influence of spin-orbit coupling the state of the atom would not remain stationary. The mean angular momenta, which are represented by $m\hbar\hat{u}$ and $m_s\hbar\hat{v}$ at $t = 0$, would precess in the course of time about their resultant $\langle\vec{j}\rangle = m\hbar\hat{u} + m_s\hbar\hat{v}$. The precession frequency is given by the energy difference between the stationary states with the quantum numbers $j = l + \frac{1}{2}$ and $j = l - \frac{1}{2}$.

14.2.2. *Energy of Spin-Orbit Coupling.* The magnetic interaction energy of the orbital and spin currents may be expressed as energy of the spin current in the magnetic field generated by the orbital motion. The spin current must be considered in a coordinate system attached to the electron. In this system, the electron's charge is at rest and does not generate any magnetic field, but the nucleus of the atom appears to be in motion and to generate a magnetic field. This field is proportional to the electron's orbital angular momentum and depends on the strength of the nuclear electric attraction upon the electron. The calculation of the coupling energy is complicated by factors mentioned below. We limit our task to presenting and discussing its result, namely,

$$E_{coup} = \left\langle \frac{Z}{r^3} \right\rangle_{nl} \frac{1}{2} \left(\frac{e}{m_e c} \right)^2 \vec{l}\cdot\vec{s}. \tag{14.13}$$

In this formula $\langle Z/r^3 \rangle_{nl}$ indicates the mean value of the effective nuclear charge in units of e divided by the cubed distance of the electron from the nucleus; the expression is averaged over the probability distribution of the electron's distance for a state with the given quantum numbers (n, l). The factor $\frac{1}{2}(e/m_e c)^2$ is the product of the gyromagnetic ratios of the orbital and spin currents.

The quantity $\langle Z/r^3 \rangle_{nl}$ in equation (14.13) and in the following pertains, strictly speaking, to an electron moving in the Coulomb field of a nucleus with charge Ze. A more general expression which takes into account the screening of other electrons and is applicable to an electron with potential energy $V(r)$ is indicated by the substitution

$$e^2 \left\langle \frac{Z}{r^3} \right\rangle_{nl} \rightarrow \left\langle \frac{1}{r} \frac{dV}{dr} \right\rangle_{nl}. \tag{14.14}$$

Equations (14.13) and (14.14) are derived by considering the *electric* field generated by the nucleus (and other electrons, if any) at the position of the electron, in the fixed-nucleus frame of reference. This field is then transformed to a frame of reference attached to the electron. The transformation yields a nonzero *magnetic* field which acts upon the electron's spin. The relative motion of the two frames of reference involves not only the electron's velocity $\vec{v}$ but also a rotation of coordinate axes. Hence the appropriate Lorentz transformation is more complex than usual and yields a magnetic field $\vec{B} \neq \vec{E} \times \vec{v}/c$. Failure to take account of this rotation of axes would remove the factor ½ *(Thomas factor)* which appears in equation (14.13).

The eigenvalues of the coupling energy (14.13) are calculated by expressing the product $2\vec{l}\cdot\vec{s}$ in terms of the quantum numbers l, s, and j of the orbital, spin, and total angular momenta. Equation (14.13) is conveniently rewritten in the form

$$E_{\text{coup}} = \left\langle \frac{Z}{r^3} \right\rangle_{nl} \left(\frac{e\hbar}{2m_e c} \right)^2 \frac{2\vec{l}\cdot\vec{s}}{\hbar^2} = \left\langle \frac{Z}{r^3} \right\rangle_{nl} \mu_B^2 \frac{2\vec{l}\cdot\vec{s}}{\hbar^2}, \tag{14.13a}$$

where the factor $(2\vec{l}\cdot\vec{s})/\hbar^2$ is dimensionless and μ_B indicates the Bohr magneton defined by equation (14.5). The vector identity

$$|\vec{j}|^2 = |\vec{l} + \vec{s}|^2 = |\vec{l}|^2 + |\vec{s}|^2 + 2\vec{l}\cdot\vec{s} \tag{14.15}$$

yields

$$\frac{2\vec{l}\cdot\vec{s}}{\hbar^2} = \frac{|\vec{j}|^2 - |\vec{l}|^2 - |\vec{s}|^2}{\hbar^2} = j(j+1) - l(l+1) - s(s+1). \tag{14.16}$$

The total energy eigenvalue of the electron is the sum of a term E_{nl} calculated in accordance with chapter 13 and of the spin-orbit energy (14.13a). The latter is obtained from (14.16) with the appropriate values of $j = l \pm \frac{1}{2}$ and of $s = \frac{1}{2}$. The eigenvalue is given by

$$E_{nlj} = E_{nl} + \left\langle \frac{Z}{r^3} \right\rangle_{nl} \mu_B^2 \times \begin{cases} l & \text{for } j = l + \frac{1}{2}, \\ -(l+1) & \text{for } j = l - \frac{1}{2}. \end{cases} \tag{14.17}$$

The difference of the two levels represented by equation (14.17) constitutes the *doublet interval* of the fine structure of alkali spectra.

For the sodium 3p state, whose transition to the ground state yields the yellow doublet, the experimental doublet interval is 2.1×10^{-3} eV. The value of μ_B^2, in appropriate units, is 5.3×10^{-5} eV Å^3. Therefore, the value of $\langle Z/r^3 \rangle_{\text{Na3p}}$ amounts to 13 Å^{-3}.

14.2.3. *Addition of Angular Momenta.* The phenomenon of weak coupling of spin and orbital motion is a prototype of a large class of atomic (and molecular) phenomena. Its essential feature is that the atom contains two or more subsystems (the orbital and spin currents in our prototype) whose energies and whose angular momenta, $\vec{j}_1$ and $\vec{j}_2$, are independent in first approximation. A weak interaction between the two subsystems applies torques on them, and therefore $\langle \vec{j}_1 \rangle$ and $\langle \vec{j}_2 \rangle$ are no longer constant. However, the interaction leaves unaffected the total angular momentum $\vec{j} = \vec{j}_1 + \vec{j}_2$. Stationary states can be constructed which are eigenstates of $|\vec{j}|^2$, of one component, j_z, and of the squared magnitudes $|\vec{j}_1|^2$ and $|\vec{j}_2|^2$. An example occurs in the interaction between electronic and nuclear currents whose effect is outlined in section S14.5.

To construct the stationary states of the combined system one starts by considering a set of $(2j_1 + 1)(2j_2 + 1)$ stationary states of its separate subsystems,

$$|j_1 \; m_1, \; j_2 \; m_2\rangle. \tag{14.18}$$

(We have not written any other quantum numbers, such as n, pertaining to characteristics of the atom that are not affected by the weak interaction.) The states (14.18) are not stationary in the presence of the interaction. Stationary states are characterized by eigenvalues of the squared total angular momentum $|\vec{j}|^2 = j(j + 1)\hbar$ and of its component $j_z = m_j\hbar$. The problem of addition of angular momenta consists of two tasks: (a) the determination of possible values of the quantum number j and (b) the determination of the coefficients of the superposition of states (14.18).

Task (a) is carried out as was done for the spin-orbit coupling in section 14.2.1, that is, by enumerating the number of possible states with each value of the quantum number m_j. There is a single state for $m_j = j_1 + j_2$, and there are two states for $m_j = j_1 + j_2 - 1$. If $j_1 \geqslant j_2$, the number of states increases successively by one until there are $2j_2 + 1$ states with $m_j = j_1 - j_2$; this is also the number of states for all smaller values of $|m_j|$. One concludes that the possible values of j range in steps of one from $j_1 + j_2$ to $|j_1 - j_2|$,

$$j = j_1 + j_2, \; j_1 + j_2 - 1, \; j_1 + j_2 - 2, \ldots, \; |j_1 - j_2|. \tag{14.19}$$

The important point, in general as for the spin-orbit coupling, is that one finds just as many eigenstates of $|\vec{j}|^2$ and j_z as there are pairs of numbers (j, m_j) for which j satisfies (14.19) and m_j satisfies $-j \leqslant m_j \leqslant j$. The total number of states is, in fact, $(2j_1 + 1)(2j_2 + 1)$ and thus equals the number of states (14.18)

Each eigenstate of $|\vec{j}|^2$ and j_z is indicated by $|j_1 \; j_2 \; j \; m_j\rangle$, sometimes simply by $|j \; m_j\rangle$. It is represented as a superposition of the states (14.18) by the general formula

$$|j_1 \; j_2 \; j \; m_j\rangle = \sum_{m_1, m_2}^{(m_1+m_2=m_j)} |j_1 \; m_1, j_2 \; m_2\rangle\langle j_1 \; m_1, j_2 \; m_2 | j_1 \; j_2 \; j \; m_j\rangle, \tag{14.20}$$

which is a generalization of equations (14.10). The probability amplitudes $\langle j_1 \; m_1, j_2 \; m_2 | j_1 \; j_2 \; j \; m_j\rangle$ which connect the two sets of eigenstates (14.18) and (14.20) are called Wigner coefficients or Clebsch-Gordan coefficients. They are functions of quantum numbers only. Their determination, which constitutes task (b) above in the addition of angular momenta, is not carried out here. Their values are given by algebraic formulae which are fairly simple when one of the quantum numbers j_1 or j_2 is small; they are also tabulated extensively.[9]

As in the case of spin-orbit coupling, the energy eigenvalue of a stationary state $|j_1\ j_2\ j\ m_j\rangle$ is proportional to the product $\vec{j}_1 \cdot \vec{j}_2$ if the interaction between the two systems can be described as the potential energy of two magnetic dipoles. In this event, the dependence of the eigenvalue on the quantum number j is obtained by a formula analogous to equation (14.16):

$$2\vec{j}_1 \cdot \vec{j}_2 = [j(j+1) - j_1(j_1+1) - j_2(j_2+1)]\hbar^2. \tag{14.21}$$

More complicated interactions—for example, electric quadrupole interaction—can often be represented as functions of $\vec{j}_1\ \vec{j}_2$ and calculated by using equation (14.21).

14.3. ANOMALOUS ZEEMAN EFFECT

Application of an external magnetic field $\vec{B}$ removes the degeneracy of fine-structure levels which have a given value of j and different values of m_j. The presence of this field, whose direction is chosen as the z axis, adds to the magnetic energy of spin-orbit coupling (14.13a) a term $-\mu_z B$. Here the magnetic moment μ_z consists of two terms due to the orbital and spin currents. These terms equal, respectively, the angular momentum components multiplied by the appropriate gyromagnetic ratios,

$$\begin{aligned} \mu_z = (\mu_z)_{orb} + (\mu_z)_{spin} &= -\frac{e}{2m_e c}\, l_z - \frac{e}{m_e c}\, s_z \\ &= \frac{e}{2m_e c}\,(l_z + 2s_z) = \mu_B \frac{j_z + s_z}{\hbar}. \end{aligned} \tag{14.22}$$

The Zeeman effect of alkali atoms is "anomalous" because the term $(\mu_z)_{spin}$ in equation (14.22) has twice the basic gyromagnetic ratio of $(\mu_z)_{orb}$. The anomaly is represented conveniently by the term s_z which appears besides j_z in the expression (14.22).

The calculation of the magnetic energy $\mu_z B$ is complicated, in the presence of spin-orbit coupling, by the fact that s_z is incompatible with the factor $2\vec{l} \cdot \vec{s}$ in equation (14.13a) and with $|\vec{j}|^2$. A calculation which takes full account of this incompatibility is given in section S14.4. However, one may proceed by a simple approximation in the usual case in which the energy $\mu_z B$ remains much

9. For a table of algebraic expressions see, for instance, E. U. Condon and G. Shortley, *Theory of Atomic Spectra* (Cambridge: Cambridge University Press, 1963), pp. 76-77 (cited hereafter as *TAS*); for numerical tables see M. Rotenberg, R. Bivins, N. Metropolis, and J. K. Wooten jr., *The 3-j and 6-j Symbols* (Cambridge, Mass.: M.I.T. Press, 1959).

smaller than the spin-orbit coupling energy. (In the example of the sodium 3p doublet, $\mu_z B$ would match the interval 2.1×10^{-3} eV of the doublet for a field strength of $\sim 4 \times 10^5$ gauss.) For a sufficiently weak field, the spin-orbit coupling remains undisturbed and the eigenstates of $|\vec{j}|^2$ and j_z are approximately stationary. The energy $-\mu_z B$ is then evaluated by setting in equation (14.22) j_z equal to its eigenvalue $m_j \hbar$ and s_z equal to its *mean value* in the state with quantum numbers j and m_j. This mean value can be obtained from the representation (14.10) of the states with $j = l + \frac{1}{2}$ (or $j = l - \frac{1}{2}$) utilizing the expression (14.12) of cos β. For $j = l + \frac{1}{2}$, expression (14.10a) shows that the state $|n \; l \; m_j - \frac{1}{2} \; \frac{1}{2}\rangle$, with $s_z = \frac{1}{2}\hbar$, has the probability $\cos^2 \beta$, while the state with $s_z = -\frac{1}{2}\hbar$ has the probability $\sin^2 \beta$. The mean values of $\langle s_z \rangle$ for both values of $j = l \pm \frac{1}{2}$ are summarized by

$$\langle s_z \rangle = \pm \tfrac{1}{2}\hbar \, (\cos^2 \beta - \sin^2 \beta) = \pm \frac{m_j \hbar}{2l + 1} = \pm \frac{j_z}{2l + 1}. \tag{14.23}$$

Combining this formula with (14.22), one obtains the mean value of the energy

$$\langle -\mu_z B \rangle = m_j \left(1 \pm \frac{1}{2l + 1}\right) \mu_B B = m_j g \mu_B B, \tag{14.24}$$

where g is called the *Landé factor*.

The Landé factor is usually evaluated by a method that does not utilize the eigenstate representation (14.10) explicitly. After showing that $\langle s_z \rangle = \vec{s} \cdot \vec{j} \, j_z / |\vec{j}|^2$, one replaces $\vec{s} \cdot \vec{j}$ by $\frac{1}{2}(|\vec{j}|^2 + |\vec{s}|^2 - |\vec{l}|^2)$. This yields

$$g = \frac{j_z + \langle s_z \rangle}{j_z} = 1 + \frac{1}{2} \, \frac{j(j + 1) + s(s + 1) - l(l + 1)}{j(j + 1)}, \tag{14.25}$$

which reduces to $1 \pm (2l + 1)^{-1}$ for $j = l \pm \frac{1}{2}$. This expression of the Landé factor is directly applicable to atoms with many electrons, as will be discussed in section 16.3.4. As anticipated in section 7.3, experimental observations of the anomalous Zeeman effect can be represented by assigning to each atomic level an empirical value of the gyromagnetic ratio. The Landé factor provides a theoretical expression for the ratio of this empirical value to the orbital gyromagnetic ratio $-e/2m_e c$. The theoretical value is fully in agreement with the experimental results in a sufficiently weak field.

In a classical treatment, the anomalous Zeeman effect was interpreted from the point of view that orbital and spin currents would precess at different rates in an external magnetic field in the absence of spin-orbit coupling. If the coupling predominates over the external field, it locks the two currents together, forcing them to precess at the same rate. The Landé factor is the ratio of this rate to the Larmor frequency of purely orbital currents. In this case again, the pre-quantummechanical interpretation is valid for the mean magnetic moment in a nonstationary state.

14.3.1. *A Remark on Stern-Gerlach Experiments.* We are now in a position to explain the cautionary remark on the Stern-Gerlach experiment in footnote 6 on p. 141. If the magnetic field is strong enough to disturb the spin-orbit coupling appreciably, the stationary states of the atom are not eigenstates of $|\vec{j}|^2$. The eigenvalues of μ_z are not equally spaced, but may be calculated in accordance with section S14.4. The gyromagnetic ratio is different for the different eigenstates of μ_z.

The field in a Stern-Gerlach magnet is usually not strong enough to disturb the spin-orbit coupling appreciably. However, it is often of the right order of magnitude to disturb the coupling of the electron currents with the nuclear currents, which is discussed in section S14.5. This coupling can also be disregarded if the magnetic field is strong enough to overcome it or sufficiently weak to leave it undisturbed. In conclusion, the design of a Stern-Gerlach experiment must take careful account of the strength of the field as compared to the energies of spin-orbit coupling and of electron-nuclear interactions.

SUPPLEMENTARY SECTIONS

S14.4. DIRECT CALCULATION OF MAGNETIC ENERGY EIGENSTATES

The qualitative analysis of spin-orbit coupling at the beginning of section 14.2 indicates that stationary states should be represented approximately as superpositions of a *finite* set of states. These states $|n\ l\ m\ m_s\rangle$ would be stationary and degenerate in the absence of magnetic interaction. Here we determine the coefficients of the superposition by the general procedure of constructing and solving the appropriate form of the Schrödinger equation. The magnetic interaction with an external field is included in the energy at the outset together with the energy of spin-orbit coupling. The results thus obtained extend the treatment of the anomalous Zeeman effect in section 14.3 to an external field of arbitrary strength. The formulation and solution of our problem are typical of a class of problems in atomic and subatomic physics.

The total magnetic energy of an atom in a field $\vec{B}$ is the sum of the energy of spin-orbit coupling (14.13a) and of the orbital and spin currents in the external field,

$$E_{magn} = \left\langle \frac{Z}{r^3} \right\rangle_{nl} \mu_B^2 \frac{2\vec{l}\cdot\vec{s}}{\hbar^2} + \mu_B \frac{l_z}{\hbar} B + 2\mu_B \frac{s_z}{\hbar} B = W \left[\frac{2\vec{l}\cdot\vec{s}}{\hbar^2} + \frac{B}{B_0} \frac{l_z + 2s_z}{\hbar} \right]. \tag{14.26}$$

Here we have set

$$W = \left\langle \frac{Z}{r^3} \right\rangle_{nl} \mu_B^2, \tag{14.27}$$

$$B_0 = \left\langle \frac{Z}{r^3} \right\rangle_{nl} \mu_B. \tag{14.28}$$

Following the procedure of section 12.1, we begin by expressing the mean value of E_{magn} for an arbitrary state $|a\rangle$ represented by the superposition

$$|a\rangle = \sum_{mm_s} |n\ l\ m\ m_s\rangle\langle n\ l\ m\ m_s|a\rangle. \tag{14.29}$$

Omitting the fixed-value indices n and l, the mean value is expressed in terms of matrices, by a formula analogous to (8.28) or (10.19),

$$\langle E_{magn}\rangle_a = W\Big\{\sum_{mm_s}\sum_{m'm'_s} \langle a|m\ m_s\rangle\langle m\ m_s|\frac{2\vec{l}\cdot\vec{s}}{\hbar^2}|m'\ m'_s\rangle\langle m'\ m'_s|a\rangle$$
$$+\frac{B}{B_0}\sum_{mm_s}\sum_{m'm'_s}\langle a|m\ m_s\rangle\langle m\ m_s|\frac{l_z+2s_z}{\hbar}|m'\ m'_s\rangle\langle m'\ m'_s|a\rangle\Big\}. \tag{14.30}$$

The explicit form of the matrices will be specified below.

The eigenvalues E_η of E_{magn} and the corresponding probability amplitudes $\langle m\ m_s|E\ \eta\rangle$ are determined by a Schrödinger equation.[10] This equation is obtained from the variational condition (12.8) in its form (12.8b). In our problem the partial derivative of the mean value $\langle E_{magn}\rangle_a$ is obtained from (14.30) and $\partial\mathcal{P}/\partial\langle a|m\ m_s\rangle$ equals $\langle m\ m_s|a\rangle$; the derivative $\partial\langle E_{magn}\rangle_a/\partial\mathcal{P}$ at $\mathcal{P}=1$ reduces to the eigenvalue E_η when $|a\rangle \to |E\ \eta\rangle$. The Schrödinger equation is then

$$W\Big\{\sum_{m'm'_s}\langle m\ m_s|\frac{2\vec{l}\cdot\vec{s}}{\hbar^2}|m'\ m'_s\rangle\langle m'\ m'_s|E\ \eta\rangle$$
$$+\frac{B}{B_0}\sum_{m'm'_s}\langle m\ m_s|\frac{l_z+2s_z}{\hbar}|m'\ m'_s\rangle\langle m'\ m'_s|E\ \eta\rangle\Big\} = \langle m\ m_s|E\ \eta\rangle E_\eta. \tag{14.31}$$

This equation is algebraic, because $\langle E_{magn}\rangle_a$ is an algebraic function of a finite number of probability amplitudes $\langle m\ m_s|a\rangle$, while the Schrödinger equation (12.9) for the motion of a particle is differential.[11]

---

10. The energy eigenstates are here labeled by η rather than by n to avoid confusion with the principal quantum number.

11. Solving an algebraic system of homogeneous eigenvalue equations is equivalent to diagonalizing the matrix of the coefficients of the system. In our case solution of the system (14.31) is equivalent to diagonalizing the matrix of the magnetic energy

$$\Big|\langle m\ m_s|W\Big\{2\frac{\vec{l}\cdot\vec{s}}{\hbar^2}+\frac{B}{B_0}\frac{l_z+2s_z}{\hbar}\Big\}|m'\ m'_s\rangle\Big|,$$

which is used in (14.30) to express $\langle E_{magn}\rangle_a$. The eigenvalues E_η of (14.31) are the elements of the diagonalized matrix. The probability amplitudes $\langle m\ m_s|E\ \eta\rangle$ are the eigenvectors of the matrix.

In the coefficients of equation (14.31), the matrix of $l_z + 2s_z$ is diagonal because the states $|m \ m_s\rangle$ are eigenstates of both l_z and s_z; we have in fact

$$\langle m \ m_s | \frac{l_z + 2s_z}{\hbar} | m' \ m_s' \rangle = (m + 2m_s) \ \delta_{mm'} \ \delta_{m_s m_s'}. \tag{14.32}$$

The matrix of $2\vec{l}\cdot\vec{s}$ is, instead, nondiagonal. Therefore (14.31) represents a system of linear equations. To limit the size of matrices and the number of equations, we *restrict* our treatment *to the example of* $l = 1$.

The elements of the matrix of $\vec{l}\cdot\vec{s}$ are obtained from those of the separate matrices of the components of $\vec{l}$ and $\vec{s}$,

$$\begin{aligned}\langle m \ m_s | \vec{l}\cdot\vec{s} | m' \ m_s' \rangle &= \langle m \ m_s | l_x s_x + l_y s_y + l_z s_z | m' \ m_s' \rangle \\ &= \langle m | l_x | m' \rangle \langle m_s | s_x | m_s' \rangle + \langle m | l_y | m' \rangle \langle m_s | s_y | m_s' \rangle \\ &+ \langle m | l_z | m' \rangle \langle m_s | s_z | m_s' \rangle.\end{aligned} \tag{14.33}$$

The last term of this expression is zero unless $m = m'$ and $m_s = m_s'$:

$$\langle m | l_z | m' \rangle \langle m_s | s_z | m_s' \rangle = m \, m_s \hbar^2 \ \delta_{mm'} \ \delta_{m_s m_s'}. \tag{14.34}$$

The matrices of s_x and s_y coincide with those given in chapter 8, equations (8.29a) and (8.29b) for angular momentum $j = 1/2$. The matrices of l_x and l_y, for $l = 1$, can be calculated utilizing the probability amplitudes (10.23).[12] The matrices are

$$\begin{matrix} & m' = 1 & 0 & -1 \end{matrix} \qquad\qquad \begin{matrix} & m' = 1 & 0 & -1 \end{matrix}$$

$$\left|\langle m | \frac{l_x}{\hbar} | m' \rangle\right| = \begin{vmatrix} 0 & \sqrt{1/2} & 0 \\ \sqrt{1/2} & 0 & \sqrt{1/2} \\ 0 & \sqrt{1/2} & 0 \end{vmatrix}, \quad \left|\langle m | \frac{l_y}{\hbar} | m' \rangle\right| = \begin{vmatrix} 0 & -i\sqrt{1/2} & 0 \\ i\sqrt{1/2} & 0 & -i\sqrt{1/2} \\ 0 & i\sqrt{1/2} & 0 \end{vmatrix}. \tag{14.35}$$

Direct multiplication[13] of the matrices of l_x and s_x yields

$$(m', m_s') = \quad (1\tfrac{1}{2}) \quad (1 -\tfrac{1}{2}) \quad (0\tfrac{1}{2}) \quad (0 -\tfrac{1}{2}) \quad (-1\tfrac{1}{2}) \quad (-1 -\tfrac{1}{2})$$

$$\left| 2\langle m | \frac{l_x}{\hbar} | m' \rangle \langle m_s | \frac{s_x}{\hbar} | m_s' \rangle \right| = \begin{vmatrix} 0 & 0 & 0 & \sqrt{1/2} & 0 & 0 \\ 0 & 0 & \sqrt{1/2} & 0 & 0 & 0 \\ 0 & \sqrt{1/2} & 0 & 0 & 0 & \sqrt{1/2} \\ \sqrt{1/2} & 0 & 0 & 0 & \sqrt{1/2} & 0 \\ 0 & 0 & 0 & \sqrt{1/2} & 0 & 0 \\ 0 & 0 & \sqrt{1/2} & 0 & 0 & 0 \end{vmatrix} \tag{14.36}$$

---

12. Part of the calculation is carried out in problem 10.2.

13. The "direct" or "Kronecker" product of two matrices A and B yields a matrix of elements each of which is the product of one element of A and one element of B. This definition of matrix product is implied in (14.33).

Similarly, we have

$$(m', m_s') = \quad (1\tfrac{1}{2}) \quad (1\,\text{-}\tfrac{1}{2}) \quad (0\tfrac{1}{2}) \quad (0\,\text{-}\tfrac{1}{2}) \quad (\text{-}1\tfrac{1}{2}) \quad (\text{-}1\,\text{-}\tfrac{1}{2})$$

$$2|\langle m\,\frac{l_y}{\hbar}\,|m'\rangle\langle m_s|\frac{s_y}{\hbar}\,|m_s'\rangle| = \begin{vmatrix} 0 & 0 & 0 & -\sqrt{\tfrac{1}{2}} & 0 & 0 \\ 0 & 0 & \sqrt{\tfrac{1}{2}} & 0 & 0 & 0 \\ 0 & \sqrt{\tfrac{1}{2}} & 0 & 0 & 0 & -\sqrt{\tfrac{1}{2}} \\ -\sqrt{\tfrac{1}{2}} & 0 & 0 & 0 & \sqrt{\tfrac{1}{2}} & 0 \\ 0 & 0 & 0 & \sqrt{\tfrac{1}{2}} & 0 & 0 \\ 0 & 0 & -\sqrt{\tfrac{1}{2}} & 0 & 0 & 0 \end{vmatrix} \tag{14.37}$$

The sum of the matrices (14.36) and (14.37) and of the matrix with the elements (14.34) yields the desired matrix

$$|\langle m \;\; m_s|\frac{2\vec{l}\cdot\vec{s}}{\hbar^2}|m' \;\; m_s'\rangle| = \begin{vmatrix} 1 & 0 & 0 & 0 & 0 & 0 \\ 0 & -1 & \sqrt{2} & 0 & 0 & 0 \\ 0 & \sqrt{2} & 0 & 0 & 0 & 0 \\ 0 & 0 & 0 & 0 & \sqrt{2} & 0 \\ 0 & 0 & 0 & \sqrt{2} & -1 & 0 \\ 0 & 0 & 0 & 0 & 0 & 1 \end{vmatrix} \tag{14.38}$$

An important simplification results from the fact that the nonzero elements of the matrix (14.38) are concentrated in the four blocks along the diagonal. While equation (14.31) appears to represent, for $l = 1$, a system of six equations corresponding to the six pairs of values of (m, m_s) with $|m| \leq 1$ and $m_s = \pm\tfrac{1}{2}$, it actually breaks up into four separate systems owing to the structure of the matrix (14.38). Each system corresponds to a fixed value of $m + m_s = m_j$ (indeed, the structure of (14.38) reflects the fact that $\vec{l}\cdot\vec{s}$ is invariant under coordinate rotations about the z axis and is accordingly compatible with j_z.) Thus we shall determine separate stationary states which are eigenstates of j_z, in accordance with sections 14.2.1 and 14.3, and we may replace their label η by the appropriate value of the quantum number m_j.

Substitution of the matrices (14.38) and (14.32) into (14.31) reduces this system to the explicit form

$$W(1 + 2\,\frac{B}{B_0}\,)\langle 1 \;\; \tfrac{1}{2}|E \;\; \tfrac{3}{2}\rangle = \langle 1 \;\; \tfrac{1}{2}|E \;\; \tfrac{3}{2}\rangle E_{3/2}, \tag{14.39a}$$

$$\left.\begin{aligned} &W[-\;\langle 1 \;-\tfrac{1}{2}|E \;\; \tfrac{1}{2}\rangle + \sqrt{2}\;\langle 0 \;\; \tfrac{1}{2}|E \;\; \tfrac{1}{2}\rangle] = \langle 1 \;-\tfrac{1}{2}|E \;\; \tfrac{1}{2}\rangle E_{1/2}, \\ &W[\sqrt{2}\;\langle 1 \;-\tfrac{1}{2}|E \;\; \tfrac{1}{2}\rangle + \frac{B}{B_0}\;\langle 0 \;\; \tfrac{1}{2}|E \;\; \tfrac{1}{2}\rangle] = \langle 0 \;\; \tfrac{1}{2}|E \;\; \tfrac{1}{2}\rangle E_{1/2}, \end{aligned}\right\} \tag{14.39b}$$

$$\left.\begin{aligned} W\Big[-\frac{B}{B_0}\langle 0\ -\tfrac{1}{2}|E - \tfrac{1}{2}\rangle + \sqrt{2}\ \langle -1\ \tfrac{1}{2}|E\ -\tfrac{1}{2}\rangle\Big] &= \langle 0\ -\tfrac{1}{2}|E\ -\tfrac{1}{2}\rangle E_{-1/2}, \\ W[\sqrt{2}\ \langle 0\ -\tfrac{1}{2}|E\ -\tfrac{1}{2}\rangle\ - \langle -1\ \tfrac{1}{2}|E\ -\tfrac{1}{2}\rangle] &= \langle -1\ \tfrac{1}{2}|E\ -\tfrac{1}{2}\rangle E_{-1/2}, \end{aligned}\right\} \tag{14.39c}$$

$$W\left(1 - 2\frac{B}{B_0}\right)\langle -1\ -\tfrac{1}{2}|E\ -\tfrac{3}{2}\rangle = \langle -1\ -\tfrac{1}{2}|E - \tfrac{3}{2}\rangle E_{-3/2}. \tag{14.39d}$$

The two single equations (14.39a) and (14.39d) show that the states $|n\ l\ m\ m_s\rangle$ with $l = 1$ and with $m + m_s = \pm\ 3/2$ are stationary not only in the presence of spin-orbit coupling—as found in section 14.2.1—but also in the presence of an external field B. The energy of these two states,

$$E_{\pm 3/2} = W\left(1 \pm 2\frac{B}{B_0}\right), \tag{14.40}$$

equals the sum of (14.17) and of (14.24) for $l = 1, j = 3/2$, and $m_j = \pm 3/2$, regardless of the magnitude of B/B_0.

The system of two equations (14.39b) has two solutions, with eigenvalues

$$E_{1/2} = \tfrac{1}{2}W\left\{-\left(1 - \frac{B}{B_0}\right) \pm \left[8 + \left(1 + \frac{B}{B_0}\right)^2\right]^{1/2}\right\}. \tag{14.41}$$

To each of these eigenvalues corresponds a pair of probability amplitudes which can be expressed in terms of a parameter β by

$$\langle 0\ \tfrac{1}{2}|E\ \tfrac{1}{2}\rangle = \cos\beta, \qquad \langle 1\ -\tfrac{1}{2}|E\ \tfrac{1}{2}\rangle = \sin\beta, \tag{14.42a}$$

or

$$\langle 0\ \tfrac{1}{2}|E\ \tfrac{1}{2}\rangle = -\sin\beta, \qquad \langle 1\ -\tfrac{1}{2}|E\ \tfrac{1}{2}\rangle = \cos\beta, \tag{14.42b}$$

respectively, as was done in (14.10). Here β is determined by the equation

$$\tan 2\beta = \frac{\sqrt{8}}{1 + B/B_0} \tag{14.43}$$

which is equivalent to equation (14.12) for $l = 1, m_j = 1/2$, and $B = 0$. The solutions of (14.39c), with $m_j = -1/2$, are obtained from those of (14.39b) by reversing the signs of B and of all the magnetic quantum numbers.

There are, all together, six eigenvalues, namely, the two values $E_{\pm 3/2}$ given by (14.40), the pair $E_{1/2}$ given by (14.41) with the alternative signs of the square root, and the corresponding pair $E_{-1/2}$. These six values are plotted against the field strength ratio B/B_0 in Figure 14.1. At $B = 0$ the curves converge to the two degenerate levels, W and $-2W$, which are given by (14.17) for $l = 1$.

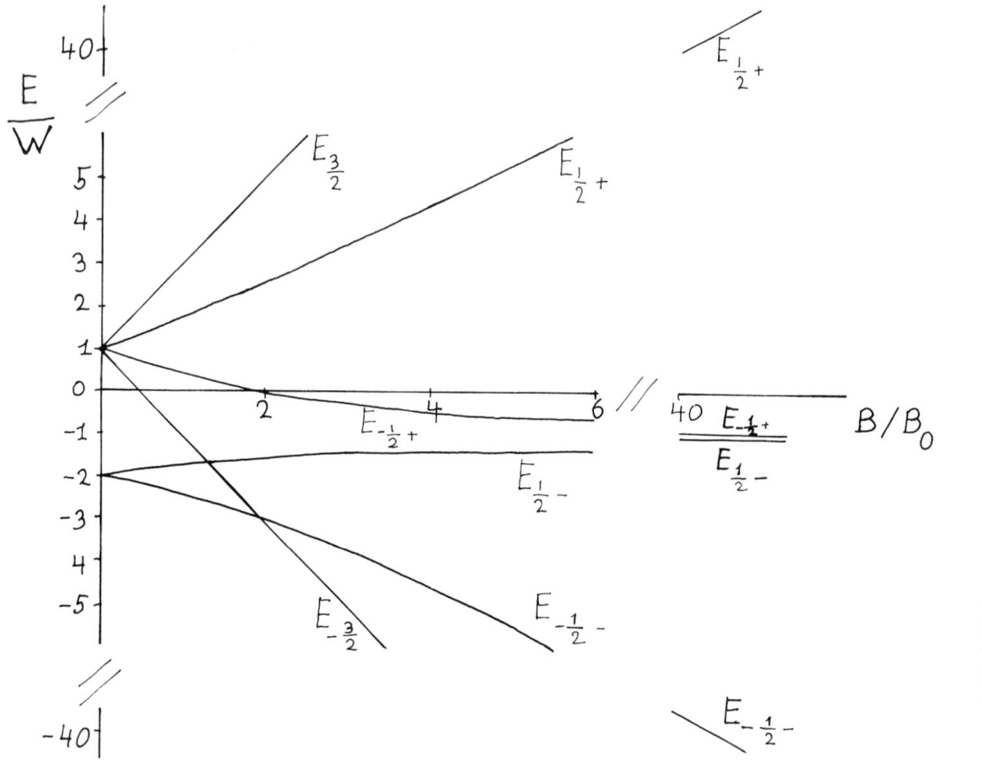

Fig. 14. 1. Magnetic energy levels of atoms with $l = 1$ as functions of the field strength B/B_0. The ± indices correspond to ± signs in equation (14. 41).

For small values of B/B_0 all plots are approximately or exactly linear with the slope predicted by equation (14. 24). For extremely large values of B/B_0 the plots are again linear because the interaction with the external field overcomes the spin-orbit coupling. Under this condition the states $|n\ l\ m\ m_s\rangle$ are approximately stationary and the orbital and spin currents remain independent. Therefore, the anomalous Zeeman effect becomes normal in the high-B limit, in which case it is called Paschen-Back effect.

S14. 5. HYPERFINE STRUCTURE.[14]

Atomic spectra show a still larger number of levels than can be accounted for by the alternative orientations of electron spin and orbital currents (fine structure). The additional *hyperfine structure* of the levels is due to alternative

---

14. See Segrè, *Nuclei and Particles*, pp. 222 ff.

mutual orientations of the electron currents and of currents within the atomic nucleus. Nuclei with nonzero angular momentum have a magnetic moment which interacts with the electron currents. Nuclei often also have an asymmetric charge distribution whose orientation has a small effect on the orbital motion of electrons. These interactions are, in general, weaker than spin-orbit coupling, primarily because the gyromagnetic ratio of nuclear currents is inversely related to the mass of nuclear particles.

The number of energy levels in the hyperfine structure depends on the magnitude of the angular momenta of the nucleus and of the electron. The angular momentum of the nucleus is usually indicated by $\vec{I}$, and the eigenvalues of its squared magnitude $|\vec{I}|^2$ are expressed in terms of a quantum number I. The combined angular momentum of nucleus and electron is indicated by $\vec{F} = \vec{I} + \vec{j}$, and its squared magnitude by a quantum number F. The stationary states of the combined energy are eigenstates of $|\vec{F}|^2$ and of one component F_z for the reasons explained in section 14.2.3.

The possible values of the quantum number F are determined by the procedure of addition of angular momenta and are given by equation (14.19). They range in steps of one from a maximum of j + I to a minimum of $|j - I|$. The energy levels of electron-nucleus interaction are proportional to $[F(F+1) - j(j+1) - I(I+1)]$, in accordance with equation (14.21), when only the magnetic dipole interaction, proportional to $\vec{I} \cdot \vec{j}$, matters.

The splitting of electronic levels which constitutes the hyperfine structure is of the order of magnitude 10^{-5}eV and therefore corresponds to frequencies in the radiofrequency to microwave range. Transitions between hyperfine structure levels of the same electronic state are observed with great precision by the methods of radiofrequency spectroscopy mentioned in section 7.3.2.

CHAPTER 14 PROBLEMS

14.1. A beam of nitrogen atoms in their ground state splits into four components when analyzed by a Stern-Gerlach magnet; the parameter μ ("magnetic moment") is found to be 2.8×10^{-20} erg/gauss (see problems 6.4 and 7.5). Calculate from these data the gyromagnetic ratio of the atoms and thus determine whether their magnetic moment stems from orbital or spin currents.

14.2. The spectral line of mercury at 1850 Å splits, under the influence of a magnetic field of 1000 gauss, into three components separated by 0.0016 Å intervals. Determine whether this Zeeman effect is normal or anomalous, and thereby determine the orbital or spin character of the current affected by the field.

14.3. Calculate the gyromagnetic ratio of the hydrogen atom in states with the following values of the quantum numbers l and j: $(0, \frac{1}{2})$, $(1, \frac{1}{2})$, $(1, \frac{3}{2})$, $(2, \frac{3}{2})$, and $(2, \frac{5}{2})$.

14.4. Boron atoms have a doublet ground state; the two levels of the doublet are $p_{1/2}$ and $p_{3/2}$, that is, they have quantum numbers ($l = 1, j = 1/2$) and ($l = 1$, $j = 3/2$). A beam of boron emerging from an oven contains atoms in all states of the doublet. Analysis of this beam by a Stern-Gerlach magnet splits it into components collected at different points along the z axis of a screen. How many components are there? What are their relative spacings?

14.5. The rubidium atom (isotope with mass number 85) has electronic ground state $s_{1/2}$ and a nucleus with spin quantum number $I = 3/2$. How many hyperfine-structure levels correspond to the electronic ground state, and what are the values of the quantum number F?

14.6. In the absence of an external field, the doublet lines of the rubidium spectrum at 12,578.96 cm^{-1} and 12,816.56 cm^{-1} are emitted in transitions to the ground state ($l = 0, j = 1/2$) from the doublet pair of excited states ($l = 1$, $j = 1/2$) and ($l = 1, j = 3/2$). (a). Calculate the magnetic field strength required to disturb the spin-orbit coupling in the $l = 1$ states to the extent of changing by 1% the energy difference between the two levels with $m_j = 1/2$. (b). Calculate the energies of all levels with $l = 1$ in a magnetic field of 5×10^6 gauss. (c). Calculate the frequencies of all Zeeman components which you would expect to observe, in the field of 5×10^6 gauss, from transitions between levels with $l = 1$ and levels with $l = 0$. Disregard the presence of nuclear spin.

14.7. Consider the magnetic energy of a hydrogen atom in its ground state, in an external field, inclusive of interaction with the magnetic moment of the proton. For a field of strength B along the z axis, the energy is represented by

$$E_{magn} = \hbar\omega_0 \frac{\vec{I}\cdot\vec{s}}{\hbar^2} + \left[g_e \frac{s_z}{\hbar} - g_p \frac{m_e}{M} \frac{I_z}{\hbar}\right] \mu_B B.$$

Here $\vec{s}$ and $\vec{I}$ are the spin angular momenta of the electron and proton, $I = 1/2$, $g_e = 2$, $g_p = 5.59$ is an empirical Landé factor for the proton, $m_e/M = 1/1836$ is the electron-proton mass ratio, and $\omega_0 = 1420$ MHz. (The hyperfine-structure frequency ω_0 is used as a standard of highest precision and its emission serves to detect hydrogen in radio astronomy.) Calculate the eigenvalues of E_{magn} following the method of section S14.4.

14.8. Calculate the electron density distribution over a spherical surface $r =$ const., for states of a hydrogen atom with $l = 1, j = l \pm 1/2$ and with each value of m_j. (The influence of spin-orbit coupling upon the orbital motion smears out the density distribution; the problem consists of evaluating this effect.)

SOLUTIONS TO CHAPTER 14 PROBLEMS

14.1. Splitting of the beam into four components implies that $j = 3/2$ (see section 6.3.1). The gyromagnetic ratio is $\gamma = \mu/j\hbar = 2.8 \times 10^{-20}/(3/2 \times 1.05 \times 10^{-27}) = 1.8 \times 10^7$ radians/sec gauss, that is, twice the gyromagnetic ratio for orbital motion given by equation (2.43). Therefore, the magnetic moment is due to spin currents.

14.2. The fractional wavelength shift of $0.0016/1850 = 1/(1.2 \times 10^6)$ due to the magnetic field implies a corresponding shift of the photon energy, which is 6.7 eV $= 10.7 \times 10^{-12}$ ergs. An energy shift of $10.7 \times 10^{-12}/(1.2 \times 10^6)$ $= 9 \times 10^{-18}$ ergs caused by a field of 1000 gauss implies a difference of 9×10^{-21} erg/gauss = 1 Bohr magneton between successive eigenvalues of μ_z; the Zeeman effect is accordingly normal and the magnetic moment is due to orbital currents.

14.3. Under conditions of the anomalous Zeeman effect, the gyromagnetic ratio is the ratio of the mean value of μ_z, given be equation (14.24) in the form $\langle\mu_z\rangle = -m_j g \mu_B$, to the eigenvalue $j_z = m_j \hbar$. Therefore, $\gamma = -g\mu_B/\hbar = -g(e/2m_e c)$. The values of g are given by equation (14.25) as follows:

l	0	1	1	2	2
j	$1/2$	$1/2$	$3/2$	$3/2$	$5/2$
g	2	$2/3$	$4/3$	$4/5$	$6/5$.

14.4. The number of components is given by the number of eigenstates of j_z with different mean values of $\langle\mu_z\rangle$, under conditions of the anomalous Zeeman effect; their spacings are proportional to the differences between their respective mean values of $\langle\mu_z\rangle$. Atoms with $j = 1/2$ have two eigenstates of j_z with $m_j = 1/2$ and $-1/2$ and $\langle\mu_z\rangle = \mp 1/2\, g\mu_B = \mp 1/2 \times 2/3\, \mu_B = \mp 1/3\, \mu_B$ according to equations (14.24) and (14.25). Atoms with $j = 3/2$ have four eigenstates of j_z with $m_j = 3/2, 1/2, -1/2$, and $-3/2$, with $\langle\mu_z\rangle = \mp m_j g\mu_B = \mp m_j \, 4/3\, \mu_B$. Thus there will be six components with deflections proportional to the following values of $\langle\mu_z\rangle$:

j	$3/2$	$3/2$	$1/2$	$1/2$	$3/2$	$3/2$
m_j	$3/2$	$1/2$	$1/2$	$-1/2$	$-1/2$	$-3/2$
$\langle\mu_z\rangle$	$-2\mu_B$	$-2/3\,\mu_B$	$-1/3\,\mu_B$	$1/3\,\mu_B$	$2/3\,\mu_B$	$2\mu_B$.

14.5. The values of F range in steps of one from $I + j = 3/2 + 1/2 = 2$ down to $|I - j| = 3/2 - 1/2 = 1$. Thus there are two hyperfine levels, with $F = 2$ and $F = 1$.

14.6. (a). The energy difference between the two fine-structure levels at zero field equals the difference between the two line frequencies, 237.60 cm^{-1}, multiplied by hc; it is 2.95×10^{-2} eV. The energy shift in an external field B is $-\langle\mu_z\rangle$ B; equations (14.24) and (14.25) give, for $l = 1$ and $m = 1/2$: $1/3\,\mu_B B$ for $j = 1/2$, and $2/3\,\mu_B B$ for $j = 3/2$. Thus we seek the value of B for which $1/3\,\mu_B B = 10^{-2} \times 2.95 \times 10^{-2}$ eV; this value is 1.5×10^5 gauss. (b). The results of section S14.4 are applicable. The energy difference of the two levels at zero field, 2.95×10^{-2} eV, is represented by $3\,W = 3\mu_B B_0$. Therefore, $B_0 =$

2.95×10^{-2} eV/$3\mu_B$ = 1.70×10^6 gauss; B = 5×10^6 gauss gives $B/B_0 = 2.95$. Equations (14.40) and (14.41), or Figure 14.1, gives

$E_{3/2}$	$E_{1/2+}$	$E_{1/2-}$	$E_{-1/2+}$	$E_{-1/2-}$	$E_{-3/2}$
6.89W	3.39W	−1.46W	−0.26W	−3.69W	−4.89W.

The scale for these levels is identified by the fact that at B = 0, $E_{3/2}$ = W. Therefore, the zero of the scale lies at 12,816.56 − W/hc = 12,737.36 cm^{-1} above the zero-field ground state. (c). The ground state $s_{1/2}$ has the Landé factor g = 2 and is split by the magnetic field into two levels with $m'_j = \pm \frac{1}{2}$ and with energies $\pm\mu_B B = \pm$ 2.95W. The frequency of the transition from $|l = 1, j = \frac{3}{2}, m_j = \frac{3}{2}\rangle$ to $|l' = 0, j' = \frac{1}{2}, m'_j = \frac{1}{2}\rangle$ is given, in cm^{-1}, by $(E_{3/2} - \mu_B B)/hc$ = 12,737 + 6.89W/hc − 2.95W/hc = 13,049 cm^{-1}. The complete array of frequencies for transitions allowed by the selection rule $|m_j - m'_j| \leq 1$ is

$l = 1,\ m_j =$		$\frac{3}{2}$	$\frac{1}{2}$	$\frac{1}{2}$	$-\frac{1}{2}$	$-\frac{1}{2}$	$-\frac{3}{2}$
$l' = 0,\ m'_j =$							
	$\frac{1}{2}$	13,049	12,773	12,389	12,484	12,212	
	$-\frac{1}{2}$		13,240	12,856	12,950	12,679	12,584

14.7. The expression of E_{magn} in this problem differs from equation (14.26) by the replacement of $\vec{l}$ and l_z with $\vec{I}$ and I_z besides changes of coefficients. The matrices of l_x and l_y in equation (14.35) must then be replaced by the matrices of I_x and I_y which have the form (8.29a) and (8.29b) because I = $\frac{1}{2}$. The analogs of matrices (14.36) and (14.38) are

$$(m'_I, m'_S) = \tfrac{1}{2},\tfrac{1}{2} \quad \tfrac{1}{2},-\tfrac{1}{2} \quad -\tfrac{1}{2},\tfrac{1}{2} \quad -\tfrac{1}{2},-\tfrac{1}{2}$$

$$2\left|\langle m_I|\frac{I_x}{\hbar}|m'_I\rangle\right|\left|\langle m_S|\frac{S_x}{\hbar}|m'_S\rangle\right| = \begin{vmatrix} 0 & 0 & 0 & \frac{1}{2} \\ 0 & 0 & \frac{1}{2} & 0 \\ 0 & \frac{1}{2} & 0 & 0 \\ \frac{1}{2} & 0 & 0 & 0 \end{vmatrix}$$

$$\left|\langle m_I\, m_S|\frac{2\vec{I}\cdot\vec{s}}{\hbar^2}|m'_I\ \ m'_S\rangle\right| = \begin{vmatrix} \frac{1}{2} & 0 & 0 & 0 \\ 0 & -\frac{1}{2} & 1 & 0 \\ 0 & 1 & -\frac{1}{2} & 0 \\ 0 & 0 & 0 & \frac{1}{2} \end{vmatrix}$$

The system of four linear equations corresponding to equation (14.27) splits into two uncoupled equations, for $m_I + m_S = \pm 1$, analogous to (14.39a and d) and one system of coupled equations. The uncoupled equations give

$E_{\pm 1} = \frac{1}{4}h\omega_0 \pm \frac{1}{2}(g_e - g_p\, m_e/M)\, \mu_B B$. The system is

$$\left[-\tfrac{1}{4}\hbar\omega_0 - \tfrac{1}{2}(g_e + g_p\, m_e/M)\, \mu_B B\right]\langle \tfrac{1}{2}\ \ -\tfrac{1}{2}|E\ \ 0\rangle + \tfrac{1}{2}\hbar\omega_0\langle -\tfrac{1}{2}\ \ \tfrac{1}{2}|E\ \ 0\rangle = \langle \tfrac{1}{2}\ \ -\tfrac{1}{2}|E\ \ 0\rangle\, E_0,$$

$$\tfrac{1}{2}\hbar\omega_0\langle \tfrac{1}{2}\ \ -\tfrac{1}{2}|E\ \ 0\rangle + \left[-\tfrac{1}{4}\hbar\omega_0 + \tfrac{1}{2}(g_e + g_p\, m_e/M)\, \mu_B B\right]\langle -\tfrac{1}{2}\ \ \tfrac{1}{2}|E\ \ 0\rangle = \langle -\tfrac{1}{2}\ \ \tfrac{1}{2}|E\ \ 0\rangle\, E_0,$$

and yields two energy eigenvalues

$$E_0 = -\tfrac{1}{4}\hbar\omega_0 \pm \left[(\tfrac{1}{2}\hbar\omega_0)^2 + \tfrac{1}{4}(g_e + g_p\, m_e/M)^2(\mu_B B)^2\right]^{1/2}.$$

Comparison of this formula with spectral data provides accurate determination of its parameters, particularly of g_p.

14.8. Each state $|j\ \ m_j\rangle$ is represented by equation (14.10) as the superposition of states with independent orbital and spin motions. Since our problem pertains only to orbital motion, it is solved by averaging over the different spin states. Therefore, equation (14.10a) implies that the electron density is given as a function of θ and ϕ by $|\langle\theta\ \ \phi|l\ \ m_j - \frac{1}{2}\rangle|^2 \cos^2\beta +$ $|\langle\theta\ \ \phi|l\ \ m_j + \frac{1}{2}\rangle|^2 \sin^2\beta$ for $j = l + \frac{1}{2}$. For $j = l - \frac{1}{2}$, the coefficients of $\cos^2\beta$ and $\sin^2\beta$ must be interchanged according to equation (14.10b). Explicit expressions of $\langle\theta\ \ \phi|l\ \ m\rangle$ are given in section 13.2 and the values of β are given by equation (14.12). Thus we find, for $l = 1$,

$$|\langle\theta\ \ \phi|j = \tfrac{3}{2}, m_j = \tfrac{3}{2}\rangle|^2 = |\langle\theta\ \ \phi|l = 1, m = 1\rangle|^2 = (3/8\pi)\sin^2\theta,$$

$$|\langle\theta\ \ \phi|j = \tfrac{3}{2}, m_j = \tfrac{1}{2}\rangle|^2 = |\langle\theta\ \ \phi|l = 1, m = 0\rangle|^2 \times \tfrac{2}{3} + |\langle\theta\ \ \phi|l = 1, m = 1\rangle|^2 \times \tfrac{1}{3} = (\tfrac{1}{8\pi})(1 + 3\cos^2\theta),$$

$$|\langle\theta\ \ \phi|j = \tfrac{1}{2}, m_j = \tfrac{1}{2}\rangle|^2 = |\langle\theta\ \ \phi|l = 1, m = 0\rangle|^2 \times \tfrac{1}{3} + |\langle\theta\ \ \phi|l = 1, m = 1\rangle|^2 \times \tfrac{2}{3} = \tfrac{1}{4\pi}.$$

Replacement of m_j by $-m_j$ does not change the density distribution. Note that the distribution is uniform for the $p_{1/2}$ state ($l = 1, j = \frac{1}{2}$), as it is for $s_{1/2}$ states ($l = 0, j = \frac{1}{2}$).

15. identical particles and the exclusion principle

The ability of condensed matter to withstand pressure is not accounted for completely by the stability of atomic electrons under nuclear attraction. Resistance to pressure also implies that atoms have the property of withstanding interpenetration. Moreover, the resistance to interpenetration must operate not only between atoms but also between the electrons of a single atom; if it didn't, the increasing nuclear charge of successive atoms along the periodic system of elements would draw their electrons into decreasing volumes, contrary to evidence.

The stability of atomic electrons under nuclear attraction is interpreted in chapter 11 as a reaction of electrons to confinement. The stability against interpenetration derives from another property of electrons formulated by Pauli and called the *exclusion principle*. In brief, once an electron is confined within a certain region of space it tends to exclude from that space any other electron whose spin current is not opposite to its own. More specifically, as we shall see, when two electrons with parallel spin orientation are forced within the same region of space, their kinetic energy increases. In this respect the reaction to interpenetration resembles the reaction to confinement.

The exclusion principle is an outstanding consequence of a more general property of systems of identical particles. Exclusion itself applies to any system of identical particles with spin angular momentum quantum number $s = 1/2$; for example, it applies to protons and neutrons. Identical particles with zero spin angular momentum—for example, α-particles and helium atoms—are not subject to the exclusion principle. Instead, the positions and motions of these identical particles are so correlated as to produce the outstanding phenomena of superfluidity of liquid helium and superconductivity. The study of these phenomena exceeds the scope of this book, but the general properties of identical particles with and without spin angular momentum are relevant to the rotation of molecules (section 20.2.4). Here we deal mostly with the exclusion principle owing to its dominant influence on the structure of matter.

Alternative formulations of the exclusion principle arise in different contexts. The simplest and most familiar formulation springs directly from the gross properties of light atoms, but it is dependent on the use of an approximate model (section 15.1). The general formulation follows from an analysis of the relative motion of two identical particles, but it is not readily applied to many-particle systems (section 15.2). Both of these formulations are embodied in a more flexible one which involves, however, an artificial mathematical representation (section 15.3).

The interpretation of the spectrum of helium was an early outstanding success of the combination of quantum mechanics and the exclusion principle. Section 15.4 outlines this interpretation as an application of the exclusion principle and as an introduction to the study of many-electron atoms.

15.1 EVIDENCE FROM LIGHT ATOMS

Evidence on the exclusion principle for electrons is provided by the trend of properties of atoms along the periodic system. It is sufficient to consider the ground state of atoms of the first five or six elements, and specifically their radii, their ionization potentials, and their magnetic properties. The radii and the ionization potentials are inversely related to one another, as it is to be expected since a small radius implies a tight binding of the electrons.

The magnetic properties of an atom depend on its orbital and spin currents and on their combinations. Orbital currents occur only in excited states of rotational motion; for example, none exists in the ground state of the hydrogen atom. However, each electron carries a spin current and angular momentum. The spin angular momenta of two or more electrons may combine into alternative eigenstates of their squared total spin momentum $\vec{S}$,

$$|\vec{s}_1 + \vec{s}_2 + \cdots|^2 = |\vec{S}|^2. \tag{15.1}$$

The possible eigenvalues of $|\vec{S}|^2$, identified by a quantum number S,[1] are determined by the procedure of addition of angular momenta outlined in section 14.2.3. Equation (14.19) yields for the addition of two electron spins $S = s_1 \pm s_2 = \frac{1}{2} \pm \frac{1}{2}$, that is, $S = 1$ or 0. Eigenstates with $S = 1$ or $S = 0$ are loosely said to have parallel or antiparallel spins; they are also often labeled by the prefixes *ortho* and *para*, respectively. The addition of more than two electron spins proceeds stepwise by finding the possible values of S first for two, then for three electrons, etc.

Consider now the properties of successive atoms in the periodic system. Compared with hydrogen, helium has a small radius and a high ionization potential, 24.6 eV against 13.6 eV. This property can be attributed to the stronger attraction by the nucleus compensated only in part by the repulsion between the two electrons. A beam of helium atoms in their ground state remains undeflected in Stern-Gerlach experiments, showing that the electron pair has $S = 0$. No state of helium with $S = 1$ (parallel spins) exists whose binding energy is comparable to that of the ground state. The lowest level with $S = 1$ has a binding energy of 4.8 eV as compared to 24.6 eV for the ground state. Its energy is only 0.8 eV lower than that of the first excited level with $S = 0$.

The three-electron atom, lithium, is larger than He and also larger than H and has a lower ionization potential, 5.4 eV. It behaves like hydrogen in Stern-Gerlach experiments; that is, it exhibits in the ground state the spin current of a single electron and no orbital current. Thus the nuclear attraction appears unable to pack three electrons into as small a space as two electrons. Also, two out of three electron spin currents cancel out in the ground state. The Li^+ ion is quite similar to the He atom, only smaller; its radius is approximately three times smaller than that of the neutral Li atom. Thus the neutral Li

---

1. Capital letters are commonly used to indicate variables and quantum numbers pertaining to a system of two or more particles.

appears to consist of a small helium-like Li^+ core and of a third loosely bound electron.

The four-electron atom, beryllium, has higher ionization potential than Li and remains undeflected in magnetic experiments just like He, thus showing zero angular momentum. Boron (five electrons) has a slightly lower ionization potential than Be. A beam of B atoms is resolved by a Stern-Gerlach magnet into two components, like a beam of H or Li. However, the gyromagnetic ratio of the ground state of B is much smaller than the value $-e/m_ec$ which pertains to an electron spin current. Its value, $-e/3m_ec$, corresponds to an orbital current with $l = 1$ coupled with a single electron spin. Thus the ground state of boron shows evidence of an electron with excited rotational motion. There is no need to continue this survey since the properties described thus far are sufficient for our purposes.

These properties are readily organized in an *independent electron approximation* which regards each electron as moving under the combined influence of the nuclear attraction and of the average repulsion by the other electrons. Each electron is initially taken to be in a stationary state with quantum numbers ($n \;\; l \;\; m \;\; m_s$), just as if it were in a hydrogen atom without consideration of spin-orbit coupling. In this approximation each electron has two possible ground states, with $n = 1, l = m = 0$, and $m_s = \pm\frac{1}{2}$. The He atom in the ground state is then loosely regarded as having one electron in each of these two states. (If both electrons had the same value of m_s, the resultant S value would be unity, contrary to evidence.[2]) Since all states of He with $S = 1$ have energies much higher than that of the ground state, one of the electrons must be excited with quantum number $n \geqslant 2$ in these states.

The larger radius and lower ionization potential of Li in its ground state imply that one of its electrons is in a state with $n > 1$. (Note that in the independent electron approximation we may assign an electron to an excited state even though the atom as a whole is in its ground state.) The absence of orbital current shows that the excited electron has $l = m = 0$ and, therefore, that its excitation is radial. The lowest excited state with these specifications has $n = 2$. The properties of Be in its ground state are accounted for by assigning to its four electrons the quantum numbers

n	l	m	m_s
1	0	0	$\frac{1}{2}$
1	0	0	$-\frac{1}{2}$
2	0	0	$\frac{1}{2}$
2	0	0	$-\frac{1}{2}$.

2. Actually, identification of the value of S requires a superposition of independent-electron states. This will be done in equation (15. 5).

The occurrence of rotational excitation in the boron ground state shows that one electron must have quantum number $l = 1$.

This observation completes the essential evidence for the formulation of the exclusion principle in the language of the independent electron approximation: "No two electrons have the same set of four quantum numbers." In the ground state of an atom the electrons have the lowest quantum numbers consistent with this rule.

15.2. STATES OF TWO IDENTICAL PARTICLES

The characteristic property of systems of identical particles emerges from the study of isolated pairs of particles. A pair is considered isolated insofar as its center of mass is not subject to external forces. Hence we do not consider here a pair of atomic electrons because the motion of their center of mass is influenced by nuclear attraction. Collisions between two identical particles, such as free electrons, helium atoms, or α-particles, are suitable for our study. So are molecules consisting of two identical atoms since the nuclei rotate freely about their center of mass and are practically undisturbed by the inertia of the electrons.

Consider first a system consisting of a pair of identical particles without any spin angular momentum. The position of the system with respect to its center of mass is identified by the interparticle distance r and by the polar coordinates Θ and Φ of the interparticle axis.[3] However, the range of Θ and Φ extends only over a half sphere since opposite directions of the interparticle axis are equivalent. Therefore, the mutual position of identical particles has, loosely speaking, half as many eigenstates as exist for a pair of distinguishable particles, or for the position of a single particle. We choose Θ to run from 0 to $\pi/2$ (upper hemisphere) while Φ runs from 0 to 2π, as usual.

Experimentally, the identity of particles has a striking influence on collision experiments. For instance, when an α-particle is scattered by a nucleus of helium gas, the nucleus recoils and two particles emerge from each collision. It is inherently undefined which of the emerging particles is the incident one and which is the recoiling one. The two particles emerge in perpendicular directions, as shown in Figure 15.1.[4] The two counters at angles θ and θ' score simultaneously. If a single counter is used, it will score at equal rates at θ and at $\theta' = 90° - \theta$. This means that a plot of the scattering cross-section is symmetric with respect to $\theta = 45°$. Since the deflection Θ in the center-of-

---

3. Directions in a center-of-mass system will be indicated by capital letters.

4. This fact results from the conservation of momentum in a center-of-mass frame, followed by transformation of the motion from the center-of-mass frame to the laboratory frame (see *BPC*, 1: 169 ff.). It also follows that the laboratory system deflection θ equals one-half the deflection Θ in the center-of-mass system.

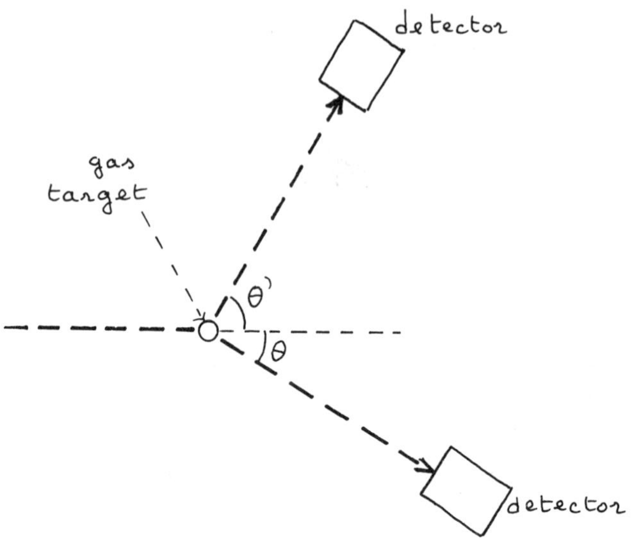

Fig. 15. 1. Schematic arrangement to observe collisions of identical particles.

mass frame is twice as large as θ,[5] the symmetry of the plot is a consequence of the equivalence of directions Θ and $\pi - \Theta$ in the center-of-mass frame. Note for comparison that the Rutherford cross-section (1. 32) for the scattering of distinguishable particles decreases steadily as Θ increases to 180°. An example of the symmetry in α-particle scattering is shown in Figure 15. 2.

The theoretical analysis of the phenomenon goes as follows. The state of a pair of colliding particles is discussed with reference to its orbital angular momentum $\vec{L}$ of rotation about the center of mass. The state of the colliding pair is an eigenstate of the component L_z in the direction of incidence, with eigenvalue $L_z = 0$ as there is no rotation about this direction. This state is, however, not an eigenstate of $|\vec{L}|^2$ but should rather be represented as a superposition of such eigenstates as noted in section S13. 6. The eigenfunctions of $|\vec{L}|^2$ with $L_z = 0$ representing rotational motion along meridians are indicated by $\langle \Theta | L\ 0 \rangle$ and are given by equation (13. 19). Each of these eigenfunctions yields equal particle density in the directions Θ and $\pi - \Theta$, which are fully equivalent for identical particles. The sign of each of the wave functions is, however, equal or opposite in these two equivalent directions depending on whether L is even or odd. The essential point is that superposition of eigenstates with even and odd values of the quantum number L would yield interference effects which are asymmetric with respect to $\Theta = 90°$ and therefore with respect to $\theta = 45°$ in the laboratory frame. Superposition of eigenstates with odd and even L is thus to be excluded as inconsistent with the indistinguishability of particles.

5. See p. 333, n. 4.

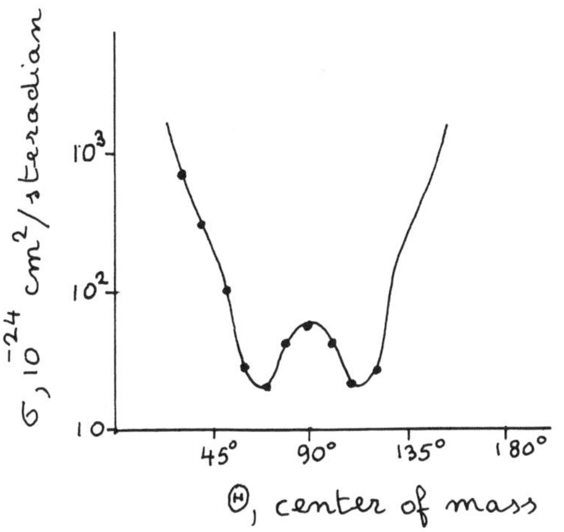

Fig. 15. 2. Cross-section, σ, for α-particle scattering on helium at 150 keV. Solid line indicates theoretical curve. [Adapted from N. P. Heydenburg and G. M. Temmer, *Phys. Rev.* 104 (1957): 123.]

Superposition of states with odd L would yield a dark fringe at $\Theta = 90°$; superposition of states with even L yields a bright fringe at $\Theta = 90°$. Figure 15. 2 clearly shows a bright fringe centered at $\Theta = 90°$ and thus shows that *only even values of L occur for a pair of α-particles*. Thus, in this case, not only the particle density but the wave function itself is identical for the equivalent values of Θ. These conclusions apply to all pairs of identical particles without any spin angular momentum, as illustrated in Figure 15. 3 for colliding pairs of helium atoms.

As regards pairs of identical particles with nonzero spin, their states are discussed not only with reference to the orbital angular momentum $\vec{L}$ but also with reference to the combined spin angular momentum $\vec{S}$ of the pair. In fact, the identity of particles comes fully to the fore for eigenstates of $|\vec{S}|^2$; control of the spin orientation of the individual particles would make them, in effect, distinguishable.[6]

The arguments concerning the symmetry of the scattering cross-section about the laboratory angle $\theta = 45°$ apply to eigenstates of $|\vec{S}|^2$ irrespective of the value of S. Experimentally, the symmetry of the cross-section plots is verified in all cases. However the experimental plots for particle pairs with nonzero spin do not show as clear a pattern of fringes as those in Figures 15. 2 and 15. 3, because different eigenstates of $|\vec{S}|^2$ usually contribute to the scattering cross-section. It is nevertheless determined that a *bright* fringe occurs at $\theta = 45°$ for all *even* values of S, a *dark* fringe for all *odd* values of S. Since

---

6. States of a pair in which the two particles have definite spin orientation are represented as superpositions of eigenstates of $|\vec{S}|^2$.

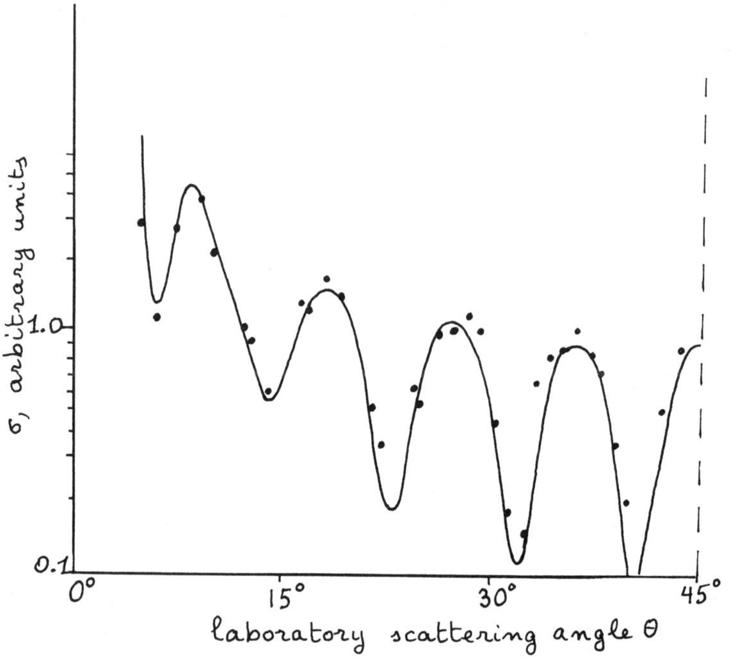

Fig. 15. 3. Cross-section for scattering of He atoms of thermal energy by helium gas. Solid line indicates theoretical curve. The oscillations of the cross-section are due to details of the interaction between pairs of He atoms. However, as in the case of α particles, the cross-section has a maximum at 45° in the laboratory frame. [Adapted from P. E. Siska et al., *J. Chem. Phys.* (**1971**).]

we know that a bright fringe at $\theta = 45°$ derives from even values of L, and a dark fringe from odd values, we conclude that even values of L and S go together and so do odd values. That is, all states of a pair of identical particles which are eigenstates of $|\vec{L}|^2$ and $|\vec{S}|^2$ obey the rule

$$L + S = \text{even}. \tag{15.2}$$

Direct evidence of this rule is provided by the experimental study of rotational energy levels of molecules with identical nuclei. The characteristics of these levels result in striking macroscopic properties of simple substances such as hydrogen, especially at very low temperatures, as will be discussed in chapter 20.

In the particular case of S = 1, which occurs for a pair of electrons with parallel spin, the rule (15.2) excludes the value L = 0 of the orbital quantum number of the pair.[7] Thus it enforces a nonzero rotational motion characterized by $L \geqslant 1$ and by a nonzero rotational kinetic energy

$$K_{rot, L} = \frac{L(L + 1)\hbar^2}{2\bar{m}r^2} \tag{15.3}$$

whose centrifugal effect keeps the electrons apart. This formula has the structure of equation (13.21); $\bar{m}$ now indicates the reduced mass of the identical particles and r their distance apart. It is this centrifugal effect which prevents the interpenetration of atoms and which provides the basis for the exclusion principle.

15.3. WAVE FUNCTIONS FOR IDENTICAL PARTICLES

For a pair of identical particles, the limitation to the eigenvalues of $|\vec{L}|^2$ and $|\vec{S}|^2$ expressed by (15.2) constitutes a formulation of the exclusion principle in terms of observable quantities. An equivalent but more flexible formulation specifies a property of the wave functions of a system of identical particles with spin quantum number $\frac{1}{2}$. It is convenient and usual to express these wave functions in a form that would be appropriate if the particles were not identical. The effects of this unrealistic assumption are subsequently compensated by applying a restriction to the wave function or other representation of a state of the system.

Let separate position coordinates $\vec{r}_1, \vec{r}_2, \ldots$, and spin orientations m_{s1}, $m_{s2}, \ldots$, be assigned to individual electrons, as though position eigenstates of a pair of electrons $|\vec{r}_1, \vec{r}_2\rangle$ and $|\vec{r}_2, \vec{r}_1\rangle$ were actually different. Thereby one increases artificially the number of different states of the system of electrons only to decrease it again at a later stage of the treatment. (This artifice is analogous to the assignment of a phase to each probability amplitude, even though only phase differences are observable; see section 8.4.) We shall first write the wave function for the motion of a pair of electrons in their center-of-mass system as though they were distinguishable. We will then single out a symmetry property of states that obey the rule (15.2). This symmetry property is equally applicable to states of any number of identical particles regardless of the motion of the center-of-mass of each pair.

Consider a state $|a\rangle$ of a pair of electrons with the following characteristics:

1) It is an eigenstate of the motion of the center of mass, $\vec{R} = \frac{1}{2}(\vec{r}_1 + \vec{r}_2)$, with momentum $\vec{P}$.

2) It is an eigenstate of the kinetic energy E_r of the relative "radial" motion of the two electrons along the line that joins them; the distance between the two electrons is represented by $r = |\vec{r}_1 - \vec{r}_2|$.

---

7. Recall that the quantum number L refers here to the motion of a pair of particles about the center of mass of the pair.

3) It is an eigenstate of the rotational motion of the pair about the center of mass with quantum numbers L and M; the direction of $\vec{r}_1 - \vec{r}_2$ is represented by polar coordinates Θ and Φ.

4) It is an eigenstate of the combined spin angular momentum of the electron pair with quantum numbers S and M_S.

The wave function of this state is represented by

$$\langle \vec{r}_1\ m_{s1}, \vec{r}_2\ m_{s2}\,|a\rangle = \langle \vec{R}\,|\,\vec{P}\rangle\ \langle r\,|E_r\rangle\ \langle \Theta\ \ \Phi\,|L\ \ M\rangle\ \langle m_{s1}\ \ m_{s2}\,|S\ \ M_S\rangle. \tag{15.3}$$

We investigate here how this wave function is modified by a permutation of the two electrons, that is, by interchanging their position coordinates $\vec{r}_1$ and $\vec{r}_2$ and their spin orientation coordinates m_{s1} and m_{s2}.

The factor $\langle \vec{R}\,|\,\vec{P}\rangle$ of equation (15.3) remains unchanged by the permutation, since $\vec{R}$ is the half-sum of $\vec{r}_1$ and $\vec{r}_2$. The factor $\langle r\,|E_r\rangle$ is also unchanged since r represents the distance of the two electrons. The factor $\langle \Theta\ \ \Phi\,|L\ \ M\rangle$, whose form is given by equations (13.23), (13.22) and (13.14), is altered by the substitutions $\Theta \to \pi - \Theta, \Phi \to \Phi + \pi$. The first of these substitutions multiplies the wave function by $(-1)^{L-M}$, the second by $(-1)^M$; together, they yield a factor $(-1)^L$.

The factor $\langle m_{s1}\ \ m_{s2}\,|S\ \ M_S\rangle$ must be written in explicit form to determine the effect of permutation. This factor is expressed in terms of single-electron spin probability amplitudes of the type $\langle m_s\,|\,\tfrac{1}{2}\rangle$. This probability amplitude equals unity for $m_s = \tfrac{1}{2}$ and zero for $m_s = -\tfrac{1}{2}$. For the largest value of M_S, namely, $M_S = 1$, both electrons have $m_s = \tfrac{1}{2}$ ("spin up"). Therefore, the electron-pair probability amplitude is

$$\langle m_{s1}\ \ m_{s2}\,|S = 1, M_S = 1\rangle = \langle m_{s1}\,|\,\tfrac{1}{2}\rangle\ \langle m_{s2}\,|\,\tfrac{1}{2}\rangle. \tag{15.4}$$

For $M_S = 0$ there are two possible combinations with $m_{s1} + m_{s2} = 0$ (one spin up and one spin down). With these two combinations one constructs two eigenstates of $|\vec{S}|^2$ with $S = 1$ and $S = 0$ by the procedure of addition of angular momenta of section 14.2.3. The two states are represented by the general formula (14.20). In our problem this equation takes the same form as (14.10) with the following replacements:

$$l \to s_1 = \tfrac{1}{2};\qquad s \to s_2 = \tfrac{1}{2};\quad m \to m_{s1} = \pm\tfrac{1}{2};$$
$$m_s \to m_{s2} = \pm\tfrac{1}{2};\quad j \to S;\qquad m_j \to M_S = 0.$$

Thereby the electron-pair eigenfunctions are given by

$$\langle m_{s1}\ \ m_{s2}\,|S = 1, M_S = 0\rangle = \langle m_{s1}\,|-\tfrac{1}{2}\rangle\langle m_{s2}\,|\,\tfrac{1}{2}\rangle\sqrt{\tfrac{1}{2}} + \langle m_{s1}\,|\,\tfrac{1}{2}\rangle\langle m_{s2}\,|-\tfrac{1}{2}\rangle\sqrt{\tfrac{1}{2}}, \tag{15.5a}$$

$$\langle m_{s1}\ \ m_{s2}\,|S = 0, M_S = 0\rangle = -\langle m_{s1}\,|-\tfrac{1}{2}\rangle\langle m_{s2}\,|\,\tfrac{1}{2}\rangle\left(-\sqrt{\tfrac{1}{2}}\right) + \langle m_{s1}\,|\,\tfrac{1}{2}\rangle\langle m_{s2}\,|-\tfrac{1}{2}\rangle\sqrt{\tfrac{1}{2}}. \tag{15.5b}$$

Finally, we have

$$\langle m_{s1}\ m_{s2}\,|S = 1, M_S = -1\rangle = \langle m_{s1}\,|-\tfrac{1}{2}\rangle\,\langle m_{s2}\,|-\tfrac{1}{2}\rangle. \tag{15.6}$$

Permutation of m_{s1} and m_{s2} leaves the eigenfunctions with $S = 1$ unchanged and reverses the sign for $S = 0$. That is, the permutation has the effect of multiplying all eigenfunctions by $(-1)^{S+1}$:

$$\langle m_{s2}\ m_{s1}\,|S\ M_S\rangle = (-1)^{S+1}\,\langle m_{s1}\ m_{s2}\,|S\ M_S\rangle. \tag{15.7}$$

Pulling all these results together, we have

$$\langle \vec{r}_2\ m_{s2}, \vec{r}_1\ m_{s1}\,|a\rangle = (-1)^{L+S+1}\,\langle \vec{r}_1\ m_{s1}, \vec{r}_2\ m_{s2}\,|a\rangle. \tag{15.8}$$

The rule (15.2), $L + S =$ even, is thus equivalent to the statement: *Permutation of the complete sets of coordinates of two electrons—including the spin coordinates—reverses the sign of the wave function,*

$$\langle \vec{r}_2\ m_{s2}, \vec{r}_1\ m_{s1}\,|a\rangle = -\,\langle \vec{r}_1\ m_{s1}, \vec{r}_2\ m_{s2}\,|a\rangle. \tag{15.9}$$

In other words, the wave function is antisymmetric under permutation of electron coordinates. The same result would have been obtained for any pair of particles with half-integer values of the spin quantum number s. All these particles are said to "obey Fermi statistics" and are called *fermions*. For any pair of particles with integer spin quantum number s, one finds a rule equivalent to equation (15.7) but with $(-1)^S$ instead of $(-1)^{S+1}$.[8] This substitution causes the negative sign of equation (15.9) to disappear thus requiring the wave function to be symmetric under permutation of all particle coordinates. These particles "obey Bose statistics" and are called *bosons*.

The antisymmetry condition (15.9) constitutes the common formulation of the exclusion principle for a pair of electrons. In particular, it requires the complete wave function to vanish wherever the position and spin coordinates for the two electrons coincide, that is, for $\vec{r}_1 = \vec{r}_2$ and $m_{s1} = m_{s2}$. The reason is that permutation of the electrons must in this case both leave the wave function unchanged and reverse its sign.

In atoms and molecules the center of mass of two electrons does not move freely. Moreover, the state of the pair is not an eigenstate of $|\vec{L}|^2$; it may, however, be an eigenstate of $|\vec{S}|^2$ provided the interaction between spin and orbital currents is negligible. These states are identified by quantum numbers S and M_S while the state of orbital motion may remain unspecified at this point and be indicated by the letter b. The eigenfunctions of these states are

---

8. The basic property of wave functions for a pair of spin orientations is that the function of the type (15.4) with $M_S = S = 2s$ is unchanged by permutation of the particles. The factor $(-1)^{S+1}$ in (15.7) can be expressed as $(-1)^{S+2s}$ for all values of s.

not factored as fully as (15.3) but have the form

$$\langle \vec{r}_1 \; m_{s1}, \vec{r}_2 \; m_{s2} | a \rangle = \langle \vec{r}_1 \; \vec{r}_2 | b \; S \rangle \langle m_{s1} \; m_{s2} | S \; M_S \rangle . \tag{15.10}$$

The complete wave function (15.10) obeys the antisymmetry condition (15.9), and its spin factor has the property (15.7). Therefore, the factor $\langle \vec{r}_1 \; \vec{r}_2 | b \; S \rangle$, which depends on position coordinates only, has the symmetry property

$$\langle \vec{r}_2 \; \vec{r}_1 | b \; S \rangle = (-1)^S \langle \vec{r}_1 \; \vec{r}_2 | b \; S \rangle . \tag{15.11}$$

This symmetry condition requires the wave function to have a node at $\vec{r}_1 = \vec{r}_2$, and thus to represent an excited state, for all states with $S = 1$.

The basic formulation (15.9) of the exclusion principle for a pair of electrons extends to a system of any number of electrons. A complete set of eigenstates of the position and spin orientation of N electrons, labeled as though the electrons were different, is represented by $| \vec{r}_1 \; m_{s1}, \vec{r}_2 \; m_{s2}, \ldots, \vec{r}_N \; m_{sN} \rangle$. This set of eigenstates can also be represented as a product of ket symbols $| \vec{r}_1 \; m_{s1} \rangle | \vec{r}_2 \; m_{s2} \rangle \ldots | \vec{r}_N \; m_{sN} \rangle$. A general state $|a\rangle$ of the system of N electrons may be represented as a superposition of these single-electron eigenstates with probability amplitudes $\langle r_1 \; m_{s1}, r_2 \; m_{s2}, \ldots, r_N \; m_{sN} | a \rangle$. Equation (15.9) requires these probability amplitudes to reverse their sign under permutation of all coordinates of *any two electrons*. This is the general formulation of the exclusion principle. Here again, the condition implies that the wave function vanishes wherever the position and spin coordinates of any two electrons coincide, that is, for $\vec{r}_i = \vec{r}_j$ together with $m_{si} = m_{sj}$.

The initial formulation of the exclusion principle given in section 15.1 in terms of quantum numbers of independent electrons can now be obtained as a special case of the general formulation. The independent electron approximation starts by considering a complete set of states $|n \; l \; m \; m_s\rangle$ of each electron. A complete set of states of an N-electron system is represented by the product of ket symbols.

$$|n_1 \; l_1 \; m_1 \; m_{s1}\rangle |n_2 \; l_2 \; m_2 \; m_{s2}\rangle \ldots |n_N \; l_N \; m_N \; m_{sN}\rangle . \tag{15.12}$$

This product of states need not be a stationary state of the system of N electrons, whereas it was assumed to be stationary in section 15.1. However, any stationary state of the system is a superposition of these ket products with probability amplitudes

$$\langle n_1 \; l_1 \; m_1 \; m_{s1}, n_2 \; l_2 \; m_2 \; m_{s2}, \ldots\ldots, n_N \; l_N \; m_N \; m_{sN} | a \rangle . \tag{15.13}$$

This probability amplitude must be antisymmetric under permutation of the sets of quantum numbers of any two electrons. Accordingly, it must vanish whenever any two such sets coincide, that is, whenever $(n_i, l_i, m_i, m_{si}) = (n_j, l_j, m_j, m_{sj})$. Product states (15.12) in which two electrons have the same set of quantum numbers are thereby excluded from the superposition.

15.4. ORTHO- AND PARA-HELIUM

Analysis of the emission spectrum of helium according to the Rydberg-Ritz combination principle (section 2.3), shows that the helium atom has two distinct sets of spectral terms. Each observed line frequency corresponds to the energy difference between two terms of one set. The spectrum appears thus to be emitted by a mixture of two substances; the two "substances" thus identified by empirical analysis of the spectrum were called *para-helium* and *ortho-helium*. The ground state of helium belongs to the para-helium set of levels and lies nearly 20 eV lower than any other level. Ortho-helium constitutes a metastable form whose conversion into para-helium occurs hardly at all by emission of radiation but rather by less easily observable mechanisms such as collisions among gas atoms or with the container's walls.

The para-helium spectrum exhibits no fine structure and shows a normal Zeeman effect. This means that the spin currents of the two electrons in the para-helium states cancel each other; that is, they combine with resultant spin quantum number S = 0. The ortho-helium spectrum exhibits a fine structure and an anomalous Zeeman effect, which indicate parallel spin currents and a spin quantum number S = 1. That "intercombination lines" emitted in transitions between the two spectra are seldom observed shows that the probability of a spin flip in a radiative transition is extremely low.

Figure 15.4 shows a diagram of some energy levels of the He atom. The

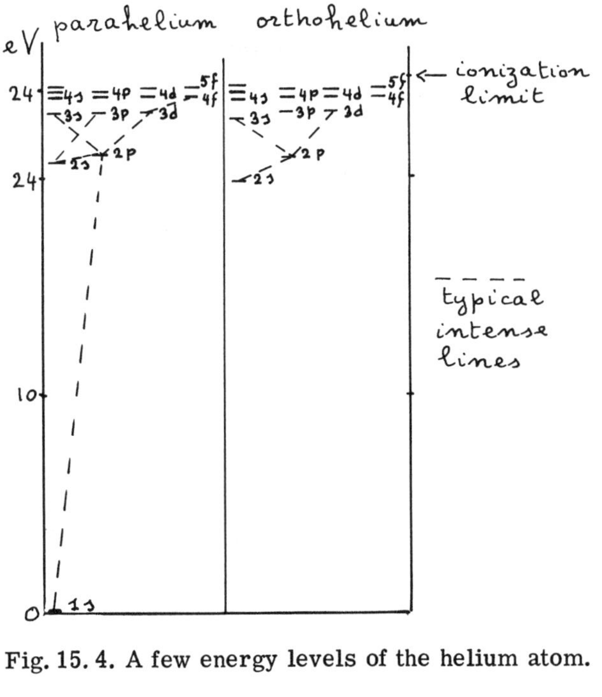

Fig. 15.4. A few energy levels of the helium atom.

dashed lines indicate transitions observed with appreciable intensity in the emission spectrum. The *para* and *ortho* systems of levels are separate, and they are similar except that the deep-lying ground level of para-helium has no counterpart in ortho-helium. In the figure the levels are arranged in columns so that dashed lines join only levels of adjacent columns. The levels of each column constitute a series which converges to the limit of ionization. The fine structure of the ortho-helium spectrum is not shown in Figure 15.4; instead, each group of levels that differ only in their spin-orbit coupling is indicated as a single level.

The subdivision of the helium levels into para and ortho systems and the main characteristics of these systems follow directly from the exclusion principle and particularly from the symmetry property (15.11) of two-electron wave functions. In particular, all ortho-helium wave functions, with $S = 1$, have a node at $\vec{r}_1 = \vec{r}_2$. It follows, in the first place, that ortho-helium has no nodeless wave function and, therefore, no low-energy ground state. Second, given that there is a node, the location of the node at $\vec{r}_1 = \vec{r}_2$ makes a close approach of the two electrons less likely than if the node were somewhere else. Since the electrons repel each other, their potential energy is reduced by increased average distance. Thus the energy of each ortho level is slightly lower than that of the corresponding para level, as seen in Figure 15.4.

Each level of ortho (or para) helium is classified by a pair of quantum numbers (n, l) which characterizes the radial motion and the rotational energy of a single electron attracted toward a center of force, as in the hydrogen atom. The classification implies that one regards the helium atom in its excited states as the combination of a single electron and an He^+ ion. Indeed, He^+ is very small, and its electron density is distributed with spherical symmetry about the nucleus as in the ground state of the hydrogen atom. An electron outside a He^+ ion is therefore attracted toward the center of the ion as though the ion were a hydrogen nucleus. The attraction becomes stronger than in the hydrogen atom when the electron is close to the nucleus of the He^+ ion; it follows that states with equal quantum number n and different l have somewhat different energies, as in the alkali atoms. Notice that the quantum number l is the same for all levels of each column of Figure 15.4 and differs by one in adjacent columns. The classification thus agrees with the selection rule stating that an electron emits light intensely only in transitions between stationary states whose quantum numbers l differ by one (section 13.5).

The description of the He atom as a combination of a He^+ ion and an outer electron can be developed into a systematic approximation theory. The quantitative results of this theory are reasonably accurate for the excited states, less accurate for the ground state. Each electron is initially assumed to move independently of the other one and to be attracted toward the nucleus by a force which is not quite the Coulomb force that prevails in a hydrogen atom. The schematized attraction is represented by a potential energy $U(r)$ which depends only on the distance r from the nucleus. Stationary states of motion of a single electron are then represented by wave functions $\langle \vec{r} | n \; l \; m \rangle$ which differ from the hydrogen wave function only in the radial factor $\langle r | n \; l \rangle$, owing to the difference in the potential energy.

We consider stationary states of the helium atom with one electron in the ground state, ($n = 1, l = m = 0$), and the other one in some excited state with unspecified quantum numbers n, l, and m. The energy of such a state of the complete atom may be indicated in the initial approximation as

$$E_{1,0} + E_{nl}\,. \tag{15.14}$$

Disregarding the spin coordinates and indicating the position coordinates by $\vec{r}_1$ and $\vec{r}_2$ as though the electrons were not identical, one may write the wave functions of two degenerate states with the energy (15.14) in which one or the other electron is excited, namely,

$$\langle\vec{r}_1|1\ 0\ 0\rangle\langle\vec{r}_2|n\ l\ m\rangle \quad \text{and} \quad \langle\vec{r}_2|1\ 0\ 0\rangle\langle\vec{r}_1|n\ l\ m\rangle. \tag{15.15}$$

Starting from the energy eigenvalues (15.14) and eigenfunctions (15.15), one considers the disturbance, or "perturbation," represented by the difference between the assumed potential energy and the actual potential energy of the two electrons. The actual energy is the sum of the attraction by the nucleus and the mutual repulsion of the electrons. The pertubation energy is

$$V(\vec{r}_1\vec{r}_2) = -\frac{2e^2}{r_1} - \frac{2e^2}{r_2} + \frac{e^2}{|\vec{r}_1 - \vec{r}_2|} - [U(r_1) + U(r_2)]. \tag{15.16}$$

The perturbation is assumed to be small in the sense that in the next approximation each stationary state is represented as a superposition of the single pair of degenerate states (15.15). This procedure is an application of the method used in section 14.2 for the treatment of spin-orbit coupling.

In the treatment of spin-orbit coupling the next step was to identify stationary states as eigenstates of $|\vec{j}|^2$ and of j_z. Here we notice that the perturbation potential energy (15.16) is invariant under permutation of the two electrons. Therefore, we can choose its eigenfunctions to be symmetric or antisymmetric under permutation of the electron positions. According to equation (15.11), this choice implies that the eigenstates of the perturbation energy are also eigenstates of $|\vec{S}|^2$ with $S = 0$ or $S = 1$. Thus theory shows that, upon removal of the degeneracy, the symmetry of the perturbation energy requires the approximate energy eigenstates to be ortho or para.

Since the wave functions (15.15) result from one another by the interchange of the electron position, their symmetric and antisymmetric superpositions are

$$\langle\vec{r}_1,\vec{r}_2|1\ 0\ 0,n\ l\ m;S\rangle = \sqrt{1/2}\ [\langle\vec{r}_1|1\ 0\ 0\rangle\langle\vec{r}_2|n\ l\ m\rangle + (-1)^S\langle\vec{r}_2|1\ 0\ 0\rangle\langle\vec{r}_1|n\ l\ m\rangle]. \tag{15.17}$$

The wave function (15.17) with either value of S yields a probability distribution of the position of the two electrons which is no longer the product of the two particle densities of the first and second electrons. Therefore this wave function describes a state of correlated, rather than independent, electrons.

The ground state is a special case in that there is only one wave function (15.15). This wave function obeys the condition (15.11) with $S = 0$, whereby the ground state is shown to be a para state.

The correlated density distribution of the two electrons can be analyzed in the following manner. For brevity, we indicate the single electron wave functions $\langle \vec{r} | 1\ 0\ 0 \rangle$ and $\langle \vec{r} | n\ l\ m \rangle$ by $u(\vec{r})$ and $v(\vec{r})$ respectively. Thereby equation (15.17) becomes

$$\langle \vec{r}_1, \vec{r}_2 | 1\ 0\ 0, n\ l\ m; S \rangle = \sqrt{\tfrac{1}{2}}\ [u(\vec{r}_1)v(\vec{r}_2) + (-1)^S\ v(\vec{r}_1)u(\vec{r}_2)]. \quad (15.17a)$$

The electron density is then represented by the square of the binomial in equation (15.17a) and thus consists of two square terms and two cross-terms,

$$|\langle \vec{r}_1, \vec{r}_2 | 1\ 0\ 0, n\ l\ m; S \rangle|^2 = \tfrac{1}{2}\{|u(\vec{r}_1)|^2 |v(\vec{r}_2)|^2 + |v(\vec{r}_1)|^2 |u(\vec{r}_2)|^2\}$$
$$+ (-1)^S\ \tfrac{1}{2}\{[u(\vec{r}_1)v^\star(\vec{r}_1)][v(\vec{r}_2)u^\star(\vec{r}_2)] + [u^\star(\vec{r}_1)v(\vec{r}_1)][v^\star(\vec{r}_2)u(\vec{r}_2)]\}. \quad (15.18)$$

Each of the squared terms represents the uncorrelated particle density of one electron in the state u and one electron in the state v. This is the density one would expect if the two electrons were distinguishable and independent. The two cross-terms represent the interference effect due to the identity of the electrons.

The contribution of the two cross-terms in equation (15.18) to the particle density is called the *exchange density* because it results from the superposition of the two independent electron states (15.15) which differ only by the interchange of the two electrons. The addition or subtraction of the exchange density, depending on the sign of $(-1)^S$, causes the average distance of the two electrons to differ in ortho and para states. To illustrate the structure of the exchange density terms in equation (15.18), one may regard each factor of the type $u(\vec{r}_1)v^\star(\vec{r}_1)$ as representing a part of the density distribution of one independent electron. This part may be called an *interference density* because it would occur if the electron's wave function were a superposition of $u(\vec{r})$ and $v(\vec{r})$.[9] Figure 15.5 shows the interference density of one electron for the superposition of 1s and 2s states.

The final step of our approximate theory of helium energy levels consists of calculating the mean value of the perturbation energy (15.16) for the states represented by the wave function (15.17):

$$\langle V(\vec{r}_1, \vec{r}_2) \rangle = \int\int d\vec{r}_1 d\vec{r}_2 V(\vec{r}_1, \vec{r}_2)\ \tfrac{1}{2} |u(\vec{r}_1)v(\vec{r}_2) + (-1)^S\ v(\vec{r}_1)u(\vec{r}_2)|^2. \quad (15.19)$$

Expansion of the squared binomial in the integrand yields two types of contributions to $\langle V \rangle$ which we call V and W. The two squared terms of the expansion contribute equally to the mean energy and together yield the amount

$$V_{1,0,0;n,l,m} = \int\int d\vec{r}_1 d\vec{r}_2 V(\vec{r}_1, \vec{r}_2) |u(\vec{r}_1)|^2 |v(\vec{r}_2)|^2. \quad (15.20)$$

---

9. Like all interference terms, the product $u(\vec{r})v^*(\vec{r})$ represents a contribution to the particle density which can be positive, negative, or even complex.

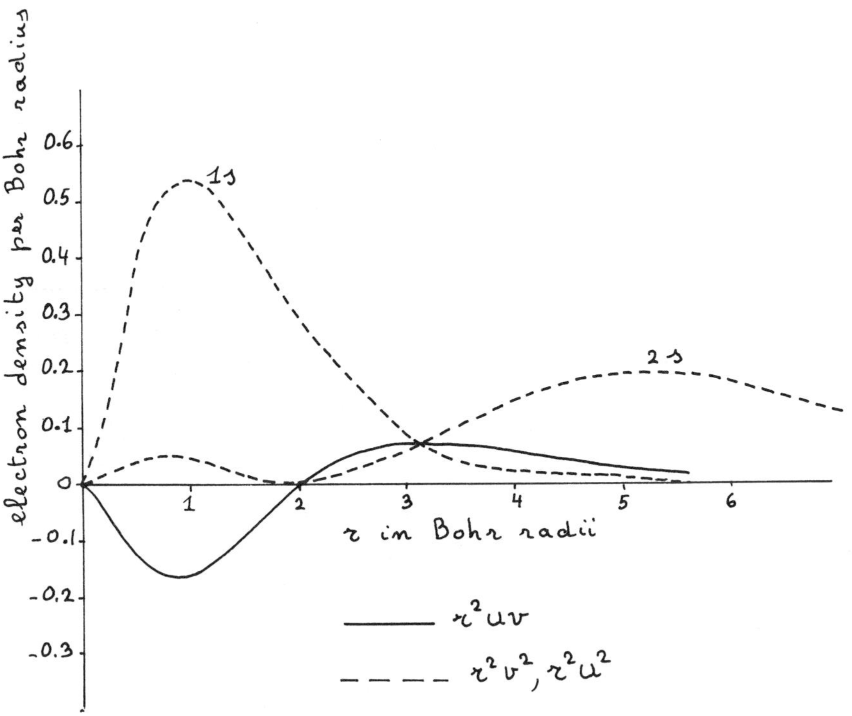

Fig. 15. 5. Interference density resulting from the superposition of 1s and 2s states with radial wave functions u and v.

The two cross-terms also contribute equal—or complex conjugate—amounts and yield together the so-called *exchange energy*

$$(-1)^s W_{1,0;n,l,m} = (-1)^s \mathrm{Re}\{\int\int d\vec{r}_1 d\vec{r}_2 V(\vec{r}_1, \vec{r}_2)[u(\vec{r}_1)v^*(\vec{r}_1)][v(\vec{r}_2)u^*(\vec{r}_2)]\}. \tag{15.21}$$

The mean energy is thus

$$\langle V(\vec{r}_1, \vec{r}_2)\rangle = V_{1,0,0;n,l,m} + (-1)^s W_{1,0,0;n,l,m}. \tag{15.22}$$

The energy difference of the para and ortho states, namely, $(V + W) - (V - W)$, equals twice the exchange energy, in this independent electron approximation.

The result (15.22), obtained by Heisenberg in 1926, constituted an early important success of quantum mechanics because the distinction of ortho and para levels had been a complete puzzle until then. Indeed, the exchange energy results from a superposition of degenerate states, a phenomenon which has no classical analog. That the superposition of degenerate states yields a stationary state of lower energy, the ortho state in our case, is a phenomenon of great importance to chemical physics where it is called *stabilization by resonance*, as we will see in later chapters.

CHAPTER 15 PROBLEMS

15.1. Consider electrons moving in a one-dimensional square-well potential similar to that of problem 12.1 with $V = \infty$ for $x < 0$, $V = 0$ for $0 < x < a$, $V = \infty$ for $x > a$. Single-electron eigenfunctions are $\langle x|n\rangle = (2/a)^{1/2} \sin(n\pi x/a)$, with $n > 0$. Disregarding the repulsion between electrons, (a) construct two-electron wave functions $\langle x_1|1\rangle\langle x_2|2\rangle$ and $\langle x_1|2\rangle\langle x_2|1\rangle$, analogous to the helium wave functions (15.15); (b) construct para and ortho wave functions $\langle x_1, x_2|1,2; S\rangle$ analogous to the wave functions (15.17); (c) sketch the positions of the nodes for the four wave functions (a) and (b) on a plane with coordinate axes (x_1, x_2).

15.2. Consider the total electron density distribution $\rho(\vec{r})$ in a helium atom, that is, the probability $\rho(\vec{r})d\vec{r}$ of finding one electron in a volume element $d\vec{r}$ at the position $\vec{r}$, irrespective of the position of the other electron. For a state represented by the wave function (15.17a), show that $\rho(\vec{r}) = |u(\vec{r})|^2 + |v(\vec{r})|^2$ for both $S = 0$ and 1, provided that $u(\vec{r})$ and $v(\vec{r})$ are solutions of the same Schrödinger equation for a single electron.

15.3. A relationship between wave functions $\langle \vec{r}_1\ m_{s1}, \vec{r}_2\ m_{s2}|a\rangle$ and $\langle \vec{r}_2\ m_{s2}, \vec{r}_1\ m_{s1}|a\rangle$ (see, e.g., eq. [15.9]) may be represented by writing $\langle \vec{r}_2\ m_{s2}, \vec{r}_1\ m_{s1}|a\rangle = \mathcal{P}_{12}\langle \vec{r}_1\ m_{s1}, \vec{r}_2\ m_{s2}|a\rangle$, where $\mathcal{P}_{12}$ indicates the *permutation operator* which interchanges all the coordinates of electrons 1 and 2. Show that this operator has only two distinct eigenvalues, namely, 1 and -1.

15.4. Two electrons are confined to the vicinity of a point by an elastic attraction represented by the potential energy $V(\vec{r}) = \frac{1}{2}k(x^2 + y^2 + z^2) = \frac{1}{2}kr^2$. (This is the potential considered in problem 12.5.) Disregard initially the repulsion between the electrons, as well as any spin-orbit coupling. Consider the para and ortho states of the electron pair, for the case of one electron in the ground state (called $\{000\}$ in the solution of problem 12.5) and the other electron in the first excited state of motion along the x axis (called $\{100\}$ in problem 12.5). (a) Write the wave functions of the para and ortho states as functions of separate coordinates $(\vec{r}_1, \vec{r}_2)$ of the two electrons. (b) Write the same wave functions in terms of the center-of-mass coordinate $\vec{R} = \frac{1}{2}(\vec{r}_1 + \vec{r}_2)$ and of the relative position $\vec{r} = \vec{r}_1 - \vec{r}_2$ of the two electrons. (c) Evaluate the energy difference of the para and ortho states as the mean value of the electric potential e^2/r between the two electrons, using the wave functions of part (b). (The dependence of the wave functions on spin quantum numbers m_{s1} and m_{s2} may be omitted here as it is omitted in eq. [15.17].)

SOLUTIONS TO CHAPTER 15 PROBLEMS

15.1. (a) $\langle x_1|1\rangle\langle x_2|2\rangle = (2/a)\sin(\pi x_1/a)\sin(2\pi x_2/a), \langle x_1|2\rangle\langle x_2|1\rangle$ $= (2/a)\sin(2\pi x_1/a)\sin(\pi x_2/a)$. (b) $\langle x_1, x_2|1,2;S=0\rangle = \sqrt{1/2}\ (2/a)2 \times$ $\sin(\pi x_1/a)\sin(\pi x_2/a)[\cos(\pi x_2/a) + \cos(\pi x_1/a)] = [(4\sqrt{2})/a]\sin(\pi x_1/a)\times$ $\sin(\pi x_2/a)\cos[\pi(x_1+x_2)/2a]\cos[\pi(x_1-x_2)/2a], \langle x_1, x_2|1,2;S=1\rangle$ $= \sqrt{1/2}\ (2/a)2\sin(\pi x_1/a)\sin(\pi x_2/a)[\cos(\pi x_2/a) - \cos(\pi x_1/a)]$ $= [(4\sqrt{2})/a]\sin(\pi x_1/a)\sin(\pi x_2/a)\sin[\pi(x_1+x_2)/2a]\sin[\pi(x_1-x_2)/2a]$.

(c)

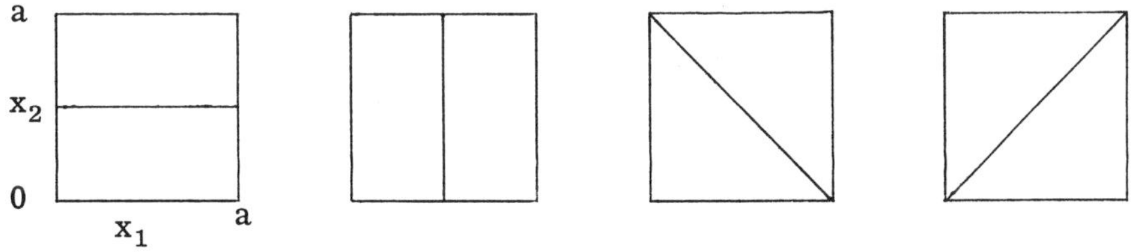

$\langle x_1|1\rangle\langle x_2|2\rangle$ $\langle x_1|2\rangle\langle x_2|1\rangle$ $\langle x_1, x_2|1,2;S=0\rangle$ $\langle x_1, x_2|1,2;S=1\rangle$

15.2. The problem statement defines $\rho(\vec{r})$ as equal to

$$[\int d\vec{r}_2|\langle\vec{r}_1,\vec{r}_2|0\ 0\ 0, n\ l\ m; S\rangle|^2]_{\vec{r}_1=\vec{r}}$$
$$+ [\int d\vec{r}_1|\langle\vec{r}_1,\vec{r}_2|1\ 0\ 0, n\ l\ m; S\rangle|^2]_{\vec{r}_2=\vec{r}}.$$

Note that the two terms of this expression are manifestly equal, owing to symmetry under permutation of $\vec{r}_1$ and $\vec{r}_2$, and substitute in them equation (15.18). This yields

$$\rho(\vec{r}) = |u(\vec{r})|^2\int d\vec{r}_2|v(\vec{r}_2)|^2 + |v(\vec{r})|^2\int d\vec{r}_2|u(\vec{r}_2)|^2$$
$$+ (-1)^S\{u(\vec{r})v^*(\vec{r})\int d\vec{r}_2 v(\vec{r}_2)u^*(\vec{r}_2) + u^*(\vec{r})v(\vec{r})\int d\vec{r}_2 v^*(\vec{r}_2)u(\vec{r}_2)\}.$$

Each of the first two integrals in this equation equals unity, since it represents an integrated probability distribution, and each of the other two integrals vanishes owing to orthogonality of the states represented by $u(\vec{r}_2)$ and $v(\vec{r}_2)$.

15.3. The square of this operator is the identity operator, because two successive permutations reproduce the initial wave function. Therefore any eigenvalue of $\mathcal{P}_{12}$ must be one of the square roots of 1, i.e., either 1 or -1.

15.4. Eigenfunctions for single electrons in the potential V(r), obtained from problem 12.5 and section 12.4 are:

$$\langle x\ y\ x|0\ 0\ 0\rangle = \langle x|0\rangle\langle y|0\rangle\langle z|0\rangle$$
$$= (\pi a^2)^{-1/4}\exp(-\tfrac{1}{2}x^2/a^2)(\pi a^2)^{-1/4}\exp(-\tfrac{1}{2}y^2/a^2)(\pi a^2)^{-1/4}\exp(-\tfrac{1}{2}z^2/a^2)$$
$$= (\pi a^2)^{-3/4}\exp(-\tfrac{1}{2}r^2/a^2),$$

$$\langle x\ y\ z|1\ 0\ 0\rangle = \langle x|1\rangle\langle y|0\rangle\langle z|0\rangle = \sqrt{2}\ (\pi a^2)^{-1/4}\ (x/a)\ \exp\ (-\tfrac{1}{2}x^2/a^2)\langle y|\ 0\rangle\langle z|0\rangle$$
$$= \sqrt{2}\ (x/a)(\pi a^2)^{-3/4}\ \exp\ (-\tfrac{1}{2}r^2/a^2),$$

where $a = (\hbar^2/km_e)^{1/4}$. (a) Substitution of these eigenfunctions in equation (15.17) gives the two-electron wave functions

$$\langle \vec{r}_1, \vec{r}_2|0\ 0\ 0, 1\ 0\ 0; S\rangle = (\pi a^2)^{-3/2}[x_2 + (-1)^S\ x_1]\ a^{-1}\ \exp[-\tfrac{1}{2}(r_1^2 + r_2^2)/a^2].$$

(b) Further substitution of $\vec{r}_1 = \vec{R} + \tfrac{1}{2}\vec{r}$ and $\vec{r}_2 = \vec{R} - \tfrac{1}{2}\vec{r}$ gives two different functions for $S = 0$ and $S = 1$:

$$\langle \vec{R}, \vec{r}|0\ 0\ 0, 1\ 0\ 0; S = 0\rangle$$
$$= (2/\pi a^2)^{3/4}(2X/a)\ \exp\ (-R^2/a^2)(2\pi a^2)^{-3/4}\ \exp\ (-\tfrac{1}{4}r^2/a^2)$$
$$= \langle 2^{1/2}\vec{R}|1\ 0\ 0\rangle\langle 2^{-1/2}\vec{r}|0\ 0\ 0\rangle,$$

$$\langle \vec{R}, \vec{r}|0\ 0\ 0, 1\ 0\ 0; S = 1\rangle$$
$$= (2/\pi a^2)^{3/4}\ \exp\ (-R^2/a^2)(2\pi a^2)^{-3/4}(-x/a)\ \exp\ (-\tfrac{1}{4}r^2/a^2)$$
$$= \langle 2^{1/2}\vec{R}|0\ 0\ 0\rangle\langle 2^{-1/2}\vec{r}|\ 1\ 0\ 0\rangle.$$

Note how in this problem the two-electron wave functions factor out in center-of-mass coordinates; in the para function the motion of the center of mass is excited, in the ortho the relative motion of the two electrons is excited. (c) Since the potential energy e^2/r depends only on the distance r between the two electrons, its calculation is straightforward in center-of-mass coordinates. (Otherwise, the separate dependence of $e^2/r = e^2/|\vec{r}_1 - \vec{r}_2|$ on $\vec{r}_1$ and $\vec{r}_2$ would have to be worked out explicitly.) Thus we have

$$\left\langle \frac{e^2}{r} \right\rangle_{000,100;\ S=0} = \int_0^\infty \frac{e^2}{r}\ \frac{\exp\ (-\tfrac{1}{2}r^2/a^2)}{(2\pi a^2)^{3/2}}\ 4\pi r^2 dr = \left(\frac{2}{\pi}\right)^{1/2} \frac{e^2}{a},$$

$$\left\langle \frac{e^2}{r} \right\rangle_{000,100;\ S=1} = \int \frac{e^2}{r}\ \frac{x^2/a^2}{(2\pi a^2)^{3/2}}\ \exp\ (-\tfrac{1}{2}r^2/a^2)d\vec{r} =$$

$$= 2\pi \int_0^\pi \cos^2\theta\ d\cos\theta \int_0^\infty \frac{e^2}{r}\ \frac{r^2/a^2}{(2\pi a^2)^{3/2}}\ \exp\ (-\tfrac{1}{2}r^2/a^2) r^2 dr$$

$$= \frac{2}{3}\left(\frac{2}{\pi}\right)^{1/2} \frac{e^2}{a}.$$

The energy of the ortho state ($S = 1$) is lower by $(\tfrac{1}{3})(2/\pi)^{1/2}e^2/a$.

16. atoms with many electrons

16.1. INTRODUCTION

Having learned to treat a single atomic electron as well as pairs of identical particles, we move on to atoms with many electrons. This chapter deals with some general theoretical methods and illustrates them with examples. The aim is to provide the student with a framework of concepts and of general information which can serve to approach specialized treatises.[1]

The most widely studied property of an atom is its spectrum of energy levels. These levels involve the excitation of outer electrons, in which case they are usually called optical levels, as well as the excitation of inner electrons which is responsible for X-ray spectra. Volumes of extremely accurate experimental data have been collected on this subject. Another important field is the study of transition probabilities between stationary—or quasi-stationary—states by emission or absorption of radiation or by a redistribution of energy among electrons. Characteristics of stationary states determine many gross physico-chemical properties of atoms, such as their chemical reactivity. In addition, singular properties of specific levels are responsible for large-scale phenomena in nature as well as in technological applications. A classical example is the use of the chromium ion in ruby crystals to produce, besides the characteristic color, the delicate chain of transitions which permits the laser action. It is remarkable how seemingly minute details of atomic structure play a major role both in nature and in technology.

Collisions between an electron and an atom—or between an electron and an ion—may be regarded as transitions between different stationary states of the electron-atom system, with transfer of energy from one to another form of motion. The study of these collisions is of increasing importance for the understanding of ionized gases in space physics, in astrophysics, and in thermonuclear plasmas.

In principle, all atomic properties can be predicted theoretically, starting from general laws. The energy of a set of electrons in the field of a nucleus can be expressed as the sum of kinetic energies, of the electrostatic potential energies between each pair of particles, and of spin-orbit couplings. There are also smaller spin-spin interactions and relativistic corrections which will not be considered here. From the energy expression one formulates a Schrödinger equation which can be solved, in principle, by numerical methods to any desired degree of accuracy. Numerical solutions have, in fact, been obtained for atoms

---

1. J. C. Slater, *Quantum Theory of Atomic Structure* (New York: McGraw Hill, 1960); Condon and Shortley, *TAS*; H. A. Bethe, and R. W. Jackiw, *Intermediate Quantum Mechanics* (2nd ed; New York: W. A. Benjamin, 1968). N. F. Mott and H. S. W. Massey, *Theory of Atomic Collisions* (3rd ed; Oxford: Clarendon Press, 1965).

with a few electrons, specifically for helium, yielding energy eigenvalues accurate to about one part in a million.

For most purposes, however, the inherent complication of a many-particle system requires analytical approximations. Firstly, the capacity of current calculators is inadequate for accurate treatment of atoms with many electrons. More importantly, analytical approximation is essential to sort out the qualitative aspects of each phenomenon and thus to permit extrapolation from one to another situation. Theory normally proceeds in successive steps of refinement, starting from schematic models. Thus the energy of the atomic electrons is often schematized at the outset by regarding each electron as subject only to a net attraction by the nucleus. On this basis one accounts for many properties, such as the shell structure and the size of atoms. Increasingly realistic expressions of the energy are then used in succession to provide increasingly detailed predictions of the energy levels and transition probabilities.

The initial schematic model developed in this chapter is the independent electron approximation, described in section 16.2. Insofar as the average attraction experienced by an atomic electron obeys approximately the Coulomb law, the energy levels of the electron depend only on the principal quantum number n, as discussed in section 13.3. The average distance of each electron from the nucleus depends mainly on n, and varies rapidly with n. Accordingly, atomic electrons are distributed in shells; the nth shell has $2n^2$ orthogonal states, namely, n^2 orthogonal states of electron motion, as noted at the end of section 13.3, and two orthogonal spin states for each of them. This shell structure would cause the periodic system of elements to have $2n^2$ elements in the nth row. However, modifications of the periodic system occur because the average attraction experienced by an atomic electron departs from the Coulomb law. Section 16.3 discusses these modifications and describes results of simple calculations that account for them.

Accounting for the spacing and angular momenta of energy levels, even only the lowest ones, requires a treatment of electron-electron interaction and spin-orbit coupling beyond the independent electron approximation. Predictions of the multiplicity of discrete levels and of their approximate relative spacings can be obtained by constructing superpositions of a finite number of independent-electron states and by solving finite algebraic systems of equations. Section 16.4 outlines the procedures for this purpose; it also describes the systems of classification of the resulting energy levels.

Accurate calculation of energy levels and of the transition probabilities requires, however, superposition of infinite sets of first-approximation states. In this case the resulting mathematical problems are more complicated and require the development of suitable approximation methods. This will be done in the following chapter. Let us note here that the superposition of a finite set of discrete states can represent only oscillatory phenomena with a finite number of discrete frequencies. The methods of this chapter are thus inadequate to represent phenomena in which an atom evolves irreversibly, for instance by emission of radiation.

16.2. INDEPENDENT ELECTRON APPROXIMATION.

The potential energy of each atomic electron at any one time and place depends on the position of all other electrons at that time. The independent electron approximation assumes, on the contrary, that the potential energy of an electron depends only on its distance r from the nucleus, as though the density of all other electrons had spherical symmetry and were independent of the position of the electron under consideration. The potential energy is therefore represented by a function U(r) which equals the nuclear potential $-Ze^2/r$ reduced by the screening action of electrons.

16.2.1 *Screened Potential.* The repulsion among electrons is represented conveniently in terms of the force acting on an electron at a distance r from the nucleus. Under the assumption of spherical symmetry, all electrons closer to the nucleus than r repel the electron under consideration as though their position coincided with the nucleus. All electrons farther than r exert no net force. The net attraction experienced by an electron then has the strength

$$F(r) = \left[Z - \int_0^r \rho(r')4\pi r'^2 dr'\right]\frac{e^2}{r^2}, \tag{16.1}$$

where ρ represents the density of electrons other than the one under consideration.

The potential energy of an electron at a distance r is obtained by integrating the force F(r) from infinity to r. The integration is carried out by parts, with $1/r^2$ taken first, and yields

$$\begin{aligned} U(r) &= -\int_r^\infty F(r')dr' \\ &= -\left[Z - \int_0^r \rho(r')4\pi r'^2 dr'\right]\frac{e^2}{r} + e^2 \int_r^\infty \rho(r')4\pi r' dr'. \end{aligned} \tag{16.2}$$

The integral in the brackets represents the influence of the electron at distances $r' < r$ and is called the *inner screening* number. This number vanishes for $r \to 0$ and approaches $Z - 1$ for $r \to \infty$ for a neutral atom. The last term on the right of equation (16.2) represents the potential energy of *outer screening*. This integral may attain about 20% of U(r) for small r and large Z. Splitting the potential U(r) according to (16.2) proves useful for approximate estimation of independent-electron wave functions and energy levels.[2] The distance r_0 for which the inner screening number reaches approximately its limiting

---

2. The wave function depends primarily on the force (16.1) in the region where the wave function is large. In fact, it is often approximated by using a hydrogen-like model with the nuclear charge diminished by an estimate of the inner screening number. On the other hand, the scale of energy levels has its zero point anchored to that of the potential energy at $r = \infty$. Therefore, the value of the outer screening serves to establish the scale of energy levels.

value $Z - 1$ may be interpreted as the radius of the atom. This radius turns out to be of the order of 1 Å for all atoms.

The electron density $\rho(r)$, the potential energy $U(r)$, and hence the atomic radius are often determined by the *Hartree self-consistent* method. Starting from a trial function $U_0(r)$, one calculates the motion of all electrons, the combined density $\rho(r)$, and hence a new estimate $U_1(r)$ from equation (16.2).[3] This estimate agrees with $U_0(r)$ if $U_0(r)$ was a good trial function; otherwise $U_1(r)$ serves as a new trial function, and the calculation is repeated until a satisfactory estimate is attained.

Once a screened potential $U(r)$ has been chosen, the Schrödinger equation for the motion of each independent atomic electron can be written

$$-\frac{\hbar^2}{2m_e}\nabla^2\langle\vec{r}\,|n\ l\ m\rangle + U(r)\langle\vec{r}\,|n\ l\ m\rangle = \langle\vec{r}\,|n\ l\ m\rangle E_{nl}. \tag{16.3}$$

This equation can be written in polar coordinates. Its variables are then separated as in the equations (13.8) for the hydrogen atom. The eigenvalues and eigenfunctions for the rotational motion are the same as those obtained for the hydrogen atom in section 13.2. The radial equation analogous to (13.8c),

$$-\frac{\hbar^2}{2m_e}\left[\frac{d^2}{dr^2}+\frac{2}{r}\frac{d}{dr}\right]\langle r\,|n\ l\rangle + \left[\frac{|\vec{l}\,|^2}{2m_e r^2} + U(r)\right]\langle r\,|n\ l\rangle = \langle r\,|n\ l\rangle\, E_{nl}, \tag{16.4}$$

must generally be solved numerically, but this problem presents no difficulty. Thus one obtains a complete set of single-electron energy eigenvalues E_{nl} and eigenfunctions $\langle r\,|n\ l\rangle\,\langle\theta\,|l\ m\rangle\,\langle\phi\,|m\rangle$ for the orbital motion. Each wave function must be complemented by an indication of the spin quantum number m_s.

The complete set of eigenvalues E_{nl} of the radial equation (16.4) includes not only discrete negative eigenvalues identified by integer values of n but also a continuous set of positive eigenvalues which correspond to ionized states. (The continuous energy spectrum of the hydrogen atom was discussed briefly in section S13.6.) In this chapter we will deal only with states of the discrete spectrum.

The eigenvalues E_{nl} and the radial wave functions differ, of course, from those of hydrogen owing to screening. For any given value of n the states with lower orbital quantum number l have lower energy eigenvalues and lower mean value of r, as noted previously for the alkalis and for helium. Quantitatively, this dependence on l is in some cases a sensitive function of the atomic number and is thus responsible for major characteristics of the periodic system of elements, as will be discussed in section 16.3.

---

3. Strictly, the density $\rho(r)$ depends on the state of the electron whose screening is being considered and therefore should be different for different states of each atomic electron. In practice, however, one often uses a single average function $\rho(r)$ for all states.

For purposes of discussion, the energy eigenvalues E_{nl} may be expressed in the form of hydrogen atom eigenvalues with screened potential parameters. To a state with radial wave function $\langle r|n\ l\rangle$ and mean radial distance $\langle r\rangle_{nl} = \int_0^\infty r|\langle r|n\ l\rangle|^2 r^2 dr$ one may attribute the average screening number $\int_0^{\langle r\rangle_{nl}} \rho(r) 4\pi r^2 dr$, the *effective atomic number*

$$(Z_{\text{eff}})_{nl} = Z - \int_0^{\langle r\rangle_{nl}} \rho(r) 4\pi r^2 dr, \tag{16.5}$$

and the effective potential energy of outer screening

$$(V_{\text{outer}})_{nl} = e^2 \int_{\langle r\rangle_{nl}}^\infty \rho(r) 4\pi r dr. \tag{16.6}$$

The eigenvalue E_{nl} may be cast in the form analogous to equation (3.17):

$$E_{nl} = -\frac{m_e (Z_{\text{eff}})_{nl}^2 e^4}{2\hbar^2 (n - \sigma_{nl})^2} + (V_{\text{outer}})_{nl}. \tag{16.7}$$

Given the values of E_{nl}, $(Z_{\text{eff}})_{nl}$, and $(V_{\text{outer}})_{nl}$, the equation (16.7) determines the value of the parameter σ_{nl}, which is called *Rydberg correction* or *quantum defect*. This parameter would vanish if the inner screening were constant over most of the range of integration of the radial equation (16.4); it does vanish approximately for low values of n, for which the inner screening is negligible and $(Z_{\text{eff}})_{nl} \approx Z$. Accordingly, the value of σ_{nl} indicates the effect of the stronger nuclear attraction which prevails near the nucleus, at $r < \langle r\rangle_{nl}$. For excited states with large values of $\langle r\rangle_{nl}$, the inner screening approaches $Z - 1$ for a neutral atom and the outer screening vanishes. In this case, the eigenvalue formula (16.7) coincides with the Bohr equation (3.17) with $Z = 1$. Under these conditions the parameter σ_{nl} becomes nearly independent of n; in fact, the sets of eigenvalues E_{nl} with large n and fixed l constitute *Rydberg series* of levels. This result accounts for the experimental evidence on series of spectral lines described in sections 2.2 and 2.3.

16.2.2. *Configurations.* We consider now the states of a system of N electrons in the screened potential field U(r) pertaining to an atom of atomic number Z. The number N of electrons equals Z, of course, for a neutral atom, but may be smaller (or larger) than Z for an ionized atom. In the initial independent electron approximation a stationary state of the atom is identified by N sets of quantum numbers (n, l, m, m_s) pertaining to the N single-electron states. No two of these sets of quantum numbers ever coincide owing to the exclusion principle. For brevity we will call "N-fold set" each particular selection of the N sets of quantum numbers (n, l, m, m_s). For instance, one excited state of beryllium is identified in this approximation by the fourfold set $(1,0,0,\frac{1}{2})$ $(1,0,0,-\frac{1}{2})$ $(2,0,0,\frac{1}{2})$ $(2,1,1,\frac{1}{2})$.

Single-electron states with the same quantum number n are said to belong to the same shell. Different shells are often identified by code letters, rather than by values of n, as follows:

n	1	2	3	4	5	...
code	K	L	M	N	O	... .

States with the same pair of quantum numbers n and l belong to the same subshell. Subshells are usually indicated by the combination of number and letter introduced at the end of section 13.3; for example, 3p means $n = 3, l = 1$.

In the independent electron approximation the energy of the atom depends only on the eigenvalues E_{nl} and on the number, N_{nl}, of electrons in the same subshell. We indicate this energy by

$$E_{Ns} = \Sigma_{nl} N_{nl} E_{nl}, \tag{16.8}$$

where "Ns" stands for the N-fold set of quantum numbers that identifies the state. States of the whole atom that differ only by magnetic quantum numbers in the N-fold set, and have therefore the same energy, are said to have the same *configuration.* A configuration is identified by the distribution of the N electrons among the different subshells, that is, by the set of numbers N_{nl}. A configuration is usually represented by a symbol which displays the code names of the different subshells, with the numbers N_{nl} as superscripts. For example, the ground state of the silicon atom has the configuration formula

$$(1s)^2(2s)^2(2p)^6(3s)^2(3p)^2, \tag{16.9}$$

meaning that $N_{1,0} = 2, N_{2,0} = 2, N_{2,1} = 6, N_{3,0} = 2, N_{3,1} = 2$. As in all ground-state configurations, the 14 electrons are distributed in this example among the lowest energy shells, that is, in such a way as to minimize the ground-state energy. Often, configuration formulae are not written out in complete form as they are in expression (16.9); instead, only the formula for the outermost subshell or subshells is given explicitly. Thus the ground-state configuration of Si is indicated by $(3p)^2$ or by $(3s)^2(3p)^2$.

Recall from section 13.2.4 that each state of a single electron in a central field has even or odd parity under inversion of coordinates, $\vec{r} \to -\vec{r}$, depending on whether the quantum number l is even or odd. For an atom with N electrons, inversion of coordinates means that $\vec{r}_i \to -\vec{r}_i$ for $i = 1,2,\ldots,N$. Since inversion of coordinates multiplies by $(-1)^l$ the wave function of each single electron with quantum number l, each N-electron configuration has even or odd parity depending on whether $\Sigma_{nl} N_{nl} l$ is even or odd. Parity is compatible with the energy of an atom, irrespective of any approximation, because the expression of the energy remains unchanged under inversion of all electron coordinates. (Inversion of position coordinates $\vec{r}_i$ implies inversion of velocities $d\vec{r}_i/dt$ and of momenta $\vec{p}_i$. On the other hand, orbital angular momenta $\vec{r}_i \times \vec{p}_i$ remain invariant under this inversion and so do the spin angular momenta.)

16.2.3. *Wave Functions.* The independent electron approximation implies at the outset that the combined wave function of all electrons in an atom is the product of separate single-electron wave functions. (An example for two electrons was given in eq. [15.15].) Electron density distributions obtained by squaring such product wave functions would also be products of factors representing the densities of separate electrons. This means that the positions of different electrons would be totally uncorrelated. Actually, we know from chapter 15 that the positions of identical particles are necessarily correlated.

In the treatment of the helium spectrum (section 15.4), product wave functions with different permutations of electron coordinates were superposed to meet the symmetry requirements of the exclusion principle. For two electrons the antisymmetric wave function can be split, as in equation (15.10), into a position factor and a spin factor, the superposition being carried out for the two factors separately. For more than two electrons, factorization of the type of (15.10) is not always possible. The required antisymmetric superposition can nevertheless be written in reasonably compact form for any number of electrons by a *Slater determinant* formulation. We shall write here such a superposition for the ground state wave function of the Li atom.

The ground state of Li has the configuration $(1s)^2(2s)^1$. Therefore, there are only two different single-electron position wave functions $\langle \vec{r}|n\ l\ m\rangle$, that is, one with $n = 1, l = m = 0$, which we call u(r), and one with $n = 2, l = m = 0$, to be called v(r). We also introduce the "spin up" symbol α for the spin wave function $\langle m_s|\tfrac{1}{2}\rangle$ and the "spin down" symbol β for $\langle m_s|-\tfrac{1}{2}\rangle$. The lithium atom has two degenerate ground states, which differ by the orientation of the spin of the 2s electron. We consider here the spin-up state, whose wave function is built with products of the three single-electron wave functions $u\alpha$, $u\beta$, and $v\alpha$. One of the product wave functions has the form $u(r_1)\alpha_1 u(r_2)\beta_2 v(r_3)\alpha_3$; five others differ by permutations of the electron indices 1, 2, and 3.[4] The superposition of these six product wave functions which changes sign under permutation of any two electron indices is represented by the determinant

$$\Psi = \frac{1}{\sqrt{6}} \begin{vmatrix} u(r_1)\alpha_1 & u(r_1)\beta_1 & v(r_1)\alpha_1 \\ u(r_2)\alpha_2 & u(r_2)\beta_2 & v(r_2)\alpha_2 \\ u(r_3)\alpha_3 & u(r_3)\beta_3 & v(r_3)\alpha_3 \end{vmatrix} . \tag{16.10}$$

Expansion of this determinant yields the wave function as the desired superposition of the six product wave functions. Permutation of all coordinates—spin as well as position—of any two electrons changes the sign of determinant (16.10), as one verifies by noticing that such a permutation interchanges two rows of the determinant.

---

4. As emphasized in section 15.3, attributing indices 1, 2, 3, ... to electrons as though they were distinguishable is a mathematical artifice compensated by accepting only antisymmetric wave functions.

For an atom with N electrons each state is identified, in our approximation, by the choice of an N-fold set of quantum numbers and can be identified briefly by $|Ns\rangle$. To each N-fold set corresponds a set of N single-electron wave functions. The state $|Ns\rangle$ has the Slater-determinant wave function constructed with this set. All elements of the determinant in the ith row pertain to the ith electron; all elements of the kth column consist of the kth wave function of the set. The expansion of the determinant has N! terms. Accordingly, it has a normalization factor $(N!)^{-1/2}$ to ensure that the integrated particle density equal unity.

The square of the expansion of a Slater determinant represents a correlated distribution of electron density, analogous to the distribution (15.18) for two electrons. Here again we have a polynomial consisting of square terms and of cross-terms. Each of the square terms represents the uncorrelated particle density with one electron in each of the single-electron states. Together, these terms contribute the density $\rho(r)$ which appears in the expression (16.2) of the screened potential. The cross-terms contribute an *exchange density* which results from the identity of electrons. The exchange density was disregarded in the determination of the screened potential. The correlation of particle positions and spins which is embodied in a Slater-determinant wave function is called the *exchange correlation.*

The screened potential (16.2) may be modified to take into account the exchange density approximately. However, the existence of exchange correlations makes it actually impossible to write a separate Schrödinger equation with an ordinary potential U(r) for each single-electron state. Instead, one must solve a system of N Schrödinger equations for the N single-electron wave functions. The system is derived by starting from an expression of the mean energy $\langle K\rangle_a + \langle V\rangle_a$ for a state $|a\rangle$ of the system of N electrons but otherwise analogous to (12.6). The exchange density is included in the calculation of $\langle V\rangle_a$ and contributes to $\langle V\rangle_a$ terms analogous to the exchange energy (15.21). This procedure, widely used in atomic physics, is called the *Hartree-Fock* method.[5]

16.2.4. *Degeneracy and Stabilization by Resonance.* As we have seen in section 16.2.2, each atomic state is identified, in the independent electron approximation, by an N-fold set of quantum numbers (n, l, m, m_s). The energy of the state of the whole atom is represented by equation (16.8) and thus depends only on the number, N_{nl}, of electrons in each subshell and not on any particular choice of the values of m and m_s in the N-fold set. That is, as noted before, all states of the atom with the same configuration are degenerate in this approximation.

A survey of the number of degenerate states of a given configuration is a prerequisite for developing better approximations. Let us first consider a single subshell with given values of n and l. In this subshell there are $2(2l + 1)$ single-electron states with different pairs of quantum numbers (m, m_s), as discussed in section 14.2. Out of these $2(2l + 1)$ states, N_{nl} must be picked

5. See, for instance, Slater, *Quantum Theory of Atomic Structure,* chap. 17; Bethe and Jackiw, *Intermediate Quantum Mechanics,* chap. 6.

for inclusion in the N-fold set. The number of different selections is represented by the symbol of combinatorial analysis, called the binomial coefficient:

$$\binom{2(2l+1)}{N_{nl}} = \frac{[2(2l+1)]!}{N_{nl}!\,[2(2l+1)-N_{nl}]!}. \tag{16.11}$$

This number reduces to unity for a full subshell, that is, when $N_{nl} = 2(2l+1)$, as well as for an empty subshell, when $N_{nl} = 0$.

The ground-state configuration of an atom consists normally of full subshells, except for the outermost one. Its number of degenerate states is therefore given by equation (16.11) for the outermost subshell only. In the example of the configuration (16.9) for the gound state of Si, the outermost subshell has $l = 1$ and $N_{3,1} = 2$. The number of degenerate states is then 15.

When there is more than one partially occupied subshell, as generally happens in excited states, the number of degenerate states is given by the product of possible selections for each subshell,

$$\Pi_{nl} \binom{2(2l+1)}{N_{nl}}. \tag{16.12}$$

For example, excitation of a 3p electron of Si to 3d yields the configuration $(3s)^2(3p)(3d)$ with 6×10 degenerate states.

An important property of the combination formula (16.11) is that it has the same value if one interchanges the number of electrons N_{nl} and the number of vacancies $2(2l+1) - N_{nl}$. For example, silicon and sulfur have ground-state configurations $(3p)^2$ and $(3p)^4$, respectively. The number of degenerate states is 15 in both cases, since sulfur has two vacancies in the 3p subshell.

The degeneracy of states for an atom with partially filled subshells is the starting point for the calculation of eigenstates and eigenvalues more realistic than those given by the independent electron approximation. New stationary states are constructed by superposition of states that are degenerate in the first approximation. This construction will be discussed in section 16.4. We note here the following general feature of the effect of superposition which has already appeared in the treatment of helium in section 15.4, as well as in connection with spin-orbit coupling in section 14.2.

The superposition of degenerate states introduces correlations between the position and spin coordinates of different electrons in addition to those introduced by the exclusion principle. The energy of states represented by these superpositions depends on the coefficients of the superposition and may be higher or lower than the first-approximation energy (16.8). The reduction of the energy of orthohelium due to the exclusion principle correlation (section 15.4) is an example of this effect. In general, the lowest energy eigenvalue lies lower than the first approximation value (16.8).

Other circumstances being equal, the larger the degeneracy of first-approximation states, the higher the degree of correlation obtained by their superposition and the lower the energy of the minimum eigenvalue. Therefore, the number of degenerate states obtained in a first approximation serves as a criterion for assessing the stability of a many-particle system.

This effect of degeneracy is often described in words derived from a somewhat incidental characteristic. Whenever a stationary state of lower energy can be represented as a superposition of first-approximation states, one can also construct nonstationary states by a different superposition. Some of these states evolve in the course of time to coincide periodically, first with one, then with another, of the degenerate first-approximation states. This periodic alternation of the nonstationary states, made possible by the first-approximation degeneracy, is called resonance, in analogy to the alternation of the motion of coupled pendulums of equal frequency. As discussed in Appendix B, the alternations of the oscillation amplitudes of coupled pendulums are associated with the existence of normal modes, one of which has lowest frequency. For this reason, the lowering of the energy derived from superposition of degenerate states is called *stabilization by resonance*.

16.3. ANALYSIS OF THE PERIODIC SYSTEM BY THE INDEPENDENT-ELECTRON MODEL

The principal features of the periodic table of elements are accounted for by an analysis of atomic properties based on the independent electron approximation:

1) The elements of each column have ground-state configurations with the same number of electrons in their outermost subshells; all these subshells have the same quantum number l. This common feature explains the similarities of spectroscopic and chemical properties.

2) Sequences of elements with increasing numbers of electrons in the same subshell have generally increasing ionization potentials (see Fig. 3.6). This increase follows from the steady increase of nuclear attraction which depresses each single-electron energy level E_{nl} as Z increases. The ionization potential drops abruptly at the start of a new subshell.

3) Stabilization by resonance accounts for the somewhat higher ionization potentials of elements with half-filled subshells. (For instance, nitrogen has a higher ionization potential than oxygen.)

The dependence of the chemical properties of atoms on their position in the periodic system will be discussed in chapter 19. However, we consider here the condition that singles out chemically inert atoms. The ground state of an inert atom should not be readily modified by external actions. The ground state must then be nondegenerate; that is, its configuration must consist only of filled subshells. Moreover, there must not exist any state with low excitation energy, that is, quasi-degenerate with the ground state. For example, He and Ne meet both conditions as their ground-state configurations consist of full shells. The alkaline-earth elements, on the other hand, have ground-state configurations with a complete subshell $(ns)^2$; they are not inert because their

$(ns)^1(np)^1$ configuration has low excitation energy, amounting only to a very few eV. The inert behavior of Ar, Kr, and Xe, whose ground-state configuration is $(np)^6$, indicates that excitation to the nd subshell requires a large amount of energy.[6]

Some major features of the periodic system, in particular the occurrence of transition elements and rare earths, can be traced to a critical dependence of single-electron energy eigenvalues and eigenfunctions upon the atomic number. For many atoms, eigenvalues E_{nl} with different pairs of indices (n, l) —for example, E_{4s} and E_{3d}—are approximately equal. In this case different configurations are approximately degenerate and the very concept of ground-state configuration tends to lose its relevance. In the next subsections we devote primary attention to such special situations.

16.3.1. *Screened Potential and Centrifugal Potential.* The net attraction experienced by an atomic electron is opposed by the centrifugal force, for $l \neq 0$. These two forces are represented in the Schrödinger equation for radial motion (16.4) by the screened potential U(r) and by the effective potential term $|\vec{l}|^2/2m_e r^2$. As we know from the study of the hydrogen atom, the attractive potential prevails at large distance from the nucleus, of the order of $l(l + 1)$ Bohr radii. At such distances the screened potential of any neutral atom is approximately the same as for hydrogen because the inner screening number is $\sim Z - 1$. At smaller distances from the nucleus, of the order of two or three Bohr radii, the centrifugal potential prevails for $l > 1$, while the screening number remains near $Z - 1$. Closer still to the nucleus, the inner screening number drops rapidly and the nuclear attraction again prevails. The net potential has, therefore, a two-valley profile as shown in Figure 16.1 for $l = 2$ and $l = 3$. An electron bound to the atom with $l \geqslant 2$ is then effectively confined in one or the other of the two valleys.

The outer valley is shallow so that any electron confined in it has potential and kinetic energies of the order of 1 eV or less; that is, its binding energy is one order of magnitude smaller than that of ground-state electrons. A state of an atom with large electron density in the region of the outer valley is therefore an excited state. The inner valley is always comparatively narrow and deep. An electron may be confined within it if the nuclear attraction is sufficiently large for the valley to contain at least one bright fringe of electron density (see section 12.4).

Figure 16.1 shows a comparison of the net potential profile for a few atoms with relatively small but important differences in atomic number. Both the depth of the inner valley and the height of the barrier between the two valleys are sensitive functions of Z in some ranges of the atomic number. These rapid variations of the net potential from one element to the next result in a rapid variation of the single-electron energy levels. Figure 16.2 shows in

---

6. This excitation energy decreases progressively from element to element down the column. In fact, it was discovered in 1962 that Xe is not quite inert.

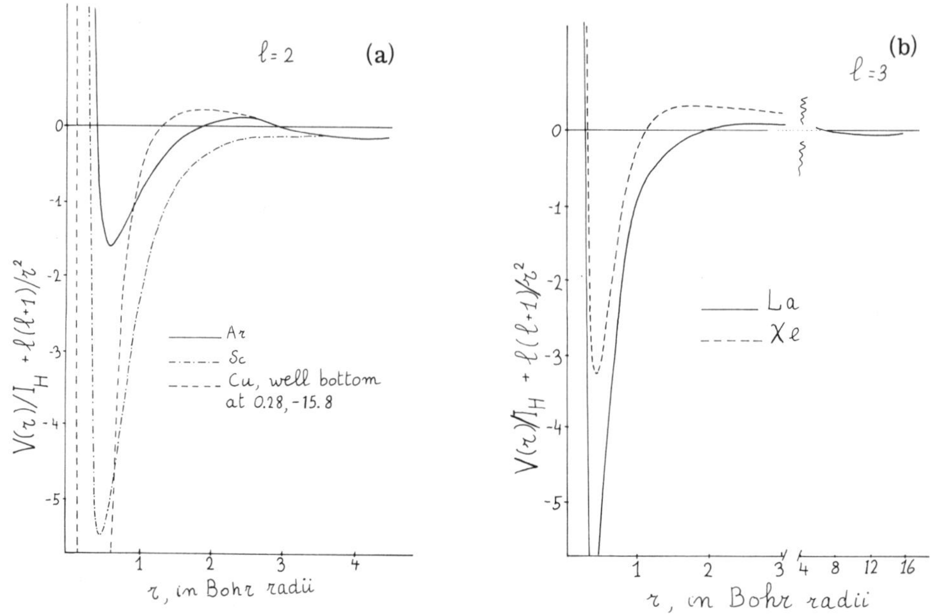

Fig. 16. 1. Examples of two-valley potential energies. (*a*) Argon, scandium, copper. (*b*) Xenon and lanthanum. Courtesy A. R. P. Rau.

particular the rapid drop of the energy of the 3d level in the range of Z between 20 and 25.[7]

The results illustrated in Figures 16. 1 and 16. 2 serve to interpret some outstanding properties of elements along the periodic system. The inert-gas behavior of Ar is explained by the large energy gap between the ground-state level E_{3p} and the next higher level E_{4s} at $Z = 18$ in Figure 16. 2. The E_{3d} level lies still higher than E_{4s} at $Z = 18$. This is because the inner valley shown for Ar in Figure 16. 1 has insufficient depth to confine an electron with $l = 2$; the 3d electron density therefore lies in the shallow outer valley. The inner valley is much deeper for scandium ($Z = 21$), and the barrier between the two valleys is lower. In fact, the inner valley is able, in this case, to confine an electron, as shown by the fact that a 3d electron is included in the ground-state configuration of Sc. Figure 16. 2 shows a very rapid drop of the energy level E_{3d} for $Z > 20$.

7. The potentials and energies shown in Figures 16. 1 and 16. 2 are calculated under simplifying assumptions and therefore details of the plots need not be realistic.

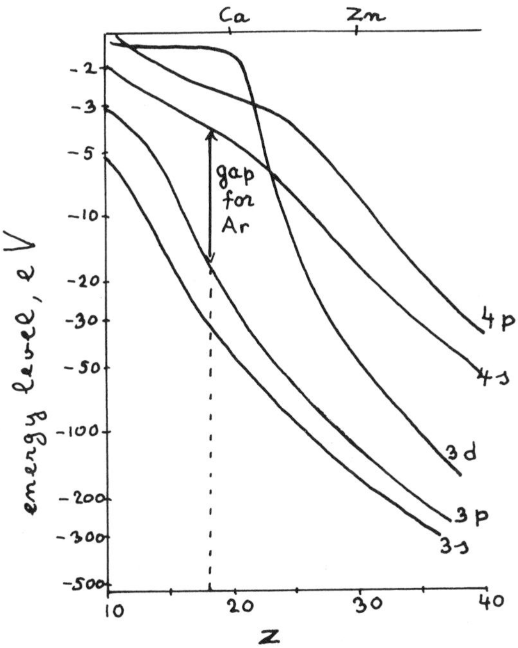

Fig. 16. 2. Typical variations of single-electron energy levels as functions of the atomic number Z. [Adapted from R. Latter, *Phys. Rev.* 99 (1955): 515.]

The group of *transition elements* between scandium and nickel (Z = 28) is characterized by the fact that their configurations with different distributions of electrons in the 3d and 4s subshells have very nearly the same energy. For this reason the transition elements do not show any clear-cut trend in their chemical properties with increasing Z. The 3d subshell is filled in the ground-state configuration of copper (Z = 29), and it shows increasingly the stability of an inner subshell in the successive elements. The cycle of properties of Ar and the following transition elements is repeated in the next rows of the periodic system and is similarly traced to the properties of 4d and 5d electrons.

The two-valley profile of the potential for $l = 3$ has a higher barrier and is thus even more distinctive than in the case of $l = 2$. So is its influence on the properties of elements whose ground-state configuration includes a partially filled 4f subshell. For all elements up to lanthanum (Z = 57) the 4f state is highly excited, with a binding energy of less than 1 eV. On the other hand, a 4f electron is included in the ground-state configuration of cerium (Z = 58). The particle density for this electron is confined within the inner valley, which lies deeper inside the atom than for d electrons. Therefore, the 4f electron of Ce does not participate in chemical reactions and Ce has chemical properties similar to those of La. The same holds for the next 13 elements, up to lutetium, which have increasing numbers of 4f electrons confined in the inner potential valley. Since La belongs to the chemical group of elements called earths, the elements from Ce to Lu which are relatively scarce in nature are called the *rare earths* or *lanthanides*. The optical spectra of the rare earths are exceedingly complicated because the atoms have nearly degenerate configurations with varying numbers of 4f, 5d, and 6s electrons; furthermore, each of these configurations is highly degenerate. The 4f states are involved in emission and absorption spectra even though they do not contribute to chemical activity.

16.3.2. *Ions.* Many of the atomic properties that vary periodically along the table of elements depend on the configuration of outer-shell electrons and hence on the number of electrons rather than on the nuclear charge. Consequently, these properties are basically common to one neutral atom and to all ions of its *isoelectronic sequence*, which have the same number of electrons but different atomic numbers. However, the screened potential becomes increasingly attractive with increasing Z along an isoelectronic sequence whereas the centrifugal potential is independent of Z. Thus the balance between these potentials shifts rapidly in favor of the attractive potential, and the total potential loses its two-valley character.

In other words, single-electron levels E_{nl} with lower n and high l lie comparatively much lower for positive ions than for neutral atoms. For example, in the isoelectronic sequence of neutral potassium, with 19 electrons, the ground-state configuration of the Sc^{++} ion includes a 3d electron, whereas in K the ground state has a 4s electron and 3d lies even higher than 4p. Similarly, in the copper ion, Cu^+, the ground-state configuration is $(3d)^{10}$ and the configuration $(3d)^8(4s)^2$ is excited by nearly 9 eV. However, for Ni, the neutral atom isoelectronic with Cu^+, the configuration $(3d)^8(4s)^2$ is more stable than $(3d)^{10}$ by nearly 2 eV. Analogous effects occur in the filling of the 4f levels for positive ions.

Negative ions have limited stability since an electron outside a neutral atom is attracted to it only by a weak polarization force (see section 18.1). More specifically, a negative ion does not exist with a single electron in an outermost subshell. On the other hand, negative ions are normally stable when the excess electron is included in a partially filled subshell and is accordingly subjected to a stronger nuclear attraction. The energy required to remove an electron from a negative ion is called the *electron affinity* of the element. The electron affinities range from a fraction of 1 eV to a maximum of nearly 4 eV for the halogens whose outer subshells have a single vacancy available for the additional electron.

16.4. SPECTRAL AND CHEMICAL EFFECTS OF ELECTRON CORRELATION

The independent electron approximation disregards correlations among electron positions by treating the potential energy of each electron as independent of the position of other electrons. Similarly, it disregards correlations of orbital and spin currents. Owing to these simplifications, each atomic configuration appears to have a single energy level; this result is correct only for configurations consisting of filled subshells. In fact, the interaction between electrons establishes correlations among electron positions. Introduction of these correlations lifts the degeneracy of levels of partially filled subshells with two (or more) electrons or vacancies.

An important result is the occurrence of a number of levels in place of a single degenerate ground state. These levels are generally rather closely spaced, with separations of the order of 1 eV or less. They are therefore easily reached, often by thermal agitation or by absorption of visible or infrared light if the transition is allowed by selection rules. It stands to reason that the existence of these low-lying levels increases the chemical reactivity of an atom; the physico-mathematical mechanism by which this occurs will be described in the following chapters. As we shall see in this section, many optical transitions between these low-lying levels are forbidden by selection rules so that the excited states are metastable with respect to light emission. Accordingly, once an atom has been excited to one of these levels it may store the excitation energy for a long time.

This section outlines improvements of the independent electron approximation which introduce a degree of correlation among electrons sufficient to yield the experimentally observed multiplicity of levels. Consider for this purpose the Schrödinger equation for a system of N atomic electrons with the electrostatic potential between all particles and the spin-orbit coupling for each electron:

$$\left[-\frac{\hbar^2}{2m_e}\sum_{i=1}^{N}\nabla_i^2-\sum_{i=1}^{N}\frac{Ze^2}{r_i}+\sum_{i=1}^{N}\sum_{j=1}^{i-1}\frac{e^2}{|\vec{r}_i-\vec{r}_j|}+E_{coup}\right]$$
$$\times\langle\vec{r}_1 m_{s1},\ldots,\vec{r}_N\ m_{sN}|E\ \eta\rangle=\langle\vec{r}_1\ m_{s1},\ldots,\vec{r}_N\ m_{sN}|E\ \eta\rangle E_\eta. \tag{16.13}$$

This equation is an extension of equation (12.9) to a system of many electrons. It is derived from the energy expression

$$E=\sum_i K_i-\sum_{i=1}^{N}\frac{Ze^2}{r_i}+\sum_{i=1}^{N}\sum_{j=1}^{i-1}\frac{e^2}{|\vec{r}_i-\vec{r}_j|}+E_{coup}, \tag{16.13a}$$

whose first term represents the kinetic energy of the N electrons, the second one their attraction by the nucleus, and the third one the mutual repulsion of all pairs of electrons. (The summations $\sum_{i=1}^{N}\sum_{j=1}^{i-1}$ are such as to count each electron pair only once.) The spin-orbit term E_{coup} will be given later. The eigenstates of (16.13) are labeled by an index η pending a more adequate classification.

Our goal is to construct approximate eigenfunctions of the equation (16.13) by superposition of independent-electron wave functions of the same configuration.[8] The point of departure is that the Slater-determinant wave function described in section 16.2.3 for a state $|Ns\rangle$ is an eigenfunction of the Schrödinger equation for an independent-electron model. This equation is obtained from equation (16.3) applied to each electron, namely,

$$\left[-\frac{\hbar^2}{2m_e}\sum_i \nabla_i{}^2 + \sum_i U(r_i)\right]\langle \vec{r}_1\ m_{s1}, \ldots, \vec{r}_N\ m_{sN}|Ns\rangle$$

$$= \langle \vec{r}_1\ m_{s1}, \ldots, \vec{r}_N\ m_{sN}\,|\,Ns\rangle\, E_{Ns}, \tag{16.14}$$

where $E_{Ns} = \sum_{nl} N_{nl} E_{nl}$ is the energy eigenvalue (16.8) of the independent electron approximation. The equation (16.14) is developed from the energy expression

$$\sum_{i=1}^{N} K_i + \sum_{i=1}^{N} U(r_i). \tag{16.14a}$$

Note at the outset that the interactions between particles included in the Schrödinger equation (16.13) exert no torque on the atom as a whole. Mathematically, the equation is invariant under coordinate rotations about the nucleus.[9] Therefore, the atom has stationary states which are eigenstates of the total squared angular momentum $|\vec{J}|^2$ and of one of its components, J_z. Accordingly, our first step is to determine superpositions of eigenstates of (16.14) which are eigenstates of $|\vec{J}|^2$ and of J_z, as was done for spin-orbit coupling in section 14.2. This requirement was sufficient to identify the desired superposition in the case of spin-orbit coupling. The same is not true when dealing with orbital and spin momenta of more than one electron, because one obtains different superpositions with the same eigenvalues of $|\vec{J}|^2$ and J_z.[10] This remaining degeneracy must be eliminated, in general, by formulating and solving a new Schrödinger equation which is derived from equation (16.13) by taking advantage of the previous solution of (16.14).[11]

This new Schrödinger equation is obtained by the variational method of section 12.1, that is, by substituting appropriate expressions of the partial derivatives in equation (12.8b). In chapter 12 we consider states represented by wave

---

8. To obtain quantitative agreement with the experimental energy levels it is generally necessary to superpose states of different configurations; this may be done by the method of section 17.1.3.

9. In this chapter we regard the nucleus as fixed and we do not distinguish between the electron's mass m_e and the reduced mass.

10. The method used for spin-orbit coupling is adequate not only for a single electron outside a complete subshell but also for a configuration with a single vacancy in an otherwise filled subshell.

11. Mathematically, the new equation is the projection of (16.13) onto the subspace of a set of degenerate eigenstates of (16.14).

functions $\langle\vec{r}|a\rangle$. Here, we seek stationary states which are eigenstates of $|\vec{J}|^2$ and J_z and are represented as superpositions of degenerate eigenstates of (16.14). We consider then states of the form

$$|a\ J\ M_J\rangle = \sum_{Ns} |Ns\rangle\langle Ns|a\ J\ M_J\rangle, \tag{16.15}$$

identified by the probability amplitudes $\langle Ns|a\ J\ M_J\rangle$. The derivative $\partial\mathcal{P}/\partial\langle a\ J\ M_J|Ns\rangle$, analogous to (12.8d), is simply $\langle Ns|a\ J\ M_J\rangle$ itself. The limit of $\partial\langle E\rangle_{aJM_J}/\partial\mathcal{P}$, analogous to (12.8e), that is the energy eigenvalue, will be here indicated by E_η. The main task for writing the Schrödinger equation is to construct an explicit expression for $\langle E\rangle_{aJM_J}$ in terms of $\langle Ns|a\ J\ M_J\rangle$ and to obtain its derivative analogous to (12.8c).

In constructing the explicit expression of $\langle E\rangle_{aJM_J}$ we take advantage of the fact that the energy (16.14a) has the value E_{Ns} for each state $|Ns\rangle$. Accordingly, we rewrite the complete energy expression (16.13a) as the sum of expression (16.14a) and a complement (*perturbation energy*),

$$E = \sum_{i=1}^{N} [K_i + U(r_i)] + V(\vec{r}_1, \vec{r}_2, \ldots, \vec{r}_N) + E_{coup}, \tag{16.16}$$

where

$$V(\vec{r}_1, \vec{r}_2, \ldots, \vec{r}_N) = -\sum_{i=1}^{N} \frac{Ze^2}{r_i} + \sum_{i=1}^{N}\sum_{j=1}^{i-1} \frac{e^2}{|\vec{r}_i - \vec{r}_j|} - \sum_i U(r_i), \tag{16.17}$$

is a generalization of the expression (15.16) for helium. The third term, E_{coup}, will be represented for our purposes as the sum of the spin-orbit coupling energies as given by equations (14.16) and (14.14) for each independent electron,

$$E_{coup} = \sum_{i=1}^{N} \left\langle \frac{1}{r}\frac{dU}{dr}\right\rangle_{n_i l_i} \frac{1}{2m_e^2c^2}\,\vec{l}_i\cdot\vec{s}_i. \tag{16.18}$$

The mean value of the energy is then represented by

$$\begin{aligned}\langle E\rangle_{aJM_J} &= E_{Ns} + \langle V\rangle_{aJM_J} + \langle E_{coup}\rangle_{aJM_J}\\ &= \sum_{Ns}\sum_{Ns'} \langle a\ J\ M_J|Ns\rangle\,[E_{Ns}\delta_{ss'} + \langle Ns|V|Ns'\rangle + \langle Ns|E_{coup}|Ns'\rangle]\\ &\quad\times \langle Ns'|a\ J\ M_J\rangle,\end{aligned} \tag{16.19}$$

as a linear combination of matrix elements $\langle Ns|V|Ns'\rangle$ and $\langle Ns|E_{coup}|Ns'\rangle$. These matrix elements must be calculated by means of the independent-electron Slater-determinant wave functions described in section 16.2.3. Explicit calculation of such matrix elements is an often laborious task of theoretical spectroscopy. Recall that the sums in equation (16.19) extend over a set of degenerate states only.

The derivative $\partial\langle E\rangle_{aJM_J}/\partial\langle a\ J\ M_J|Ns\rangle$, analogous to (12. 8c), is now obtained from equation (16. 19) and has algebraic form since no derivative appears in the mean value (16. 19). Accordingly, the Schrödinger equation has the form of an algebraic system analogous to equation (14. 31), namely,

$$E_{Ns}\,\langle Ns|E\ \eta\ J\ M_J\rangle + \sum_{Ns'}[\langle Ns|V|Ns'\rangle + \langle Ns|E_{coup}|Ns'\rangle]\,\langle Ns'|E\ \eta\ J\ M_J\rangle$$
$$= \langle Ns|E\ \eta\ J\ M_J\rangle\,E_\eta. \tag{16.20}$$

Solution of this finite system is a routine numerical problem and yields the more accurate energy eigenvalue E_η and the corresponding eigenstate representation

$$\sum_{Ns}|Ns\rangle\langle Ns|E\ \eta\ J\ M_J\rangle. \tag{16.21}$$

We abandon here the general procedure and show only how simpler approximate solutions can be obtained in various cases of interest.

16. 4. 1. *LS Coupling.* The magnetic energy of spin-orbit coupling, E_{coup}, is generally much smaller than the electrostatic potential energy of electrons, as noted in section 14. 2. Even in heavy atoms the value of E_{coup} for a single electron is at least one order of magnitude smaller than the energy eigenvalue E_{nl}. Accordingly, the mean magnetic energy $\langle E_{coup}\rangle_a$ is often, though not always, much smaller than the mean electrostatic perturbation energy $\langle V\rangle_a$. A simplifying approximation disregards initially $\langle E_{coup}\rangle_a$. This approximation is particularly appropriate for the treatment of ground-state or low-excitation configurations.

The electrostatic perturbation energy (16. 17) is independent of the mutual orientation of orbital and spin currents. Therefore, eigenstates of V can also be eigenstates of the squared total spin angular momentum $|\vec{S}|^2$ and of the total squared orbital angular momentum $|\vec{L}|^2 = |\sum_i \vec{l}_i|^2$.[12] Accordingly, one selects superpositions of independent-electron states $|Ns\rangle$ which are characterized by quantum numbers S and L, as well as by J and M_J,

$$|a\ S\ L\ J\ M_J\rangle = \sum_{Ns}|Ns\rangle\langle Ns|a\ S\ L\ J\ M_J\rangle. \tag{16.22}$$

The addition of spin and orbital angular momenta performed by this superposition is called *LS coupling* or also *Russell-Saunders coupling*. As we know, the requirement that a superposition represent an eigenstate of J and M_J normally identifies the state $|a\ J\ M_J\rangle$ of a given configuration only if the configuration has a single electron (or a single vacancy). The additional requirement of LS coupling identifies the state $|a\ S\ L\ J\ M_J\rangle$ in a larger class of

---

12. Here the orbital momentum $\vec{L}$ is the moment of momentum about the center of mass of the atom, whereas in section 15. 2 we dealt with the moment of momentum about the center of mass of a pair of particles.

cases. In particular, it does so for all configurations with two electrons (or two vacancies) in addition to filled subshells. When a state is thus completely identified, the label "a" is no longer necessary in equation (16.22).

The values of S and L for the states of a given configuration are determined by the general procedure of addition of angular momenta (section 14.2.3). First one combines the orbital angular momenta of two electrons with their spin momenta, then one combines the resultant with the momenta of a third electron, and so on. The restrictions placed by the exclusion principle upon the quantum numbers of electrons in the same subshell limit the possible values of the quantum numbers S and L which result from the addition process. We only give without proof one example of these limitations. The S and L quantum numbers of two electrons (or two vacancies) in the same subshell are subject to the restriction $L + S = \text{even}$. Note that this rule has the same form as equation (15.2) even though the meaning of L is different in the two cases.

When the quantum numbers L, S, J, and M_J identify the state, the average perturbation energy $\langle V \rangle_{SL}$ of the state $|S\ L\ J\ M_J\rangle$ is already an eigenvalue of the electrostatic perturbation energy V itself. This value is independent of the orientation of the angular momenta $\vec{S}$ and $\vec{L}$ and therefore independent of the quantum numbers J and M_J. However, $\langle V \rangle_{SL}$ does depend on the spin quantum number S because this quantum number determines the exchange correlation of electron positions and thereby the sign of the exchange energy terms in the expression of $\langle V \rangle_{SL}$. In particular, the values of $\langle V \rangle_{SL}$ usually follow *Hund's rule* stating that the larger the value of S, the lower the value of $\langle V \rangle_{SL}$. A particular case of this rule applies to the helium spectrum where each ortho level lies below the para level with the same orbital quantum number. Another application occurs for configurations with a half-filled subshell. Here, there is one state of lowest energy with all spins parallel in the half-filled subshell, that is, with $S = l + \frac{1}{2}$, and with $L = 0$. The electron positions in this state are correlated so as to maximize their average distances. For instance, the ground state of the nitrogen atom with $(2p)^3$ configuration has $S = \frac{3}{2}$ and is particularly stable. The particular stability of atoms with half-filled outer subshells is apparent in the plot of ionization potentials in Figure 3.6.

Next to the exchange correlation, the mutual orientation of orbital angular momenta is important. This last effect obeys often the empirical *second Hund rule* stating that, for a given value of S, the mean value of $\langle V \rangle_{SL}$ is lower for larger L.

Emission or absorption of radiation normally does not disturb the coupling of electron spins. Hence it occurs in transitions between states with the same value of S; these are allowed transitions. Accordingly, the sets of levels of each atom with a given value of S tend to form distinct spectra, as in the example of ortho and para spectra of helium. This distinction becomes less marked in the spectra of elements with larger values of Z, where the LS coupling approximation gets poorer.

Optical transitions between two levels with different values of S are called intercombination lines. A well-known example of intercombination lines in heavy elements is afforded by the line of Hg at $\lambda = 2537$ Å. This line is emitted in a transition between a state with L = 1 and S = 1 of the configuration (6s)(6p) and the ground state $(6s)^2$, L = 0, S = 0. This line is quite intense; nevertheless, it is some 30 times less intense than the companion, fully allowed, line at λ = 1850 Å, which is emitted in the transition from (6s)(6p), L = 1, S = 0 to the ground state.

We have seen that eigenstates of $|\vec{S}|^2$ and $|\vec{L}|^2$ are still degenerate in the initial Russell-Saunders approximation. This degeneracy of states with the same values of S and L but different J is removed by taking into account the effect of spin-orbit coupling, that is, by considering the interaction E_{coup} represented by equation (16.18). This interaction depends on the mutual orientation of the spin and orbital current of *each electron*; accordingly, it is not compatible with the electrostatic perturbation energy V which depends, instead, on the coupling of all spins with each other and of all orbital momenta with each other. In the LS coupling approximation one evaluates not the eigenvalues of E_{coup} but only its mean value for each state $|S\ L\ J\ M_J\rangle$. The energy of these states is then represented by

$$E_{Ns} + \langle V\rangle_{SL} + \langle E_{coup}\rangle_{SLJ}, \tag{16.23}$$

where E_{Ns} is the independent-electron energy (16.8). The energy (16.23) thus depends on J, while $E_{Ns} + \langle V\rangle_{SL}$ does not.

The evaluation of $\langle E_{coup}\rangle_{SLJ}$ starts from the expression (16.18) of E_{coup}. The mean value of each orbital momentum $\vec{l}_i$ in this expression is proportional to $\vec{L}$; similarly, the mean value of $\vec{s}_i$ is proportional to $\vec{S}$. Thus one writes, in analogy to equations (14.13a), (14.16) and (14.26),

$$\langle E_{coup}\rangle_{SLJ} = W\frac{2\vec{L}\cdot\vec{S}}{\hbar^2} = W[J(J+1) - L(L+1) - S(S+1)], \tag{16.24}$$

where W is a proportionality constant determined by detailed calculations. Generally, W is much smaller than the perturbation energy $\langle V\rangle_{SL}$ insofar as spin-orbit coupling is weak.

Levels with the same values of L and of S, but with different values of J, are generally very close to one another, owing to the small value of W, and are said to form a *multiplet*. The number of levels of a multiplet is 2S + 1 or 2L + 1, whichever is smaller. A multiplet is called a *singlet, doublet, triplet, quartet*, etc., depending on whether S = 0, ½, 1, 3/2, etc., even though 2L + 1 may be smaller than 2S + 1. Thus, for example, a single level with S = 1 and L = 0 is called a triplet. This convention derives from the fact that sets of levels with the same value of S form distinct spectra.

For example, consider again the Hg levels with the configuration (6s)(6p). There are four such levels, all with L = 1. Three of them form a triplet with S = 1 and J = 0, 1, and 2 and have energies of 4.67, 4.89, and 5.46 eV, respectively, above the ground-state energy. The fourth level is a singlet with S = 0 and has higher energy, according to Hund's rule, namely, 6.70 eV above the ground state. The line at λ = 2537 Å is emitted in a transition from the triplet level with J = 1. Transitions to the ground state from the triplet levels with J = 0 and J = 2 are not observed in optical spectra because they are forbidden by the selection rules on the variations of J.[13]

Energy levels and eigenstates identified by quantum numbers L, S, and J are usually labeled by spectroscopic code symbols. The central symbol is a capital letter representing the value of L according to the code

L	0	1	2	3	4	5	...
Symbol	S	P	D	F	G	H	... .

The value of S is indicated by a numerical superscript on the left-hand side equal to the *multiplicity* 2S + 1 of the level. The value of J is indicated as a subscript on the right-hand side. A superscript,°, is added on the right to the symbol of odd-parity states.[14] Thus, for example, the ground state level of the nitrogen atom is represented by $(2p)^3\ {}^4S^{\circ}_{3/2}$; this symbol reads "2p cubed, quartet, S, odd, three halves." The Hg levels mentioned above are indicated by $(6s)(6p)\ {}^3P^{\circ}_0$, $(6s)(6p)\ {}^3P^{\circ}_1$, $(6s)(6p)\ {}^3P^{\circ}_2$, and $(6s)(6p)\ {}^1P^{\circ}_1$, in order of increasing energy.

As an additional example we describe the set of levels of the ground-state configuration of carbon, $(2p)^2$, and of oxygen, $(2p)^4$. According to equation (16.11) each of these configurations has 15 independent-electron degenerate states. Superposition of these 15 states yields 15 different LS coupling states. There are nine ${}^3P$ states, represented by ${}^3P_0$, ${}^3P_1$, and ${}^3P_2$ with different values of M_J; five states ${}^1D_2$, and one ${}^1S_0$. According to Hund's rule the triplet states are energetically lowest. The state ${}^3P_0$ is the ground state for carbon, for which W in equation (16.24) is positive. However, for oxygen the multiplet is *inverted*, meaning that W is negative as usual for configurations with more than half-filled subshells; accordingly, ${}^3P_2$ is the ground state level. The level separations within the multiplet are of the order of 0.01 eV for both carbon and oxygen. The ${}^1D_2$ and ${}^1S_0$ levels lie successively higher for both atoms and have separations of the order of 1 eV. These levels are metastable; that is, they do not decay readily by radiation emission, since the decays are forbidden by selection rules. These rules would require a change of parity, no change of quantum number S, and a change of L not exceeding unity. The metastability of ${}^1S_0$ and ${}^1D_2$ states of oxygen permits oxygen atoms in the upper atmosphere to store for hours excitation energy received from the sun.

16.4.2. *jl and jj Coupling.* The theory of energy levels according to the LS coupling approximation proves unrealistic under various circumstances which reduce the influence of the electrostatic perturbation energy V. This occurs particularly when an electron is highly excited, and the atom may be regarded

---

13. A complete statement and derivation of all selection rules is beyond the scope of this book; it may be found, for instance, in Condon and Shortley, *TAS*.

14. Each energy eigenstate of an atom is an eigenstate of parity. If it is represented by a superposition of states with different configurations, the superposition includes only configurations of the same parity.

as consisting of an ionic core with an electron loosely attached to it. In this event, the influence of the electrostatic perturbation energy V upon the motion of the excited electron is reduced. On the other hand, the spin-orbit coupling maintains its normal strength within the ionic core. Therefore, the level intervals within the multiplet ground states of the core are larger than the level separations of the excited electron. In this case the theoretical analysis must determine first the energy levels of the ionic core and characterize each of them by its quantum number J. The effect of the electrostatic perturbation V between the core and the orbital motion of the excited electron is taken into account in a second step of the calculation. This procedure is called *jl coupling*. The spin-orbit coupling for the excited electron is usually small and is taken into account as a last step of the procedure, after the jl coupling. However, the spin-orbit coupling of the excited electron is sometimes taken into account in parallel with the spin-orbit coupling of the ionic core before considering the perturbation between ionic core and electron. The procedure is then called *jj coupling*.

16.4.3. *X-Ray Levels*. The spectral effects of electron correlations which we have been discussing are important for configurations with two or more electrons or vacancies in an unfilled subshell. They are not of qualitative importance for configurations with a single electron or vacancy in an unfilled subshell, as occur in the alkalis. Situations substantially analogous to that of the alkalis occur for configurations with inner-shell vacancies which are involved in X-ray spectra.

Atoms emit characteristic X-rays when a vacancy in an inner shell is filled by an electron from a more external shell which is usually not the outermost one (see section S3.7). The transition shifts the vacancy from one subshell to another. Thus the initial, as well as the final, state of the transition belongs to a configuration with a single vacancy in an inner shell. The degeneracy of either the initial or the final configuration is then removed by taking into account the effect of spin-orbit coupling if one disregards the minor correlations between inner- and outer-shell vacancies. The level splitting is then calculated by the method of section 14.2 and may amount even to a few keV for L-shell levels of heavy atoms. Accordingly, each of the levels involved in an X-ray transition is identified by the quantum numbers (n, l, j). Traditionally, a special code designation is given to these levels. The designation consists of the letter K, L, M, . . . , which corresponds to the value of n, as indicated in section 16.2, and of a Roman numeral subscript which designates a pair of values (l, j) as follows:

l	0	1	1	2	. . .
j	$1/2$	$1/2$	$3/2$	$3/2$	. . .
code	I	II	III	IV	. . . .

The classification and calculation of levels becomes more complicated in the presence of additional vacancies in the inner shells or when the interaction with partially filled outer shells is appreciable.

16.4.4. *Zeeman Effect.* The experimental observation of the Zeeman effect in optical spectra provides important checks on the classification of levels and on the accuracy of different coupling approximations. In the first place, transitions between levels which are properly characterized as singlets ($S = 0$) in the LS approximation should exhibit a normal Zeeman effect. The Zeeman effect is anomalous for all other levels. A good illustration of the Zeeman effect is afforded by the Hg spectrum. This spectrum includes lines due to singlet-singlet transitions, with the normal Zeeman effect, as well as triplet-triplet and intercombination lines with the anomalous Zeeman effect.

When the Zeeman effect is anomalous, analysis of the line pattern serves to determine experimentally the J values of the levels. The magnitude of the level splitting determines the gyromagnetic ratio for each level and the equivalent Landé factor $g = \gamma/(-e/2m_e c)$. In LS coupling approximation, the Landé factor results from an average of the spin and orbital gyromagnetic ratios, just as it does for a single electron. It is in fact given by the same equation (14.25) used for a single electron with substitution of the appropriate values of L, S, and J. Comparison of theoretical and experimental values of the Landé factor provides a test of the approximation used in the theory.

16.4.5. *Transition Probabilities.* Theory can predict not only the frequencies of spectral lines but also their intensities. The intensity of a line depends on the amplitude of oscillation of the electron density arising from interference between the initial and final states of the transition. In the independent electron approximation the oscillations of the electron density are calculated as described for the hydrogen atom in section 13.5. When the stationary eigenstates are represented as superpositions of independent-electron states as in equation (16.21), the density oscillations are likewise represented as superpositions of independent-electron densities.

Comparison of theoretical and experimental results on spectral intensities provides a more critical test of theory than the comparison of frequencies because the calculation of intensities is much more sensitive to inaccuracy in the approximation. In fact, the theoretical results on absolute intensities have been quite poor until recent years. Theoretical predictions of relative intensities have been far more satisfactory in many cases.

CHAPTER 16 PROBLEMS

16.1. The ground-state configuration of the N atom, $(1s)^2(2s)^2(2p)^3$, has three "terms," namely, $^4S^\circ$, $^2D^\circ$, and $^2P^\circ$, with energies of 0, 2.4, and 3.6 eV. (a) How many distinct levels has this configuration in the absence of external fields, and what are their spectroscopic symbols? (b) How many distinct (Zeeman) levels does it have in a weak magnetic field? (c) What is the degeneracy of the ground state in the independent electron approximation?

16. 2. Determine the parity of the ground states of the following elements: Na, Mg, Al, Si, P, S, Cl.

16. 3. Determine the quantum numbers of the ground state level of the following atoms: Al$[(1s)^2(2s)^2(2p)^6(3s)^2(3p)]$, S$[(1s)^2(2s)^2(2p)^6(3s)^2(3p)^4]$, Mn$[(1s)^2(2s)^2(2p)^6(3s)^2(3p)^6(3d)^5(4s)^2]$, Fe$[(1s)^2(2s)^2(2p)^6(3s)^2(3p)^6(3d)^6(4s)^2]$. Use the first Hund rule and the fact that the multiplets of configurations with more than half-filled subshells are inverted.

16. 4. An oven at ~500° C can produce atomic beams of the following elements (with indicated outer-subshell configurations): P $(3p)^3$, Cr $(3d)^5(4s)^1$, S $(3p)^4$. Predict the number of component beams to be observed by Stern-Gerlach analysis of each element, and the Landé factors of the components.

16. 5. The energy levels of the ground-state ^3P term of the C atom lie at 0, 16. 4, and 43. 5 cm^{-1}, those of the ^3P term of the O atom at 0, 158. 5, and 226. 5 cm^{-1}. Verify the extent of the agreement between equation (16. 24) and the experimental data, and determine approximate values of the spin-orbit coupling factor W for each atom.

16. 6. The Lα "line" of the characteristic X-ray spectra of heavy atoms consists of several components of different frequencies corresponding to the various allowed transitions from levels with $n = 3$ to levels with $n = 2$. Predict the number of different frequencies to be observed, on the basis of the selection rules $\Delta l = \pm 1$ and $\Delta j = 0$ or ± 1.

16. 7. List the N-fold sets of quantum numbers of the degenerate independent-electron states of the ground-state configuration of the C atom, $(1s)^2(2s)^2(2p)^2$. [It is actually sufficient to list the 15 twofold sets of magnetic quantum numbers (m, m_s) for the 2p electrons, because all other quantum numbers are identical in all 15 sets.] (a) Classify these sets according to their values of $M_J = \sum_i (m_i + m_{si})$, of $M_L = \sum_i m_i$, and of $M_S = \sum_i m_{si}$. (b) Show that the number of N-fold sets for each value of M_J coincides with the number of LS-coupled states $|S\ L\ J\ M_J\rangle$ with the same value of M_J for the $(2p)^2$ configuration. (c) Show that the four Slater determinants with $|M_J| = 2$ represent LS-coupled wave functions. (The construction of the other LS-coupled wave functions requires superposition of two or more Slater determinants.)

SOLUTIONS TO CHAPTER 16 PROBLEMS

16. 1. (a) Five levels, of which two pairs form doublets: $^4S^\circ_{3/2}$, $^2D^\circ_{3/2}$, $^2D^\circ_{5/2}$, $^2P^\circ_{1/2}$, $^2P^\circ_{3/2}$.

(b)

Zero-field level:	$^4S^\circ_{3/2}$	$^2D^\circ_{3/2}$	$^2D^\circ_{5/2}$	$^2P^\circ_{1/2}$	$^2P^\circ_{3/2}$
Number of Zeeman components:	4	4	6	2	4

Total number: 20.

(c) Equation (16.11) with $l = 1$ and $N_{nl} = 3$ gives 20-fold degeneracy. [The eigenfunctions of the 20 levels obtained in (b) are approximated by 20 orthogonal superpositions of the 20 Slater determinants of the independent-electron states.]

16.2. These elements belong to a row of the periodic table and have a common neon-like core $(1s)^2(2s)^2(2p)^6$ with even parity. Each additional 3s electron contributes even parity, each 3p electron odd parity. Therefore the parities are

Na	Mg	Al	Si	P	S	Cl
even	even	odd	even	odd	even	odd.

16.3. Al: $^2P^\circ_{1/2}$; S: $^3P_2$; Mn: $^6S_{5/2}$; Fe: $^5D_4$.

16.4. The oven temperature corresponds to $k_BT = 0.07$ eV; it is sufficient only to excite levels of the ground-state multiplet. According to the first Hund rule the ground-state terms are $^4S^\circ$ for P, ^7S for Cr, and ^3P for S; there is just one level for $^4S^\circ$ and ^7S, but three levels for ^3P. Since $L = 0$ for P and Cr, their atoms have $g = 2$ and $J = 3/2$ and 3, respectively, and split into four and seven components. For sulfur, equation (14.25) gives $g = 3/2$ for all levels of the ground-state multiplet. Therefore, the three components of $^3P_1$ atoms, with $M_J = 1, 0$, and -1, are not separated from those of $^3P_2$ with the same M_J; the $^3P_0$ atoms are undeflected, as other atoms with $M_J = 0$. Thus sulfur atoms are split into five equidistant components with $g = 3/2$.

16.5. Relative values of the energy levels are represented by 0, 2W, and 6W for $J = 0, 1$, and 2 according to equation (16.24); their intervals should be in a ratio 1:2. The experimental ratios are $16.4/27.1 = 1{:}1.65$ for C and $68.0/158.5 = 1{:}2.23$ for O whose triplet is inverted. The constant W can thus be determined only to approximately 10-20% accuracy, at $W \sim 7\ \text{cm}^{-1}$ for C and $W \sim -37\ \text{cm}^{-1}$ for O.

16.6. The levels with $n = 3$ and $n = 2$ and the allowed (+) and forbidden (×) lines are:

n = 2		n = 3: l	0	1	1	2	2
l	j	j	1/2	1/2	3/2	3/2	5/2
0	1/2		×	+	+	×	×
1	1/2		+	×	×	+	×
1	3/2		+	×	×	+	+

There are seven allowed transitions.

16.7. (a) The 15 twofold sets of different pairs (m, m_S), with $|m| \leq 1$ and $|m_S| = 1/2$ are listed below, with their values of M_L, M_S, M_J:

Twofold sets

$(m,$	$m_S)$	$(m,$	$m_S)$	M_L	M_S	M_J
1	1/2	0	1/2	1	1	2
1	1/2	1	–1/2	2	0	2
1	1/2	–1	1/2	0	1	1
1	1/2	0	–1/2	1	0	1
1	–1/2	0	1/2	1	0	1
0	1/2	–1	1/2	–1	1	0
1	1/2	–1	–1/2	0	0	0
1	–1/2	–1	1/2	0	0	0
0	1/2	0	–1/2	0	0	0
1	–1/2	0	–1/2	1	–1	0
0	1/2	–1	–1/2	–1	0	–1
0	–1/2	–1	1/2	–1	0	–1
1	–1/2	–1	–1/2	0	–1	–1
–1	1/2	–1	–1/2	–2	0	–2
0	–1/2	–1	–1/2	–1	–1	–2

(b) The states $|p^2\ S\ L\ J\ M_J\rangle$ are listed below for each value of M_J:

$M_J = 2$: $^1D_2$ $^3P_2$

$M_J = 1$: $^1D_2$ $^3P_2$ $^3P_1$

$M_J = 0$: $^1D_2$ $^3P_2$ $^3P_1$ $^3P_0$ $^1S_0$

$M_J = -1$: $^1D_2$ $^3P_2$ $^3P_1$

$M_J = -2$: $^1D_2$ $^3P_2$

(c) Slater-determinant wave functions $(1s)^2(2s)^2(2p)^2$ with $M_S = \pm 1$ represent triplet states because singlets have $M_S = 0$; Slater determinants with $M_L = \pm 2$ represent D states because P and S have $|M_L| < 2$. Part (b) of this problem lists only one triplet and one D state with $M_J = \pm 2$. The following coincidences between states listed in (a) and (b) are thus established: $(M_L = 1, M_S = 1) \equiv {}^3P_2$, $(M_L = 2, M_S = 0) \equiv {}^1D_2$, $(M_L = -1, M_S = -1) \equiv {}^3P_2$, $(M_L = -2, M_S = 0) \equiv {}^1D_2$.

17. polarization, decay, and collision. perturbation methods

In the preceding chapters we have represented the energy of atomic electrons disregarding one or another interaction among electrons or with external fields. For instance, in chapter 13 the hydrogen atom was discussed without reference to the electron spin. The study of atoms with many electrons in chapter 16 is based on the independent electron approximation which considers only the average interaction between an atomic electron and all its mates. Upon refining of the initial models we have obtained new stationary states by superposing a discrete set of stationary states of the initial model. This is what was done in chapter 14 when we took into account the spin-orbit coupling of a single electron, and again in section 16.4 when we considered the effects of electron correlation upon the set of states of a single configuration.

In this chapter we extend the range of model refinements to deal with phenomena which must be represented by superposing complete, and therefore infinite, sets of stationary states of the initial model. A complete set of states of motion of an atomic electron includes ionized states of the continuous spectrum, as noted briefly in sections S13.6 and 16.2.1. Whereas previous chapters dealt with phenomena involving only discrete states, states of the continuous spectrum play an important role in the phenomena to be considered here.

Often a superposition of states of the initial model represents observed phenomena adequately even though the probability amplitudes of all but one (or a few) of the superposed states are small. These small probability amplitudes can then be calculated by procedures of successive approximation called *perturbation methods*. The change of state thus obtained, though quantitatively small, may represent drastically new features of the system under consideration.

We shall consider here a few examples chosen both for their intrinsic interest and because their treatment serves as a prototype for countless applications. In fact, resort to perturbation methods is necessary to treat a large fraction of atomic phenomena. The first example in this chapter deals with the electric polarization of atoms, that is, with the production of a dipole moment under the influence of an external field. In this problem one may still identify in good approximation a discrete set of energy levels of the polarized atom. More profound changes result, however, in the case of degeneracy between states of the discrete and continuous spectra of the initial model. In this event a discrete state of the initial model does not go over into a stationary state; the decay of excited states of atoms by radiation emission or by the Auger effect is a typical occurrence of this type. A third type of phenomenon is represented by collision processes in which the incident particle energy has continuous eigenvalues, as has the energy after collision.

17.1. ELECTRIC POLARIZATION OF ATOMS

In the presence of an electric field $\vec{F}$ an atom is expected to acquire a mean electric dipole moment $\langle\vec{d}\rangle$. As in chapter 2, we consider primarily fields sufficiently weak for the disturbance of the atom to lie within the elastic limit. Under this condition we should have

$$\langle\vec{d}\rangle = \alpha\vec{F}. \tag{17.1}$$

The proportionality constant α, called the *polarizability* of the atom, is a scalar for atoms with spherical symmetry but is generally a symmetric tensor. We outline here the quantum-mechanical calculation of α for a single atom.

17.1.1. *Perturbation treatment for discrete eigenstates.* Consider an unpolarized atom whose complete infinite set of energy eigenvalues E_s and eigenstates $|U \; s\rangle$ is regarded as known from previous solution of the Schrödinger equation for zero field strength. We shall see later that only limited accuracy may be required in the preliminary calculation of E_s and $|U \; s\rangle$. Here s represents a complete set of quantum numbers adequate to identify an eigenstate. We intend to formulate and solve the Schrödinger equation for the atom polarized by a field of strength F, whose direction serves as the z axis of coordinates. Eigenstates of this equation can be eigenstates of the total angular momentum component J_z, with eigenvalue $M_J\hbar$, because the field exerts no torque about the z axis. These energy eigenstates of the polarized atom will be indicated by $|P \; n \; M_J\rangle$.

Following the general procedure for deriving the Schrödinger equation (section 12.1), we seek an expression of the mean energy of an unspecified state $|a\rangle$ of the atom in the field $\vec{F}$. As a result of the calculation the state $|a\rangle$ must be so determined that it coincides with an energy eigenstate $|P \; n \; M_J\rangle$ of the polarized atom. We start by representing the state $|a\rangle$ as a superposition of the complete set of initial states $|U \; s\rangle$:

$$|a\rangle = \sum_s |U \; s\rangle\langle U \; s|a\rangle. \tag{17.2}$$

The Σ_s in this case is understood to include an integration over the energy of the states $|U \; s\rangle$ of the continuous spectrum. The mean energy is represented by

$$\langle E\rangle_a = \sum_s \langle a|U \; s\rangle E_s\langle U \; s|a\rangle + eF\,\langle z\rangle_a, \tag{17.3}$$

where $z = \Sigma_i z_i$ is the sum of the z coordinates of all electrons, with the origin of coordinates set at the nucleus. The charge of each electron has been entered in equation (17.3) as $-e$; the nuclear charge does not contribute to the mean energy because its position is chosen as the point of zero potential of the field F. The mean value $\langle z\rangle_a$ can be expressed in terms of matrix elements of the type (10.20), whereby equation (17.3) becomes

$$\langle E\rangle_a = \sum_s \langle a|U \; s\rangle E_s\langle U \; s|a\rangle + eF\sum_s\sum_{s'} \langle a|U \; s\rangle\langle U \; s|z|U \; s'\rangle\langle U \; s'|a\rangle. \tag{17.4}$$

If we choose, as usual, a set of states $|U\ s\rangle$ of the unpolarized atom which are eigenstates of J_z, the matrix elements $\langle U\ s|z|U\ s'\rangle$ vanish unless the magnetic quantum numbers of $|U\ s\rangle$ and $|U\ s'\rangle$ are equal. (This selection rule follows from the same argument utilized for the single-electron equation [13. 37a].) With this choice, the mean value (17. 4) splits into a sum of terms each of which stems from the superposition of a subset of states $|U\ s\rangle$ with the same value of the quantum number M_J. Since we are seeking energy eigenstates of the polarized atom which are eigenstates of J_z with quantum number M_J, we can restrict our consideration to states $|a\rangle$ constructed by superposing one subset of states $|U\ s\rangle$, which is, however, still infinite.

From the algebraic expression (17. 4) of the mean energy $\langle E\rangle_a$ one obtains a Schrödinger equation just as the Schrödinger equation (16. 20) was obtained from the mean energy (16. 19). Here again, the Schrödinger equation takes the form of an algebraic system, namely,

$$E_s\langle U\ s|P\ n\ M_J\rangle + eF \sum_{s'} \langle U\ s|z|U\ s'\rangle\langle U\ s'|P\ n\ M_J\rangle \\ = \langle U\ s|P\ n\ M_J\rangle\mathcal{E}_n\,, \tag{17.5}$$

where $\mathcal{E}_n$ indicates the energy eigenvalue to be determined. While this equation is formally similar to equations (14. 31) and (16. 20), substantive differences should be noted.

The interactions considered in sections S14. 4 and 16. 2 have the primary effect of removing the degeneracy among a finite number of states. In the present problem we deal with no degenerate states because we have selected for treatment only eigenstates of J_z with the same value of M_J. We cannot even single out a finite subset of quasi-degenerate eigenstates because the energy differences between successive levels of the spectrum of an atom are comparable; indeed, these differences converge to zero and vanish in the continuum. Therefore, we expect the electric field to change each initial unpolarized state $|U\ s\rangle$ into a superposition of an infinite set of states with different values of the unperturbed energy E_s. This problem is therefore more complicated; on the other hand, we can treat the field strength F as a small quantity. Accordingly, we solve the Schrödinger equation by a method of successive approximations, called the *Rayleigh-Schrödinger perturbation method.*

We represent the unknown probability amplitudes and eigenvalues in equation (17. 5) as power series in F. For brevity we omit the symbol M_J from the probability amplitudes and write

$$\langle U\ s|P\ n\rangle = \langle U\ s|P\ n\rangle_0 + \langle U\ s|P\ n\rangle_1 F + \langle U\ s|P\ n\rangle_2 F^2 + \cdots, \tag{17.6}$$

and

$$\mathcal{E}_n = \mathcal{E}_{n0} + \mathcal{E}_{n1}F + \mathcal{E}_{n2}F^2 + \cdots. \tag{17.7}$$

The two power series are substituted in equation (17. 5) and terms of each order in F are required to satisfy the equation separately. Thus equation

(17.5) splits into the system

$$E_s\langle U\ s|P\ n\rangle_0 = \langle U\ s|P\ n\rangle_0 \mathcal{E}_{n0}, \tag{17.8a}$$

$$E_s\langle U\ s|P\ n\rangle_1 F + eF\sum_{s'}\langle U\ s|z|U\ s'\rangle\langle U\ s'|P\ n\rangle_0$$
$$= \langle U\ s|P\ n\rangle_1 F\mathcal{E}_{n0} + \langle U\ s|P\ n\rangle_0 \mathcal{E}_{n1}F, \tag{17.8b}$$

$$E_s\langle U\ s|P\ n\rangle_2 F^2 + eF\sum_{s'}\langle U\ s|z|U\ s'\rangle\langle U\ s'|P\ n\rangle_1 F$$
$$= \langle U\ s|P\ n\rangle_2 F^2\mathcal{E}_{n0} + \langle U\ s|P\ n\rangle_1 F\mathcal{E}_{n1}F + \langle U\ s|P\ n\rangle_0 \mathcal{E}_{n2}F^2, \tag{17.8c}$$

. . . .

The equations (17.8) still constitute an infinite system, but one that can be unraveled by first solving (17.8a), then the next equation, and so on. The unraveling can be carried as far as necessary to obtain adequate results. The zeroth-order equation (17.8a) has a nonzero solution $\langle U\ s|P\ n\rangle_0$ only if $\mathcal{E}_{n0}$ coincides with one of the zero-field eigenvalues E_s. Thus, as expected, each state $|n\rangle$ must coincide to zeroth order with one state $|s\rangle$, as indicated by

$$\langle U\ s|P\ n\rangle_0 = \delta_{sn} \tag{17.9a}$$

$$\mathcal{E}_{n0} = E_n. \tag{17.10a}$$

Upon substitution of these results, the sum over s' in the first order equation (17.8b) reduces to the single term with $s' = n$. The equation (17.8b) leads now to separate results for $s = n$ and $s \neq n$. For $s = n$, the terms with $\langle U\ s|P\ n\rangle_1$ cancel out and equation (17.8b) reduces to

$$\mathcal{E}_{n1}F = eF\langle U\ n|z|U\ n\rangle. \tag{17.10b}$$

This first-order perturbation energy normally vanishes because, in the unperturbed state $|U\ n\rangle$, the mean position of all electrons coincides with the nucleus and therefore $\langle U\ n|z|U\ n\rangle = 0$. The value of $\langle U\ s|P\ n\rangle_1$ for $s = n$ remains undetermined and is conveniently set to zero.[1] For $s \neq n$ the last term of equation (17.8b) vanishes owing to (17.9a) and the rest of the equation yields

$$\langle U\ s|P\ n\rangle_1 F = -eF\frac{\langle U\ s|z|U\ n\rangle}{E_s - E_n}, \quad \text{for } s \neq n. \tag{17.9b}$$

Application of the same procedure to equation (17.8c) with $s = n$ yields the value of $\mathcal{E}_{n2}F^2$,

---

1. Any other choice, e.g., $\langle U\ s|P\ n\rangle_1 = b$, would merely cause the probability amplitude $\langle U\ s|P\ n\rangle$ calculated to all orders to be multiplied by $1 + bF$. This factor would then be eliminated in the eventual normalization process.

$$\mathcal{E}_{n2}F^2 = -e^2F^2 \sum_{s \neq n} \frac{\langle U\ n|z|U\ s\rangle\langle U\ s|z|U\ n\rangle}{E_s - E_n}. \qquad (17.10c)$$

The equations (17.8c) with $s \neq n$ would yield the value of $\langle U\ s|P\ n\rangle_2 F^2$. However, the results obtained at this stage of the calculation, namely, the nonzero value (17.9b) of the first-order probability amplitude and the nonzero value of the second-order energy (17.10c), are sufficient for us to discuss the effect of the polarizing field $\vec{F}$. Accordingly we carry the calculation no further.

17.1.2. *Discussion of Results.* The main effect of the polarizing field is represented by the first-order term in the power expansion of the probability amplitude, namely, by equation (17.9b). This term gives the approximate amplitude of the unperturbed state $|U\ s\rangle$ in the superposition indicated by equation (17.2). The matrix element $\langle U\ s|z|U\ n\rangle$ in equation (17.9b) may be interpreted as the mean value of z for a nonstationary state constructed by superposing the zero-field states $|U\ s\rangle$ and $|U\ n\rangle$ with equal amplitudes. The probability amplitude (17.9b) equals then the ratio of the mean polarization energy of this nonstationary state to the energy difference of $|U\ s\rangle$ and $|U\ n\rangle$. The magnitude of this ratio is estimated by considering that the matrix element $\langle U\ s|z|U\ n\rangle$ is generally of the order of atomic dimensions (~ 1 Å) and that $E_s - E_n$ usually is of the order of a few eV. Therefore, the first-order probability amplitude (17.9b) will indeed be a small number (as implied when we truncated the expansion [17.6]) unless the field strength F approaches the order of 1 volt/Å = 10^8 volts/cm.

The energy shift (17.10c) produced by the polarization may be interpreted as the average of the mean polarization energies $eF\langle U\ n|z|U\ s\rangle$ weighted by the ratio (17.9b). The value of (17.10c) is clearly negative when $|P\ n\rangle$ is the ground state of an atom, in which case all E_s are larger than E_n in equation (17.10c). Thus electrons in the ground state are pulled by the field to positions of lower potential energy. For excited states the energy shift (17.10c) can be of either sign. The dependence of the energy eigenvalues $\mathcal{E}_n$ upon the field strength F, represented by equation (17.7), is called the *Stark effect.* Normally the Stark effect is quadratic because the coefficient $\mathcal{E}_{n1}$ of the linear term of (17.7) vanishes.

The excited stationary states of hydrogen and of the isoelectronic ions He^+, Li^{++}, etc., are exceptional in their response to an electric field because they need not be eigenstates of $|\vec{l}|^2$ and $|\vec{J}|^2$ at zero field (see section 13.4). In this case the matrix element $\langle U\ n|z|U\ n\rangle$ in (17.10b) does not vanish for zero-field states of the type (13.32) with off-center density distribution, and the Stark effect is *linear.*

Consider now the dipole moment induced by the electric field. The field has the effect of replacing each state $|U\ s\rangle$ of the unpolarized set by a polarized stationary state $|P\ n\rangle$ with $n \equiv s$. The dipole moment of the atom in this stationary state is represented by

$$\langle d_z\rangle_{Pn} = -e\langle z\rangle_{Pn} = -e\sum_{ss'} \langle P\ n|U\ s\rangle\langle U\ s|z|U\ s'\rangle\langle U\ s'|P\ n\rangle, \qquad (17.11)$$

where the probability amplitudes $\langle P\ n|U\ s\rangle$ and $\langle U\ s'|P\ n\rangle$ are given by the expansion (17.6). Limiting this expansion to first order in F, we have

$$\langle d_z\rangle_{Pn} = -e\langle U\ n|z|U\ n\rangle + 2e^2\sum_{s\neq n}\frac{\langle U\ n|z|U\ s\rangle\langle U\ s|z|U\ n\rangle}{E_s - E_n}F. \quad (17.12)$$

The first term, which is nonzero only for excited states of the hydrogen isoelectronic sequence, represents a zero-field moment. The coefficient of F in the second term represents the polarizability defined by equation (17.1), namely,

$$\alpha = 2e^2\sum_{s\neq n}\frac{\langle U\ n|z|U\ s\rangle\langle U\ s|z|U\ n\rangle}{E_s - E_n} = 2e^2\sum_{s\neq n}\frac{|\langle U\ s|z|U\ n\rangle|^2}{E_s - E_n}. \quad (17.13)$$

An essential feature of this result is that the polarizability is *inversely proportional* to the energy difference $E_s - E_n$ between the stationary state $|U\ n\rangle$ which is being polarized and the states $|U\ s\rangle$ for which $\langle U\ s|z|U\ n\rangle$ is large. Polarizability measurements thus provide an estimate of this difference.

For purposes of numerical calculation of the polarizability α it is important that α itself does not depend on every detail of the wave functions of the unpolarized states $|U\ s\rangle$. All that is needed is sufficient data to evaluate adequately the matrix elements $\langle U\ s|z|U\ n\rangle$. Again, only as much accuracy is required in the calculation of these matrix elements and of the corresponding energy difference $E_s - E_n$ as permits to evaluate the Σ_s in equation (17.13).

The expression (17.13) of α may be compared with the corresponding expression (2.29a) in the semiempirical treatment of chapter 2. It was assumed there that elastic displacements of atomic charges can be represented by superposition of a set of normal modes with characteristic frequencies ω_r. These frequencies can now be identified with the atomic frequencies of oscillation by establishing the correspondence

$$\omega_r \leftrightarrow \frac{E_s - E_n}{\hbar}, \quad (17.14)$$

which relates each normal mode to one pair of stationary states $|U\ s\rangle$ and $|U\ n\rangle$. (Macroscopic observations usually deal with atoms in their ground state, in which case $|U\ n\rangle$ indicates the ground state.) On this basis, comparison of equations (2.29a) and (17.13) establishes the correspondence

$$b_r \leftrightarrow 2e^2\frac{E_s - E_n}{\hbar}|\langle U\ s|z|U\ n\rangle|^2, \quad (17.15)$$

or, in terms of oscillator strengths,

$$f_r \leftrightarrow \frac{2m_e(E_s - E_n)}{\hbar^2}|\langle U\ s|z|U\ n\rangle|^2. \quad (17.15a)$$

This formula is used for the theoretical calculation of oscillator strengths from the matrix elements $\langle U \; s | \; z | U \; n\rangle$. The calculation of the matrix elements themselves with good accuracy is no simple task, except for the hydrogen atom; equation (17.15a) provides a sensitive test of this calculation against measured oscillator strengths.

Reference may be made, in closing, to the phenomenon of field emission of electrons from metals or from single atoms described briefly in section 12.3. This phenomenon is *not* included in our perturbation treatment which describes polarization as a small modification of the states $|U \; s\rangle$. Field emission, however low its probability, constitutes a qualitative change of the states $|U \; s\rangle$; this change results from lack of convergence of the expansions (17.6) and (17.7).

17.1.3. *Generalization.* The above treatment of the effect of an electric field upon an atom may serve as a prototype for the perturbation treatment of different atomic problems. At the start, an atomic system is represented by a schematic model, for which a complete set of stationary states $|U \; s\rangle$ and their energy eigenvalues E_s are regarded as known. The perturbation treatment is designed to improve upon this model. Usually the model includes an incomplete expression of the energy; the remainder, disregarded in the model, is called the *perturbation energy* and is often indicated by V. (In the polarization problem the perturbation energy is eFz; the electrostatic potential energy represented by eq. [16.17] is another typical example of perturbation energy.)

The Rayleigh-Schrödinger perturbation treatment aims at constructing a new set of perturbed states $|P \; n\rangle$ which are eigenstates of the energy inclusive of V. The states $|P \; n\rangle$ are to be identified by probability amplitudes $\langle U \; s | P \; n\rangle$ which are the solutions of a Schrödinger equation analogous to equation (17.5), namely,

$$E_s \langle U \; s | P \; n\rangle + \sum_{s'} \langle U \; s | V | U \; s'\rangle\langle U \; s' | P \; n\rangle = \langle U \; s | P \; n\rangle \mathcal{E}_n. \qquad (17.16)$$

The desired probability amplitudes $\langle U \; s | P \; n\rangle$ and the eigenvalues $\mathcal{E}_n$ are obtained by expansion into powers of the matrix elements $\langle U \; s | V | U \; s'\rangle$ of the perturbation energy. The expression of $\langle U \; s | P \; n\rangle$ expanded through first-order terms is analogous to equations (17.9a) and (17.9b) combined,

$$\langle U \; s | P \; n\rangle = \delta_{sn} - \frac{\langle U \; s | V | U \; n\rangle}{E_s - E_n}(1 - \delta_{sn}) + \cdots. \qquad (17.17)$$

The energy eigenvalue expanded through second-order terms is

$$\mathcal{E}_n = E_n + \langle U \; n | V | U \; n\rangle - \sum_{s \neq n} \frac{|\langle U \; s | V | U \; n\rangle|^2}{E_s - E_n} + \cdots, \qquad (17.18)$$

and is analogous to the results (17.10).

An important qualitative result is implied by equations (17.17) and (17.18).

The effect of a perturbation upon the nth eigenstate and eigenvalue of a system is inversely proportional to the energy differences $E_s - E_n$ between other unperturbed eigenvalues and the nth one. In other words, the closer the unperturbed energy levels of a system, the more subject the system is to perturbations.

The perturbation method is appropriate when the ratios $\langle U\ s|V|U\ n\rangle/(E_s - E_n)$ are small for all pairs of states $|U\ s\rangle$ and $|U\ n\rangle$. Note that if E_n is the lowest eigenvalue of the model, the Σ_s in equation (17.17) is certainly positive and thus has the effect of lowering the perturbed eigenvalue $\mathcal{E}_n$. Thereby the perturbation stabilizes the ground state of the system. More generally, the perturbation has the effect of making the intervals between successive levels of the spectrum more nearly uniform. Consider, in particular, two successive unperturbed levels E_n and $E_{n'}$ with $E_{n'} > E_n$. The expression (17.18) of the perturbed energy $\mathcal{E}_n$ includes a negative term $-|\langle U\ n'|V|U\ n\rangle|^2/(E_{n'} - E_n)$, which lowers $\mathcal{E}_n$. Conversely, $\mathcal{E}_{n'}$ includes a positive term of equal magnitude obtained by interchanging $E_{n'}$ and E_n. The upper one of the two levels is thereby raised. Colloquially one says that the perturbation causes any two levels to "repel" each other.

The expansions (17.17) and (17.18) clearly diverge whenever the perturbation energy V has nonzero matrix elements $\langle U\ s|V|U\ n\rangle$ between two different degenerate states for which $E_s - E_n = 0$. If V has nonzero matrix elements between degenerate states, the effect of the perturbation upon each subset of degenerate states must be worked out prior to application of the perturbation treatment, thus removing the degeneracy in a preliminary step. This is exactly what was done for spin-orbit coupling in section 14.2; the set of degenerate states $|n\ l\ m\ m_s\rangle$ was replaced there by the set $|n\ l\ j\ m_j\rangle$ whose energies depend on j. Again, in section 16.4 we considered the effect of the electric and magnetic perturbation energies V and E_{coup} upon the degenerate states of each atomic configuration; the perturbation treatment may be applied as a further refinement after the degeneracy is removed by solving equation (16.20). The preliminary removal of degeneracies was made unnecessary in our polarization problem by initial selection of states $|U\ s\rangle$ which are eigenstates of J_z. This selection was appropriate because J_z is compatible with the energy in an electric field and there are normally no two states $|U\ s\rangle$ which are degenerate and have the same value of M_J. The selection rule (13.37a) requires the matrix elements $eF\langle U\ s|z|U\ s'\rangle$ to vanish for any pair of states $|U\ s\rangle$ and $|U\ s'\rangle$ with different values of M_J.

17.2. DIPOLE OSCILLATIONS OF ATOMS

It is often important to study how the state of an atom evolves in the course of time under the influence of a perturbation. We consider first the nonstationary state of an atom which is subject to a constant electric field F, as in section 17.1; here, the atom is assumed to be unpolarized at the time $t = 0$. That is, we assume that the nonstationary state of the atom, $|a(t)\rangle$, coincides at a particular instant of time with an unperturbed state $|U\ n\rangle$ which would be stationary at zero field. We call the particular time $t = 0$ and thus set $|a(0)\rangle \equiv |U\ n\rangle$. The evolution of $|a(t)\rangle$ under the influence of the field could be represented by a superposition of the perturbed stationary states $|P\ n\rangle$ constructed in section 17.1. We shall follow, instead, a more direct procedure.

17.2.1. *Time-dependent Perturbation Treatment.* The procedure utilizes a time-dependent Schrödinger equation which derives from the time-independent equation (17.5) in the same way as equations (11.22) and (12.13) derive from equations (11.17) and (12.9), respectively. Let us multiply both sides of equation (17.5) by the still unspecified probability amplitudes

$$\langle P\ n\ M_J|a(t)\rangle = \exp(-i\mathcal{E}_n t/\hbar)\langle P\ n\ M_J|a(0)\rangle, \tag{17.19}$$

and sum over all values of the index n. On the right side we have, among other factors, the product $\mathcal{E}_n\langle P\ n\ M_J|a(t)\rangle$ which can be replaced by $i\hbar\partial\langle P\ n\ M_J|a(t)\rangle/\partial t$. After this replacement the summation over n can be carried out and yields the time-dependent equation

$$E_s\langle U\ s|a(t)\rangle + eF\sum_{s'}\langle U\ s|z|U\ s'\rangle\langle U\ s'|a(t)\rangle = i\hbar\frac{\partial\langle U\ s|a(t)\rangle}{\partial t}. \tag{17.20}$$

This equation must be solved by starting with the initial condition that $|a(0)\rangle$ coincides with $|U\ n\rangle$, that is,

$$\langle U\ s|a(0)\rangle = \delta_{sn}, \tag{17.21}$$

and by assuming that $\langle U\ s|z|U\ s'\rangle$ vanishes for $s' = s$.

Equation (17.20) is simplified by expressing the probability amplitude in the form

$$\langle U\ s|a(t)\rangle = \exp(-iE_s t/\hbar)b_s(t), \tag{17.22}$$

where b_s is a new dependent variable. Substitution of equation (17.22) into (17.20) reduces this equation to

$$i\hbar\frac{db_s}{dt} = eF\sum_{s'\neq s}\langle U\ s|z|U\ s'\rangle\exp[i(E_s - E_{s'})t/\hbar]\ b_{s'}(t). \tag{17.23}$$

In first approximation the perturbation treatment considers only variations of b_s proportional to F. Therefore, $b_{s'}(t)$ is replaced in equation (17.23) by its value *at zero field* which is given, in accordance with (17.21), by $b_{s'}(t) = \delta_{s'n}$. Integration of equation (17.23) now yields

$$b_s(t) = \delta_{sn} - eF\langle U\ s|z|U\ n\rangle\frac{\exp[i(E_s - E_n)t/\hbar] - 1}{E_s - E_n}(1 - \delta_{sn}), \tag{17.24}$$

and hence

$$\langle U\ \ s|a(t)\rangle = \delta_{sn}\exp(-iE_nt/\hbar)$$
$$-eF\langle U\ \ s|z|U\ \ n\rangle\frac{\exp(-iE_nt/\hbar)-\exp(-iE_st/\hbar)}{E_s-E_n}(1-\delta_{sn}). \qquad (17.25)$$

This equation contains the main result of the calculation.

The dipole moment of the atom, induced by the field F, is calculated as in equation (17.11), with appropriate replacement of probability amplitudes.

$$\langle d_z\rangle_{a(t)} = -e\langle z\rangle_{a(t)} = -e\sum_{ss'}\langle a(t)|U\ \ s\rangle\langle U\ \ s|z|U\ \ s'\rangle\langle U\ \ s'|a(t)\rangle. \qquad (17.26)$$

Only terms through first order in the field strength F will be retained, as was done in equation (17.12). Thus we find

$$\langle d_z\rangle_{a(t)} = -e\langle U\ \ n|z|U\ \ n\rangle + e^2\sum_{s\neq n}\frac{\langle U\ \ n|z|U\ \ s\rangle\langle U\ \ s|z|U\ \ n\rangle}{E_s-E_n}$$
$$\times\{1-\exp[-i(E_s-E_n)t/\hbar]+1-\exp[i(E_s-E_n)t/\hbar]\}F$$
$$= -e\langle U\ \ n|z|U\ \ n\rangle$$
$$+\ 2e^2\sum_{s\neq n}\frac{|\langle U\ \ s|z|U\ \ n\rangle|^2}{E_s-E_n}\left[1-\cos\left(\frac{E_s-E_n}{\hbar}t\right)\right]F. \qquad (17.27)$$

As one might have expected, this result bears a close resemblance to equation (17.12). It differs from (17.12) by the insertion of the oscillating factor $1-\cos[(E_s-E_n)t/\hbar]$ which multiplies the contribution of each term to the Σ_s. This factor vanishes at $t = 0$, in accordance with the initial conditions of our problem; its average value over a period of oscillation equals 1. In fact, the dipole moment of the nonstationary state $|a(t)\rangle$ reduces to the dipole moment (17.12) of the stationary state $|P\ \ n\rangle$ when its oscillations are averaged out.

The influence of the field F is often described by saying that the field induces the atom to perform transitions ("quantum jumps") from the initial state $|U\ \ n\rangle$ to one or the other of the states $|U\ \ s\rangle$. (See the discussion on this subject in section 7.1.2.) Suppose that the field F were suddenly removed at a time t. This could be done by having the atom exit from the space between condenser plates where the field is present. The states $|U\ \ s\rangle$ remain stationary after the exit time t. One may then determine the probability $P_{sn} = |\langle U\ \ s|a(t)\rangle|^2$ that the atom, having entered the field at $t = 0$ in the state $|U\ \ n\rangle$, emerges from the field in the state $|U\ \ s\rangle$; the excitation may be observed through the subsequent emission of radiation. From the expression (17.25) of the probability amplitudes we obtain for $s \neq n$,

$$P_{sn} = e^2F^2|\langle U\ \ s|z|U\ \ n\rangle|^2\, 2\,\frac{1-\cos[(E_s - E_n)t/\hbar]}{(E_s - E_n)^2}$$

$$= e^2F^2|\langle U\ \ s|z|U\ \ n\rangle|^2\,\frac{4\sin^2[(E_s - E_n)t/2\hbar]}{(E_s - E_n)^2}\,. \tag{17.28}$$

17.2.2. *Polarization by an Alternating Field.*[2] The time-dependent treatment of polarization may be applied to calculate the polarizability of atoms under the influence of an oscillating external field. To facilitate comparison with the macroscopic treatment of chapter 2, we represent the field strength F by

$$F = F_0 e^{-i\omega t}. \tag{17.29}$$

We substitute this expression of F in equation (17.23) and set, as before, $b_{s'}(t) = \delta_{s'n}$. Integration then yields a modified form of equation (17.24), namely,

$$b_s(t) = \delta_{sn} - eF_0\langle U\ \ s|z|U\ \ n\rangle\,\frac{\exp\,[i(E_s - E_n)t/\hbar]e^{-i\omega t} - 1}{E_s - E_n - \hbar\omega}\,(1-\delta_{sn}). \tag{17.30}$$

The corresponding probability amplitude is

$$\langle U\ \ s|a(t)\rangle = \delta_{sn}\exp(-iE_n t/\hbar)$$

$$- eF_0\langle U\ \ s|z|U\ \ n\rangle\frac{\exp[-i(E_n + \hbar\omega)t/\hbar] - \exp(-iE_s t/\hbar)}{E_s - E_n - \hbar\omega}\,(1-\delta_{sn}). \tag{17.31}$$

The induced dipole moment $\langle d_z\rangle_{a(t)}$ is calculated as before by equation (17.26). In this case, however, the probability amplitude $\langle a(t)|U\ \ s\rangle$ is not quite the complex conjugate of $\langle U\ \ s|a(t)\rangle$, because we have used the complex representation (17.29) of the external field.[3] To calculate the inverse probability amplitude $\langle a(t)|U\ \ s\rangle$, one must reverse the sign of i in the equation corresponding to (17.23) everywhere except in the expression of F. This prescription may be restated by saying that one may reverse the sign of i throughout the calculation of $\langle a(t)|U\ \ s\rangle$ provided one also reverses the sign of ω. Therefore, we have

$$\langle a(t)|U\ \ s\rangle = \delta_{ns}\exp(iE_n t/\hbar)$$

$$-\frac{\exp[i(E_n - \hbar\omega)t/\hbar] - \exp(iE_s t/\hbar)}{E_s - E_n + \hbar\omega}\,eF_0\langle U\ \ n|z|U\ \ s\rangle(1-\delta_{sn}). \tag{17.32}$$

---

2. This section, while following logically the preceding one, is not necessary for further developments.

3. As explained in chapter 2, only the real part of the complex field F has physical significance. The complex representation of the field is analytically simpler to handle and may be removed at the end of the calculation.

The induced dipole moment is then calculated substituting (17.31) and (17.32) into (17.26) and retaining, as usual, only terms through first order in F,

$$\langle d_z\rangle_{a(t)} = -e\langle U\ n|z|U\ n\rangle \tag{17.33}$$

$$+ e\sum_{s\neq n}\Big\{\exp(iE_n t/\hbar)\langle U\ n|z|U\ s\rangle eF_0\langle U\ s|z|U\ n\rangle$$

$$\times \frac{\exp[-i(E_n+\hbar\omega)t/\hbar] - \exp(-iE_s t/\hbar)}{E_s - E_n - \hbar\omega}$$

$$+ \frac{\exp[i(E_n-\hbar\omega)t/\hbar] - \exp(iE_s t/\hbar)}{E_s - E_n + \hbar\omega} eF_0\langle U\ n|z|U\ s\rangle\langle U\ s|z|U\ n\rangle$$

$$\times \exp(-iE_n t/\hbar)\Big\}.$$

The two terms in the braces can be summed to yield

$$\langle d_z\rangle_{a(t)} = -e\langle U\ n|z|U\ n\rangle$$

$$+ e^2\sum_{s\neq n}\Big\{\frac{2(E_s-E_n)|\langle U\ s|z|U\ n\rangle|^2}{(E_s-E_n)^2-(\hbar\omega)^2}\left[e^{-i\omega t} - \cos\left(\frac{E_n-E_s}{\hbar}t\right)\right]$$

$$-i\frac{2\hbar\omega|\langle U\ s|z|U\ n\rangle|^2}{(E_s-E_n)^2-(\hbar\omega)^2}\sin\left(\frac{E_n-E_s}{\hbar}t\right)\Big\}F_0. \tag{17.34}$$

The alternating-field polarizability $\alpha(\omega)$ is represented by the coefficient of $F_0e^{-i\omega t}$ in equation (17.34), that is, by

$$\alpha(\omega) = 2e^2\sum_{s\neq n}\frac{(E_s-E_n)|\langle U\ s|z|U\ n\rangle|^2}{(E_s-E_n)^2-(\hbar\omega)^2}. \tag{17.35}$$

This result reduces to (17.13) for $\omega = 0$. It may be compared with the result of the macroscopic model in chapter 2, that is, with the ratio

$$\frac{\sum_r \vec{p}_{or}}{\vec{E}_0} = \sum_r \frac{b_r}{\omega_r^2 - \omega^2 - i\Gamma_r\omega} \tag{17.35a}$$

obtained from equation (2.17). The two equations (17.35) and (17.35a) are made to coincide by the correspondences (17.14) and (17.15) and by setting $\Gamma_r = 0$. Our present treatment of the atom leads to $\Gamma_r = 0$ because it has not considered any damping effects on atomic oscillations. The oscillating term of equation (17.34) proportional to $\cos[(E_n - E_s)t/\hbar]$ ensures that the dipole moment has zero value at the initial time $t = 0$. It represents a normal-mode oscillation of the atom induced by the transient at $t = 0$, an effect disregarded in chapter 2. The imaginary term proportional to $\sin[(E_n - E_s)t/\hbar]$

is a mathematical artifact due to the complex representation (17.29) of the external field **F**.

17.3. DECAY OF EXCITED ATOMS

The perturbation method has been introduced as applicable insofar as each matrix element of the perturbation energy, $\langle U\ s | V | U\ n\rangle$, is much smaller than the energy difference $E_s - E_n$. The method may break down when this energy difference becomes very small because in this case the perturbed probability amplitudes diverge. However, the divergence does not actually occur in the time-dependent expressions (17.25) and (17.28) of the probability amplitudes and of their square magnitudes. In the limit $E_s - E_n \to 0$ these expressions remain finite. It is thus possible to treat by the perturbation method an important class of problems without removing degeneracy by a preliminary treatment.

This class of problems occurs when the initial model of an atomic system has a discrete energy level E_n degenerate with levels E_s of a continuous spectrum. As an example, consider an atom, in the independent electron approximation, with an inner electron in an excited state. If the excitation energy is transferred to an electron of an outer shell, this electron usually acquires sufficient energy to escape from the atom (the Auger effect; see section S3.7.6). For instance, in the case of the lithium atom the excited configuration $(1s)^1(2s)^2$ has an excitation energy $E_{2s} - E_{1s}$; transfer of this energy to a 2s electron, whose initial energy level is E_{2s}, permits it to escape with kinetic energy $\epsilon = 2E_{2s} - E_{1s}$ (see Fig. 17.1). Thus the $(1s)^1(2s)^2$ configuration is degenerate with the configuration $(1s)^2(\epsilon s)^1$ which belongs to the continuous spectrum of the ionized atom.

Still more important is the degeneracy of the combined system of atom plus radiation field. Thus far we have disregarded the interaction between atoms and radiation in our atomic models. In fact, any excited state of an atom in the absence of radiation is degenerate with lower levels of the atom combined with a photon of the radiation field. The interaction between atom and radiation field acts as a perturbation which causes transitions with de-excitation of the atom and excitation of a photon.

As we shall see, the first-order perturbation treatment yields a total transition probability, that is, the probability of finding the system in any state of the continuous spectrum, which increases linearly with time if the system is taken to be initially in a discrete state. The linear increase holds only in the lowest order of the perturbation treatment, because this approximation does not ensure that the total probability of finding the system in any possible state equals unity. This requirement is met by higher-order approximations which also calculate the decrease of the probability of finding the system in the initial state. Under simple conditions this probability decays exponentially in accordance with the semimacroscopic considerations of section 3.4. The lowest-order perturbation treatment calculates, in effect, the linear term of a power expansion of the exponential decay. The whole perturbation treatment

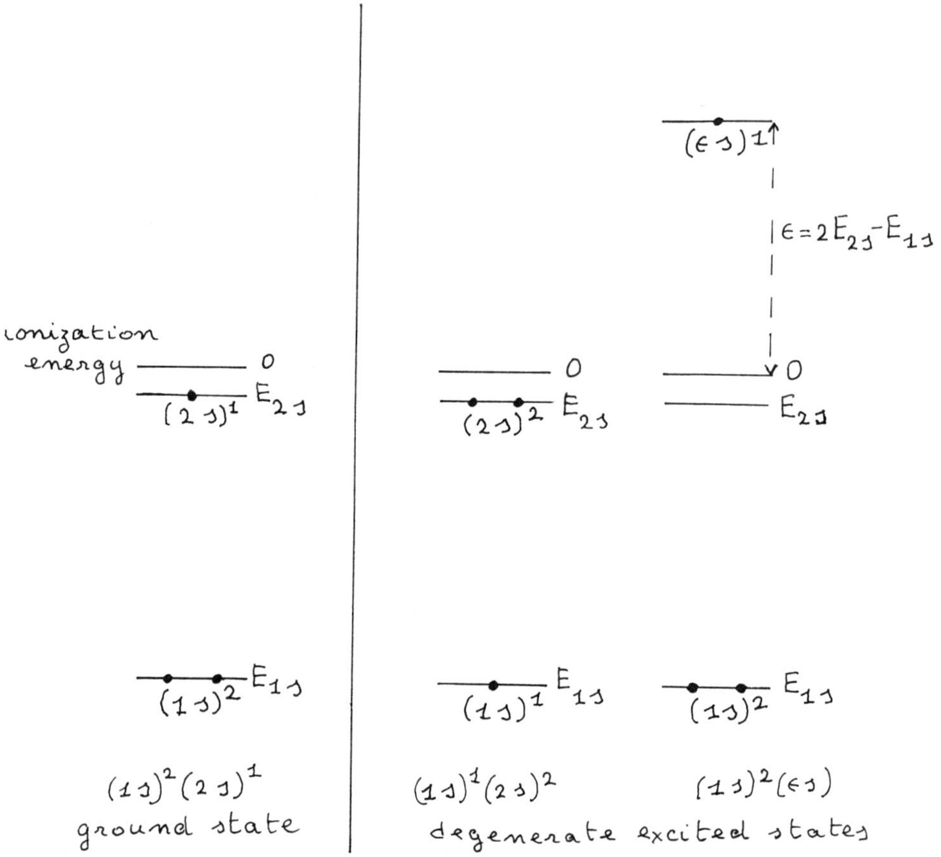

Fig. 17. 1. Configurations of lithium atom, showing example of Auger transition between degenerate states of the independent electron approximation.

of systems with continuous spectra holds insofar as the transition probabilities remain sufficiently small.

17. 3. 1 *Auger Effect (Autoionization)*. The main perturbation energy which causes transitions of an atom from a state of an excited configuration to a state of an ionized configuration is the electrostatic energy V represented by equation (16. 17). Occasionally, the relevant perturbation is the spin-orbit coupling energy. In chapter 16 we discussed primarily how these interactions reduce the degeneracy between states belonging to the same configuration. Here we discuss how the same interactions cause transitions between states of different configurations.

Degenerate states of a given configuration have been labeled in chapter 16 by an index Ns, which stands for the N-fold set of discrete quantum numbers of all electrons. Labeling the states of an ionized configuration requires an index ϵ which represents the energy of the ionized electron and an index which we call σ, which stands for all the discrete quantum numbers of all electrons. Thus σ indicates the spin orientation and the orbital quantum numbers l and m of the ionized electron, as well as all the quantum numbers of the electrons that remain in the ion. The energy of the state $|\epsilon\ \sigma\rangle$ will be indicated by $\epsilon + E_\sigma$. Matrix elements of the electric interaction V (or of the magnetic interaction E_{coup}) will be indicated by $\langle\epsilon\ \sigma|V|s\rangle$; here and in the following we reduce the label Ns to s. These matrix elements may be calculated by using the independent-electron Slater-determinant wave functions described in section 16.2.3. Their numerical calculation is generally laborious but quite feasible.

We consider then a nonstationary state which at the time $t = 0$ coincides with a state $|s\rangle$ of a discrete configuration—for example, the configuration $(1s)^1(2s)^2$ of lithium. The probability that this state coincides after the time t with an ionized state $|\epsilon\ \sigma\rangle$ will be indicated by $P_{\sigma s}(\epsilon)$. The value of this probability, calculated by the time-dependent perturbation method, is given by equation (17.28) with appropriate changes of symbols:

$$P_{\sigma s}(\epsilon) = |\langle\epsilon\ \sigma|V|s\rangle|^2 \frac{4\sin^2[(\epsilon + E_\sigma - E_s)t/2\hbar]}{(\epsilon + E_\sigma - E_s)^2}. \tag{17.36}$$

The quantity of interest is the integral of $P_{\sigma s}(\epsilon)$ over all values of ϵ. Mathematically, it is apparent that the main contribution to this integral arises from values of ϵ which differ from $E_s - E_\sigma$ by an amount of the order of $\hbar/t$. Otherwise the sine factor oscillates rapidly as a function of ϵ and the denominator increases. Physically, this means that the main contribution to the total probability arises from transitions to final states whose energy $\epsilon + E_\sigma$ is confined to a narrow range $E_s \pm \hbar/t$.[4] The variations of the squared matrix element over this narrow range can then be disregarded in equation (17.36); accordingly, the matrix element may be evaluated at $\epsilon = E_s - E_\sigma$ and factored out of the integral. The remaining integral is of the standard type $\int_{-\infty}^{\infty}(\sin x/x)^2dx = \pi$, with x related to ϵ by the substitution $\epsilon = E_s - E_\sigma + (2\hbar/t)x$. Thus one finds the total probability

$$P_{\sigma s}^{tot} = \int_0^{\infty} d\epsilon P_{\sigma s}(\epsilon) = \frac{2\pi}{\hbar}|\langle E_s - E_\sigma, \sigma|V|s\rangle|^2 t. \tag{17.37}$$

This formula is so important that it is often colloquially called the *golden rule*.

---

4. Note that this limitation to the energy range results from the superposition of Fourier components that oscillate with different frequencies. The longer the time interval t, the narrower the spectral range that contributes to the integral of $P_{\sigma s}(\epsilon)$. This mathematical effect is analogous to the one encountered in the inversion of the Fourier expansion (section A.3.5). The departure from energy conservation, $E_\sigma + \epsilon - E_s$, at any particular time t is of the order of $\hbar/t$ in accordance with the uncertainty relation (11.30a).

Many textbooks give it with an additional normalization factor

$$P_{\sigma s}^{tot} = \int_0^\infty d\epsilon P_{\sigma s}(\epsilon) = \frac{2\pi}{\hbar} |\langle E_s - E_\sigma, \sigma | V | s\rangle|^2 \rho(E_s - E_\sigma)t. \tag{17.37a}$$

The factor ρ equals unity when the wave functions of the ionized atom are normalized in accordance with equation (10.12) as we have assumed here. Specifically, if the wave function is indicated by $\langle\tau | \epsilon\ \sigma\rangle$, where τ stands for all position and spin coordinates of all electrons, equation (10.12) becomes $\int d\tau \langle\epsilon'\ \sigma' | \tau\rangle\langle\tau | \epsilon\ \sigma\rangle = \delta(\epsilon' - \epsilon)\delta_{\sigma'\sigma}$. The wave functions normalized with this convention may be inconvenient, as discussed in footnote 9 on p. 218. For this reason, other conventions are often used and the difference is compensated by the factor ρ which is called the *density of states* in the continuous spectrum.

Numerical evaluation of $P_{\sigma s}^{tot}$ gives results in fair agreement with experimental evidence. The rate of occurrence of the Auger (autoionization) transition, $dP_{\sigma s}^{tot}/dt$, is often of the order of 10^{14} sec^{-1}, and is largely independent of the energy $E_s - E_\sigma$ of the ejected electron. Auger transitions thus proceed approximately two or more orders of magnitude more slowly than electron oscillations. This means that an atom with excitation sufficient to produce the Auger effect may still experience a large number of oscillations before decaying.

17.3.2. *Radiation Emission.* So far we have studied the motion of atomic electrons disregarding their interaction with the electromagnetic field of radiation. Consider states of an atom which are stationary in the presence of all intra-atomic forces as well as of interactions with constant external and magnetic fields. These states are no longer stationary when one takes also into account the coupling of electrons with radiation fields.

This coupling may be represented for our purposes by a perturbation energy formally similar to the interaction with a constant field,

$$\mathcal{V} = -e\sum_i \vec{\mathcal{F}}\cdot\vec{r}_i = -e\vec{\mathcal{F}}\cdot\vec{r}. \tag{17.38}$$

Here $\vec{r}_i$ represents the position coordinates of the ith electron, and $\vec{r}$ the resultant of all $\vec{r}_i$. The radiation field $\vec{\mathcal{F}}$ is now no longer a macroscopic variable (as was the field strength F in sections 17.1 and 17.2), but it is a quantum-mechanical variable.[5]

We will apply the time-dependent perturbation treatment to a process of radiation emission by an atom. Specifically, we consider a state of the atom coinciding at $t = 0$ with an excited state $|s\rangle$ which would be stationary in the absence of the interactions $\mathcal{V}$. At $t = 0$ we also assume that the radiation field is in a state that contains no photon; in the language of section 12.5 we say that

5. The expression (17.38) of the interaction energy $\mathcal{V}$ is valid under the assumption that the radiation field $\vec{\mathcal{F}}$ is practically uniform over the whole atom. For short-wavelength radiation, such as X-rays, the simple dipole form of the interaction (17.38) must be replaced by a more accurate one.

the amplitude Q of each Fourier component of the radiation is in the ground state of oscillation with $n = 0$. We shall calculate the probability that at a time t the atom is in a state $|s'\rangle$ different from $|s\rangle$ and that the radiation field is in a state that contains one photon. For simplicity we consider only states $|s'\rangle$ with the same value of the quantum number M_J as the state $|s\rangle$. We also consider initially only a single set of Fourier components of the radiation field, indicated by σ, which travel in a direction $\hat{k}_\sigma$ and whose linear polarization has the electric field $\mathcal{F}$ in the direction $\hat{u}_\sigma$.[6] The frequency ω_σ of these Fourier components is left initially unspecified. Each component has amplitude Q_σ and energy levels identified by n_σ. We indicate by E_s the energy at $t = 0$ of the complete system of atom plus radiation, with the atom in the state $|s\rangle$ and with no radiation photon. The energy of the final state of the complete system is then indicated by $E_{s'} + \hbar\omega_\sigma$.

Following the treatment of section 17.3.1, one calculates the probability $P_{\sigma s', s}(\omega_\sigma)$ analogous to equation (17.36) that at time t a transition has occurred from the atomic state $|s\rangle$ to the state $|s'\rangle$ with emission of one photon of energy $\hbar\omega_\sigma$. The important quantity is the total probability of emission of one photon of any frequency ω_σ, that is, $\int_0^\infty d\omega_\sigma P_{\sigma s', s}(\omega_\sigma)$. Here, as in section 17.3.1, the main contribution to the integral arises from values of ω_σ such that the final state energy $E_{s'} + \hbar\omega_\sigma$ coincides approximately with E_s. This condition of energy conservation determines the value of ω_σ. Thus one obtains the probability expression analogous to equation (17.37a):

$$P^{tot}_{\sigma s', s} = \int_0^\infty d\omega_\sigma P_{\sigma s', s}(\omega_\sigma) = \frac{2\pi}{\hbar^2}\left|\langle n_\sigma = 1, s'|V|n_\sigma = 0, s\rangle\right|^2 \rho(\omega_\sigma)t. \qquad (17.39)$$

Here $\rho(\omega_\sigma)$ is a normalization factor discussed below and evaluated at $\omega_\sigma = (E_s - E_{s'})/\hbar$; the factor $\hbar^{-2}$ appears in equation (17.39) in place of $\hbar^{-1}$ in (17.37a) because $P^{tot}_{\sigma s', s}$ is an integral over frequencies ω_σ rather than over energies.

In the calculation of the matrix element in (17.39), only the z component of $\vec{r}$ gives a nonzero contribution, owing to the selection rule (13.38a), because $|s\rangle$ and $|s'\rangle$ have the same value of M_J. The product $\vec{\mathcal{F}}\cdot\vec{r}$ in (17.38) then reduces to $\mathcal{F} z\, \hat{u}_\sigma\cdot\hat{z}$, and we have

$$\langle n_\sigma = 1, s'|V|n_\sigma = 0, s\rangle = -e\langle n_\sigma = 1|\mathcal{F}|n_\sigma = 0\rangle\langle s'|z|s\rangle\, \hat{u}_\sigma\cdot\hat{z} \qquad (17.40)$$

To obtain the matrix element of $\mathcal{F}$ we must adapt the treatment of radiation in section 12.5 to the case of radiation in free space.

To this end we represent the electric field of radiation by a superposition of Fourier components

$$\vec{\mathcal{F}} = \sum_\sigma Q_\sigma \hat{u}_\sigma \exp(i\vec{k}_\sigma\cdot\vec{r}). \qquad (17.41)$$

---

6. The index σ pertains here, as in section 17.3.1, to states of a system with a continuous spectrum; however, here we deal with states of radiation whereas in the preceding section we dealt with states of an ionized atom.

(Note that Q_σ now indicates the amplitude of the electric field whereas in section 12.5 Q referred to the magnetic field.) We consider the Fourier expansion over a very large but finite volume L^3; hence the possible values of each Cartesian component of $\vec{k}_\sigma$ vary in steps of $2\pi/L$, much as the frequencies in equation (A.22) vary in steps of π/T. On this basis the energy of the σth Fourier component of the radiation, analogous to (12.51), is

$$E_\sigma = \frac{1}{2}\frac{L^3}{4\pi c^2 k_\sigma^2}\left(\frac{dQ_\sigma}{dt}\right)^2 + \frac{1}{2}\frac{L^3}{4\pi}Q_\sigma^2 . \qquad (17.42)$$

The eigenvalues of E_σ are $(n_\sigma + \frac{1}{2})\hbar\omega_\sigma$, with $\omega_\sigma = ck_\sigma$, where n_σ is the number of photons. The matrix element of $\mathcal{F}$ in (17.40) now coincides with that of the harmonic-oscillator amplitude Q_σ. It can therefore be taken from equation (12.46) with $n' = 1$ and $n = 0$. The value of a in (12.46) is taken from (12.43); the values of k and m to be entered in this equation are obtained by comparing the energy expressions (12.48) and (17.42). Thus we find

$$\langle n_\sigma = 1|\mathcal{F}|n_\sigma = 0\rangle = \left(\frac{2\pi\hbar\omega_\sigma}{L^3}\right)^{1/2} . \qquad (17.43)$$

There remains the calculation of the normalization factor $\rho(\omega_\sigma)$ of equation (17.39). Recall that the magnitude of each component of the wave vector $\vec{k}_\sigma$ varies in steps of $2\pi/L$, L being very large. Specifically, the wave vector $\vec{k}_\sigma$ is represented by $(h_\sigma\hat{x} + l_\sigma\hat{y} + m_\sigma\hat{z})(2\pi/L)$, where h_σ, l_σ, and m_σ are integers. Since L is very large, a summation over $(h_\sigma, l_\sigma, m_\sigma)$ is equivalent to a volume integration over $\vec{k}_\sigma$. The appropriate differential volume element is

$$\left(\frac{L}{2\pi}\right)^3 dk_{\sigma x}dk_{\sigma y}dk_{\sigma z} = \left(\frac{L}{2\pi}\right)^3 k_\sigma^2 d\hat{k}_\sigma dk_\sigma = \left(\frac{L}{2\pi c}\right)^3 \omega_\sigma^2 d\hat{k}_\sigma d\omega_\sigma. \qquad (17.44)$$

The coefficient of $d\omega_\sigma$ in this formula constitutes the normalization factor $\rho(\omega_\sigma)$ for Fourier components within an infinitesimal bundle of directions $d\hat{k}_\sigma$.

We can now substitute in the probability expression (17.39) the results (17.40), (17.43), and (17.44) and obtain

$$P^{tot}_{\sigma s',s} = \frac{1}{\hbar}\, e^2 |\langle s'|z|s\rangle|^2 \,\frac{(\hat{u}_\sigma\cdot\hat{z})^2}{2\pi}\left(\frac{\omega_\sigma}{c}\right)^3 d\hat{k}_\sigma\, t. \qquad (17.45)$$

This formula may be extended by summation over all the Fourier components σ, that is, over their possible polarizations $\hat{u}_\sigma$ and directions of propagation $\hat{k}_\sigma$. The sum of $(\hat{u}_\sigma\cdot\hat{z})^2$ over two orthogonal polarizations $\hat{u}_\sigma$, perpendicular to $\hat{k}_\sigma$, equals $1 - (\hat{k}_\sigma\cdot\hat{z})^2$. The integration of this factor over $d\hat{k}_\sigma$ yields $8\pi/3$ and thereby

$$P^{tot}_{s's} = \sum_\sigma P^{tot}_{\sigma s',s} = \frac{4}{3\hbar}\, e^2 |\langle s'|z|s\rangle|^2 \left(\frac{\omega_\sigma}{c}\right)^3 t = \frac{4}{3}\frac{e^2}{\hbar c}\, |k_\sigma\langle s'|z|s\rangle|^2 \omega_\sigma t. \qquad (17.46)$$

This probability can be discussed by considering its value for a time interval t of the order of the period of oscillation of the radiating atom, that is, for $\omega_\sigma t \approx 1$. The ratio $e^2/\hbar c$ equals 1/137; the product $k_\sigma\langle s'|z|s\rangle$ is usually of the order of 1/100. Accordingly, the total probability of emission by the atom within this time interval is of the order of 10^{-6}. This is why atomic excited states $|s\rangle$ are considered stationary for many purposes.

17.3.3. *Natural Line Width.* In the study of the Auger effect and of radiation emission we have calculated the probability of the process as an integral over all possible values of the energy ϵ of the escaping electron or of the energy $\hbar\omega_\sigma$ of the emitted photon. In both cases the main contribution to the integral was seen to arise from a narrow energy range centered at the "energy conservation" values, which are, for the two problems, $\epsilon \sim E_s - E_\sigma$ and $\hbar\omega_\sigma \sim E_s - E_{s'}$, respectively. It was also noted that the width of this energy range, namely, $\hbar/t$, is inversely proportional to the time elapsed from the beginning of the process. Here, we stress the important point that the duration of the process is effectively limited since the probability of the initial state decays to zero and the probability of all the final states saturates to unity. Therefore, the range of energies of the escaping electron or photon does not converge to zero with increasing length of observation. The limiting value of this range, called *natural line width,* is determined by observing the distribution of energies with a spectrometer which collects all the electrons or photons emitted by a population of identical isolated atoms in a given excited state.[7]

The existence of a nonzero natural line width derives from the fact that the initial state of the atom is not rigorously stationary; its energy E_s is defined only within a preliminary model that excludes the mechanism of decay. In the presence of this mechanism the state is not stationary and thus has no energy eigenvalue. The center of the observed line identifies the mean energy of the initial state.

Detailed calculation of the energy distribution of the emitted electrons or photons requires a high-order treatment of the process of perturbation and decay. We state here only that the results of a complete analysis of the phenomenon agree with those of the macroscopic model of chapter 2. In particular, the energy spectrum of an emitted photon is represented by the Fourier transform of the damped dipole oscillation (2.36). The damping parameter Γ_r in equation (2.36) represents the rate of decay of the initial state. If the initial state decays only by radiation emission, then the relevant value of Γ_r is that which we had called $\Gamma_{r,rad}$, given by equation (2.22); in this case $\Gamma_{r,rad}$ represents the natural line width.

In general, the energy of any "stationary state" of an atomic system is defined only to within a natural line width $\hbar\Gamma_r$, where Γ_r represents the total rate of decay of the state by any process whatsoever. An adequate theory of any phenomenon which serves to measure the energy of a state must include all interactions which lead to the decay of the state; the combined effect of all interac-

---

7. The resolving power of the spectrometer must, of course, be higher than the natural line width.

tions is represented by the decay constant Γ_r. For example, the theory of absorption spectra leads back to the macroscopic formula (2.18).

17.4. COLLISION THEORY IN BORN APPROXIMATION

When an electron of sufficiently high velocity collides with an atom, the electric forces between the electron and the atomic particles are strong only for a very short time and therefore have a small probability of causing any observable effect. Under these circumstances, the effect of a collision may be calculated by the time-independent perturbation treatment of section 17.1.3.[8] However, any collision phenomenon involves an intrinsic degeneracy since the incident electron may be deflected in an infinity of directions with equal energy. Treatment of this degeneracy requires a considerable effort to adapt the formulas obtained in section 17.1. The perturbation treatment of collision processes is called the Born approximation.

The theory of the collision can be formulated in the manner outlined in section 6.4. Initially, one considers an unperturbed state of the system of atom plus incident electron. We call this state $|U\ \vec{p}_0\ n_0\rangle$, where $|\vec{p}_0\rangle$ is a momentum eigenstate of the incident electron and $|n_0\rangle$ a state of the atom (usually the ground state) which is stationary apart from interaction with other systems. In the presence of the interaction energy V between the incident electron and the atom, the state $|U\ \vec{p}_0\ n_0\rangle$ is replaced by a perturbed stationary state $|P\ \vec{p}_0\ n_0\rangle$. This state can be represented as a superposition of a complete set of unperturbed states

$$|P\ \vec{p}_0\ n_0\rangle = \sum_n \int d\vec{p}\,|U\ \vec{p}\ n\rangle\langle U\ \vec{p}\ n|P\ \vec{p}_0\ n_0\rangle, \tag{17.47}$$

analogous to (17.5). The probability that the colliding electron is scattered in a direction $\hat{p}$, leaving the atom in a stationary state n, is proportional to the squared probability amplitude $|\langle U\ \vec{p}\ n|P\ \vec{p}_0\ n_0\rangle|^2$.

17.4.1. *Probability Amplitudes*. Our perturbation treatment consists of solving a Schrödinger equation of the form (17.16) to lowest order in the interaction energy. We also disregard the presence of electron spin and treat the incident electron as distinguishable from the atomic electrons. Note at the outset that in our problem the energy eigenvalue of the perturbed state coincides with the energy of the unperturbed state $|U\ \vec{p}_0\ n_0\rangle$, because the interaction operates over an infinitesimal fraction of the infinite space over which the electron moves.[9] We indicate the coincidence of the eigenvalues by writing

---

8. The time-dependent treatment of section 17.3 could also be applied, and would lead to equivalent results.

9. The treatment could be carried out formally by assuming the atom and the incident electron to be confined in a very large box. In this case the eigenvalue $\mathcal{E}_{\vec{p}_0 n_0}$ would be calculated by an equation analogous to (17.18), but it would reduce to (17.48) in the limit where the box becomes infinitely large.

$$\mathcal{E}_{\vec{p}_0 n_0} = \frac{p_0^2}{2m_e} + E_{n_0}. \tag{17.48}$$

Equation (17.16) takes now the form

$$\left(\frac{p^2}{2m_e} + E_n\right)\langle U\ \vec{p}\ n|P\ \vec{p}_0\ n_0\rangle + \Sigma_{n'}\int d\vec{p}'\langle U\ \vec{p}\ n|V|U\ \vec{p}'\ n'\rangle\langle U\ \vec{p}'\ n'|P\ \vec{p}_0\ n_0\rangle$$
$$= \langle U\ \vec{p}\ n|P\ \vec{p}_0\ n_0\rangle\left(\frac{p_0^2}{2m_e} + E_{n_0}\right). \tag{17.49}$$

The calculation of the matrix element of V will be described in section 17.4.4.

The solution of equation (17.49) is obtained as a sum of terms of successively higher order in the interaction energy V, similar to the equations (17.6) and (17.17),

$$\langle U\ \vec{p}\ n|P\ \vec{p}_0\ n_0\rangle = \langle U\ \vec{p}\ n|P\ \vec{p}_0\ n_0\rangle_0 + \langle U\ \vec{p}\ n|P\ \vec{p}_0\ n_0\rangle_1 + \dots . \tag{17.50}$$

The zeroth-order term represents the state in the absence of interaction and is simply

$$\langle U\ \vec{p}\ n|P\ \vec{p}_0\ n_0\rangle_0 = \delta(\vec{p} - p_0)\delta_{nn_0}. \tag{17.51}$$

The equation that determines the first-order term is obtained by substituting equations (17.50) and (17.51) into equation (17.49) and retaining only first-order terms, in analogy to equation (17.8b). The integral equation (17.49) reduces then to the algebraic form

$$\left(\frac{p^2}{2m_e} + E_n\right)\langle U\ \vec{p}\ n|P\ \vec{p}_0\ n_0\rangle_1 + \langle U\ \vec{p}\ n|V|U\ \vec{p}_0\ n_0\rangle$$
$$= \langle U\ \vec{p}\ n|P\ \vec{p}_0\ n_0\rangle_1\left(\frac{p_0^2}{2m_e} + E_{n_0}\right). \tag{17.52}$$

The solution of equation (17.52) would be trivial, were it not for the degeneracy which occurs when the energy of the state $|U\ \vec{p}\ n\rangle$, namely, $p^2/2m_e + E_n$, coincides with the eigenvalue (17.48), that is, with $p_0{}^2/2m_e + E_{n_0}$. In this event solving equation (17.52) involves division by zero, an undefined operation which introduces a singularity in the solution $\langle U\ \vec{p}\ n|P\ p_0\ n_0\rangle_1$. One gets around this difficulty by writing a formal solution of (17.52) and specifying how the singularity is treated when $\langle U\ \vec{p}\ n|P\ \vec{p}_0\ n_0\rangle_1$ is used to construct the superposition (17.47). The formal solution may be written in the form

$$\langle U\ \vec{p}\ n|P\ \vec{p}_0\ n_0\rangle = \mathfrak{P}\frac{\langle U\ \vec{p}\ n|V|U\ \vec{p}_0\ n_0\rangle}{p_0^2/2m_e + E_{n_0} - p^2/2m_e - E_n}$$
$$+ \delta\left(\frac{p_0^2}{2m_e} + E_{n_0} - \frac{p_0^2}{2m_e} - E_n\right) C(\vec{p}, n). \tag{17.53}$$

Here $\mathfrak{P}$ indicates that the "principal part" is to be taken in any integration over the singularity and C is a still undetermined integration parameter which may depend on $\vec{p}$ and n.[10] One verifies that the first-order probability amplitude (17.53) satisfies equation (17.52) by substituting this amplitude in the equation and noting that the δ-function term does not contribute to (17.52) since $x\delta(x) = 0$.

The presence of the integration parameter C in the first-order probability amplitude (17.53) derives from the intrinsic degeneracy of the states of the colliding electron which is not confined in space. This is the same kind of degeneracy we met in chapter 12 when dealing with the one-dimensional reflection and transmission of a particle interacting with a barrier. The general solution of the Schrödinger equation for those problems (for example, eqs. 12.26 and 12.27) contained undetermined parameters. The values of the parameters were then chosen so that the wave function represented a particle flux incident from one direction toward the barrier. In the present collision problem the incident electron flux is represented by the zeroth-order probability amplitude (17.51). Therefore, the first-order probability amplitude must represent no flux incident from infinity toward the atom. Enforcement of this condition will determine the value of C. We must then work out explicitly the electron flux at large distances from the atom; this flux is represented implicitly by the probability amplitude (17.53). To this end we construct a wave-function representation of this probability amplitude.

17.4.2. *The Wave Function.* The representation (17.47) of the perturbed state includes momentum eigenstates $|\vec{p}\rangle$ of the colliding electron, with $\vec{p} \neq \vec{p}_0$, because the electron interacts with the atom. The state $|\vec{p}\rangle$ itself is represented by the plane wave (11.6), $\langle \vec{r}\,|\vec{p}\rangle = h^{-3/2} \exp(i\vec{p}\cdot\vec{r}/\hbar)$. The wave function of the atom may be simply indicated at this point by $u_n(\tau)$, where τ represents all coordinates of atomic electrons and of the nucleus. (We shall actually regard the nucleus as having infinite mass fixed at the origin of coordinates.) The combined wave function of the system of atom plus colliding electron is then

$$\begin{aligned}\Psi(\vec{r}, \tau) &= \Psi_0(\vec{r}, \tau) + \Psi_1(\vec{r}, \tau) + \dots \qquad (17.54)\\ &= h^{-3/2} \exp(i\vec{p}_0\cdot\vec{r}/\hbar)u_{n_0}(\tau)\\ &\quad + \sum_n \int d\vec{p}\, h^{-3/2} \exp(i\vec{p}_0\cdot\vec{r}/\hbar)u_n(\tau)\langle U\ \vec{p}\ n\,|\,P\ \vec{p}_0\ n_0\rangle_1 + \dots .\end{aligned}$$

To evaluate the electron flux of interest, we need to carry out explicitly the integration over $\vec{p}$ for electron positions $\vec{r}$ at large distances from the atom. We represent $\vec{p}$ in polar coordinates with axis $\hat{r}$. Thus we set $\vec{p}\cdot\vec{r} = pr\cos\theta$

---

10. The principal part integration is defined by

$$\mathfrak{P}\int_a^b f(x)dx/x = \lim_{\epsilon=0}\left\{\int_a^{-\epsilon} f(x)dx/x + \int_\epsilon^b f(x)dx/x\right\},$$

thus excluding the singular point.

and $d\vec{p} = p^2\, dp\, d(\cos\theta)d\phi$. We shall use the property that the integral

$$\int_0^{\pi} d\cos\theta\, e^{ipr\cos\theta/\hbar} = \frac{\hbar}{ipr}\left[e^{ipr/\hbar} - e^{-ipr/\hbar}\right] \tag{17.55}$$

is inversely proportional to the large distance r. Repeated application of integration by parts over $\cos\theta$ to the integral in (17.54) yields then an expansion of $\Psi_1(\vec{r}, \tau)$ into powers of $1/r$. We retain only the first-order terms of this expansion, namely, we set

$$\Psi_1(\vec{r}, \tau) \approx h^{-3/2} \sum_n u_n(\tau) \int_0^{\infty} p^2 dp \int_0^{2\pi} d\phi \frac{\hbar}{ipr} \{e^{ipr/\hbar} \langle U\; p\hat{r}\;\; n \,|\, P\;\; \vec{p}_0\;\; n_0 \rangle_1 - e^{-ipr/\hbar} \langle U\; -p\hat{r}\;\; n \,|P\;\; \vec{p}_0\;\; n_0\rangle_1\}. \tag{17.56}$$

Here $p\hat{r}$ represents the momentum $\vec{p}$ at the limit of integration $\theta = 0$, while $-p\hat{r}$ represents $\vec{p}$ at the limit $\theta = \pi$.

Equation (17.56) represents the wave function $\Psi_1(r, \tau)$ as the superposition of a spherical wave $\exp(ipr/\hbar)$ which propagates away from the atom and of an ingoing wave $\exp(-ipr/\hbar)$. The outgoing wave yields an outward flux; the ingoing wave would yield a flux toward the atom, which must be set to zero, since $\Psi_1(\vec{r}, \tau)$ must contain no flux from infinity toward the atom.

How this condition determines the parameter C will be apparent after performing the integration over ϕ and p in (17.56). The integrand no longer depends on ϕ since p is now either parallel or antiparallel to $\hat{r}$. Therefore, the integration over ϕ yields a factor 2π. The integration over p is conditioned by the rapid oscillation of the factors $\exp(\pm ipr/\hbar)$. Here, as in the integrals of section 17.3, the only contribution to the integral derives from values of p within a very narrow range, of width $\sim\hbar/r$, near the singularity of the probability amplitude (17.53). The singularity occurs at a value of p, to be called p_n, for which the denominator of (17.53) vanishes. This condition sets the kinetic energy of the electron scattered with momentum $\vec{p}$ at the value required by energy conservation, namely,

$$\frac{p_n^2}{2m_e} = \frac{p_0^2}{2m_e} - (E_n - E_{n_0}). \tag{17.57}$$

We omit here the details of the integration and give only its result

$$\Psi_1(r, \tau) = h^{-3/2} \sum_n u_n(\tau) \frac{\hbar}{ir} 2\pi m_e \times \{\exp(ip_n r/\hbar)[-i\pi\langle U\;\; p_n\hat{r}\;\; n\,|V|\,U\;\; \vec{p}_0\;\; n_0\rangle + C(p_n\hat{r}, n)] - \exp(-ip_n r/\hbar)[i\pi\langle U\;\; -p_n\hat{r}\;\; n\,|V|\,U\;\; \vec{p}_0\;\; n_0\rangle + C(-p_n\hat{r}, n)]\}. \tag{17.58}$$

The requirement that Ψ_1 contain only components propagating away from the atom in any direction $\hat{r}$ is met by setting

$$C(\vec{p}, n) = -i\pi(U \ \vec{p} \ \ n|V|U \ \vec{p}_0 \ \ n_0). \tag{17.59}$$

Thereby the coefficient of $\exp(-ip_n r/\hbar)$ vanishes in equation (17.58), and the final expression of the wave function Ψ_1 for large r becomes

$$\Psi_1(\vec{r}, \tau) = -\sum_n u_n(\tau) \frac{2\pi m_e}{h^{1/2} r} \exp(ip_n r/\hbar)\langle U \ \ p_n\hat{r} \ \ n|V|U \ \vec{p}_0 \ \ n_0\rangle. \tag{17.60}$$

17.4.3. *Scattering Cross-section.* Recall from section 1.3.3 that the experimental data on the frequency of elementary collision processes of a given type are expressed conveniently in terms of cross-sections. A cross-section is the area σ of a mock target associated with each atom, which would be hit by incident particles as often as the scattered particles reach a specified detector. The representation we have obtained of the state $|P \ \vec{p}_0 \ \ n_0\rangle$ of an electron interacting with an atom contains all the necessary information to obtain a theoretical expression of cross-sections for electron-atom scattering.

Consider an ideal detector of area A placed at a position $\vec{r}$ very far from the atom with its surface perpendicular to $\hat{r}$. The detector scores all and only electrons that traverse its surface with kinetic energy $p_n^2/2m_e$ defined by equation (17.57). The energy of these electrons indicates that as a result of the collision the atom is in a state $|n\rangle$, with energy E_n, which may or may not coincide with the initial state $|n_0\rangle$. The state of electrons to be scored by the detector is represented by the nth component of the wave function Ψ_1 given by equation (17.60); in fact, this component represents a state of the joint system "scattered electron plus atom in the state $|n\rangle$."

The probability that the detector scores a count per unit time is the product of the area A and the electron flux at $\vec{r}$. The value of the relevant flux $\vec{\Phi}_n(\vec{r})$ is calculated by applying equation (11.16) to the nth component of Ψ_1. However, the factor $u_n(\tau)$ in the component of Ψ_1 is excluded from the calculation because it pertains to the internal coordinates of the atom which are not relevant to the flux of scattered electrons. Thus we find

$$\begin{aligned}\vec{\Phi}_n(\vec{r}) &= \frac{(2\pi m_e)^2}{hr^2} |\langle U \ \ p_n\hat{r} \ \ n|V|U \ \vec{p}_0 \ \ n_0\rangle|^2 \frac{\hbar}{m_e}\frac{p_n}{\hbar}\hat{r} \\ &= \frac{4\pi^2 m_e}{hr^2} |\langle U \ \ p_n\hat{r} \ \ n|V|U \ \vec{p}_0 \ \ n_0\rangle|^2 p_n\hat{r}.\end{aligned} \tag{17.61}$$

Consider now a mock detector of area σ_n, attached to the atom, which scores incident particles. The probability that this detector scores a count per unit time is the product of σ_n and the flux Φ_0 calculated for the wave function that represents the incident electron. This is the zeroth-order wave function Ψ_0 given by (17.54). The flux $\vec{\Phi}_0$ is

$$\vec{\Phi}_0 = \frac{1}{h^3 m_e} \vec{p}_0. \tag{16.62}$$

The cross-section σ_n is obtained by equating the two probabilities of scoring a count

$$A\Phi_n(\vec{r}) = \sigma_n \Phi_0 . \tag{17.63}$$

We have then

$$\sigma_n = (2\pi m_e h)^2 |\langle U \ p_n \hat{r} \ n | V | U \ \vec{p}_0 \ n_0 \rangle|^2 \frac{p_n}{p_0} \frac{A}{r^2} . \tag{17.64}$$

As stressed in section 1.3.3, the quantity independent of incidental details is the differential cross-section (or effective cross-section per unit solid angle), namely,

$$\frac{d\sigma_n}{d\Omega} = \lim_{A \to 0} \frac{\sigma_n}{A/r^2} = (2\pi m_e h)^2 |\langle U \ p_n \hat{r} \ n | V | U \ \vec{p}_0 \ n_0 \rangle|^2 \frac{p_n}{p_0} . \tag{17.65}$$

The ratio p_n/p_0 reduces to unity for an elastic collision; otherwise it represents the ratio of velocities of the outgoing and the incident electron. The essential part of the cross-section is represented by the squared matrix element of the interaction energy V, which we now proceed to study.

17.4.4. *Matrix Element of Interaction Energy.* The main interaction between the incident electron and the atom is the electrostatic potential energy

$$V = -\frac{Ze^2}{r} + \sum_i \frac{e^2}{|\vec{r} - \vec{r}_i|} . \tag{17.66}$$

The first term of V is due to attraction by the nucleus with atomic number Z, at the origin of the coordinates. The other terms represent the repulsion between the incident and the atomic electrons whose positions are called $\vec{r}_i$. The contribution of magnetic interaction to V will be disregarded; this contribution is important only when the incident electron's velocity approaches the velocity of light.

The calculation of the matrix elements of the interaction V is simplified by representing the Coulomb potentials by Fourier integrals

$$\frac{1}{r} = \frac{1}{2\pi^2} \int d\vec{k} \, \frac{e^{i\vec{k}\cdot\vec{r}}}{k^2} , \tag{17.67a}$$

$$\frac{1}{|\vec{r} - \vec{r}_i|} = \frac{1}{2\pi^2} \int d\vec{k} \, \frac{\exp[i\vec{k}\cdot(\vec{r} - \vec{r}_i)]}{k^2} . \tag{17.67b}$$

The simplification derives from the fact that each plane wave of the Fourier representation is the product of a function of the position of the incident electron and a function of the position of one atomic particle. Therefore, in the calculation of the matrix element, integration over the position of each particle can be carried out separately.

The wave function of each state $|U\ \vec{p}\ n\rangle$ is represented by $h^{-3/2}\exp(i\vec{p}\cdot\vec{r}/\hbar)\times u_n(\tau)$ as in section 17.4.2; the symbol τ stands for the set of position coordinates $\vec{r}$ of all atomic electrons, as well as for the spin coordinates which are unaffected by the interaction. The explicit form of the matrix element of V is

$$\langle U\ \vec{p}\ n|V|U\ \vec{p}_0\ n_0\rangle$$

$$= \int d\vec{r}\int d\tau\ h^{-3/2}e^{-i\vec{p}\cdot\vec{r}/\hbar}\ u_n^*(\tau)\left[-\frac{Ze^2}{r}+\sum_i \frac{e^2}{|\vec{r}-\vec{r}_i|}\right]h^{-3/2}\exp(i\vec{p}_0\cdot\vec{r}/\hbar)u_{n_0}(\tau)$$

$$= \frac{e^2}{2\pi^2h^3}\int\frac{d\vec{k}}{k^2}\int d\vec{r}\int d\tau\ \exp[i(\vec{p}_0-\vec{p})\cdot\vec{r}/\hbar]e^{i\vec{k}\cdot\vec{r}}\left[-Z+\sum_i \exp(-i\vec{k}\cdot\vec{r}_i)\right]$$

$$\times u_n^*(\tau)\ u_{n_0}(\tau). \tag{17.68}$$

The integration over the position $\vec{r}$ of the colliding electron involves only a product of three plane waves and thus reduces to a δ-function (see eq. [A.46])

$$\int d\vec{r}\exp[i(\vec{p}_0-\vec{p})\cdot\vec{r}/\hbar]e^{i\vec{k}\cdot\vec{r}} = (2\pi)^3\delta\left(\vec{k}+\frac{\vec{p}_0-\vec{p}}{\hbar}\right). \tag{17.69}$$

Substitution of this result into the expression (17.68) shows that a single Fourier component of the Coulomb interaction contributes to the matrix element. This is the component with $\vec{k}=(\vec{p}-\vec{p}_0)\hbar$. Thereby the matrix element reduces to

$$\langle U\ \vec{p}\ n|V|U\ \vec{p}_0\ n_0\rangle = \frac{4\pi e^2}{h^3}\frac{\hbar^2}{|\vec{p}-\vec{p}_0|^2}\int d\tau\left\{-Z+\sum_i \exp[i(\vec{p}_0-\vec{p})\cdot\vec{r}_i/\hbar]\right\}$$

$$\times\ u_n^*(\tau)\ u_{n_0}(\tau). \tag{17.70}$$

The integral in equation (17.70) depends only on atomic wave functions. The term $-Zu_n^*(\tau)u_{n_0}(\tau)$ averages out to zero, owing to orthogonality of the wave functions, except for $n=n_0$, that is, for elastic scattering. The other term represents a Fourier coefficient of the electron density function $u_n^*(\tau)u_{n_0}(\tau)$. For $n=n_0$ this coefficient coincides with the form factor defined by equation (4.21); for $n\neq n_0$, the coefficient is called a *generalized form factor* and is indicated by

$$F_{nn_0}\left(\frac{\vec{p}_0-\vec{p}}{\hbar}\right) = \int d\tau\sum_i \exp[i(\vec{p}_0-\vec{p})\cdot\vec{r}_i/\hbar]u_n^*(\tau)u_{n_0}(\tau). \tag{17.71}$$

The matrix element (17.70) can thus be written in the form

$$\langle U\ \vec{p}\ n|V|U\ \vec{p}_0\ n_0\rangle = -\frac{e^2}{\pi h|\vec{p}-\vec{p}_0|^2}\left[Z\delta_{nn_0}-F_{nn_0}\left(\frac{\vec{p}_0-\vec{p}}{\hbar}\right)\right]. \tag{17.72}$$

Substituting this result in the differential cross-section (17.65), we obtain the final result

$$\frac{d\sigma_n}{d\Omega} = \left(\frac{2m_e e^2}{|\vec{p}_0 - \vec{p}|^2}\right)^2 \left|Z\delta_{nn_0} - F_{nn_0}\left(\frac{\vec{p}_0 - \vec{p}}{\hbar}\right)\right|^2 \frac{p_n}{p_0} . \qquad (17.73)$$

17.4.5. *Discussion.* The differential cross-section (17.73) has the same structure as the Rutherford cross-section in its form (1.33). When comparing the two cross-sections, one should set $z = -1$ in (1.33), to represent the charge of the colliding electron. The substantive difference consists of the replacement of the nuclear charge Z by $Z\delta_{nn_0} - F_{nn_0}[(\vec{p}_0 - \vec{p})/\hbar]$. One may interpret this result by stating that only a single Fourier component of the distribution of atomic charges contributes to electron scattering. This component depends on the momentum change $\vec{p} - \vec{p}_0$ experienced by the colliding electron. This interpretation substantiates the surmise of section S4.4.1, namely, that the number of electrons scattered with a given momentum transfer is proportional to the squared Fourier coefficient of the electrostatic potential generated by the scatterer atom.[11]

The cross-section (17.73) applies both to elastic and inelastic scattering, whereas section S4.4.1 dealt only with the elastic case. The extension to inelastic processes was, however, implied in section S4.5 which dealt with the simultaneous transfers of momentum and energy resulting from charge oscillations within a material. We have learned now how to represent atomic charge oscillations in terms of wave functions of stationary states (sections 10.2.1 and 13.5). The charge density which oscillates with frequency $(E_n - E_{n_0})/\hbar$ and can transfer the energy $E_n - E_{n_0}$ to the colliding electron is indeed proportional to the product of wave functions $u_n^*(\tau)u_{n_0}(\tau)$ whose Fourier coefficient is represented by the generalized form factor F_{nn_0} in equation (17.72).

The treatment of collisions by lowest-order perturbation theory presupposes that the wave function of the colliding electron is modified but little even within the atom. Without examining this modification in detail, we take as a criterion for the validity of the method that the scattering cross-section be much smaller than the cross-sectional area of the atom. This condition is met for colliding electrons much faster than the atomic electrons, in accordance with the limitation set forth at the beginning of section 17.4. For high velocity and large deflection angles, $|\vec{p}_0 - \vec{p}|$ is comparable to p_0 itself, and the ratio $2m_e e^2/|\vec{p}_0 - \vec{p}|^2$ represents a length much smaller than the radius of the atom. In the opposite limit of small deflections, the factor $Z\delta_{nn_0} - F_{nn_0}$ of the cross-section is very small. (This factor vanishes for $\vec{p}_0 - \vec{p} = 0$, since it represents in this case the integral over a charge density that averages out to zero; for nonzero values of $\vec{p}_0 - \vec{p}$ the integral is small if the exponentials vary but little over the atomic volume.) A more detailed analysis of equation (17.73) shows

11. Indeed, the matrix element (17.70) can be expressed as the matrix element of the energy of the colliding electron subjected to the potential $V(\vec{q}) \exp(i\vec{q}\cdot\vec{r})$, with $V(\vec{q})$ given by equation (4.27).

that the cross-section is small not only in limiting cases but under all conditions provided the incident electron is sufficiently fast.

Equation (17.73) is in satisfactory agreement with experimental results for collisions of sufficiently fast charged particles against isolated atoms and molecules. Relativity corrections become important only when the velocity of the incident particle is within a few percent of the light velocity.

CHAPTER 17 PROBLEMS

17.1. Consider the one-dimensional motion of an electron confined in a potential well $\frac{1}{2}kz^2$ with $k = 1$ ev/Å^2, and subjected also to a perturbing electric field of strength F. Defining the dipole moment of the system as $-e\langle z\rangle_n$ for each stationary state, calculate the polarizability α: (a) by the perturbation method, using the matrix elements and energy levels given in section 12.4.3; (b) by considering the exact energy eigenstates of the system in the presence of the field.

17.2. Consider an electron confined in the square potential well of problem 12.1, with width $a = 1$ Å. Calculate the electric polarizability of this system—defined as in problem 17.1—for the ground state ($n = 0$) and the first excited state ($n = 1$), when the electric field is parallel to the x axis. Calculate numerically to two significant figures only.

17.3. Calculate the linear Stark effect of the excited state of the H atom with $n = 2$. That is, determine which superpositions of these degenerate states are approximate energy eigenstates in the presence of an electric field $F\hat{z}$ and calculate their energies. Disregard electron spin and spin-orbit coupling.

17.4. Collisions with small deflection angles θ and small energy transfer $E_n - E_{n_0}$, i.e., with small momentum transfer $|\vec{p}_0 - \vec{p}_n|^2$, are most frequent when fast electrons (or other charged particles) traverse matter. Obtain the limiting expression for small $|\vec{p}_0 - \vec{p}_n|$ of the Born-approximation cross-section, equation (17.73), for collision of an electron with a neutral atom. Show in particular that (a) the cross-section is nonzero for excitations $|n_0\rangle \to |n\rangle$ that are allowed by spectroscopic selection rules (qualitatively, the passage of a fast electron applies to an atom a pulsed electric field akin to a pulse of white light; hence the connection between electron collisions and light absorption); (b) the cross-section (a) is inversely proportional to the square of the incident electron's velocity $v_0^2 = (p_0/m_e)^2$ and to $[(E_n - E_{n_0})^2/p_0^2v_0^2 + \theta^2]$; thus it is largest at small angles $\theta < (E_n - E_{n_0})/p_0v_0$. Calculate the numerical parameters of this cross-section for an incident electron energy of 10 keV and for $E_n - E_{n_0} = 10$ eV. (c) Show also that the cross-section for elastic collisions or for transitions forbidden by selection rules is constant at small deflection angles.

Hints. Expand the exponential in equation (17.71) to second order in powers of the exponent. Expand equation (17.57) in the form $p_n \sim p_0 - (dp_0/dE_0)_{p_0} \times (E_n - E_{n_0}) = p_0[1 - (E_n - E_{n_0})/p_0 v_0]$ and represent the vector p_n in the form $[1 - (E_n - E_{n_0})/p_0 v_0]\vec{p}_0 + p_0 \theta \hat{\gamma}$, where $\hat{\gamma}$ is perpendicular to $\vec{p}_0$. Introduce a vector $\hat{\delta}$ parallel to $\vec{p}_n - \vec{p}_0$.)

SOLUTIONS TO CHAPTER 17 PROBLEMS

17.1. (a) Evaluation of the polarizability by equation (17.13) requires values of the matrix elements $\langle U \; s|z|U \; n\rangle$ and energy differences $E_s - E_n$. According to equation (12.46), the matrix element $\langle U \; s|z|U \; n\rangle$ for $s = n + 1$ is $[\frac{1}{2}(n + 1)]^{1/2}a$, with $a = [\hbar^2/km_e]^{1/4}$; it is $(\frac{1}{2}n)^{1/2}a$ for $s = n - 1$, and zero otherwise. For $s = n \pm 1$ we also have $E_s - E_n = \pm\hbar\omega = \pm(\hbar^2 k/m_e)^{1/2}$. Hence $\alpha = 2e^2 \frac{1}{2}(\hbar^2/km_e)^{1/2}[(n + 1)(\hbar^2 k/m_e)^{-1/2} - n(\hbar^2 k/m_e)^{-1/2}] = e^2/k = (14.4 \text{ eV Å})/(1 \text{ eV Å}^{-2}) = 14.4 \text{ Å}^3$, irrespective of the value of n. (b) The electron energy in the presence of the field, $\mathcal{E} = \frac{1}{2}m_e v^2 + \frac{1}{2}kz^2 + eFz$, can be written as $\mathcal{E} = \frac{1}{2}m_e v^2 + \frac{1}{2}k(z + eF/k)^2 - \frac{1}{2}e^2F^2/k$. The last expression represents the energy of an unperturbed oscillator with its equilibrium position shifted from $z = 0$ to $z = -eF/k$ and with its total energy diminished by the constant amount $-\frac{1}{2}e^2F^2/k$. The energy shift affects all energy levels equally while the shift of the equilibrium position yields a dipole moment $(-e)(-eF/k) = e^2F/k$; this dipole moment may be represented in terms of a polarizability $\alpha = e^2/k$.

17.2. Problem 12.1 gives the energy-level differences $E_s - E_n = (h^2/8m_e a^2) \times [(s + 1)^2 - (n + 1)^2]$ to be entered in equation (17.13) and it gives the eigenfunctions, but the matrix elements $\langle U \; s|z|U \; n\rangle$ must be calculated with these wave functions. For n even and s odd we have $\langle U \; s|x|U \; n\rangle = (-1)^{(s+n-1)/2} \times (2/a)\int_{-a/2}^{a/2} dx\, x \cos[(n + 1)\pi x/a] \sin[(s + 1)\pi x/a] = 8a(s + 1)(n + 1)\pi^{-2} \times [(n + 1)^2 - (s + 1)^2]^{-1}$. The same result is obtained for n odd and s even, while the matrix element is zero when n and s are both even or both odd. Equation (17.13) then gives

$$\alpha = 2e^2 \frac{512 m_e a^4}{\hbar^2 \pi^4} \sum_s^{n-s \text{ odd}} \frac{(s + 1)^2(n + 1)^2}{[(s + 1)^2 - (n + 1)^2]^3}$$

$$= 4\left(\frac{2}{\pi}\right)^6 \frac{a^4}{a_B} \sum_s^{n-s \text{ odd}} \frac{(s + 1)^2(n + 1)^2}{(s - n)^3(s + n + 2)^3}.$$

Note the rapid convergence of the sum with increasing values of $|s - n|$. With $a = 1$ Å and $a_B = \hbar^2/m_e e^2 = 0.53$ Å, we find

$$\alpha_{n=0} = 0.50\left[\frac{4}{27} + \frac{16}{3375} + \ldots\right] = 0.077\ \text{Å}^3$$

$$\alpha_{n=1} = 0.50\left[-\frac{4}{27} + \frac{36}{125} + \frac{100}{9261} + \ldots\right] = 0.076\ \text{Å}^3.$$

17.3. A complete orthogonal set of states $|n\ l\ m\rangle$ with $n = 2$ consists of $|2\ 0\ 0\rangle$, $|2\ 1\ 1\rangle$, $|2\ 1\ 0\rangle$, and $|2\ 1\ -1\rangle$. An arbitrary superposition of these states is represented by equation (17.2) if the sum over states $|U\ s\rangle$ extends over these four states. In equation (17.4) the zero-field energy E_s coincides now with E_2 for all four states. This equation also includes 16 matrix elements $\langle U\ s'|z|U\ s\rangle$ which are to be calculated by the use of the wave functions of chapter 13. However, owing to selection rules, all but two of these matrix elements are zero; the two remaining ones are equal: $\langle 2\ 1\ 0|z|2\ 0\ 0\rangle =$

$\langle 2\ 0\ 0|z|2\ 1\ 0\rangle = 3a_B$. Equation (17.4) is now

$$\langle E\rangle_a = E_2 \sum_s |\langle U\ s|a\rangle|^2 + eF3a_B[\langle a|2\ 1\ 0\rangle\langle 2\ 0\ 0|a\rangle + \langle a|2\ 0\ 0\rangle\langle 2\ 1\ 0|a\rangle].$$

Separate Schrödinger equations (17.5) are obtained from $\langle E\rangle_a$ for $m = 1, m = -1$ and $m = 0$. The first two are trivial and show that $|2\ 1\ 1\rangle$ and $|2\ 1\ -1\rangle$ remain unperturbed, with unperturbed energy eigenvalues E_2. For $m = 0$, equation (17.5) has the explicit form

$$E_2\langle 2\ 1\ 0|P, m = 0\rangle + 3eFa_B\langle 2\ 0\ 0|P, m = 0\rangle = \langle 2\ 1\ 0|P, m = 0\rangle\mathcal{E}_{m=0},$$
$$E_2\langle 2\ 0\ 0|P, m = 0\rangle + 3eFa_B\langle 2\ 1\ 0|P, m = 0\rangle = \langle 2\ 0\ 0|P, m = 0\rangle\mathcal{E}_{m=0}.$$

The determinant of this system is $(E_2 - \mathcal{E}_{m=0})^2 - (3eFa_B)^2 = 0$. Thus we have the two eigenvalues $\mathcal{E}_{m=0} = E_2 \pm 3eFa_B$, for which $\langle 2\ 1\ 0|P, m = 0\rangle/\langle 2\ 0\ 0|P, m = 0\rangle = \pm 1$. The eigenstate with the + sign has the wave function (13.32).

17.4. Expansion of the exponential in equation (17.71) gives

$$F_{nn_0}\left(\frac{\vec{p}_n - \vec{p}_0}{\hbar}\right) = Z\delta_{nn_0} + i\,\frac{|\vec{p}_n - \vec{p}_0|}{\hbar}\sum_i \langle n|\hat{\delta}\cdot\vec{r}_i|n_0\rangle$$
$$- \frac{1}{2}\,\frac{|\vec{p}_n - \vec{p}_0|^2}{\hbar^2}\sum_i \langle n|(\hat{\delta}\cdot\vec{r}_i)^2|n_0\rangle + \ldots .$$

The term $Z\delta_{nn_0}$ cancels the identical term of equation (17.73). Either one of the next two terms of the expansion of F_{nn_0} must vanish for any given transition $|n_0\rangle \to |n\rangle$. This is because $\hat{\delta}\cdot\vec{r}_i$ is odd under coordinate reflections and thus $\langle n|\hat{\delta}\cdot\vec{r}_i|n_0\rangle$ vanishes if the product of atomic eigenfunctions $u_n{}^*(\tau)u_{n_0}(\tau)$ is even; conversely $\langle n|(\hat{\delta}\cdot\vec{r}_i)^2|n_0\rangle$ vanishes if $u_n^* u_{n_0}$ is odd. In our problem $|\vec{p}_0 - \vec{p}_n|^2 \sim p_0{}^2[(E_n - E_{n_0})^2/p_0{}^2v_0{}^2 + \theta^2]$.

(a) The matrix element $\Sigma_i \langle n|\hat{\delta}\cdot\vec{r}_i|n_0\rangle$ coincides with the matrix element $\langle s'|z|s\rangle$ in the radiation emission formula (17.46) if we take a z axis in the direction $\hat{\delta}$. Therefore it has nonzero values under the same selection rules as apply to transitions in spectroscopy, including the requirement that u_n^* and u_{n_0} have opposite parity; this requirement excludes the elastic case $n = n_0$.

(b) When u_n^* and u_{n_0} have opposite parity, the last term in the expansion of F_{nn_0} vanishes. Substitution in equation (17.73) gives

$$\frac{d\sigma_n}{d\Omega} \sim \left(\frac{2e^2}{\hbar v_0}\right)^2 \frac{|\Sigma_i\langle n|\hat{\delta}\cdot\vec{r}_i|n_0\rangle|^2}{(E_n - E_{n_0})^2/p_0^2v_0^2 + \theta^2} \frac{p_n}{p_0}.$$

For the given data we have $2e^2/\hbar v_0 = 7.4 \times 10^{-2}$ and $(E_n - E_{n0})/p_0 v_0 = 5 \times 10^{-4}$ radian. (Note that the validity of the approximation method is not impaired if the differential cross-section exceeds the cross-sectional area of the atom over a small solid angle.)

(c) When u_n^* and u_{n_0} have the same parity, the linear term of F_{nn_0} vanishes. Substitution into equation (17.73) eliminates $|\vec{p}_0 - \vec{p}_n|^4$ from the denominator of that formula and gives

$$\frac{d\sigma_n}{d\Omega} \sim \left|\frac{\langle n|\Sigma_i(\hat{\delta}\cdot\vec{r}_i)^2|0\rangle}{a_B}\right|^2 \frac{p_n}{p_0}$$

since $m_e^2 e^4/\hbar^4 = 1/a_B^2$.

Part IV

AGGREGATES OF ATOMS

introduction

Parts I to III have considered properties of single atoms isolated or under the action of controlled well-defined external fields. This part begins the study of atoms interacting with one another and forming aggregates of different kinds. The aim is to show how properties of matter in bulk can be understood and predicted with various degrees of accuracy from knowledge of atomic properties and laws.

Various specific kinds of interactions among atoms provide for the chemical bonds which hold together molecules, crystals, and other aggregates. Less specific electric forces are responsible for a degree of cohesion among all kinds of atoms, particularly through the mechanism of van der Waals forces. Magnetic interactions have a minor influence on the gross structure of matter, but the study of their effects has great theoretical and practical importance.

Aggregates of atoms have a large variety of excited stationary states. Many of these excitations require only a modest amount of energy, readily supplied by thermal agitation. The enormous variety of material properties of technological interest is due largely to the existence of these low-energy excited states. Systematic theoretical study of these excitations has been practicable only for aggregates of a few atoms or under simplifying circumstances which are often rooted in symmetry. For instance, the symmetry of crystalline aggregates underlies the advanced development of solid-state physics.

The study of aggregates of many electrons and nuclei rests on the laws of quantum mechanics but also requires methods of approximation. These methods must be designed to provide a basis for useful qualitative and semiquantitative concepts such as interatomic potentials, bond energies, etc. The development of appropriate concepts and calculational procedures remains an open field of research. The treatment of matter in bulk utilizes, besides the methods of quantum theory, concepts and formulae from thermal physics, which are summarized in Appendix E.

The extension of atomic theory to include the approximation methods required for the treatment of atomic aggregates is carried out in this book for typical small molecules. The approach is developed in sufficient detail to serve as an introduction to molecular spectroscopy. By contrast, the atomistic interpretation of macroscopic properties, such as cohesion, density, paramagnetism, and heat capacity, is conducted on a qualitative or phenomenological basis. Illustrative data are taken from experiments or calculations which are not described in detail.

18. electric and magnetic polarization

The formation and stability of atomic aggregates depends, to a considerable extent, on the response of atoms to the influence of their neighbors. Neighboring atoms influence one another mainly by electric forces and, to a lesser extent, by magnetic forces. These forces may distort the electron density distribution within individual atoms and may also change their orientation.

Electric forces among neighboring atoms are generally much stronger than magnetic forces because the charges within the atoms have speeds much lower than the speed of light. The electrostatic field generated by the presence of ions or of asymmetric charge distributions within molecules contributes substantially to the cohesion of many materials. This field is orders of magnitudes stronger than external electric fields normally produced in the laboratory. On the other hand, the internal magnetic fields due to the motion of charges within a material are relatively weak and their effects are generally smaller than those produced by external magnetic fields. Only in the special phenomenon of ferromagnetism are the effects of internal and external magnetic fields comparable.

The response of a material to external electric or magnetic fields—that is, the susceptibility of the material—is important as an indicator of atomic structure and properties. Experimental data on magnetic susceptibility are more directly related to the properties of individual atoms (or molecules) than are electric susceptibility data, just because interatomic magnetic interactions are weak.[1]

The electric polarizability of atoms studied in section 17.1 causes any atom to be attracted by an ion or any other system that generates a nonuniform electric field (section 18.1). It also makes it possible and energetically convenient for a pair of neutral atoms or molecules to undergo correlated deformations which yield a net mutual attraction called the *van der Waals force* (section 18.2). In this chapter we deal with the electric polarization due to deformation of electron density within an atom. Effects due to change of orientation of molecular groups will be considered in chapter 20.

The magnetic polarization of atoms or small groups of atoms by external fields is responsible for the macroscopic phenomena of diamagnetism and paramagnetism.[2] Diamagnetism (section 18.3) arises from the induction of Larmor precession currents; paramagnetism (section 18.4) arises mostly from the orientation of orbital and spin currents. These two effects are characterized experimentally by their opposite signs and are often concurrent. Paramagnetic effects are particularly important as an analytical tool and can be measured very accurately in resonance experiments (section S18.5). The conspicuous phenomenon of ferromagnetism is characteristic of certain large aggregates of atoms and will be discussed briefly in chapter 22.

---

1. Even the phenomenon of ferromagnetism is due primarily to electrostatic interactions, as will be noted at the end of chapter 22.

2. See *BPC*, vol. 2, chap. 10.

18.1. ELECTRIC POLARIZABILITY AND ITS COHESION EFFECT

The response of a material to an external electric field is characterized macroscopically by its electric susceptibility χ_e.[3] In chapter 2 we have considered low-density materials whose susceptibility equals the product of the polarizability α of a single atom and of the number N of atoms per unit volume. In denser materials, each atom is influenced not only by the external field but also, to an appreciable extent, by the polarization of surrounding atoms. In this case the relationship among χ_e, α, and N is no longer linear. It is represented for many cases of interest, in particular for dense gases, by[4]

$$\chi_e = \frac{N\alpha}{1-(4\pi/3)N\alpha}. \tag{18.1}$$

The electric susceptibility χ_e of a material is defined, and can be measured, as the ratio of the magnitude of the electric dipole moment, P, induced in a unit volume of material to the inducing electric field strength F,[5]

$$\chi_e = P/F. \tag{18.2}$$

With regard to energy, a dipole moment $\vec{P}$ in a field $\vec{F}$ has energy $-\vec{P}\cdot\vec{F}$. The energy increment of a unit volume of the material for an increment $d\vec{F}$ of the field is $-\vec{P}\cdot d\vec{F} = -\chi_e \vec{F}\cdot d\vec{F}$. Accordingly, the polarization energy, E_{pol}, of a unit volume of material is given by

$$E_{pol} = -\int_0^\infty \chi_e F\, dF = -\tfrac{1}{2}\chi_e F^2. \tag{18.3}$$

Conversely, the susceptibility can be defined, and also measured, in terms of the dependence of the energy of a material on the applied field strength,

$$\chi_e = -\left(\frac{d^2E_{pol}}{dF^2}\right)_{F=0}. \tag{18.4}$$

Note that an equivalent relationship applies to the energy $\mathcal{E}_n$ of a single atom in an electric field, calculated in section 17.1. Equation (17.7) gives $(d^2\mathcal{E}_n/dF^2)_{F=0} = 2\mathcal{E}_{n2}$; the value of $-2\mathcal{E}_{n2}$ obtained from equation (17.10c) coincides with the polarizability α of the atom given by equation (17.12). Recall that the coefficient of the linear term of equation (17.7) is zero for the ground

---

3. See *BPC*, 2: 323.

4. See *BPC*, 2: 337; limitations of this formula are also discussed on that page.

5. Here, as in chapter 17, the electric field strength is indicated by F rather than E, to avoid confusion with energy. This is at variance with the notation of chapter 2.

state of an atom. Disregarding powers of F larger than 2, we can rewrite (17.7) in the form analogous to (18.3)

$$E_{pol} = \mathcal{E}_n - E_n = -\tfrac{1}{2}\alpha F^2. \qquad (18.5)$$

This formula is convenient for representing the change of energy of a neutral atom in the electric field due to its neighbors.

Consider now that an ion carrying a unit charge e generates at a distance r in the surrounding space a field of strength $F = e/r^2$. Any neutral atom in the neighborhood of the ion is polarized by this field and experiences an energy change $-\tfrac{1}{2}\alpha F^2$ if the distance r is much larger than an atomic radius. At closer distances the results of the perturbation treatment serve only as a guide for estimation because the ionic field can no longer be taken as weak and uniform, nor can the polarized atom be regarded as a simple dipole.[6] The energy $-\tfrac{1}{2}\alpha F^2 = -\tfrac{1}{2}\alpha e^2/r^4$ may be regarded as a potential energy of the neutral atom due to attraction by the ion. This energy is inversely proportional to the fourth power of the ion-atom distance r, and the corresponding force varies as r^{-5}.

The magnitude of the attractive potential between ion and atom may be estimated by inserting in equation (18.5) the field strength $F = e/r^2$ and the expression (17.13) of α and by regrouping the factors in the form

$$E_{pol} = -\frac{e^2}{r} \sum_{s \neq n} \frac{e^2/r}{E_s - E_n} \left| \frac{\langle U\ s|z|U\ n\rangle}{r} \right|^2. \qquad (18.6)$$

The right-hand side of equation (18.6) is factored into a potential energy factor, e^2/r, and a sum of dimensionless ratios. The factor e^2/r is of the order of 1 eV for $r \sim 10$ Å. The first dimensionless ratio, the ratio of e^2/r to an excitation energy of the neutral atom, is of the order of 1/10 for $r \sim 10$ Å. The last factor, the squared ratio of a dipole matrix element to the distance r, is of the order of $(1/10)^2$ for $r \sim 10$ Å. Thus the net energy of attraction at this distance may be three orders of magnitude smaller than e^2/r; however, it increases rapidly at shorter distances.

The attraction exerted by an ion on surrounding neutral atoms provides one of the mechanisms for the cohesion of matter. As will be mentioned in chapter 22, the same mechanism applies to electric fields generated by molecular groups that carry no net charge but possess an average dipole moment. These groups not only attract neutral groups but are themselves oriented by the field of other charges.

---

6. The effect of a stronger field may be evaluated by extending the calculation to higher powers of F in the expansions (17.6) and (17.7). The effects of non-uniformity of the field can also be treated by the general procedure of section 17.1.3.

18.2. VAN DER WAALS FORCE

A cohesive force operates between any two neutral atoms or groups of atoms. Its strength is rather low and decreases rapidly as the distance between the atoms increases. The existence of this force was emphasized by van der Waals a century ago, but its interpretation in terms of atomic mechanics came only with quantum physics.

To trace the origin of the van der Waals force, consider two neutral atoms first as independent isolated systems and then as a single system. The statement that each isolated atom in its ground state has zero dipole moment means that the mean position of all its electrons, averaged over their density distribution, coincides with the position of the nucleus. Now consider the two atoms simultaneously, though still treating them in an independent atom approximation analogous to the independent electron approximation for single atoms. In this approximation the probability distributions of the electrons of the two atoms are still independent and their mean dipole moments are still zero. Actually, the positions of the electrons of the two atoms are correlated by electrostatic repulsion between electrons which must be considered in a better approximation. For example, it is unlikely for electrons of both atoms to be concentrated in the space between the two nuclei. The most likely distributions of the electrons are those that minimize the potential energy. Examples of likely and unlikely distributions are shown in the diagram of Figure 18.1. More specifically, a calculation given below shows that the potential energy of the system of two atoms is a minimum when the electrons of both atoms are polarized in the same direction along the internuclear axis. The mean value of the dipole moment of each atom remains zero, but the product of the two dipole moments of the atoms has nonzero mean value.[7]

The correlation of electron positions in the two atoms results in a net attraction because two aligned dipoles of equal orientation attract each other. Therefore, the ground-state energy eigenvalue of a two-atom system is lower when the atoms are close to one another than when they are far apart.

The total energy of a two-atom system may be regarded as consisting of two parts. One part is the kinetic and potential energy of the particles within each atom, which does not concern us at the moment. The other part is the electrostatic potential energy between each particle of one atom and each particle of the other. We consider for simplicity only one pair of unit charges in each atom, carried by one nucleus and one electron. The contribution of additional pairs of charges can be treated separately.

Call $\vec{R}'$ the position of the nucleus of one atom and $\vec{r}' = \vec{R}' + \vec{\rho}'$ the position of the electron of the same atom; call $\vec{R}''$ and $\vec{r}'' = \vec{R}'' + \vec{\rho}''$ the corresponding

---

7. This is a typical quantum effect. The mean dipole moment of an atom is zero though its mean square dipole moment is nonzero, much as the mean velocity of an atomic electron may be zero while its mean square velocity is not.

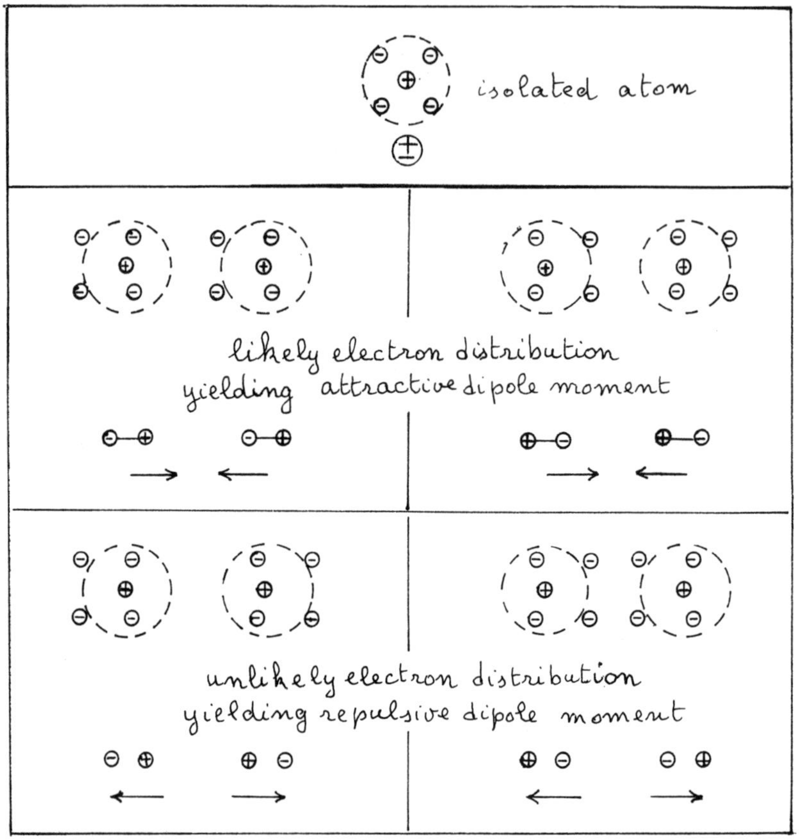

Fig. 18. 1. Examples of likely and unlikely electron distributions.

positions for the other atom. The relative position of the two nuclei will be indicated by $\vec{R} = \vec{R}' - \vec{R}''$. The potential energy of one atom with respect to the other is

$$
\begin{aligned}
V &= e^2\left\{\frac{1}{|\vec{R}' - \vec{R}''|} + \frac{1}{|\vec{r}' - \vec{r}''|} - \frac{1}{|\vec{r}' - \vec{R}''|} - \frac{1}{|\vec{r}'' - \vec{R}'|}\right\} \\
&= e^2\left\{\frac{1}{R} + \frac{1}{|\vec{R} + \vec{\rho}' - \vec{\rho}''|} - \frac{1}{|\vec{R} + \vec{\rho}'|} - \frac{1}{|\vec{R} - \vec{\rho}''|}\right\}.
\end{aligned} \tag{18.7}
$$

This energy would vanish if the electron-nucleus distance within each atom, ρ' and ρ'', were altogether negligible as compared to the internuclear distance

R. The nonvanishing part of the interaction energy V is then calculated by expanding (18.7) into powers of ρ' and ρ''. The expansion yields

$$V = e^2 \left\{ \frac{1}{R} + \left[\frac{1}{R} + \frac{(\vec{\rho}\,' - \vec{\rho}\,'') \cdot \hat{R}}{R^2} + \frac{3/2[(\vec{\rho}\,' \overset{\downarrow}{-} \vec{\rho}\,'') \cdot \hat{R}]^2 - 1/2 |\vec{\rho}\,' \overset{\downarrow}{-} \vec{\rho}\,''|^2}{R^3} + \dots \right] \right.$$

$$- \left[\frac{1}{R} + \frac{\vec{\rho}\,' \cdot \hat{R}}{R^2} + \frac{3/2(\vec{\rho}\,' \cdot \hat{R})^2 - 1/2 \rho'^2}{R^3} + \dots \right]$$

$$\left. - \left[\frac{1}{R} - \frac{\vec{\rho}\,'' \cdot \hat{R}}{R^2} + \frac{3/2(\vec{\rho}\,'' \cdot \hat{R})^2 - 1/2 \rho''^2}{R^3} + \dots \right] \right\}$$

$$= -e^2 \left[\frac{3\vec{\rho}\,' \cdot \hat{R}\, \vec{\rho}\,'' \cdot \hat{R} - \vec{\rho}\,' \cdot \vec{\rho}\,''}{R^3} + \dots \right]. \qquad (18.8)$$

The right-hand side of this formula consists of all the terms that do not cancel out in the expansion, up to third order in $1/R$.[8] These terms originate from the cross-terms in the squared binomials marked by the arrows. Equation (18.8) can represent the interaction between neutral atoms with any number of electrons N' and N'' if $\vec{\rho}\,'$ and $\vec{\rho}\,''$ are redefined as

$$\vec{\rho}\,' = \sum_{i=1}^{N'} \vec{\rho}_i{}' \quad \text{and} \quad \vec{\rho}\,'' = \sum_{i=1}^{N''} \vec{\rho}_i{}'' . \qquad (18.9)$$

The interaction energy **V** has the effect of modifying the energy eigenvalues of the two-atom system. The modification is calculated by the perturbation method of section 17.1.3. The input data for the calculation are changed as follows. Instead of the stationary states $|U\ s\rangle$ of a single unpolarized atom, we consider here stationary states $|U\ s'\ s''\rangle$ of an unperturbed pair of atoms. (Here primed symbols refer to one atom; double-primed symbols, to the other.) The energy eigenvalues of the unperturbed pair are given by the sum of separate energies $E'_{s'} + E''_{s''}$. The perturbation energy V is given here by (18.8). We consider energy eigenvalues $\mathcal{E}_n$ for the perturbed pair of atoms in their ground state, $n \equiv (s' = 0, s'' = 0)$, expanded through the second order term as given by equation (17.18). The first-order perturbation energy vanishes in the present problem (as in many others) because each single radial vector $\vec{\rho}\,'$ or $\vec{\rho}\,''$ has mean value zero in an unperturbed stationary state. The result of interest consists of the second-order correction to the energy,

$$\mathcal{E}_{00} - (E'_0 + E''_0) = -\sum_{s's''} \frac{\langle U\ 0\ 0|V|U\ s'\ s''\rangle\langle U\ s'\ s''|V|U\ 0\ 0\rangle}{E'_{s'} + E''_{s''} - E'_0 - E''_0}. \qquad (18.10)$$

---

8. Any interaction with the same mathematical structure as the right-hand side of (18.8) is called a dipole-dipole tensor interaction.

Substitution of the expression of the interaction energy V from (18.8) yields

$$\mathcal{E}_{00} - (E_0' + E_0'') = -\frac{e^4}{R^6}\sum_{s's''} \frac{|\langle U\ s'\ s''|3\vec{\rho}'\cdot\hat{R}\,\vec{\rho}''\cdot\hat{R} - \vec{\rho}'\cdot\vec{\rho}''|U\ 0\ 0\rangle|^2}{E_{s'}' + E_{s''}'' - E_0' - E_0''}. \tag{18.11}$$

An important result contained in this formula is that for any pair of atoms the interaction energy is negative—that is, attractive—and inversely proportional to the sixth power of the internuclear distance. The order of magnitude of the van der Waals interaction energy (18.11) may be estimated by the method used for the polarization energy of a single atom (18.6). This energy is of the order of 10^{-5} to 10^{-4} eV for atoms about 10 Å apart. In some cases this interaction energy is sufficient to keep two atoms close together. For instance, at low temperature, it holds two argon atoms at a distance of approximately 4 Å; at this distance the interaction amounts to a little more than 10^{-2} eV.

18.3. DIAMAGNETISM

Diamagnetism is the effect of atomic currents induced by the onset of a magnetic field. According to the Lenz law, the direction of these currents is such as to oppose the inducing field. The currents persist as long as the field is present because no resistance dissipates the energy of intra-atomic currents. Macroscopic experiments measure the magnetic susceptibility χ_m of materials[9] which results from the combined opposite effects of diamagnetism and of paramagnetism; the latter is discussed in the following section. Diamagnetism contributes to the measured susceptibility χ_m a negative amount to be indicated by χ_d. For an assembly of N atoms (or small molecules) per unit volume the macroscopic magnetic susceptibility is represented accurately by the product of N and the polarizability of each atom (or molecule).[10]

Diamagnetism derives from the induction of orbital currents; spin currents do not contribute significantly for reasons that will become apparent shortly. The theory of diamagnetism for single atoms does not require an explicit modification of the wave functions of their stationary states. It is sufficient to consider the motion of electrons in a frame of reference rotating with the Larmor velocity $\omega_L = \gamma B$ (see section 2.4). In this frame, the wave functions coincide to first approximation with the zero-field wave functions. However, the kinetic energy of the electrons is different in the rotating frame, and we consider here the resulting shift of atomic energy levels.

The change of kinetic energy of the ith atomic electron is obtained by adding to the velocity $\vec{v}_i$ of this electron the vector product of the Larmor velocity

---

9. See *BPC*, vol. 2, chap. 10.

10. This statement means that the magnetic equation analogous to (18.1) reduces to an approximation linear in B because the dimensionless values of the susceptibility χ_m are much smaller than unity.

$\vec{\omega}_L$ and of the electron's position vector $\vec{r}_i$.[11] The vector $\vec{\omega}_L$ has direction opposite to $\gamma\vec{B}$ (see eq. [2.54]), and the value of γ is $-e/2m_ec$ for the orbital motion of electrons. Accordingly, the velocity $\vec{v}_i$ is replaced by

$$\vec{v}_i \rightarrow \vec{v}_i + \vec{\omega}_L \times \vec{r}_i = \vec{v}_i + \frac{e}{2m_ec}\vec{B} \times \vec{r}_i. \tag{18.12}$$

The kinetic energy of the electron becomes

$$K_i = \tfrac{1}{2}m_ev_i^2 + m_e\vec{\omega}_L \times \vec{r}_i \cdot \vec{v}_i + \tfrac{1}{2}m_e|\vec{\omega}_L \times \vec{r}_i|^2$$

$$= \tfrac{1}{2}m_ev_i^2 + \frac{e}{2m_ec}\,Bl_{iz} + \frac{e^2}{8m_ec^2}\,B^2(x_i^2 + y_i^2). \tag{18.13}$$

Here we have chosen the $\hat{z}$ axis in the direction of $\vec{B}$ and indicated the orbital angular momentum $m_e\vec{r}_i \times \vec{v}_i$ by $\vec{l}_i$, as usual. The factor $x_i^2 + y_i^2$ represents the square of the component of $\vec{r}_i$ perpendicular to $\vec{B}$.

The first term of equation (18.13) represents the zero-field kinetic energy. The second term represents the Zeeman-effect energy of an orbital current, such as was considered in section 14.3. This term averages out for an assembly of atoms of random orientation, that is, with random values of the magnetic quantum number. The third term is independent of current circulation at zero field and thus represents the effect of diamagnetic induction. This term, which is quadratic in the field strength B, has been disregarded in the treatment of isolated atoms in previous chapters because it is negligible compared with the linear Zeeman-effect term. However, the diamagnetic term is important for an assembly of atoms for which the linear term averages out even only approximately. The diamagnetic term may be regarded as a magnetic potential energy of each electron since it depends explicitly only on the electron position.

Consider now how the diamagnetic energy of each electron affects the stationary states of a whole atom. Eigenstates of the total angular momentum component J_z remain stationary if the field $\vec{B}$ is directed along the z axis. However, the energy eigenvalue of each zero-field stationary state $|n\ J\ M_J\rangle$ is shifted by the diamagnetic effect. In a first and adequate approximation the energy shift equals the mean value of the sum of the diamagnetic energies of all electrons, calculated for the state $|n\ J\ M_J\rangle$. This shift is thus represented by

$$\Delta E_{nJM_J} = \frac{e^2}{8m_ec^2}\,B^2\,\langle\textstyle\sum_i (x_i^2 + y_i^2)\rangle_{nJM_J}. \tag{18.14}$$

---

11. This procedure is straightforward only when applied to single atoms because they possess an axis of symmetry through the nucleus which also serves as the origin of coordinates $\vec{r}_i$.

The energy shift ΔE_{nJM_J}, called the *quadratic Zeeman effect*, is proportional to the mean cross-sectional area occupied by the electron density, that is, to the magnetic flux linked with the induced current. Consequently, this effect acquires increasing importance for extended orbital motions. In fact, the approximate theory breaks down for excited states of an atom with $n \geqslant 30$ (orbital radii of the order of 1000 Å) in a field of ~25,000 gauss, because the diamagnetic energy shift (18.14) becomes comparable to the separation of levels with successive values of n. The diamagnetic effect is also large for the ground state of atomic aggregates, like graphite, where the orbital motion of electrons extends throughout the aggregate (see chapter 21). The diamagnetic effect becomes extreme in the superconducting state of matter, in which electrons move over macroscopic distances. It is also extreme for magnetic fields of astronomic strength (of the order of 10^{12} gauss) which occur in pulsars. Vice versa, the diamagnetic effect is minute for the spin currents which are narrowly confined and thus linked to a very small magnetic flux.

Barring extreme conditions under which the approximation (18.14) breaks down, measurement of the diamagnetic susceptibility provides an evaluation of the area encompassed by the orbital motion of electrons. The diamagnetic susceptibility is measured as part of the total magnetic susceptibility of a material, $\chi_m = \chi_d + \chi_p$, where χ_p is the paramagnetic susceptibility.[12] The magnetization energy per unit volume of a material in a field $\vec{B}$ is represented by $-\frac{1}{2}\chi_m B^2$. The diamagnetic contribution to this energy for an assembly of N atoms per unit volume in states $|n\ J\ M_J\rangle$ equals $N\Delta E_{nJM_J}$. Accordingly, we have an expression of χ_d analogous to (18.4), namely,

$$\chi_d = -N\left(\frac{\partial^2 \Delta E_{nJM_J}}{\partial B^2}\right)_{B=0} = -N\frac{e^2}{4m_e c^2}\langle\sum_i (x_i^2 + y_i^2)\rangle_{nJM_J}. \tag{18.15}$$

For atoms in a nondegenerate ground state with spherical symmetry, that is, with $n = J = M_J = 0$, the diamagnetic susceptibility χ_d depends only on the mean square distance of the electrons from the nucleus. In this case the mean value of $\Sigma_i (x_i^2 + y_i^2)$ equals two-thirds of the mean value of $\Sigma_i r_i^2$. It follows that

$$\chi_d = -N\frac{e^2}{6m_e c^2}\langle\sum_i r_i^2\rangle_{0,0,0}. \tag{18.16}$$

This expression relates conveniently the measured values of χ_d to the size of atoms. For example, susceptibility measurements of rare gases in their ground state yield the following values (in angstroms):

	He	Ne	Ar	Kr	Xe
$\langle\sum_i r_i^2\rangle_{0,0,0}^{1/2}$	0.57	0.54	0.85	1.03	1.44.

12. Experimentally, the paramagnetic susceptibility is sorted out in most cases through its temperature dependence discussed in section 18.4.

These values agree satisfactorily with theoretical determinations from calculated electron wave functions.

Note that the diamagnetic susceptibility (18.15) is negative for all states of atoms whereas the electric polarizability discussed in section 17.2 is positive for the ground state and for some, though not all, other states. Diamagnetism originates from an increase in electron kinetic energy, whereas electric polarization consists of a shift of electron density to a region of lower potential energy.[13]

18.4. PARAMAGNETISM

Paramagnetism is the partial orientation of atomic currents under the influence of an external magnetic field. Thermal agitation tends to randomize the magnetic quantum numbers of atoms with nonzero angular momentum. However, in the presence of a magnetic field $\vec{B}$, states with lower magnetic energy are favored by having a larger Boltzmann factor. Accordingly, the material acquires a nonzero mean magnetic moment. This mean magnetic moment depends on temperature because it results from a balance between the selecting action of the field and the randomizing action of thermal agitation. Paramagnetism is observed experimentally as a positive temperature-dependent contribution to the magnetic susceptibility of a material.

We calculate first the mean magnetic moment induced in one atom of an ensemble[14] by a magnetic field of strength B parallel to the z axis. This will be done using formulas of thermal physics summarized in Appendix E. An atom with total angular momentum quantum number J and gyromagnetic ratio $\gamma = -g\mu_B/\hbar$ has, in the magnetic field, $2J + 1$ nearly degenerate stationary states with energies $g\mu_B BM_J$. Here, as in chapter 14, g is the Landé factor and μ_B the Bohr magneton. The relative probabilities of these states are given by the Boltzmann factor $\exp(-M_J g\mu_B B/k_B T)$. For brevity we introduce the parameter

$$y = \frac{g\mu_B B}{k_B T} = \beta g\mu_B B, \tag{18.17}$$

---

13. The magnetic analog of electric polarization results from a second-order effect of the main interaction between field and current. This interaction, represented by the second *(ordinary Zeeman)* term of (18.13), has no higher-order effect on the stationary states of atoms which are eigenstates of l_z. However, it distorts the current distribution for stationary states that are not even approximately eigenstates of l_z, particularly in molecules. This effect is calculated by the method of section 17.1.3 with the perturbation energy $-(e/2m_e c)l_z B$. The result is a negative energy shift analogous to (18.6) and proportional to B^2. This shift yields a positive, temperature-independent contribution to the magnetic susceptibility, and is called *Van Vleck paramagnetism*.

14. The concept of ensemble is introduced in *BPC*, 5:56.

whereby the Boltzmann factor becomes exp $(-M_J y)$. The partitition function Z is calculated in accordance with equation (E.3) as the sum of a geometric progression

$$Z = \sum_{M_J=-J}^{J} \exp(-M_J y) = \frac{\sinh (J + \frac{1}{2})y}{\sinh \frac{1}{2}y}. \qquad (18.18)$$

The mean magnetic moment component, $\langle \mu_z \rangle$, is obtained from the derivative of the mean energy $-\langle \mu_z \rangle B$ with respect to the field strength. According to statistical mechanics the mean energy expression appropriate to an atom of an ensemble at constant temperature is the free energy

$$F = -k_B T \ln Z. \qquad (18.19)$$

We have then[15]

$$\langle \mu_z \rangle = -\frac{\partial F}{\partial B} = g\mu_B [(J + \tfrac{1}{2}) \coth (J + \tfrac{1}{2})y - \tfrac{1}{2} \coth \tfrac{1}{2}y]. \qquad (18.20)$$

This result simplifies in the limits of a very strong and of a very weak field. In the strong-field limit $y \to \infty$, the coth functions reduce to unity and

$$\lim_{B=\infty} \langle \mu_z \rangle = g\mu_B J = \mu, \qquad (18.21)$$

where μ is the largest eigenvalue of μ_z as defined in section 6.3. In this case the atom is said to be fully oriented along the field. This *saturation* limit is approached only for extremely strong fields or for very low temperatures, since $\mu_B/k_B = 6.7 \times 10^{-5}$ °K/gauss.

For the values of y which usually obtain, one utilizes the expansion

$$\coth x = \frac{1}{x} + \frac{x}{3} - \frac{x^3}{45} + \dots, \qquad (18.22)$$

which yields

$$\begin{aligned} \langle \mu_z \rangle &= g\mu_B \left[(J + \tfrac{1}{2}) \frac{(J + \frac{1}{2})y}{3} - \dots - \tfrac{1}{2} \frac{\frac{1}{2}y}{3} + \dots \right] \\ &\approx (g\mu_B)^2 \, J(J+1) \frac{B}{3k_B T} = \frac{|\vec{\mu}|^2}{3k_B T} B. \end{aligned} \qquad (18.23)$$

15. The mean magnetic moment can also be calculated by the seemingly more direct method of averaging the different eigenvalues of $\mu_z = -M_J g\mu_B$, with the weights exp $(-M_J y)$. The two calculations actually coincide because the product of $-M_J g\mu_B$ and of exp $(-M_J y)$ can be expressed as $k_B T \, \partial \exp(-M_J y)/\partial B$.

Here $|\vec{\mu}|^2$ indicates the squared magnetic moment $\gamma^2 |\vec{J}|^2 = (g\mu_B)^2 J(J + 1)$.[16] This theoretical result verifies the *Curie law* according to which the paramagnetic effect is inversely proportional to the Kelvin temperature.

For the purpose of direct comparison with experiments one wants to obtain the paramagnetic susceptibility χ_p for a macroscopic assembly of N atoms per unit volume. If the atoms are independent of one another, the total free energy of the assembly is just N times the free energy (18.19) of each atom. The susceptibility is then given by a formula analogous to (18.15), that is

$$\chi_p = -N\left(\frac{\partial^2 F}{\partial B^2}\right)_{B=0} = N\frac{|\vec{\mu}|^2}{3k_B T} = \frac{C}{T}, \qquad (18.24)$$

where C is called the *Curie constant.*

Paramagnetism is a property of any system with nonzero squared angular momentum, whether the angular momentum belongs to electronic—spin or orbital—or to nuclear motions. The effect is, of course, proportional to the gyromagnetic ratio. Atomic vapors of most elements are paramagnetic; the main exceptions occur for atoms whose electron configurations consist of filled subshells. Conversely, the molecules of most chemically stable substances show hardly any paramagnetism because their electronic currents yield no total angular momentum, as will be shown in the next chapter. (Oxygen is an outstanding exception among molecular substances; see section 19.6.3.)

The paramagnetism of crystalline solids will not be discussed here. There are, however, important types of crystals whose paramagnetism can be traced to single-atom behavior. These are crystals containing atoms of the transition elements and of the rare earths, with partially filled d or f subshells. As discussed in section 16.3, the electrons of these subshells are confined within a small radius and remain comparatively isolated from surrounding atoms. In fact, susceptibility measurements serve to study the interaction of these electrons with their surroundings.

Sample data from such a study are shown in Table 18.1. The table compares measured values of the paramagnetic susceptibility of crystals containing various ions with theoretical values obtained from different models. The data are expressed as "effective magneton numbers," $[(\chi_p)_{exper}\, 3k_B T/N\mu_B^2]^{1/2}$, where N is the concentration of ions under consideration. For the ions of the lanthanide group the data are matched well, in accordance with equation (18.23),

---

16. A similar result is obtained by the macroscopic Langevin theory which considers a continuous distribution of values of $\mu_z = \vec{\mu}\cdot\hat{B}$. Here $\vec{\mu}$ represents a macroscopic vector that may take all possible orientations. This theory yields the partitition function $Z = (k_B T/\mu B)\sinh(\mu B/k_B T)$ and

$$\langle\mu_z\rangle = \mu[\coth(\mu B/k_B T) - k_B T/\mu B],$$

an expression that coincides with (18.20) in both limits, for large and for small y.

TABLE 18.1

Sample Data on Effective Magneton Numbers

for Ions in Crystals

(From C. Kittel, *Introduction to Solid State Physics*)

Ion	Configuration	Zero-Field Level	$\left[\frac{(\chi_p)\ \text{exper}\ 3k_BT}{N\mu_B^2}\right]^{1/2}$	$g[J(J+1)]^{1/2}$	$2[S(S+1)]^{1/2}$
Ce^{3+}....	$(4f)^1$	$^2F_{5/2}$	2.4	2.54	1.73
Pr^{3+}....	$(4f)^2$	$^3H_4$	3.5	3.58	2.83
Gd^{3+}....	$(4f)^7$	$^8S_{7/2}$	8.0	7.94	7.94
Yb^{3+}....	$(4f)^{13}$	$^2F_{7/2}$	4.5	4.54	1.73
Ti^{3+}....	$(3d)^1$	$^2D_{3/2}$	1.8	1.55	1.73
Cr^{3+}....	$(3d)^3$	$^4F_{3/2}$	3.8	0.77	3.87
Fe^{3+}....	$(3d)^5$	$^6S_{5/2}$	5.9	5.92	5.92
Fe^{2+}....	$(3d)^6$	$^5D_4$	5.4	6.70	4.90

by the value of $g[J(J + 1)]^{1/2} = [|\vec{\mu}|^2]^{1/2}/\mu_B$ appropriate to orbital and spin current in LS coupling (section 16.4.4). For the ions of the iron group, the experimental effective magneton numbers generally differ from $g[J(J + 1)]^{1/2}$ and lie much closer to the values $2[S(S + 1)]^{1/2}$ predicted on the basis of spin currents only. The contribution of orbital currents to paramagnetism thus appears to be suppressed (or *quenched*). The quenching is attributed to lack of axial symmetry of the interaction of the orbital motion with the surrounding atoms in the crystal, the interaction being much stronger for the d-electrons ($l = 2$) of the iron group than for the more narrowly confined f-electrons of the lanthanides. In the absence of axial symmetry the orbital current no longer flows around the direction of the magnetic field. The energy eigenstates are no longer eigenstates of the orbital momentum component L_z, nor are they represented by superpositions of states with $\langle L_z \rangle \neq 0$; they are superpositions of states with $\langle L_z \rangle = 0$.

The orientation of nuclear currents is treated in the same way as the orientation of electron currents. The gyromagnetic ratio, indicated by $\gamma = g\mu_B/\hbar$ in the equations of this section, takes, of course, different and much smaller values. Similarly, the quantum number J is replaced by the quantum number I (see section 14.5), or by F when nuclear and electronic currents are coupled.

Electronic and nuclear paramagnetism has a remarkable application to the cooling of materials at extremely low temperatures, through the phenomenon of *adiabatic demagnetization*. In this phenomenon the magnetization of electrons or nuclei is saturated by the application of a very strong field after preliminary cooling to 1° K or less by another method. The selection of states with highest μ_z thus achieved is equivalent, thermodynamically, to the compression or liquefaction of a gas in an ordinary refrigerator. Sudden removal of the magnetic field under adiabatic conditions, that is, after the material has been thermally isolated, reduces the temperature of the material as the gas expansion does in a refrigerator. Knowledge of the entropy of the spin system is obtained from the partition function (18.18) and provides design data for cooling devices.

SUPPLEMENTARY SECTION

S18.5 **MAGNETIC RESONANCE**

Magnetic resonance is a phenomenon used to detect and measure electric and magnetic interactions among atomic and subatomic particles within a macroscopic material. The phenomenon rests on the paramagnetic orientation of electronic or nuclear currents by an external field and on the Larmor precession of these currents about the field direction. The frequency of this precession is proportional to the magnetic field strength prevailing at the position of the precessing electron or nucleus. When neighboring particles contribute to the local magnetic field, their contribution is measured through a shift of the precession frequency. An additional shift of the precession frequency may be contributed by nonuniform electric fields generated by neighboring particles.

Magnetic resonance is observed through variations of the magnetic moment $\vec{M}$ of a macroscopic sample of material driven by an external field. The vector $\vec{M}$ is the sum of the average moments $\langle\vec{\mu}\rangle$ of all constituent atomic systems in the sample; normally the observed variations of $\vec{M}$ are due to the precession of the moments $\langle\vec{\mu}\rangle$ of a single type of constituent—for instance, the nuclei of hydrogen atoms.

The mean magnetic moment $\langle\vec{\mu}\rangle$ of an atomic system which results from paramagnetic orientation is normally parallel to the local field $\vec{B}_0$ which we regard as constant. Therefore, $\langle\vec{\mu}\rangle$ does not precess about $\vec{B}_0$ unless it is deflected by a disturbing action. If $\langle\vec{\mu}\rangle$ is deflected, it precesses with the frequency γB_0, where γ is the gyromagnetic ratio associated with $\langle\vec{\mu}\rangle$; the value of γ is assumed to be known from other experiments. Deflection of $\langle\vec{\mu}\rangle$ is achieved by application of an oscillating cross-field of strength $B_1 \cos \omega t$, if ω coincides with the precession frequency γB_0 (see section 7.3). This matching of frequencies constitutes the condition of magnetic resonance. The occurrence of precession is observed through the magnetic field generated by the variation of $\langle\vec{\mu}\rangle$ or, more commonly, through the power loss by the oscillating crossfield.

Magnetic resonance experiments serve to map the field strengths within a material at the location of the currents whose resonance is observed. For example, a typical experiment detects the resonance of proton spin currents in an organic material and thereby determines the magnetic field strengths that prevail at the nuclei of the various hydrogen atoms. If the field strengths at the locations of the different nuclei were the same as the strength of the externally applied constant field, resonance would be detected at a single frequency ω. A uniform difference between the internal and the external field causes a uniform shift of the resonance frequency ω. Variations of the internal field strength from location to location result in the occurrence of resonance at different frequencies.

In essence, the experimental arrangement consists of a macroscopic specimen placed between the poles of a magnet and surrounded by a radio frequency coil or microwave cavity (Fig. 18.2). The extremely high precision of tuning of high-frequency equipment and its sensitivity to power loss are main assets of magnetic resonance technology. In the normal experimental procedure the oscillation frequency ω of the cross-field is kept fixed and resonance is searched by sweeping the field strength B_0 and thus varying slowly the precession frequency γB_0. The field sweep is displayed on the abscissa of an oscilloscope while the cross-field power loss is proportional to the component of $\vec{M}$ which oscillates out of phase with the driving field $\vec{B}_1 \cos \omega t$; alternatively, the oscilloscope may display the component of $\vec{M}$ in phase with the cross-field (Fig. 18.3). Figure 18.4 shows typical results of magnetic resonance analysis.

The analytical representation of the precession starts from the macroscopic equation of the Larmor precession (2.55), replacing the macroscopic angular momentum $\vec{J}$ by the mean magnetic moment $\langle\vec{\mu}\rangle$:

$$\frac{d\langle\vec{\mu}\rangle}{dt} = -\gamma\vec{B} \times \langle\vec{\mu}\rangle. \tag{18.25}$$

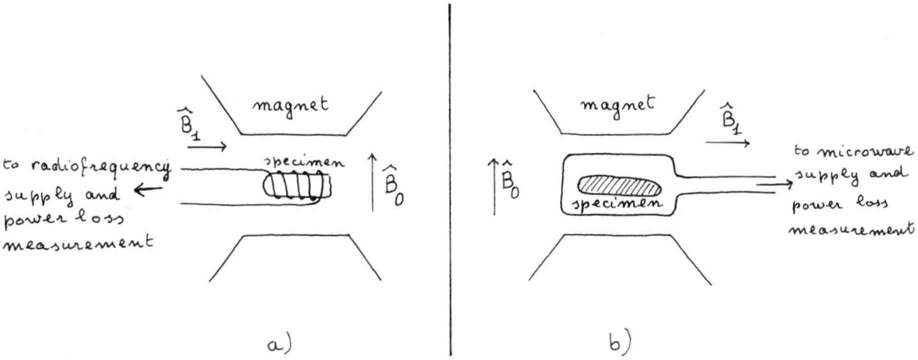

Fig. 18. 2. Schematic arrangement for magnetic resonance experiment. Resonance is achieved in different frequency ranges for: (*a*) nuclear spin with low γ, and (*b*) electron spin with larger γ.

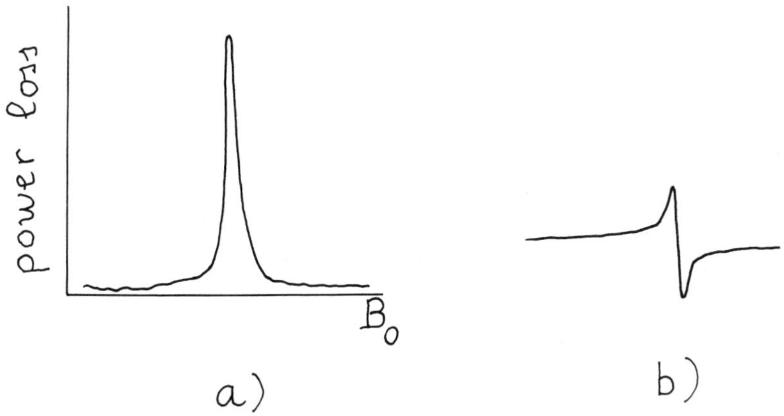

Fig. 18. 3. Magnetic resonance signals of proton in liquid H_2. (*a*) Power loss, (*b*) component of $\vec{M}$ in phase with cross-field.

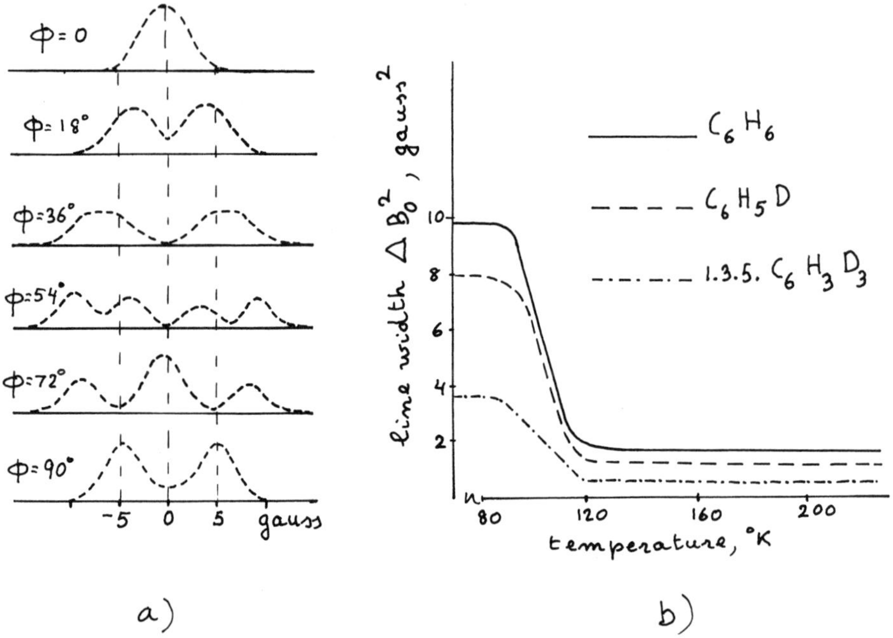

Fig. 18. 4. Results of magnetic resonance analysis.

(*a*) Power loss versus external field modulation, due to proton resonance in crystal of $CaSO_4 2H_2O$ for different angles ϕ between the field $\vec{B}_0$ and one of the lattice vectors. The occurrence of a number of resonances shows the existence of a number of nonequivalent proton locations with different local field strengths. The abscissa of a resonance peak measures the component parallel to $\vec{B}_0$ of the field generated by the neighboring protons at one location. For example, at $\phi = 90°$ this component equals $\pm$ 5 gauss at two nonequivalent locations.

Analysis of the complete pattern determines the positions of hydrogen nuclei in a crystal cell. The protons of H_2O molecules are the only particles with nonzero magnetic moment in the sample. [Adapted from G. E. Pake, *J. Chem. Phys.* 16 (1948): 327]

(*b*) Line width of proton resonance in benzene crystals versus temperature. Resonances due to protons at nonequivalent locations are not resolved. The resonance width represents the mean square deviation ΔB_0^2 of the field strength at various locations. Substitution of deuterium reduces ΔB_0^2. The resonance narrowing at ~ 100° K reveals a phase transition which permits rapid changes of molecular orientation; the field thus becomes more nearly uniform on a time average. [Adapted from E. R. Andrew and R. G. Eades, *Proc. Roy. Soc.* (Lond.), Ser. A 218 (1953): 537]

When the magnetic field $\vec{B}$ is constant and directed along the z axis, that is, when $\vec{B} = B_0\hat{z}$, the mean magnetic moment is also directed along the z axis, $\langle\vec{\mu}\rangle = \mu_0\hat{z}$, where μ_0 coincides with the value of $\langle\mu_z\rangle$ given by the paramagnetism equation (18.20). The oscillating cross-field $B_1\hat{x}\cos\omega t$ drives $\langle\vec{\mu}\rangle$ away from $\mu_0\hat{z}$, but thermal agitation opposes this action. To obtain a realistic description of magnetic resonance we must incorporate this effect of thermal agitation into (18.25).

When a magnetic field is switched off, thermal agitation causes the mean magnetic moment $\langle\vec{\mu}\rangle$ induced by the field to decay exponentially as indicated by

$$\frac{d\langle\vec{\mu}\rangle}{dt} = -\frac{\langle\vec{\mu}\rangle}{T}, \tag{18.26}$$

where T is called the *relaxation time*. In the presence of the constant field $B_0\hat{z}$, components of $\langle\vec{\mu}\rangle$ parallel and perpendicular to $\hat{z}$ relax to their equilibrium values at different rates, represented respectively by T_1 and T_2. Often T_1 is much larger than T_2. The values of these longitudinal and transverse relaxation times depend on the detailed interactions of each particle with its surroundings and are usually obtained from experiments. In fact, the experimental determination of T_1 and T_2 is one of the objectives of magnetic resonance studies.

In the presence of the constant field $B_0\hat{z}$, of the oscillating cross-field $B_1\hat{x}\cos\omega t$, and of relaxation, the time variations of the various components of $\langle\vec{\mu}\rangle$ are represented by

$$\frac{d\langle\mu_x\rangle}{dt} = \gamma B_0\langle\mu_y\rangle - \frac{\langle\mu_x\rangle}{T_2}, \tag{18.27a}$$

$$\frac{d\langle\mu_y\rangle}{dt} = -\gamma B_0\langle\mu_x\rangle + \gamma B_1\cos\omega t\,\langle\mu_z\rangle - \frac{\langle\mu_y\rangle}{T_2}, \tag{18.27b}$$

$$\frac{d\langle\mu_z\rangle}{dt} = -\gamma B_1\cos\omega t\,\langle\mu_y\rangle - \frac{\langle\mu_z\rangle - \mu_0}{T_1}. \tag{18.27c}$$

These equations will be solved here in the weak cross-field approximation in which $\langle\mu_x\rangle$ and $\langle\mu_y\rangle$ remain much smaller than $\langle\mu_z\rangle$. As will be seen, this approximation implies that $\gamma B_1 T_2 \ll 1$. In this approximation both γB_1 and $\langle\mu_y\rangle$ are small in equation (18.27c) and therefore $\langle\mu_z\rangle$ does not depart significantly from μ_0. Accordingly, $\langle\mu_z\rangle$ may be replaced by μ_0 in (18.27b), and this equation, together with (18.27a), is decoupled from (18.27c).

The system of the two equations (18.27a) and (18.27b) can be replaced by two separate equations with the variables $\langle\mu_x \pm i\mu_y\rangle$. The equations are

$$\frac{d\langle\mu_x + i\mu_y\rangle}{dt} + i\gamma B_0\langle\mu_x + i\mu_y\rangle + \frac{\langle\mu_x + i\mu_y\rangle}{T_2} = i\gamma B_1\mu_0\cos\omega t, \tag{18.28}$$

and its complex conjugate. A solution to equation (18.28) is obtained by representing cos ωt as the sum of $\frac{1}{2}$ exp ($i\omega t$) and $\frac{1}{2}$ exp ($-i\omega t$). The solution is then a superposition of two terms proportional to exp ($\pm i\omega t$),

$$\langle\mu_x + i\mu_y\rangle = \tfrac{1}{2}\gamma B_1\mu_0\left(\frac{e^{i\omega t}}{\omega + \gamma B_0 - i/T_2} + \frac{e^{-i\omega t}}{-\omega + \gamma B_0 - i/T_2}\right). \quad (18.29)$$

The first term in the parentheses is always negligible since the sum of ω and γB_0 is large.[17] The second term becomes quite appreciable when ω nearly coincides with γB_0, that is, at or near magnetic resonance. Discarding the first term and separating the real and imaginary parts of the second one, we have

$$\langle\mu_x\rangle = \frac{\frac{1}{2}\gamma B_1 T_2\mu_0}{[1 + (\gamma B_0 - \omega)T_2^2]^{1/2}} \cos(\omega t - \phi), \quad (18.30a)$$

$$\langle\mu_y\rangle = -\frac{\frac{1}{2}\gamma B_1 T_2\mu_0}{[1 + (\gamma B_0 - \omega)^2 T_2^2]^{1/2}} \sin(\omega t - \phi), \quad (18.30b)$$

where the phase lag ϕ is given by

$$\cot\phi = (\gamma B_0 - \omega)T_2. \quad (18.31)$$

Equations (18.30) show that the ratio of the transverse component of $\langle\vec{\mu}\rangle$ to its longitudinal component μ_0 attains a maximum of $\sim\gamma B_1 T_2$. This maximum is attained at resonance, more specifically for a departure from resonance such that $|(\gamma B_0 - \omega)T_2| \lesssim 1$. The phase lag, which characterizes the energy dissipation, is proportional to the reciprocal relaxation time $1/T_2$—below resonance—as one might have anticipated. The power dissipated by the oscillating field, averaged over a period of oscillation, is

$$P = \frac{\omega}{2\pi}\int_0^{2\pi/\omega} B_1 \cos\omega t \frac{d\langle\mu_x\rangle}{dt}\,dt = \tfrac{1}{4}\frac{\omega\gamma B_1 T_2\mu_0 B_1}{[1 + (\gamma B_0 - \omega)^2 T_2^2]^{1/2}} \sin\phi$$

$$= \tfrac{1}{4}\frac{\omega\gamma B_1^2 T_2\mu_0}{1 + (\gamma B_0 - \omega)^2 T_2^2}. \quad (18.32)$$

CHAPTER 18 PROBLEMS

18.1. Consider ferric and ferrous ions, Fe^{3+} and Fe^{2+}, in solution. Their ground states have the spectroscopic symbols $3d^5\ {}^6S_{5/2}$ and $3d^6\ {}^5D_4$, respectively. (a) Calculate their Larmor precession frequencies in a field B of

---

17. This statement implies that $\gamma > 0$. If γ is negative, one should interchange γ and $-\gamma$ and "first term" with "second term" in the following discussion and formulae.

1000 gauss. (b) Calculate the mean magnetic moment of 1 mg of Fe^{3+} ions in a field of 1000 gauss and at 27° C.

18.2. An H^+ ion and an H atom in the ground state lie 10 Å apart. (a) Calculate the force between them, in eV/Å, given that the polarizability of the H atom is $(9/2)a_B{}^3 = 0.67$ Å^3. (b) Calculate what fraction of this polarizability is contributed by the term of equation (17.13) with $s \equiv n = 2$ and $n = 1$.

18.3. Using the data from problem 18.2(b), estimate the van der Waals energy of two H atoms in their ground states at 10 Å, which is contributed by the states $2p_z$ of the two atoms, that is, by $|s'\rangle \equiv |2\ 1\ 0\rangle$, $|s''\rangle \equiv |2\ 1\ 0\rangle$, with the z axis in the direction $\hat{R}$. (Hint: since each ground state has spherical symmetry, $3\vec{\rho}'\cdot\hat{R}\vec{\rho}''\cdot\hat{R} - \vec{\rho}'\cdot\vec{\rho}''$ may be substituted by $2z'z''$ in eq. [18.11].)

18.4. Consider the diamagnetic polarizability of the H atom in the following three states $|n\ l\ m\rangle$: $|1\ 0\ 0\rangle$, $|2\ 1\ 0\rangle$, $|2\ 1\ 1\rangle$, disregarding initially the electron spin. (a) Without calculation, rank these three states in order of increasing diamagnetic polarizability. (b) Calculate the diamagnetic polarizability of each state. (c) Suppose that the atom is in one of the states $|1\ 0\ 0\rangle$, $|2\ 1\ 0\rangle$, or $|2\ 1\ 1\rangle$ with spin state $|m_s = -\tfrac{1}{2}\rangle$ at the time $t = 0$ and that the spin-orbit coupling operates thereafter. Indicate for each of the states whether its diamagnetic polarizability is going to increase, decrease, or remain unchanged in the course of time. Disregard any paramagnetic effect.

SOLUTIONS TO CHAPTER 18 PROBLEMS

18.1 (a) Since Fe^{3+} has $L = 0$, it has $g = 2$ and $\gamma = -2(e/2m_e c) = -1.76 \times 10^7$ radians/gauss sec; at $B = 1000$ gauss, $\omega = \gamma B = -1.76 \times 10^{10}$ radians/sec. For Fe^{2+}, equation (14.25) gives $g = 3/2$; hence $\omega = \gamma B = -1.32 \times 10^{10}$ radians/sec. (b) The mean magnetic moment is given by equation (18.23) provided the parameter y is small. In our problem, $y = 2 \times 0.927 \times 10^{-20}$ (ergs/gauss) $\times$ 1000 gauss/(1.38×10^{-16} (ergs/° K) $\times$ 300° K) $= 4.5 \times 10^{-4}$. Equation (18.33) gives $\langle\mu_z\rangle = \tfrac{1}{3} g\mu_B J(J+1)y = (35/6)\mu_B \times 4.5 \times 10^{-4} = 2.4 \times 10^{-23}$ ergs/gauss. The number of ions in 1 mg of Fe is $N = 1.1 \times 10^{19}$, hence $N\langle\mu_z\rangle = 2.6 \times 10^{-4}$ ergs/gauss.

18.2 (a) By equation (18.5), the polarization energy is $E_{pol} = -\tfrac{1}{2}\alpha F^2$ with $F = [e/(10\text{Å})^2]$. The force between ion and atom is $-dE_{pol}/dr = 2\alpha e^2/r^5$. Since $e^2/(10\text{Å}) = 1.44$ eV, we find $(-dE_{pol}/dr)_{r=10\text{Å}} = 2(9/2)a_B^3 e^2/(10\text{Å})^5 = 9 \times [a_B/(10\text{Å})]^3\ 1.44\ \text{eV}/10\text{Å} = 1.9 \times 10^{-4}$ eV/Å. (b) For the H atom (without reference to spin) the only nonzero matrix element $\langle n = 2, 1\ m|z|n = 1, 0\ 0\rangle$ is $\langle n = 2, 1\ 0|z|n = 1, 0\ 0\rangle = (2^{3/2}/3)^5 a_B$. Substitution into equation (17.13) reduces the first term of α to $2e^2(2^{3/2}/3)^{10} a_B{}^2/(3/8)\,(e^2/a_B) = (16/3)\,(2^{3/2}/3)^{10} \times a_B{}^3 = 3.0 a_B{}^3$, i.e., $\sim(2/3)\,\alpha$.

18.3 With the substitutions indicated, the single term of equation (18.11) becomes $-[e^4/(10\text{Å})^6]|\langle n=2,1\ 0|z|n=1,0\ 0\rangle|^4/[2\times \frac{3}{8}(e^2/a_B)] = -(4/3)\times [e^2 a_B/(10\text{Å})^6](2^{3/2}/3)^{20} a_B{}^4 = -(4/3)(2^{3/2}/3)^{20}(a_B/10\text{Å})^5 \times 1.44\text{ eV} = -2.4\times 10^{-7}$ eV.

18.4. (a) The polarizability is lowest for the ground state $|1\ 0\ 0\rangle$ because the electron has by far the lowest mean distance from the nucleus and from the z axis as well. The states $|2\ 1\ 0\rangle$ and $|2\ 1\ 1\rangle$ have the same mean distance from the nucleus, but their electron densities are proportional to $\cos^2\theta$ and $\sin^2\theta$, respectively; hence the mean value of $x^2 + y^2$ is lower for $|2\ 1\ 0\rangle$ than for $|2\ 1\ 1\rangle$. (b) Equation (18.16) applies to the ground state, which has spherical symmetry,

$$(\chi_d/N)_{1,0,0} = -\frac{e^2}{6m_e c^2}\int_0^\infty r^2 dr\, r^2 (4/a_B)^3 \exp(-2r/a_B) = -\frac{e^2 a_B{}^2}{2m_e c^2}\,.$$

For the 2p states, equation (18.15) applies, with $x^2 + y^2 = r^2\sin^2\theta$, and we have

$$(\chi_d/N)_{2,1,0} = -\frac{3e^2 a_B{}^2}{m_e c^2},\quad (\chi_d/N)_{2,1,1} = -\frac{6e^2 a_B{}^2}{m_e c^2}\,.$$

(c) The ground state has spherical symmetry and no spin-orbit coupling; its polarizability remains unaffected by the electron spin. Spin-orbit coupling changes the state $|2\ 1\ 0\rangle|m_s = -\frac{1}{2}\rangle$ into a nonstationary superposition of that state and of $|2\ 1\ -1\rangle|m_s = \frac{1}{2}\rangle$ which has the same value of $m + m_s$. The latter state has the same electron density and the same polarizability as $|2\ 1\ 1\rangle$ which is larger than that of $|2\ 1\ 0\rangle$. Hence the spin-orbit coupling increases the mean value of the polarizability above its value at $t = 0$. The corresponding effect reduces the initial polarizability of $|2\ 1\ 1\rangle$.

19. chemical bonds

19.1. INTRODUCTION

The aggregation of atoms into macroscopic bodies results in part from nonspecific mechanisms of cohesion and in part from the formation of more specific chemical bonds. As we have seen in section 18.2, all species of atoms attract each other by van der Waals forces. These forces are weak, but they are sufficient to overcome the resistance of atoms to confinement. This resistance is thousands of times smaller than the expansive force of electrons because it is inversely proportional to the mass of the confined particle (see, for instance, eq. [11.32]). The main obstacle to aggregation by van der Waals forces arises from thermal agitation.

To the van der Waals cohesion is added in many cases the effect of electric polarization by ions or other groups of atoms (see section 18.1). Polarization effects are generally stronger than van der Waals forces and have often a major influence on the structure of matter as, for instance, in the case of water (chap. 22). Yet, these cohesion effects amount generally to a fraction of 1 eV per atom ($\lesssim 10^4$ cal/mole), whereas several eV per atom are usually released in the formation of stable chemical compounds.

The mechanism of the strong *heteropolar bonding* of salts was understood in its essence by Berzelius in the early nineteenth century. In modern language we say that it takes rather little energy to transfer an electron from a metal atom like Na to a halogen atom. Since, for example, Cl has an electron affinity of 3.7 eV, and Na has an ionization potential of 5.1 eV, the energy required for the transfer would be 1.4 eV for the two atoms at infinite distance. The process is, however, exoergic if the two atoms are less than 10 Å apart. The pair of ions Cl^- and Na^+ at the distance of 2.8 Å, where their electron shells begin to interpenetrate, has 3.9 eV less energy than a neutral pair (Na, Cl) at infinite distance.

Yet, the ionic bond mechanism cannot possibly account for the formation of molecules from identical, or electrically similar, atoms. In particular, it cannot account for the formation of the very stable molecules of familiar gases like N_2, O_2, H_2, etc. Indeed, the formation of *homopolar (covalent)* bonds remained a main puzzle of chemistry until the advent of quantum mechanics. The break came with the realization that the resistance of atoms to interpenetration operates fully only between atoms with complete subshells, according to the exclusion principle. Upon approach of two atoms with incomplete outer subshells, the electrons of both atoms form a single system whose energy depends on the distance of the nuclei. The stability of a bond between two (or more) atoms must accordingly be studied by determining the stationary state of minimal energy of the combined system.

Atoms with incomplete subshells thus have a *valence*—ability to form bonds—which becomes saturated when the complete electronic system of the molecule has the characteristics of a filled subshell.[1] Molecules with saturated valences have a characteristic feature common to single atoms with filled outer subshells, namely, that the ground state of the electronic system has zero magnetic moment. Therefore, chemically stable matter is normally nonparamagnetic.

Molecules whose electronic system does not partake of the stability of complete shells—that is, molecules with unsaturated valences—do exist, but they are normally very reactive. They are called *free radicals* and are found in small concentrations in many materials. Free radicals are typically paramagnetic, and their presence is detected by magnetic resonance. Two exceptional small molecules are chemically stable and yet paramagnetic. These are O_2 and NO; they will be discussed at the end of this chapter. Also exceptional in this respect are substances containing transition elements and lanthanides whose d and f electrons are confined within a small radius and do not participate in chemical bond formation.

The simplest covalently bonded structure is the molecule ion H_2^+. The theoretical study of this ion is the counterpart in molecular physics to the study of the H atom in atomic physics and will accordingly be developed in some detail (section 19.2). The classification of the stationary states of H_2^+ provides a pattern for the discussion of the states of other molecules. However, the independent electron approach is quantitatively less successful when applied to molecules than it is for atoms.

In principle, one may write a complete Schrödinger equation for the motion of all electrons and nuclei of any molecular system, much as one does for an atom with many electrons. Here again, there is a choice between direct numerical solution of the complete Schrödinger equation and reliance on a variety of semianalytical approximations. Numerical solution has produced agreement with experimental results for the ground state of the H_2 molecule to five-digit accuracy. A continuing effort is devoted to calculations, primarily of the ground states of molecular systems with increasing numbers of particles.

Semianalytic approximations provide results of limited accuracy but permit a qualitative analysis of molecular energy levels, transition probabilities, etc., as they do for atoms. The variety of approximation methods is greater for molecules than for atoms in relation to the complexities of the problems, and the accuracy of results is lower. Note that a bond energy is obtained as a small difference between the much larger energy eigenvalue of a molecule and the energy of its separate component atoms.

The comparative characteristics of different approaches can be illustrated by a discussion of the simplest neutral molecule, H_2, given in section 19.3. The rest of the chapter sketches qualitatively the bonding mechanisms of a few simple molecules. The descriptions are based on experimental evidence inter-

---

1. An empirical rule for this effect had been worked out by G. N. Lewis, as an "octet formation principle," about 1920, prior to quantum mechanics.

preted according to the concepts and formulae developed for the H_2^+ ion. They also afford initial examples of the symmetry considerations which are essential for the analysis of diatomic and polyatomic molecules. The analytical development in the first part of this chapter and the qualitative descriptions in the second may serve as a preliminary introduction to the field of quantum chemistry.[2] An introduction to the study of excited molecules and of their spectra follows in chapter 20. Most of the accurate experimental data on molecules are obtained from spectroscopic observations, though basic data on the energy released upon formation of molecules derive from thermochemistry. The study of chemical reactions and of chemical equilibria is altogether beyond our scope.

19.2. THE HYDROGEN MOLECULE ION

The study of the hydrogen molecule ion presents two types of problems which we have not yet met when dealing with atoms; both of them are common to the study of all molecules or larger aggregates of atoms. The first problem is the interaction between the motion of the electrons and that of the nuclei. The complications resulting from this interaction can be sorted out by taking advantage of the smallness of the ratio m_e/M of the electron to the nuclear mass. The procedure used for this purpose is called the *Born-Oppenheimer approximation*. The second problem is due to the lack of central symmetry of the forces which attract the electron toward the nuclei. However, in the H_2^+ ion the forces still possess axial symmetry which makes it possible to adapt many of the analytical procedures used in the treatment of atoms. Axial symmetry is common to all diatomic molecules with which we shall be mostly concerned throughout this chapter and the next.

We begin by considering only electrostatic interactions, disregarding spin-orbit coupling and other minor perturbations. We indicate the particle positions by relating them to the midpoint $\vec{R}_0$ of the two nuclei. The electron position with respect to this point is indicated by $\vec{r}$, and the relative positions of the two muclei by $\vec{R}$. Thus we have:

Positions of H nuclei: $\vec{R}_0 + \frac{1}{2}\vec{R}$, $\vec{R}_0 - \frac{1}{2}\vec{R}$;
Electron position: $\vec{R}_0 + \vec{r}$.

The particles' velocities are indicated by corresponding symbols $\vec{V}_0 \pm \frac{1}{2}\vec{V}$ and $\vec{V}_0 + \vec{v}$. The mass of one proton is called M. The sum of the kinetic and electrostatic potential energies of all three particles is then

---

2. This field is treated, for example, by C. A. Coulson, *Valence* (Oxford: University Press, 1961); J. C. Slater, *Quantum Theory of Matter* (New York: McGraw-Hill, 1968); and F. L. Pilar, *Elementary Quantum Chemistry* (New York: McGraw-Hill, 1968).

$$E_{tot} = \tfrac{1}{2}M|\vec{V}_0 + \tfrac{1}{2}\vec{V}|^2 + \tfrac{1}{2}M|\vec{V}_0 - \tfrac{1}{2}\vec{V}|^2 + \tfrac{1}{2}m_e|\vec{V}_0 + \vec{v}|^2$$
$$- \frac{e^2}{|\vec{r} - \tfrac{1}{2}\vec{R}|} - \frac{e^2}{|\vec{r} + \tfrac{1}{2}\vec{R}|} + \frac{e^2}{R}. \tag{19.1}$$

Note that the potential energy is independent of $\vec{R}_0$.

As a second step, we separate from E_{tot} the energy of the center-of-mass motion by introducing the

$$\text{Center-of-mass velocity } \vec{V}_{CM} = \vec{V}_0 + \frac{m_e}{2M + m_e}\vec{v}.$$

Thus we can express $\vec{V}_0$ in equation (19.1) by means of $\vec{V}_{CM}$ and $\vec{v}$ and reduce (19.1) to the form

$$E_{tot} = E_{CM} + E = \tfrac{1}{2}(2M + m_e)V_{CM}^2 + \tfrac{1}{2}(\tfrac{1}{2}M)V^2$$
$$+ \tfrac{1}{2}\frac{2Mm_e}{2M + m_e}v^2 - \frac{e^2}{|\vec{r} - \tfrac{1}{2}\vec{R}|} - \frac{e^2}{|\vec{r} + \tfrac{1}{2}\vec{R}|} + \frac{e^2}{R}. \tag{19.1a}$$

This equation involves separately the total mass of the molecule, $2M + m_e$, the reduced mass of the two nuclei, $\tfrac{1}{2}M$, and the reduced mass of the electron and nuclei, $\overline{m}_e = 2Mm_e/(2M + m_e)$.

To determine the energy levels and the eigenfunction of the electron and the two nuclei within the molecule, we need not consider E_{CM}. A Schrödinger equation can be constructed by the procedure of section 12.1, starting from an expression of the mean value of the energy of relative motion E. For a state $|a\rangle$ represented by probability amplitudes $\langle\vec{R}\ \vec{r}|a\rangle$, this mean value is

$$\langle E\rangle_a = \int d\vec{R}\int d\vec{r}\langle a|\vec{R}\ \vec{r}\rangle\left[-\frac{\hbar^2}{M}\nabla_R^2 - \frac{\hbar^2}{2\overline{m}_e}\nabla^2 - \frac{e^2}{|\vec{r} - \tfrac{1}{2}\vec{R}|} - \frac{e^2}{|\vec{r} + \tfrac{1}{2}\vec{R}|}\right.$$
$$\left. + \frac{e^2}{R}\right]\langle\vec{R}\ \vec{r}|a\rangle. \tag{19.2}$$

Here the symbol $\vec{\nabla}_R$ indicates the gradient operator with respect to $\vec{R}$ and $\vec{\nabla}$ means, as usual, the same operator with respect to the electron's coordinate $\vec{r}$. For purposes of accurate calculations the full Schrödinger equation is derived from (19.2) and solved by numerical methods. We follow here an approximation procedure.

19.2.1. *The Born-Oppenheimer Approximation.* The key idea in this approximation is that the electron moves much faster than the nuclei owing to its small mass. Accordingly, the initial model treats the electron's motion as though the nuclei were fixed at a distance R. This is done by determining the eigenstates of the terms of the energy expression (19.1a) that depend on the electron motion only, namely, the third, fourth, and fifth terms in the equation. Eigenstates of the complete system, inclusive of the nuclei, are then constructed in a second step of the calculation using the eigenstates of the electron alone.

Consider then the motion of an electron under the attraction of two fixed nuclei. We represent the position $\vec{r}$ of the electron in a system of coordinates with the z axis along $\vec{R}$. The Schrödinger equation for this problem is

$$\left[-\frac{\hbar^2}{2\overline{m}_e}\nabla^2 - \frac{e^2}{|\vec{r} - \frac{1}{2}R\hat{z}|} - \frac{e^2}{|\vec{r} + \frac{1}{2}R\hat{z}|}\right]\langle\vec{r}\,|n\;\vec{R}\rangle = \langle\vec{r}\,|n\;\vec{R}\rangle\,\mathcal{E}_n(R). \tag{19.3}$$

The index n indicates one among a complete set of stationary states. The internuclear distance $\vec{R}$ has been written in the probability amplitude to emphasize that the eigenstates depend on the magnitude and direction of $\vec{R}$; however, the eigenvalues depend only on the magnitude R. The solution of equation (19.3) will be developed in the next section. We assume here that a complete set of solutions has been obtained for each value of $\vec{R}$, and we proceed to use these solutions to simplify the expression (19.2) of the mean energy of all three particles in the molecule.

To this end we expand the wave function $\langle\vec{R}\;\;\vec{r}\,|a\rangle$, for each value of $\vec{R}$, into a superposition of the complete set of wave functions $\langle\vec{r}\,|n\;\;\vec{R}\rangle$,

$$\langle\vec{R}\;\;\vec{r}\,|a\rangle = \sum_n \langle\vec{r}\,|n\;\vec{R}\rangle\,\langle\vec{R}\,|n\;a\rangle. \tag{19.4}$$

Upon substitution of this expansion and its complex conjugate, equation (19.2) takes the form

$$\begin{aligned}\langle E\rangle_a = \int d\vec{R}\int d\vec{r}\,\sum_{n'n}\,\langle n'\;\;a\,|\vec{R}\rangle\,\langle n'\;\;\vec{R}\,|\vec{r}\rangle\Bigg[-\frac{\hbar^2}{M}\nabla_R^2 \\ -\frac{\hbar^2}{2\overline{m}_e}\nabla^2 - \frac{e^2}{|\vec{r} - \frac{1}{2}R\hat{z}|} - \frac{e^2}{|\vec{r} + \frac{1}{2}R\hat{z}|} + \frac{e^2}{R}\Bigg]\langle\vec{r}\,|n\;\;\vec{R}\rangle\,\langle\vec{R}\,|n\;\;a\rangle.\end{aligned} \tag{19.5}$$

The fact that $\langle\vec{r}\,|n\;\;\vec{R}\rangle$ is an eigenfunction of equation (19.3) serves now to replace the three terms $-(\hbar^2/2\overline{m}_e)\nabla^2 - e^2/|\vec{r} - \frac{1}{2}R\hat{z}| - e^2/|\vec{r} + \frac{1}{2}R\hat{z}|$ in the brackets of equation (19.5) by the eigenvalue $\mathcal{E}_n(R)$. Furthermore, the orthogonality of the eigenfunctions, $\int d\vec{r}\langle n'\;\;\vec{R}\,|\vec{r}\rangle\langle\vec{r}\,|n\;\;\vec{R}\rangle = \delta_{n'n}$, serves to eliminate these eigenfunctions altogether from the last four terms of equation (19.5), which thus reduces to

$$\begin{aligned}\langle E_a\rangle = \int d\vec{R}\int d\vec{r}\,\sum_{n'n}\langle n'\;\;a\,|\vec{R}\rangle\,\langle n'\;\;\vec{R}\,|\vec{r}\rangle\left[-\frac{\hbar^2}{M}\nabla_R^2\right]\langle\vec{r}\,|n\;\;\vec{R}\rangle\,\langle\vec{R}\,|n\;\;a\rangle \\ + \int d\vec{R}\,\sum_n\,\langle n\;\;a\,|\vec{R}\rangle\left[\mathcal{E}_n(R) + \frac{e^2}{R}\right]\langle\vec{R}\,|n\;\;a\rangle.\end{aligned} \tag{19.5a}$$

The orthogonality of the electron eigenfunctions cannot be applied directly to the first term of (19.5a) which contains the gradient operator ∇_R^2. This operator introduces derivatives of the electron wave functions $\langle \vec{r} | n\ \vec{R}\rangle$ because these wave functions depend on $\vec{R}$. The Born-Oppenheimer approximation consists of disregarding terms containing these derivatives because they give a small contribution to the total value of $\langle E\rangle_a$. In this approximation equation (19.5a) reduces further to

$$\langle E\rangle_a = \int d\vec{R} \sum_n \langle a\ n | \vec{R}\rangle \left[-\frac{\hbar^2}{M} \nabla_R^2 + \mathcal{E}_n(R) + \frac{e^2}{R} \right] \langle \vec{R} | n\ a\rangle. \tag{19.5b}$$

The variational treatment of section 12.1 can now be applied to equation (19.5b). The calculation of $\partial\langle E\rangle_a/\partial\langle n\ a|\vec{R}\rangle$ and of the other derivatives, $\partial\langle E\rangle_a/\partial\mathcal{P}$ and $\partial\mathcal{P}/\partial\langle n\ a\ |\vec{R}\rangle$, is straightforward. As in previous applications—for instance, in section 16.4—we call $|n\ \eta\rangle$ the eigenstates identified by the variational procedure. In the limit $|n\ a\rangle \to |n\ \eta\rangle$ we obtain the Schrödinger equation for the nuclear motion

$$\left[-\frac{\hbar^2}{M} \nabla_R^2 + \mathcal{E}_n(R) + \frac{e^2}{R} \right] \langle \vec{R} | n\ \eta\rangle = \langle \vec{R} | n\ \eta\rangle E_{n\eta}. \tag{19.6}$$

Equation (19.6) treats the nuclear motion as separated from the electron motion. This separation is the core of the Born-Oppenheimer approximation. The approximation reduces the Σ_n in equation (19.4) to a single term for each energy eigenstate of the whole molecule. In equation (19.6) the electron energy eigenvalue $\mathcal{E}_n(R)$ represents a potential energy of the nuclei due to the electron motion, much as the rotational energy of the electron in the hydrogen atom contributes a centrifugal potential to the radial equation (13.8c). The potential energy $\mathcal{E}_n(R)$ represents the force which holds together the two nuclei, that is, the chemical bond. The calculation of $\mathcal{E}_n(R)$ is presented in the next section; the solution of the Schrödinger equation for the nuclear motion will follow.

Corrections to the Born-Oppenheimer approximation can be worked out by retaining the derivatives which have been disregarded in the expression of $\langle E\rangle_a$. When the derivatives are retained, one no longer finds independent equations (19.6) for different values of n; instead, one finds a system of interlinked equations. The interlinkage terms can be treated as perturbation energies by the method of chapter 17. The treatment shows that the corrections are inversely related to the energy differences $\mathcal{E}_n(R) - \mathcal{E}_{n'}(R)$ between energy levels of the electron motion. The initial Born-Oppenheimer approximation is good to the extent that the correlations between the nuclear and electron motion contribute to the mean energy $\langle E\rangle_a$ amounts smaller than the electron level separations.[3] Examples of phenomena involving a breakdown of the initial Born-Oppenheimer approximation will be given in the following chapters.

19.2.2. *Electron Motion.* The motion of the electron in the H_2^+ ion can be resolved, as for the H atom, into three distinct partially independent types of

3. Closer mathematical analysis beyond our scope shows that the Born-Oppenheimer method can be cast as an expansion of the Schrödinger equation into powers of the parameter $(m_e/M)^{1/4} \sim 0.15$.

motion. The motion around the internuclear axis, which we have chosen as the z axis, is independent of nuclear attraction and has the same eigenstates as for the hydrogen atom. Whereas in the hydrogen atom one considers the motion along meridian circles centered at the nucleus, here one considers the motion along meridian ellipses with foci at the two nuclei. In place of the radial motion of the hydrogen atom one considers here motion along hyperbolas with foci at the nuclei.

Mathematically, the analysis is carried out by representing the electron position $\vec{r}$ in elliptical coordinates ξ, η, ϕ defined by

$$\tan\phi = \frac{y}{x}, \quad \xi = \frac{|\vec{r} + \frac{1}{2}R\hat{z}| + |\vec{r} - \frac{1}{2}R\hat{z}|}{R}, \quad \eta = \frac{|\vec{r} + \frac{1}{2}R\hat{z}| - |\vec{r} - \frac{1}{2}R\hat{z}|}{R}. \qquad (19.7)$$

Figure 19.1 shows, in a plane of constant ϕ, one line of constant ξ and lines of

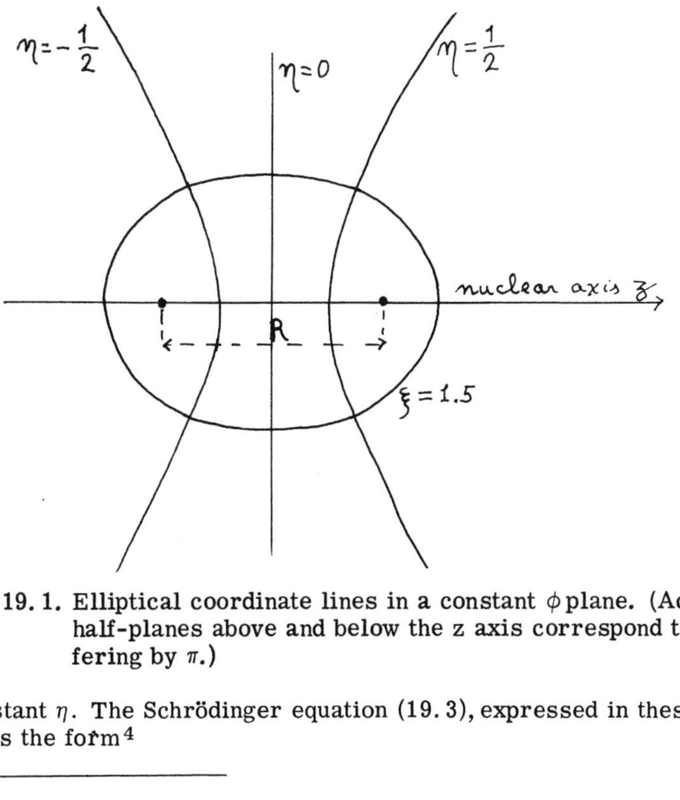

Fig. 19.1. Elliptical coordinate lines in a constant ϕ plane. (Actually, the two half-planes above and below the z axis correspond to ϕ values differing by π.)

constant η. The Schrödinger equation (19.3), expressed in these coordinates, takes the form[4]

---

4. For the expression of ∇^2 in orthogonal curvilinear coordinates see, e.g., Kaplan, *Advanced Calculus*, p. 154; Boas, *Mathematical Methods in Physical Sciences*, p. 44. The transformation of the potential energy is given by

$$|\vec{r} + \tfrac{1}{2}R\hat{z}|^{-1} + |\vec{r} - \tfrac{1}{2}R\hat{z}|^{-1} = 4\xi/(\xi^2 - \eta^2)R.$$

$$\left[\frac{\partial}{\partial\xi}(\xi^2-1)\frac{\partial}{\partial\xi}+\frac{\partial}{\partial\eta}(1-\eta^2)\frac{\partial}{\partial\eta}+\left(\frac{1}{\xi^2-1}+\frac{1}{1-\eta^2}\right)\frac{\partial^2}{\partial\phi^2}\right.$$
$$\left.+\ ½\,\frac{\overline{m}_e R^2}{\hbar^2}\,\mathcal{E}_n(R)(\xi^2-\eta^2)+2\frac{R}{a_B}\,\xi\right]\langle\xi\ \eta\ \phi\,|\,n\ \vec{R}\rangle=0, \tag{19.8}$$

where a_B is the Bohr radius.[5] One can verify, as for the hydrogen-atom equation (13.6), that equation (19.8) is separable, that is, that it has eigenfunctions of the form

$$\langle\xi\ \eta\ \phi\,|\,n\ \vec{R}\rangle=\langle\xi\,|\,n_\xi\ \ \lambda\ \vec{R}\rangle\langle\eta\,|\,n_\eta\ \ \lambda\ \vec{R}\rangle\langle\phi\,|\,\lambda\rangle. \tag{19.9}$$

Here, n_ξ and n_η are quantum numbers indicating the number of nodes of the ξ and η wave functions and therefore the number of dark fringes in the electron density. These fringes are, respectively, ellipsoids with constant ξ and hyperboloids with constant η. The motion along ϕ, that is, along parallel circles on the ellipsoids, is quite the same as the motion along ϕ for the hydrogen atom.

Indeed, the wave function $\langle\phi\,|\,\lambda\rangle=(2\pi)^{-1/2}\exp(i\lambda\phi)$ coincides with the wave function (13.14) except that, conventionally, the magnetic quantum number m is replaced by λ when dealing with molecules.

Wave functions with λ values of equal magnitude and opposite sign represent, of course, degenerate states. Here again, these wave functions may be replaced by wave functions $\langle\phi\,|\,\lambda^2\pm\rangle$ of the form (13.17). Stationary states with either type of these functions are conventionally indicated by Greek letters according to the code

$\|\lambda\|$	0	1	2	...
Code	σ	π	δ	....

The electron motion around the z axis can be separated provided only that the molecule has axial symmetry. Accordingly, independent-electron wave functions for any molecule with this symmetry include a factor $\langle\phi\,|\,\lambda\rangle$.

On the other hand, the motion of the electron along ξ and η separates out only for the H_2^+ ion. In this case the eigenfunctions $\langle\xi\,|\,n_\xi\ \ \lambda\ \vec{R}\rangle$ and $\langle\eta\,|\,n_\eta\ \ \lambda\ \vec{R}\rangle$ are obtained by numerical solution of two separate single-variable differential equations. Figure 19.2 shows the energy eigenvalues obtained by this procedure, plotted as functions of the internuclear distance R. The labeling of levels in the figure indicates the values of the quantum numbers n_ξ and n_η

---

5. Strictly speaking, the value of a_B to be used in this section should be obtained by replacing the electron mass m_e in equation (3.21) by the reduced mass $\overline{m}_e$. Actually, the difference of the two values is negligible within the Born-Oppenheimer approximation.

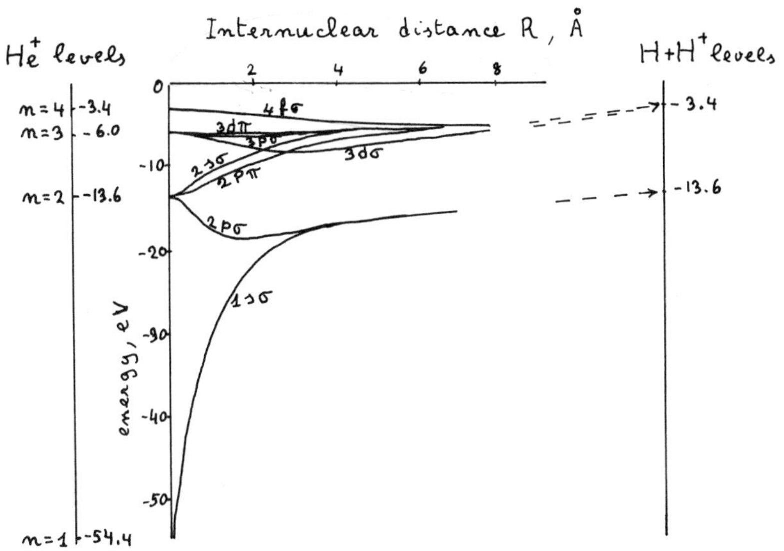

Fig. 19. 2. Energy levels $\mathcal{E}$ (R) for H_2^+ molecule ion. [Adapted from E. Teller, *Z. Physik* 61 (1930): 458].

by a numerical index which is the value of the principal quantum number $n = n_\xi + n_\eta + |\lambda| + 1$, and by a letter determined by the code

$n_\eta + \|\lambda\|$	0	1	2	3	...
Code	s	p	d	f	....

Thus, for example, 3dσ means $|\lambda| = 0, n_\xi = 0, n_\eta = 2$; 3d$\pi$ means $|\lambda| = 1, n_\xi = 0, n_\eta = 1$.

The energy eigenvalues $\mathcal{E}$(R), plotted in Figure 19. 2, provide the essential information to be entered in the nuclear motion equation (19. 5). We will see in the next section how this information serves to predict the formation and stability of the H_2^+ molecule ion. Basically, a decrease of $\mathcal{E}$ (R) with decreasing R represents a force that draws the nuclei together. Here we discuss the general trend of dependence of the eigenvalues $\mathcal{E}$ (R) and of the corresponding wave functions on the internuclear distance R. The purpose is to disentangle the qualitative features of the electron state which lead to the formation of molecular bonds.

The characterization of each stationary state of H_2^+ by the quantum numbers n_ξ and n_η is independent of the internuclear distance R. Thus the topological pattern of dark fringes of electron density remains constant as R varies, even though the shape of the individual fringes changes. Each stationary state is characterized in the two opposite limiting cases $R = 0$ and $R = \infty$. In the limit $R = 0$, called the *united-atom* limit, the positions of the two nuclei coincide and the eigenvalues and eigenfunctions coincide with those of the He^+ ion. In the opposite *(separate-atom)* limit, $R = \infty$, the molecule ion is split into an H^+ ion and an H atom; its energy levels and eigenfunctions are readily derived from those of the H atom. The character of each stationary state at intermediate R is studied conveniently starting from the separate-atom limit and considering how the state evolves with diminishing R to coincide with the united-atom limit at $R = 0$.

In the separate-atom limit each stationary state is degenerate because the electron has pairs of wave functions equal in all respects except that they are centered on one or the other of the nuclei. For instance, we have two degenerate wave functions for the ground state of the electron in the system $H + H^+$ at large fixed internuclear distance R, namely,

$$u(\vec{r} + \tfrac{1}{2}R\hat{z}) = \left(\frac{1}{\pi a_B{}^3}\right)^{1/2} \exp(-|\vec{r} + \tfrac{1}{2}R\hat{z}|/a_B), \tag{19.10a}$$

$$u(\vec{r} - \tfrac{1}{2}R\hat{z}) = \left(\frac{1}{\pi a_B{}^3}\right)^{1/2} \exp(-|\vec{r} - \tfrac{1}{2}R\hat{z}|/a_B). \tag{19.10b}$$

As the internuclear distance R decreases from infinity to a finite value, the charge of the bare proton perturbs the state of the electron centered on the other proton. As we know, the first step in the treatment of a perturbation is to remove the degeneracy of the unperturbed states, that is, to identify those superpositions of states (19.10) which remain approximately stationary under the influence of the perturbation. Recall that in section 14.2 the relevant superpositions of states represented eigenstates of the squared angular momentum. In the present problem the system has no spherical symmetry, but it has axial symmetry about the z axis and it has symmetry under reflection at the plane $z = 0$. Both wave functions (19.10) have axial symmetry about z; two superpositions of the pair (19.10) remain approximately stationary, one symmetric and one antisymmetric under reflection at the plane $z = 0$. The symmetric superposition has no node at $z = 0$, or anywhere else; therefore, it represents the limit of the $1s\sigma$ state $n_\xi = n_\eta = \lambda = 0$, and is

$$\sqrt{\tfrac{1}{2}}\,[u(\vec{r} + \tfrac{1}{2}R\hat{z}) + u(\vec{r} - \tfrac{1}{2}R\hat{z})] = \lim_{R=\infty} \langle \xi|0\ \ 0\ \ \vec{R}\rangle\langle\eta|0\ \ 0\ \ \vec{R}\rangle\langle\phi|0\rangle, \quad 1s\sigma. \tag{19.11a}$$

The antisymmetric superposition has a single node at $z = 0$; it is the limit of $2p\sigma$ state and is given by

$$\sqrt{\tfrac{1}{2}}\,[u(\vec{r} + \tfrac{1}{2}R\hat{z}) - u(\vec{r} - \tfrac{1}{2}R\hat{z})] = \lim_{R=\infty} \langle \xi|0\ \ 0\ \ \vec{R}\rangle\langle\eta|1\ \ 0\ \ \vec{R}\rangle\langle\phi|0\rangle, \quad 2p\sigma. \tag{19.11b}$$

As the nuclei begin to approach, each wave function $u(\vec{r} \pm \tfrac{1}{2}R\hat{z})$ is deformed by electric polarization due to the field of the other nucleus, and the energy decreases for both $1s\sigma$ and $2p\sigma$ states (in accordance with section 18.1). Upon further approach of the nuclei, the electron density at the midpoint, $\vec{r} = 0$, begins to build up for the state $1s\sigma$, but remains at zero (dark fringe) for $2p\sigma$ (Figs. 19.3 and 19.4). Therefore, the $2p\sigma$ state acquires the character of an excited state, as its energy, $\mathcal{E}_{2p\sigma}$ (R), begins to increase with decreasing R, whereas the ground-state energy $\mathcal{E}_{1s\sigma}$ (R) keeps decreasing. For this reason the $1s\sigma$ state

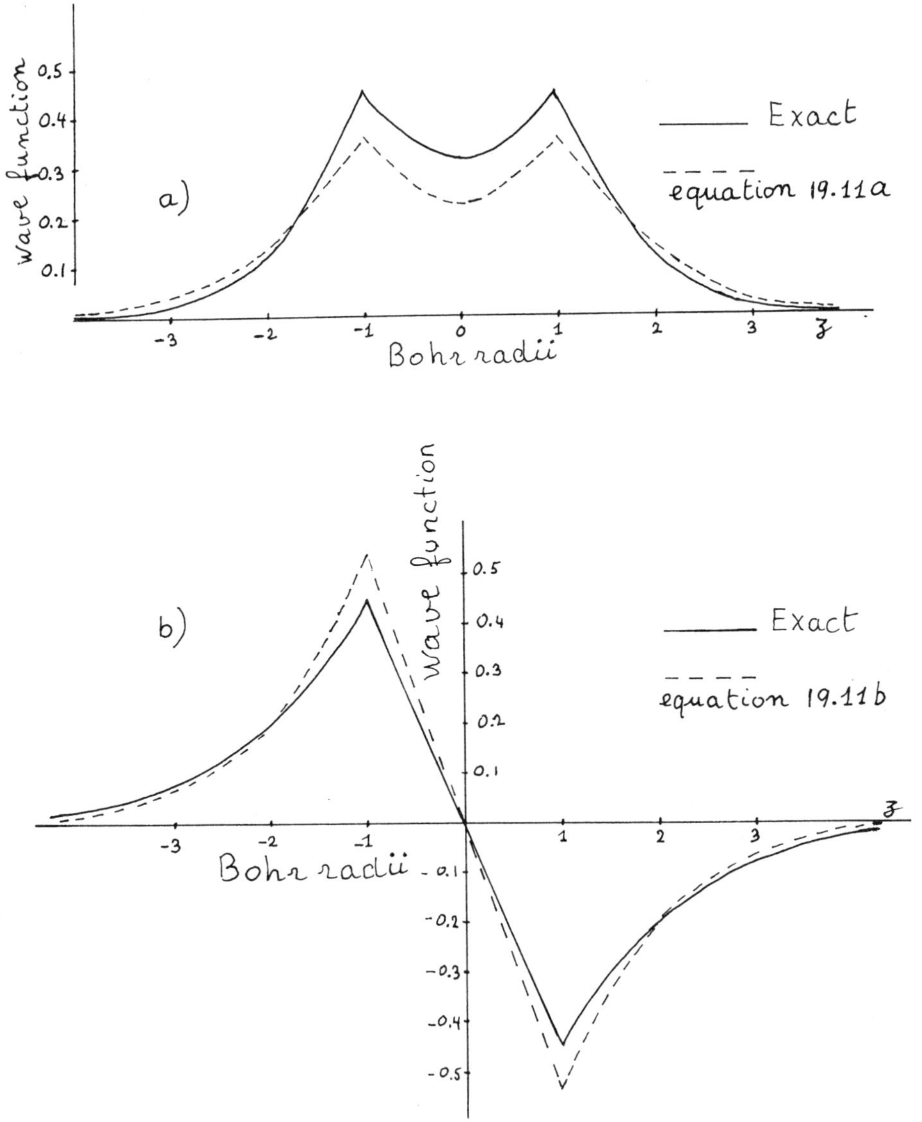

Fig. 19.3. Plot of H_2^+ wave functions along the z axis, for R = 2 Bohr radii. (*a*) $1s\sigma$, (*b*) $2p\sigma$.

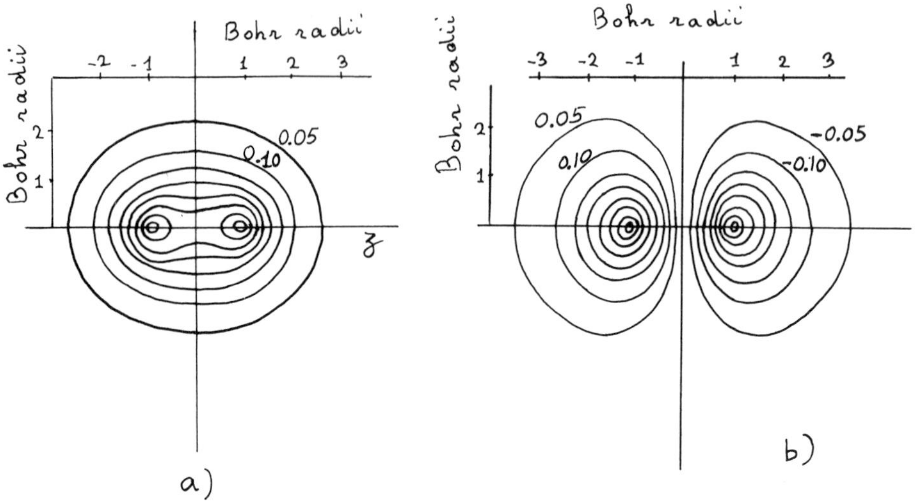

Fig. 19. 4. Contour diagram of H_2^+ wave functions in (x, z) plane for R = 2 Bohr radii. (*a*) 1sσ, (*b*) 2pσ. [Adapted from C. A. Coulson, *Valence*.]

is said to be *bonding*, while the 2pσ state is called *antibonding*. The bonding effect may be interpreted in terms of the accumulation of electron density in the region of attraction by both nuclei. The antibonding results from the electrons being squeezed out of that region. In the limit of $R \to 0$ the 1sσ wave function reduces to the 1s of He^+ and the 2pσ reduces to the $2p_z$ (i.e., 2p with $m = 0$) of He^+.

The wave functions of other excited states of $H_2{}^+$ are similarly derived from the hydrogen-atom wave functions in the separate-atom limit. As an illustration we describe here the construction of $H_2{}^+$ wave functions derived from the $n = 2$ level of the hydrogen atom. This level has fourfold degeneracy in a single atom and, therefore eightfold degeneracy for the pair $H + H^+$. The suitable set of four orthogonal wave functions with $n = 2$ of the hydrogen atom consists of v_+, v_-, w, and $\overline{w}$, given by

$$v_\pm(\vec{r}) = \mp\left(\frac{1}{16\pi a_B^3}\right)^{1/2} \exp\left(\frac{-r}{2a_B}\right) \frac{r}{2a_B} \sin\theta \, e^{\pm i\phi}, \qquad (19.12a)$$

$$w(\vec{r}) = \left(\frac{1}{16\pi a_B^3}\right)^{1/2} \exp\left(\frac{-r}{2a_B}\right) \left(\frac{r+z}{2a_B} - 1\right), \qquad (19.12b)$$

and

$$\overline{w}(\vec{r}) = \left(\frac{1}{16\pi a_B^3}\right)^{1/2} \exp\left(\frac{-r}{2a_B}\right) \left(\frac{r-z}{2a_B} - 1\right). \qquad (19.12c)$$

The wave functions $v_\pm(\vec{r})$ are the basic wave functions with $l = 1$ and $m = \pm 1$. The wave function w coincides with (13.32) and yields the off-center electron density shown in Figure 13.9; $\bar{w}$ is a mirror image of $\overline{w}$. The four wave functions are selected because they remain approximately stationary in an electric field parallel to the z axis.

By using the hydrogen-atom wave functions $v_+(\vec{r} + \frac{1}{2}R\hat{z})$ and $v_+(\vec{r} - \frac{1}{2}R\hat{z})$, one obtains two superpositions, one symmetric and one antisymmetric, both of them representing π states, with $\lambda = 1$. They are

$$\sqrt{\tfrac{1}{2}}\,[v_+(\vec{r} + \tfrac{1}{2}R\hat{z}) + v_+(\vec{r} - \tfrac{1}{2}R\hat{z})] = \lim_{R=\infty} \langle \xi\,|0\ \ 1\ \ \vec{R}\rangle\langle \eta\,|0\ \ 1\ \ \vec{R}\rangle\langle \phi\,|1\rangle,\quad 2p\pi \tag{19.13a}$$

$$\sqrt{\tfrac{1}{2}}\,[v_+(\vec{r} + \tfrac{1}{2}R\hat{z}) - v_+(\vec{r} - \tfrac{1}{2}R\hat{z})] = \lim_{R=\infty} \langle \xi\,|0\ \ 1\ \ \vec{R}\rangle\langle \eta\,|1\ \ 1\ \ \vec{R}\rangle\langle \phi\,|1\rangle,\quad 3d\pi. \tag{19.13b}$$

The wave function (19.13a) has no node in any constant ϕ plane and hence it has $n_\xi = n_\eta = 0$ and bonding character. The wave function (19.13b) is antibonding with the nodal plane $z = 0$, that is, $\eta = 0$; accordingly, it has $n_\xi = 0$ and $n_\eta = 1$. In the united atom limit ($R = 0$) these two wave functions reduce respectively to the He^+ wave functions 2p and 3d with $m = 1$. A similar pair of wave functions with $\lambda = -1$ and degenerate with (19.13) is constructed by using $v_-(\vec{r} + \frac{1}{2}R\hat{z})$ and $v_-(\vec{r} - \frac{1}{2}R\hat{z})$. The wave functions w and $\overline{w}$ must be used together to construct superpositions symmetric or antisymmetric with respect to the $z = 0$ plane of the $H_2{}^+$ molecule ion. Thus we have

$$\sqrt{\tfrac{1}{2}}\ [w(\vec{r} + \tfrac{1}{2}R\hat{z}) + \overline{w}(\vec{r} - \tfrac{1}{2}R\hat{z})] = \lim_{R=\infty} \langle \xi\,|0\ \ 0\ \ \vec{R}\rangle\langle \eta\,|2\ \ 0\ \ \vec{R}\rangle\langle \phi\,|0\rangle,\quad 3d\sigma; \tag{19.14a}$$

$$\sqrt{\tfrac{1}{2}}\ [w(\vec{r} + \tfrac{1}{2}R\hat{z}) - \overline{w}(\vec{r} - \tfrac{1}{2}R\hat{z})] = \lim_{R=\infty} \langle \xi\,|0\ \ 0\ \ \vec{R}\rangle\langle \eta\,|3\ \ 0\ \ \vec{R}\rangle\langle \phi\,|0\rangle,\quad 4f\sigma; \tag{19.14b}$$

and

$$\sqrt{\tfrac{1}{2}}\ [\overline{w}(\vec{r} + \tfrac{1}{2}R\hat{z}) + w(\vec{r} - \tfrac{1}{2}R\hat{z})] = \lim_{R=\infty} \langle \xi\,|1\ \ 0\ \ \vec{R}\rangle\langle \eta\,|0\ \ 0\ \ \vec{R}\rangle\langle \phi\,|0\rangle,\quad 2s\sigma; \tag{19.15a}$$

$$\sqrt{\tfrac{1}{2}}\ [\overline{w}(\vec{r} + \tfrac{1}{2}R\hat{z}) - w(\vec{r} - \tfrac{1}{2}R\hat{z})] = \lim_{R=\infty} \langle \xi\,|1\ \ 0\ \ \vec{R}\rangle\langle \eta\,|1\ \ 0\ \ \vec{R}\rangle\langle \phi\,|0\rangle,\quad 3p\sigma. \tag{19.15b}$$

Figure 19.5 shows a diagram of the $2s\sigma$ superposition (19.15a).

Quite generally, the antibonding character of a state is associated in the separate-atom limit with a minus sign of the wave function, and for any value of R

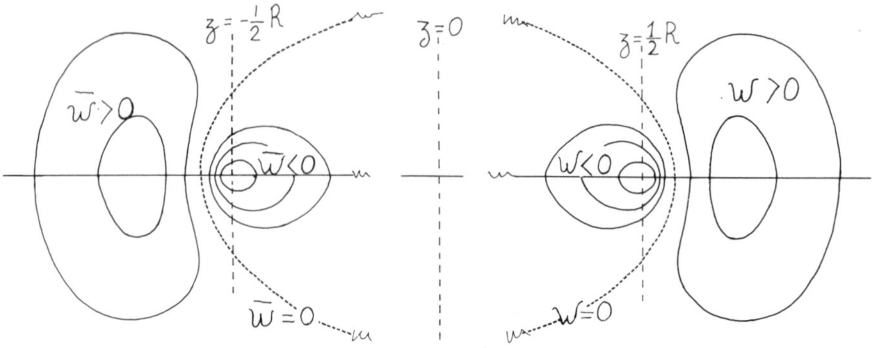

Fig. 19. 5. Contour diagram of H_2^+ wave function $2s\sigma$ in (x, z) plane for large R. Compare with electron density map for hydrogen atom in Figure 13. 9.

with a dark fringe of electron density across the middle of the internuclear axis. This dark fringe occurs whenever n_η is odd. As R decreases, the n_ξ ellipsoidal fringes of each wave function become progressively rounder, and they become spherical in the limit $R = 0$. In the same limit, the n_η hyperboloidal fringes sharpen into cones identical with those of the hydrogen atom. Indeed, as shown in Figure 19. 2, all stationary states of $H_2{}^+$ coincide at $R = 0$ with those of the hydrogen-like ion He^+ and $n_\eta + |\lambda|$ reduces to the orbital quantum number l.

A diagram connecting each level of the united atom with the corresponding level of the separate atom is called a *correlation diagram*. The plot of energy levels in Figure 19. 2 serves as a correlation diagram. These diagrams are particularly useful when studying many-electron molecules in the independent electron approximation (section 20. 4). In molecules other than $H_2{}^+$ the pattern of single-electron wave functions may change with varying internuclear distance. Each single-electron state cannot be readily classified on the basis of the quantum numbers n_ξ and n_η. A classification is then established by connecting each state to its united-atom or separate-atom limit in a correlation diagram. In any event each single-electron state can be labeled by the quantum numbers belonging to its united-atom limit.

19. 2. 3. *Nuclear Motion.* We return now to the study of nuclear motion in the $H_2{}^+$ ion, that is, to the solution of the Schrödinger equation (19. 6). The potential energy of this motion, U(R), is a function of the internuclear distance R, and it equals the sum of the internuclear repulsive potential and the electron energy

$$U(R) = \frac{e^2}{R} + \mathcal{E}_n(R). \qquad (19.16)$$

Figure 19. 6 shows graphs of this function obtained by adding the proton-proton

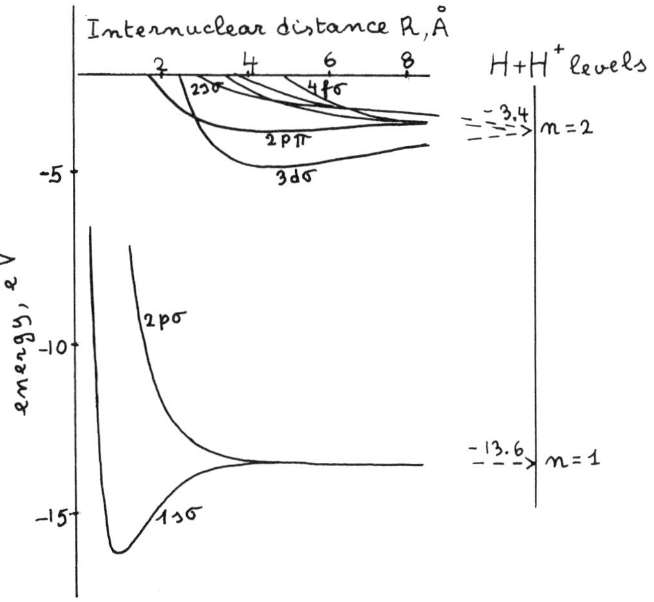

Fig. 19. 6. Potential energy U(R) of the nuclear motion for H_2^+. [Adapted from E. Teller, *Z. Physik* 61 (1930): 458].

repulsion to the functions $\mathcal{E}_n(R)$ plotted in Figure 19. 2 which represents the bonding (or antibonding) effect of the electron. The function U(R) has a minimum for the bonding states, and thus constitutes a potential well. The molecule ion holds together when the internuclear motion is confined within this well For the antibonding states, U(R) does not form a potential well sufficiently deep to confine the internuclear motion. Detailed experimental evidence on the potential function U(R) is obtained indirectly from analysis of spectral, collision, and thermal data. The results provide a check on the theory.

Since the potential energy U(R) is independent of the direction of the internuclear axis $\vec{R}$, the motion of the pair of nuclei is equivalent to the motion of a single particle in a spherically symmetrical potential. Therefore, the rotational motion of the pair of nuclei *about* their center of mass may be treated before treating the motion *along* the internuclear axis $\vec{R}$, as was done for the hydrogen atom in section 13. 2. The rotational energy is then entered as a centrifugal potential in the Schrödinger equation for the motion of the nuclei along the axis $\vec{R}$. The motion along $\vec{R}$ is called the *vibrational motion* of the two nuclei.

Traditionally, the rotational and the vibrational motion of the nuclei are assumed at first to be altogether independent. This approximation rests on the fact that the centrifugal force generated by the rotational motion is much weaker than the force $|dU/dR|$ which confines the motion along the internuclear axis.

One also approximates the function U(R) near its point of minimum by a parabola. In this case the internuclear motion along R coincides with that of a harmonic oscillator, hence the name "vibrational motion." The energy levels thus obtained are then improved by various corrective terms. Here we proceed without disregarding centrifugal forces at the outset.

The rotational energy is given by a formula analogous to (13.21), that is, by the squared angular momentum of nuclear rotation divided by $2(\frac{1}{2}M)R^2$.[6] Here, however, the angular momentum of rotation of the pair of nuclei, $\vec{\mathcal{R}}$, is not a constant of the motion.[7] The constant is the sum of all angular momenta, spin and orbital, of both nuclei and of the electron,

$$\vec{F} = \vec{\mathcal{R}} + \vec{l} + \vec{s} + \vec{T}, \tag{19.17}$$

where $\vec{l}$ and $\vec{s}$ are the orbital and spin momenta of the electron and

$$\vec{T} = \vec{I}_1 + \vec{I}_2 \tag{19.17a}$$

is the total spin of the two nuclei. Since in H_2^+ the electron spin $\vec{s}$ and the total spin $\vec{T}$ of the two nuclei are weakly coupled with the other angular momenta, the rotational energy is practically independent of their orientation and depends only on the squared magnitude of

$$\vec{N} = \vec{\mathcal{R}} + \vec{l}. \tag{19.18}$$

We consider then states of the molecule which are approximately stationary and are eigenstates of $|\vec{N}|^2$.[8] To find the rotational energy of each of these states we must evaluate $|\vec{\mathcal{R}}|^2 = |\vec{N} - \vec{l}|^2 = |\vec{N}|^2 + |\vec{l}|^2 - 2\vec{N}\cdot\vec{l}$. The states of electron motion treated in section 19.2.2 are not eigenstates of $|\vec{l}|^2$ or of $\vec{N}\cdot\vec{l}$. In the Born-Oppenheimer approximation the nuclear motion does not influence the electron motion. Accordingly, the rotational energy of the whole ion depends on the mean values $\langle|\vec{l}|^2\rangle$ and $\langle\vec{N}\cdot\vec{l}\rangle$ for the electron state $\langle\vec{r}|n\,\vec{R}\rangle$. The mean value $\langle|\vec{l}|^2\rangle_n$ has no simple expression and will be indicated by $\hbar^2\langle l(l+1)\rangle_n$. The mean value $\langle\vec{N}\cdot\vec{l}\rangle$ is shown to equal $\lambda^2\hbar^2$, where λ is the quantum

---

6. Recall that we deal here with the reduced mass of the nuclear pair, namely, $\frac{1}{2}M$.

7. The angular momentum of nuclear rotation is usually called $\vec{R}$; we use the symbol $\vec{\mathcal{R}}$ here to avoid confusion with the internuclear distance vector.

8. The identity of the two nuclei enforces a relationship between the quantum numbers corresponding to the magnitudes of $\vec{T}$ and $\vec{N}$. This relationship will be treated in section 20.2.4.

number of the electron motion introduced in section 19.2.2.[9] The rotational energy of the ion is thus given by

$$E_{rot} = \frac{|\vec{N}|^2 + \langle|\vec{l}|^2\rangle - 2\langle\vec{N}\cdot\vec{l}\rangle}{MR^2} = \frac{\hbar^2}{MR^2}\,[N(N+1) + \langle l(l+1)\rangle_n - 2\lambda^2]. \tag{19.19}$$

The coefficient $\hbar^2/MR^2$ is conventionally indicated by[10]

$$\frac{\hbar^2}{MR^2} = B(R) \tag{19.19a}$$

Note that MR^2 is twice the moment of inertia of the two nuclei.

The stationary states of rotation of the ion are represented by wave functions $\langle\Theta\ \Phi|N\ \lambda\ M\rangle$, where Θ and Φ are the polar coordinates of the internuclear axis $\hat{R}$ in a laboratory frame with polar axis Z. For brevity we do not write or solve here the Schrödinger equation which determines these wave functions; we merely describe its solution. The motions along Θ and Φ are separated as in the wave functions of the electron in the hydrogen atom, by writing

$$\langle\Theta\ \Phi|N\ \lambda\ M\rangle = \langle\Theta|N\ \lambda\ M\rangle\langle\Phi|M\rangle. \tag{19.20}$$

Here the factor $\langle\Phi|M\rangle$ is given by the hydrogen-atom equation (13.14),

$$\langle\Phi|M\rangle = \frac{1}{(2\pi)^{1/2}}\,e^{iM\Phi}. \tag{19.20a}$$

The remaining factor of the wave function, $\langle\Theta|N\ \lambda\ M\rangle$, reduces to the hydrogen-atom wave function $\langle\theta|l\ \ m\rangle$ when $\lambda = 0$. For $\lambda \neq 0$, the electron motion about the internuclear axis influences the rotation of the axis itself. The Schrödinger equation for the motion along Θ differs from the hydrogen atom's equation (13.18) only in that m^2 is replaced by $M^2 + \lambda^2 - 2\lambda M\cos\Theta$. [Whereas, in section 13.2, m^2 is the eigenvalue of $l_z^2/\hbar^2$, $M^2 + \lambda^2 - 2\lambda M\cos\Theta$ is the eigenvalue of $|(\vec{N}\cdot\hat{Z})\hat{Z} - (\vec{l}\cdot\hat{R})\hat{R}|^2/\hbar^2$.] The wave functions are polynomials of

---

9. The stationary states of electron motion considered in section 19.2.2 are eigenstates of the orbital angular momentum component $\vec{l}\cdot\hat{R}$ with eigenvalues $\lambda\hbar$. The arguments of section 6.3.4, showing that $\langle\vec{\mu}\rangle = \mu_z\hat{z}$, yield in the present problem $\langle\vec{l}\rangle = \lambda\hbar\hat{R}$. Thus we have $\langle\vec{N}\cdot\vec{l}\rangle = \langle\vec{N}\cdot\hat{R}\rangle\lambda\hbar$. Substitution of (19.18) yields then $\langle\vec{N}\cdot\vec{l}\rangle = (\vec{\mathcal{R}}\cdot\hat{R} + \vec{l}\cdot\hat{R})\lambda\hbar = \vec{\mathcal{R}}\cdot\hat{R}\lambda\hbar + \lambda^2\hbar^2$. The factor $\vec{\mathcal{R}}\cdot\hat{R}$ is zero because $\vec{\mathcal{R}} = \vec{R}\times\vec{P}$, where $\vec{P}$ is the momentum of one nucleus in a frame attached to the other nucleus.

10. In spectroscopy literature B often indicates the energy $\hbar^2/MR^2$ expressed in wavenumbers, that is, the quantity $\hbar/2\pi McR^2$.

degree N in sin Θ and cos Θ whose general form is given by

$$\langle\Theta|N\ \lambda\ M\rangle = \left[\frac{(2N+1)!(N+M)!}{(N+\lambda)!(N-\lambda)!(N-M)!}\right]^{1/2} \frac{(1+\cos\Theta)^{(\lambda-M)/2}}{(1+\cos\Theta)^{(\lambda+M)/2}}$$

$$\times\left[\left(\frac{d}{d\cos\Theta}\right)^{N-M}(1-\cos\Theta)^{N-\lambda}(1+\cos\Theta)^{N+\lambda}\right]. \qquad (19.20b)$$

These polynomials coincide within a normalization factor with the functions $d^{(N)}_{\lambda M}(\Theta)$ introduced in equation (10.22) and given in Table 10.1 for N = 1.[11]

We come now to the vibrational motion of the two nuclei along the internuclear axis, that is, to the solution of the Schrödinger equation for the dependence of the wave function on the internuclear distance R. The equation is obtained from (19.6) by expressing the complete nuclear wave function $\langle\vec{R}|n\ \eta\rangle$ as the product $\langle\Theta\ \Phi|N\ \lambda\ M\rangle\langle R|E\ N\ v\rangle$.[12] Upon substitution of this product into (19.6), this equation separates like the Schrödinger equation for the hydrogen atom (13.6). The rotational wave function factors out, and the terms of (19.6) with derivatives with respect to the coordinates Θ and Φ are replaced by the rotational kinetic energy (19.19). The Schrödinger equation is then analogous to (13.8c), namely,

$$\left\{-\frac{\hbar^2}{M}\left(\frac{d^2}{dR^2}+\frac{2}{R}\frac{d}{dR}\right)+\frac{e^2}{R}+\mathcal{E}_n(R)+B(R)\,[N(N+1)\right.$$

$$\left.+\langle l(l+1)\rangle_n - 2\lambda^2\right\}\langle R|E\ N\ v\rangle = \langle R|E\ N\ v\rangle E_{Nv}, \qquad (19.21)$$

where v indicates the quantum number of vibrational motion. This equation is reduced to simpler form by introducing the modified potential function

$$\overline{U}(R) = \frac{e^2}{R} + \mathcal{E}_n(R) + B(R)\,[\langle l(l+1)\rangle_n - 2\lambda^2] \qquad (19.22)$$

---

11. These two functions coincide because the problem of the rotation of the H_2^+ ion, or even of a single particle, is substantially equivalent to the problem of selection of angular momentum eigenstates considered in chapters 6-10. We know that the molecule ion has angular momentum $\lambda\hbar$ about the internuclear axis $\hat{R}$ and angular momentum $M\hbar$ about the laboratory axis $\hat{Z}$. The probability amplitude of a given orientation of the axis $\hat{R}$ is then essentially equivalent to the probability amplitude $\langle\hat{Z}\ M|\hat{R}\ \lambda\rangle$.

12. In this expression the electronic quantum number n has been dropped for brevity; the quantum number η, replaced by N, M, and v, is not to be confused with the electron coordinate η.

and the modified vibrational wave function[13]

$$\chi_{Nv}(R) = R\langle R|E\ N\ v\rangle. \tag{19.23}$$

Equation (19.21) now takes the form

$$-\frac{\hbar^2}{M}\frac{d^2}{dR^2}\chi_{Nv}(R) + [\overline{U}(R) + B(R)N(N+1)]\,\chi_{Nv} = \chi_{Nv}E_{Nv}. \tag{19.24}$$

For bound states of the H_2^+ ion the motion of the nuclei along R is confined within the potential well $\overline{U}(R)$. The vibrational character of this motion (see page 445-6) is demonstrated by expanding $\overline{U}(R)$ into a power series in the vicinity of the *equilibrium distance* R_e at which $\overline{U}(R)$ has a minimum value. Limiting the expansion to the second power yields the parabolic approximation

$$\overline{U}(R) \approx \overline{U}(R_e) + \tfrac{1}{2}k(R - R_e)^2, \tag{19.25}$$

with

$$k = (d^2\overline{U}/dR^2)_{R_e}. \tag{19.25a}$$

This approximation, together with $B(R) \approx B(R_e)$, reduces equation (19.24) to the Schrödinger equation (12.15) for a particle with mass $\tfrac{1}{2}M$ and with the harmonic-oscillator potential energy (12.41). The energy eigenvalues are then obtained from (12.44) in the form

$$E_{Nv} = \overline{U}(R_e) + B(R_e)N(N+1) + (v + \tfrac{1}{2})\hbar\omega, \tag{19.26}$$

where the vibrational frequency is

$$\omega = 2\pi c\nu = \left(\frac{k}{\tfrac{1}{2}M}\right)^{1/2}. \tag{19.26a}$$

The vibrational wave functions are obtained from (12.45). The value of $\hbar\omega$ of the H_2^+ ion in its electronic ground state is 0.28 eV ($\nu = 2297\ \text{cm}^{-1}$), and the vibration extends about $R_e = 1.06$ Å over an approximate range given by $a = (\hbar/\tfrac{1}{2}M\omega)^{1/2} \approx 0.18$ Å, for $v = 0$. The *rotational constant* $B(R_e)$ is 3.7×10^{-3} eV.

The approximation of the vibrational potential by that of a harmonic oscillator independent of the rotational quantum number N is reasonably good for molecules with nuclei much heavier than hydrogen. For the H_2^+ ion the approximation is not accurate because the vibrational motion extends over

---

13. This modification is the same as was introduced for the radial wave equation of the H atom as "change of variable 3" on p. 290.

most of the width of the potential well even for v = 0. Accurate values of the vibrational energy levels must be obtained by numerical solution of equation (19.24).

Short of numerical solution, one may improve the initial approximation by expanding the potential $\overline{U}(R) + B(R)N(N+1)$ to higher powers of $R - R_e$ and treating the additional terms as perturbations, by the method of section 17.1.3. Improved approximations of $\overline{U}(R)$ reproduce the decreasing slope of this potential with increasing R, shown in Figure 19.7. These approximations yield *anharmonicity corrections* represented by an energy term proportional to $(v + \frac{1}{2})^2$. This term is negative and brings about a decrease of the level separation for successive values of v. The dependence of B on R introduces two corrections to the energy levels (19.26). One correction reflects the centrifugal stretching of the molecule and is proportional to $[N(N+1)]^2$; the

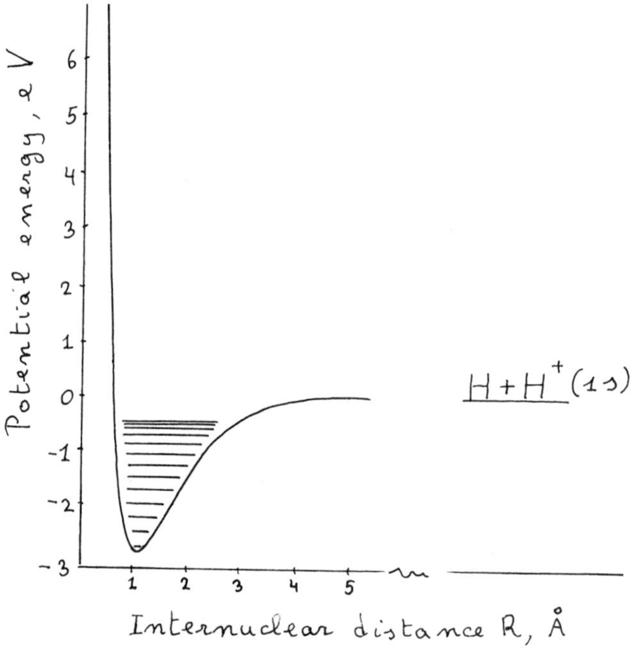

Fig. 19.7. Potential function U(R) and vibrational levels for H_2^+ in the ground state. Note that the value of U(R) for $R = \infty$ is taken here as the zero point of the ordinates.

other represents a vibrational-rotational coupling and is proportional to $(v + \frac{1}{2})N(N+1)$.

19.3. THE HYDROGEN MOLECULE

The H_2 molecule is the simplest molecular structure with fully saturated covalent bonds. In the study of molecules it plays a role analogous to that of helium in atomic physics. In fact, both He and H_2 have their electrons in the ground state orbital with opposite spins and are thus particularly stable.

From the point of view of the independent-electron and Born-Oppenheimer approximations the ground state of the molecule has a $(1s\sigma)^2$ configuration.

The repulsion between the two electrons may be taken into account approximately by introducing a screened potential in the Schrödinger equation. The resulting wave function of each electron is then no longer exactly factorable in the form (19.9) applicable to H_2^+; only the rotational motion about the internuclear axis, represented by $\langle\phi\,|\,\lambda\rangle$, is factorable. The internuclear repulsion is the same as it is in H_2^+, but in the neutral molecule two electrons exert a bonding action. Therefore, the equilibrium distance between the nuclei, R_e, is about 30% smaller in the H_2 molecule than it is in H_2^+ and equals ~ 0.74 Å for H_2. At this short internuclear distance the electron states somewhat resemble those of the united-atom limit, in that the electron density for the molecule is not greatly elongated in the direction of the nuclear axis. The modified internuclear potential function $\overline{U}(R)$ forms a well deeper and narrower than for H_2^+ and leads to a high vibrational frequency with $\hbar\omega$ approximately 0.54 eV (Fig. 19.8).

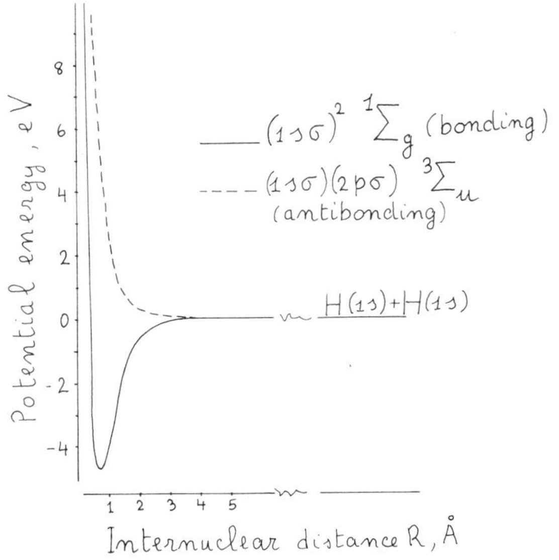

Fig. 19.8. Potential function U(R) for two electronic states of H_2.

Quantitatively, however, the independent electron approximation serves only for purposes of initial estimation. Even with the best self-consistent potential one accounts only for approximately 75% of the stability of the H_2 molecule. The lowest energy eigenvalue calculated with this approximation lies only 3.4 to 3.5 eV lower than the energy of two separate H atoms, while the experimental value of the dissociation energy is 4.48 eV. On the other hand, elaborate numerical calculations have been carried out, whose results agree with the experimental energy to about 1 part in 10^5. These calculations solve the Schrödinger equation with wave functions that include correlations between the positions of both electrons and nuclei.

The correlated wave functions are no longer characterized by the configuration formula $(1s\sigma)^2$ but by the symbol $^1\Sigma_g$. The superscript 1 indicates here, as for atomic states, that the state is a singlet with total spin angular momentum $S = 0$. The symbol Σ means that the electron pair has zero angular momentum about the internuclear axis. The subscript g indicates that the electron wave function is unchanged under reflection of coordinates with respect to the center of mass.[14]

From the beginning of this section we have regarded the H_2 molecule as a single system and we have been concerned with the energy and wave function of its ground state. From this point of view the motion of each electron extends over the whole molecule and is thus described by a *molecular orbital*. The state in which the molecule is dissociated into two separate atoms was considered only as a point of reference for the energy scale of Figure 19.8. An alternative point of view starts by considering the two separate H atoms and how the combined energy of their two electrons changes when the atoms approach and the electron densities overlap significantly. Historically, this approach was used in the first calculation of a covalent bond by Heitler and London in 1927. The Heitler-London (or *valence bond*) method proves reasonably accurate for values of the internuclear distance larger than the equilibrium distance R_e. It has the advantage, particularly for molecules more complex than H_2, of utilizing the knowledge of atomic wave functions, which is generally superior to that of molecular orbitals.

19.3.1. *The Valence-Bond Treatment*. Heitler and London patterned their treatment of the electron pair in the H_2 molecule after the treatment of the para and ortho states of He (section 15.4). In the initial approximation one considers the wave functions of the two electrons as if each electron belonged to a separate atom. The nuclei of the two atoms lie along the z axis at $\pm\frac{1}{2}R\hat{z}$, respectively. Utilizing the symbols of the two equations (19.10a, b), one writes initial wave functions of two degenerate states in the product form analogous to (15.15)

---

14 The letter g stands for *gerade*, which is German for "even." The alternative symbol is u, which stand for *ungerade*, or "odd". The English words "even" and "odd" usually refer to reflection of all coordinates, as discussed in section 20.2.2.

$$u(\vec{r}_1 + \tfrac{1}{2}R\hat{z})u(\vec{r}_2 - \tfrac{1}{2}R\hat{z}), \qquad u(\vec{r}_2 + \tfrac{1}{2}R\hat{z})u(\vec{r}_1 - \tfrac{1}{2}R\hat{z}). \tag{19.27}$$

Superposition of these wave functions with alternative symmetry under permutation of electron coordinates yields wave functions analogous to (15.17), for alternative values of the combined spin quantum number S. However, the cases of He and H_2 differ in that the two single-electron wave functions for He are orthogonal, since one corresponds to the ground state and one to an excited state of electrons in the same central field. In the H_2 case the single-electron wave functions (19.10a) and (19.10b) pertain to atoms with different centers; hence they are not orthogonal but they have a nonzero *overlap integral*

$$\Delta(R) = \int d\vec{r}\ u(\vec{r} - \tfrac{1}{2}R\hat{z})\ u(\vec{r} + \tfrac{1}{2}R\hat{z}). \tag{19.28}$$

Accordingly, the symmetrized and normalized two-electron wave functions analogous to (15.17) are

$$\{2[1 + (-1)^S\Delta(R)^2]\}^{-1/2}\ [u(\vec{r}_1 + \tfrac{1}{2}R\hat{z})u(\vec{r}_2 - \tfrac{1}{2}R\hat{z}) + (-1)^S u(\vec{r}_2 + \tfrac{1}{2}R\hat{z})u(\vec{r}_1 - \tfrac{1}{2}R\hat{z})]. \tag{19.29}$$

These wave functions yield a correlated density of the two electrons analogous to (15.18). The correlation results in an increased density of both electrons in the space between the two nuclei when $S = 0$. Conversely, in the triplet state with $S = 1$ the density of the two electrons is diminished in the space between the nuclei and increased in the outer region.

The wave functions (19.29) serve to estimate the two lowest eigenvalues of the exact Schrödinger equation for electron motion. The energy of the system of two electrons is represented by

$$\mathcal{E} = \frac{p_1^2}{2m_e} - \frac{e^2}{|\vec{r}_1 + \tfrac{1}{2}R\hat{z}|} - \frac{e^2}{|\vec{r}_1 - \tfrac{1}{2}R\hat{z}|} + \frac{p_2^2}{2m_e} - \frac{e^2}{|\vec{r}_2 + \tfrac{1}{2}R\hat{z}|} - \frac{e^2}{|\vec{r}_2 - \tfrac{1}{2}R\hat{z}|} + \frac{e^2}{|\vec{r}_1 - \vec{r}_2|}, \tag{19.30}$$

where $\vec{p}_1$ and $\vec{p}_2$ are the momenta of the two electrons. The two lowest eigenvalues are obtained by evaluating the mean value of $\mathcal{E}$ for each of the states represented by (19.29). The calculation can be carried out analytically, though laboriously, and gives

$$\langle\mathcal{E}\rangle = \left[-2 + \frac{Q + (-1)^S A}{1 + (-1)^S\Delta^2}\right] I_H. \tag{19.31}$$

Here $I_H = 13.6$ eV indicates the ionization energy of each separate H atom.

Therefore, the term $-2I_H$ represents the energy of the pair of atoms at infinite distance; the remaining terms represent the bonding or antibonding effects of the electronic motion. In particular, the term A represents the exchange energy of the electron pair, analogous to the energy W in equation (15.22) for the He atom, while Q is analogous to V; both A and Q are functions of R. The denominator $1 + (-1)^S\Delta^2$ stems from the normalization coefficient of the wave function (19.29).

The interaction energy, U(R), between two hydrogen atoms can be expressed as the sum of the second term in the brackets in equation (19.31) and a term representing the repulsion of the two nuclei,

$$U(R) = \frac{Q + (-1)^S A}{1 + (-1)^S \Delta^2} I_H + \frac{e^2}{R}. \tag{19.32}$$

To this expression one should still add the Van der Waals energy (18.11), which actually predominates for $R \gg a$, since it decreases as R^{-6} while (19.32) decreases exponentially.

The exchange energy term A in equation (19.32) has a decisive influence on the formation of a covalent bond. This term represents the potential energy of the exchange density of the electron pair. Part of this energy is due to the attraction by the nuclei and part to the mutual repulsion of the electrons. Only this second part is present in the analogous calculation for the He atom, in which case the energy W (corresponding to A) is positive. Here, instead, A is negative and yields the main contribution to bond formation in the *singlet state* (S = 0). For the triplet state (S = 1) the term $(-1)^S A$ is positive and thus yields an antibonding effect. The trend of U(R) in the antibonding state is shown in Figure 19.8. From the point of view of the independent electron approximation the state with S = 1 has the configuration $(1s\sigma)^1 (2p\sigma)^1$, with one bonding and one antibonding electron. Without reference to the state of individual electrons, this state is classified as $^3\Sigma_u$.

Numerical evaluation of U(R) from equation (19.32) yields, as anticipated, results of fair accuracy for internuclear distances of the order of a few Bohr radii. However, the calculation yields a minimum of U(R) at a distance $R = 1.65\, a_B$, that is, nearly 20% in excess of the experimental value of the equilibrium distance R_e. The calculated minimum of U(R) is only about two-thirds of the experimental value.

The valence-bond approximation may be improved by generalizing the wave functions (19.29). In this equation the single-atom wave functions $u(\vec{r})$ were assumed to have the form $(\pi a^3)^{-1/2} \exp(-r/a_B)$ of the hydrogen atom. One generalization consists of replacing a_B by a variable parameter. One then determines which value of this parameter minimizes the mean energy $\langle \mathcal{E} \rangle$ given by equation (19.31) for each given value of the internuclear distance R. Finally, one finds the value of R for which the energy U(R) given by (19.32) is a minimum. This method yields a value of R_e very near the experimental value and a minimum of U(R) equal to over 85% of the experimental value. The optimum value of the variable parameter turns out to be $a_B/1.17$.

The valence-bond and molecular-orbital approximations may be compared by considering their wave functions for large internuclear distances. To this end, we construct a wave function of the $(1s\sigma)^2$ configuration in the independent-electron molecular-orbital approximation. For each $1s\sigma$ orbital we use the H_2^+ wave function (19.11a) applicable to large values of R. We improve the normalization of this wave function by introducing the overlap integral Δ which does not vanish for finite values of R. The resulting wave function is

$$\frac{1}{2[1+\Delta(R)]}\,[u(\vec{r}_1+\tfrac{1}{2}R\hat{z})+u(\vec{r}_1-\tfrac{1}{2}R\hat{z})][u(\vec{r}_2+\tfrac{1}{2}R\hat{z})+u(\vec{r}_2-\tfrac{1}{2}R\hat{z})]$$

$$=\frac{1}{2[1+\Delta(R)]}\,\{u(\vec{r}_1+\tfrac{1}{2}R\hat{z})u(\vec{r}_2+\tfrac{1}{2}R\hat{z})+u(\vec{r}_1-\tfrac{1}{2}R\hat{z})u(\vec{r}_2-\tfrac{1}{2}R\hat{z})$$

$$+u(\vec{r}_1+\tfrac{1}{2}R\hat{z})u(\vec{r}_2-\tfrac{1}{2}R\hat{z})+u(\vec{r}_1-\tfrac{1}{2}R\hat{z})u(\vec{r}_2+\tfrac{1}{2}R\hat{z})\}. \qquad (19.33)$$

The last two terms in the braces coincide, aside from normalization, with the wave function (19.29) for $S = 0$. The first two terms represent an extension of the superposition to include states with both electrons in the same atom. These states correspond to the ionic formula H^+H^-; this formula is unrealistic for large internuclear distances because it takes 13.6 eV to rip off an electron from one H atom, while only 0.75 eV are gained in the formation of H^-. However, the molecular-orbital description becomes more realistic at low values of the internuclear distance where the electrons roam more freely and independently throughout the molecule.

19.4. BONDS BETWEEN MONOVALENT ATOMS

The treatment of the covalent bond in H_2^+ and H_2 requires some adaptation when extended to bonds between other pairs of monovalent atoms. One modification is needed to account for the influence of electrons that do not participate in the bond. A more substantial effect occurs when the molecule consists of two unlike atoms.

The influence of nonbonding electrons is apparent in the example of the Na_2 molecule. A complete study would involve the description of the motion of all 22 electrons in the two atoms. However, we know that 10 electrons in each atom form a neon-like closed-shell core. This core does not contribute to bond formation. Rather, the two cores repel each other strongly at short internuclear distances. In the Schrödinger equation for the motion of the nuclei, analogous to (19.6), the repulsive energy between the two complete shells can be separated out from the total energy eigenvalue $\mathcal{E}_n$ of all electrons. In other words, $\mathcal{E}_n$ can be regarded as the sum of two terms, one representing the energy of the 20 electrons in the two ionic cores and the other representing the energy of the two bonding electrons. The energy of the two cores may be estimated from the size of the separate Na^+ ions without any detailed calculation; it rises sharply when the internuclear distance becomes too small. The energy of the bonding electrons is calculated much as for the H_2 molecule. However, the internuclear equilibrium distance remains much larger than in the H_2 molecule or in the H_2^+ ion; this is due in part to the core repulsion but mostly to the fact that each outer electron is weakly bound to either atom. Consequently, the Na_2 bond energy amounts to only 0.73 eV, and the internuclear distance is unusually large, namely, 3.08 Å.

In general, bonds formed between atoms with rather large closed-shell cores are similarly weak. In the condensed state, these covalent molecular bonds are replaced by a different mechanism of aggregation, that is, by the *metallic bond* (see chap. 21).

For purposes of semiquantitative analysis of molecular bonds one represents the state of each bonding electron by a molecular orbital wave function constructed as a superposition of *atomic orbitals,* that is, of wave functions of the valence electrons of each atom. For a pair of identical atoms this superposition is analogous to the H_2^+ wave function (19.11a). Naturally, the atomic wave function $u(\vec{r})$ of each atom must represent an independent-electron state for the atom under consideration rather than for hydrogen. Furthermore, the superposition is normalized in accordance with the value of the overlap integral $\Delta(R)$, given by equation (19.28). The molecular wave function thus obtained is

$$\{2[1 + \Delta(R)]\}^{-1/2}\,[u(\vec{r} + \tfrac{1}{2}R\hat{z}) + u(\vec{r} - \tfrac{1}{2}R\hat{z})]. \tag{19.34}$$

This type of wave function is called an *MO-LCAO,* which stands for *molecular orbital, linear combination of atomic orbitals.*

For a pair of unlike atoms X and Y the function (19.34) must be generalized first by introducing different wave functions $u_X(\vec{r})$ and $u_Y(\vec{r})$ for the two atoms. Furthermore, the coefficients of the two wave functions u_X and u_Y need not be equal because the density of the bonding electrons need not be distributed evenly between the two atoms. We set the origin of coordinates at the center of mass of the two atoms; therefore, the positions of the two nuclei are represented respectively by $-(M_Y/M)R\hat{z}$ and $(M_X/M)R\hat{z}$, where $M = M_X + M_Y$ is the mass of the whole molecule. The MO-LCAO wave function is then

$$\{c_X^2 + c_Y^2 + 2c_Xc_Y\Delta_{XY}(R)\}^{-1/2}\,[c_Xu_X(\vec{r} + \frac{M_Y}{M}R\hat{z}) + c_Yu_Y(\vec{r} - \frac{M_X}{M}R\hat{z})]. \tag{19.35}$$

The ratio of coefficients c_X/c_Y is a variational parameter which is to be determined to minimize the energy of the bonding electrons for each value of the internuclear distance R.

In any molecule consisting of two different atoms the electron density is naturally concentrated, to a larger or lesser extent, on the side of the atom with higher electron affinity. The optimum value of the ratio c_X/c_Y indicates the magnitude of the electron density shift toward one atom. This means that the bond between unlike atoms always partakes of heteropolar (or ionic) character. If the ordinary chemical-bond symbol X-Y is understood to represent a covalent bond, one must say that the molecule has a structure intermediate between those represented by X-Y and by the ionic formula X^+Y^- (or X^-Y^+). The ability of the electron density to shift with continuity to yield a minimum energy contributes increased stability to bonds between unlike atoms. These bonds are said to be *stabilized by resonance* between the covalent and ionic structure. The effect of stabilization is particularly large when one of the atoms is strongly electronegative—that is, when it belongs in the upper right-hand corner of the periodic table—and the other atom is electropositive. Bond formation in this case is called combustion.

The ionic character of a bond between unlike atoms causes the resulting molecule, or more generally, any pair of atoms linked by a bond, to possess an electric dipole moment $\vec{\mu}$. For an aggregate of N electrons at positions $\vec{r}_i$ and of nuclei with atomic numbers Z_k at positions $\vec{R}_k$, one defines the mean electric dipole moment by

$$\langle\vec{\mu}\rangle = Ne\left\langle \sum_k \frac{Z_k \vec{R}_k}{N} - \sum_i \frac{\vec{r}_i}{N} \right\rangle . \qquad (19.36)$$

The measurement of the electric dipole moments of molecules is discussed in section 20.5. If a chemical bond were fully ionic, it would have a dipole moment equal to the product of the electron charge e and of the internuclear distance R_e. Table 19.1 gives examples of dipole moments of molecules, or molecular groups, with partially ionic bonds.

TABLE 19.1

Dipole Moments of Molecular Groups

Molecular Group	Dipole Moment $\mu(10^{-18}$esu-cm)	Internuclear Distance $R(10^{-8}$cm)	μ/eR_e
HF	1.91	0.92	0.43
HCl	1.05	1.27	0.17
HBr	0.80	1.41	0.11
HI	0.42	1.62	0.05
KF	7.3	2.55	0.60
CsF	7.6	2.34	0.68
HN (in NH_3)	0.63	1.01	0.13
HO (in H_2O)	1.51	0.97	0.32

19.5. MOLECULES WITH ONE POLYVALENT ATOM

This section deals with the structure of molecules consisting of one polyvalent atom bonded to a number of monovalent atoms. We shall assume for simplicity that the molecule is of the type XY_n, having in mind, for example, the methane molecule CH_4, or the tetrafluoride CF_4.

The wave function of each bonding electron may be represented initially by the superposition (19.35) of one atomic orbital, u_X, of the polyvalent atom X, and one atomic orbital of a monovalent atom Y. We set the origin of coordinates at the nucleus of atom X and call $\vec{R}_k$ the position of the nucleus of the kth atom Y. The atomic orbital of the k-th atom Y is then indicated by $u_Y(\vec{r} - R_k)$. This orbital is superposed with a particular orbital of the atom X, which we shall call $u_{Xk}(\vec{r})$. This superposition yields the wave function

$$\psi_k(\vec{r}) = \frac{1}{[c_{Xk}^2 + c_Y^2 + 2c_{Xk}c_Y\Delta_{XYk}(R_k)]^{1/2}} [c_{Xk}u_{Xk}(\vec{r}) + c_Y u_Y(\vec{r} - \vec{R}_k)], \tag{19.37}$$

where k is an integer between 1 and n. The characteristics of the set of orbitals u_{Xk} entering in the n wave functions determine not only the geometric structure of the molecule—in particular, the equilibrium positions $\vec{R}_{ke}$ of the various nuclei of the Y atoms—but also the energy of the complete molecule.

The atomic wave functions u_{Xk} need not represent stationary states of the isolated atom X; rather, they must be such that the whole molecule is in a stationary state of minimum energy. The wave functions u_{Xk} must be orthogonal to one another. Each of them may be represented as a superposition of stationary states of the isolated atom X. Such a superposition is called, in the context of bond formation, a *hybrid orbital*. The mean energy of a hybrid orbital may be larger than the energy of an atomic orbital in the ground state of the atom, provided this energy increase is overcompensated by the energy released in bond formation. Overcompensation occurs typically in the bonds formed by carbon with other atoms. The hybrid orbitals u_{Xk} of carbon belong to the excited configuration $(2s)^1(2p)^3$ whose energy lies about 5 eV above that of the ground-state configuration $(2s)^2(sp)^2$. Four orthogonal orbitals u_{Xk} may be constructed from the $(2s)^1(2p)^3$ configuration, whereas only two can be formed from the ground-state configuration. Furthermore, the configuration $(2s)^1(2p)^3$ includes orbitals akin to the hydrogen atom orbitals (13.34) with off-center electron density. These orbitals form stronger bonds than those of the $(2s)^2(2p)^2$ configuration whose electron density is centered at the nucleus. In fact, the energy of each C-H bond in methane amounts to ~4 eV after subtraction of ~1 eV, the energy needed per electron to excite the $(2s)^1(2p)^3$ configuration.

The four orthogonal hybrid orbitals u_{Xk} constructed by superposition of 2s and 2p states have the characteristic geometrical pattern of the wave functions (13.34)-(13.34b) of hydrogen. Their electron densities extend along four rays pointing from the central carbon atom toward the vertices of a regular tetrahedron and thus forming angles of 109°28' with one another. For methane, CH_4, the equilibrium positions $\vec{R}_{ke}$ of the four hydrogen nuclei lie at 1.09 Å from the carbon nucleus along these rays. The valence of the C atom is saturated by the formation of the four bonds because carbon has four electrons in the incomplete shell with n = 2, and because there are just four orbitals 2s and 2p giving rise to the four hybrid orbitals u_{Xk}. (Actually, more accurate single-electron wave functions are constructed that include in the superposition higher excited states of the unperturbed C atom, with small coefficients.)

The MO-LCAO wave function ψ_k has been introduced with reference to one single bond of the CH_4 molecule, without any consideration of the symmetry of the molecule. The four wave functions ψ_k corresponding to the four bonds are meant to serve as starting points for constructing approximate eigenfunctions for electrons subject to the average electrostatic potential of the whole molecule. A main feature of this potential is its symmetry with respect to the four bonds. On the other hand, the four states represented by the ψ_k have no such symmetry and are degenerate. The four individual ψ_k must then be replaced at the outset by four superpositions which reflect the symmetry of the molecule. One such *symmetry-adapted* superposition, namely,

$$\tfrac{1}{2}(\psi_1 + \psi_2 + \psi_3 + \psi_4), \tag{19.38}$$

remains invariant under all rotation and reflection operations that interchange the positions of any of the atoms Y. Three additional symmetry-adapted superpositions, orthogonal to (19.38) and to one another, are

$$\tfrac{1}{2}(\psi_1 + \psi_2 - \psi_3 - \psi_4), \tag{19.39a}$$

$$\tfrac{1}{2}(\psi_1 - \psi_2 + \psi_3 - \psi_4), \tag{19.39b}$$

$$\tfrac{1}{2}(\psi_1 - \psi_2 - \psi_3 + \psi_4). \tag{19.39c}$$

Each of these wave functions is an eigenfunction, with eigenvalue 1 or -1, of three 180° coordinate rotations which interchange two pairs of atoms Y. For instance, one such rotation interchanges atom 1 with atom 2 and atom 3 with atom 4; this rotation leaves the wave function (19.39a) unchanged and reverses the sign of wave functions (19.39b) and (19.39c). Each of the superpositions (19.39) represents a wave function with a nodal plane that contains the X atom and is equidistant from the four Y atoms; the Y atoms 1 and 2 lie on one side of the nodal plane of (19.39a), and atoms 3 and 4 on the other side. No such node is introduced in the construction of the totally symmetric wave function (19.38). Therefore, the symmetric wave function constitutes an initial approximation to a single-electron state whose energy is lower than the energy of the three states (19.39).

In general, the identification of symmetry-adapted wave functions is an important step in the study of the electron states of polyatomic molecules. The systematic analysis of the symmetry properties of molecular eigenstates applies the theory of point groups.

To complement the description of the tetrahedral bonds formed by carbon, we mention briefly the characteristics of molecules XY_n formed by other elements in the first row of the periodic system. Such are the fluorides BeF_2 and BF_3 and the hydrides NH_3 and H_2O.

The Be atom has a ground-state configuration $(1s)^2(2s)^2$ consisting of filled subshells, but is not chemically inert because it is easily excited to the $(2s)^1(2p)^1$ configuration. This excited configuration yields two orthogonal hybrid orbitals u_{Xk} akin to the hydrogen orbital (13.32). Here again, the electron density is not centered at the nucleus but extends along two rays pointing from the central Be atom in opposite directions. (Figure 13.9 illustrates the electron density for the orbital [13.32]; the other orbital extends along the −z axis.) Accordingly, the valence of Be is saturated by the formation of two bonds, and the BeF_2 molecule has a linear structure. Similarly, the ground-state configuration of the boron atom, namely, $(2s)^2(2p)^1$, is readily excited to $(2s)^1(2p)^2$. Here, we have three orthogonal hybrid orbitals u_{xk} analogous to the hydrogen wave functions (13.33) and (13.33a). Their off-center electron densities extend along three rays in a *trigonal arrangement*, that is, forming angles of 120°. The boron valence is therefore saturated by formation of BF_3.

The ammonia and water molecules are isoelectronic with methane, with neon, and also with HF. They all have 10 electrons, two of which belong in a 1s shell. The molecule NH_3 has the shape of a regular triangular pyramid, the angle between the axes of any two N-H bonds is known experimentally to be about 107°, that is, only a little smaller than the tetrahedral angle of methane. The electronic configuration of NH_3 can accordingly be described, starting from that of CH_4. One may consider a deformation of CH_4 in which one of the hydrogen nuclei is made to approach the carbon nucleus until the two coalesce in the united-atom limit. Note, on the other hand, that nitrogen could form three bonds in directions perpendicular to each other, starting from the ground-state configuration $(2s)^2(2p)^3$. The fact that the bond angles of NH_3 are only a bit smaller than the tetrahedral angles indicates that the electronic configuration of the N atom in NH_3 is closer to $(2s)^1(2p)^4$ than to $(2s)^2(2p)^3$. The interpretation of the bond angles also takes into account that each N-H bond is appreciably heteropolar (see Table 19.1). The resulting average positive charge of the hydrogen atoms forces these atoms apart and thus tends to increase the bond angles. The water molecule may be discussed qualitatively along the same lines as for NH_3. The bond angle H-O-H is approximately 105°.

Polyvalent atoms of successive rows of the periodic system form molecules of the type XY_n much in the same way as the corresponding atoms of the first row. The d electrons in the ground-state configurations of transition elements may take a lesser part in bond formation as they lie deep inside the atom (see section 16.3.1). On the other hand, excited d orbitals are often included to a substantial extent in the hybrid orbitals $u_{xk}(\vec{r})$; the valence of an atom is thereby increased. For example, sulfur has the ground-state configuration $(3s)^2$ $(3p)^4$ homologous to that of oxygen, but it has a d orbital in the same shell as the ground-state orbitals; for this reason sulfur behaves as hexavalent in the molecule SF_6. The six wave functions u_{xk} of this molecule yield off-center electron densities extending from the central atom of sulfur toward the vertices of a regular octahedron. Superposition of s, p, and d wave functions could yield, in principle, nine orthogonal orbitals u_{xk}. However, molecules XY_n with $n > 6$ do not exist because of the repulsion among the Y atoms on close approach.

19.6. MULTIPLE BONDS

Polyvalent atoms may be bonded to one another by sharing more than one pair of electrons. The multiple bonds thus formed contribute through several mechanisms to increase the variety of properties of molecules and of macroscopic aggregates. We describe in this section a few examples illustrating different characteristics of multiple bonds.

19.6.1. *The Double Bond of Ethylene.* This molecule consists of two CH_2 groups. Each carbon atom is thus joined to three atoms—namely, two hydrogens and the other carbon. The three bonds are formed in the trigonal arrangement described in section 19.5 for the BF_3 molecule. The trigonal bonds utilize only three of the four available valence electrons of each carbon.

The electronic structure of the molecule is usually described by choosing the x and y coordinate axes in a plane that contains the equilibrium positions of all six atoms. (The planar structure of the molecule is an important consequence of the double bond, as explained below.) The wave functions of the electrons forming the trigonal bonds are then obtained by hybridization, that is, by superposition, of the 2s, $2p_x$, and $2p_y$ orbitals of each carbon atom.[15] There remains, for each carbon atom, one electron in the $2p_z$ wave function whose distribution is indicated schematically in Figure 19.9. The second C—C bond results from the sharing of these two electrons in a singlet state. The molecular orbital

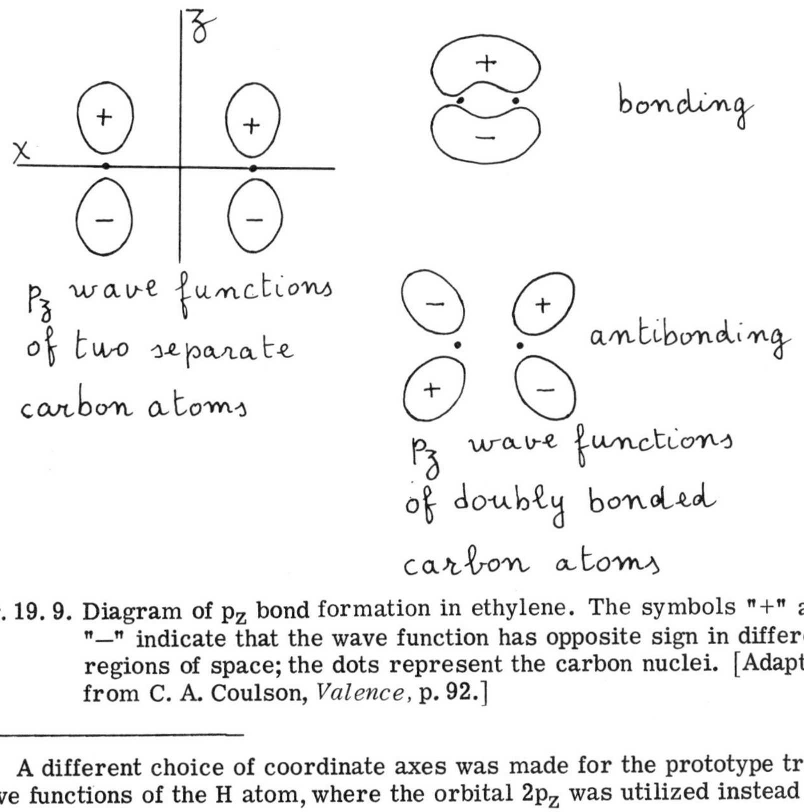

Fig. 19.9. Diagram of p_z bond formation in ethylene. The symbols "+" and "—" indicate that the wave function has opposite sign in different regions of space; the dots represent the carbon nuclei. [Adapted from C. A. Coulson, *Valence*, p. 92.]

---

15. A different choice of coordinate axes was made for the prototype trigonal wave functions of the H atom, where the orbital $2p_z$ was utilized instead of $2p_y$.

occupied by this electron pair may be represented approximately by superposition of the $2p_z$ orbitals of the carbon atoms.

Notice that the electron densities of the $2p_z$ orbitals extend from each carbon atom along rays perpendicular to the (x, y) plane and thus are not directed toward the other carbon atom. The overlap of the $2p_z$ orbitals is therefore lower than for ordinary bonds, as is the energy released by the formation of this second bond. Consequently, this second bond is fairly susceptible to breakage and replacement by bonds with other atoms. For this reason, ethylene is called an *unsaturated* hydrocarbon, even though the double bond does saturate the valence electrons. An index of the strength of the double bond is the reduction of the equilibrium distance of the two carbon atoms from 1.54 Å in ethane (single bond) to 1.34 Å in ethylene.

Any departure of the two CH_2 groups from coplanarity would reduce the overlap of the $2p_z$ orbitals and thus reduce the strength of the double bond. Therefore, the formation of the double bond constricts the equilibrium position of the molecule into a planar configuration. However, the two halves of the molecule can twist about the C—C axis by small angles or bend the axis in the course of nuclear vibration.

The wave functions of the four electrons forming the double bond must have characteristic symmetry properties irrespective of whether they are represented as superpositions of atomic orbitals. The two electrons of the trigonal C—C bond have a density distribution with approximate axial symmetry about the bond axis and zero angular momentum about this axis. Their state is classified as σ according to the code introduced in section 19.2.2 for the states of the electron in H_2^+. This classification implies that the mean square angular momentum of the electron about the C—C axis is approximately zero. The two electrons forming the second C—C bond have zero density on the (x, y) plane and wave functions of opposite sign above and below this plane. The existence of this nodal plane implies that their state is approximately an eigenstate of the squared angular momentum component along the bond axis with eigenvalue $\hbar^2$. The state is accordingly a π state.

19.6.2. *Nitrogen and Related Molecules*. The N_2 molecule is exceedingly stable owing to the triple bond between the two nitrogen atoms. This stability is shared to various degrees by a number of related molecules, several of which contain 14 electrons like nitrogen itself.

The description of the electronic structure of these molecules may begin with the acetylene molecule H—C ≡ C—H. The equilibrium positions of the four nuclei lie on a straight line which we take as the z axis; the molecule has full symmetry about this axis. Each carbon atom is connected to each of its neighbors by a bond similar to those of divalent Be in the BeF_2 molecule. The wave functions for these bonds are constructed by utilizing the 2s and $2p_z$ atomic orbitals of each carbon atom. These wave functions represent a molecular orbital with zero angular momentum about the z axis, that is, a σ state. Two additional bonds between the two carbon atoms are formed by sharing the two remaining pairs of electrons of each carbon atom. The two wave functions for the second and third bond may be constructed, starting from the $2p_x$ and $2p_y$

orbitals of each carbon atom, much as the second C—C bond of ethylene is constructed from the p_z orbital. These wave functions are classified as π states. Since the molecule is linear, its many-electron stationary states are eigenstates of angular momentum about the molecular axis. The ground state has zero angular momentum because the momenta of its π electrons cancel out, and is therefore classified as Σ. The molecule is symmetrical with respect to reflection at its center of mass, and therefore its stationary states are classified as g or u. The ground state of acetylene is $^1\Sigma_g$. The equilibrium internuclear distance between the triply bonded carbons is 1.20 Å, that is, substantially shorter than for doubly bonded ethylene.

The nitrogen molecule in its ground state has electronic structure $^1\Sigma_g$ analogous to that of acetylene, with each nitrogen atom corresponding to a C—H group. Recall how the electronic structure of NH_3 was derived in section 19.5 from that of CH_4 as the united-atom limit in which a CH bond is shortened to the point where the H and C nuclei coalesce. The same limiting consideration applies to the relationship of HC ≡ CH and N ≡ N, as well as to numerous organic molecules in which an N atom may replace a CH group. The electron pairs which form the CH bonds in acetylene remain essentially inert in each N atom in the N_2 molecule, and they are often called nonbonding pairs.

The equilibrium distance of the N nuclei in the N_2 molecule is only 1.09 Å. As an index of the strength of the bond we may consider that the vibrational frequency ω of the two nuclei corresponds to an energy $\hbar\omega$ of nearly 0.3 eV. This energy is a little larger than that of the H_2^+ ion whose nuclear masses are 14 times smaller. Accordingly, the elastic force holding the N nuclei near their equilibrium position is approximately 14 times larger than in the case of H_2^+. The very short internuclear distance and high vibrational frequency of N_2 are exceptional. They are matched only by molecules involving the lightest elements, H and He, and are approached by molecules, such as CO, whose electronic structure is similar to that of N_2.

The CO molecule has the same number of electrons as N_2. One may regard the electronic structure of the two molecules as differing only by the greater attraction of electrons to the O nucleus and lesser attraction to the C nucleus. The CO electron density is not symmetrical with respect to the center of mass (see Fig. 19.10), and thus the molecule has a nonzero dipole moment. The equilibrium internuclear distance is 1.13 Å, only a bit larger than for N_2 and shorter than the distance 1.21 Å of the doubly bonded $>C = O$ group in organic molecules. Carbon monoxide thus partakes to a considerable extent of the triple-bond character of N_2. The H—C ≡ N molecule and the $(C \equiv N)^-$ ion have electronic structure similar to those of C_2H_2 and CO and have analogous triply bonded character.

19.6.3. *Two Stable "Free Radicals": O_2 and NO.* It was noted in the introduction to this chapter that atomic aggregates whose valences are saturated are normally in a singlet state, and are accordingly nonparamagnetic. Paramagnetism is characteristic of free radicals with unsaturated valences. Two small molecules are outstanding exceptions. Their stability, in spite of the presence of unpaired electrons, can be traced to the exceptional stability of the triply bonded electronic structure of N_2.

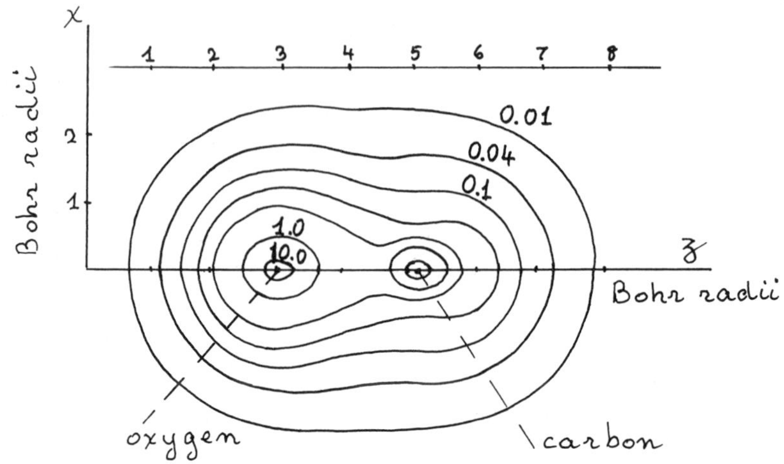

Fig. 19. 10. Total electron density contours for CO (electrons per cubic Bohr radius). [Adapted from W. M. Huo, *J. Chem. Phys.*, 43 (1965): 624].

The simpler of these two molecules is NO, which differs from N_2 by the addition of one nuclear charge and one electron. This is the only neutral and stable small molecule with an odd number of electrons. This molecule can be regarded as consisting of an NO^+ core, isoelectronic to N_2 and CO, and of an additional electron. The molecular orbital of lowest energy which is available for this electron is an antibonding π orbital. Its wave function is represented in the LCAO method by equation (19. 35) with the following interpretation. The subscripts X and Y pertain to the oxygen and nitrogen atoms, each wave function u represents a $2p_x$ or $2p_y$ orbital or a combination thereof, and the coefficients c_X and c_Y are of *opposite* sign to yield an antibonding effect. This effect reduces the stability of the molecule only by a moderate amount, as indicated by the value 1. 15 Å of the equilibrium internuclear distance.

The oxygen molecule may be similarly regarded as consisting of an O_2^{++} core, partaking of the N_2 stability, and two additional electrons. Here again, the wave functions of these two electrons have the same antibonding π character as that of the unpaired electron of NO. Since the O_2 molecule has a center of symmetry, the π antibonding orbitals have a definite parity with respect to reflection of coordinates; this parity is "g," and the states are designated at π_g.[16]

---

16. Reflection of coordinates at the center of the molecule may be regarded as the result of two successive operations, namely, reflection at the molecular axis and reflection on a plane across the molecular axis. An antibonding π wave function is odd under each of these reflections; it is therefore even (gerade) under the combined operations.

The configuration of this electron pair is degenerate because there are two degenerate π antibonding orbitals. For example, these two orbitals can be constructed with $2p_x$ or with $2p_y$ atomic orbitals; alternatively, they may have quantum number $\lambda = \pm 1$. Allowing for alternative spin orientations, there are four π_g states, and the $(\pi_g)^2$ configuration is sixfold degenerate.

The electrostatic repulsion between the two electrons removes this degeneracy. The ground-state level has the highest possible spin quantum number, in accordance with the first Hund rule (see section 16.4.1). For this reason the oxygen molecule is paramagnetic in its ground state. Owing to the exclusion principle the independent-electron wave functions of the two electrons with parallel spins must have different quantum numbers—for example, $\lambda = 1$ and $\lambda = -1$. Hence, the combined orbital angular momentum of the two electrons about the internuclear axis vanishes, and their joint state has $^3\Sigma$ character.

The two-electron wave function of the O_2 ground state affords an example of a type of molecular symmetry which we have not yet encountered. According to the form (15.11) of the exclusion principle, the wave function of a triplet state of two electrons must reverse its sign under permutation of the two electron positions. If we represent these positions by means of cylindrical coordinates ρ_1, z_1, ϕ_1 and ρ_2, z_2, ϕ_2, with $\hat{z}$ as the internuclear axis, the joint wave function of the two antibonding π electrons must be symmetric with respect to ρ_1, ρ_2 and z_1, z_2 and must depend on ϕ_1 and ϕ_2 through the antisymmetric factor $\sin(\phi_1 - \phi_2)$. This wave function then also has the property that it changes its sign under reflection of the coordinates at any plane containing the z axis, since such a reflection reverses the sign of $\phi_1 - \phi_2$. Odd parity of the wave function under this reflection is indicated by a minus sign in the state symbol, which is then written as $^3\Sigma_g^-$.

The removal of degeneracy of the ground-state configuration of O_2 yields, besides the ground-state level of the molecule, two singlet levels with excitation energies of approximately 1 and 2 eV. The lower of these two states has the electrons with parallel orbital angular momenta and is classified as $^1\Delta_g$. The higher one is a $^1\Sigma_g^+$ state. Both of these states are metastable with respect to radiation emission, for reasons analogous to those that apply to the oxygen atom in the states ^1D and ^1S (see p. 369). The existence of metastable states of O_2 is also important in atmospheric physics.

CHAPTER 19 PROBLEMS

19.1. Consider the following molecules, free radicals, and ions in their electronic ground states, ignoring nuclear spins: N_2, NO, NH_3, NH_4^+, CO, CH_3, CH_3Cl, OH^-, Cl_2.
(a) Which of them is paramagnetic? (b) Which of them is polar, i.e., has a nonzero mean electric dipole moment in the frame of reference where the nuclei are at rest?

19.2. On the basis of general properties of molecular bonds, predict whether the formaldehyde molecule, $\genfrac{}{}{0pt}{}{H}{H}\!\!>\!C = 0$, is planar and estimate its approximate bond angles and lengths. Is this molecule polar?

19.3. On the basis of the LCAO approximate wave functions for H_2^+, (19.14a), (19.14b), and (19.15b) and of the model in Figure 19.5, sketch the nodes of the 3dσ, 4fσ, and 3pσ wave functions.

19.4. The electronic states of diatomic molecules can be classified by symbols analogous to the $^1\Sigma$ and $^3\Sigma$ used for H_2 in section 19.4, from a knowledge of their separate atom limits. Classify the states of LiH and HF, starting from the separate-atom limit in which both atoms are in their ground-state configuration. Which of these molecular states is the ground state?

19.5. The potential energy curve U(r) of the hydrogen molecule shown in Figure 19.8 is the same for the different isotopes of hydrogen, to a good approximation. However, different isotopes have different vibrational (and rotational) energy levels, and therefore different dissociation energies, owing to their different reduced masses. Given that the ordinary H_2 isotope has a dissociation energy of 4.48 eV and a vibrational level interval of 0.54 eV, and assuming that the vibrational levels are approximated by those of a harmonic oscillator, calculate the dissociation energy of the D_2 (deuterium) molecule and of HD. (Accurate experimental values of the dissociation energies of these three molecules are 4.476, 4.554, and 4.511 eV.)

SOLUTIONS TO CHAPTER 19 PROBLEMS

19.1. (a) Paramagnetic are the atomic systems with an odd number of electrons: NO and CH_3.

(b) Potentially polar are all molecular groups with bonds between different atoms; this class includes all the systems listed except N_2 and Cl_2. However, the dipole moments of different bonds cancel out in molecular groups of sufficient symmetry; for this reason the tetrahedral ion NH_4^+ and the trigonal free radical CH_3 are not polar.

19.2. The double bond between the C and O atoms is similar to the double bond in ethylene. Accordingly, three bonds joining the C atom to each of the other atoms should be formed by electron pairs in σ orbitals in an approximately trigonal arrangement. That is, the bonds should be in a plane at angles of ~120°. The second bond between C and O should be weaker and should have a π character. The CO bond is strongly polar, owing to the electronegativity of oxygen. The C—H bond lengths should not depart much from their value in CH_4, which is stated in section 19.5 to be 1.09 Å. The length of the CO double bond in organic molecules is 1.21 Å according to section 19.6.2. The experimental values of the bond angles and distances are: 126° for the H—C—H angle (and hence 117° for H—C = 0), 1.06 Å for the CH distance, and 1.23 Å for CO.

19. 3.

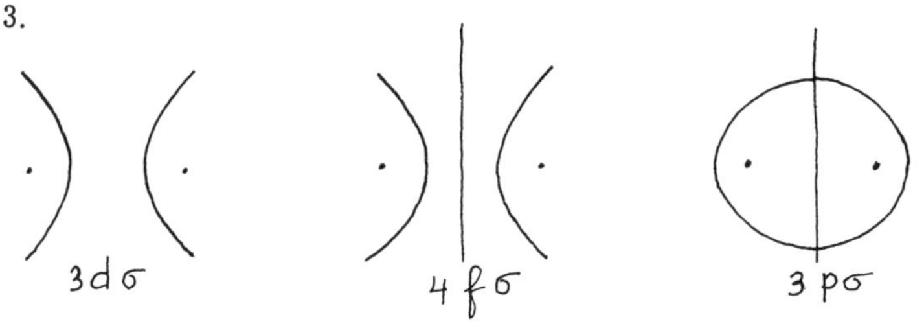

19. 4. Lithium has the ground-state configuration $(1s)^2(2s)^1$ with a single level ^2S analogous to that of H. From the separate-atom limit Li(^2S) + H(^2S), the same molecular states are formed as for H_2, namely, $^1\Sigma$ (bonding, ground state) and $^3\Sigma$ (antibonding). Fluorine has the ground-state configuration $(1s)^2(2s)^2(2p)^5$ with one doublet term ^2P. LCAO molecular orbitals formed from this doublet can have σ or π character depending on the m quantum number of the state, while H in the ground state can only form σ bonds. Therefore, Σ and Π molecular states can be formed, but Π states will have one electron excited. The ground state will be $^1\Sigma$, with both electrons in the molecular orbital of lowest energy.

19. 5. The dissociation energy for a molecule in its lowest energy level of rotational and vibrational motion (N = 0, v = 0) equals $-E_{N=0,\ v=0}$. Equation (19. 26) gives $E_{0,0} = \overline{U}(R_e) + \frac{1}{2}\hbar\omega$, and equation (19. 26a) gives $\omega = (k/\frac{1}{2}M)^{1/2}$, where $\frac{1}{2}M$ represents the reduced mass of the nuclei. Given that $\hbar\omega = 0.54$ eV for H_2 and that the reduced masses of D_2 and HD are 2 and 4/3 times the reduced mass of H_2, the values of $\hbar\omega$ for D_2 and HD are $(\sqrt{\frac{1}{2}}) \times 0.54 = 0.38$ eV and $(\frac{1}{2}\sqrt{3}) \times 0.54 = 0.46$ eV, respectively. Therefore, the dissociation energies of D_2 and HD exceed that of H_2 by $\frac{1}{2}(0.54 - 0.38) = 0.08$ eV and $\frac{1}{2}(0.54 - 0.46) = 0.04$ eV, respectively.

20. molecular excitations

20.1. INTRODUCTION

Structural properties of molecules are studied by observing excitations to states in which the molecule holds together, that is, stationary states of the discrete spectrum. Ionization, dissociation, and other chemical transformations of molecules involve excited states of the continuous spectrum. Both types of excitations are studied through the absorption and emission of radiation; the transformations and dissociation of molecules are now studied increasingly through collision experiments.

The study of excited states of the continuum is potentially most important because it underlies a detailed description of the elementary processes involved in chemical reactions. However, this study is still at its beginning, whereas the spectroscopy of discrete states has been developed extensively. Spectroscopy has provided not only voluminous and often very accurate data on molecular properties, but also the framework of concepts which serve to describe the dynamics of molecular transformations.

Stationary states of molecules are classified to a large extent according to the constants of motion of the whole molecular system, such as angular momenta, parities, and other symmetries. The treatment of symmetry is more elaborate for molecules than it is for atoms in relation to the complexity of the geometrical structure. The effects of symmetries include selection rules which limit the symmetry character of the final state of a molecule on the basis of the symmetry of the initial state and of an external agent such as radiation.

The main approach to the study of molecular excitations rests on the separation of the motion of electrons and of nuclei, in accordance with the Born-Oppenheimer approximation (section 19.2.1). The description of the electronic states as functions of the internuclear distances makes it possible to follow how the state of a complex system evolves as two component atoms (or groups of atoms) approach and as they separate. A chemical reaction has occurred when the outgoing groups are different from the incoming ones. However, the very change of electronic state which generally accompanies a reaction implies a partial breakdown of the Born-Oppenheimer approximation.

Another partial breakdown of this approximation occurs in the conversion of electronic excitation into excitation of nuclear motion, a phenomenon which often takes place within a single molecule. The conversion constitutes a degradation of energy since the energy of a single electronic excitation is sufficient to provide a number of different nuclear excitations. This kind of process, as well as the evolution of a molecular rearrangement, is still hardly amenable to detailed quantitative description.

This chapter treats first in some detail the phenomena of molecular excitation (rotational, vibrational, and electronic) which give rise to the discrete spectra. This is done within the Born-Oppenheimer approximation and with primary reference to small molecules, mostly diatomic. The emphasis shifts then progressively to a qualitative description of phenomena which are characteristic of larger aggregates of atoms.

20.2. ROTATIONAL MOTION

Emission or absorption of radiation by a molecule is generally accompanied by a change of angular momentum. The kinetic energy of free rotation of the molecule about its center of mass is therefore also changed. Analysis of the spectrum of the absorbed or emitted radiation makes it possible to sort out the changes of rotational energy from other energy changes. The determination of rotational energies in turn provides accurate data on the moments of inertia of molecules and thereby on internuclear distances.

When the rotational quantum number changes by one unit, the change of rotational energy is expressed in terms of a parameter B inversely proportional to the moment of inertia of the molecule. This parameter was introduced in equation (19.19) for the H_2^+ molecular ion, in which case $B(R) = \hbar^2/MR^2$ and the moment of inertia is $\frac{1}{2}MR^2$. The energy change is of the order of magnitude of 0.01 eV for H_2, the smallest and lightest molecule. For CO_2, whose moment of inertia is much larger, the energy change is of the order of 10^{-4} eV, which corresponds to a radiation wavelength of 1 cm. Accordingly, transitions between molecular levels that differ only in rotational energy are observed in the far-infrared or, more commonly, in the microwave region of the spectrum. A transition involving changes of vibrational as well as rotational energy appears in the infrared range of the spectrum. When electronic excitation is also involved, transitions appear in the ultraviolet or visible range.

Transitions without any change of electronic state emit or absorb radiation intensely only if the molecule has a net average dipole moment—that is, if the average positions of the electronic and nuclear charges do not coincide. This is to say that infrared and microwave spectra are not observed for simple homonuclear molecules like H_2 and N_2. For these molecules, changes of rotational energy are observed only together with electronic transitions which give rise to a dipole moment. Rotational changes are also observed for these molecules in the Raman spectra (see section 4.5), that is, through frequency shifts of scattered light; these spectra will not be discussed in this book.

The selection rules for absorption or emission of radiation require that the initial and final states have angular momentum quantum numbers differing by no more than one unit and have opposite parity. The required parity change generally occurs only when the angular momentum quantum number changes by one unit, as in the simple example of the hydrogen atom in section 13.5. This quantum number may, however, remain unchanged when a parity change is provided by electronic transition, as discussed below.

We begin by considering the simple case of diatomic molecules for which the dependence of the rotational energy on the angular momentum quantum number is represented by $B(R)N(N + 1)$ in equation (19.19). For diatomic molecules the parameter B is given by $B(R) = \hbar^2/2\overline{M}R^2$, where $\overline{M}$ is the reduced mass of the nuclei and R is the internuclear distance; the dependence of B on R must be averaged over the state of vibrational motion. The quantum number N identifies the molecular angular momentum exclusive of electron or nuclear spin. (The influence of spin will be discussed later.)

The change of rotational energy in a transition can be analyzed roughly in two parts. One part is due to the change of the moment of inertia represented by a change of the rotational constant B. (A substantial change in the moment of inertia occurs only as a result of structural changes which accompany an electronic transition.) The second part is due to simple speeding up or slowing down of the rotation and corresponds to a change of the quantum number N. The two contributions may be separated by writing the rotational energy change for a transition $N \rightarrow N'$ in the form

$$\begin{aligned} \Delta E_{rot} &= B'(R)\, N'(N'+1) - B(R)N(N+1) \\ &= [B'(R) - B(R)]N(N+1) + B'(R) \times \begin{cases} 2(N+1) & \text{for } N' = N+1, \\ 0 & \text{for } N' = N, \\ -2N & \text{for } N' = N-1. \end{cases} \end{aligned} \quad (20.1)$$

This formula is particularly simple for transitions in the infrared and microwave spectra. For these transitions the moment of inertia remains essentially constant and the coefficient $B'(R) - B(R)$ vanishes; furthermore, the possibility $N' = N$ is excluded by the parity selection rule. Consequently, ΔE_{rot} results only from a change of rotational speed and is a linear function of N. Accordingly, the spectra show bands consisting of equally spaced lines corresponding to different initial values of N. Actually, a progressive decrease in the spacing of the lines is often observed for increasing N since the mean value of B(R) decreases by *rotational stretching*.

In the microwave absorption spectra of diatomic molecules one observes a single set of lines with $N' = N + 1$, corresponding to an increase of rotational energy. In the infrared spectra the change of rotational energy causes only a minor shift of the frequency of emission or absorption which is determined primarily by a change in vibrational energy. The various possible values of the change in rotational energy cause the basic vibrational frequency to be replaced by two sets of frequencies; one set corresponds to $N' = N + 1$ with all possible different values of N, the other set corresponds to $N' = N - 1$ (see Fig. 20.1). The relative intensity of the different lines depends on the fraction of molecules in initial states with different values of N; therefore, it is normally a calculable function of temperature. Observation of the intensity distribution within each band of a molecular gas spectrum serves to determine the gas temperature—for example, in astrophysics. Observed departures from the distribution expected on the basis of thermal equilibrium may indicate unusual characteristics of the state of the gas.

In optical spectra the rotational energy change is an even smaller fraction of the whole transition energy. Each basic frequency due to electronic and vibrational transition is split by the change of rotational energy into a large number of lines, that is, into a band. There is a branch corresponding to $N' = N + 1$ with different values of N, a branch $N' = N - 1$, and often also a third branch

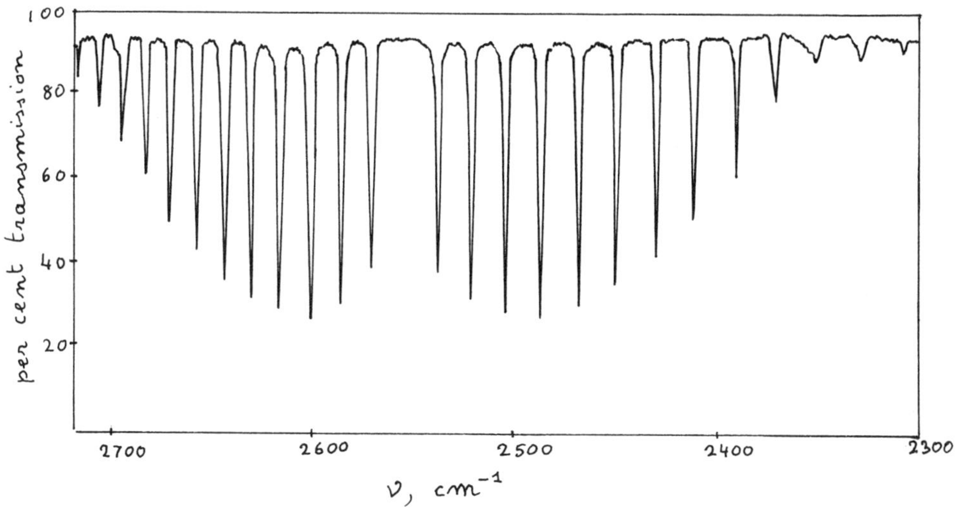

Fig. 20. 1. Absorption band of gaseous HBr in the infrared spectral region. [Adapted from G. M. Barrow, *The Structure of Molecules* (New York: Benjamin, 1964).]

with $N' = N$ when the parity change is provided by the electronic transition. The rotational energy change depends in this case quadratically on N, according to the full formula (20. 1). It follows that the rotational lines crowd into the portion of a band for which $d\Delta E_{rot}/dN \sim 0$; this portion is called the *band head*.

20. 2. 1. *Coupling of Electron Spin.* The coupling of electron spin with other molecular currents is complicated by the fact that the spin-orbit interaction is comparable in magnitude to the rotational energies. This complication is avoided in the microwave and infrared spectra of most molecules because their electronic ground state is a singlet with no net spin current. Otherwise, different types of couplings arise, depending on the relative magnitudes of spin-orbit and rotational energies. The different types have been classified by Hund and are commonly referred to as Hund's case a, b, c, etc. Here, we mention briefly only two extreme cases, a and b.

In Hund's case b, the total spin angular momentum of all electrons, $\vec{S}$, couples directly with the angular momentum of rotational and electronic motion, $\vec{N}$, to yield the total angular momentum $\vec{J} = \vec{N} + \vec{S}$. This case obtains when the spin-orbit coupling is extremely weak. The component of the orbital momentum along the internuclear axis of a linear molecule, $\Sigma_i \vec{l}_i \cdot \vec{R} = \Lambda\hbar$, then interacts more strongly with the rotational angular momentum $\vec{\mathcal{R}}$ than with $\vec{S}$. Hund's case b also obtains for all Σ states of linear molecules, that is, in the absence of net orbital currents. The spin-orbit coupling is very small only for the lightest elements, that is, for the very ones whose rotational energy is largest.

For this reason the spin-orbit coupling was not even mentioned in the discussion of H_2^+. In this molecule each rotational level splits into a spin doublet with extremely small separation. Another typical example of Hund's case b occurs for the $^3\Sigma$ ground state of O_2; here the net electron spin interacts only, and weakly, with the molecular rotation. This interaction is weak because the gyromagnetic ratio of the molecular rotation vanishes to a good approximation, that is, insofar as the electrons rotate rigidly with the nucleus.

In the opposite extreme case (Hund's case a) the spin-orbit interaction takes precedence over the rotational motion. The stationary states of linear molecules are then characterized by a quantum number Ω which indicates the projection of the total (spin and orbit) angular momentum of all electrons along the molecular axis $\hat{R}$,

$$|\textstyle\sum_i(\vec{l}_i + \vec{s}_i)\cdot\hat{R}| = |(\vec{L} + \vec{S})\cdot\hat{R}| = \Omega\hbar \tag{20.2}$$

The quantum number Ω replaces the quantum number Λ which is used in case b. Similarly, the quantum number J which pertains to the total squared angular momentum

$$|\vec{J}|^2 = |\vec{\mathcal{R}} + \vec{L} + \vec{S}|^2 \tag{20.3}$$

replaces in case a the quantum number N. As an example, the spin-orbit coupling prevails over rotation in the ground state of NO. Here, the doublet splitting between the levels $^2\Pi_{1/2}$ and $^2\Pi_{3/2}$, corresponding to $\Omega = 1/2$ and $\Omega = 3/2$, amounts to 0.015 eV, whereas the rotational constant $B(R_e)$ is only 2×10^{-4} eV. The rotational energy levels depend on J rather than on N whenever Hund's case a applies; the treatment of rotational motion in section 19.2.3 and at the beginning of this section must be adjusted accordingly.

20.2.2. *Hyperfine Structure and Isotopic Effect.* The rotational motion of a molecule is influenced by three types of nuclear effects: (1) The coupling of nuclear spin to the rotational angular momentum, which gives rise to hyperfine structure; (2) the effects of isotopic differences which shift the rotational energies; and (3) the effect of the indistinguishability of identical nuclei, which will be discussed in section 20.2.4.

The hyperfine structure of molecular spectra is analogous to that observed for atoms (see section S14.5). It is, however, much more conspicuous in microwave spectra where the hyperfine splittings of rotational levels are comparatively large and accessible to accurate measurement. These measurements have provided much detailed information on the magnetic and electric fields prevailing at the location of the nuclei, and thus on the distribution of currents and charges within molecules.

Isotopic shift is the displacement of spectral lines for atoms or molecules which differ only in their isotopic constitution. This effect is very small in atoms, whose reduced mass depends very little on the nuclear mass, but is rather large for the rotational and vibrational spectra of molecules. For instance, the constant B(R) for the deuterium molecule is half as large as for H_2;

as another example, this constant differs by about 3% in the oxygen molecules $^{16}O^{17}O$ and $^{16}O^{16}O$. The rotational spectra of substances with mixed isotopic composition appear as two (or more) superposed spectra whose relative intensities reflect the relative abundance of different isotopes.

20.2.3. *Eigenstates of Parity*. The energy of a molecule is unchanged by reflection of the position coordinates of all electrons and nuclei with respect to the center of mass. Therefore, stationary states can be classified according to their parity under this transformation. This symmetry property has an important consequence on the rotational wave functions of diatomic and other linear molecules for the following reason. Reflection of all coordinates changes the sign of the component of the electronic angular momentum along the nuclear axis. That is, reflection changes the sign of $\vec{L}\cdot\hat{R} = \Lambda\hbar$, or of $(\vec{L} + \vec{S})\cdot\hat{R} = \Omega\hbar$, because it changes $\hat{R}$ into $-\hat{R}$ but leaves $\vec{L}$ and $\vec{S}$ unchanged.

Following the treatment of H_2^+ we would classify a stationary state of a linear molecule as an eigenstate of the angular momentum component $\vec{L}\cdot\hat{R}$ and as a state of rotational motion with quantum numbers N, Λ, and M. However, such a state is not an eigenstate of parity because reflection of coordinates changes the sign of its eigenvalue of $\vec{L}\cdot\hat{R}$. An eigenstate of parity is constructed by superposing two eigenstates of $\vec{L}\cdot\hat{R}$ with eigenvalues of equal magnitude and opposite sign. The complete wave function of an eigenstate of parity is built with three factors, for the electronic, vibrational, and rotational motions. The electronic wave function will be indicated by $\langle \vec{r}_i | n \;\; \Lambda\rangle$, where $\vec{r}_i$ stands for the coordinates of all electrons, n is a principal quantum number, and Λ indicates an eigenvalue of $\vec{L}\cdot\hat{R}$. The vibrational wave function will be indicated by $\langle R | v\rangle$, where v is the vibrational quantum number. The rotational wave function is given by $\langle \Theta \;\; \Phi | N \;\; \Lambda \;\; M\rangle$, as in section 19.2.3, in Hund's case b. An eigenstate of parity is represented by the complete wave function

$$\langle \vec{r}_i \;\; R \;\; \Theta \;\; \Phi | n \;\; \Lambda^{\pm} \;\; v \;\; N \;\; M\rangle = \sqrt{1/2}\, [\langle \vec{r}_i | n \;\; \Lambda\rangle\langle \Theta \;\; \Phi | N \;\; \Lambda \;\; M\rangle \pm \langle \vec{r}_i | n \; -\Lambda\rangle \langle \Theta \;\; \Phi | N \; -\Lambda \;\; M\rangle]\langle R | v\rangle . \quad (20.4)$$

The form of this wave function is such that reflection of the coordinates interchanges the two terms in the brackets. (This is because the reflection changes $\langle \vec{r}_i | n \;\; \Lambda\rangle$ into $(-1)^{\Lambda} \langle \vec{r}_i | n \; -\Lambda\rangle$ and $\langle \Theta \;\; \Phi | N \;\; \Lambda \;\; M\rangle$ into $\langle \pi - \Theta \;\; \Phi + \pi | N \;\; \Lambda \;\; M\rangle = (-1)^{N} \langle \Theta \;\; \Phi | N \; -\Lambda \;\; M\rangle$.) Thus reflection of the coordinates multiplies the entire wave function (20.4) by a factor $\pm(-1)^{N}$. Therefore, this wave function represents an eigenstate of parity.

The + or − sign, which appears as a superscript of Λ in the wave function symbol on the left of equation (20.4) and in the brackets on the right side, pertains to the change of the wave function under reflection of the electrons' coordinates with respect to any plane containing the internuclear axis. This reflection reverses the sign of $\vec{L}\cdot\hat{R}$ and is included in a reflection of all coordinates (nuclear and electronic) at the center of mass. The ± sign in equation 20.4 has the same effect on the parity of the complete wave function as the ± symbol introduced in section 19.6.3 for the ground state wave function $^3\Sigma^-$ of the oxygen molecule and for this reason is indicated by the same superscript.

However, this superscript represents a parity of the electronic wave function only in the case of Σ states. For these states the wave function (20.4) reduces to a single term, since $\Lambda = 0$,

$$\langle \vec{r}_i \; R \; \Theta \; \Phi | n \; 0^{\pm} \; v \; N \; M \rangle = \langle \vec{r}_i | n \; 0^{\pm} \rangle \langle R | v \rangle \langle \Theta \; \Phi | N \; 0 \; M \rangle, \quad \text{for } \Lambda = 0 \tag{20.4a}$$

The parity of this wave function is the same as for (20.4), namely, $\pm(-1)^N$.

Two stationary states of electron motion, Σ^+ and Σ^-, generally have quite different energies because the electrons have different correlations in the two types of states. On the other hand, pairs of states Π^+ and Π^-(or Δ^+ and Δ^-) have very nearly equal energies because the electronic wave functions $\langle \vec{r}_i | n \; \Lambda \rangle$ and $\langle \vec{r}_i | n \; -\Lambda \rangle$ represent degenerate states which differ only by the direction of rotation of all electrons about $\hat{R}$. The energy difference between Π^+ and Π^- states is quite small since it is due only to a weak coupling between electron motion and the rotation of the whole molecule. This weak coupling is disregarded in the Born-Oppenheimer approximation, and its effect is called Λ *doubling*.

20.2.4. *Homonuclear Diatomic Molecules*. The energy of a molecule with two identical nuclei remains unchanged not only under simultaneous reflection of the position coordinates of all electrons and nuclei but also under separate reflection at the center of mass of either the electronic or the nuclear coordinates. Therefore, each stationary state of such a molecule has two separate classification labels. Parity under reflection of electron coordinates is indicated by the subscript g or u introduced in section 19.3. Reflection of the nuclear coordinates amounts to an interchange of their positions and is indicated by s or a, meaning *symmetric* or *antisymmetric*. States classified as gs or ua are even under the reflection of all position coordinates; ga or us states are odd. This classification of parity for homonuclear molecules by an electronic and a nuclear quantum number must coincide with the parity $\pm(-1)^N$ which also combines an electronic ($\pm$) and a nuclear $(-1)^N$ property for any linear molecule. In particular, for a given electronic character (g or u, or + or −), the s or a character of the nuclear wave function depends on whether N is even or odd, as follows:

	g+	g−	u+	u−
N even:	s	a	a	s
N odd:	a	s	s	a .

A further restriction on the possible combinations of quantum numbers results from taking into account the identity of nuclei, in accordance with chapter 15. The states of relative motion of two identical particles were classified in chapter 15 by quantum numbers L and S subject to the rule (15.2), L + S = even. Now, the symmetry s of a nuclear wave function for a homonuclear diatomic molecule coincides with the symmetry of an even-L wave function (15.3). In both cases, interchange of the position coordinates of the two particles leaves the wave function unchanged. Similarly, antisymmetry a and odd values of L yield sign reversal of the wave function under interchange of positions. The quantum number that corresponds to S for a pair of molecular nuclei has been called T in equation (19.17a). Therefore, the rule "L + S = even" takes the form "s for even T, a for odd T." On this basis the consistency rule of nuclear and electronic quantum numbers can be expressed, for homonuclear diatomic molecules, on the basis of whether N + T is even or odd:

electronic:	g+	g–	u+	u–
N + T:	even	odd	odd	even .

Consider, for example, oxygen molecules consisting of the ordinary ^{16}O isotope with zero nuclear spin; their electronic ground state has the character $^3\Sigma_g^-$ (section 19.6.3). The g– character requires that N be odd, since T = 0; the ground state of rotation has N = 1 and nonzero energy. Each rotational state is then split into a triplet with J = (N – 1, N, N + 1) in accordance with section 20.2.1.

An important application of the consistency requirement of quantum numbers occurs for hydrogen at very low temperature. The electronic ground state of the hydrogen molecule has the classification $^1\Sigma_g^+$. The *para-hydrogen* states with T = 0 are necessarily symmetric and, being gs states, they have even parity. Since they are + states, they must have even N. Conversely, *ortho-hydrogen,* with T = 1, has only rotational states with odd values of N. At low temperatures, of the order of 20° K, the molecules of liquid hydrogen become concentrated in the lowest rotational level with N = 0. This, however, can only occur as the spins of the two nuclei take opposite orientations, that is, as the concentration of para-hydrogen increases. The equilibrium shift between the concentrations of ortho- and para-hydrogen takes a long time, of the order of weeks, in the absence of magnetic disturbances. A subsequent, reasonably rapid, increase of temperature then produces hydrogen gas in its para form. Conversion to the ordinary mixture of para- and ortho-hydrogen then takes place slowly, at a rate which depends on the temperature and may be increased by magnetic catalysts, such as oxygen. Note that the equilibrium ratio of ortho-para concentration at room temperature is 3 : 1, since the state with T = 1 is triply degenerate. The statistical predominance of ortho-hydrogen is detected in rotational spectra, where the states with odd N values appear to be more densely populated.

20.2.5. *Nonlinear Molecules.* Here the rotational motion of the whole molecule is often complicated by interaction with vibrational and electronic motion, even though these motions may have no axis of symmetry and therefore no constant component of angular momentum. For purposes of orientation one considers the molecular rotation irrespective of this interaction as though all internuclear distances were rigidly fixed.

In this *rigid rotator* model the rotational parameter B(R) introduced in equation (19.19a) is fixed, but one must, in general, consider three separate parameters, B_x, B_y, and B_z, pertaining to the three principal axes of inertia of the molecule and inversely proportional to the respective moments of inertia I_x, I_y, I_z. These three constants are equal for molecules with high symmetry, such as methane. In this case the energy levels of the rigid rotator are represented by

$$E_{rot} = BN(N + 1), \tag{20.5}$$

with

$$B = \frac{\hbar^2}{2I}. \tag{20.5a}$$

Two of the constants are equal for molecules, such as NH_3, with sufficiently high symmetry about one axis, which we take as $\hat{z}$. In this case (*symmetric top*) we have $B_y = B_x$, and the rotational motion has stationary states which are eigenstates of the angular momentum component N_z. The energy levels are then given by the sum of the energies of rotation about z and about any axis x perpendicular to z,

$$E_{rot} = B_z N_z^2 + B_x[N(N + 1) - N_z^2] = B_x N(N + 1) + (B_z - B_x)N_z^2. \tag{20.6}$$

The corresponding wave function has the form $\langle \Theta \ \Phi | N \ N_z \ M \rangle$ which is the same as that given in equation (19.20) but has the quantum number N_z in place of λ.

In the more complicated case of an *asymmetric top*, that is, of a molecule with $B_x \neq B_y \neq B_z$, energy eigenfunctions are superpositions of wave functions with different N_z and the energy levels must be calculated numerically.

20.3. VIBRATIONAL MOTION

The vibrational motion of all diatomic molecules has essentially the same characteristics as that of H_2^+. This motion can be considered initially as a harmonic oscillation, but this is only a rough approximation. The level intervals are equal for a harmonic oscillator, but unequal intervals are observed for vibrational levels, as shown in Table 20.1 for the example of HCl.

TABLE 20.1

Vibrational Level Intervals for HCl

$E_1 - E_0$	$E_2 - E_1$	$E_3 - E_2$	$E_4 - E_3$	
2885.9	2782.1	2678.9	2576.1	cm^{-1}.

Similarly, transitions with $\Delta v > 1$ are observed in infrared spectra, even though they violate the selection rule for the harmonic oscillator represented by equation (12.46). As shown schematically in Figure 19.7 for H_2^+ and in Figure 19.8 for H_2, the vibrational levels converge to the limiting value of U(R) for $R \to \infty$, which is called the *dissociation limit.*

Above the dissociation limit there exists a continuous spectrum of levels of the relative motion of the two atoms. This motion is subject to forces due to the bonding action of the electrons, but it has sufficient energy to escape confinement. The states of relative motion of the two atoms in the dissociated molecule are analogous to the states of motion of an electron in an ionized atom. Dissociated states of a molecule can be formed by excitation of a discrete bound state of vibrational motion. They also occur when two atoms approach in the course of a collision forming a temporary molecule; the relative motion of the atoms is influenced by the same potential U(R) which would confine the vibrational motion at lower energy. The wave functions of dissociated states have the form of standing waves which extend to infinite values of the interatomic distance R. These wave functions are often represented approximately by the WKB method of section 12.2.4. [In our molecular problem the "repulsive potential" which causes the total reflection is the portion of U(R) which rises rapidly at very short interatomic distance.] A continuous spectrum of levels of the relative motion of two atoms exists also when their joint electronic state is antibonding and the potential U(R) is repulsive for all values of R up to several angstroms, as shown in Figure 19.8 for the antibonding state $^3\Sigma_u^+$ of the hydrogen molecule.

Spectroscopic observations are the primary source of data on the discrete spectrum of vibrational motion. The absorption spectrum of a gas at ordinary temperature shows primarily transitions from the lowest vibrational state with $v = 0$ because the population of states with $v > 0$ is small. The final states of these transitions, however, may have a large value of v if the transitions are in the visible or ultraviolet range and thus involve electronic excitation. Quite generally, states with large values of v are readily attained through electronic excitation, whether by absorption of light or by collisions in a discharge or hot gas. The excitation of an electron usually reduces its bonding action and therefore increases the equilibrium internuclear distance R_e. For example, electronic excitation of the H_2 molecule to the lowest state with $^1\Pi_u$ character increases the equilibrium distance R_e by nearly 40% from 0.74 to 1.03 Å. The effects of the increase of R_e on the vibrational motion are treated in the following way.

In an optical transition the radiation interacts with the motion of electrons; the nuclear vibrations are controlled by this motion. Consider the example of the vibrational transition of H_2 from the ground state with vibrational wave function $\langle R \,|\, ^1\Sigma_g, v = 0\rangle$ to a state $\langle R \,|\, ^1\Pi_u \;\; v'\rangle$. In first, and rather good, approximation the probability of this transition is proportional to the squared overlap integral[1]

$$\left| \int_0^\infty dR \langle ^1\Pi_u \;\; v' \,|\, R\rangle\langle R \,|\, ^1\Sigma_g, v = 0\rangle \right|^2. \tag{20.7}$$

1. The expression (20.7) is but one factor of the complete transition probability.

Owing to the large difference of R_e for the two electronic states, this integral is very small for small values of v'. That is, those vibrational states v' will be excited preferentially whose wave functions extend just far enough from the equilibrium distance of $^1\Pi_u$ to maximize the overlap integral with the ground-state vibrational wave function.

This prediction of the relative probability of the different vibrational transitions which may accompany an electronic process is an application of the *Franck-Condon principle*. Franck first formulated the principle in the language of classical mechanics. He considered that an electronic transition takes place rapidly as compared to nuclear motion. Following the electronic transition, the nuclei may find themselves far from their new equilibrium positions and they start vibrating with large amplitude. Spectral analysis based on the Franck-Condon principle has yielded much information on the difference between the values of R_e for different electronic states and also on the profile of the potential curves $U(R)$.

20.3.1. *Normal Modes of Polyatomic Molecules*. The vibrational motion of all nuclei in a polyatomic molecule must be considered simultaneously because the motion of any single nucleus would shift the center of mass of the molecule and would generally alter more than one bond length or bond angle. The potential energy of the nuclear motion, which was indicated by $U(R)$ for a diatomic molecule, is now a function $U(R_1, R_2, \ldots, R_k)$ of all internuclear distances. For each bonding stationary state of electronic motion the function U has a minimum for a set of equilibrium values $R_{1e}, R_{2e}, \ldots, R_{ke}$ of the internuclear distances, and can be expanded into powers of the deviations of these distances from equilibrium, $R_k - R_{ke}$. Truncation of the expansion after the quadratic term yields the usual elastic approximation.

Determination of the normal modes of vibration in the elastic approximation for each molecule in each electronic state constitutes a standard problem of classical mechanics (see Appendix B). Solution of this problem reduces the expression of the potential and kinetic energies of all nuclei to the form

$$E = E_{rot} + E_{vibr} = E_{rot} + \sum_i [\tfrac{1}{2}M_i(dQ_i/dt)^2 + \tfrac{1}{2}k_iQ_i^2], \tag{20.8}$$

where Q_i represents a normal coordinate—a linear function of the deviations $R_k - R_{ke}$—and where M_i and k_i represent the relevant inertial and elastic constants. Each ith term of the vibrational energy has the form of the harmonic-oscillator energy given in equation (12.48). Therefore, the eigenvalues of vibrational energy are represented in this approximation as

$$E_{vibr} = \sum_i (v_i + \tfrac{1}{2})\hbar\omega_i \tag{20.9}$$

with[2]

$$\omega_i = (k_i/M_i)^{1/2} = 2\pi c \nu_i. \tag{20.9a}$$

---

2. We follow here the practice of expressing vibrational frequencies in cm^{-1}; this practice requires the factor c in (20.9a).

The wave functions are similarly obtained from section 12.4.3. The treatment becomes much more complicated in high approximations.

In molecules with several atoms, vibrational excitation may be partitioned among different normal modes in many different ways. There results a very large number of closely spaced levels. The density of levels increases rapidly with increasing vibrational excitation to the point where the levels can no longer be resolved.

Diagrams of the normal modes of three simple types of molecules are sketched in Figure 20.2. Notice the symmetry characteristics of these normal modes.

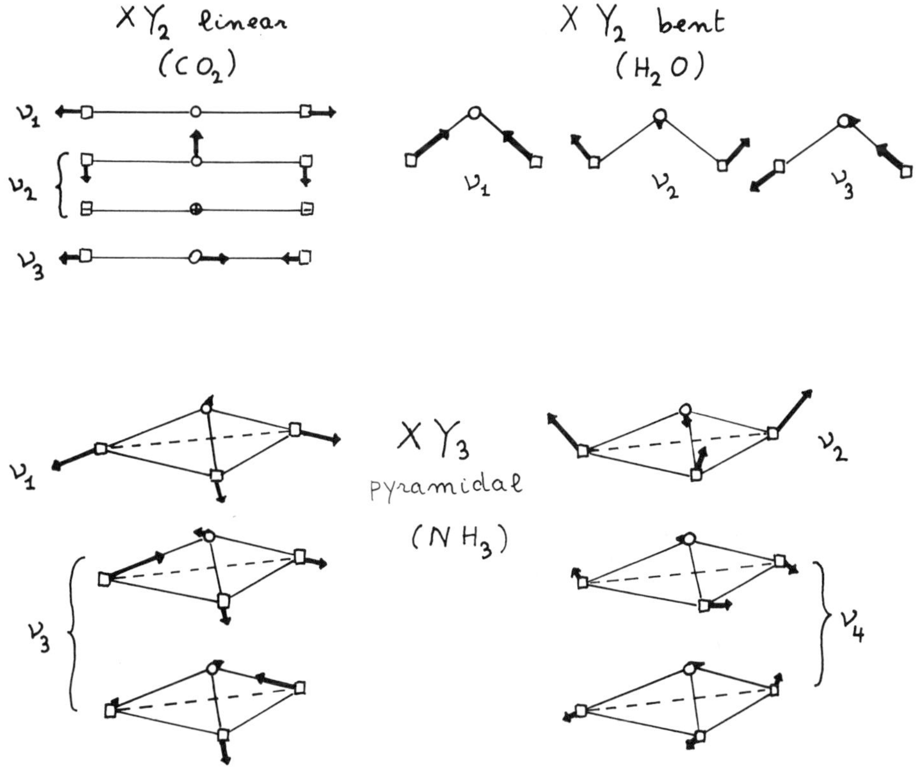

Fig. 20.2. Schematic diagram of normal modes of vibration of XY_n molecules. (Degenerate modes are grouped together by braces.) [Adapted from Herzberg, *Infrared and Raman Spectra*.]

For example, both the CO_2 and H_2O molecules have a plane of symmetry through the central atom and perpendicular to the plane of the drawing. Two of the normal modes are symmetric under reflection of all coordinates with respect to this plane, the third normal mode is antisymmetric. Many of the normal modes of molecular vibration can be identified by symmetry considerations. In this event the stationary states of vibrational motion can be classified irrespective of the elastic approximation. The systematic exploitation of molecular symmetry utilizes the theory of point groups.

A remarkable example of very low frequency vibration occurs in NH_3, but is not shown in Figure 20.2. The vibrational distortion with frequency ν_2 shown in the figure can actually extend until the N nucleus moves across the plane of the H nuclei. Thus the pyramid becomes "inverted," with the N nucleus "below" the H nuclei. This motion proceeds through a region of high distortion of bonds, that is, through a region of high potential energy. The two-valley potential is shown schematically in Figure 20.3. Inversion of the pyramid proceeds by tunneling through the barrier (see section 12.3), at a comparatively very slow rate. The frequency of inversion, Δ, is of the order of 1 cm^{-1}, while the frequency ν_2 is of the order of 10^3 cm^{-1}. The occurrence of the tunneling process splits each vibrational level of the ν_2 mode into two closely spaced levels whose wave functions are even and odd under reflection at the center of mass, as indicated in Figure 20.3. Transitions between the two levels of such a pair are induced by radiation at the microwave frequency Δ; the radiation absorption constitutes the *inversion spectrum* of ammonia.

The distortion of the molecule in the inversion phenomenon is represented by a time-dependent wave function obtained by the superposition of the stationary-

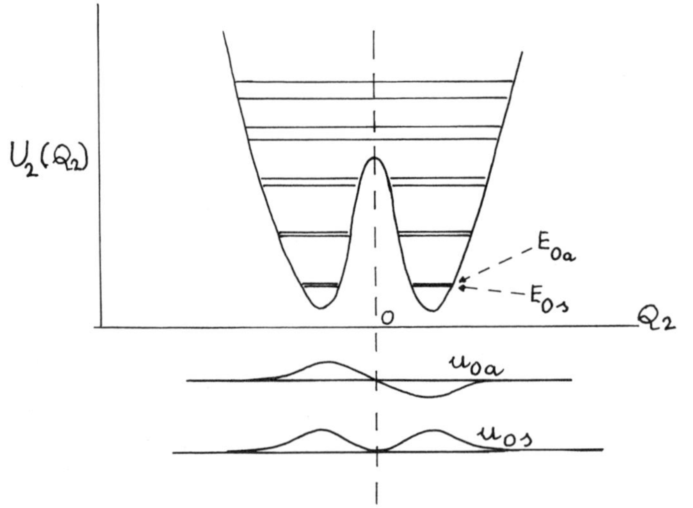

Fig. 20.3. Sketch of two-valley potential and wave functions for the inversion spectrum of NH_3. The coordinate Q_2 is the distance of the N atom from the H_3 plane. The separation $hc\Delta$ between E_{0a} and E_{0s} is much exaggerated. [Adapted from Herzberg, *Infrared and Raman Spectra*, p. 222.]

state wave functions indicated by u_{0s} and u_{0a} in Figure 20.3,

$$\psi(Q_2, t) = \sqrt{\tfrac{1}{2}}\left[u_{0s}(Q_2)e^{i\pi c\Delta t} + u_{0a}(Q_2)e^{-i\pi c\Delta t}\right]\exp(-i\pi c\nu_2 t). \qquad (20.10)$$

The two wave functions u_{0s} (Q_2) and u_{0a} (Q_2) interfere constructively, at $t = 0$, for negative values of Q_2 and destructively for $Q_2 > 0$, thus representing the molecule with the N nucleus "below" the H nuclei. The interference is reversed after each time interval $t = 1/(2c\Delta)$. Similar time-dependent wave functions can be constructed with other pairs of energy eigenfunctions, u_{1s} and u_{1a}, u_{2s} and u_{2a}, etc., corresponding to excited states of the ν_2 vibration.

20.4. ELECTRONIC EXCITATION

Stationary states of electronic excitation of diatomic molecules are classified according to their angular momentum and to their symmetry which has been discussed in previous sections. The properties that are classified include the orbital angular momentum component about the internuclear axis; the total spin quantum number S, and, for certain coupling cases, its projection along the internuclear axis; the parity (+ or −) under reflection on a plane containing the internuclear axis; and, for homonuclear molecules, the parity (g or u) under reflection at the center of mass. A summary of the various quantum numbers is given in Appendix F.

A large amount of experimental information on excited electronic states of numerous diatomic molecules has been accumulated and is tabulated in standard references.[3] Theoretical calculation of energy levels and wave functions usually starts from the independent-electron molecular-orbital approximation with various improvements. Applications have been limited by the complications of the problems.

An important aspect of electronic states is their evolution as the internuclear distance varies from the united-atom limit to the separate-atom limit. Schematic representations of such evolution are called *correlation diagrams*. Figure 20.4a shows a correlation diagram for single-electron molecular orbitals of a diatomic molecule; such a diagram serves as a basis for correlating many-electron states of the whole molecule. Figure 20.4b shows a partial correlation diagram for levels of a whole molecule.

While correlation diagrams serve for purposes of qualitative discussion, the real goal is to obtain plots of the energy of each electronic level of the whole molecule as a function of internuclear distance. Such plots are of the type of Figure 19.2 or 19.5 for H_2^+, depending on whether the energy includes the internuclear repulsion. Plots of this type have been prepared for several molecules with various degrees of accuracy.

---

3. See G. Herzberg, *Molecular Spectra and Molecular Structure. I. Spectra of Diatomic Molecules; II. Infrared and Raman Spectra of Polyatomic Molecules; III. Electronic Spectra and Electronic Structure of Polyatomic Molecules* (Princeton, N. J.: Van Nostrand, 1950, 1954, 1966).

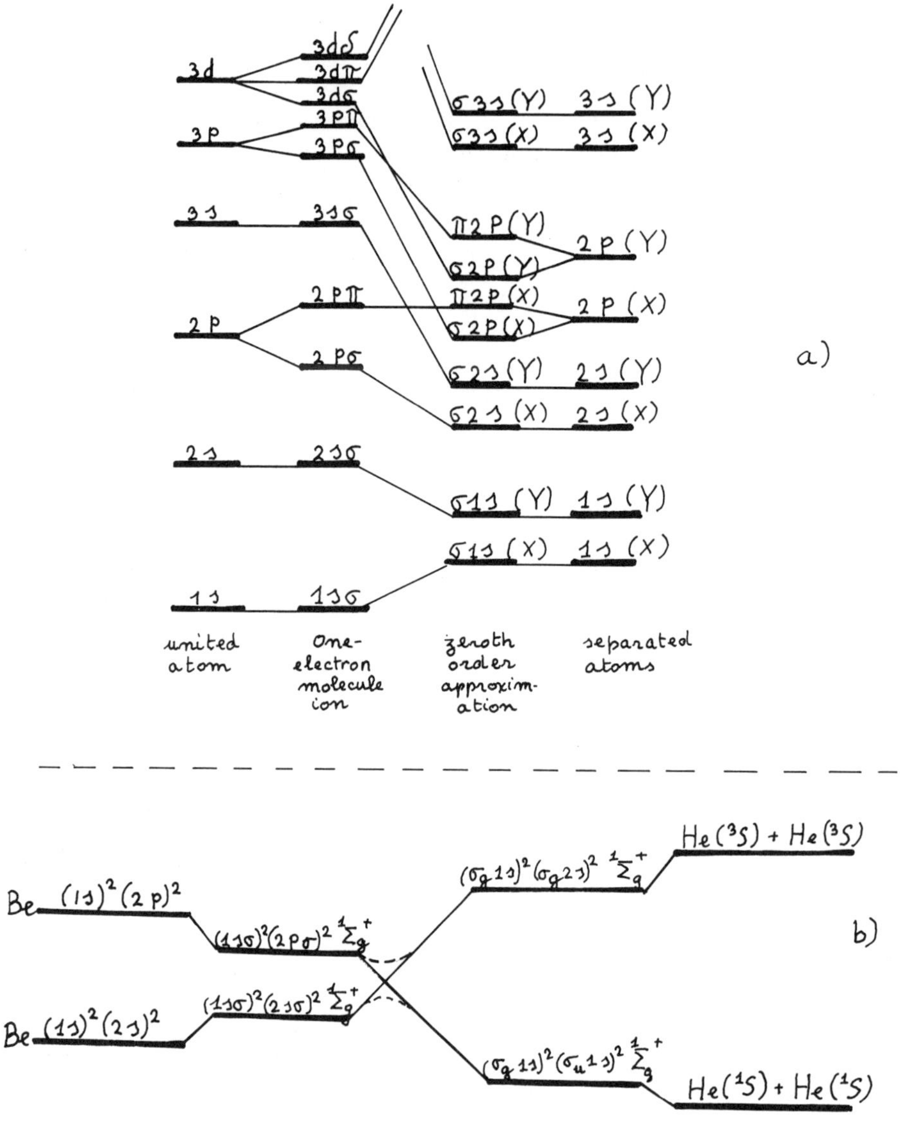

Fig. 20. 4. Correlation diagrams. Symbols of the type $\sigma_g 2s$ indicate a molecular orbital σ_g formed from 2s atomic orbitals of separate atoms.

(*a*) Single-electron levels for XY molecule. [Adapted from Pilar, *Elementary Quantum Chemistry*.]

(*b*) Levels for a four-electron XX molecule. [Adapted from Herzberg, *Spectra of Diatomic Molecules*.]

In the construction of correlation diagrams or detailed plots of energy levels against R, the following point is important. As we know, degenerate stationary states are normally distinguished by being different eigenstates of some physical quantity other than energy, that is, by having different values of some quantum number. States which are characterized by the same set of values of all quantum numbers should accordingly have different energies. The plots of their energies against the internuclear distance should, therefore, not cross *(no-crossing rule)*.

Note, in this connection, that some of the quantum numbers used to characterize a state are defined only within a given approximation. For instance, in the diagram of Figure 20.4b, the two states of the whole molecule are both $^1\Sigma_g$; however, one of these states belongs to a configuration with two u independent-electron states, the other one to an independent-electron configuration including only g states. The two $^1\Sigma_g$ levels of the whole molecule could cross if the independent electron approximation were strictly applicable. Actually, when the interatomic distance reaches a critical value, the independent electron approximation breaks down and the levels nearly coincide but do not actually cross. This effect is called an *avoided crossing* and is shown schematically in Figure 20.4b by the dotted lines.

In polyatomic molecules the electronic states are classified, to whatever extent is possible, by symmetry properties. These properties are worked out by considering the various coordinate transformations which leave the mutual positions of all nuclei unchanged. As in the case of nuclear vibrations, the systematic treatment of these symmetries is based on the theory of point groups. For polyatomic molecules the energy levels are functions of many internuclear distances, and therefore the energy level plots and correlation diagrams become multidimensional.

For all molecules with any symmetry whatsoever, intense emission or absorption of radiation is restricted to transitions which obey certain selection rules. For example, the angular momentum quantum number Λ for diatomic molecules can change by no more than one unit. In homonuclear diatomic molecules, allowed transitions occur only between a *gerade* (g) and an *ungerade* (u) state. These rules are analogous to the selection rules for atoms limiting the change of magnetic quantum number and enforcing the change of parity.

In the electronic excitation of atoms we have singled out the excited states which belong to the ground-state configuration, that is, those in which no electron is shifted out of the ground-state subshell. One could similarly single out excited states belonging to configurations with every electron in the same shell, if not subshell, as the ground-state configuration. Excitation does not greatly increase the size of an atom as long as no electron moves to an outer shell. On the other hand, if one electron is pushed into an outer shell, the atom takes the character of an ionic core with a loosely attached electron.

Similarly, electronic excitation of molecules may be distinguished depending on whether an electron is removed from the others so as to remain loosely attached to the molecule ion. In this case the resulting excited state of the whole molecule is usually called a *Rydberg state*.[4] Conversely, the molecule is in a state of *valence excitation* when all electrons remain approximately confined within the volume of the molecule in its ground state. In molecules, especially polyatomic molecules, a large variety of molecular orbitals can be constructed by superposition of atomic orbitals of the same shell. Many of these molecular orbitals are antibonding, but excitation of a single electron to an antibonding state does not necessarily cause dissociation. Increasing size of molecules affords increasing opportunity for electronic excitations of the valence type.

As the energy of electronic excitation increases, the separation of successive levels decreases, partly because of the large variety of available modes of excitation. As we know, small level separation also occurs at avoided crossings. Under these circumstances, one faces a breakdown of the Born-Oppenheimer approximation which has been the basis of our treatment so far. Specifically, the excitation energy can no longer be regarded as clearly apportioned between electronic and vibrational motion. Therefore, energy received by electrons through light absorption or from another external agent is readily converted into vibrational energy and eventually degraded into heat. Processes of energy exchange between electronic and nuclear motion have been known and understood qualitatively for a long time, but they have not been easily accessible to quantitative experimental or theoretical study. Yet chemical reactions consist just of such elementary processes of exchange, and an increasing effort has been devoted recently to this study. In the remaining part of this section we will outline simple examples of energy transfer between electronic and nuclear motion.

20.4.1. *Fluorescence and Internal Conversion.* Molecules which absorb light in the visible or near-ultraviolet range often reemit it in the process of fluorescence with reduced photon energy. The energy difference is dissipated into nuclear vibration through some or all of the following processes. In the first place, radiation absorption by electronic excitation also gives rise to vibrational excitation, according to the Franck-Condon mechanism mentioned in section 20.3. Second, before the excited molecule reemits radiation, the vibrational excitation is usually dissipated to various modes of nuclear motion of the molecule or in collisions with other molecules. Finally, the process of fluorescent reemission is likely to leave the molecule in the ground state of electronic motion but in an excited state of vibrational motion. The cycle of radiation absorption, degradation, and fluorescence is shown schematically in Figure 20.5.

The fluorescence process is often sidetracked to some extent into a long-delayed emission called *phosphorescence*. This occurs particularly when the lowest electronic excited state is a triplet—a common occurrence which we first met in the case of helium (see section 15.4). The radiation absorption normally leads from the singlet ground state to a singlet excited state with

---

4. One calls a *Rydberg series* a sequence of excited states in which a single electron has increasing values of the principal quantum number (see sections 2.3 and 16.2.1).

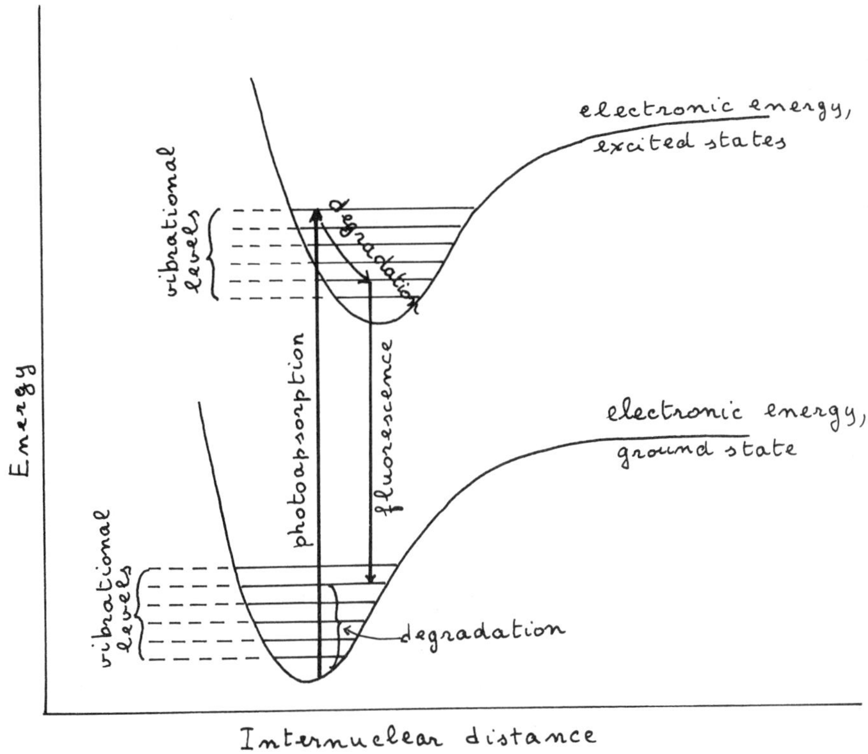

Fig. 20. 5. Diagram of cycle of radiation absorption, degradation, and fluorescence.

considerable vibrational excitation. The spectrum of vibrationally excited *singlet* states overlaps considerably with that of vibrationally excited *triplet* states. In a sufficiently large molecule there generally exist many pairs of singlet and triplet states which are degenerate or quasi-degenerate. Following excitation by photon absorption and in the course of vibrational degradation, the spin-orbit coupling often causes the singlet state to change into a quasi-degenerate triplet state. By the time vibrational degradation is completed, the molecule is in the lowest triplet state. Transition from this state to the singlet ground state is forbidden in that it can occur only with concurrent action of the weak spin-orbit interaction; hence this transition proceeds very slowly. Phosphorescent emission often has a half-life of the order of seconds or longer.

The phenomena of fluorescence and phosphorescence with emission of lower-energy photons afford direct, though partial, evidence of the general phenomenon of *internal conversion* of energy. This conversion occurs in all but the simplest molecules following electronic excitation of a molecule by light absorption or by collision. Recall that the representation of molecular energy levels in the

Born-Oppenheimer approximation disregards the influence of nuclear motion upon the motion of the electrons. Specifically, in this approximation the Schrödinger equations for the electronic and nuclear motions in H_2^+ were derived disregarding the derivative $\vec{\nabla}_R$ of the electronic wave functions $\langle \vec{r} | n\, \vec{R} \rangle$ in the mean energy expression (19.5a). When these derivatives are taken into account, the resulting states, stationary or not, are superpositions of Born-Oppenheimer stationary states with different energies. Application of this treatment to molecules other than H_2^+ is completely analogous. The internal conversion of energy occurs when different stationary states of the Born-Oppenheimer approximation are nearly degenerate. In this case, taking full account of the gradient operator $\vec{\nabla}_R$ in the mean energy expression interlinks different Born-Oppenheimer wave functions in the complete Schrödinger equation.

The complete Schrödinger equation is unraveled by the perturbation methods of chapter 17. One considers usually a nonstationary state which coincides initially with one Born-Oppenheimer stationary state and evolves in the course of time by *radiationless transitions* into a superposition of different Born-Oppenheimer states. This superposition generally develops in time to include a larger admixture of states of increasingly lower electronic excitation and higher nuclear vibration. A chain of radiationless transitions with attendant energy degradation may continue until the electronic motion has returned to the ground state, or it may stop when the electronic motion is at a low level of excitation, thus resulting in fluorescence or phosphorescence.[5]

The exchange of energy between electronic and nuclear motion may also occur with energy transfer from the nuclei to the electrons. When two atoms or groups of atoms collide, part of the nuclear kinetic energy of the colliding systems may be transferred to electronic excitation by a radiationless transition. The electronic energy is later reemitted in the form of light.

20.4.2. *Predissociation.* One type of radiationless transition has been particularly accessible to experimental and theoretical study because of its simplicity. This is the process which changes a bonding state of a molecule into an antibonding one. In the antibonding state the molecule is dissociated; the nuclear motion is no longer confined and has a continuous energy spectrum. Dissociation also results from a radiationless transition to an electronic bonding state when the energy transferred to the vibrational motion of the nuclei exceeds the dissociation limit. One calls predissociation the excitation of a molecule to any electronic bonding state which then evolves slowly by a radiationless transition into a dissociated state. Evidence of predissociation appears in optical spectra as a broadening of discrete lines—that is, as an increase of the natural line width in proportion to the probability of radiationless transition (see section 17.3.3).

---

5. Experimental analysis of the degradation process becomes possible as instrumentation is perfected to observe phenomena over time intervals of the order of 10^{-11} to 10^{-12} sec.

The probability of dissociation by radiationless transition is usually calculated by the time-dependent perturbation method of section 17.3. Recall that this method serves generally to calculate the probability of transitions from a state which is stationary if one disregards some terms of the total energy to a final state which belongs to a continuous spectrum. The Auger effect (section 3.7) and the emission of radiation belong to the class of phenomena treated by this method. The probability of transition is proportional to the squared matrix elements of the energy terms disregarded initially.

20.5. INFLUENCE OF EXTERNAL FIELDS

The influence of external electric and magnetic fields on molecules is in some ways analogous to the influence on single atoms; in some ways it is peculiar to aggregates of two or more atoms. Basically, the influence on the electronic motion (Zeeman effect, Stark effect, etc.) is largely independent of whether or not atoms are bound into molecules. Only a few remarks will be made on these effects at the beginning of this section. The typical effects occur in molecules with polar bonds. Here the rotation of the nuclei with respect to each other is strongly influenced by an external electric field and thus gives a major contribution to the polarizability of the molecule.

An external magnetic field produces on the optical spectra of molecules a Zeeman effect analogous to that produced on atomic spectra. As for atoms, the effect is normal for singlet states, anomalous for doublets and for states of higher spin multiplicity. The magnetic properties of molecules in their ground states have been very successfully analyzed by Stern-Gerlach techniques. In particular, the magnetic analysis of molecules has produced fundamental information on nuclear moments. The study of nuclei is often simpler in a molecular than in an atomic environment because most stable molecules have zero electronic angular momentum and are therefore nonparamagnetic.

The influence of an external electric field on the electronic spectra of molecules is called Stark effect, as for atoms. The electric field shifts the positions of spectral lines by an amount which is generally proportional to the square of the field strength (see section 17.1.1). The reason is that stationary states, for molecules as well as for atoms, are eigenstates of parity and therefore have no average electric dipole moment in the absence of an external electric field. The electric field spoils the symmetry of the molecule under reflection at the center of mass but produces a linear effect only when its strength is sufficient to change appreciably the zero-field eigenstates. An appreciable change does occur when zero-field energy eigenstates of opposite parity are degenerate or quasi-degenerate. Quasi-degeneracy occurs particularly for diatomic molecules with $\Lambda \neq 0$, since the energy split of rotational levels due to Λ-doubling is extremely small. Thus, polar molecules in states with $\Lambda \neq 0$ show a linear Stark effect. On the other hand, the ground state of most polar diatomic molecules has $^1\Sigma$ character and consequently exhibits only a quadratic Stark effect.

The electric polarizability α of a molecule in a particular stationary state coincides with the coefficient of the quadratic Stark effect for that state, according to equation (18.5). However, the electric susceptibility of the substance in bulk depends on the average value of α for all states populated at the prevailing temperature and thus generally depends on temperature like the magnetic susceptibility. The distinction vanishes for a monatomic vapor at normal temperature because excited states lie at energies much larger than $k_B T$ and therefore are not populated in single atoms (apart from fine-structure effects). The contribution to electric polarizability due to the perturbation of electronic motion in molecules is also temperature independent insofar as the electron motion is independent of the excitation of nuclear rotation and vibration.

Temperature dependence of the average polarizability α results therefore from perturbation of the rotational or, possibly, vibrational motion. The external electric field can produce such a distortion only if the mean position of the electrons does not coincide with the center of charge of the nuclei, that is, if the molecule has a nonzero mean dipole moment (see eq. [19.36]). Temperature dependence of the electric polarizability is, accordingly, the hallmark of polar molecules. The possibility of experimentally sorting out polar from nonpolar molecules has played a great role in the qualitative and quantitative study of molecular structure. Among the most elementary applications one may mention the case of CO_2. The nonpolar character of this molecule shows that the polar C=O bonds have opposite directions and, therefore, that the molecule has linear structure. Conversely, the large dipole moment of H_2O shows that its two O—H bonds are far from collinear.

Quantitatively, the Stark effect of an external electric field upon each rotational (or vibrational) energy level of a molecule can be calculated by the perturbation method of section 17.1. The effect is very large because it is inversely proportional to the small spacing of energy levels. However, this effect is positive for some levels, negative for others, in analogy to the linear Zeeman effect in paramagnetic systems (see section 18.4). In both cases the effect is largely canceled by the random distribution of the molecules among different levels, under the influence of thermal agitation. The result of perturbation treatment of the Stark effect followed by averaging over the population of states at thermal equilibrium yields a moderate value of the average polarizability, in accordance with experimental data. Nearly equivalent results are obtained starting from a macroscopic model which disregards the level structure but considers each molecule as a macroscopic electric dipole under the joint influence of the external field and of thermal agitation.

Following the perturbation method, we express the polarizability α_{rot} of a molecule in a rotational state $|N\ M\rangle$ by equation (17.13),[6]

$$\alpha_{rot}(N, M) = 2e^2 \sum_{N'} \frac{|\langle N'\ M|z|N\ M\rangle|^2}{E_{N'} - E_N}. \tag{20.11}$$

---

6. Here the vibrational motion is regarded as unperturbed by the external field.

The numerator of (20.11) is of the order of magnitude of the square of the dipole moment (19.36); therefore, it is comparable to the corresponding factor in a calculation of the polarizability due to perturbation of electronic motion. The denominator of (20.11), however, is of the order of the rotational constant B discussed in section 20.2, and is therefore thousands of times smaller than the energy difference of electronic levels. Accordingly, α_{rot} is far larger than the electronic polarizability. However, when one proceeds to average α_{rot} over the population of states, it can be shown that $\langle \alpha_{rot} \rangle$ cancels out upon averaging over the values of M, except for $N = 0$. That is, $\langle \alpha_{rot} \rangle$ cancels for all rotational states with nonzero squared angular momentum. The net value of $\langle \alpha_{rot} \rangle$ stems entirely from the small fraction of molecules which, at normal temperature, are in states with zero squared angular momentum. According to equation (E2) in Appendix E this fraction is inversely proportional to the partition function; at sufficiently high temperatures the partition function is proportional to the Kelvin temperature, as discussed in the next section. As noted above, calculation of the temperature-dependent electric polarizability can also be carried out by the same classical Langevin theory which applies to atomic paramagnetism (section 18.4). The results of all theories coincide in the high-temperature limit, in which they reduce to the Curie law expression

$$\alpha_{rot} = \frac{|\langle \vec{\mu} \rangle|^2}{3k_B T}, \tag{20.12}$$

where $\langle \vec{\mu} \rangle$ represents now the electric dipole moment of equation (19.36) rather than a magnetic moment.

20.6. THERMAL PROPERTIES

Rotational and vibrational excitation of molecules is attained to a substantial extent by thermal agitation. Therefore, the study of the thermal properties of simple molecules affords a comparatively simple example of the relation of thermal properties to the spectrum of excited levels; it may thus serve as an introduction to the study of larger aggregates. As summarized in Appendix E, thermal properties such as heat capacity are derived from the partition function $Z(\beta) = Z(1/k_B T)$.

The partition function incorporates the effects of rotational, vibrational, and electronic excitation. In the approximation where each energy eigenvalue is the sum of independent terms arising from the three different kinds of motion, equations (E5) and (E6) apply; the partition function then factors into a product of contributions from the separate motions

$$Z \approx Z_{rot} \times Z_{vib} \times Z_{el}. \tag{20.13}$$

These contributions are additive in the calculation of thermal properties which depend linearly on the logarithm of Z.

In the limit of zero temperature there is no molecular excitation, and the partition function is equal to the degeneracy of the ground state. With increasing temperature, Z_{rot} starts increasing when k_BT reaches the order of magnitude of the rotational parameter B. The vibrational contribution, Z_{vib}, steps in when k_BT approaches the energies $hc\nu_i$ of equation (20.9). At room temperature and for small molecules Z_{rot} is much larger than unity and Z_{vib} is beginning to be appreciable.

The factor Z_{rot} can be evaluated analytically for diatomic molecules in the approximation in which the rotational levels are given by BN(N + 1) and the rotational constant B is averaged over all internuclear distances. Under these assumptions we have

$$Z_{rot} \approx \sum_{N=0}^{\infty} (2N + 1)\, e^{-\beta BN(N+1)}. \tag{20.14}$$

Here, 2N + 1 represents the number of different values of the magnetic quantum number M, that is, the degeneracy, or *statistical weight*, of the Nth level. When the exponential factor is appreciable for many values of N, the summation can be replaced approximately by integration,

$$\sum_{N=0}^{\infty} (\ldots) \approx \int_0^{\infty} dN(\ldots). \tag{20.15}$$

It is particularly convenient here that $(2N + 1)\, dN = d\,[N(N + 1)] = (\beta B)^{-1} \times d\,[\beta BN(N + 1)]$. The integral thus reduces to unity, to within the factor $(\beta B)^{-1}$, and we have

$$Z_{rot} \approx \frac{1}{\beta B} = \frac{k_BT}{B} \quad (\text{for } T \to \infty). \tag{20.16}$$

These general considerations show that Z_{rot} is constant at very low temperature and increases at higher temperatures in proportion to T.

As noted in Appendix E, proportionality of Z to a power of the absolute temperature leads to the *equipartition of energy*. In equation (20.16) Z_{rot} is proportional to the first power of T. Consequently, at sufficiently high temperatures, the internal energy $U_{rot}(T)$ amounts to k_BT per molecule (two degrees of freedom) and the heat capacity $C_{V rot}$ has the constant value k_B. In conclusion, the heat capacity of rotational motion vanishes in the low-temperature limit, where Z_{rot} is constant, and rises to its value k_B for values of $k_BT \gtrsim B$.

Analogous results hold for the vibrational motion. The calculation of Z_{vib} can be carried out analytically irrespective of temperature with the single approximation that the vibrational motion be harmonic and that its energy be represented by the normal-mode formula (20.9). In this case, Z_{vib} factors further into a product of contributions from the separate normal modes and each contribution is given by the sum of a geometric series,

$$Z_{vib} = \Pi_i\, Z_{vib,i} = \Pi_i \sum_{v_i=0}^{\infty} \exp(-\beta v_i hc\nu_i) = \Pi_i \frac{1}{1 - \exp(-\beta hc\nu_i)}. \tag{20.17}$$

Here the energy eigenvalue with vibrational quantum number v_i is represented by $v_i hc\nu_i = (v_i + \frac{1}{2})hc\nu_i - \frac{1}{2}hc\nu_i$. This energy is thus expressed in a scale whose zero-point lies at the ground state, $v_i = 0$, in accordance with the definition of partition function in Appendix E. Recall also that ν_i is expressed in cm^{-1} following the practice of infrared spectroscopy.

At low temperatures, the product in (20.17) departs from unity only by exponentially small terms of the form $\exp(-\beta hc\nu_i)$. At high temperatures expansion of the exponential to first order in β yields

$$Z_{vib} \sim \Pi_i \frac{k_B T}{hc\nu_i} \qquad \text{(for } T \to \infty). \tag{20.18}$$

It follows that the heat capacity has the characteristic quantum behavior of approaching zero exponentially for low temperatures and of approaching a constant for high temperatures. This behavior of the heat capacity, which was first derived by Einstein, is shown in Figure 20.6 for a single normal mode. The total heat capacity for a set of normal modes is obtained by substituting the partition function (20.17) in the general formula (E14),

$$C_{Wib} = \frac{1}{k_B T^2} \frac{d^2 \ln Z_{vib}}{d\beta^2} = k_B \sum_i \left(\frac{hc\nu_i}{k_B T}\right)^2 \frac{\exp(-hc\nu_i/k_B T)}{[1 - \exp(-hc\nu_i/k_B T)]^2}. \tag{20.19}$$

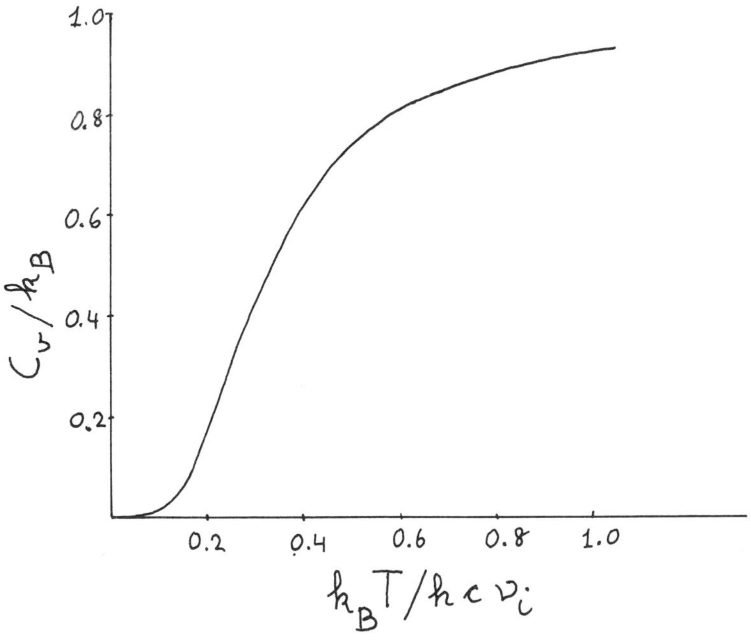

Fig. 20. 6. Heat capacity due to a single vibrational mode of frequency ν_i cm^{-1}.

This formula represents the heat capacity function $C_v(T)$ as a superposition of identical functions of the ratio ν_i/T. The function $C_v(T)$ can be calculated for any molecule whose frequency spectrum is known. Equation (20.19) serves also to calculate the heat capacity of most materials, that is, of all aggregates of atoms whose heat capacity depends primarily on the vibration of their atoms. It applies even to the radiation of the blackbody mentioned in section S3.5. The information required for the calculation is the spectrum of vibrational frequencies.

CHAPTER 20 PROBLEMS

20.1. The successive rotational lines of the HCl molecule, in a spectrum analogous to that shown in Figure 20.1 for HBr, are separated by 21.2 cm^{-1}. Calculate the equilibrium distance of the H and Cl nuclei, assuming that the spectrum pertains to molecules of the ^{35}Cl isotope. Would the difference between the ^{35}Cl and ^{37}Cl isotopes be apparent at the level of significance of your information?

20.2. The NH_3 molecule has the shape of a pyramid with an equilateral triangle of three H atoms as its base. Show that two of the principal moments of inertia of the molecule are equal, thus verifying that the levels of its rotational spectrum are given by the "symmetric top" formulae.

20.3. Calculate the rotational energy levels of the CH_4 molecule, considering that the nuclei of its H atoms lie approximately at the vertices of a regular tetrahedron at 1.09 Å from the C nucleus.

20.4. The potential energy U of the nuclear motion of a CO_2 molecule may be expressed in terms of the Cartesian coordinates of the nuclear positions by

$$U = \tfrac{1}{2}k_s\,(z_{02} - z_C - R_e)^2 + \tfrac{1}{2}k_s\,(z_C - z_{01} - R_e)^2$$
$$+ \tfrac{1}{2}k_b\,(x_{01} + x_{02} - 2x_C)^2 + \tfrac{1}{2}k_b\,(y_{01} + y_{02} - 2y_C)^2,$$

where k_s and k_b are the stretching and bending force constants. The center of mass can be fixed at the origin, which permits elimination of $x_C, y_C,$ and z_C. (The figure shows the equilibrium positions of the nuclei.) (a) Express this potential energy in terms of the normal-mode coordinates

$$Q_s = z_{02} - z_{01} - 2R_e \quad , \quad Q_a = z_{01} + z_{02},$$
$$Q_x = x_{01} + x_{02} \quad , \quad Q_y = y_{01} + y_{02}.$$

(b) Express the kinetic energy K of the nuclei as a sum of terms which depend separately on the four normal-mode coordinates Q and on the rotational coordinates $x_{01} - x_{02}$ and $y_{01} - y_{02}$. (c) Determine the force constants k_s and k_b from knowledge of the nuclear masses and of the experimental vibrational frequencies $\nu_1 = 1337$ cm^{-1}, $\nu_2 = \nu_x = \nu_y = 667$ cm^{-1}, $\nu_3 = 2349$ cm^{-1}, which correspond to those indicated in Figure 20.2. (You will find two somewhat inconsistent values of k_s owing to the inadequacy of the simple harmonic model.)

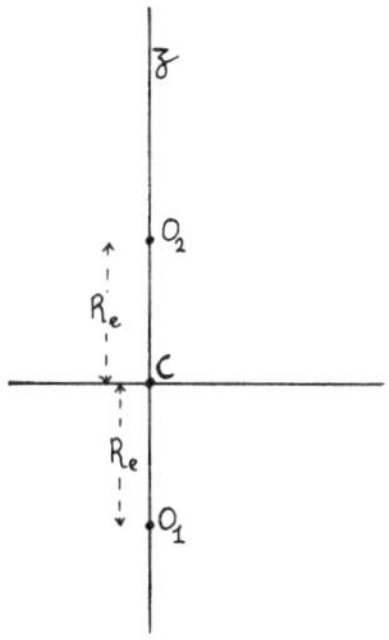

20.5. An analytically solvable model for the tunneling in the NH_3 inversion phenomenon can be formulated and treated by combining the square-well and square-barrier problems of chapter 12. Consider a particle in a potential V consisting of a pair of wells of width a separated by a barrier of height V' and width 2b and represented by $V(x) = \infty$ for $|x| > a + b$; $V(x) = 0$ for $b < |x| < a + b$; $V(x) = V'$ for $|x| < b$. Find the equation which determines the energy eigenvalues and eigenfunctions and discuss it in the case of a thick barrier, $[2m(V' - E)]^{1/2}\, b/\hbar \gg 1$.

20.6. Calculate the contribution to the heat capacity of the NO molecule due to excitation of its electronic spin doublet, and plot the result as a function of temperature. The energy separation of the doublet levels is $\Delta = 0.015$ eV. Disregard any interlinkage of this excitation with rotational or vibrational excitations, as well as the existence of other electronic levels.

20.7. (a) Evaluate and plot the partition function of the rotational energy of para-hydrogen by numerical calculation of the contributions of successive rotational levels between 0° and 300° K. Calculate to three significant figures only. The value of the constant B for H_2 is 7.5×10^{-3} eV. (b) Calculate the mean internal rotational energy U_{rot} of 1 gram of para-hydrogen at 200° K.

SOLUTIONS TO CHAPTER 20 PROBLEMS

20.1. According to equation (20.1) the lines corresponding to successive values of N are separated by 2B, where B and B′ are equal in a vibrational-rotational transition. Therefore, the average value of B, namely, $\sim B(R_e)$, is given in cm^{-1} by $B = \hbar/4\pi\overline{M}cR_e^2 = \frac{1}{2}\, 21.2 = 10.6\ cm^{-1}$ and we have $R_e = [\hbar/4\pi\overline{M}c\,(10.6\ cm^{-1})]^{1/2}$. The reduced mass of the H and ^{35}Cl nuclei is, within 1% accuracy, $(35/36) \times 1.66 \times 10^{-24}\ g = 1.61 \times 10^{-24}$ g. This yields $R_e = 1.28$ Å; the value given in standard references taking into account all corrections is 1.27460 Å. For the isotope ^{37}Cl we would obtain a value of R_e differing in the ratio $[(37/38)/(35/36)]^{-1/2} \sim 1 - (36.5)^{-2}$; which does not depart from 1 significantly to our level of accuracy.

20.2. If the molecule has an axis z of inertial symmetry, this axis would be the height of the pyramid. Therefore, it is sufficient to show that the three H atoms yield equal moments of inertia with respect to any two orthogonal axes x and y lying on their plane. Take these axes as in the figure and call M the mass of each H atom. We have $I_y = M[(\frac{1}{2}d)^2 + (\frac{1}{2}d)^2] = \frac{1}{2}Md^2$;
$I_x = M[(\frac{1}{2}\sqrt{\frac{1}{3}}d)^2 + (\frac{1}{2}\sqrt{\frac{1}{3}}d)^2 + (\sqrt{\frac{1}{3}}d)^2] = \frac{1}{2}Md^2 = I_y$. Q.E.D.

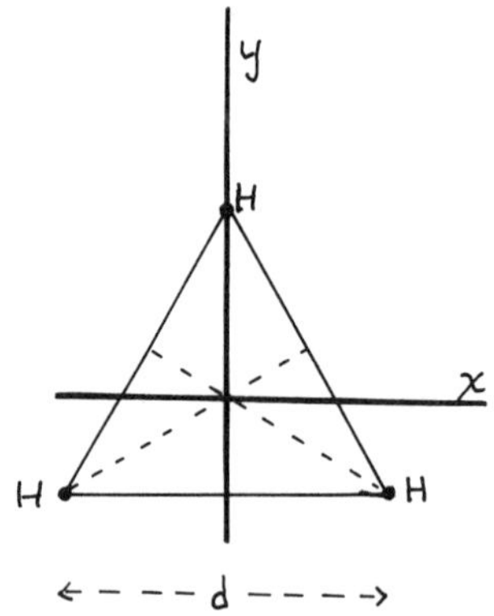

20.3. The principal moments of inertia of a regular tetrahedron are all equal. It is sufficient to calculate the moment of inertia of three H nuclei, at the vertices of an equilateral triangle, with respect to an axis perpendicular to the triangle at its center. The distance of each H nucleus from this center is $\frac{1}{3}\sqrt{8} \times 1.09$ Å $= 1.03$ Å. Therefore, the moment of inertia is $I = 3 \times 1.67 \times 10^{-24} \times (1.03)^2 \times 10^{-16} = 5.3 \times 10^{-40}$ g cm², and $B = \hbar^2/2I = 1.04 \times 10^{-15}$ ergs $= 6.5 \times 10^{-4}$ eV. The energy levels are given by $E_N = 6.5 \times 10^{-4} \times N(N+1)$eV.

20.4. With the center of mass at the origin we have $x_c = -M_0Q_x/M_c$, $y_c = -M_0Q_y/M_c$, $z_c = -M_0Q_a/M_c$ where M_0 and M_c are the oxygen and carbon masses. Set $\gamma = 1 + 2M_0/M_c = 3.67$. (a) Substitution of x_c, y_c, and z_c and of $z_{01} + R_e$, $z_{02} - R_e$, etc., by normal coordinates gives $U = \frac{1}{4}k_s(Q_s^2 + \gamma^2 Q_a^2) + \frac{1}{2}\gamma^2 k_b(Q_x^2 + Q_y^2)$. (b) Indicating the time derivatives by dots and performing the same changes of variables as in part a, we have

$$K = \tfrac{1}{4}M_0[\gamma(\dot{Q}_x^2 + \dot{Q}_y^2 + \dot{Q}_a^2) + \dot{Q}_s^2 + (\dot{x}_{01} - \dot{x}_{02})^2 + (\dot{y}_{01} - \dot{y}_{02})^2].$$

(c) The combined expression of K and U represents the sum of the rotational kinetic energy and of the energies of four harmonic oscillators with frequencies $\omega_1 = (k_s/M_0)^{1/2}$, $\omega_3 = (\gamma k_s/M_0)^{1/2}$, $\omega_2 = \omega_x = \omega_y = (2\gamma k_b/M_0)^{1/2}$. In this approximation, theory predicts $\omega_3/\omega_1 = \gamma^{1/2} = 1.92$, whereas experimentally $\nu_3/\nu_1 = 1.76$. The values of k_s and k_b in terms of ν_1, ν_2, and ν_3 are

$$k_s = (2\pi c\nu_1)^2 M_0 \quad = 1.69 \times 10^6 \text{ ergs/cm}^2 = 105 \text{ ev/Å}^2,$$

$$k_s = (2\pi c\nu_3)^2 M_0/\gamma \quad = 1.42 \times 10^6 \text{ ergs/cm}^2 = 89 \text{ eV/Å}^2,$$

$$k_b = (2\pi c\nu_2)^2 M_0/2\gamma = 5.72 \times 10^4 \text{ ergs/cm}^2 = 3.6 \text{ eV/Å}^2.$$

20.5. Indicate the wave function by ψ, ψ', and ψ'' in three separate regions, as in equation (12.66). As in problem 12.1, the condition $\psi(x) = 0$ at $x = \pm(a + b)$ can be enforced by suitable alternative representations of ψ, ψ', and ψ'' which are symmetric (+) or antisymmetric (—) under reflection at $x = 0$. We set $k_0 = (2mE)^{1/2}\hbar$, and $k' = i\kappa = [2m(E - V')]^{1/2}/\hbar$, as in chapter 12, and

$-(a+b) < x < -b$	$\psi_+(x) = A_+ \sin[k_0(a+b+x)]$	$\psi_- = A_- \sin[k_0(a+b+x)]$
$-b < x < b$	$\psi_+'(x) = B_+ \cos k'x$	$\psi_- = -B_- \sin k'x$
$b < x < a+b$	$\psi_+''(x) = A_+ \sin[k_0(a+b-x)]$	$\psi_-'' = -A_- \sin[k_0(a+b-x)]$.

Owing to symmetry it is sufficient now to enforce continuity of $\psi_\pm$ and of its derivative at $x = -b$,

$$A_+ \sin k_0 a = B_+ \cos k'b \quad , \quad A_- \sin k_0 a = B_- \sin k'b,$$

$$k_0 A_+ \cos k_0 a = k' B_+ \sin k'b \quad , \quad k_0 A_- \cos k_0 a = -k' B_- \cos k'b.$$

The condition for these systems to have a solution is obtained by eliminating the A and B coefficients,

$$\tan k_0 a = \frac{k_0}{k'} \cot k'b \quad = -\left(\frac{E}{V' - E}\right)^{1/2} \frac{1 + e^{-2\kappa b}}{1 - e^{-2\kappa b}} \sim -\left(\frac{E}{V' - E}\right)^{1/2} (1 + 2e^{-2\kappa b});$$

$$\tan k_0 a = -\frac{k_0}{k'} \tan k'b = -\left(\frac{E}{V' - E}\right)^{1/2} \frac{1 - e^{-2\kappa b}}{1 + e^{-2\kappa b}} \sim -\left(\frac{E}{V' - E}\right)^{1/2} (1 - 2e^{-2\kappa b}).$$

The roots of these equations occur for values of $k_0 a$ a little lower than a multiple of π, for $E/(V' - E) \ll 1$; that is, $k_0 a$ is a little lower than an eigenvalue of problem 12.1, to the extent that V' is finite and the barrier can be penetrated. The two equations for symmetric and antisymmetric wave functions differ by the sign of $\exp(-2\kappa b)$; it follows that the value of $k_0 a$ for each symmetric eigenfunction is a little lower than the root for the corresponding antisymmetric eigenfunction. The wave-function amplitudes under the barrier are very small for $\kappa b \gg 1$, since

$$\frac{B_+}{A_+} = \frac{\sin k_0 a}{\cos k'b} \quad , \quad \frac{B_-}{A_-} = \frac{\sin k_0 a}{\sin k'b} \quad .$$

20.6. According to appendix E the doublet contributes to the partition function, under the stated conditions, a factor $Z_{doub} = 1 + \exp(-\Delta/k_B T)$. Substitution into equation (E.14) gives

$$C_{V\,doub} = k_B \left(\frac{\Delta}{k_B T}\right)^2 \frac{e^{-\Delta/k_B T}}{[1 + \exp(-\Delta/k_B T)]^2}.$$

Note that $C_{V\,doub}$ vanishes both for $T \to 0$ and $T \to \infty$, in contrast to $C_{V\,vib}$ (eq. [20.19]). Mathematically, the different behavior is due to the sign difference in the denominator; physically, to the fact that the doublet excitation absorbs no further heat once the temperature is sufficiently high to equalize the probabilities of the two levels. Conversely, the vibration or rotation of a molecule can be excited to higher and higher levels as the temperature increases.

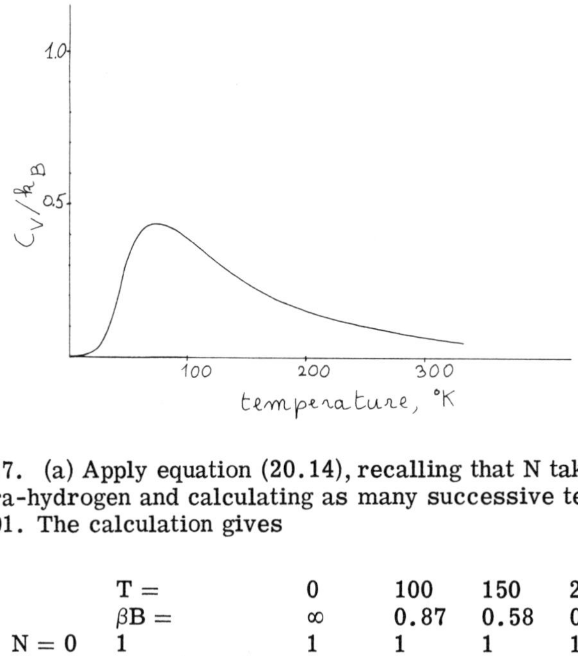

20.7. (a) Apply equation (20.14), recalling that N takes only even values for para-hydrogen and calculating as many successive terms as are greater than 0.01. The calculation gives

	$T =$	0	100	150	200	250	300 °K
	$\beta B =$	∞	0.87	0.58	0.44	0.35	0.29
$N = 0$	1	1	1	1	1	1	1
2	$5\exp(-6\beta B)$	0.00	0.03	0.15	0.37	0.62	0.88
4	$9\exp(-20\beta b)$	0.00	0.00	0.00	0.0015	0.01	0.03
	Z	1	1.03	1.15	1.37	1.63	1.91

(b) Apply equation (E11), initially to a single molecule, using the data in part a: $(U_{rot})_{T=200°} = [Z^{-1}\Sigma_N BN(N+1)(2N+1)\exp(-\beta BN(N+1))]_{T=200°} = \frac{B}{1.37}[6 \times 0.37 + 20 \times 0.0015] = 1.23 \times 10^{-2}$ eV. One gram of para-hydrogen contains 3.01×10^{23} molecules and thus has a rotational energy of $3.01 \times 10^{23} \times 1.23 \times 10^{-2}$ eV = 594 joules = 142 cal.

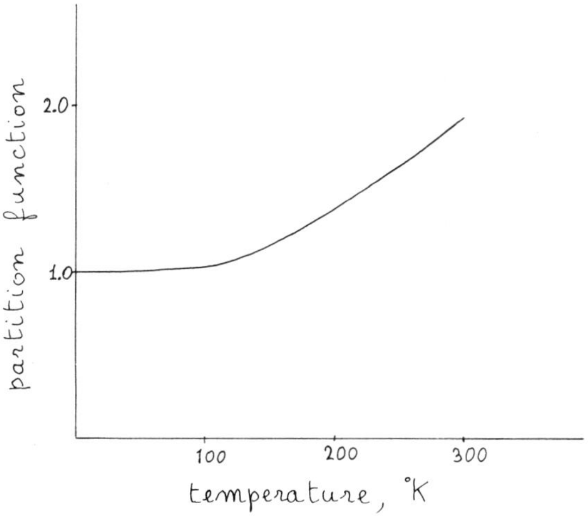

21. nonlocalized bonds

The chemical bonds described in chapter 19 link any one atom of a molecule to atoms immediately adjacent to it. Thereby, in a polyatomic molecule, distant atoms are linked only indirectly through a chain of bonds and any disturbance of equilibrium is transmitted through the molecule stepwise from one atom to the next. The concept of a localized chemical bond is emphasized by the ordinary structure formulae of chemistry which represent each bond by a dash between the symbols of two elements. This concept goes a long way toward a realistic description of chemical structures but fails to account for the properties of many important aggregates.

Basically, bond formation results from the merging of the electronic motion of different atoms into a single system. This mechanism is not exploited fully by considering the motion of electron pairs extending only over two adjacent atoms. Additional stability accrues to many molecules when a number of electron pairs spread smoothly over a number of atoms, thus welding all atoms together by a single diffuse bond. In these cases, relaxing the restriction that electrons may move only over adjacent atoms allows the system of *all* electrons to settle in a stationary state of lower energy. This effect is analogous to the gain of stability which results from bonding with intermediate homopolar and ionic character (section 19.4).

The blending of many electrons into a single system gives molecules certain characteristic properties besides increased stability. In the first place, electrons moving over an extended volume readily transmit the effects of external disturbances throughout that volume. The spacing of electronic energy levels decreases as the electrons move over larger volumes; since a system is the more sensitive to external disturbances the closer are its energy levels (section 17.1.3), the motion of electrons throughout a molecule gives the molecule some of the properties of a conductor. Molecules with nonlocalized bonds have a key role in biochemistry owing to their sensitivity to disturbances and to their ability to transmit them.

Chemical structure formulae are not suited to represent nonlocalized bonds. In quantum-mechanical language the state of a molecule with a nonlocalized bond may be represented by a superposition of different states each of which corresponds to a different structure formula. In chemical language the molecule is described by assigning to it a number of alternative structure formulae and by stating that the molecule is stabilized by resonance among all of them.

This chapter gives a series of examples of molecular aggregates in which bonding electrons move over increasing distances ranging from a few atoms to macroscopic distances. The first examples deal with small molecules (section 21.1) and are followed by examples of motion along chains and plane networks of atoms (section 21.2). Metals may be regarded as limiting cases of large molecules with unrestricted range of electronic motion (section 21.3). The study of bonding and of vibrations in metals and other crystal lattices is the core of solid-state physics and is not attempted here. In this chapter we intend only to show how concepts and formulae developed for molecules extend to infinite lattices. A sample treatment of electron wave functions for the two-dimensional graphite layer is developed as an extension of the treatment of the benzene molecule in supplementary section S21.4.

21.1. SMALL MOLECULES

21.1.1. *Carbon Dioxide.* This molecule is often described by the structure formula O=C=O, with two localized double bonds. However, the experimental value of the CO internuclear distance in CO_2 is 1.15 Å, as compared to about 1.21 Å for the doubly bonded CO group in organic molecules such as aldehydes or ketones. Also, the total bond energy of the CO_2 molecule exceeds by 1.4 eV the sum of the energies of two localized C=O double bonds. As noted in section 20.5, the nonpolar character of CO_2 shows that this molecule has linear structure.

The electronic structure of CO_2 can be described, as we did for doubly bonded structures in section 19.6, starting from the formation of two localized bonds which connect the carbon atom to each of the two oxygen atoms. These bonds are formed by hybrid orbitals with σ character extending from the carbon and oxygen nuclei along the internuclear axis; we take this axis as the z axis, as we did for H–C≡C–H. The remaining bonds of CO_2 are formed by π orbitals, constructed from the p_x and p_y orbitals of carbon and oxygen, as in the acetylene and nitrogen molecules (section 19.6).

Eight electrons are available for the formation of π orbital bonds, three from each oxygen and two from carbon.[1] Molecular orbitals corresponding to the ordinary doubly bonded formula O=C=O can be constructed, for example, by superposing the p_x orbital of carbon with the p_x orbital of oxygen atom #1, and the p_y orbital of carbon with the p_y orbital of oxygen atom #2. Each of these two molecular orbitals would be occupied by a pair of bonding π electrons. There would remain two pairs of π electrons, one assigned to the p_x orbital of oxygen atom #2, and one to p_y of oxygen atom #1. Clearly this localized bond structure would lack any symmetry with respect to reflection at the carbon atom.

A more symmetric and realistic structure is obtained by constructing MO-LCAO orbitals which extend over the whole molecule. Such wave functions are superpositions of p orbitals of all three atoms. We take, as usual, the origin of coordinates at the mid-point of the linear molecule, and indicate the positions of the oxygen nuclei as $\pm R\hat{z}$. A first set of single-electron p orbitals of the three atoms is then indicated by $u_O(\vec{r} - R\hat{z})$, $u_C(\vec{r})$, $u_O(\vec{r} + R\hat{z})$. A second set, orthogonal to the first one, is indicated by $v_O(\vec{r} - R\hat{z})$, $v_C(\vec{r})$, and $v_O(\vec{r} + R\hat{z})$.

---

1. The CO_2 molecule has all together 22 electrons. The six K-shell electrons of the three atoms are nonbonding. Four electrons form the two localized bonds in σ orbitals. Two more electrons of each oxygen may be assigned to nonbonding σ orbitals.

(The wave functions u and v may be p_x and p_y, respectively, or any other degenerate orthogonal pair such as the p orbitals with $m = \pm 1$.) The MO-LCAO superpositions may then be represented by

$$\frac{c_{01}u_0(\vec{r} - R\hat{z}) + c_C u_C(\vec{r}) + c_{02}u_0(\vec{r} + R\hat{z})}{[c_{01}^2 + c_C^2 + c_{02}^2 + 2(c_{01} + c_{02})c_C\Delta]^{1/2}}, \tag{21.1}$$

and by an analogous expression with wave functions v. The coefficients c remain to be determined and Δ indicates the overlap integral for adjacent atoms, analogous to (19.28).

The coefficients can be determined by working out the mean energy of an electron with the wave function (21.1) and by applying the variational procedure of section 12.1. There results a system of three linear algebraic equations analogous to the system obtained in sections S14.4 and 16.4. The lowest-energy eigenvalue, E_0, corresponds to a molecular orbital, π_0, in which all three coefficients have the same sign and $c_{01} = c_{02}$; this orbital is symmetric with respect to reflection at a plane perpendicular to the molecular axis at its midpoint. Analogy to the H_2^+ orbitals (19.11) indicates that this lowest-energy orbital has bonding character all over the molecule. The next higher energy eigenvalue, E_1, corresponds to an antisymmetric wave function, π_1, with $c_{01} = -c_{02}$ and $c_C = 0$. This wave function has essentially nonbonding character. The highest eigenvalue, E_2, corresponds to another symmetric wave function, π_2, in which the sign of c_C is opposite to that of $c_{01} = c_{02}$. This wave function is antibonding. Each of the energy eigenvalues thus determined is degenerate since it corresponds not only to a wave function (21.1) but also to a second wave function with the same coefficients and with atomic orbitals of the second set v.

The eight π electrons of the CO_2 molecule are distributed among the states with the lowest possible energy (as they are in the ground state configuration of atoms) as follows: one pair to the state π_0 constructed with atomic orbitals u; one pair to the state π_0 constructed with atomic orbitals v; the remaining two pairs to states π_1 constructed with u and v. The antibonding state π_2 remains unoccupied. The total energy of the eight electrons equals $4E_0 + 4E_1$. In this approximation the special stability of the nonlocalized bond arises from the particularly low value of E_0.

In chemical language the stability of CO_2 is accounted for by regarding the state of the molecule as a combination of the structures

$$O{=}C{=}O \qquad O^-{-}C{\equiv}O^+ \qquad O^+{\equiv}C{-}O^-. \tag{21.2}$$

This combination gives a partial triple bond character to each CO group and is stabilized by resonance. Resonance among the three structures represented in (21.2) involves a shift of electrons from one to the other end of the molecule. The stability of CO_2 is shared by two molecules with the same electronic structure, namely, N_2O and N_3^-. Their structure formulae analogous to O=C=O are, respectively, $N^-{=}N^+{=}O$ and $N^-{=}N^+{=}N^-$.

21.1.2. *The BO_3^{---}, CO_3^{--}, and NO_3^- Molecule Ions.* These three ions form a sequence isoelectronic with the BF_3 molecule described in section 19.5. Their electronic structure may be described starting from the chemical structure formulae

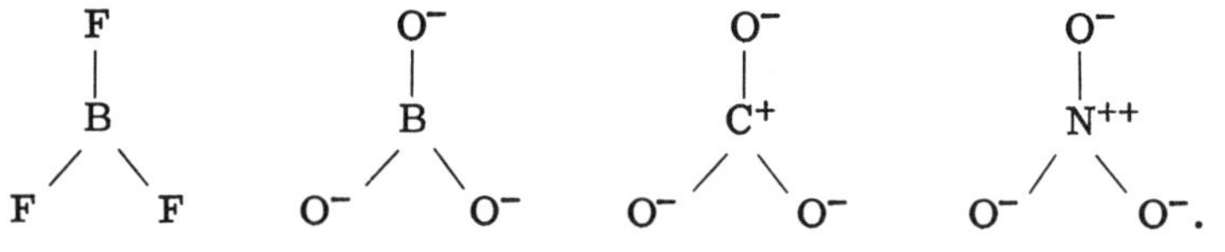

(21.3)

All these structures have three localized single bonds, with the central atom contributing hybrid orbitals in trigonal arrangement. These structures do not utilize the orbital of the central atom which extends in a direction perpendicular to the plane of the molecule; if we take this direction as the z axis, then the unused orbital is p_z.

The singly bonded structures of equation (21.3) are increasingly unrealistic proceeding from BF_3 to NO_3^-. The electron pairs in the p_z orbitals of the O^- ions need not remain confined within the ion to which they are assigned by the structure formulae, but they can expand into the p_z orbital of the central atom. This effect is increasingly strong with increasing atomic number of the central atom.

In the MO-LCAO approximation, one constructs for each molecule four symmetry-adapted wave functions which are superpositions of the p_z orbitals of the four atoms. The state with the lowest energy eigenvalue has full triangular symmetry and exerts a bonding action throughout the molecule. (This bond has different degrees of ionic character depending on the atomic number of the central atom.) Two other wave functions are orthogonal and degenerate and contribute no bonding. The fourth one has antibonding character. The six p_z electrons, all of which are assigned to the O^- ions in the formulae of (21.3), are assigned in the MO-LCAO approximation as follows. One pair belongs in the bonding state of lowest energy, and contributes added stability to the molecule. The two remaining pairs belong in the degenerate nonbonding orbitals. The antibonding orbital remains empty.

In chemical language the stability of these molecular ions is attributed to stabilization by resonance among different structures. One structure is represented by (21.3) and the others by formulae involving a double bond at alternative locations.

21.2. CONJUGATED DOUBLE BONDS

Very important classes of substances contain nonlocalized bonds between carbon atoms. These bonds may be regarded as resulting from the merger of several double bonds of the type described for ethylene in section 19.6.1. Most of these substances consist of organic molecules, but pure carbon in the form of graphite is also held together by nonlocalized carbon-carbon bonds.

21.2.1. *Butadiene*. The smallest molecule of this class is butadiene, whose basic structure formula is indicated by $CH_2{=}CH{-}CH{=}CH_2$, and more explicitly, in the alternative *cis* and *trans* forms, by

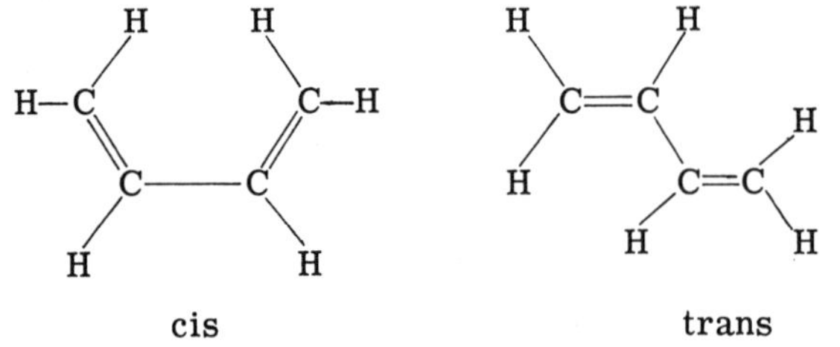

(21.4)

The cis and trans forms differ by a 180° rotation of one-half of the molecule around the central carbon-carbon bond. The fact that the two forms have different spectra and have planar structure shows that the rotation about the central bond is not free and, therefore, that this bond is not an ordinary single bond.[2] In fact, the internuclear C—C distance is 1.46 Å for this bond, as compared to 1.54 Å for a single bond and to 1.34 Å for the double bond in ethylene. Moreover, the measured total bond energy of butadiene exceeds by about 0.15 eV the energy calculated on the basis of the structures (21.4). These facts indicate that the central C—C bond has a partial double-bond character. This character results, as we shall see, from the existence of two adjacent double bonds. Double bonds which are separated by a single C—C bond are called *conjugated double bonds*.

According to the structure formulae (21.4) each of the conjugated double bonds has the structure of the ethylene double bond. Recall from section 19.6.1 that one of the electron pairs in the double bond of ethylene is assigned to a π orbital; this molecular orbital may be represented as a superposition of p_z atomic orbitals when the z axis is perpendicular to the plane of the molecule. Notice now that in butadiene the p_z orbitals of the two middle atoms of the carbon chain overlap each other as well as the orbitals of the end atoms. A nonlocalized bond is thus formed from the p_z orbitals of all carbon atoms in the chain.

Four π molecular orbitals can be formed starting from the p_z atomic orbitals of the four carbon atoms. The orbital of lowest energy has bonding action throughout the molecule. The molecular orbital with second lowest energy has a node between the two central carbon atoms and is therefore antibonding in this region. The other two orbitals have higher energy and remain unoccupied in the ground-state configuration of the molecule.

In chemical language the butadiene molecule is regarded as stabilized by re-

---

2. The cis and trans forms of butadiene convert into each other rapidly at room temperature, but the rate of conversion is much slower than the frequencies of vibration. Therefore, separate frequencies appear in the spectra of the two forms.

sonance between structures represented by $CH_2{=}CH{-}CH{=}CH_2$ and $CH_2{-}CH{=}CH{-}CH_2$. The central C—C bond has only weak double-bond character because the structure $CH_2{-}CH{=}CH{-}CH_2$ is incompletely bonded and participates in the resonance with small amplitude.

21.2.2. *Benzene*. The benzene molecule, C_6H_6, consists of a closed ring of six CH groups with three conjugated double bonds. These groups are welded into a regular hexagon by a nonlocalized bond whose electrons are distributed uniformly all around the ring. The total bond energy of benzene exceeds by 1.5 eV the value calculated on the basis of three single and three double bonds. The stability of benzene results from its complete hexagonal symmetry which is explained in turn by a bonding arrangement distributed uniformly all over the ring.

The symmetry of the molecule simplifies greatly the calculation of wave functions in the MO-LCAO approximation. In this approximation the wave function of each π electron may be indicated by

$$\psi = \sum_{k=1}^{6} c_k u(\vec{r} - \vec{R}_k), \tag{21.5}$$

where $u(\vec{r} - \vec{R}_k)$ indicates the p_z orbital of the carbon atom whose nucleus lies at $\vec{R}_k$. (Here we assume for simplicity that the wave functions have been modified to be orthogonal to those of the adjacent atoms; that is, such that the overlap integrals vanish.) We want to write the mean value of the energy of an electron with the wave function (21.5) in a form analogous to equation (16.19) in order to derive from it a Schrödinger equation. The coefficients c_k of (21.5) correspond to the probability amplitudes $\langle Ns \mid a \;\; J \;\; M_J\rangle$ of the state representation (16.15). The energy E_0 of a p_z electron in a single carbon atom corresponds to the energies E_{Ns} in equation (16.19). The perturbation energy is represented here by the energy $-V$ which binds each atom to its neighbors on either side. Thus we write

$$\begin{aligned}\langle \mathcal{E} \rangle &= \sum_{kk'} c_k^* [E_0 \, \delta_{kk'} - V(\delta_{k+1,k'} + \delta_{k-1,k'})] c_{k'} \\ &= E_0 \sum_k |c_k|^2 - V \sum_k c_k^* \, (c_{k+1} + c_{k-1}).\end{aligned} \tag{21.6}$$

Note that $-V$ is in essence the contribution of the second bond to the energy of ethylene and that a negative value of the product $c_k^* \, c_{k+1}$ yields a positive contribution to the energy (21.6) and thus represents an antibonding effect. Owing to the ring shape of the molecule, the index k is defined modulo 6, that is, for $k = 6$, $c_{k+1} \equiv c_1$.

The Schrödinger equation derived from (21.6) and analogous to (16.20) is

$$\frac{\partial \langle \mathcal{E} \rangle}{\partial c_k^*} = E_0 c_k - V(c_{k+1} + c_{k-1}) = c_k \mathcal{E}. \tag{21.7}$$

The cyclic symmetry of the molecule is reflected in the invariance of the equation under the substitution of the index k by k + m, where m is any integer.

It follows (in accordance with Appendix A) that (21.7) is solved by taking c_k to be an exponential function of k,

$$c_k = \sqrt{1/6}\, e^{ik\alpha}. \tag{21.8}$$

Here the coefficient $\sqrt{1/6}$ is required by the normalization condition $\Sigma_{k=1}^{6} |c_k|^2 = 1$. The fact that the index k is defined modulo 6 requires that

$$\alpha = \frac{2\pi n}{6}, \tag{21.9}$$

where n is an integer and serves as a quantum number.

The coefficients (21.8) satisfy the Schrödinger equation (21.7) with

$$\mathcal{E} = E_0 - 2V\cos\alpha. \tag{21.10}$$

The condition (21.9) upon the values of α sets the energy eigenvalues at

$$\mathcal{E}_n = E_0 - 2V\cos(1/3\,\pi n). \tag{21.11}$$

Thus the Schrödinger equation (21.7) has six independent eigenfunctions, characterized by the six different values of n:

$$n = 0, \pm 1, \pm 2, 3. \tag{21.12}$$

Other values of n yield the same eigenvalues and eigenfunctions. The eigenfunctions (21.5) with the coefficients (21.8) are examples of symmetry adapted wave functions. They are real for n = 0 and n = 3, but complex for n = ±1 and n = ±2. The electron flux $\vec{\Phi}$ defined by equation (11.16) is accordingly nonzero for n = ±1 or ±2 and runs around the ring in opposite directions depending on the sign of n.

Equation (21.11) represents a set of four different energy levels, namely,

$$\begin{aligned} \mathcal{E}_0 &= E_0 - 2V, \\ \mathcal{E}_1 = \mathcal{E}_{-1} &= E_0 - V, \\ \mathcal{E}_2 = \mathcal{E}_{-2} &= E_0 + V, \\ \mathcal{E}_3 &= E_0 + 2V, \end{aligned} \tag{21.13}$$

two of which are doubly degenerate. Since a total of six electrons belong in the π orbitals, the ground-state configuration includes one pair of electrons in the $\mathcal{E}_0$ level, and two pairs in the degenerate level $\mathcal{E}_{\pm 1}$. The total energy of the six electrons amounts then to

$$\mathcal{E}_0 + \mathcal{E}_1 + \mathcal{E}_{-1} = 6E_0 - 8V, \tag{21.14}$$

in this approximation. This result may be compared with the energy $6(E_0 - V)$ which would result from three nonconjugated ethylene π bonds. The difference, $-2V$, equals the ethylenic π bond strength of an additional pair of electrons in this approximation and represents the extra bond strength due to conjugation.

21.2.3. *Chains.* Chains of conjugated double bonds occur with various lengths in numerous organic molecules. Chains of CH groups represented by the general formula $-(CH{=}CH)_n-$ are called *polymethine* groups. The nonlocalized bonds which weld these chains together become increasingly uniform as the chain length increases; that is, the chains have no spot as weak as the one in the middle of butadiene. As the chain length increases, the density of electrons forming the nonlocalized bond can oscillate with lower frequencies; this trend is common to all mechanical systems of increasing size. Specifically, polymethine chains with more than four conjugated double bonds have characteristic electronic absorption bands in the visible or infrared spectrum. Many organic dyes consist of molecules with such polymethine chains.

The single-electron energy levels and wave functions of polymethine chains can be calculated in the MO-LCAO approximation by the same method used for benzene. The Schrödinger equation may be cast in the form (21.7), which remains invariant under replacement of k by k + 1, except when k has its lowest or its highest value at the beginning or at the end of the chain.[3] The eigenvalues maintain the general form (21.10), but the possible values of α are determined by the boundary conditions at the two ends of the chain as well as by the length of the chain. For a chain of 2N carbon atoms, with N conjugated double bonds, there are 2N energy eigenvalues corresponding to 2N values of α. All their eigenfunctions have zero flux since the chain does not form a circuit. For this reason no energy level is degenerate.

The spacing of the energy levels decreases with increasing length of the chain, since an increasing number of levels lie in the range from $E_0 - 2V$ to $E_0 + 2V$. The visible or infrared absorption spectrum of long chains is due to transitions between such closely spaced levels. The close spacing of levels also makes the molecule sensitive to external disturbances.

In the ground-state configuration, the 2N π electrons occupy in pairs the lower half of the set of 2N energy levels. It should be noted that a finite energy gap between the lower and upper half of the spectrum often persists even for very long chains, whereas the gap does not appear in our approximation. When there is such a gap, the electronic absorption spectrum does not shift indefinitely toward low frequency with increasing length of the chain. This means that conjugation has not completely wiped out the difference between successive bonds of the chain.

21.2.4. *Lattices.* Conjugated double bonds are also a main feature of the polycyclic aromatic hydrocarbons. These molecules consist of two or more benzene rings welded into a plane lattice with common sides. The structures

---

3. This analytical model actually oversimplifies the treatment of end effects. Its results are not correct in all respects.

of some substances of this type are shown schematically in Figure 21. 1. In these diagrams, vertices common to different rings indicate the positions of C atoms; vertices belonging to a single ring—that is, at the edge of the lattice—indicate the positions of CH groups. Each atom of the lattice is linked to the adjacent atoms by ordinary covalent localized bonds. All carbon atoms are further welded into a single system by a nonlocalized bond formed by π electrons.

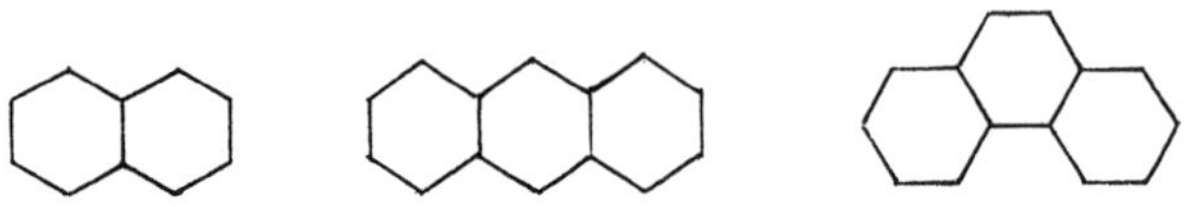

naphthalene anthracene phenanthrene

Fig. 21. 1. Schematic structure of some carbon lattices.

Conjugated bonds also occur in a great variety of other polycyclic molecules. Often, CH groups are replaced by nitrogen atoms, and rings may consist of five atoms only. There are also combinations of rings and of polymethine chains again welded into single systems by nonlocalized bonds. Figure 21. 2 shows, as an example, the structure formula of tetrapyrrole. This basic ring, with an iron or magnesium ion added in the middle and with different minor modifications, constitutes the pigment and key functional element of hemoglobin and of chlorophyll.

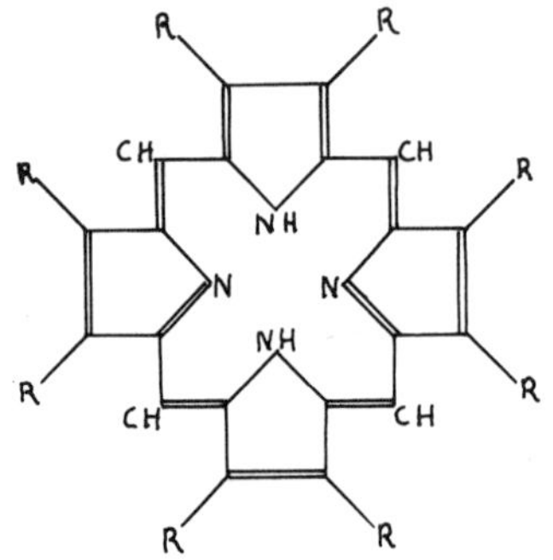

Fig. 21. 2. Structure formula of tetrapyrrole ring. Numerous substances have this structure with various radicals at positions R. Carbon atoms lie at unmarked corners of the network of conjugated bonds.

Lattices of hexagonal carbon rings of any size exist up to macroscopic dimensions. A large lattice consists almost entirely of carbon atoms since only the carbons at the edge of the lattice have a free valence for bonding with hydrogen or other atoms. In the limit of macroscopic size, a lattice of carbon atoms constitutes a layer of a graphite crystal. For our purposes, we will regard such a layer as infinitely extended over a plane. The stacking up of layers will be discussed in chapter 22.

The properties of a graphite layer can be derived in part from our previous discussion of chains of conjugated double bonds. The different energy levels of π electrons become increasingly close as the number of these electrons in the system increases. Indeed, the separation of the levels vanishes in an infinite lattice with infinitely many π electrons. Accordingly, the system acquires the ability to absorb radiation of low frequency down to zero in transitions between adjacent energy levels. Absorption of energy from a static (zero frequency) field characterizes the system as a conductor. Graphite actually behaves as a conductor at ordinary temperatures.

Closer examination of the properties of graphite, especially at low temperatures, shows that it has some of the characteristics of a semiconductor. The spectrum of electronic energy levels of a semiconductor is continuous over large energy ranges which are, however, separated by critical gaps. Calculation of the energy levels of π electrons in graphite (section S21.4) shows that these levels are effectively separated into two bands. Each of the two bands contains as many eigenstates as there are π electrons in the lattice; the states of the lower band have bonding character, those of the upper band are antibonding. In the limit of low temperature, where thermal agitation is disregarded, the system of π electrons is in its ground-state configuration. The states of the lower band are filled, those of the upper band are unoccupied; graphite is not a conductor here. As the temperature increases, thermal agitation supplies sufficient energy to raise an appreciable fraction of electrons to the upper band. In this temperature range the conductivity of graphite increases with increasing temperature.

21.3. THE METALLIC BOND

In the examples described so far, nonlocalized bonds are subsidiary to an underlying system of localized covalent, or partially ionic, bonds. The localized bonds determine the kind, number, and relative positions of two or more adjacent atoms among which a nonlocalized bond can be formed. On the other hand, nonlocalized bonds also occur without any undergirding by localized bonds; in particular, they provide the mechanism of aggregation of metal atoms.

A metal consists of a three-dimensional lattice of atoms whose valence electrons form a single system and move throughout the lattice in all directions. Owing to the sharing of all the valence electrons by the lattice atoms, the energy of the system is lower than the sum of the energies of the separated atoms. This reduction of energy constitutes the *metallic bond*. The bond energy of a metal lattice is of the order of a few eV per pair of valence electrons. Conjugated double bonds are intermediate between covalent and metallic bonds. The characteristic conductivity of metals is accounted for in the independent electron approximation on the basis of the continuous spectrum of energy levels of single electrons. As for graphite, the energy levels merge into a continuum because each electron can move over an infinitely large distance.

The metallic bond links together evenly an indefinitely large number of atoms with equal, or similar, electrical and chemical properties. (Atoms with different electrochemical properties form ionic bonds.) Typically, most of the chemical elements are metals; in other words, equal atoms of most elements form aggregates held together by a metallic bond under normal conditions of pressure and temperature. By contrast, a limited number of important elements, such as nitrogen and oxygen, are stable when they form small molecules with covalent bonds. To illustrate the conditions that favor the formation of metallic bond one may compare two elements, like Na and O, both of which form diatomic molecules in their gaseous state.

As noted at the beginning of section 19.4, Na atoms form weakly bonded molecules because they consist of large ionic cores with loosely attached electrons. Conversely, oxygen atoms have small ionic cores and form tightly bonded molecules. Upon condensation, the opportunity of sharing electrons with many neighbors is energetically advantageous for the Na atom because the bond with any one neighbor would be weak anyhow. The opposite circumstance prevails for oxygen. Indeed, nonmetallic elements have small ionic cores and belong in the upper right-hand corner of the periodic table. Heavier elements have larger ionic cores and more weakly bound valence electrons.

The wave functions of electrons participating in a metallic bond can be studied in the independent electron approximation by solving an appropriate Schrödinger equation. The potential in this equation is the sum of the potentials representing the attraction of the electron by all ionic cores. The resulting potential is periodic because the positions of the ionic cores which yield greatest stability form a regular crystal lattice. Treatment of the Schrödinger equation with a periodic potential by the method of Appendix C and section S12.8 shows that each of its eigenfunctions has the form of a plane wave whose amplitude is modulated with the periodicity of the lattice. That is, the eigenfunctions have the form

$$\psi_{E\vec{K}}(\vec{r}) = \exp(i\vec{k}\cdot\vec{r})\, v_{\vec{K}}(\vec{r}) \tag{21.15}$$

with the condition

$$v_{\vec{K}}(\vec{r} + m\vec{a} + n\vec{b} + p\vec{c}) = v_{\vec{K}}(\vec{r}). \tag{21.16}$$

Here $\vec{k}$ is a wave vector which serves, much as a quantum number, to characterize the wave function; $\vec{a}$, $\vec{b}$, and $\vec{c}$ are the lattice vectors which define the size and shape of a crystal cell, and m, n, and p are arbitrary integers as in equation (4.10). The relationship between the metal wave function (21.15) and the wave functions of the electrons of conjugated double bonds is described in section S21.4.

Each energy eigenvalue is a function of the wave vector $\vec{k}$, much as the energy eigenvalues (21.11) for benzene are functions of the quantum number n. As explained in section S21.4, all eigenfunctions can be obtained considering only a limited range of values of $\vec{k}$, such that $|\vec{k}\cdot\vec{a}|$, $|\vec{k}\cdot\vec{b}|$, and $|\vec{k}\cdot\vec{c}|$ remain smaller than π. Apart from the values of $\vec{k}$, the energy levels and their wave functions can be classified to some extent with reference to a separate-atom limit. One may consider a crystal lattice of atoms in the limit of very large internuclear distances. In this limit a single electron has degenerate levels in which it is attached to one or another atom in an energy eigenstate with given quantum numbers. As the internuclear distance is reduced, each degenerate level splits into a band of levels with different vectors $\vec{k}$. Different bands correspond to different quantum numbers of the separate-atom limit.

The energy ranges spanned by several bands usually overlap. In the ground-state configuration of the metal the valence electrons occupy in pairs the lowest available energy levels, whether these levels belong to one band or to different bands. The energy of the metallic bond is obtained, in this approximation, by comparing the total energy of all valence electrons in the ground-state configuration with the sum of the energies of the valence electrons in the separate atoms.

SUPPLEMENTARY SECTION

S21.4. ELECTRON WAVE FUNCTIONS IN LATTICES

This section presents calculations based on the very rudimentary model used for benzene. The purpose is to show in the simplest form the effects of symmetry upon energy eigenstates.

21.4.1. *LCAO Treatment of Graphite Layers*. The plane lattice of a graphite layer consists of hexagonal benzene-type rings. A primitive cell of the lattice consists of two carbon atoms whose relative position is represented by the vector $\vec{d}$ in Figure 21.3. The mutual position of adjacent cells in the plane lattice is defined by two fundamental translation vectors $\vec{a}$ and $\vec{b}$, whereas three vectors are needed for a crystal (section 4.2.2). Each primitive cell is identified by a pair of integers (m, n); the positions of its two atoms may be indicated by $\vec{R}_{m,n} = m\vec{a} + n\vec{b}$ and $\vec{R}_{m,n} + \vec{d}$.

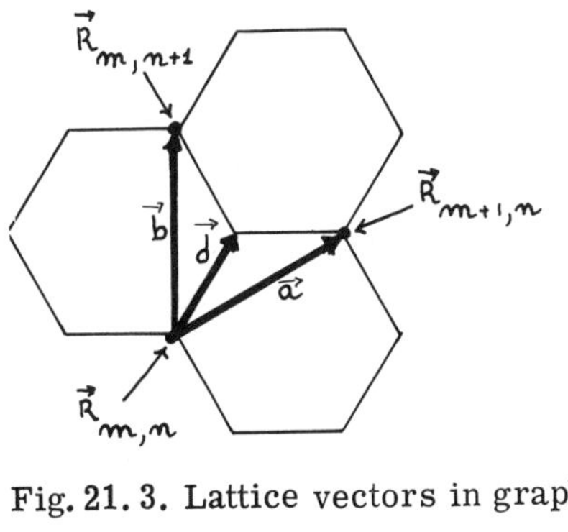

Fig. 21.3. Lattice vectors in graphite.

An LCAO π wave function analogous to the benzene wave function (21.5) has the form

$$\psi = \sum_{m,n} [c_{m,n}\, u(\vec{r} - \vec{R}_{m,n}) + c'_{m,n}\, u(\vec{r} - \vec{R}_{m,n} - \vec{d})]. \tag{21.17}$$

Here $c_{m,n}$ is the coefficient of the atomic orbital centered at $\vec{R}_{m,n}$, and $c'_{m,n}$ is the coefficient of the orbital centered at $\vec{R}_{m,n} + \vec{d}$. The mean energy of an electron with this wave function is represented by an expression analogous to (21.6), namely,

$$\begin{aligned} \langle \mathcal{E} \rangle = {} & E_0 \sum_{m,n} [|c_{m,n}|^2 + |c'_{m,n}|^2] \\ & - V \sum_{m,n} [c^*_{m,n} (c'_{m,n} + c'_{m-1,n} + c'_{m,n-1}) + c'^*_{m,n} \\ & \quad (c_{m,n} + c_{m+1,n} + c_{m,n+1})]. \end{aligned} \tag{21.18}$$

Each term of the type $-Vc^*_{m,n}\, c'_{m,n}$, or $-Vc^*_{m,n}\, c'_{m-1,n}$, stems from the interaction of two adjacent atoms in the lattice. The Schrödinger equation derived from (21.18) and analogous to (21.7) has the form of a system of equations

$$\begin{aligned} \frac{\partial \langle \mathcal{E} \rangle}{\partial c^*_{m,n}} &= E_0 c_{m,n} - V\,(c'_{m,n} + c'_{m-1,n} + c'_{m,n-1}) = c_{m,n}\, \mathcal{E}, \\ \frac{\partial \langle \mathcal{E} \rangle}{\partial c'^*_{m,n}} &= E_0 c'_{m,n} - V\,(c_{m,n} + c_{m+1,n} + c_{m,n+1}) = c'_{m,n}\, \mathcal{E}. \end{aligned} \tag{21.19}$$

The translational symmetry of the lattice is reflected in the invariance of the system (21.19) under substitution of the pair of indices (m, n) by any other pair of integers. Consequently, the system (21.19) is solved by taking $c_{m,n}$ and $c'_{m,n}$ to be exponential functions of m and n. We write in analogy to (21.8)

$$c_{m,n} = N^{-1/2}\, c e^{i(m\alpha + n\beta)}, \quad c'_{m,n} = N^{-1/2}\, c' e^{i(m\alpha + n\beta)}, \tag{21.20}$$

where c and c′ remain to be determined and N indicates the number of lattice cells, to be taken as ∞ at the end of the calculation. The parameters α and β are analogous to the parameter α in the benzene equation (21.8). Substitution of (21.20) in the system (21.19) reduces the Schrödinger equation to the form

$$\begin{aligned} E_0 c - V\,(1 + e^{-i\alpha} + e^{-i\beta}) c' &= c\, \mathcal{E}, \\ E_0 c' - V\,(1 + e^{i\alpha} + e^{i\beta}) c &= c'\, \mathcal{E}. \end{aligned} \tag{21.21}$$

The energy eigenvalues $\mathcal{E}$ are determined as functions of the parameters α and β by the condition that the determinant of the system (21.21) vanish, $(E_0 - \mathcal{E})^2 - V^2 |1 + \exp(i\alpha) + \exp(i\beta)|^2 = 0$, that is,

$$\mathcal{E} = E_0 \pm V |1 + e^{i\alpha} + e^{i\beta}|. \tag{21.22}$$

The energy $\mathcal{E}$ is a doubly periodic function of α and β, which may be mapped over the range $-\pi < \alpha \leq \pi$ and $-\pi < \beta \leq \pi$. For a finite lattice of N cells, boundary conditions would single out a discrete set of N pairs of values (α, β) analogous to (21.9).

Irrespective of boundary conditions, the values of $\mathcal{E}$ range over a pair of bands corresponding to bonding and antibonding states. The spectrum of bonding levels ranges from $E_0 - 3V$ (for $\alpha = \beta = 0$) to E_0. The antibonding levels range from E_0 to $E_0 + 3V$. The value E_0 of the energy is reached only for special pairs of values $\alpha = \pm 2\pi/3, \beta = \mp 2\pi/3$; otherwise the two bands of the pair are separated by a gap. The two bands are in contact only at isolated points of the (α, β) plane, in this approximation. This accounts for the semiconductor property of graphite mentioned at the end of section 21.2.

21.4.2 *Connection with Bloch Wave Functions*. Substitution of the expression (21.20) of the coefficients $c_{m,n}$ and $c'_{m,n}$ into the wave function (21.17) yields

$$\psi = N^{-1/2} \sum_{m,n} e^{i(m\alpha + n\beta)} [cu(\vec{r} - \vec{R}_{m,n}) + c'u(\vec{r} - \vec{R}_{m,n} - \vec{d})]. \quad (21.23)$$

We want to express the wave function ψ, insofar as possible, as a single periodic function of the position $\vec{r}$ of the electron with the periodicity of the lattice. The function $\Sigma_{m,n} u(\vec{r} - \vec{R}_{m,n})$ possesses this periodicity, but the coefficient $\exp[i(m\alpha + n\beta)]$ does not. This coefficient will then be expressed in suitable form.

To this end we interpret the parameters α and β as the components of a vector $\vec{k}$ along the fundamental translation vectors; that is, we set

$$\alpha = \vec{k}\cdot\vec{a}; \qquad \beta = \vec{k}\cdot\vec{b}. \quad (21.24)$$

With this definition of $\vec{k}$ the exponent in (21.23) becomes

$$m\alpha + n\beta = m\vec{k}\cdot\vec{a} + n\vec{k}\cdot\vec{b} = \vec{k}\cdot\vec{R}_{m,n}. \quad (21.25)$$

With this substitution the wave function (21.23) becomes

$$\begin{aligned}\psi &= N^{-1/2} \sum_{m,n} \exp(i\vec{k}\cdot\vec{R}_{m,n}) [cu(\vec{r} - \vec{R}_{m,n}) + c'u(\vec{r} - \vec{R}_{m,n} - \vec{d})] \\ &= N^{-1/2} e^{i\vec{k}\cdot\vec{r}} \sum_{m,n} \exp[-i\vec{k}\cdot(\vec{r} - \vec{R}_{m,n})] [cu(\vec{r} - \vec{R}_{m,n}) \\ &\qquad + c'u(\vec{r} - \vec{R}_{m,n} - \vec{d})].\end{aligned} \quad (21.26)$$

The last summation is a periodic function of $\vec{r}$, because it resumes the same value whenever $\vec{r}$ is shifted by a multiple of either lattice vector $\vec{a}$ or $\vec{b}$. We call this summation $N^{1/2} v_{\vec{k}}(\vec{r})$, whereby (21.26) takes the form of a *Bloch wave function*[4]

$$\psi = e^{i\vec{k}\cdot\vec{r}} v_{\vec{k}}(\vec{r}). \quad (21.27)$$

---

4. Functions of the type (21.26) are called instead *Wannier wave functions*.

The range of variation of the vector $\vec{k}$, that is, of the parameters α and β, is illustrated conveniently by mapping it in the space of the reciprocal lattice.

The base vectors of the reciprocal lattice have been defined for a three-dimensional lattice by equation (4.13). Here, we have a two-dimensional space of reciprocal vectors with the base vectors $\vec{A}$ and $\vec{B}$ defined by

$$\vec{A}\cdot\vec{a} = 2\pi \quad , \quad \vec{A}\cdot\vec{b} = 0, \qquad \vec{B}\cdot\vec{a} = 0 \quad , \quad \vec{B}\cdot\vec{b} = 2\pi, \tag{21.28}$$

and shown in Figure 21.4b. When the vector $\vec{k}$ is represented by an arrow as in Figure 21.4b, its tip may lie anywhere within the hexagon shown in the figure. Any vector $\vec{k}'$ whose tip lies outside the hexagon yields a value of $\exp(i\vec{k}'\cdot\vec{R}_{m,n})$ equal to the value obtained from a vector $\vec{k}$ within the hexagon which differs from $\vec{k}'$ by a multiple of either $\vec{A}$ or $\vec{B}$ or both. For this reason the vectors $\vec{k}$ within the hexagon yield all possible values of the coefficients $\exp(i\vec{k}\cdot\vec{R}_{m,n})$ in equation (21.26). The hexagon about the origin of the reciprocal lattice space, shown in Figure 21.4b, is called the first Brillouin zone.

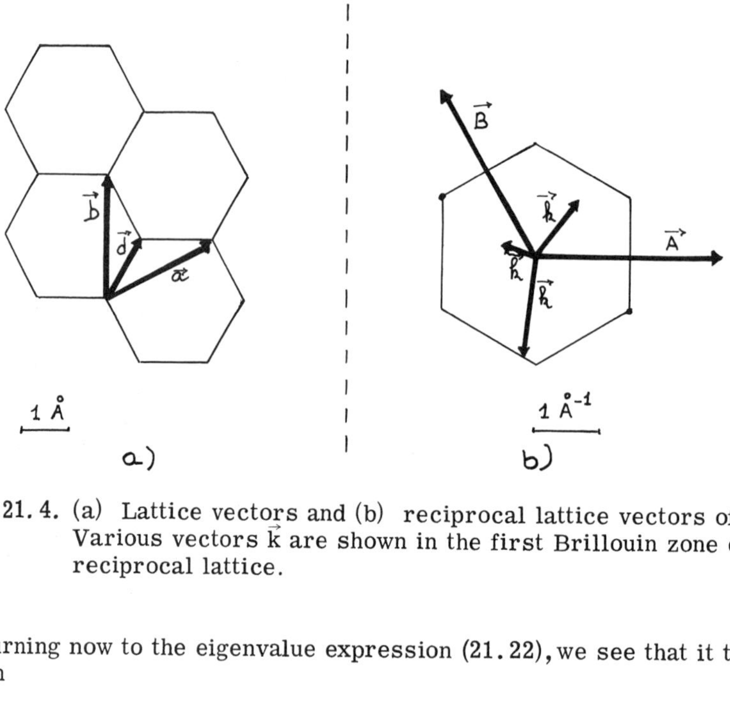

Fig. 21.4. (a) Lattice vectors and (b) reciprocal lattice vectors of graphite. Various vectors $\vec{k}$ are shown in the first Brillouin zone of the reciprocal lattice.

Returning now to the eigenvalue expression (21.22), we see that it takes the form

$$\mathcal{E} = E_0 \pm V\,|1 + e^{i\vec{k}\cdot\vec{a}} + e^{i\vec{k}\cdot\vec{b}}|. \tag{21.29}$$

The coefficient of V vanishes in this formula only when the tip of the vector $\vec{k}$ coincides with one of the vertices of the Brillouin zone in Figure 21.4b.

For any other value of $\vec{k}$, equation (21.29) yields two distinct energy levels which correspond respectively to an antibonding and to a bonding state. The highest and lowest energy levels of the pair of bands correspond to $\vec{k} = 0$.

21.4.3. *Generalization.* In the study of molecules the LCAO wave functions have been used for orientation purposes as crude approximations to independent-electron molecular orbitals. By the same token, the Bloch wave functions (21.27) obtained from the LCAO wave functions (21.26) represent only crude approximations to independent-electron wave functions in a crystal lattice. A better approximation is obtained by considering a more general type of wave function. Thus far we have used the same atomic orbitals u for all energy levels of the band. The wave functions can be generalized by selecting an appropriate atomic orbital u for each wave vector $\vec{k}$. This amounts to calculating the periodic factor $v_{\vec{k}}(\vec{r})$ of the Bloch wave function by writing and solving a Schrödinger equation for the lattice electrons without starting from a single given carbon atomic orbital u. This is, in fact, the procedure outlined for metals in section 21.3.

Independent-electron wave functions for crystal lattices, three-dimensional as well as two-dimensional, nonmetal as well as metal, are generally calculated without use of the LCAO approximation. The fact that the eigenfunctions of the Schrödinger equation have the Bloch form (21.27) follows from the treatment of equations with periodic potentials outlined in Appendix C. The vector $\vec{k}$ of the Bloch wave function ranges over a Brillouin zone whose shape is determined by the crystal symmetry and which is a polyhedron for three-dimensional lattices. The factor $v_{\vec{k}}(\vec{r})$ of the Bloch wave function is generally calculated by Fourier expansion, that is, by expressing the Schrödinger equation as a system of linear equations analogous to, but naturally more complicated than, equation (C.13).

The internuclear distances in a crystal, and in particular the magnitudes of the lattice vectors $\vec{a}$, $\vec{b}$, and $\vec{c}$, play the same role as the internuclear distance $\vec{R}$ does for diatomic molecules. The solution of the Schrödinger equation for the electronic motion is carried out treating $\vec{a}$, $\vec{b}$, and $\vec{c}$ as parameters. In the limit of large magnitudes of these vectors, the calculation reduces to a problem for separate atoms and yields discrete energy levels. Recall that for a homonuclear diatomic molecule each level of the separate atoms splits when the atoms approach into a pair of levels of the molecule; one level usually corresponds to a bonding and one to an antibonding state. Similarly, for crystal lattices the discrete energy levels of separate atoms split upon shortening of the internuclear distances. Since a crystal consists of infinitely many atoms, each separate atom level splits into a continuous band of crystal levels.

CHAPTER 21 PROBLEMS

21.1. The structure formula H—C(=O)—O—H represents a molecule of formic acid. This substance is an acid because it tends to dissociate, in water solution, into H^+ and a negative radical ion $HCOO^-$. By contrast, the structure

formula

$$\underset{\underset{H}{|}}{\overset{\overset{H}{|}}{H-C-O-H}}$$

represents a molecule of methanol, a substance with no

acid character—that is, which does not split off an H^+ ion when in aqueous solution. The acid character of the $HCOOH$ molecule is attributed in the language of chemistry to resonance between two structure formulae for the $HCOO^-$ ion, that is, to formation of a particularly stable nonlocalized bond, which would not occur for H_3CO^-. Which are the resonating structure formulae?

21.2. Adapt to butadiene the treatment of the nonlocalized bond formed by p_z orbitals in benzene.

SOLUTIONS TO CHAPTER 21 PROBLEMS

21.2. The resonating formulae for the $HCOO^-$ molecule ion are

$$\text{H-C}\overset{\displaystyle\nearrow^{O}}{\underset{\displaystyle\searrow_{O^-}}{}} \quad \text{and} \quad \text{H-C}\overset{\displaystyle\nearrow^{O^-}}{\underset{\displaystyle\diagdown_{O}}{}}.$$

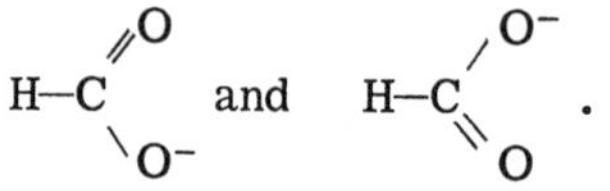

Each of these formulae shows an asymmetry between the two CO bonds whereas both O are bonded to C directly and to no other atom. Note that the symmetry of the two O is spoiled by addition of an H^+ to form a bond with either of them. In the language of the independent electron approximation, we say that the $HCOO^-$ ion is stabilized by formation of a nonlocalized bond, with a molecular orbital that extends over the OCO chain. No such bond formation could occur in a H_3CO^- molecule ion.

21.2 The benzene equations (21.6) and (21.7) apply to butadiene, with the following modifications. The Σ_k extends only over $k = 1, \ldots, 4$; the terms with c_{k-1} are missing for $k = 1$, and the terms with c_{k+1} are missing for $k = 4$. Adaptation of the benzene treatment is achieved by the analytical device of regarding the Σ_k as extending to all values of k while enforcing the condition that the coefficients c_k vanish for $k = 0$ modulo 5. Exponential solutions of the form (21.8), $\exp(ik\alpha)$, do not meet this condition, but a super-position of $\exp(\pm ik\alpha)$ does. Set $c_k = N \sin k\alpha$, where N is a normalization coefficient and α is a multiple of $\pi/5$ to ensure that $c_0 = c_5 = 0$. Substitution of $c_k = N \sin(kn\pi/5)$ satisfies the equation (21.7), with the eigenvalue $E_n = E_0 - 2V \cos(n\pi/5)$. There are four nondegenerate eigenvalues, with $n = 1, 2, 3, 4$. The levels with $n = 1, 2$ are occupied by the four π electrons of the butadiene molecule.

This problem can also be solved directly by writing explicitly the four equations analogous to (21.7):

$$E_0c_1 - Vc_2 = c_1\mathcal{E},$$

$$-Vc_1 + E_0c_2 - Vc_3 = c_2\mathcal{E},$$

$$-Vc_2 + E_0c_3 - Vc_4 = c_3\mathcal{E},$$

$$-Vc_3 + E_0c_4 = c_4\mathcal{E}.$$

The eigenvalues $\mathcal{E}$ are determined by setting to zero the determinant of coefficients, $(E_0 - \mathcal{E})^4 - 3V^2(E_0 - \mathcal{E})^2 + V^4 = 0$. Solution of this equation yields in a first step $(E_0 - \mathcal{E})^2 = \frac{1}{2}(3 \pm \sqrt{5})V^2 = [\frac{1}{2}(1 \pm \sqrt{5})V]^2$. A second step gives the four roots corresponding to the four combinations of sign in the formula $\mathcal{E} = E_0 \pm \frac{1}{2}(1 \pm \sqrt{5})V$. For each of these eigenvalues the system is solved completely. Thus one finds the set of eigenvalues and eigenfunction coefficients

n	$\mathcal{E}$	c_1	c_2	c_3	c_4
1	$E_0 - 1.62V$	0.37	0.60	0.60	0.37
2	$E_0 - 0.62V$	0.60	0.37	−0.37	−0.60
3	$E_0 + 0.62V$	0.60	−0.37	−0.37	0.60
4	$E_0 + 1.62V$	0.37	−0.60	0.60	−0.37

Note the increasing number of alternations of sign, with increasing n, showing the increasingly antibonding character of the state.

22. macroscopic aggregates

22.1. CONNECTION BETWEEN MACROSCOPIC AND ATOMIC PROPERTIES

The description of nonlocalized bonds in chapter 21 completes the survey of the basic physical mechanisms that hold atoms and molecules together in condensed matter. A tremendous number of different materials with a countless variety of properties is formed from a limited number of chemical elements. This variety is made possible by the combined play of a few mechanisms of aggregation in different configurations and in different proportions.

The study of the structure and properties of materials is an endless task, subdivided in the many branches of chemical and solid-state physics. As noted in the introduction to Part IV, a detailed theory has been developed only for simple materials with high structural symmetry, primarily for crystals. In general, the analysis takes a phenomenological approach, expressing experimental properties in terms of semiempirical parameters. Atomic and molecular theory must then account for these parameters. This chapter gives a few illustrative examples of such a phenomenological approach.

The effects of forces among atoms and molecules vary greatly, depending on the state of condensation of a material. They are small for a rarefied gas of neutral molecules. In this case the properties of the gas are accounted for as properties of single molecules and by gas kinetic theory. With increasing density, the properties of single molecules are altered by collisions whose effects increase with continuity to the limit of condensation. On the other hand, in ionized gases (*plasmas*), major effects of interaction among electrons and molecules occur even at very low density, owing to the long-range effect of electric and magnetic forces.

Among the gross properties of condensed materials, the density is determined by the equilibrium internuclear distance R_e of nearest neighbors and by the steric arrangement of the atoms. The cohesive energy released upon formation of the aggregate is the sum of all bond energies within the aggregate, and is given essentially by the value of the interatomic potential function U(R) at $R = R_e$.[1] The curvature of the function U(R) at R_e accounts for the compressibility in the zero temperature limit. The compressibility is the reciprocal of the *bulk modulus*

$$B = -V \frac{dp}{dV}, \tag{22.1}$$

where V is the volume and p is the pressure. In the low-temperature limit, where the entropy is negligible, we have

1. An exact expression of the cohesive energy should take into account, besides $U(R_e)$, the lowest-energy eigenvalue $\frac{1}{2}\hbar\omega_i$ of all vibrational modes of the system. This correction is small for all but the lightest elements.

$$p = -\frac{dU(R)}{dV} = -\frac{dU(R)}{dR}\frac{dR}{dV}, \tag{22.1a}$$

and

$$B = V\left(\frac{d^2U}{dR^2}\right)_{R=R_e}\left(\frac{dR}{dV}\right)^2_{R=R_e} \tag{22.1b}$$

since $dU(R)/dR$ vanishes at $R = R_e$.

Examples of the evaluation of these properties for simple crystals are given in section 22.2. Section 22.3 gives a few examples of structures in which different bonding mechanisms coexist. It also indicates how the presence of polar molecules, particularly water, influences greatly the aggregation of ionic structures both in crystals and in liquids. Finally, section 22.4 indicates how some outstanding thermal and electrical properties of materials relate to molecular processes.

22.2. BONDING OF SIMPLE CRYSTALS

The structural properties of some simple crystals can be accounted for in terms of a single bonding mechanism. The basic mechanisms are: the covalent bond, the ionic bond, the metallic bond, and the van der Waals force, in addition to the more complex interactions of polar groups.

22.2.1. *Covalent Crystals.* Diamond is an outstanding example of structure with purely covalent bonds. In this case the basic data for determining the crystal properties can be obtained from the study of single C-C bonds in small molecules. The data are: bond energy 3.6 eV, internuclear distance $R_e = 1.54$ Å, and tetrahedral arrangements of bond directions around each atom as in methane (section 19.5). In this arrangement each carbon atom is at the center of a tetrahedron and is surrounded by four nearest neighbors at the vertices of the tetrahedron. One can verify that a cube with edge $4 \times \sqrt{1/3} \times R_e = 3.6$ Å contains eight atoms, and, therefore, a mass of $8 \times 12 \times 1.66 \times 10^{-24}$g. Consequently, 1 cm^3 of diamond contains 1.75×10^{23} atoms and $2 \times 1.75 \times 10^{23}$ bonds. It has a mass of 3.52 g and a cohesive energy of $3.50 \times 10^{23} \times 3.6$ eV $= 2.0 \times 10^{12}$ erg $= 48$ kcal. These results agree with the experimental properties of pure diamond, showing that the purely covalent bond between carbon atoms maintains its characteristics in the crystal.[2]

22.2.2. *Ionic Crystals.* Bonds which are partially heteropolar in small molecules, like NaCl, become essentially ionic in the crystal. This shift in bonding character permits an arrangement in which each ion is surrounded by a number of nearest neighbors all of the opposite charge. Such ionic bonds

---

2. It must be noted that there are few covalent structures as simple as diamond.

prevail in salts and oxides. In this case, the experimental properties of a crystal cannot be predicted from the bond data on small molecules. Instead, many crystal properties are correlated with remarkable success by a rather crude model.

Each ion is regarded as an almost rigid sphere of charge e or a small multiple of e. The spheres are packed against each other; the nearest neighbor distance, R, is a parameter of the model. The net Coulomb potential energy of a single ion due to the interaction with all its neighbors (near and far) is represented by $-\alpha e^2/R$. The coefficient α, called the *Madelung constant,* serves to represent the contribution of different ions at different distances. In the example of NaCl, each ion has six nearest neighbors of opposite charge at distance R, 12 next nearest of the same charge at distance $\sqrt{2}$ R, etc. The Madelung constant is then given by the series

$$\alpha = \frac{6}{1} - \frac{12}{\sqrt{2}} + \frac{8}{\sqrt{3}} - \frac{6}{2} + \cdots . \tag{22.2}$$

This type of series converges poorly; its successful calculation for the different types of ionic structures has required a nontrivial effort.

The Coulomb attraction which tends to reduce the distance R is resisted by the rigidity of the ionic spheres. Each ion of the crystal has complete-shell electron configuration. The rigidity of the spheres is understood as an effect of antibonding which increases rapidly as the internuclear distance R decreases. The antibonding effect is represented, in the language of chapter 19, by the potential function U(R). In the model of ionic crystals one represents U(R) for a pair of nearest neighbors by

$$U(R) = \lambda e^{-R/\rho} - e^2/R, \tag{22.3}$$

where λ and ρ are empirical parameters. While these parameters are fitted to experimental data for each specific crystal, the resulting function U(R) in the form (22.3) must be, and is, consistent with the predictions of molecular bond theory for the electronic structure of the interacting ions.

The total cohesive energy for a crystal consisting of N pairs of monovalent ions is given by

$$U_{tot}(\lambda, \rho, R) = N\left(z\lambda e^{-R/\rho} - \frac{\alpha e^2}{R}\right), \tag{22.4}$$

where z is the number of nearest neighbors. Other measurable quantities are the density and the bulk modulus of the crystal. The values of these three quantities determine the values of the parameters R, λ, and ρ. As a check of the theory one must verify that the potential energy has a minimum at the equilibrium distance R_e, that is,

$$\left(\frac{dU_{tot}}{dR}\right)_{R = R_e} = 0. \tag{22.5}$$

This condition is verified satisfactorily for a large number of crystals. Furthermore, it is possible to assign to each of the common ions an effective radius which remains the same independently of the combination in which the ion takes part. The equilibrium internuclear distances in different crystals are represented within a few percent by the sum of the effective radii of the ions. For example, the equilibrium distance for NaCl is 2.820 Å, as observed by X-ray diffraction. This value can be compared with the value 1.05 + 1.80 = 2.85 Å, which represents the sum of effective radii of Na^+ and Cl^- ions.

22.2.3. *Crystals of Inert Gases.* Van der Waals forces, though much weaker than chemical bonds, may be sufficient to hold together aggregates of inert atoms or molecules. The atoms of inert gases and small molecules, such as N_2 and O_2, form crystals at very low temperatures. Large molecules are held together even at normal temperature in aggregates such as paraffin.

The interatomic potential U(R) for pairs of rare gas atoms is known fairly accurately from atomic collision experiments. It is also known to some extent from calculations of the electronic energy of a pair of atoms, whose state must be antibonding because the atoms have a closed-shell configuration. The potential function U(R) is represented adequately by the *Lennard-Jones* analytic form

$$U(R) = 4\epsilon\left[\left(\frac{\sigma}{R}\right)^{12} - \left(\frac{\sigma}{R}\right)^{6}\right] \tag{22.6}$$

with the semiempirical parameters ϵ and σ. The attractive term of this potential has the same dependence on R as the van der Waals potential (18.11), and the constants in the two expressions must coincide. The repulsive term predominates at short distances and is analogous to the exponential term in the expression (22.4) for ionic crystals. The parameter σ represents the value of R at which the repulsive and attractive potentials cancel out. The minimum of the potential lies at $R = 1.122\,\sigma$, where $U(R) = -\epsilon$.

Starting from the Lennard-Jones potential, one calculates the cohesive energy, that is, the energy of all pairs of atoms in a crystal as a function of R. The minimum value of the cohesive energy corresponds to the internuclear equilibrium distance, R_e. This calculated distance is $1.09\,\sigma$ for all rare gases if one disregards the vibrational kinetic energy, as we have done for the ionic crystals. Actually, the existence of a nonzero mean kinetic energy of the nuclei increases the equilibrium distance by about 5% for neon. The experimental values of R_e range from 3.1 Å for neon to 4.4 Å for xenon, and the cohesive energy ranges from 0.02 to 0.17 eV per atom, in good agreement with the calculated values. The cohesive energy increases with atomic weight because the heavier atoms have lower excitation energies and are therefore more easily polarized.

22.3. AGGREGATES WITH MIXED BONDING

Molecular structures in the form of chains or sheets often stack into crystalline aggregates held together by van der Waals forces. A typical example is graphite whose sheetlike structure has been discussed in section 21.2.4. The interlayer distance is 3.35 Å. It is thus comparable to the interatomic equilibrium distance in inert gas crystals and is ~2.4 times as large as the C—C distance within each layer. The cumulative action of van der Waals forces over the whole macroscopic surface of each layer is sufficient to keep the crystalline aggregate stable at ordinary temperature.

Van der Waals forces operate universally between any two molecular aggregates. Molecular groups possessing a dipole moment exert much stronger forces upon other groups. The occurrence of polar forces is not universal but is, nevertheless, very common, particularly because of the widespread presence of water. The bonding action due to the electrostatic field of polar groups approaches in strength that of valence and ionic bonds. The polar action of water molecules is rather complicated in pure water, whether liquid or crystalline, but its role can be described more readily in the structure of some ionic crystals.

For purposes of orientation we may begin by estimating the electric potential due to the field of a single O—H group. At a distance r, along the axis of the O—H bond, this potential equals $2|\langle\vec{\mu}\rangle|/r^2$, where $|\langle\vec{\mu}\rangle|$ is the dipole moment given in Table 19.1. At a distance r = 2 Å, the potential amounts to ½ volt.

The binding between ions and the polar groups of water stabilizes many ionic solutions with respect to precipitation. An estimate of the bond between an ion and the surrounding water can be obtained through macroscopic considerations. The electric field strength $E = ze/r^2$ at distance r from an ion of charge z is reduced in water to $ze/\epsilon r^2$, where $\epsilon \sim 80$ is the dielectric constant of water. The bond energy between an ion of radius r_0 and the surrounding water (*hydration energy*) may be estimated as the difference of the energy of the ion's electric field in vacuum and in water. The vacuum energy is

$$\int_{r_0}^{\infty} \frac{E^2}{8\pi} 4\pi r^2 dr = \tfrac{1}{2}(ze)^2 \int_{r_0}^{\infty} \frac{dr}{r^2} = \tfrac{1}{2}\,\frac{z^2e^2}{r_0}\,. \qquad (22.7)$$

Practically, this energy is the hydration energy since the energy of the ion in water is only 1/6400 of it. In the example of Cl^-, the value $r_0 = 1.8$ Å, derived from the equilibrium distance in ionic crystals, yields a value of 4.0 eV for the expression (22.7). This value coincides with the experimental hydration energy of Cl^- within 10%. Similar results are obtained for other negative ions. Estimates for positive ions obtained from (22.7) are very large because of the small values of r_0. However they generally exceed the experimental values by nearly a factor of 2. For example, (22.7) yields 38 eV for Mg^{++} with $r_0 = 0.75$ Å, whereas the experimental value is 20 eV. Presumably, the macroscopic model applies poorly when extended to the small radii of positive ions. Recall that the macroscopic analysis of polarization energy assumes a linear response of the atom to the field; the atomic perturbation theory (sections 17.1 and 18.1) makes the same assumption when carried out to lowest order. Neither treatment is expected to be accurate for the very strong field of a positive ion at very short range.

Among other limitations, the macroscopic model does not allow for the difference in size of the hydrogen and oxygen atoms in the water molecule. Since the hydrogen atoms are very small, their average positive charge approaches other groups more closely than does the negative charge of the larger oxygen ions. Consequently, negative ions are more tightly bound to hydrogen than positive ones are to oxygen. The preferential binding of polar groups on the hydrogen side is called *hydrogen bond*.

Bearing in mind these estimates of the binding of ions to water, we shall now describe the role of water molecules in a simple example, the ionic crystals of $MgCl_2$. This substance has an anhydrous crystalline form, in which each Mg^{++} is surrounded symmetrically by 8 Cl^- ions arranged at the vertices of a cube. However, the substance crystallizes from a water solution in the hydrated form $[Mg(H_2O)_6]^{++}(Cl)_2{}^{--}$. In this ionic structure each magnesium ion remains at the center of a cube and is surrounded by the six oxygens of the water molecules, in directions pointing toward the vertices of an octahedron, that is, toward the centers of a cube's faces. The twelve hydrogen atoms lie in pairs on the faces of a cube. The eight chlorine ions, surrounding the $[Mg(H_2O)_6]^{++}$ complex, are attracted to its positive charge and occupy positions particularly close to the hydrogen atoms. Since there are twelve hydrogens to eight chlorines, the symmetry of the crystal is not exactly cubic; the crystal is stabilized by optimizing the hydrogen bond.

The calculation of the bonding energy of the hydrated crystal can be approached as a problem of electrostatic interaction among the ionic charges and the polar charge distribution within the water molecules. This calculation is analogous to the calculation of the Madelung constant for purely ionic crystals, though of course it is more complicated. We do not pursue this calculation since our aim is simply to indicate how water enters into an aggregate. In the specific example of $MgCl_2$, the role of water may also be filled by NH_3. The anhydrous form absorbs ammonia to form a complex $Mg(NH_3)_6Cl_2$.

Whether a positive ion is surrounded by water or other polar molecules, the effect of the association is to substitute a small metal ion by a much larger aggregate with the same net charge. This larger aggregate can then pack tightly with negative ions with the help of the hydrogen bond mechanism. Furthermore, variations of the orientation, or even of the number, of the polar molecules make the shape and size of the positive aggregate exceedingly flexible and thus capable of forming a stable network of bonds with negative ions of various sizes and shapes.

22.4. THERMAL, ELECTRIC, AND MAGNETIC PROPERTIES

In a crystalline aggregate the nuclear motion is entirely vibrational, as a rule. The vibrational motion can be described, in first approximation, as a superposition of independent normal modes, as was done for molecules in chapter 20. The frequency spectrum of these normal modes extends toward zero frequency as the crystalline lattice attains macroscopic size. The heat capacity of the aggregate is primarily attributable to this vibrational motion and is therefore represented by the same equation (20.19) which applies to molecules. According to this equation, the contribution to the heat capacity of each vibrational motion with frequency ν_i cm^{-1} increases rapidly over the temperature range in which k_BT is comparable to the vibrational quantum $hc\nu_i$. Thus measurement of the heat capacity as a function of temperature provides information on the frequency spectrum of vibrational modes. At sufficiently high temperature all normal modes contribute and the heat capacity of a material attains the value $3N_Ak_B$ per mole of atoms, where N_A is Avogadro's number.

For noncrystalline materials some of the heat capacity accrues from translational and rotational motion. Electronic excitation does not normally contribute appreciably to heat capacity because its energy levels are too high to be reached by thermal agitation. In metals there exist extremely low electronic excitation levels within the reach of thermal agitation. However, excitation to these levels involves only a small fraction of all electrons. This is apparent in the independent electron approximation, where only the electrons belonging in the highest occupied levels can be excited by thermal agitation into vacant levels. For this reason the electronic contribution to the heat capacity of a macroscopic material is not very significant at ordinary temperatures even in metals.[3]

Whereas the concept of heat capacity applies to small molecular aggregates as well as to macroscopic ones, thermal conductivity is a property of large aggregates only. This is because thermal conductivity is the ratio of heat flow to temperature gradient, and temperature gradient is not defined within a molecule. In a crystalline aggregate we have regarded thermal agitation as the excitation of independent vibrational normal modes each of which extends over the whole macroscopic lattice. However, this schematization is inconsistent with the existence of a temperature gradient. The phenomenon of heat conduction can be treated only in a higher approximation which takes into account the coupling of different vibrational modes. In fact, heat conduction, being a transport phenomenon, must be described in terms of nonstationary states.

The conduction of electricity by metals is also a transport phenomenon which is foreign to the physics of small molecules and must be treated in the context of nonstationary states. Indeed, the concept of a potential difference applied to the terminals of a conductor is incompatible with the concept of a stationary state of the whole electronic system of the conductor. In a nonstationary treatment, an external electric field keeps exciting the valence electrons of a metal to states of higher and higher momentum in the direction of the field. This effect of the field is compensated by the transfer of electronic excitation to nuclear vibration. The interaction between electronic and nuclear motion, which is disregarded in the Born-Oppenheimer approximation, plays the same role in ohmic conduction as the interaction among normal modes of nuclear vibration plays in thermal conduction. In general, the treatment of

---

3. The lack of appreciable contribution to heat capacity by conduction electrons remained a puzzle until 1927, when the spectrum of electronic energy levels was understood.

transport properties requires an extension of the conceptual frame developed in this book.

Ferromagnetism is another phenomenon whose treatment requires an extension of concepts developed so far. In chapter 18 we considered the action of an external magnetic field upon circulating currents within atoms or small groups of atoms. The extension of this treatment to macroscopic aggregates does not meet any difficulty for ordinary materials in which neither spin currents nor orbital currents of different electrons normally have any net resultant in the electronic ground state or in states of low excitation. Recall that the cancellation of spin angular momenta is normally required in the formation of chemical bonds, whether covalent, metallic, or ionic. However, this rule does not necessarily apply to transition elements and elements of the rare-earth group in which the electrons of d and f subshells do not really participate in bond formation. For these substances the effects of repulsion among electrons may prevail over the effects of chemical bonding; parallel spin orientation in the ground state may then occur for an aggregate of atoms as it does for single atoms with incomplete shells (see Hund rule, section 16.4.1). In this case parallel orientation of the spins of an appreciable fraction of electrons yields a large-scale magnetization of the material; that is, it makes the material ferromagnetic. The assembly of parallel spins behaves then as a single system under the influence of an external magnetic field.

APPENDIX A

LINEARITY, INVARIANCE, AND FOURIER ANALYSIS

This Appendix presents mathematical formulae and techniques which have extensive application in physics. They are developed here in the context of examples selected for purposes of illustration. We stress here the correspondence between the characteristics of physical systems and those of the mathematical models that describe them in order that one may appreciate the suitability of particular mathematical techniques for dealing with particular problems. The linearity of equations corresponds to superposition of different process (section A1); constancy of parameters under translation in space or time singles out a class of phenomena represented by exponential or sinusoidal functions (section A2). The combination of linearity and constant parameters permits the exploitation of Fourier analysis techniques (section A3).

A1. SUPERPOSITION

Suppose that an electromagnetic disturbance at a point P results from two different causes, e.g., from the onset of currents i_1 and i_2 in two separate wires. The electric and magnetic fields at the point **P**, $\vec{E}(P)$ and $\vec{B}(P)$, can be represented respectively as the sums of two components, $\vec{E} = \vec{E}_1 + \vec{E}_2$ and $\vec{B} = \vec{B}_1 + \vec{B}_2$, each of which would occur alone if only one of the currents, i_1 or i_2, had been started. This is a consequence of the *principle of superposition*, stating that two or more disturbances can propagate at the same time through the same region of space without affecting each other.

Mathematically, the superposition principle results from the linearity of Maxwell's equations:

$$\vec{\nabla} \times \vec{E} = -\frac{1}{c}\frac{\partial \vec{B}}{\partial t}, \tag{A1a}$$

$$\vec{\nabla} \times \vec{B} = \frac{1}{c}\frac{\partial \vec{E}}{\partial t} + \frac{4\pi}{c}\vec{j}, \tag{A1b}$$

$$\vec{\nabla} \cdot \vec{E} = 4\pi\rho, \tag{A1c}$$

$$\vec{\nabla} \cdot \vec{B} = 0. \tag{A1d}$$

If one calculates $\vec{E}_1$ and $\vec{B}_1$ by solving these equations with the current density[1] $\vec{j} = \vec{j}_1$ and $\rho = 0$ and then one calculates $\vec{E}_2$ and $\vec{B}_2$ by solving for $\vec{j} = \vec{j}_2$, the sums $\vec{E} = \vec{E}_1 + \vec{E}_2$ and $\vec{B} = \vec{B}_1 + \vec{B}_2$ represent the solution calculated from

---

1. The current intensity in a wire, i, is the integral of the current density $\vec{j}$ over each cross-section of the wire.

$\vec{j} = \vec{j}_1 + \vec{j}_2$ and $\rho = 0$. (If $\rho \neq 0$, one can calculate additional contributions to $\vec{E}$ and $\vec{B}$ by solving eqs. [A1] with $\vec{j} = 0$ and with the given ρ.)

The superposition principle applies to all physical phenomena governed by linear equations. In particular, it applies to the currents driven by radiation in matter, within the limits of the linear approximation discussed in chapter 2.

The separation in space of sources assumed in the preceding example is, in fact, unnecessary for applications of the superposition principle. One can regard any given source current density $\vec{j}$ as the sum of components, $\vec{j}_1 + \vec{j}_2$, with any desirable characteristic (e.g., direction of flow or monochromaticity) and study independently the radiation originating from each component. Similarly, one can consider the effects of radiation on matter by superposing the effects of convenient radiation components.

Notice, however, that superposition applies to the radiation *fields* and to the initial or induced current densities but not to the energy density or energy flow associated with them. For example, the energy density of the electric field in space is proportional to the squared strength $|\vec{E}|^2$ which differs from the sum of energy densities of two components, $|\vec{E}_1|^2 + |\vec{E}_2|^2$, by the cross-term $2\vec{E}_1 \cdot \vec{E}_2$; this cross term can be positive or negative.

A2. EXPLOITING INVARIANCE IN SPACE AND TIME

Radiations generated by identical (or equivalent) sources in different regions of space and/or at different times propagate through empty space in the same manner, insofar as the properties of space are uniform and constant in time. The uniformity of these properties is represented mathematically by the fact that the Maxwell equations (A1) have *constant coefficients*. The same uniformity permits the existence of special types of radiation—represented by plane monochromatic waves—which maintain uniform intensity, direction, and frequency throughout all space and in the course of time. Whereas radiation fields of such uniformity constitute only an idealized limiting case, application of the superposition principle permits treatment of any radiation as a superposition of these idealized radiations.

Centering attention on phenomena with a high degree of uniformity is a common procedure in physics, with practical as well as conceptual significance. For example, electric signals are often described in terms of their frequency spectrum, i.e., as superpositions of steady sinusoidal oscillations whose separate propagation is most easily studied. Conversely, the properties of complex electric systems are often studied by observing their responses to sinusoidal drives of various frequencies. Similarly the structure of matter is typically studied by observing its reaction to monochromatic and collimated radiation. Moreover, quantum physics reveals previously unsuspected connections between uniformities in time or space and the conservation of energy or momentum.

The study of invariances in space and time utilizes mathematical techniques centered on Fourier analysis. Space and time together involve four independent coordinates. We begin this review of relevant physico-mathematical procedures by an example that involves only the time variable.

A2.1. *Oscillations of Electric Charge.* Consider the familiar example of electric charge variations, Q(t), on a condenser of capacity C shorted through an inductance L and a resistance R (see Fig. A1). In the absence of potential sources the charge obeys the equation

$$L\frac{d^2Q}{dt^2} + R\frac{dQ}{dt} + \frac{Q(t)}{C} = 0. \tag{A2}$$

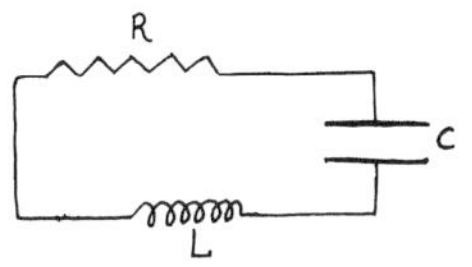

Fig. A1. Simple oscillating circuit

Since the coefficients L, R, and C are constant, we seek a solution whose characteristics are as invariant as possible in the course of time. Requiring that Q(t) is constant leads to the uninteresting solution Q = 0. A lesser but more realistic requirement is that a shift of the time variable merely multiplies Q(t) by a constant factor, as represented by the condition

$$Q(t + \tau) = a(\tau)Q(t). \tag{A3}$$

The property (A3) is the characteristic of exponential functions, i.e., of

$$Q(t) = Q_0 e^{Kt}, \tag{A4}$$

where K is a constant; with this choice of Q(t), the coefficient $a(\tau)$ in (A3) becomes $\exp K\tau$. Substitution of (A4) into (A2) permits the factorization of Q(t) itself out of the differential equation, thus reducing (A2) to the algebraic form

$$LK^2 + RK + \frac{1}{C} = 0. \tag{A5}$$

This characteristic equation has a pair of solutions which may be real or complex conjugate, namely,

$$K = -\frac{1}{2}\frac{R}{L} \pm \left(\frac{1}{4}\frac{R^2}{L^2} - \frac{1}{LC}\right)^{1/2} = -\kappa + i\omega, \tag{A6}$$

where $-\kappa$ and ω indicate the real and imaginary parts of K.

For any set of initial conditions on Q(t), e.g., for given values of Q and dQ/dt at t = 0, the solution Q(t) is constructed as a superposition of two terms of the form (A4) with the alternative values of K. The solution is particularly simple in the absence of energy dissipation, that is, for R = 0, in which case (A6) reduces to

$$K = i\omega = \pm i(LC)^{-1/2}. \tag{A7}$$

The procedure outlined above is standard for the solution of ordinary homogeneous differential equations with constant coefficients. The theory of these equations states that their general solution is a *superposition*, that is, a linear combination of special solutions of the form (A4). Each of the special solutions is an exponential whose exponent has a coefficient K equal to one root of a characteristic equation analogous to (A5). Extensions of this well-known theory underlie the applications we have in mind, including the treatment of driven oscillations and of problems with many variables.

A2.2. *Exponential and Sinusoidal Variations*. In physical applications, the coefficients of a differential equation and of the corresponding characteristic equation are normally real, but the roots of the characteristic equation nevertheless often occur in complex-conjugate pairs. One then constructs real solutions Q(t) of physical interest by superposing pairs of complex-conjugate exponentials. These superpositions contain sinusoidal factors cos ωt or sin ωt. For this reason one often regards sinusoidal oscillating functions rather than exponentials as the basic solutions of linear differential equations with constant coefficients. Actually, both exponential and sinusoidal functions can serve to construct solutions Q(t).

The relations between complex exponentials and sinusoidal functions and some of their relevant properties are displayed in Table A1 for the simple case (A7), where the root, K, of the characteristic equation is purely imaginary. Notice that each of the three transformations indicated in the table, that is, $f \to f(t+\tau)$, $f \to df/dt$, $f \to d^2f/dt^2$, simply *multiplies* exp $(i\omega t)$ *by a constant factor*. However, the first two transformations change cos ωt and sin ωt into different sinusoidal functions.

TABLE A1

Comparative Properties of Exponential and Sinusoidal Functions

Function			
f(t)	$\exp(i\omega t)$	$\cos\omega t$	$\sin\omega t$
	$= \cos\omega t + i\sin\omega t$	$= \frac{1}{2}[\exp(i\omega t) + \exp(-i\omega t)]$	$= -i\frac{1}{2}[\exp(i\omega t) - \exp(-i\omega t)]$
$f(t+\tau)$	$\exp[i\omega(t+\tau)]$	$\cos[\omega(t+\tau)]$	$\sin[\omega(t+\tau)]$
	$= \exp(i\omega\tau)\exp(i\omega t)$	$= \cos\omega\tau\cos\omega t - \sin\omega\tau\sin\omega t$	$= \cos\omega\tau\sin\omega t + \sin\omega\tau\cos\omega t$
df/dt	$i\omega\exp(i\omega t)$	$-\omega\sin\omega t$	$\omega\cos\omega t$
d^2f/dt^2	$-\omega^2\exp(i\omega t)$	$-\omega^2\cos\omega t$	$-\omega^2\sin\omega t$

Mathematically, one describes this fact by saying that exp $(i\omega t)$ is an *eigenfunction* of all three transformations, but cos ωt and sin ωt are eigenfunctions of d^2/dt^2 only. The broader properties of invariance of the complex exponential often make it expedient to utilize this function, instead of sines and cosines.

Theoretical results which directly represent relationships among measurable quantities must, of course, be expressed in real form. This requirement is met at the end of a calculation with complex functions by constructing a real superposition, or other appropriate combination, of complex-conjugate solutions.

A2.3 *Plane Monochromatic Radiation Waves*. Returning now to radiation in free space, we note that the Maxwell equations involve derivatives with respect to three space coordinates (x, y, z) and to time, all with constant coefficients. Radiation with high invariance characteristics has electric and magnetic fields that depend on x, y, z, and t, respectively, through four exponential factors analogous to the factor exp (Kt) in (A4). We write these factors in the form

$$\exp(ik_x x)\exp(ik_y y)\exp(ik_z z)\exp(-i\omega t) = \exp[i(k_x x + k_y y + k_z z - \omega t)]. \tag{A8}$$

The combination $k_x x + k_y y + k_z z$ in (A8) can be regarded as the scalar product of a vector $\vec{k} = (k_x, k_y, k_z)$ and of a space coordinate vector $\vec{r}$. Hence (A8) is written more compactly as

$$e^{i(\vec{k}\cdot\vec{r}-\omega t)}. \tag{A9}$$

The exponential function (A9) or (A8) is called a *plane monochromatic wave*: a "monochromatic wave" because its real and imaginary parts are sinusoidal functions of space and time variables, a "plane" wave because its complex phase, $\vec{k}\cdot\vec{r} - \omega t$, has the same value at each point of any plane perpendicular to $\vec{k}$. This phase varies in space with a gradient equal to $\vec{k}$ and in the course of time at the rate of $-\omega$ radians/sec. The phase remains constant at any point that moves in the direction of $\vec{k}$ with the speed of ω/k cm/sec. (The factor ω has been entered in [A8] with a *sign opposite* to k_x, k_y and k_z to ensure this phase propagation.[2]) For this reason $\vec{k}$ is called the *propagation vector* of the wave and ω/k the *phase velocity*. The wavelength of (A9) equals $2\pi/k$ and its frequency equals $\omega/2\pi$.

A plane monochromatic radiation wave is represented by electric and magnetic fields proportional to (A9), namely, by

$$\vec{E} = \vec{E}_0 e^{i(\vec{k}\cdot\vec{r}-\omega t)}, \qquad \vec{B} = \vec{B}_0 e^{i(\vec{k}\cdot\vec{r}-\omega t)}. \tag{A10}$$

2. This sign convention is now widely but not universally followed in physics literature; some authors use exp $[i(\omega t - \vec{k}\cdot\vec{r})]$, others exp $[i(\omega t + \vec{k}\cdot\vec{r})]$. Care is warranted to avoid misunderstandings.

The constant vectors $\vec{E}_0$ and $\vec{B}_0$, the propagation vector $\vec{k}$ and the angular frequency ω are not independent of one another but subject to relationships implied by the Maxwell equations. Indeed, substitution of (A10) into (A1), with $\vec{j} = \rho = 0$ as pertains to empty space, permits factoring out of the exponential (A9). The differential Maxwell equations are thereby reduced to algebraic form, i.e., to a system of characteristic equations analogous to (A5),

$$i\vec{k} \times \vec{E}_0 = i\frac{\omega}{c}\vec{B}_0, \tag{A11a}$$

$$i\vec{k} \times \vec{B}_0 = -i\frac{\omega}{c}\vec{E}_0, \tag{A11b}$$

$$i\vec{k}\cdot\vec{E}_0 = 0, \tag{A11c}$$

$$i\vec{k}\cdot\vec{B}_0 = 0. \tag{A11d}$$

Equations (A11c) and (A11d) require the fields $\vec{E}$ and $\vec{B}$ to be orthogonal to the propagation vector $\vec{k}$ (transverse wave condition). Equation (A11a) requires $\vec{B}$ and $\vec{E}$ to be mutually orthogonal. The constraint on the propagation vector, $\vec{k}$, and frequency, ω, of the plane wave is obtained from equation (A11b) by the following sequence of transformations: (1) Rewrite (A11b) as $\vec{k}\times\vec{B}_0 + (\omega/c)\vec{E}_0 = 0$. (2) Multiply this equation by (ω/c). (3) Substitute $(\omega/c)\vec{B}_0$ by $\vec{k} \times \vec{E}_0$, in accordance with (A11a). (4) Utilize the vector identity $\vec{k} \times (\vec{k} \times \vec{E}_0) = (\vec{k}\cdot\vec{E}_0)\,\vec{k} - k^2\vec{E}_0$. (5) Set $\vec{k}\cdot\vec{E}_0 = 0$, in accordance with (A11c). The result, namely,

$$\left[\left(\frac{\omega}{c}\right)^2 - k^2\right]\vec{E}_0 = 0, \tag{A12}$$

implies the basic relationship between k and ω,

$$k^2 = \omega^2/c^2, \tag{A13}$$

which completes the set of constraints on a plane monochromatic radiation wave in empty space.

A2.4 *Extension to Driven Oscillations*. Oscillations of currents in electric circuits, of the fields in electromagnetic radiation, and of many other physical variables occur frequently as the result of external drives. The variables of interest obey then *inhomogeneous* linear equations in which the inhomogeneity represents the external drive. For example, a nonzero current density $\vec{j}$ in (A1b) represents a source of radiation; similarly a time-variable potential source V(t) inserted in an oscillating circuit (see Fig. A2) changes equation (A2) to the inhomogeneous form

$$L\frac{d^2Q}{dt^2} + R\frac{dQ}{dt} + \frac{Q(t)}{C} = V(t). \tag{A14}$$

The linearity of (A1) and (A14) and the constancy of their coefficients can still

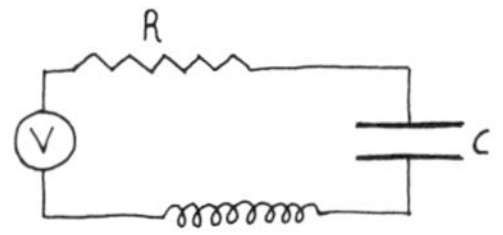

Fig. A2. Oscillating circuit with time-variable potential source V.

be exploited when the source variations in time and/or space have the characteristic invariance of exponential or sinusoidal functions. For example, when $V(t) = V_0 \exp(-i\omega t)$ in (A14), the charge variations may be represented by[3]

$$Q(t) = Q_0 e^{-i\omega t}. \tag{A15}$$

Here, ω indicates an arbitrary frequency and no longer represents a root of the characteristic equation. Upon substitution of (A15) into (A14), the factor $\exp(-i\omega t)$ drops out as it does for homogeneous equations; thereby (A15) reduces to the algebraic form

$$\left[-L\omega^2 - iR\omega + \frac{1}{C}\right] Q_0 = V_0. \tag{A16}$$

The solution[4] of this algebraic equation determines the coefficient

$$Q_0 = C \frac{V_0}{1 - LC\omega^2 - iRC\omega}. \tag{A17}$$

Notice that the expression in the brackets in (A16), which appears also in the denominator of (A17), equals the left-hand side of the characteristic equation (A5) to within the substitution $\omega \to iK$. Therefore, (A17) diverges when $i\omega$ is a root of (A5). We have here a fundamental property of all linear systems: the solution of an inhomogeneous problem becomes singular for those values of the parameters that permit a solution of the corresponding homogeneous problem.[5]

---

3. Equation (A15) is a particular solution of the inhomogeneous equation (A14); the general solution consists of a superposition of (A15) and of a solution of the homogeneous equation (A2). However, the oscillations represented by the solution of the homogeneous equation are damped and thus represent only a transient state of the system.

4. The complex coefficient Q_0 is expressed in terms of amplitude and phase by $Q_0 = CV_0[(1-LC\omega^2)^2 + (RC\omega^2)^2]^{-1/2} e^{i\delta}$, where $\delta = \arctan [RC\omega/(1 - LC\omega^2)]$. The angle δ represents the *phase lag* of Q(t) with respect to V(t), as apparent by substituting (A17) in (A15).

5. In practice circuits have a nonzero resistance; therefore, the characteristic frequencies of free oscillation given by the roots of (A6) are complex and the denominator of (A17) never vanishes for real values of the driving frequency ω.

As a second problem, consider the solutions of Maxwell's equations (A1) with an idealized distribution of currents and with zero charge density. We set

$$\rho = 0, \quad \vec{j} = \vec{j}_0 \exp\,[i(\vec{k}\cdot\vec{r} - \omega t)], \qquad \text{with } \vec{j}_0\cdot\vec{k} = 0, \tag{A18}$$

where the condition $\vec{j}_0\cdot\vec{k} = 0$ requires the currents to be transverse to the propagation vector.[6] The radiation field produced by this source is still represented by the plane monochromatic wave (A10). Substitution of (A10) and (A18) into (A1) permits elimination of the exponential factor and reduction of the equations to algebraic form. This form coincides with (A11) except for addition of $4\pi\vec{j}_0/c$ to the right-hand side of (A11b). Thereby (A12) is replaced by

$$\left[\left(\frac{\omega}{c}\right)^2 - k^2\right]\vec{E}_0 = -4\pi\,\frac{i\omega\vec{j}_0}{c^2} \tag{A19}$$

whose solution, analogous to (A17), is

$$\vec{E}_0 = 4\pi\,\frac{i\omega\vec{j}_0}{c^2k^2 - \omega^2}. \tag{A20}$$

The oscillating circuit equation (A14) and the radiation field equations (A1) with the idealized source terms (A18) constitute mathematical models which would be *altogether unrealistic, if taken by themselves*. Notice in particular that the exponential variation represented by (A18) is understood to extend *indefinitely* throughout space and time, which is indeed an extreme idealization. The models become realistic in the context of systematic *application of the superposition principle*.

Realistic source terms in (A14) or (A1) may be represented as sums of terms each of which has an idealized exponential dependence on time and/or space coordinates. One can thus represent sources that can be built experimentally and are designed to approximate ideal sources. The equations are then solved *by calculating first* the effect of each idealized source term, by means of (A17) or of (A20) or of an analogous equation, and *then by summing* the idealized exponential oscillations of charges or fields to obtain a realistic result.

This procedure is powerful in that it exploits basic invariances of physical laws and it can provide, in essence, ready-made solutions to complex problems. On the other hand it obscures cause-effect relationships by the very act of replacing differential by algebraic equations. The differential equations describe the connection between the values of fields or of charges at successive instants of time and/or at adjacent points of space. The corresponding algebraic equations relate, instead, *rate parameters* of standardized variations in time and/or space, such as $\vec{k}$ and ω, to one another and to physical constants;

---

6. If the current distribution is not transverse, i.e., if $\vec{j}_0\cdot\vec{k} \neq 0$, we must also have $\rho = \vec{k}\cdot\vec{j}/\omega$, and there results an additional longitudinal wave of no interest to us.

they also relate the magnitude of standard solutions to the strength of idealized sources. To compensate this loss of contact with the differential point of view, one should keep in sight the connection between rate parameters and point-to-point variations at each successive stage of analysis or calculation. (For example, high frequency means fast variation in time.) One should also acquire understanding and experience in the operation of Fourier analysis, that is, of the procedures for resolving realistic sources as well as field and charge distributions into idealized ones and for recombining ideal exponential components of a distribution into their resultant.

A3. FOURIER ANALYSIS AND SUPERPOSITION [7]

Consider a function f(t)—e.g., the driving potential V(t) in (A14)—defined within a very large but finite time interval $-T < t < T$. We wish to represent this function as a sum of complex exponential functions of t. For this purpose the exponentials must be periodic functions of t with the period 2T, i.e., they must have the form

$$\exp(i\pi nt/T), \tag{A21}$$

where n is a positive or negative integer. The oscillation frequency of the exponential (A21) is expressed as n/2T cycles/sec or $n\pi/T$ radians/sec. Under certain conditions which are met in all cases of interest to us, the function f(t) can actually be represented by the infinite sum[8] (Fourier expansion)

$$f(t) = \sum_{n=-\infty}^{\infty} a_n \exp(i\pi nt/T). \tag{A22}$$

A formula for calculating the coefficients a_n, if the function f(t) is described analytically or numerically, will be given below. It will become apparent that the sum in (A22) can often be approximated by a finite number of terms. In general, for any given accuracy, functions f(t) which are rapidly varying are approximated by a larger number of Fourier components than slowly varying functions.[9]

---

7. For additional examples and problems concerning the material in this section see *BPC*, vol. 3, chap. 6; also A. Papoulis, *The Fourier Integral and Its Applications* (New York: McGraw-Hill Book Co., 1962).

8. Henceforth in this Appendix Σ_n will mean that the range of n is $-\infty < n < \infty$; otherwise the limits will be stated explicitly.

9. One may notice that because the exponentials on the right-hand side of equation (A22) are periodic, the function f(t) on the left-hand side is also implied to be periodic in the sense that $f(t + m2T) = f(t)$, where m is an integer. This periodicity contradicts our initial statement that f(t) is confined to a finite range of t, but the contradiction does not matter as we intend to consider always the limit $T = \infty$, i.e., to consider values of T so large that values of $|t| > T$ are altogether irrelevant.

The pairs of exponentials exp $(i\pi nt/T)$ with $n = \pm |n|$ are complex conjugate to one another. Accordingly, if the function f(t) is real, the coefficients of any two such exponentials are complex conjugate to one another,

$$a_{-n} = a_n^*. \tag{A23}$$

This condition ensures that each pair of terms of (A22) with $n = \pm|n|$ can be replaced by a pair of real terms

$$f(t) = \sum_{n=0}^{\infty} [c_n \cos(\pi nt/T) + d_n \sin(\pi nt/T)], \tag{A22a}$$

where $c_n = a_n + a_{-n}$ and $d_n = i(a_n - a_{-n})$ are real, or by a single term with appropriate phase constant

$$f(t) = \sum_{n=0}^{\infty} (c_n^2 + d_n^2)^{1/2} \cos[\pi nt/T - \arctan(d_n/c_n)]. \tag{A22b}$$

The real representation (A22b) has a direct physical interpretation. For example, if f(t) is a driving potential V(t), one can construct an assembly of a.c. potential sources $V_0, V_1, \ldots, V_n$, with peak strengths and phases adjusted so that

$$V_n(t) = (c_n^2 + d_n^2)^{1/2} \cos[\pi nt/T - \arctan(d_n/c_n)].$$

Addition of these sources, which is achieved by connecting them in series, reproduces the potential V(t).

A3.1. *Amplitude Modulation and Bandwidth.* In practice, coefficients a_n of successive terms in (A22) often have similar magnitude. Accordingly, it is worthwhile to describe the results of the superposition of two or more exponentials with equal coefficients. For example, the sum of two terms can be expressed as

$$\exp(i\pi mt/T) + \exp(i\pi nt/T) \\ = 2\cos[\pi(m-n)t/2T]\exp[i\pi(m+n)t/2T]. \tag{A24}$$

[This formula is obtained by representing m and n as $\frac{1}{2}(m + n) \pm \frac{1}{2}(m - n)$ and factoring out the last exponential.] The right-hand side may be described as an exponential function of t with the parameter $\frac{1}{2}(m + n)$, the average of m and n, and with the coefficient $2\cos[\pi(m - n)t/2T]$ which varies much more slowly than the exponential if m and n are nearly equal. This variable coefficient is called the *amplitude modulation* of $\exp[i\pi(m + n)t/2T]$; it is also described as a *beat*, i.e., as an effect of interference between two monochromatic functions of time. (The word "interference" generally indicates any effect characteristic of the superposition of two or more components but absent in any individual component.) Physically, one may say that the two terms on the left-hand side of (A24) represent oscillations proceeding at somewhat different rates so as to be now "in step" and now "out of step" with one another. When they are in step, their amplitudes add to yield a large resultant amplitude; but when they are out of step, their amplitudes cancel, at least approximately (see Fig. A3).

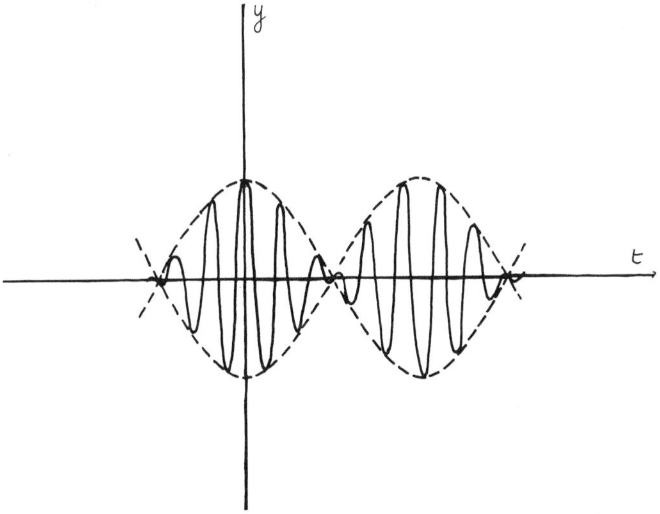

Fig. A3. Interference of two oscillations.
$y = \cos(9\pi t/T) + \cos(11\pi t/T) = 2\cos(\pi t/T)\cos(10\pi t/T)$.

Another instructive example of constructive and destructive interference is afforded by the superposition of 2N + 1 successive exponentials (A21) with equal coefficients $a_n = 1$. Setting $n = m + p$ with p ranging from $-N$ to N, the sum over n in (A22) becomes

$$\sum_{p=-N}^{N} \exp[i\pi(m+p)t/T] = \frac{\sin[\pi(2N+1)t/2T]}{\sin(\pi t/2T)} \exp(i\pi m t/T). \qquad \text{(A25)}$$

The right-hand side of (A25) is obtained by considering that its left-hand side is a geometric progression with ratio $\exp(i\pi t/T)$ and applying the standard formula for the sum of a progression. The right-hand side consists, as in (A24), of an exponential function of t—with the parameter m equal to the arithmetic mean of the parameters $m + p$ on the left-hand side—and of an amplitude modulation factor. Here, the modulation is much sharper than in the case of two components only. The $2N + 1$ components are "in step" at $t \sim 0$ but remain in step only for $|t| << 2T/(2N+1)$. The amplitude modulation factor drops to zero at $t = \pm 2T/(2N+1) \sim T/N$; then it rises again to successive maxima, but these maxima become rapidly much lower than the central one at $t = 0$ (Fig. A4). Big maxima occur again wherever $\sin(\pi t/2T)$ vanishes; but they lie outside the range $|t| \leq T$, i.e., at infinite distance in the limit $T = \infty$.

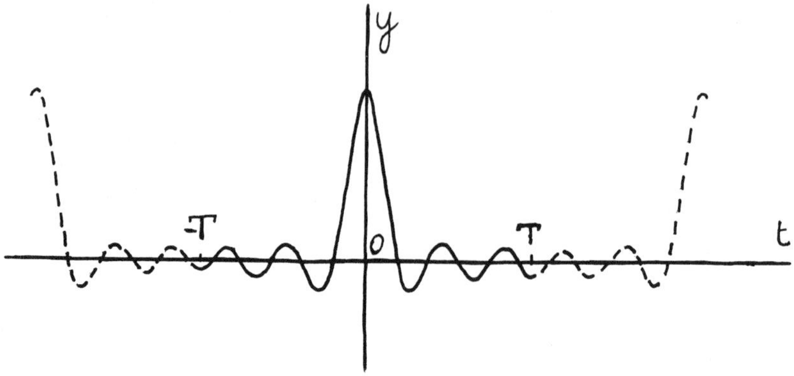

Fig. A4. Amplitude modulation factor for equation (A25) for N = 5

$$y = \frac{\sin(11\pi t/2T)}{\sin(\pi t/2T)}.$$

Notice that the profile of the modulation factor does *not* depend on the mean oscillation frequency m/2T, but only on the range of frequencies included in the superposition. This range is represented in our case by the difference

$$\Delta\nu = \frac{m+N}{2T} - \frac{m-N}{2T} = \frac{N}{T} \tag{A26}$$

between the largest and the smallest frequency in (A25). This range is called the *bandwidth* of the superposition. Note that $\Delta\nu$ is approximately the *reciprocal of the time*, $T/(N + \frac{1}{2})$, *required for the amplitude modulation factor to drop* from its main maximum to zero. A short time interval between peak and zero of the amplitude modulation factor requires a broad bandwidth of frequencies in the superposition.

Equation (A25) can also be utilized to illustrate the effect of superposing a set of plane waves of the type (A9), with slightly different wave numbers and frequencies,

$$\sum_{p=-N}^{N} \exp i[(\vec{k} + p\Delta\vec{k})\cdot\vec{r} - (\omega + p\Delta\omega)t]$$
$$= \frac{\sin[(N + \frac{1}{2})(\Delta\vec{k}\cdot\vec{r} - \Delta\omega t)]}{\sin\frac{1}{2}(\Delta\vec{k}\cdot\vec{r} - \Delta\omega t)} \exp[i(\vec{k}\cdot\vec{r} - \omega t)]. \tag{A27}$$

The result of the superposition consists of the plane wave with mean wave vector $\vec{k}$ and mean frequency ω, which propagates with the phase velocity ω/k in the direction of $\vec{k}$, and of an amplitude modulation factor. This factor—and hence the whole *wave packet*—also propagates in the course of time, in the direction of the wave-vector increments $\Delta\vec{k}$ and with the *group velocity* $\Delta\omega/\Delta k$.[10]

A3.2. *Calculation of Fourier Coefficients.* Since the Fourier expansion (A22) of a given function f(t) requires suitable values of the coefficients a_n, the determination of these values can be regarded as a procedure for fitting the sum of terms to f(t). Accordingly, we shall consider the right-hand side of (A22) initially as a function of undetermined parameters a_n. The parameters must be fitted by minimizing the difference between the two sides of (A22). More specifically, the quantity to be minimized is the squared magnitude of this difference, namely, the quantity

$$\left|f(t) - \sum_n a_n e^{i\pi nt/T}\right|^2. \tag{A28}$$

We proceed by least-squares fitting.

To achieve a fit for all values of t, one minimizes the *integral* of (A28) over the whole range of t or, equivalently, its mean value

$$M = \frac{1}{2T}\int_{-T}^{T} |f(t) - \sum_n a_n e^{i\pi nt/T}|^2\,dt. \tag{A29}$$

If numerical values of f(t) are assigned for a certain set of values of t $(t_1, t_2, .., t_i, ...)$, the integral (A29) reduces to a sum and resembles closely the familiar expression for least-squares fit. One can evaluate M by expanding the binomial in the integrand and by using the relation[11]

$$\frac{1}{2T}\int_{-T}^{T} \exp\,[i\pi(n-m)t/T]dt = \delta_{nm} \equiv \begin{cases} 1 & \text{if } n = m \\ 0 & \text{if } n \neq m. \end{cases} \tag{A30}$$

Thereby (A29) becomes

$$M = \frac{1}{2T}\int_{-T}^{T} |f(t)|^2 dt - \sum_n a_n \frac{1}{2T}\int_{-T}^{T} f^*(t)\exp\,(i\pi nt/T)dt$$
$$- \sum_n a_n^* \frac{1}{2T}\int_{-T}^{T} f(t)\exp\,(-i\pi nt/T)dt + \sum_n a_n^* a_n. \tag{A31}$$

---

10. The concept and the derivation of group velocity are treated in greater detail in *BPC*, vol. 3, chap. 6, and also particularly well in Appendix XI of Born's *Atomic Physics*.

11. The relation (A30) is true because the exponential equals unity for n = m and otherwise consists of a cosine and a sine to be integrated over a whole number of periods. The symbol δ_{nm}, defined by the right-hand side of (A30), with alternative values 1 and 0, is called the *Kronecker delta*. Equation (A30) and its generalized form (A30a) on p. 542 are called *orthogonality relations* (for $n \neq m$) because of analogy to the equation $a_x b_x + a_y b_y + a_z b_z = 0$ among the components of orthogonal vectors.

The fitting is now achieved by requiring M to be a minimum as a function of a_n and a_n^*. It is sufficient to set[12]

$$\frac{\partial M}{\partial a_n^*} = -\frac{1}{2T}\int_{-T}^{T} f(t)\exp(-i\pi nt/T)dt + a_n = 0, \qquad \text{(A32)}$$

which provides the value of each a_n,

$$a_n = \frac{1}{2T}\int_{-T}^{T} f(t)\exp(-i\pi nt/T)dt. \qquad \text{(A33)}$$

The integral in this formula can be evaluated numerically, if necessary, for any given f(t). The set of coefficients a_n defined by (A33) is called the *Fourier transform* of the function f(t).

Notice that substitution of (A33) into (A31) reduces M to the form

$$M = \frac{1}{2T}\int_{-T}^{T} |f(t)|^2 dt - \sum_n |a_n|^2, \qquad \text{(A34)}$$

which should vanish if *all* a_n had been calculated appropriately. Conversely, whenever a finite set of a_n has been evaluated, the residual value of M obtained by entering in (A34) the finite sum over $|a_n|^2$ provides an estimate of the error in the approximation used. The sum of squared amplitudes $\Sigma_{n=-N}^{N} |a_n|^2$ can never exceed the mean value of $|f(t)|^2$ since the integrand of (A29) is nonnegative.

A3.3. *Example: Analysis of an Oscillation of Limited Duration.* In practice one often deals with oscillating forces, currents, etc., that are sinusoidal only for a certain time interval. Let us see how such a variable can be represented by a superposition of exactly sinusoidal functions. Set, e.g.,

$$f(t) = e^{i\omega t} \text{ for } -\tau \leqslant t \leqslant \tau, \qquad f(t) = 0 \text{ for } |t| > \tau, \qquad \text{(A35)}$$

and substitute this expression in (A33). The result of the integration is

$$a_n = \frac{\tau}{T}\frac{\sin(\omega - \pi n/T)\tau}{(\omega - \pi n/T)\tau}. \qquad \text{(A36)}$$

This coefficient equals τ/T when $\pi n/T = \omega$ and then oscillates with an amplitude that decreases inversely to $(\omega - \pi n/T)\tau$ as $|\omega - \pi n/T|\tau$ becomes $\gg 1$. That is, the important values of $\pi n/T$ are concentrated around ω within a *bandwidth* reciprocal to the time duration τ of the oscillation. Notice the analytical similarity of (A36), regarded as a function of $\omega - \pi n/T$, to the amplitude modulation factor in (A25), regarded as a function of t.

---

12. The additional equation $\partial M/\partial a_n = 0$ yields the complex conjugate of equation (A33).

A3.4 *Example: Power Dissipation in Driven Oscillations.* Let us return to equation (A14) for the charge variations on the condenser of a circuit driven by a potential V(t) and let us represent this potential by its Fourier expansion

$$V(t) = \sum_n V_n e^{i\pi nt/T}. \tag{A37}$$

The charge variations Q(t) may be similarly represented by[13]

$$Q(t) = \sum_n Q_n e^{i\pi nt/T}. \tag{A38}$$

Solution of (A14) by application of the superposition principle shows that each Q_n is given by (A17), with $\omega = -\pi n/T$,

$$Q_n = \frac{CV_n}{1 - LC(\pi n/T)^2 + iRC\pi n/T}. \tag{A39}$$

Thereby (A14) is solved, at least formally, by the procedure of Fourier analysis.

The magnitude of a Fourier coefficient Q_n becomes very large whenever its denominator nearly vanishes. Thus Q_n peaks sharply for values of $(-\pi n/T)$ near the characteristic frequencies (A6) of the circuit without any driving potential. For $R << 2(L/C)^{1/2}$, these *resonances* occur for $n \simeq \pm\, T/\pi(LC)^{1/2}$. The evaluation of Q(t) for each value of t utilizing the coefficients (A39) is complicated by this peaking. We shall not elaborate this procedure here, but confine ourselves to the illustrative calculation of an average property of the driven charge oscillation, namely, of the power dissipated in the circuit.

Power dissipation at each time instant t is the product of the applied potential V(t) and of the current $i(t) = dQ/dt$. Therefore, its average value is

$$\langle P\rangle = \frac{1}{2T}\int_{-T}^{T} \sum_n V_n \exp\left(\frac{i\pi nt}{T}\right)\sum_{n'} Q_{n'}\frac{i\pi n'}{T}\exp\left(\frac{i\pi n't}{T}\right)dt \tag{A40}$$

$$= \sum_n V_n Q_{-n}\frac{-i\pi n}{T} = \sum_n V_n V_{-n}\frac{-iC\pi n/T}{1 - LC(\pi n/T)^2 - iRC\pi n/T},$$

owing to (A30) and (A39). This result can be simplified by noticing that the two terms of Σ_n with $n = \pm|n|$ are complex conjugates. Moreover, $V_n V_{-n}$ equals $|V_n|^2$ since V(t) is real. Therefore, their imaginary parts cancel, and (A40) reduces to

$$\langle P\rangle = \sum_n \frac{|V_n|^2/R}{1 + [1 - LC(\pi n/T)^2]^2/(RC\pi n/T)^2}. \tag{A41}$$

13. The additional terms which may have to be added to Q(t) as indicated in note 3 on p. 531, decay rapidly in the course of time; hence they usually fail to contribute appreciably to power dissipation.

Two important and well-known results are contained in this formula. First, cross-terms in the products of the expansions of V(t) and dQ/dt yield no average contribution to power dissipation. That is, different Fourier components contribute to the *average* power dissipation independently, even though cross-terms do contribute to the power *at each instant* as noted for the electric energy density in section A1. Second, each Fourier component contributes to power dissipation the amount $|V_n|^2/R$, which would obtain if the potential source were shorted through the resistance R, divided by the factor in the denominator of (A41). This factor is very large except near resonance where

$$\left|\frac{1}{LC} - \left(\frac{\pi n}{T}\right)^2\right| \leqslant \frac{R}{L}\left|\frac{\pi n}{T}\right|. \tag{A42}$$

A3. 5. *Inversion of Fourier Expansion.* We examine here, first, how superposition of Fourier components with the coefficients a_n given by (A33), actually reconstructs a given function f(t), and second, what approximation is obtained by limiting the sum of components to a finite range $-N \leqslant n \leqslant N$. The integration variable in (A33) will be indicated by t′, to distinguish it from t in the phase of the nth Fourier term. Consider then the expression which *would* yield the complete Fourier expansion (A22) in the limit $N \to \infty$, namely,

$$\sum_{n=-N}^{N} \exp\left(\frac{i\pi nt}{T}\right) a_n = \sum_{n=-N}^{N} \exp\left(\frac{-i\pi nt}{T}\right) \int_{-T}^{T} \exp\left(\frac{-i\pi nt'}{T}\right) f(t') \frac{dt'}{2T}. \tag{A43}$$

The Σ_n in this equation is equivalent to the summation performed in (A25). Therefore, equation (A43) reduces to

$$\sum_{n=-N}^{N} \exp\left(\frac{i\pi nt}{T}\right) a_n = \int_{-T}^{T} \frac{\sin[\pi(2N+1)(t-t')/2T]}{\sin[\pi(t-t')/2T]} f(t') \frac{dt'}{2T}, \tag{A43a}$$

where the integrand is the product of f(t′) and of the amplitude modulation factor which appears in (A25).

As we know from the discussion of (A25), the amplitude modulation factor in this equation is very large for $|t - t'| << T/N$ and becomes rapidly small outside this interval. Accordingly, only a very small range of values of t′ contributes significantly to the integral in (A43a), the range being centered at $t' = t$ and of width $\sim T/N$. This width can be made as small as desired by increasing the value of N. The goal of interest is to make the width so small that $|f(t') - f(t)|$ remains negligible throughout it, so that f(t′) can be effectively replaced by f(t) in the integral of (A43a). In this integral f(t) is a constant which can be factored out. Under the condition which makes $f(t') \sim f(t)$, that is, for $t' \sim t \pm T/N$, the denominator $\sin[\pi(t - t')/2T]$ can be replaced by the first term of its power expansion. By utilizing these approximations, the integral in (A43a) reduces to

$$\int_{-T}^{T} \frac{\sin[\pi(2N+1)(t-t')/2T]}{\sin[\pi(t-t')/2T]} f(t') \frac{dt'}{2T}$$

$$\sim f(t) \int_{-T}^{T} \frac{\sin[\pi(2N+1)(t-t')/2T]}{\pi(t-t')/2T} \frac{dt'}{2T}. \tag{A44}$$

The last integral has the structure of the standard $\int_{-\infty}^{\infty} [\sin(\pi x)/\pi x]dx = 1$ and, in fact, is brought to this form by the substitution $(2N + 1)(t - t')/2T = x$ and by extending the integration limits to infinity. Thereby, the expressions (A43a) and (A44) are seen to equal f(t), as we wished to show.

In this evaluation of (A43) and (A44) the limiting process $T \to \infty, N \to \infty$ has been postponed to the end. In a more familiar procedure the limit is taken when evaluating the Σ_n in (A43). At this stage, since both the values of n and T are understood to be very large, one replaces the integers n by the new variable

$$\nu = \frac{n}{2T}. \qquad \text{(A45)}$$

The $\exp[i\pi n(t - t')/2T]$ in (A43) becomes now $\exp[i2\pi\nu(t - t')]$. The change of ν when n is replaced by n + 1 is infinitesimal in the limit $T \to \infty$. Hence the Σ_n from $-N$ to N is replaced in the double limit $(T \to \infty, N \to \infty)$ by $2T\int d\nu$ with the integral extended from $-\infty$ to ∞. In particular, application of the limiting process to (A43) replaces the sum of $\exp[i\pi n(t - t')/T]$ by the integral of $\exp[i2\pi\nu(t - t')]$ over ν. This integral constitutes a well-known improper function of $(t - t')$ called the *Dirac delta*. The relationships between the various functions involved in this limiting process are indicated by the set of equations

$$\begin{array}{ccc} \dfrac{1}{2T}\sum_{n=-N}^{N} \exp[i\pi n(t - t')/T] & = & \dfrac{\sin[\pi(2N + 1)(t - t')/2T]}{2T \sin[\pi(t - t')/2T]} \\ \downarrow & \overset{T\ \ N}{\downarrow\downarrow}{}_{\infty} & \downarrow \\ \int_{-\infty}^{\infty} d\nu \exp[i2\pi\nu(t - t')] & = & \delta(t - t'). \end{array} \qquad \text{(A46)}$$

The Dirac δ-function is characterized by the twin properties of vanishing at all points where $t' \neq t$ and of yielding upon integration

$$\int_{-\infty}^{\infty} \delta(t - t')f(t')dt' = f(t). \qquad \text{(A47)}$$

(Dirac's definition of the δ-function, essentially along the lines followed here, remained somewhat controversial until the mathematician Laurent Schwartz developed the concept of *improper functions*, a class which includes the Dirac δ.)

A3.6 *Complementarity of Time and Frequency Determination.* The frequency of an oscillation is defined rigorously only for an oscillation that lasts indefinitely in the course of time, as stressed in this Appendix. In practice, signals of standard frequency, broadcast by certain radio stations, last only for a limited length of time τ. Therefore, the frequency of such a signal is defined only insofar as one can identify the midpoint of a frequency distribution of the type represented by equation (A36).

Conversely, any sharp time signal broadcast for accurate setting of clocks is obtained by amplitude modulation of a carrier wave. Irrespective of the frequency of the carrier wave, the sharpness of the signal is limited by the bandwidth of frequencies transmitted and received. This bandwidth is reciprocal to the duration of the signal in accordance with equation (A26).

This type of restriction extends to the problem of representing the difference between two values of a function, f(t) and f(t′), by Fourier analysis. According to section A. 3. 5, this difference is reproduced by superposition of Fourier components only if the superposition includes a bandwidth $\Delta\nu \gtrsim 1/(t-t')$. Use of a smaller bandwidth has the effect of obliterating the difference between f(t) and f(t′).

The same type of restriction applies, of course, not only to functions of time but also to functions of space coordinates. In this case sharp localization of a position in space implies superposition of Fourier components with a broad range of wave vectors. Conversely, the wave vector of a plane wave is defined only insofar as the wave extends over a large region of space. Whenever a functional dependence upon any one variable, ξ, is represented by Fourier superposition, the superposition involves a parameter complementary to ξ and related to it as frequency relates to time and wave vector relates to position.

A3. 7. *Generalization.* The Fourier analysis constitutes an example of a very general procedure. As we have seen, use of exponential and sinusoidal functions is appropriate for representing processes whose laws are invariant under translation of one or of several independent variables. Invariance under rotations about one point of three-dimensional space leads to a different situation and hence to another type of functions called *spherical functions*. Many other types of invariance also occur with correspondingly different types of characteristic functions. In each of these cases one considers a set of functions $u_n(x)$ which play a role quite analogous to that of $\exp(i\pi nt/T)$ in (A22). The set has properties equivalent to (A30) and (A46), namely, the orthogonality relation

$$\int u_m^*(x)\, u_n(x)\, dx = \delta_{nm} \tag{A30a}$$

and the *completeness* relation

$$\sum_n u_n^*(x')u_n(x) = \delta(x - x'). \tag{A46a}$$

The coefficients of an expansion $f(x) = \Sigma_n a_n u_n(x)$ are calculated by the generalization of (A33):

$$a_n = \int f(x)u_n^*(x)dx. \tag{A32a}$$

These procedures and formulae apply also to functions of several variables, to variables ranging over finite or infinite intervals and to sets of functions identified by continuous indices ν rather than by discrete indices n. [14]

---

[14] The process of generalization whereby these results are regarded as extensions of vector geometry is outlined in an essay by J. T. Schwartz in *The Mathematical Sciences*, p. 72.

APPENDIX B

NORMAL MODES

In many physical systems two or more variables may oscillate simultaneously but not independently. Familiar examples are a set of coupled pendulums or an electrical circuit in which currents oscillate in several loops. In these examples, as in countless other systems of interest, forces are represented as linear functions of the variables while the energy is a quadratic function, at least near equilibrium. The equations of motion take the form of systems of linear equations with constant coefficients. These systems of equations are solved conveniently by replacing the original variables by new ones which are called *normal coordinates*. These coordinates are so chosen that the equations of the system become uncoupled and each normal coordinate corresponds to an independent degree of freedom. The energy is then reduced to a sum of terms each one depending on a single coordinate. The oscillatory variation of a normal coordinate is called a *normal mode* of oscillation of the physical system.

In the example of two identical coupled pendulums swinging in parallel planes, (see Fig. B1) the normal modes are clearly those in which both pendulums oscil-

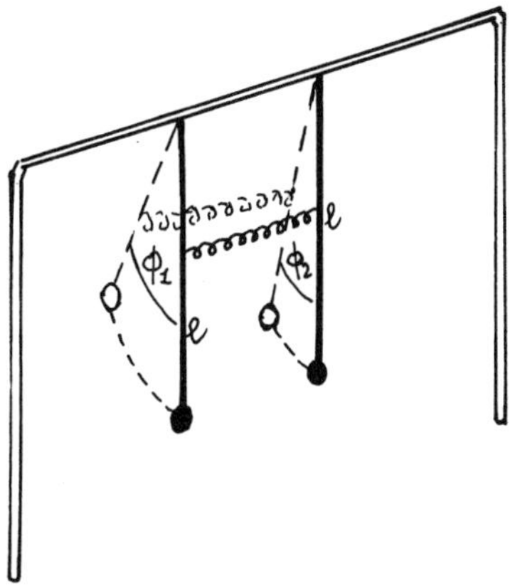

Fig. B1. Two coupled pendulums.

late with equal amplitude either in phase or with opposite phases. In this and in other simple problems, symmetry considerations suffice to identify the normal modes. Otherwise, the normal coordinates are determined by solving the algebraic problem. Often, the difficulty of a new problem lies in recognizing that it can be analyzed in terms of normal coordinates.

B1. TWO-VARIABLE SYSTEMS

A prototype of two-variable systems is afforded by the oscillation of a particle gliding near the bottom of an elliptical bowl. Let us represent the surface of the bowl by the equation

$$z = ax^2 + 2bxy + cy^2. \tag{B1}$$

The gravitational potential energy, V, of a particle of mass m at any point of this surface equals mgz. For small displacements of the particle from the bottom point of the bowl the component of gravity perpendicular to the bottom is compensated approximately by the reaction of the bowl, and the force is represented approximately by the two tangential components

$$F_x = -mg\frac{\partial z}{\partial x} = -2mg(ax + by),$$
$$F_y = -mg\frac{\partial z}{\partial y} = -2mg(bx + cy). \tag{B2}$$

The motions along x and y are coupled because a displacement from the bottom point along x introduces in general a nonzero value of the force component F_y. An initial displacement along x thus gives rise to a complicated motion in which the particle oscillates about the bottom successively in different directions. (The particle's path is called a Lissajous figure.) The oscillation maintains a fixed direction only if the particle's initial displacement lies along one of the axes of the ellipse. In this case the vector force (F_x, F_y) is and remains parallel to the displacement and proportional to it. The mathematical determination of the axes of the ellipse starting from the coefficients a, b, and c serves as a prototype for the identification of normal coordinates.

If we call $\hat{\xi}$ and $\hat{\eta}$ the axes of the ellipse and indicate displacements along these axes by (x_ξ, y_ξ) and (x_η, y_η), the parallelism of the force to a displacement along ξ is represented by the system of equations

$$F_x = -2mg(ax_\xi + by_\xi) = -k_\xi x_\xi, \quad F_y = -2mg(bx_\xi + cy_\xi) = -k_\xi y_\xi. \tag{B3a}$$

The parallelism of the force to a displacement along η is represented by

$$F_x = -2mg(ax_\eta + by_\eta) = -k_\eta x_\eta, \quad F_y = -2mg(bx_\eta + cy_\eta) = -k_\eta y_\eta. \tag{B3b}$$

Each system represents the force vector as resulting from a linear transformation of a displacement vector (x, y); the system also requires each component of the force to be proportional to the corresponding component of the displacement. The proportionality coefficients k_ξ and k_η remain to be determined. All displacements (x, y) that are solution sets of equations (B3) are called *eigenvectors* of the transformation. The proportionality coefficients k_ξ and k_η are called *eigenvalues*.

Normal coordinates are obtained in this problem by a coordinate rotation that replaces the coordinate axes $\hat{x}$ and $\hat{y}$ by $\hat{\xi}$ and $\hat{\eta}$. Calling α the rotation angle, we have

$$\xi = x\cos\alpha + y\sin\alpha, \quad \eta = -x\sin\alpha + y\cos\alpha, \tag{B4}$$

where α is given by

$$\tan 2\alpha = \frac{2b}{a - c}. \tag{B4a}$$

The potential energy and the force are now represented by

$$V = mgz = ½ \left(k_\xi \xi^2 + k_\eta \eta^2\right) \tag{B5}$$

and

$$F_\xi = -k_\xi \xi, \quad F_\eta = -k_\eta \eta, \tag{B6}$$

where k_ξ and k_η are the roots of the *secular equation*

$$\begin{vmatrix} 2mga-k & 2mgb \\ 2mgb & 2mgc-k \end{vmatrix} = 0. \tag{B7}$$

The time variations of the coordinates ξ and η obey the separate equations

$$m \frac{d^2\xi}{dt^2} = -k_\xi \xi, \quad m \frac{d^2\eta}{dt^2} = k_\eta \eta. \tag{B8}$$

Many physical phenomena are mathematically equivalent to the motion of a particle in an elliptical bowl and can be solved by mapping their variables on the (x, y) coordinates of the bowl. In the example of two coupled pendulums of Figure B1, we may call x and y the parallel displacements $l\phi_1$ and $l\phi_2$ of the two pendulums from the equilibrium positions. The energy of the coupling spring may be represented by $½\kappa(l\phi_1 - l\phi_2)^2$, where κ is the force constant of the spring. The energy of the system and its equations of motion coincide then with those of the particle in the bowl if we set

$$a = c = ½\left(\frac{1}{l} + \frac{\kappa}{mg}\right), \quad b = -\frac{\kappa}{2mg}. \tag{B9}$$

(When, as is usual, the coupling between the two pendulums is very weak, the parameters κ and b are very small and the representative bowl is nearly round.)

As an electrical example consider the two-loop circuit shown in Figure B2. Call $Q_1(t)$ and $Q_2(t)$ the variable charges on the capacitors C_1 and C_2. The variations of these charges obey the systems of equations

$$\begin{aligned} L \frac{d^2Q_1}{dt^2} + \frac{Q_1}{C_1} + \frac{Q_1 + Q_2}{C} &= 0, \\ L \frac{d^2Q_2}{dt^2} + \frac{Q_2}{C_2} + \frac{Q_1 + Q_2}{C} &= 0. \end{aligned} \tag{B10}$$

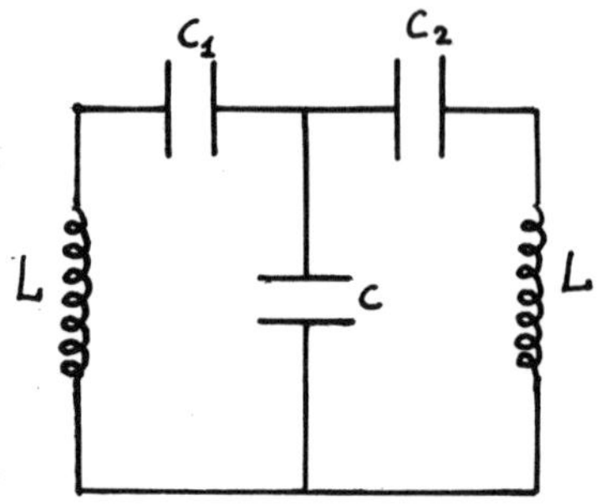

Fig. B2. Two-loop circuit.

This system of equations coincides with the system for the two coordinates of the particle in the bowl with the following substitutions:

$$Q_1 \rightarrow x, \quad Q_2 \rightarrow y, \quad L \rightarrow m,$$
$$\frac{1}{C_1} + \frac{1}{C} \rightarrow 2mga, \quad \frac{1}{C_2} + \frac{1}{C} \rightarrow 2mgc, \quad \frac{1}{C} \rightarrow 2mgb. \tag{B11}$$

(The correspondence of variables is more complicated when the two inductances are different or in the presence of a mutual inductance.) The normal coordinates are the linear combinations of charges Q_1 and Q_2 given by equation (B4) with the substitutions (B11).

B2. MANY-VARIABLE SYSTEMS

Consider a set of N identical particles of mass m, each of which can move along one coordinate $x_i (i = 1, 2, \cdots, N)$. All particles are interconnected by elastic forces, such that the potential energy of the whole set is a quadratic function of all coordinates x_i

$$V = \tfrac{1}{2} \sum_{i=1}^{N} \sum_{j=1}^{N} k_{ij} x_i x_j \tag{B12}$$

with $k_{ij} = k_{ji}$. (A set of N coupled pendulums fits this description.) The equations of motion of the particles form the system

$$m \frac{d^2 x_i}{dt^2} = -\frac{\partial V}{\partial x_i} = -\sum_{j=1}^{N} k_{ij} x_j. \tag{B13}$$

Since this is a system of linear equations with constant coefficients, it has, according to section A2, a set of exponential solutions of the form

$$x_i(t) = x_{i0} e^{-i\omega t}. \tag{B14}$$

Substitution of (B14) reduces the system of differential equations (B13) to the algebraic form

$$-m\omega^2 x_{i0} = -\sum_{j=1}^{N} k_{ij} x_{j0}. \tag{B15}$$

This system of homogeneous linear equations has nonzero solutions only if the determinant of its coefficients equals zero,

$$\begin{vmatrix} k_{11} - m\omega^2 & k_{12} & \dots & k_{1N} \\ k_{21} & k_{22} - m\omega^2 & \dots & k_{2N} \\ \dots & \dots & \dots & \dots \\ k_{N1} & k_{N2} & \dots & k_{NN} - m\omega^2 \end{vmatrix} = \mathrm{Det}\,|k_{ij} - m\omega^2\delta_{ij}| = 0. \tag{B16}$$

This secular equation is of degree N in the frequency ω^2 and has N real roots, which we call $\omega_r^2 (r = 1, 2, \dots, N)$. To each root ω_r^2 corresponds a set of values of x_{i0} which satisfy (B15), but this set is identified only within an arbitrary multiplicative factor. That is, the system of equations determines the ratios among the x_{i0}. Therefore it is convenient to choose, for each root ω_r^2, a base set of x_{i0} whose magnitude is *normalized* by appropriate choice of the multiplicative factor. We shall call this base set u_{ir}. Barring an exceptional case discussed below, the u_{ir} are determined by the joint requirements that they satisfy the system

$$\sum_j k_{ij} u_{jr} = m\omega_r^2 u_{ir} \tag{B17}$$

and the normalization condition

$$\sum_j u_{jr}^2 = 1. \tag{B18}$$

The general solution of (B15) with $\omega = \omega_r$ is then represented by

$$x_{i0} = u_{ir} Q_{r0}, \tag{B19}$$

where the coefficient Q_{r0} represents the amplitude of the rth normal mode of oscillation. The general solution of the equation of motion (B13) is a superposition of all the N normal modes, represented by

$$x_i(t) = \sum_{r=1}^{N} u_{ir} Q_{r0} \exp(-i\omega_r t) = \sum_r u_{ir} Q_r(t). \tag{B20}$$

The N variables

$$Q_r(t) = Q_{r0} \exp(-i\omega_r t) \tag{B21}$$

perform sinusoidal oscillations and constitute the *normal coordinates* of the system of particles.

The normal coordinates $Q_r(t)$ are represented in terms of the initial coordinates x_i by a linear transformation inverse to (B20). Recall that the set of N^2 numbers u_{ir} (i, r = 1, 2, ..., N) which satisfies (B17) and (B18) for all values of **r**, constitutes an orthogonal matrix with the properties

$$\sum_i u_{ir}u_{is} = \delta_{rs}, \quad \sum_t u_{ir}u_{jr} = \delta_{ij}. \tag{B22}$$

(Eq. [A30] is an example of this type of relationship.) Therefore, the coefficients of the coordinate transformation inverse to (B20) constitute a matrix which is the transpose of u_{ir}. We write

$$Q_r = \sum_i u^{\dagger}_{ri} x_i \qquad (u^{\dagger}_{ri} = u_{ir}). \tag{B23}$$

Note that in the simple case of N = 2 one can always express the orthogonal matrix $u^{\dagger}_{ri}$ in the form

$$|u^{\dagger}_{ri}| = \begin{vmatrix} \cos\alpha & \sin\alpha \\ -\sin\alpha & \cos\alpha \end{vmatrix}. \tag{B24}$$

With this expression of $u^{\dagger}_{ri}$, equation (B4) is seen to be a special case of (B23).

As anticipated at the beginning of this appendix, introduction of normal coordinates has the effect of decoupling the equations of the system (B13). Substitution of $x_i = \Sigma_r u_{ir}Q_r$ from (B20) into (B13) and application of (B17) yields

$$m \sum_r u_{ir} \frac{d^2Q_r}{dt^2} = -\sum_j k_{ij} \sum_r u_{jr}Q_r = -\sum_r m\omega_r^2 u_{ir}Qr. \tag{B25}$$

Further multiplication of this equation by $u^{\dagger}_{si}$ and summation over i gives the equation

$$\frac{d^2Q_s}{dt^2} = -\omega_s^2 Q_s \tag{B26}$$

which holds separately for each normal coordinate Q_s. Thus one verifies that each normal coordinate performs free harmonic oscillations with eigenfrequency ω_s.

Independence of normal mode oscillations holds also in the presence of a driving force. One verifies readily that application of an oscillating force $F_{i0} \times \exp(-i\omega t)$ to the particles of the system drives the normal modes of oscillation independently of one another. The same procedure that transforms (B13) into (B26) yields now

$$m \frac{d^2Q_s}{dt^2} = -\, m\omega_s^2 Q_s - F_{s0}e^{-i\omega t}. \tag{B27}$$

Here

$$F_{so} = \sum_i u^{\dagger}_{si} F_{io} \tag{B28}$$

represents the component of the force along the axis of the normal coordinate Q_s.

Many physical systems have symmetries which cause different normal modes to be equivalent, with the same frequency ω_r. In this case the secular equation (B16) has coincident (*degenerate*) roots and the conditions (B17) and (B18) no longer determine the coefficients u_{ir} and the normal coordinates Q_r uniquely. For $\omega_r^2 = \omega_s^2$, if the conditions (B16) and (B17) are satisfied by two sets of coefficients u_{ir} (r fixed, i = 1, 2, . . . N) and u_{is}, they are also satisfied by the set $u_{ir} \cos \beta + u_{is} \sin \beta$ where β is an arbitrary parameter.

An example of this *degeneracy* phenomenomenon occurs in a four-variable system whose potential energy (B12) is represented by

$$V = \tfrac{1}{2}k(x_1^2 + x_2^2 + x_3^2 + x_4^2) + \tfrac{1}{2}\kappa \left[(x_1 - x_2)^2 + (x_2 - x_3)^2 + (x_3 - x_4)^2 + (x_4 - x_1)^2\right], \tag{B29}$$

with only two different force constants, k and κ. Figure B3 shows a model of such a system. The secular equation has then the four roots

$$m\omega_1^2 = k, \qquad m\omega_2^2 = m\omega_3^2 = k + 2\kappa, \qquad m\omega_4^2 = k + 4\kappa. \tag{B30}$$

A set of solutions of the equations (B17) and (B18) is

i	u_{i1}	u_{i2}	u_{i3}	u_{i4}
1	$\frac{1}{2}$	$\sqrt{\frac{1}{2}}$	0	$\frac{1}{2}$
2	$\frac{1}{2}$	0	$\sqrt{\frac{1}{2}}$	$-\frac{1}{2}$
3	$\frac{1}{2}$	$-\sqrt{\frac{1}{2}}$	0	$\frac{1}{2}$
4	$\frac{1}{2}$	0	$-\sqrt{\frac{1}{2}}$	$-\frac{1}{2}$

(B31)

In each of the two degenerate modes, r = 2 and r = 3, two of the particles are at rest and the other two oscillate with opposite phase, thus exerting equal and opposite forces on the particles at rest. An alternative pair of degenerate modes has the form $u_{i2} \cos \beta + u_{i3} \sin \beta$, with $\beta = \pm 45°$, and is given by

i	1	2	3	4
$\beta = 45°$	$\frac{1}{2}$	$\frac{1}{2}$	$-\frac{1}{2}$	$-\frac{1}{2}$
$\beta = -45°$	$\frac{1}{2}$	$-\frac{1}{2}$	$-\frac{1}{2}$	$\frac{1}{2}$

(B31a)

The introduction of normal coordinates proves generally useful in all phenomena governed by linear equations. In particular, the treatment given above

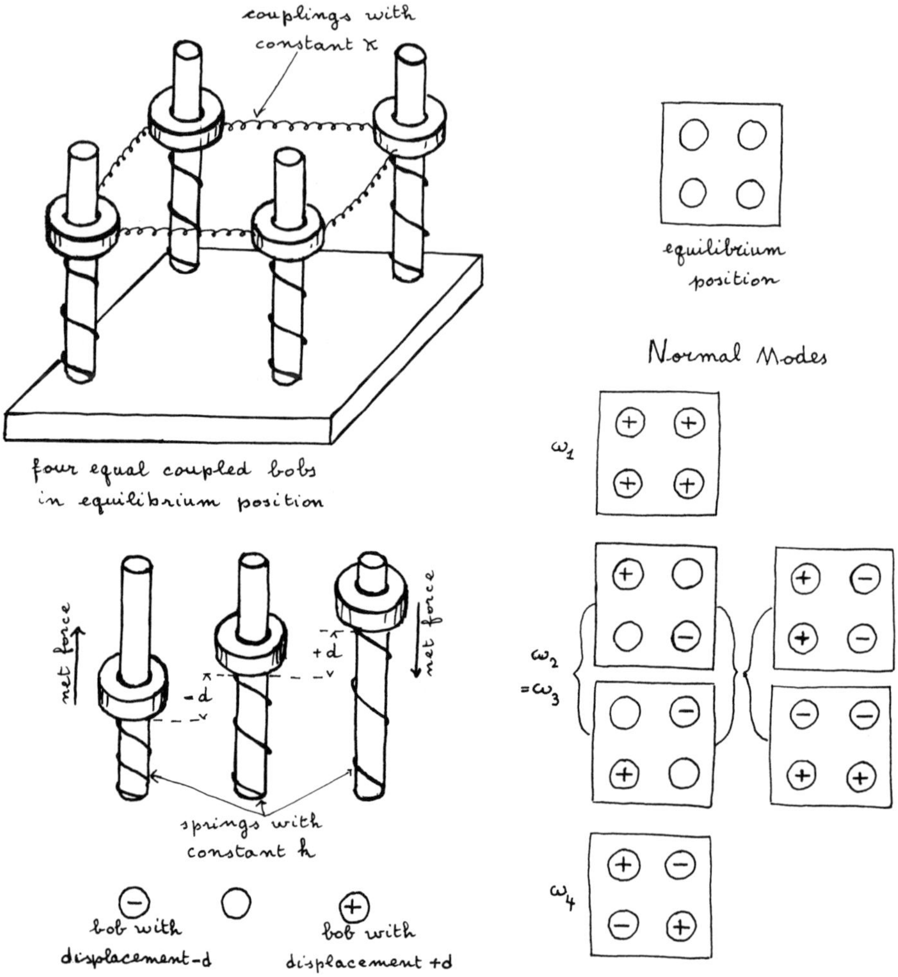

Fig. B3. System of four coupled bobs which can oscillate up and down along their pegs.

can be extended to systems whose oscillations are damped. It can also be extended to systems with infinitely many variables, or even to continuous systems such as a vibrating string. These extensions, as well as additional details, are found, for example, in J. C. Slater and N. H. Frank, *Mechanics* (New York: McGraw-Hill Book Co.), 1947, chapters 7 ff.

APPENDIX C

FOURIER ANALYSIS OF SYSTEMS WITH VARIABLE PROPERTIES

As emphasized in section A2, fields or charges driven by sources with sinusoidal variations also vary sinusoidally, with the same frequency (or propagation vector) as the source. This match of response to drive stems from the invariance of the parameters which determine the response of the system. Accordingly, this match should be lost or at least distorted when the same parameters vary in the course of time (and/or from point to point in space).

The Fourier analysis procedure is desirable for the treatment of constant-parameter systems subjected to nonsinusoidal drives, because it affords a ready calculation of the separate influence of each Fourier component of the drive. The same procedure still proves useful for the study of systems with variable parameters. Consider, for example, a circuit with condenser plates subjected to a mechanical vibration. If the circuit is driven by a sinusoidal potential source, the condenser charge will no longer simply oscillate sinusoidally with the driving frequency. However, one can usefully relate the Fourier analysis of the charge variations to the frequency of the mechanical disturbance. We shall first treat this example, pointing out features that have broader applicability, and then we will consider a vibrating string loaded with a mass distribution which varies along the string.

C1. CIRCUIT WITH A VARIABLE PARAMETER

Consider the charge oscillations in the circuit with inductance L, resistance R, and capacity C corresponding to equation (A14), assuming now that the distance between the condenser plates oscillates by a factor $1 + \alpha \cos \eta t$, the coefficient α being a small fraction of unity. Accordingly, the term $Q(t)/C$ of (A14) is multiplied by this factor. The driving potential oscillations $V(t)$ are represented by a complex exponential, as in section A2.4. The charge oscillations are then governed by the equation

$$L \frac{d^2Q}{dt^2} + R \frac{dQ}{dt} + [1 + \alpha \cos \eta t] \frac{Q(t)}{C} = V_0 \exp(-i\omega t). \tag{C1}$$

The equation (A14) without the term $\alpha \cos \eta t$ was reduced to a time-independent algebraic equation by setting $Q(t) = Q_0 \exp(-i\omega t)$, because the time-dependent factor $\exp(-i\omega t)$ could be removed from each term of the equation. Reduction of (C1) to algebraic form proceeds less rapidly owing to the presence of the time-dependent coefficient $\alpha \cos \eta t$. However, this coefficient is a periodic function of time, with period $2\pi/\eta$. This observation suggests that $Q(t)$ be represented as the product of $\exp(-i\omega t)$ and of a periodic amplitude modulation factor with the same period as $\cos \eta t$. Accordingly, we set

$$Q(t) = \left[\sum_n Q_n e^{-in\eta t}\right] e^{-i\omega t} = \sum_n Q_n e^{-i(\omega + n\eta)t}. \tag{C2}$$

Substitution of this expression of the charge variation in equation (C1) again permits the removal of the factor exp(−iωt). The resulting equation,

$$\sum_n \left[-(\omega + n\eta)^2 L - i(\omega + n\eta)\, R + \frac{1}{C} \right] Q_n e^{-in\eta t} + \alpha \cos \eta t \frac{1}{C} \sum_n Q_n e^{-in\eta t} = V_0, \qquad \text{(C3)}$$

is now no longer time independent; instead, all of its terms are periodic functions of time. Reduction to algebraic form will be achieved by expanding each term of (C3) into a Fourier series. To this end we work out the expansion of the product of two periodic functions

$$\cos \eta t \sum_n Q_n e^{-in\eta t} = \tfrac{1}{2} \sum_n Q_n [e^{-i(n-1)\eta t} + e^{-i(n+1)\eta t}] = \tfrac{1}{2} \sum_n (Q_{n+1} + Q_{n-1}) e^{-in\eta t}. \qquad \text{(C4)}$$

Substitution of this expansion in (C3) and multiplication by C reduce (C3) to the form

$$\sum_n \left\{ [1 - i(\omega + n\eta)RC - (\omega + n\eta)^2 LC] Q_n + \tfrac{1}{2}\alpha (Q_{n+1} + Q_{n-1}) - CV_0 \delta_{n0} \right\} e^{-in\eta t} = 0, \qquad \text{(C5)}$$

where δ_{n0} is the Kronecker symbol defined by equation (A30).

This time-dependent equation will be fulfilled for all values of t if and only if the coefficient of exp(−inηt) vanishes for each value of n. Reduction to the algebraic form

$$[1 - i(\omega + n\eta)\, RC - (\omega + n\eta)^2 LC]\, Q_n + \tfrac{1}{2}\alpha(Q_{n+1} + Q_{n-1}) - CV_0\, \delta_{n0} = 0 \qquad \text{(C6)}$$

has thus been achieved. Notice that (C6) represents a system of infinitely many equations rather than a single equation, because the coefficient α interlinks each Fourier coefficent Q_n of the variable charge Q(t) with Q_{n+1} and Q_{n-1}. This interlinkage disappears in the limit where the amplitude α of the condenser-plate oscillation vanishes. The assumption (C2) has led to reduction of the differential equation (C1) first to the periodic form (C3) and then to the time-independent algebraic form (C6). We conclude that (C2) represents the time dependence of the charge Q(t) correctly.

In general, when the coefficients of a linear differential equation oscillate sinusoidally with frequency η and the inhomogeneous term varies according to exp(−iωt), the Fourier representation of the solution contains only terms with frequencies that *differ from* ω *by a multiple of* η. Conversely, the experimental detection of *side frequencies* $\omega + n\eta$ in the frequency analysis of the response of a variable driven with the single frequency ω constitutes evidence that some parameter of the system oscillates with the frequency η. The Fourier components with frequencies $\omega + n\eta$ may be described as due to a modulation of the primary effect of the drive by the variations of the parameter. Should the relevant parameter (or parameters) oscillate with a multiple periodicity, i.e., in proportion to $1 + \Sigma_r \alpha_r \cos \eta_r t$, components with all frequencies $\omega + \Sigma_r n_r \eta_r$ would appear in the Fourier analysis of the driven variable.

Replacement of the differential equation (C1) by the algebraic system (C6) might be of little help were it not that in practice α is often sufficiently small to permit an adequate solution of (C5) by expansion of Q_n into powers of the coefficient α which interlinks the different equations in the system (C6). We utilize here an important procedure of algebra. Whenever the equations of a system are interlinked by terms proportional to a coefficient α, expansion into powers of α, by setting $Q_n = \Sigma_s Q_n^{(s)} \alpha^s$, unravels the system. Often an appropriate solution can be obtained by calculating only the coefficients of the first few powers of α.

In our case, substitution of the expansion of Q_n into (C6) transforms this equation itself into a power series in α. Equation (C6) is then satisfied provided the coefficient of each power of α is set to zero. (This procedure is analogous to what we have done for the Fourier series [C5].) Setting to zero only the coefficient of the zeroth power of α shows that, to zeroth order in α, Q_0 is given by (A17) and all other Q_n vanish. Setting to zero the coefficient of the first power of α, in the expansion of (C6), shows that, to first order in α,

$$Q_{\pm 1} = \frac{-\frac{1}{2}\alpha Q_0}{1 - i(\omega \pm \eta) RC - (\omega \pm \eta)^2 LC} = \frac{-\frac{1}{2}\alpha CV_0}{[1 - i\omega RC - \omega^2 LC][1 - i(\omega \pm \eta) RC - (\omega \pm \eta)^2 LC]}, \tag{C7}$$

Q_0 is unchanged and all Q_n with $|n|>1$ still vanish. Continuing the procedure to higher order, one finds that the expansion of each Q_n begins with the term $\alpha^{|n|}$.

Substituting into the Fourier series (C2) the expansion thus obtained for each coefficient Q_n, one finds that when α^2 is negligible *only the first pair* of side frequencies, $\omega \pm \eta$, occurs in the Fourier analysis of Q(t). The coefficients Q_1 and Q_{-1} of these frequencies are proportional to α, according to (C7). Conversely, experimental measurement of Q_1 and Q_{-1} determines the magnitude of the parameter α. This result underlies very extensive applications to the experimental analysis of unknown systems.

C2. STRING WITH VARIABLE LOAD

The usual problem of transverse vibrations of a string of infinite length along an x axis, with mass density ρ and subjected to a tension T, leads to the equation

$$T \frac{\partial^2 u}{dx^2} = \rho \frac{\partial^2 u}{\partial t^2}, \tag{C8}$$

where u(x, t) represents the transverse displacement at the abscissa x and time t. Owing to invariance in space and time, (C8) has solutions analogous to (A9),

$$u(x,t) = u_0 \exp[i(kx - \omega t)], \tag{C9}$$

where u_0 is arbitrary and the wave vector[1] k and frequency ω are subject to the condition $Tk^2 = \rho\omega^2$; this condition sets the propagation velocity ω/k at

$$V = (T/\rho)^{1/2}. \tag{C10}$$

[As noted in section A2. 2, it is the real part of u(x, t) which represents a traveling wave.]

For a string with a sinusoidally variable mass density, equation (C8) is replaced by the equation analogous to (C1) — though homogeneous—

$$T\frac{\partial^2 u}{\partial x^2} = (1 + \alpha \cos \eta x)\rho \frac{\partial^2 u}{\partial t^2}. \tag{C11}$$

Analogy to (C2) suggests a solution of the form

$$u(x,t) = \left[\sum_n u_n \exp(in\eta x)\right] \exp[i(kx - \omega t)] = \sum_n u_n \exp\{i[(k + n\eta)x - \omega t]\}, \tag{C12}$$

which represents a superposition of waves with equal frequencies and with wave vectors differing by multiples of η. Substitution into (C11) yields an equation analogous to (C3), which can be transformed first to a form analogous to (C5) and, finally, to the analog of (C6),

$$\left[\frac{V^2}{\omega^2}(k + n\eta)^2 - 1\right] u_n - \tfrac{1}{2}\alpha(u_{n+1} + u_{n-1}) = 0. \tag{C13}$$

Here, again, the case where α is a small number leads to an approximate solution. To first order in α we can take $k = \omega/V$ as in (C10), u_0 arbitrary,

$$u_{\pm 1} = \tfrac{1}{2}\alpha \frac{k^2}{(k \pm \eta)^2 - k^2} u_0, \tag{C14}$$

and all other $u_n = 0$. Each u_n with $|n| > 1$ differs from zero only to order

---

1. No vector symbol is needed here for k because we are dealing with propagation parallel to the x axis; however, we must distinguish propagation toward +x and toward −x by the sign of k.

$\alpha|n|$. Substitution of these results in (C12) yields, to first order in α,

$$u(x,t) = u_0 \left[e^{ikx} + \tfrac{1}{2}\alpha \frac{k^2}{(k+\eta)^2 - k^2} e^{i(k+\eta)x} + \tfrac{1}{2}\alpha \frac{k^2}{(k-\eta)^2 - k^2} e^{i(k-\eta)x} \right] e^{-i\omega t}. \quad \text{(C15)}$$

This formula shows that the inhomogeneity of the string has induced Fourier components with wave vectors $k \pm \eta$. The amplitude of these components remains small, of order α, as long as neither denominator $(k \pm \eta)^2 - k^2$ approaches zero.

Approach to zero occurs when the wave vectors k and $k \pm \eta$ have approximately the same magnitude, that is, when k and $k \pm \eta$ differ only in sign, so that the corresponding waves travel in opposite directions and have approximately equal wavenumbers. This situation occurs when $k = \pm\tfrac{1}{2}\eta$, in which case $V|k \mp \eta| = Vk = \omega$, that is, one of the induced components is driven at resonance. The amplitude of the resonant component becomes then infinite, in the approximation of (C14) and (C15).

The infinite value of the amplitude of this secondary wave indicated by (C14) is, of course, a mathematical artifact due to the failure of the approximation in this special circumstance. A more careful treatment of (C13) is given in the next section to obtain a realistic solution whenever $|k \pm \eta| \sim |k|$. We only stress here that sinusoidal variations of the properties of a medium can alter drastically the propagation of a disturbance, even though the variations are very small.

C3. HINDERED PROPAGATION AND FREQUENCY GAPS

Consider the special case where $k \sim \tfrac{1}{2}\eta$ for which $|k - \eta| \sim k$. The amplitude u_{-1} of the Fourier component with wave vector $k - \eta \sim -k$ is large in this case because it is driven near resonance; this secondary wave arises, in effect, by reflection of the wave with amplitude u_0 and wave vector k. The coefficient of u_n in the system of equations (C13) then approaches zero for both $n = 0$ and $n = -1$. Accordingly, a more accurate treatment requires that the two equations with $n = 0$ and $n = -1$ be solved simultaneously in the initial approximation. The two equations are

$$\left[\frac{V^2k^2}{\omega^2} - 1\right] u_0 - \tfrac{1}{2}\alpha u_{-1} - \tfrac{1}{2}\alpha u_1 = 0,$$

$$\left[\frac{V^2(k-\eta)^2}{\omega^2} - 1\right] u_{-1} - \tfrac{1}{2}\alpha u_0 - \tfrac{1}{2}\alpha u_{-2} = 0, \quad \text{(C16)}$$

where k/ω no longer need be equal to V, but V is still given by (C10). We disregard here the amplitudes u_1 and u_{-2} which are still expected to remain small, and we concentrate on u_0 and u_{-1}, both of which are of zeroth order in α when $k \sim \tfrac{1}{2}\eta$ and $\omega \sim Vk \sim V\tfrac{1}{2}\eta$.

When k is exactly equal to $\frac{1}{2}\eta$ and u_1 and u_{-2} are disregarded, the equations (C16) have the alternative simultaneous solutions

$$u_{-1} = \pm u_0 \tag{C17}$$

for frequency values fulfilling

$$\frac{1}{4}\,\frac{V^2\eta^2}{\omega^2} - 1 = \pm\tfrac{1}{2}\alpha. \tag{C18}$$

The required frequencies are

$$\omega_\pm = \tfrac{1}{2}\,\frac{V\eta}{(1 \pm \frac{1}{2}\alpha)^{1/2}}\,. \tag{C19}$$

The essential point is that one of these frequencies lies below $Vk = \frac{1}{2}V\eta$ and the other one above it. This frequency difference originates from the fact that the complete wave (C12) is, in this approximation, one of the standing waves

$$u_+(x,t) \propto \cos(\tfrac{1}{2}\eta x)\exp(-i\omega_+ t),$$
$$u_-(x,t) \propto \sin(\tfrac{1}{2}\eta x)\exp(-i\omega_- t). \tag{C20}$$

The wave u_+ has its antinodes where the string density $\rho\,(1 + \alpha\cos\eta x)$ is largest; therefore, the vibrational motion is concentrated in the heavier parts of the string and the frequency ω_+ is accordingly reduced. The opposite occurs for the wave u_- whose antinodes lie where the string density is smallest. Note that at $k = \frac{1}{2}\eta$ the frequency no longer has a single value for a given value of the primary wave vector k.

One can also see how the splitting of the frequencies originates as k approaches $\frac{1}{2}\eta$ from above or below. Let us indicate the fractional departure of k from $\frac{1}{2}\eta$ by δ, that is, $k = \frac{1}{2}\eta(1 \pm \delta)$ in our original approximation (C14). We find then that u_{-1} is approximately $\mp(\alpha/8\delta)u_0$, so that the complete wave becomes

$$u(x,t) \approx \left\{\exp\left[i\tfrac{1}{2}\,\eta(1 \pm \delta)x\right] \mp \frac{\alpha}{8\delta}\exp\left[-i\tfrac{1}{2}\,\eta(1 \pm \delta)x\right]\right\} e^{-i\omega t}. \tag{C21}$$

Even while δ is still much larger than α, the amplitude of $u(x,t)$ begins to build up where the string density is small, and down where the string density is large. Therefore, the frequency rises as the amplitude increases (and drops as the amplitude decreases) with respect to its normal value Vk. The resulting behavior of ω as a function of k is illustrated by Figure C1, a very important diagram which applies to all phenomena of wave propagation in periodic media.

No wave appears to propagate freely along the string with frequency lying in the *gap* between ω_+ and ω_-. What happens, then, if a disturbance with a frequency ω *within this gap* is forced upon the string by an external agent? The answer is that a wave generated by an external drive can propagate with

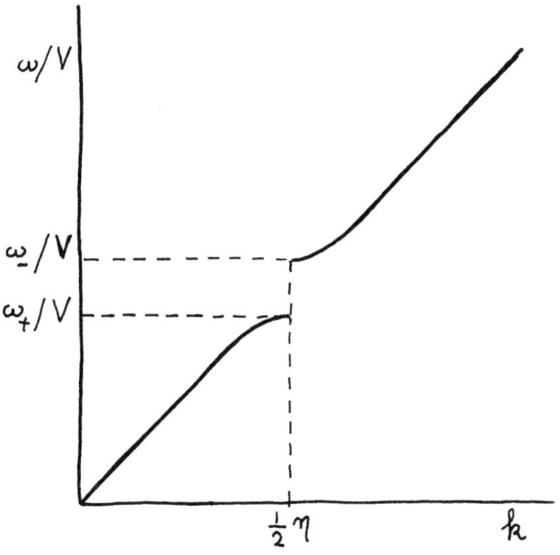

Fig. C1 Frequency gap.

characteristics that we had excluded under the unstated assumption of dealing with a string of periodic properties but of *infinite* extension. Under this assumption only waves with real wavenumbers can be accepted as representing a realistic physical phenomenon; waves with complex wavenumbers have infinitely large amplitudes at infinite distance either in the direction of propagation or in the opposite direction. However, the point of application of a drive is a singularity in the string which divides it into two semi-infinite parts. The mathematical representation of a wave on one side of this singularity does not hold for the other side. The restriction on the mathematical representation of a wave is now merely that it remain finite on one side of the point of application of the drive.

We set the origin of coordinates at the point of application and consider the solution of the system of equations (C16) in the region $x > 0$. The complex wavenumber

$$k = \tfrac{1}{2}\eta(1 + i\kappa) \tag{C22}$$

with $\kappa \geqslant 0$ is now acceptable. Substitution of this wavenumber in (C17), discarding, as before, the amplitudes u_1 and u_{-2}, yields

$$\left[\frac{V^2(\tfrac{1}{2}\eta)^2}{\omega^2}(1 - \kappa^2 + 2i\kappa) - 1\right] u_0 - \tfrac{1}{2}\alpha u_{-1} = 0, \tag{C23}$$

$$\left[\frac{V^2(\tfrac{1}{2}\eta)^2}{\omega^2}(1 - \kappa^2 - 2i\kappa) - 1\right] u_{-1} - \tfrac{1}{2}\alpha u_0 = 0.$$

This system is solved by introducing a parameter β which ranges from zero to π. We set

$$u_0 = \tfrac{1}{2}e^{-i\beta/2}, \qquad u_{-1} = \tfrac{1}{2}e^{i\beta/2} \tag{C24}$$

and find that (C23) is satisfied, to lowest order in α, by

$$\omega = \frac{\tfrac{1}{2}V\eta}{(1 + \tfrac{1}{2}\alpha \cos \beta)^{1/2}}, \tag{C25a}$$

$$\kappa = \tfrac{1}{4}\alpha \sin \beta. \tag{C25b}$$

As β increases from zero to π, the value of ω rises across the gap from ω_+ to ω_-, and the value of κ rises from 0 to $\tfrac{1}{4}\alpha$ and then returns to zero. Substitution of (C22), (C24), and (C25b) in (C12) yields

$$u(x, t) = \cos [\tfrac{1}{2}(\eta x - \beta)] \exp [-\tfrac{1}{8}\eta\alpha(\sin \beta)x - i\omega t]. \tag{C26}$$

This solution represents a wave which is driven into the string starting from $x = 0$ toward positive x with a frequency in the gap at $\omega \sim \tfrac{1}{2}V\eta$. The wave penetrates only to a distance of the order of $4/(\alpha \sin \beta)$ wavelengths because its amplitude decays exponentially. There results a phenomenon of reflection giving rise to the formation of a standing wave, represented by (C25). This is a one-dimensional analog of Bragg reflection.

APPENDIX D

NOTES ON STATISTICAL DISTRIBUTIONS

Statistics serves to describe and analyze phenomena in which the result of any single experiment is not predictable in full detail but regularities are nevertheless observed in the combined results of numerous experiments. In the physical sciences one regards the experiments as performed on an "ensemble" of "similar" systems, that is, of systems prepared and treated in identical manner. This basic physical concept is developed and illustrated in chapter 2 of *BPC*, volume 5. Here we review definitions and formulae in the context of simple examples.

D1. PROBABILITY DISTRIBUTIONS

The predictions about a series of throws of a pair of dice derive from the initial statement that all faces of each die have equal probability of turning upward in a throw. (The basis of this statement is discussed in section 5.2.) Since each die has six faces, each face of a die has the probability $1/6$ of turning up in any one throw. There are 36 combinations of faces of two dice, all equally probable; therefore, each combination has probability $1/36$ of turning up in any one throw. Out of the 36 combinations, the combination of the two faces with the number 1, which we call (1, 1), yields a point count of 2. Two combinations, namely, (2, 1) and (1, 2), yield a point count of 3, etc. Thus a 4-point throw has a total probability $3/36$ because it can result from 3 combinations, each with equal probability $1/36$. The number of combinations for each point count and the corresponding probabilities are:

Point count	n	2	3	4	5	6	7	8	9	10	11	12
Combinations	C_n	1	2	3	4	5	6	5	4	3	2	1
Probability	p_n	$1/36$	$2/36$	$3/36$	$4/36$	$5/36$	$6/36$	$5/36$	$4/36$	$3/36$	$2/36$	$1/36$.

The set of values p_n, given in this table as a function of the point count n, is called the *probability distribution* of an n-point throw. In a long series of N throws, an n-point throw occurs a number of times ν_n such that the ratio ν_n/N, called the experimental *frequency* of n-point throws, approaches the probability p_n very closely.

In quantum physics, the probability distribution $p_1, p_2, \ldots, p_r, \ldots$ of alternative events $E_1, E_2, \ldots, E_r, \ldots$ is determined either by experimental observation of the frequency of occurrence of events, or by theory, though not necessarily from an initial statement of equal probability.

D2. MEAN AND MEAN DEVIATIONS

An important quantity in a series of dice throws is the arithmetic mean of the point counts observed in the series. If a 2-point throw occurred ν_2 times, a 3-point throw ν_3 times, etc., then the mean point count is

$$\frac{2\nu_2 + 3\nu_3 + \ldots + 12\nu_{12}}{\nu_2 + \nu_3 + \ldots + \nu_{12}} = \frac{\sum_n n\nu_n}{\sum_n \nu_n} = \frac{\sum_n n\nu_n}{N}. \tag{D1}$$

Since the frequency of occurrence of an n-point throw, namely, the ratio ν_n/N, approaches the probability p_n, the mean point count approaches the quantity

$$\langle n\rangle = \sum_n np_n = 2\frac{1}{36} + 3\frac{2}{36} + \ldots + 12\frac{1}{36} = 7. \tag{D2}$$

This quantity is called the *average* or the *mean* point count in a throw, whether determined by experimental observation of frequencies or by calculation of probabilities. It is a parameter—that is, a numerical characteristic—of the probability distribution. In the problem of dice throwing, a 7-point throw is also the most probable result because point counts larger and smaller than 7 by equal amounts are equally probable. However, the probability distributions of other processes may not possess this symmetry and their mean does not, in general, coincide with the most probable value. The probability distribution of the dice point count has many characteristic parameters beside the mean, such as the mean square count $\langle n^2\rangle = \Sigma_n n^2p_n$, the mean cube $\langle n^3\rangle = \Sigma_n n^3p_n$, etc.

It is often more interesting to characterize the distribution of the deviation of the count from its mean, that is, the distribution of $n - \langle n\rangle$, than the distribution of n itself. The mean deviation vanishes, of course, because deviations by excess or defect cancel out on the average. The mean square deviation is the most interesting parameter, next to the mean of n itself, and is related to the mean square by

$$\langle (n - \langle n\rangle)^2\rangle = \sum_n p_n(n - \langle n\rangle)^2 = \langle n^2\rangle - \langle n\rangle^2. \tag{D3}$$

This *mean square deviation* is also indicated by Δn^2. In the example of dice throws we have

$$\Delta n^2 = \langle (n - \langle n\rangle)^2\rangle = \frac{1}{36}(2-7)^2 + \frac{2}{36}(3-7)^2 + \ldots + \frac{1}{36}(12-7)^2 = 5.83. \tag{D4}$$

The *root mean square deviation* Δn, equal to 2.41 in our example, indicates the order of magnitude of the deviations from the mean which are likely to occur.

D3. REPEATED TRIALS

Consider an experiment in which a certain event has a probability p to occur; for example, a dice throw in which the 4-point event has the probability $3/36 = 1/12$. In a series of N experiments the number of times, ν, in which the event occurs approaches the product of N and of the probability p, as noted above. If various trial series of N experiments each are performed, the value of ν varies in general from one trial series to the next. We consider here the distribution of ν in successive trials.

The probability $P(\nu)$ that the event occurs exactly ν times in a trial series can be calculated as follows: In any one series of N experiments, ν events may be distributed differently in the sequence (for instance, three events may be the first, second, and third, or the fifth, seventh, and eleventh). The number of alternative distributions of ν events in a series of N experiments is $N(N-1)\cdots(N-\nu+1)/\nu!$. Each experiment has probability p of yielding the event, and probability $1-p$ of not yielding it. The probability that the event occurs in ν specified experiments of the series is p^{ν}; the probability that it does not occur in any of the $N-\nu$ remaining experiments is $(1-p)^{N-\nu}$. The joint probability that the event occurs just in the ν specified experiments and in none of the others is $p^{\nu}(1-p)^{N-\nu}$. This probability multiplied by the number of alternative distributions of the ν events yields

$$P(\nu) = \frac{N(N-1)(N-2)\cdots(N-\nu+1)}{\nu!}\, p^{\nu}(1-p)^{N-\nu}. \tag{D5}$$

This probability, as a function of ν, is called the *binomial distribution*. The mean value of ν is

$$\langle\nu\rangle = \sum_{\nu} \nu P(\nu) = Np, \tag{D6}$$

as expected. The mean square deviation of ν is

$$\langle(\nu-\langle\nu\rangle)^2\rangle = \sum_{\nu}(\nu-Np)^2\, P(\nu) = Np(1-p). \tag{D7}$$

If both Np and ν are large numbers, the probability distribution (D5) is represented approximately by the *Gaussian distribution*

$$P(\nu) \approx [2\pi Np(1-p)]^{-1/2} \exp\left[-\tfrac{1}{2}\frac{(\nu-Np)^2}{Np(1-p)}\right]. \tag{D8}$$

The approximation is surprisingly good when Np and ν are only as large as 5 or 10.

D4. POISSON DISTRIBUTION

The study of the time distribution of atomic events may be regarded as a study of repeated trials. A period of observation of T seconds may be regarded as a trial consisting of a very large number N of experiments, each of them lasting for a very small interval $\Delta T = T/N$. The probability p of occurrence of an event in one experiment is therefore extremely small, but the mean number of events in a trial, $\langle\nu\rangle = Np$, need not be small. Under these conditions, ν is negligible as compared to N, and p is conveniently represented in the form $\langle\nu\rangle/N$. Substitution of this expression of p in (D5) yields

$$P(\nu) = 1\left(1-\frac{1}{N}\right)\left(1-\frac{2}{N}\right)\cdots\left(1-\frac{\nu-1}{N}\right)\frac{\langle\nu\rangle^{\nu}}{\nu!}\left(1-\frac{\langle\nu\rangle}{N}\right)^{N-\nu}. \tag{D9}$$

Upon taking the limit $N \to \infty$, all terms with N in their denominator can be set at zero, except in the factor $(1 - \langle\nu\rangle/N)^N$ where N appears in the exponent. Since the limit of this factor is $\exp(-\langle\nu\rangle)$, $P(\nu)$ reduces to the Poisson formula

$$P(\nu) = \frac{\langle\nu\rangle^\nu}{\nu!}\, e^{-\langle\nu\rangle}. \tag{D10}$$

The mean of this distribution is, of course, $\langle\nu\rangle$, and its mean square deviation reduces to $Np = \langle\nu\rangle$. This last result is usually applied to estimate the precision of counting experiments in atomic and subatomic physics.

The Poisson distribution formula can also be established by considering how $P(\nu)$ varies as a function of $\langle\nu\rangle$. The mean value $\langle\nu\rangle$ itself is proportional to the time interval T over which atomic events are scored, $\langle\nu\rangle = T/\tau$. If T is increased by ΔT, the probability $P(\nu)$ increases by the product of $P(\nu - 1)$—the preexisting probability of having observed $\nu - 1$ events—and of $\Delta T/\tau$ the probability of occurrence of νth event during the time ΔT. The probability $P(\nu)$ also decreases during ΔT by $P(\nu)\Delta T/\tau$, since any additional event raises the total number from ν to $\nu + 1$. The variation of $P(\nu)$ is thus represented by the

$$\Delta P(\nu) = P(\nu - 1)\,\frac{\Delta T}{\tau} - P(\nu)\,\frac{\Delta T}{\tau}. \tag{D11}$$

Dividing by $\Delta T/\tau = \Delta\langle\nu\rangle$ and taking the limit for $\Delta T \to 0$ reduces (D11) to the differential equation

$$\tau\,\frac{dP(\nu)}{dT} = \frac{dP(\nu)}{d\langle\nu\rangle} = P(\nu - 1) - P(\nu). \tag{D12}$$

Equation (D12) actually represents a system of differential equations, since it interlinks $P(\nu)$ with $P(\nu - 1)$. Solution of this system with the initial condition $P(\nu) = \delta_{\nu 0}$ at $\langle\nu\rangle = 0$ yields (D10).

D5. **A CORRELATION TEST**

Consider a set of balls, contained in a box, which are of several colors—for example, white, red, and blue—and are marked with different numbers of spots—for example, one, two, three, or four spots. One may wish to determine whether the number of spots and the color of each ball are correlated. For this purpose one may extract a sample of N balls from the box and classify them according to color and number of spots. Call the total number of balls of each color N_w, N_r, and N_b, respectively, the total number of balls with each number of spots N_1, N_2, N_3, or N_4 and, for example, ν_{r3} the number of red balls with 3 spots. All these numbers may be arranged in rows and columns as in the following *contingency table:*

	Number of Spots				
Color	1	2	3	4	Total
White	ν_{w1}	ν_{w2}	ν_{w3}	ν_{w4}	N_w
Red	ν_{r1}	ν_{r2}	ν_{r3}	ν_{r4}	N_r
Blue	ν_{b1}	ν_{b2}	ν_{b3}	ν_{b4}	N_b
Total	N_1	N_2	N_3	N_4	N

An analysis of possible correlations between color and spot number starts usually from the initial assumption that there is no correlation. One regards then the N balls as distributed among the 12 boxes of the table with equal probability, subject only to the condition that the totals for the boxes in each row and column have the observed values $N_w, N_r, N_b, N_1, N_2, N_3$, and N_4. The probability of any set of values of $\nu_{w1}, \nu_{w2}, \cdots, \nu_{b4}$ can then be calculated, and from the distribution of this probability the mean of each number ν is obtained. The mean number, for example, of red balls with three spots equals the number N of balls in the sample multiplied by the fraction of balls which are red, namely, N_r/N, and by the fraction of balls which have three spots, namely, N_3/N,

$$\langle \nu_{r3} \rangle = N \frac{N_r}{N} \frac{N_3}{N} = \frac{N_r N_3}{N}. \tag{D13}$$

Any correlation between color and spot number brings about systematic deviations of the numbers in the table from their mean values (D13) calculated on the basis of no correlation. In the absence of correlation there are deviations due to sampling accidents. An index of the expected magnitude of these deviations for each number, like ν_{r3}, is provided by the mean square deviation according to the Poisson distribution. A convenient index of the aggregate deviation of all numbers in the table from their means is found to be the expression

$$\chi^2 = \frac{(\nu_{w1} - \langle \nu_{w1} \rangle)^2}{\langle \nu_{w1} \rangle} + \frac{(\nu_{w2} - \langle \nu_{w2} \rangle)^2}{\langle \nu_{w2} \rangle} + \cdots + \frac{(\nu_{b4} - \langle \nu_{b4} \rangle)^2}{\langle \nu_{b4} \rangle}. \tag{D14}$$

The value of this expression depends on sampling accidents, besides possible correlations. The sums of the deviations of the numbers ν in each row and in each column of the table must vanish, so that only six out of the 12 numbers in the table are independent. A value of the index χ^2 comparable to the number of independent deviations (6 in our example) is likely to arise from sampling accidents alone. A much larger value is unlikely. The number of independent variations is called the *number of degrees of freedom* in the table. Statistical tables give the probability that χ^2 exceeds any given value, owing to sampling accidents only, for each number of degrees of freedom. If this probability is very small, one concludes that a systematic correlation exists.

D6. CORRELATION COEFFICIENT

When two variables, like the color and number of spots of balls in the preceding example, are correlated, it is useful to describe the kind and magnitude of the correlation by some index, to signify, for example, that the blue color is preferentially associated with four spots. The general procedure to investigate and describe correlations is called *analysis of variance*. Here we consider the correlation coefficient which is defined only for pairs of variables represented by numerical indices.[1]

Consider then two variables x and y, whose values can be determined by a single observation but which vary in successive observations with some probability distribution, and consider their deviations $x - \langle x \rangle$ and $y - \langle y \rangle$. A correlation of these variables may be such that positive deviations of both variables ($x - \langle x \rangle > 0, y - \langle y \rangle > 0$) are frequently found in single observations and so are negative deviations ($x - \langle x \rangle < 0, y - \langle y \rangle < 0$), whereas simultaneous occurrence of a positive and a negative deviation (for example, $x - \langle x \rangle > 0$, $y - \langle y \rangle < 0$) is infrequent. The correlation is then called positive. This qualitative characteristic of the correlation is represented by the sign of the mean product of the deviations

$$\Delta xy = \langle (x - \langle x \rangle)(y - \langle y \rangle) \rangle. \qquad \text{(D15)}$$

The magnitude of this mean product is also a characteristic of the correlation but acquires a meaning only when compared to the expected magnitude of the separate deviations $x - \langle x \rangle$ and $y - \langle y \rangle$.

A correlation coefficient should then be suitably defined as a combination of Δxy and of the root mean square deviations Δx and Δy. The definition

$$r = \frac{\Delta xy}{\Delta x \Delta y} \qquad \text{(D16)}$$

is convenient, because, as shown below,

$$r^2 \leqslant 1, \qquad \text{(D17)}$$

so that the magnitude of r constitutes an absolute index of the strength of the correlation. A value of $r = \pm 1$ indicates that pairs of deviations $x - \langle x \rangle$ and $y - \langle y \rangle$ are in constant ratio; it represents a complete correlation.

To prove equation (D17), one may consider the mean value of the combination of deviations $[(\Delta y^2)(x - \langle x \rangle) - (\Delta xy)(y - \langle y \rangle)]$. The square of this expression

---

1. The color of a ball does not fall in this class even though it may be represented by a numerical index according to an arbitrary code, for example, blue = 1, white = 2, red = 3; it would then be generally meaningless to consider an 'average' value of the color index.

cannot be negative, and neither can its mean value. We therefore write

$$0 \leq \langle[(\Delta y^2)(x - \langle x\rangle) - (\Delta xy)(y - \langle y\rangle)]^2\rangle = \Delta y^4\langle(x - \langle x\rangle)^2\rangle + (\Delta xy)^2\langle(y - \langle y\rangle)^2\rangle$$
$$- 2(\Delta y^2)(\Delta xy)\langle(x - \langle x\rangle)(y - \langle y\rangle)\rangle$$
$$= \Delta y^4 \Delta x^2 - \Delta y^2(\Delta xy)^2 = \Delta y^2[\Delta x^2 \Delta y^2 - (\Delta xy)^2]. \tag{D18}$$

Since Δy^2 is itself nonnegative, it follows from (D18) that

$$(\Delta xy)^2 \leq \Delta x^2 \Delta y^2, \tag{D19}$$

as implied by (D17).

APPENDIX E

FORMULAE FROM STATISTICAL MECHANICS

This Appendix summarizes formulae from thermal physics which are utilized in Part IV of this book. Background references are *BPC*, volume 5, particularly pages 169 and following, and F. Reif. *Fundamentals of Statistical and Thermal Physics,* especially pages 213 and following. Recall from *BPC*, volume 5, chapter 2, that all statements of probabilities and mean values of statistical physics pertain to an ensemble of "similar" systems, that is, of systems prepared and treated in identical manner.

Consider a system characterized by a complete set of states $|n\rangle$ which are stationary when the system is isolated. To each state corresponds an energy eigenvalue E_n. Degenerate states are labeled here by different indices n and n′ even though their eigenvalues coincide.[1] For the purposes of thermal physics, only the set of eigenvalues E_n matters; other characteristics of stationary states are generally immaterial. The set of eigenvalues may be discrete, continuous, or partly discrete and partly continuous. The system under consideration may be either very small (e.g., a single electron) or as large as a macroscopic body.

Thermal physics deals usually with a system in contact with an unspecified thermal bath which is characterized only by its Kelvin temperature T. Interaction between the system and the bath causes transitions among the stationary states of the system, accompanied by energy exchanges with the bath. The system is said to be in thermal equilibrium with the bath at the temperature T when, upon analysis in stationary states, the probability of each state $|n\rangle$ is proportional to the Boltzmann factor

$$\exp\,(-E_n/k_B T), \tag{E1}$$

where k_B, equal to 1.38×10^{-16} ergs per degree Kelvin, is the Boltzmann constant.[2] The absolute probability of the state $|n\rangle$ is given by the ratio of the relative probability (E1) to the sum of the corresponding relative probabilities for all states of the system

$$P_n(T) = \frac{\exp\,(-E_n/k_B T)}{\sum_n \exp\,(-E_n/k_B T)}. \tag{E2}$$

The quantity in the denominator of (E2) has a central role in the thermal physics of the system. It is called the *partition function* or *sum over states,*

---

1. The conventions with regard to the labeling of degenerate states differ in different textbooks.

2. "Thermal equilibrium" also implies that the states $|n\rangle$ are superposed *incoherently*.

and it is usually indicated by the letter Z. The function Z represents the total number of states, weighted by the probability of each state being excited at the temperature T; thus Z may be finite even though the number of states is infinite. It is often convenient to express the partition function Z as a function of the reciprocal temperature $\beta = 1/k_B T$, that is, to write

$$Z(\beta) = \sum_n \exp(-\beta E_n). \tag{E3}$$

Note that the value of the partition function tends to a finite nonzero limit when the temperature approaches absolute zero only if the zero point of the scale of energy eigenvalues is set to coincide with the lowest energy eigenvalue of the system. Thus, in the definition (E3) of Z, E_n will be understood to indicate the difference in energy between the state $|n\rangle$ and the ground state.

One often considers a system that may be regarded as consisting of separate, effectively independent, subsystems. For example, the system may consist of weakly interacting particles, or also of independent spin and orbital motions of the same particle. Independence of the subsystems implies first that each state symbol $|n\rangle$ can be expressed as a product of ket symbols of the different subsystem

$$|n\rangle = |n_1\rangle|n_2\rangle|n_3\rangle\ldots, \tag{E4}$$

and, second, that

$$E_n = E_{n_1} + E_{n_2} + E_{n_3} + \cdots. \tag{E5}$$

It follows that the sum over states (E3) can be carried out independently over the different subsystems, and the partition function is the product of separate partition functions

$$Z(\beta) = Z_1(\beta)Z_2(\beta)Z_3(\beta)\cdot\cdots. \tag{E6}$$

Note that in this case the logarithm of the partition function results from adding the logarithms of the partition functions for the component systems. Hence for an extended homogeneous system ln Z is an additive function of the size of the system. For this reason ln Z serves to represent the free energy and other extensive properties of a system.

When the system under consideration has a continuous—or practically continuous—distribution of eigenvalues with density $\rho(E)$ the partition function is represented by

$$Z = \int_0^\infty \rho(E) \exp(-\beta E) dE. \tag{E7}$$

Many systems of interest have spectral density functions $\rho(E)$ that follow a power law

$$\rho(E) = \frac{E^{\gamma-1}}{E_0^{\gamma}}. \tag{E8}$$

This distribution of energy eigenvalues occurs, for example, for the motion of a free particle, in which case $\gamma = 3/2$. The frequency distribution of radiation in free space follows the same law with $\gamma = 3$. Systems with the spectral density (E8) have the partition function

$$Z(\beta) = \frac{(\gamma - 1)!}{(\beta E_0)^\gamma}. \tag{E9}$$

The *free energy* F of a system is represented by ln Z to within a proportionality factor

$$F = -k_B T \ln Z. \tag{E10}$$

The *internal energy*, U, of a system, that is, the mean value of E_n weighted by the probability (E2), is given by

$$U = \sum_n E_n \frac{\exp(-\beta E_n)}{Z} = \frac{d \ln Z}{d\beta} \tag{E11}$$

When the partition function is given by (E9), the expression (E11) of the internal energy reduces to the *equipartition formula*

$$U = \gamma k_B T. \tag{E12}$$

The *entropy*, S, is represented, according to equations (E10) and (E11) by

$$S = \frac{U - F}{T} = k_B \left[\ln Z - \beta \frac{d \ln Z}{d\beta} \right]. \tag{E13}$$

The *heat capacity* at constant volume, C_V, is then given by

$$C_V = \left(\frac{\partial U}{\partial T} \right)_V = \frac{1}{k_B T^2} \frac{d^2 \ln Z}{d\beta^2}. \tag{E14}$$

In the special case where (E12) holds, the heat capacity C_V is temperature independent,

$$C_V = \gamma k_B. \tag{E15}$$

APPENDIX F

QUANTUM NUMBERS AND SPECTROSCOPIC NOTATION

Quantum numbers serve to identify discrete eigenvalues of physical quantities as well as geometrical characteristics of quantum-mechanical states. Quantum numbers pertaining to quantities that are compatible with energy are colloquially called *good quantum numbers* and serve to identify energy eigenstates. Each energy eigenstate is classified according to the quantum numbers of a maximal set of quantities compatible with each other and with energy. Quantum numbers pertaining to quantities compatible with an approximate expression of the energy are used to classify sets of states that are approximately stationary.

This Appendix lists the quantum numbers used in different problems and with various approximations. It also lists code names and symbols which often replace quantum numbers, particularly in spectroscopic notation. The list is arranged in alphabetical order, with Greek letters and mathematical symbols entered in accordance with the spelling of their English name; for instance, " + " is entered as "plus." The set of numbers immediately under each entry indicates the page where the entry was first introduced.

Symbol	*Page*	*Description*
a	474	Code name for homonuclear diatomic molecule state antisymmetric under permutation of position coordinates of nuclei
α	355	Code symbol for electron spin with $m_s = 1/2$
β	355	Code symbol for electron spin with $m_s = -1/2$
d	287	Code name for state with $l = 2$
D	369	Code name for state with $L = 2$ (for combinations with superscripts and subscripts, see L)
$d_{3/2}, d_{5/2}$		Code name for state with $l = 2$ and $j = 3/2$ or $j = 5/2$
δ	438	Code name for state of linear molecule with $\lambda = 2$
Δ	465	Code name for state of linear molecule with $\Lambda = 2$ (for combinations with superscripts and subscripts, see Λ)
f	287	Code name for state with $l = 3$

F	325 446	$\lvert\vec{F}\rvert^2 = F(F+1)\hbar^2 = \begin{cases}\lvert\vec{J}+\vec{I}\rvert^2 \text{ for atom}\\ \lvert\vec{J}+\vec{T}\rvert^2 \text{ for molecule}\end{cases}$ Hyperfine-structure quantum number of atom or molecule; good quantum number for weak or zero external fields
F	369	Code name for state with $L = 3$ (for combinations with superscripts and subscripts, see L)
$f_{5/2}, f_{7/2}$		Code name for state with $l = 3$ and $j = 5/2$ or $j = 7/2$
g	452n	Subscript code symbol for homonuclear diatomic molecule state symmetric (gerade) under reflection of all electron coordinates at the center of mass
I	325	$\lvert\vec{I}\rvert^2 = I(I+1)\hbar^2$ Nuclear-spin quantum number; good quantum number in atomic and molecular phenomena
j	144 162-3	$\lvert\vec{J}\rvert^2 = j(j+1)\hbar^2$ Angular momentum quantum number of unspecified system
j	313	$\lvert\vec{j}\rvert^2 = \lvert\vec{l}+\vec{s}\rvert^2 = j(j+1)\hbar^2;\ j = \{l + 1/2, l - 1/2\}$ Total angular momentum quantum number for single particle; good quantum number for single electron in central field or for many-electron atom in the independent electron approximation
J	364	$\lvert\vec{J}\rvert^2 = \lvert\sum_i \vec{l}_i + \sum_i \vec{s}_i\rvert^2 = J(J+1)\hbar^2$ Total angular momentum quantum number for many-electron atom; good quantum number for weak coupling with nucleus (condition generally satisfied)
J	472	$\lvert\vec{J}\rvert^2 = \lvert\vec{R} + \sum_i \vec{l}_i + \sum_i \vec{s}_i\rvert^2 = J(J+1)\hbar^2$ Total rotational quantum number of molecule; for linear molecule in Hund's case a, $J = \{\Omega, \Omega+1, \ldots\}$; for molecule in Hund's case b, $J = \{N+S, N+S-1, \ldots \lvert N-S\rvert\}$; good quantum number for molecule with weak nuclear coupling (condition generally satisfied)

K	354	Code name for electron shell with n = 1
l	284 287	$\lvert\vec{l}\rvert^2 = \lvert\vec{r} \times \vec{p}\rvert^2 = l(l+1)\hbar^2$ Orbital quantum number of single electron $l = \{0, 1, 2, 3, \ldots\}$ code $\{s, p, d, f, \ldots\}$ Good quantum number for electron in central field with weak spin-orbit coupling and for many-electron atom in the independent electron approximation; identifies single-electron state of molecule in united-atom or separate-atom limit (page 439-40).
L	334 366	$\lvert\vec{L}\rvert^2 = \lvert\sum_i \vec{l}_i\rvert^2 = L(L+1)\hbar$ Orbital quantum number of many-electron system $L = \{0, 1, 2, 3, \ldots\}$ Code $\{S, P, D, F, \ldots\}$ $L_{max} = \sum_i l_i$ Good quantum number for atom with weak spin-orbit coupling
L	354 370	Code name for electron shell with n = 2

	n	l	j
L_I	2	0	1/2
L_{II}	2	1	1/2
L_{III}	2	1	3/2

$L_0, L_{1/2}, L_1$	368-9	$L_0 = \{S_0, P_0, D_0, \ldots\}$ $L_{1/2} = \{S_{1/2}, P_{1/2}, D_{1/2}, \ldots\}$ $L_1 = \{S_1, P_1, D_1, \ldots\}$ .... Symbol for state of many-electron atom with quantum number L and with J = 0, 1/2, 1, ...
L°	369	$L^\circ \quad \{S^\circ, P^\circ, D^\circ, \ldots\}$ Symbol for state of atom with quantum number L and with odd parity under coordinate reflection at center of mass
$^1L, ^2L, ^3L$	369	$^1L = \{^1S, ^1P, ^1D, \ldots\}$ $^2L = \{^2S, ^2P, ^2D, \ldots\}$ $^3L = \{^3S, ^3P, ^3D, \ldots\}$ ....

Symbol for singlet, doublet, triplet, . . . state of atom with quantum number L and S = 0, ½, 1, . . .

λ — 438

$|\vec{l} \cdot \hat{R}| = \lambda\hbar$

Projection of single-electron orbital angular momentum along axis of linear molecule

$\lambda = \{0, 1, 2, 3, \ldots\}$

Code $\{\sigma, \pi, \delta, \phi, \ldots\}$

Good quantum number for weak spin-orbit coupling and for many-electron molecule in the independent electron approximation

Λ — 452, 471

$|\vec{L} \cdot \hat{R}| = \Lambda\hbar$

Projection of many-electron orbital angular momentum along axis of linear molecule

$\Lambda = \{0, 1, 2, 3, \ldots\}$

Code $\{\Sigma, \Pi, \Delta, \Phi, \ldots\}$

$\Lambda_{max} = \sum_i \lambda_i$

Good quantum number for weak spin-orbit coupling (Hund's case b)

λ_g, λ_u

$\lambda_g = \{\sigma_g, \pi_g, \delta_g, \ldots\}$

$\lambda_u = \{\sigma_u, \pi_u, \delta_u, \ldots\}$

Symbol for single-electron homonuclear diatomic molecule state with even (gerade) or odd (ungerade) parity under reflection of electron coordinates at center of mass

Λ_g, Λ_u — 452

$\Lambda_g = \{\Sigma_g, \Pi_g, \Delta_g, \ldots\}$

$\Lambda_u = \{\Sigma_u, \Pi_u, \Delta_u, \ldots\}$

Symbol for many-electron homonuclear diatomic molecule state with even (gerade) or odd (ungerade) parity under reflection of all electron coordinates at center of mass

$\Lambda_0, \Lambda_{1/2}, \Lambda_1$ — 472

$\Lambda_0 = \{\Sigma_0, \Pi_0, \Delta_0, \ldots\}$

$\Lambda_{1/2} = \{\Sigma_{1/2}, \Pi_{1/2}, \Delta_{1/2}, \ldots\}$

$\Lambda_1 = \{\Sigma_1, \Pi_1, \Delta_1, \ldots\}$

. . . .

Symbol for many-electron linear molecule state with quantum number Λ and with $\Omega = 0, \frac{1}{2}, 1, \ldots$

$^1\Lambda, {}^2\Lambda, {}^3\Lambda,$ 452

$$^1\Lambda = \{^1\Sigma, {}^1\Pi, {}^1\Delta, \ldots\}$$
$$^2\Lambda = \{^2\Sigma, {}^2\Pi, {}^2\Delta, \ldots\}$$
$$^3\Lambda = \{^3\Sigma, {}^3\Pi, {}^3\Delta, \ldots\}$$
$$\cdots\cdots\cdots\cdots\cdots\cdots$$

Symbol for singlet, doublet, triplet, ... state of linear molecule with quantum number Λ and with spin quantum number $S = 0, 1/2, 1, \ldots.$

Λ^+, Λ^- 473

$$\Lambda^+ = \{\Sigma^+, \Pi^+, \Delta^+, \ldots\}$$
$$\Lambda^- = \{\Sigma^-, \Pi^-, \Delta^-, \ldots\}$$

Symbol for linear molecule state with quantum number Λ and with even or odd parity index

m 144, 163

$$J_z = m\hbar$$

Magnetic quantum number of unspecified system

$$m = \{-j, -j + 1, \ldots j\}$$

m 282

$$l_z = m\hbar$$

Magnetic quantum number of single electron

$$m = \{-l, -l + 1, \ldots l\};$$

Good quantum number for electron in field with axial symmetry about z axis and in absence of spin-orbit coupling

M (or M_L) 338

$$L_z = \textstyle\sum_i l_{iz} = M\hbar$$

Magnetic orbital quantum number of many-electron atom

$$M = \{-L, -L + 1, \ldots L\};$$

Good quantum number for atom in field with axial symmetry about z axis and in absence of spin-orbit coupling

M 354, 370

Code name for electron shell with $n = 3$

	n	l	j
M_I	3	0	1/2
M_{II}	3	1	1/2
M_{III}	3	1	3/2
M_{IV}	3	2	3/2
M_V	3	2	5/2

M	447	$N_z = M\hbar$ Magnetic rotational quantum number for molecule with rotational quantum number N $M = \{-N, -N + 1, \ldots N\};$ Good quantum number for molecule in field with axial symmetry about z axis in absence of coupling with electron or nuclear spin
M_I		$I_z = M_I\hbar$ Magnetic nuclear spin quantum number $M_I = \{-I, -I + 1, \ldots I\}$ Good quantum number in field with axial symmetry about z axis and in absence of coupling with atomic or molecular angular momentum
m_j	312	$j_z = m_j\hbar$ Total magnetic quantum number $j_z = \{-j, -j + 1, \ldots j\}$ Good quantum number for single electron or independent electron in field with axial symmetry about z axis
M_J	366	$J_z = \sum_i l_{iz} + \sum_i s_{iz} = M_J\hbar$ Total magnetic quantum number for atom with many electrons $M_J = \{-J, -J + 1, \ldots J\}$ Good quantum number for atom in field with axial symmetry about z axis and in absence of coupling with nuclear spin
M_J		$J_z = \mathcal{R}_z + \sum_i l_{iz} + \sum_i s_{iz} = M_J\hbar$ Total magnetic quantum number for molecule $M_J = \{-J, -J + 1, \ldots J\}$ Good quantum number for molecule in field with axial symmetry about z axis and in absence of coupling with nuclear spin
m_s	308	$s_z = m_s\hbar$ Magnetic spin quantum number of single electron $m_s = \{1/2, -1/2\}$

		Good quantum number in absence of spin-orbit coupling or of magnetic field component orthogonal to z axis
M_S	338	$S_z = \sum_i s_{iz} = M_S\hbar$ Magnetic spin quantum number for many-electron system $M_S = \{-S, -S + 1, \ldots S\}$ Good quantum number in absence of spin-orbit coupling
—	218	Code symbol for states with odd parity under various coordinate reflections
—	465 473	Superscript code symbol for linear molecule state pertaining to reflection of electron coordinates at any plane containing nuclear axis
n	291	$n = n_r + l + 1$ Principal quantum number for electron in atom; identifies single-electron state of molecule in united-atom or separate-atom limit (section 19. 2. 2)
N	354	Code name for electron shell with n = 4
N	447 469	$\lvert\vec{N}\rvert^2 = \lvert\vec{\mathcal{R}} + \sum_i \vec{l}_i\rvert^2 = N(N + 1)\hbar^2$ Rotational quantum number of molecule; for linear molecule $N = \{\Lambda, \Lambda + 1, \ldots\}$ Good quantum number for molecule with spin-orbit coupling weaker than rotational energy (Hund's case b)
n_r	291	Radial quantum number; number of spherical nodal surfaces of particle density $n_r = \{0, 1, 2, \ldots\}$ Good quantum number for electron in central field with weak spin-orbit coupling and for many-electron atom in the independent electron approximation
Ω	472	$\lvert(\vec{L} + \vec{S}) \cdot \hat{R}\rvert = \Omega$ Projection of many-electron total angular momentum along axis of linear molecule $\Omega = \{0, 1/2, 1, \ldots\}$

		Good quantum number in Hund's case a
p	287	Code name for state with $l = 1$
P	369	Code name for state with $L = 1$ (for combinations with superscripts and subscripts, see L)
$p_{1/2}, p_{3/2}$		Code name for state with $l = 1$ and $j = 1/2$ or $3/2$
p_x, p_y	287	Code name for single-electron state with $l = 1$ and angular momentum component $l_x = 0$ or $l_y = 0$
p_z	287	Code name for single-electron state with $l = 1$ and $m = 0$
Parity	218	$\{1, -1\}$ Coefficient by which wave function is multiplied under various coordinate reflections. No standard symbol; alternative names: $\{1, -1\}$ {even, odd}, $\{+, -\}$ {positive, negative} for molecules {gerade, ungerade} for molecular electrons; unless otherwise stated, pertains to reflection of all particle coordinates at center of mass $(-1)^l$ for single electron in atom or in united-atom limit of molecule (section 13.2.4) $(-1)^{\Sigma_i l_i}$ for state of many-electron atom derived from independent electron approximation (page 354) $\pm(-1)^N$ for linear molecule (page 473) +1 for gs and ua in homonuclear diatomic molecules −1 for us and ga in homonuclear diatomic molecules
+	218	Code symbol for states with even parity under various coordinate reflections
+	465 473	Superscript code symbol for linear molecule state pertaining to reflection of electron coordinates at any plane containing nuclear axis
π	438	Code name for state with $\lambda = 1$

Π		Code name for state with $\Lambda = 1$ (for combinations with superscripts and subscripts, see Λ)
s	287	Code name for state with $l = 0$
s	308	$\lvert\vec{s}\rvert^2 = s(s+1)\hbar^2 = \frac{3}{4}\hbar^2$ Spin quantum number of a single electron $s = \{\frac{1}{2}\}$ Good quantum number
s	474	Code name for homonuclear diatomic molecule state symmetric under permutation of position coordinates of nuclei
S	331	$\lvert\vec{S}\rvert^2 = \lvert\sum_i \vec{s}_i\rvert^2 = S(S+1)\hbar^2$ Spin quantum number of many-electron system $S = \{0, \frac{1}{2}, 1, \ldots\}$ Code {singlet, doublet, triplet, . . .}; integer for even number of electrons; half-integer for odd number of electrons $S_{max} = \frac{1}{2}$ of number of electrons Good quantum number for system with weak spin-orbit coupling
S	369	Code name for state with $L = 0$ (for combinations with superscripts and subscripts, see L)
$s_{1/2}$		Code name for state with $l = 0, j = \frac{1}{2}$
σ	438	Code name for state with $\lambda = 0$
Σ	452	Code name for state with $\Lambda = 0$ (for combinations with superscripts and subscripts, see Λ)
T	446	$\lvert\vec{T}\rvert^2 = \lvert\sum_i \vec{I}_i\rvert^2 = T(T+1)\hbar^2$ Total nuclear spin quantum number of molecule
u	452n	Subscript code symbol for homonuclear diatomic molecule state odd (ungerade) under reflection of electron coordinates at center of mass
v	448	Vibrational quantum number of molecule.

BIBLIOGRAPHY

Background Reference Works

Berkeley Physics Course. 5 vols. New York: McGraw-Hill, 1965-71.

Boas, M. L. *Mathematical Methods in the Physical Sciences*. New York: Wiley, 1966.

Kaplan, W. *Advanced Calculus*. Reading, Mass.: Addison-Wesley, 1952.

Physical Science Study Committee. *Physics*. Boston: D.C. Heath, 1960.

Reif, F. *Fundamentals of Statistical and Thermal Physics*. New York: McGraw-Hill, 1965.

Weinberg, G. H., and Schumaker, J. A. *Statistics: An Intuitive Approach*. Belmont, Calif.: Wadsworth, 1962.

Works Comparable to the Present Text in Level and Scope

Born, M. *Atomic Physics*. 8th ed. Darien, Conn.: Hafner, 1970 (7th ed., 1962).

Harnwell, C. P., and Livingood, J. J. *Experimental Atomic Physics*. New York: McGraw-Hill, 1933.

Richtmyer, F. K.; Kennard, E. H.; and Cooper, J. N. *Introduction to Modern Physics*. New York: McGraw-Hill, 1969.

Works on Quantum Theory and Special Fields

Bethe, H. A., and Jackiw, R. W. *Intermediate Quantum Mechanics*. 2d ed. New York: Benjamin, 1968.

Condon, E. U., and Shortley, G. *Theory of Atomic Spectra*. Cambridge: At the University Press, 1963.

Coulson, C. A. *Valence*. London: Oxford University Press, 1961.

Herzberg, G. *Molecular Spectra and Molecular Structure*. I. *Spectra of Diatomic Molecules*. Princeton, N. J.: Van Nostrand, 1950.

———. *Molecular Spectra and Molecular Structure*. II. *Infrared and Raman Spectra of Polyatomic Molecules*. Princeton, N. J.: Van Nostrand, 1954.

———. *Molecular Spectra and Molecular Structure*. III. *Electronic Spectra and Electronic Structure of Polyatomic Molecules*. Princeton N. J.: Van Nostrand, 1966.

Kittel, C. *Introduction to Solid State Physics*. New York: Wiley, 1967.

Landau, L. D., and Lifschitz, E. M. *Quantum Mechanics*. London: Pergamon, 1958.

Merzbacher, E. *Quantum Mechanics*. New York: Wiley, 1970.

Mott, N. F., and Massey, H. S. W. *Theory of Atomic Collisions*. Oxford: Clarendon Press, 1965.

Pilar, F. L. *Elementary Quantum Chemistry*. New York: McGraw-Hill, 1968.

Rutherford, E.; Chadwick, J.; and Ellis, C. D. *Radiations from Radioactive Substances*. New York: Macmillan, 1930.

Segrè, E. *Nuclei and Particles*. New York: Benjamin, 1964.

Slater, J. C. *Quantum Theory of Atomic Structure*. New York: McGraw-Hill, 1960.

Index

Greek letters and mathematical symbols are entered in accordance with the spelling of their English names. Quantum numbers are listed in Appendix F.